Frontiers and Tasks of Catalysis Towards the Next Century

Frontiers and Tasks of CATALYSIS TOWARDS THE NEXT CENTURY

PROCEEDINGS OF THE INTERNATIONAL SYMPOSIUM
IN HONOUR OF PROFESSOR TOMOYUKI INUI

Editors:
S. Uemura, T. Mitsudo and M. Haruta

///VSP///
Utrecht, The Netherlands, 1998

VSP BV
P.O. Box 346
3700 AH Zeist
The Netherlands

Tel: +31 30 692 5790
Fax: +31 30 693 2081
E-mail: vsppub@compuserve.com
Home Page: http://www.vsppub.com

First published in 1998

ISBN 90-6764-284-3

Printed in The Netherlands by Ridderprint bv, Ridderkerk.

CONTENTS

Professor Tomoyuki Inui

FORWORD

This special issue is dedicated to Professor Tomoyuki Inui of Kyoto University to celebrate his retirement in March 1998. The 41 contributions constitute either invited talks or invited posters, which will be presented at the International Catalysis Symposium on Frontiers and Tasks of Catalysis towards the Next Century. The Symposium has been organized to honor Professor Inui, with the help of financial assistance from the Ministry of Education, Science, Sports and Culture, Japan and will be held at the Kyoto International Conference Hall on the 20th and 21st March 1998.

Professor Tomoyuki Inui graduated from the Department of Fuel Chemistry, Faculty of Engineering, Kyoto University in 1957 and finished his M.S. and Ph.D. courses at the same university in 1959 and 1962, respectively. Since 1962, he has worked for the same department as a Research Associate until 1978 and as an Associate Professor until 1983, after which he became a full Professor. From 1986 to 1994 he also served as a Professor of the International Corporation Center for Science and Technology, Tokyo Institute of Technology.

Professor Inui has long been widely admired for his distinguished mastery of experimental techniques and his innovative and skillful approach to catalysis research, which have made great contributions to the development of the petroleum chemical industry and to the solutions of energy resource and global environmental problems. He is recognized not only for outstanding research achievements, but also for his superb educational talents and his caring attitude which have resulted in the nurturing of many students into established, respected figures in academia and industry.

Professor Inui's research has always focused on the synthesis of novel catalysts and their practical performance. These contributions include the selective oxidation of lower hydrocarbons, the high-rate synthesis of chemical feedstocks and fuels from non-petroleum resources, the high-rate transformation of carbon dioxide to important chemical intermediates, the purification of exhaust gases, catalytic combustion, effective separation of mixed gases, high-grade purification, and the elucidation of the structure and dynamic behavior of heterogeneous catalysts. In addition, his recent research has shown the importance and validity of non-linear phenomena in catalysis in the analysis of the reaction kinetics and in the improvement of catalytic performance. He is the author of more than 310 technical papers and more than 100 review articles and books and also holds many patents for his work in heterogeneous catalysts.

During his career, Professor Inui has received many honors: the Progress

Award in the Fuel Society of Japan (1983), the Chemical Society of Japan Award (1989), Catalysis Society of Japan Award (1992), the Japan Institute of Energy Award in Academic Division (1994), and the Japan Petroleum Institute Award (1995). He has served on the editorial boards of Applied Catalysis, Catalysis Letters, Zeolites, and Catalysis Surveys form Japan. In addition, he has served as President of the Catalysis Society of Japan, and the Japan Association of Zeolites, and as a member of many important Advisory Committees of the Ministry of Education, Science, Sports and Culture and the Ministry of Industrial Trade and Industry.

It is our greatest pleasure and honor to publish this special issue as Guest Editors for Research on Chemical Intermediates and we are indebted to the editors of the journal, Professors M. Anpo, M. C. Depew, and J. K. S. Wan who kindly gave us this opportunity. We would be very grateful if this issue could help us recognize Professor Inui's outstanding contributions to catalysis science and technology, including a successful research and teaching career in Kyoto University, and his friendly and attractive personality. Most importantly we would like to thank very much all of the authors for their significant contributions in the production of this special issue.

Masatate Haruta
Osaka National Research Institute
AIST, MITI

Takeaki Mitsudo
Sakae Uemura
Department of Energy and
Hydrocarbon Chemistry
Graduate School of Engineering
Kyoto University

Waste Water Purification

Two recent UN sponsored reports [7,8] warn that the world must act now to protect its water. In many regions of the world, people take for granted the real value of the water which we use, while in many other regions water is a limited natural resource because of supply or contamination. We must not just develop new processes for treating water, but also for the processes which use and affect the water we use. This means more efficient use of water, especially in view of the continued global warming which will increase irrigation needs even further. In addition certain types of energy production consume more water than others. Coal gasification or use of oil shales require 20-50 times more water to process these fuel sources than the use of oil or natural gas.

Another factor in the quality of our drinking water is the presence of nitrates and nitrates caused by tremendous amounts of runoff from soil where nitrates are used as fertilizers. Some have advocated that a surcharge should be placed on the cost of nitrate fertilizers to reflect the added cost they impose on the local regions for the eventual removal of these nitrates. The use of nitrates as fertilizers also leads to enhanced levels of N_2O, another undesirable emission [8]. This issue is complicated by the fact that the use of NH_3 and nitrate fertilizers is critical for our survival in the production of essential amino acids, urea, DNA, and proteins. It is estimated that one third of the protein responsible for nourishing mankind depends upon the use of synthetic fertilizer. Unfortunately a problem arises from the runoff of nitrates which then accumulates in the water supply.

In attempting to treat waste water we need to appreciate that there are multiple parts to this problem. Not only are nitrates and nitrites present but also many different types of undesirable organic compounds in low levels. Many of the organic compounds cannot be removed by a single, simple, cost effective catalyst. Further, the removal of the organic compound still leaves the nitrates and nitrites in the water. In treating the waste water one has to appreciate that it must be done at low temperatures [preferably <100 °C] because one doesn't want to spend a lot of energy to vaporize water needlessly. Further, such treatment must be done in the presence of an excellent ligating material, water itself. In using catalysts for such operations, they must not only perform efficiently in the presence of large amounts of water, but catalyst erosion and leaching is unacceptable. Thus the problem is not only removing dilute levels of impurities, but also the presence of many different types of pollutants further complicated by the presence of nitrates and nitrites which may still have to be removed by some alternative approach.

Removal of Harmful Gases

Auto emission controls. B. Cooper has summarized some of the opportunities in auto emissions control. These include improved thermal durability of the catalysts, developing fast light off catalysts, developing thin wall substrate systems, improving methods of dispersion and stabilizing key catalytic components, substituting nickel with other components in order to suppress H_2S emissions, and developing new lean NOx catalysts. Further the problems are different but even more complicated for diesel engine emissions control [9].

NOx removal. Much has already been said about means to remove NOx [1,4], so I don't believe there is a need to repeat much of what has already been written. Let me just summarize some key thoughts in this regard. Additional research on this topic should not focus only on ideal atmospheres. There is a need to operate in the presence of what may seem, but are typically present, large levels of water vapor [>10%]. If one chooses to use a hydrocarbon as a reducing agent, large excesses of hydrocarbons are not acceptable, since they then must be removed. Ethane, ethylene, propane or propylene may be useful model hydrocarbons, but much more work is needed to be done with compounds similar to the components of gasoline or diesel fuels. Mobile vehicle manufacturers do not want to install two fuel systems into new vehicles, rather it is preferred to use the current fuel to treat NOx emission. In evaluating NOx removal catalysts, one needs to add SO_2 or H_2S at ppm levels since some form of sulfur is often present with emissions of NOx. Further, NOx will be present not only at dilute levels but in large volumes of gas at very high space velocities. The means to remove NOx must be demonstrated at very high GHSV, in excess of 30,000. In addition, one should not assume that removal of NOx at percent or thousands of ppms levels can extrapolate to conventional NOx emission levels at the 50-200 ppm level. Any announced breakthroughs in technology for NOx removal must be balanced and benchmarked against sustained long lived activity for hours, not just minutes. Benchmarking of new catalysts becomes increasingly important; reports of new catalysts should be compared against well established materials already reported in order to give readers a scale for comparison. Further, I believe it is necessary for us to refocus efforts away from overly popular catalysts such as Cu-ZSM-5. It is clear that this catalyst needs extensive modification because of its known hydrothermal sensitivity. Finally, NOx decomposition will always be a much more acceptable alternative in the long run because it avoids the use of another reagent, in most cases the hydrocarbons discussed above.

SO_2 removal. The removal of SO_2 is not an easy problem by any means. Many have tried in the past, but with incomplete impact upon the problem which is focused at the extremely high levels of sulfur oxides which are emitted worldwide. These sulfur oxides amount to much more [10] than the level of NOx currently emitted [Figure 1], but certainly the level of research on SOx is small compared to

that on NOx removal.

Sulfur in an exhaust or product stream is often eliminated as SO_2 or converted to dilute sulfuric acid. This means that one has then to find a market for dilute sulfuric acid or to transport the acid to another site for possible conversion to concentrated sulfuric acid which has more value. Alternatively, the oxides of sulfur may be precipitated by injected lime to make gypsum, but the amount of gypsum this would make far exceeds the world's need for this alternative form of sulfur.

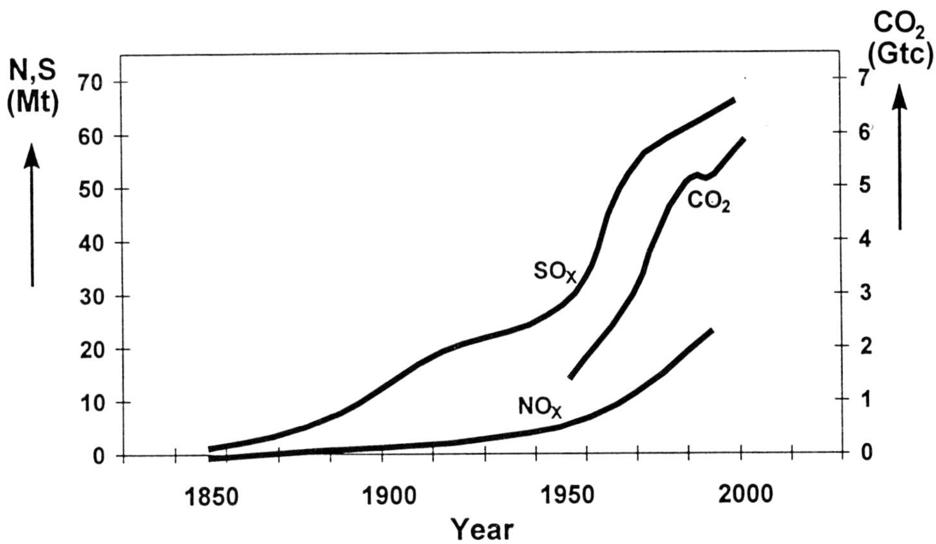

Figure 1: Relative emissions of NOx, SO_2, and CO_2 (N, S Mt = millions of tons of N in NOx or S in SO_2; Gtc= gigatons of C in CO_2) [10].

The Claus Process does offer an effective means of removing SO_2 by converting it to elemental sulfur, but again there is a limited need for elemental sulfur produced by this approach, since there are plentiful sources of sulfur currently produced by subsurface drilling. Thus, the question remains, what are we going to do with all the potential forms of by-product sulfur produced by the elimination of more than 40 million metric tons per year of SO_2?

Large scale conversion of CO_2. Earlier articles have pointed to the limited value of using the production of chemicals from CO_2 as a reasonable approach to significantly impacting CO_2 emissions [3]. In addition CO_2 will need to be purified and activated to make it an acceptable feedstock for most chemical reactions. Currently, the use of H_2 to control CO_2 emissions is impractical considering the fact that most H_2 is produced today with CO_2 as a by-product, and H_2 has a lot of value for other purposes with greater immediate and cost effective impact on reducing other pollutants. Perhaps the greatest impact upon CO_2 emissions will come from improved energy efficiency, since power plants are a major source of CO_2.

Alternative routes to H_2. Opportunities exist for new routes for the production of H_2. Currently steam reforming of hydrocarbons is the most popular approach to H_2 production with partial oxidation of hydrocarbons as a second method. Both approaches result in substantial levels of CO_2 coproduction. However, while steam reforming is a mature, energy intensive technology, there is some fertile ground. One has to keep in mind that one needs to carry out such studies at >10 atm with substantial catalyst life [current catalysts operate for > 5 years] because that is where the customer needs the H_2. There are a number of potential alternative routes, such as selective oxidation of methane to syngas, oxidative dehydrogenation, electrolysis with fuel cells, use of solar energy, and the use of more energy efficient approaches to reforming hydrocarbons. CO_2 reforming [equation 2]

$$CO_2 + CH_4 \rightarrow 2CO + 2H_2 \qquad \textbf{(2)}$$

is a popular approach today, but researchers must appreciate that the key problem is catalyst life at high pressures. Again customers need H_2 at >10 atm, and compression of large volumes of product gas is not often very acceptable. A key part of the catalyst life issue is the avoidance of carbon formation which can lead to unacceptable plugging of the reformer tubes.

Potentially large quantities of non-fossil fuel based H_2 are available, especially in the large volumes of water available on our earth, if only we could figure how to efficiently recover the H_2. Alternative approaches must avoid the coproduction of large levels of CO_2. Clearly breakthroughs are needed.

Solid acids to replace liquid acids. There is a lot of interest in replacing HF or H_2SO_4 with a solid acid in order to reduce the difficulty in handling materials like HF or in the disposal of the waste acids. Most of the current focus of publications is on acid strength, but it is important to remember that the number of acid sites is very important as well. An example of such comparisons between acid site activity vs acid sites/gm can be seen in a recent publication by Okuhara, *et al.* [11]. At this time solid acids are usually a dilute form of liquid acids, hence one needs to focus on increasing acid strength density. If those features can be obtained in a solid acid, then such a material must perform in the 100 °C range without serious deactivation. To be acceptable replacements to the current liquid acids, solid acids must offer better overall yields with similar RON (research octane number) at comparable space velocities of > 2 hr^{-1}. In testing potential solid acids for sufficient RON, the paraffin/olefin ratio is very important because recycle of more than 80% of a process stream is felt to be unacceptable [12].

CONCLUSION

Areas of opportunity for more R&D which could offer more impact on our environment include increased energy efficiency [as it impacts not only CO_2 emissions but also process simplification], waste minimization, SO_2 removal, NOx removal, other economical approaches to H_2 as a potentially clean fuel, and waste water purification. The CO_2 problem probably cannot be "solved" only by producing valued chemicals via reduction with H_2 until we have a non-fossil fuel source of cheap H_2.

Acknowledgments
I wish to thank Air Products and Chemicals, Inc. for the permission to publish this work.

REFERENCES

1. J. N. Armor, *Appl. Catal. B*, **1,** 221 (1992).
2. *Emissions of Greenhouse Gases in the United States: 1987-1992*, DOE/EIA, Report # 0573, October, 1994, US Government Printing Office, Washington, DC.
3. J. N. Armor, *Catalysis Today*, 1998, proceedings of CO_2 utilization meeting, Kyoto, 1997 accepted for an Elsevier publication.
4. J. N. Armor, *Catalysis Today*, **26**, 99 (1995).
5. M. Carson, *Oil & Gas Journal*, Aug 11, 37 (1997).
6. J-P Lange, K. deJong, J. Ansorge, and P. Tijm, *Studies in Surf. Sci. & Catal.*, **107**, 81 (1997).
7. *Chemistry & Industry*, 3 February, 80 (1997).
8. V. Smil, *Scientific American*, July, 76 (1997).
9. B.J. Cooper, *Plat. Met. Rev.* **38** 2 (1994).
10. B. Eliasson, in: *Carbon Dioxide Chemistry:Environmental Issues,* J. Paul and C-M. Pradier, (Eds.), 1994, p.5
11. T. Okuhara, M. Kimura, and T. Nakato, *Appl. Catal. A*, L9, 155 (1997).
12. P. Rao and S. Vatcha, *Oil and Gas J.*, September, 56 (1996).

Res. Chem. Intermed., Vol. 24, No. 2, pp. 115-122 (1998)

HIGHLY SELECTIVE REDUCTION OF NO IN EXCESS OXYGEN THROUGH THE INTERMEDIATE ADDITION OF REDUCTANT BETWEEN OXIDATION AND REDUCTION CATALYSTS

MASAKAZU IWAMOTO[1]*, TETSU ZENGYO[1], and ANGEL M. HERNANDEZ[2]

[1]*Catalysis Research Center, Hokkaido University, Sapporo 060, Japan*

[2]*Dpto. IPII, Universidad Autonoma Metropolitana-Iztapalapa, 09340 Mexico D.F., Mexico*

Received 13 October, 1997; accepted 14 November, 1997

Abstract-- Intermediate addition of reductant (named an IAR method) into a NO+O_2 stream between an oxidation catalyst of NO to NO_2 and a reduction catalyst of NOx to N_2 has been evidenced to be a very effective method to selectively reduce NO to N_2 in the presence of excess oxygen. Here Pt-MFI has been employed as an oxidation catalyst, Na-, Mg-, Ca-, Ba-, Zn- and In-MFI have been examined as the reduction catalysts and the reductant was ethene. The combination of Pt- and Zn-MFI was the most effective. The efficiency of ethene, namely the molar ratio of the amount of NO reduced to that of ethene consumed, reached 1.9 and was much higher than 1.4 of conventional reduction on Cu-MFI.

INTRODUCTION

The selective catalytic reduction of NO with hydrocarbons (SCR-HC) in the presence of excess oxygen is a potential method to remove NO_X from various exhausts [1]. Various metal oxides, metals, and zeolites etc. have been suggested to be active for the SCR-HC and there are many discussions on the reaction mechanism of the SCR-HC. In the latter studies, the concept that NO is oxidized to NO_2 with oxygen and then the resulting NO_2 reacts with hydrocarbons to form an active intermediate is widely suggested on various catalysts [2-4]. Several efforts have been devoted to separate the functions of catalysts; for example, platinum and Mn_2O_3 are reported active for the oxidation of NO to NO_2 in the reaction systems of NO+O_2+CH_4 on Pt/In/MFI [5] and of NO+O_2+C_3H_6 on Mn_2O_3+Ce-MFI, [6] respectively. One of the most significant problems in these systems is that the oxidation catalyst for NO is often active for oxidation of the hydrocarbons, which would result in decrease in efficiency of the reductant. In case of CH_4 as the reductant, Pt can be used for the NO oxidation because of the low oxidation activity of Pt to CH_4 [5]. With C_3H_6, Mn_2O_3 or CeO_2 should be employed because Pt or the other oxide catalysts are too active for the C_3H_6 oxidation [6].

It is well known that the diesel exhausts contain very small amounts of unburned hydrocarbons. The addition of fuels or hydrocarbons into the diesel exhausts is required to reduce NO by the SCR-HC method. On the basis of the above reaction mechanism and the necessity of hydrocarbons addition, we have proposed a novel strategy to reduce NO selectively, that is, the intermediate addition of a reductant (IAR method) into the NO- containing exhaust between an oxidation catalyst of NO (reaction 1) and a reduction catalyst to N_2 (reaction 2) [7].

$$2NO + O_2 \rightarrow 2NO_2 \quad \textbf{(1)}$$

$$2NO_2 + C_2H_4 + O_2 \rightarrow N_2 + 2CO_2 + 2H_2O \quad \textbf{(2)}$$

In the previous paper [7] we have used Pt- and In-MFI as the oxidation and reduction catalysts, respectively. Here Pt-MFI has again been employed as the oxidation catalyst. Several kinds of metal ion-exchanged MFI zeolites such as Na-, Mg-, Ca-, Ba-, and Zn-MFI have been examined as the reduction catalysts because these zeolites have been reported to be not so active for the usual selective catalytic reduction of NO with hydrocarbons [8]. More active catalytic systems than the previous combination [7] will be reported here. Finally the efficiency of hydrocarbons were compared with those of the conventional reaction on Cu-MFI alone.

EXPERIMENTAL SECTION

Catalyst Preparation

Platinum and metal ion-exchanged MFI zeolites were prepared by a conventional ion exchange method with an aqueous tetra-ammineplatinum (II) nitrate and respective metal acetate solutions. The Si/Al molar ratio of the parent MFI zeolite was 11.7. The amounts of cations were determined by ICP analysis after the samples were dissolved in a HF solution. The ion exchange levels of Pt, Na, Mg, Ca, Ba, Zn, and In-MFI prepared were 101, 97, 55, 84, 62, 97 and 83 %, respectively, on the assumption of ion exchange of two (or three) Na ions for one divalent (or trivalent) cation. The zeolites were pressed, crushed, and sieved to obtain particles of 250-425 μm.

Reaction Procedures

The catalytic reduction was carried out in a conventional flow system with twin reactors made of quartz. The flow diagram is shown in Figure 1. The twin reactors were placed in a single furnace. The Pt-MFI catalyst (0.4 g) was mounted in the first reactor and the other Metal ion-MFI (0.4 g) in the second one. The mixture of NO, O_2, and He (balance) (total flow rate, 124 cm^3 min^{-1}) was introduced into the first reactor. The reductant, C_2H_4, diluted in He (26 cm^3 min^{-1}) was then added into the $NO+O_2$ mixture between the first and the second reactors. One can change the flow route of the $NO+O_2$ mixture to the bypath from the first reactor to evaluate the effect of the first catalyst bed. Before the catalytic reaction, the catalysts were pretreated in situ at 773 K for 3 h in a He flow. The reactants were 1000 ppm of NO, 857 ppm of C_2H_4, 2 % of O_2 (helium balance) and the total flow rate of gases was 150 cm^3 min^{-1}. A gas chromatograph was employed to measure the respective concentrations of ethene, nitrogen, nitrous oxide, carbon dioxide, and carbon monoxide. A NOx analyzer was used to determine the degree of oxidation of NO to NO_2.

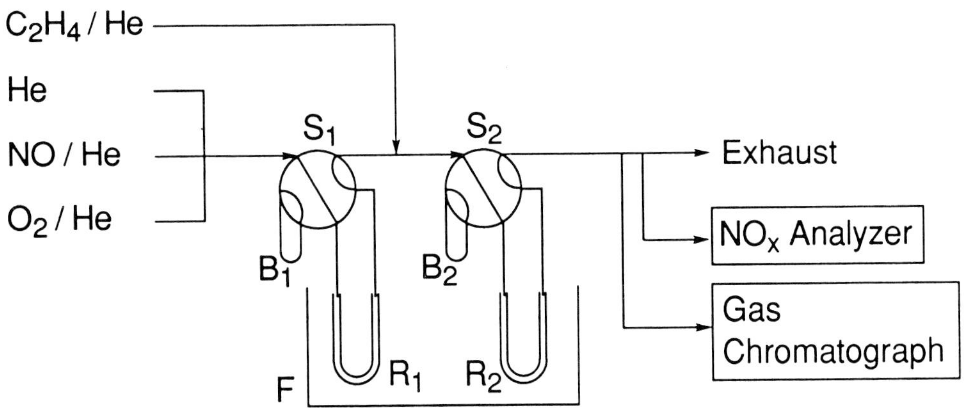

Figure 1. Flow system for the SCR-HC by the IAR method. B1 and B2, by-paths of the reactors; F, furnace; R1 and R2, reactors; S1 and S2, six-way valves.

RESULTS AND DISCUSSION

Oxidation of NO on Pt-MFI

First, the effectiveness of Pt-MFI as the oxidation catalyst of NO has been investigated. The results are plotted in Figure 2. The conversion of NO to NO_2 on Pt-MFI has been maximized at 523 K. The decrement in the conversion at higher temperatures is due to the limitation of the equilibrium as clearly shown in Figure 2. Without the Pt-MFI catalyst a very small amount of NO_2 was produced under the present experimental conditions. One can therefore recognize that the Pt-MFI zeolite is a very active catalyst for the oxidation of NO to NO_2. The catalytic activity is much higher than those reported previously [6,9]. After the pretreatment, the X-ray diffraction analysis of the Pt-MFI catalyst showed the formation of Pt particles on the MFI zeolite, which is in good agreement with the previous report

[10]. The reproducibility of the results of Figure 2 was very good within the temperature range 473-873 K and no deactivation of the Pt-MFI catalyst for the NO_2 formation has been observed over 300 h.

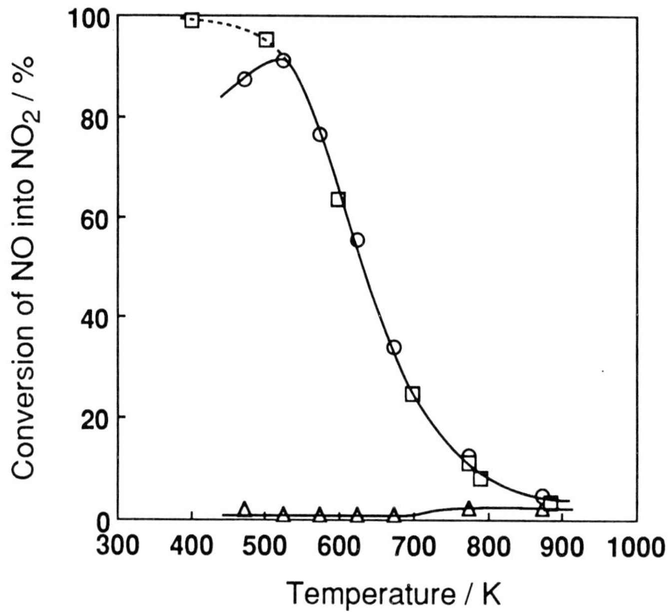

Figure 2. NO_2 formation in the NO + O_2 reaction on Pt-MFI (○). NO to NO_2 conversions without Pt-MFI (△) and under equilibrium (❑) are also shown.

Combination of Pt-MFI and In-MFI

In Figure 3, the activities of four kinds of catalytic systems have been depicted; the series combination of the Pt-MFI catalyst on the first reactor and the In-MFI catalyst on the second reactor (abbreviated as Pt|In), the physically mixed catalysts of Pt-MFI (0.4 g) and In-MFI (0.4 g) on the second reactor (Pt+In), the In-MFI catalyst alone on the second reactor without the Pt-MFI catalyst (In), and the series combination of the Pt-MFI catalyst on the first reactor and the Na-MFI catalyst on the second reactor (Pt|Na). Figure 3 clearly shows the great difference between the results of the Pt|In and the In systems. With Pt-MFI as the oxidation catalyst the catalytic activity of In-MFI for the NO reduction was greatly enhanced at 473-623 K. At the same time the conversion levels of C_2H_4 were also increased. It is evident that the increments of NO and C_2H_4 conversions are due to the formation of NO_2 on Pt-MFI. Figure 3 also indicates that there is the formation of N_2O at 473-623 K. This suggests the easy formation of N_2O in the presence of NO_2, though the reaction mechanism is not clear yet.

The results on the Pt|In and In systems at 773-873 K were approximately the same. Taking the small amount of NO_2 formed at 773-873 K (Figure 2) into consideration, the activity at this temperature range would result from the catalysis of In-MFI itself to the $NO+O_2+C_2H_4$ mixture.

The results of the Pt|In system can be compared with those of the Pt+In system. The degree of conversion of NO to N_2 on the latter was much lower than that of the former as shown in Figure 3 and almost the same as that of the Pt-MFI alone reported in the previous paper [10]. It follows that only Pt-MFI in the mixture of Pt-MFI and In-MFI is an active phase and In-MFI is little used for the reaction. In addition, the results on the Pt|Na system reveal that the indium ions are more effective than the sodium ions to the progress of catalytic reaction of NO_2 with hydrocarbons.

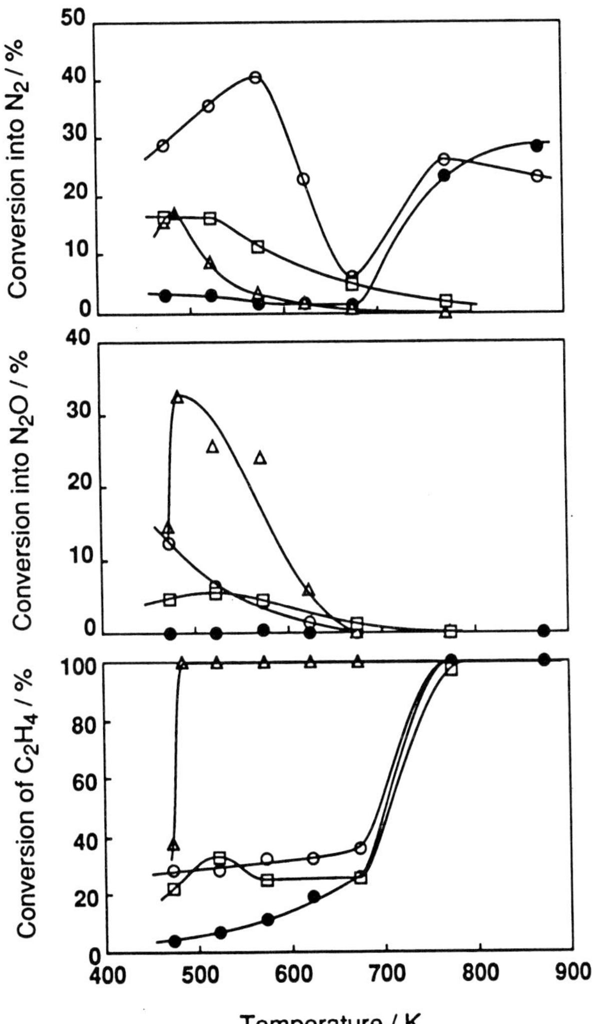

Figure 3. Reduction of NO on Pt|In (○, IAR), In (●), Pt+In (△, physically mixed), and Pt|Na (□) systems. The detailed catalyst systems and the experimental conditions are described in the text.

Another interesting phenomenon should be noted in Figure 3. The conversion levels of ethene on the Pt|In system were 20-40 % in the temperature range 473-673 K. In contrast, those on the Pt+In system were always 100 %. This indicates that the IAR method is effective for little wasting of reductant due to the combustion with oxygen, or for utilization of hydrocarbon in the selective catalytic reduction of NO in excess oxygen. This will be discussed later.

Comparison of the activities of various systems

The conversions to N_2 on the physical mixture of Pt-MFI and metal ion-exchanged MFI zeolites were almost the same as that on Pt-MFI alone, as evidently shown in Figure 3. The conversions to N_2 on the Pt|metal ion systems, therefore, are compared with those on the metal ion-exchanged MFI alone in Table 1. The reaction temperatures listed were those at which the maximum conversions to N_2 were observed. Table 1 shows the great increment in the conversions to N_2 by employing the IAR method on all of these catalysts. The order of the catalytic activity of metal ion-exchanged MFI combined with Pt-MFI is Zn > In > Mg > Ba > Ca > Na, though the reason is not clear yet.

The efficiency of hydrocarbons, that is, the E_{HC} value calculated by the following equation is also summarized in Table 1.

$$E_{HC} = \frac{\textit{the amount of NO reduced to } N_2 \ (\textit{in mol})}{\textit{the amount of hydrocarbons consumed} \ (\textit{in mol})}$$

Table 1

Conversion of NO to N_2 and efficiency of ethene for reduction of NO (E_{HC}).

Metal	Metal ion-MFI alone			Pt|Metal ion IAR system		
	T / K[a]	Conv. to N_2 / %	E_{HC}	T / K[a]	Conv. to N_2 / %	E_{HC}
Na	673	3.6	0.23	473	16.6	0.87
Mg	773	7.3	1.15	573	37.9	1.20
Ca	773	14.4	0.32	573	32.4	1.09
Ba	673	8.3	1.48	523	36.6	1.16
Zn	773	11.2	1.05	573[b]	54.1[b]	1.81[b]
In	773	23.4	0.27	573	40.5	1.46

[A] The temperature at which each maximum conversion of NO to N_2 was observed.

[B] The conversion level of NO to N_2 and the E_{HC} value at 673 were 36.0 % and 1.94.

The E_{HC} values were increased except the Ba system. The maximum E_{HC} value is 1.9 or approximately 2 on the Pt|Zn combination, which supports the selective progress of reactions 1 and 2 on the IAR system. Namely, ethene can selectively react with NO_2 on Zn-MFI and is not oxidized with O_2. On the other hand, the maximum efficiency of ethene on Cu-MFI alone under the usual flow system, where the mixture of $NO+O_2+C_2H_4$ was introduced on the catalyst, was about 1.4. The E_{HC} value of 1.9 on the Pt|Zn was much greater than that on Cu-MFI alone. It follows that the waste of hydrocarbons can be prevented by using the IAR method. In case of methane as a reductant the E_{HC} value of ca. 2 has already been reported by Kikuchi *et al.* [11] which is almost the same as the present value.

The results conclude the great effectiveness of the IAR method to reduce NO with hydrocarbons in the presence of excess oxygen. Although the effect of water addition to the system should be clarified in the future, it is clear that we don't need to develop or search the catalyst which does oxidize NO to NO_2 but does not oxidize the reductant, if we employ this method. There lie many possibilities for the proper combination of the oxidation and reduction catalysts, depending on the conditions of exhausts and the kinds of reductants. At last it should be noted that a few papers have reported the decomposition of NO_2 to NO and oxygen [12] or the reaction of NO_2 with hydrocarbons to give NO and oxidized compounds [3], which would result in decrease in not only the efficiency of the reductants but also the effectiveness of the first catalyst bed. Such reaction should be minimized by selecting the suitable second catalyst.

Acknowledgments

This work was supported by a Grant-in-aid for Scientific Research from the Ministry of Education, Science, Culture, and Sports of Japan.

REFERENCES

1. M. Iwamoto and H. Yahiro, *Catal. Today*, **22**, 5 (1994).
2. H. Hamada, Y. Kintaichi, M. Sasaki, T. Ito, and M. Tabata, *Appl. Catal.*, **70**, L15 (1991).
3. C. Yokohama and M. Misono, *J. Catal.*, **150**, 9 (1994).
4. Y. Li and J. N. Armor, *J. Catal.*, **150**, 376 (1994); M. Shelef, C. M. Montreuil, and H. W. Jen, *Catal. Lett.*, **26**, 277 (1994); D. B. Lukyanov. G. Still, J. L. d'Itri, and W. K. Hall, *J. Catal.*, **153**, 265 (1995); T. Beutel, B. J. Adelman, G. -D. Lei, and W. M. H. Sachtler, *Catal. Lett.*, **32**, 83 (1995); K. A. Bethke, C. Li. M. C. Kung, and H. H. Kung, *Catal. Lett.*, **31**, 287 (1995).
5. M. Ogura, S. Hiromoto, and E. Kikuchi, *Chem. Lett.*, 1135 (1995); E. Kikuchi, M. Ogura, N. Aratani, Y. Sugiura, S. Hiromoto, and K. Yogo, *Catal. Lett.*, **27**, 35 (1996).
6. C. Yokohama and M. Misono, *Catal. Lett.*, **29**, 1 (1994).

7. M. Iwamoto, A. M. Hernandez, and T. Zengyo, *J. Chem. Soc., Chem. Commun.*, 37 (1997).
8. M. Iwamoto and N. Mizuno, *J. Automobile Eng.*, **207**, 23 (1993).
9. A. Yu. Stakheev, C. W. Lee, S. J. Park, and P. J. Chong, *Catal. Lett.*, **38,** 271 (1996).
10. H. K. Shin, H. Hirabayashi, H. Yahiro, M. Watanabe, and M. Iwamoto, *Catal. Today*, **26**, 13 (1995).
11. E. Kikuchi and K. Yogo, *Catal. Today*, **22**, 73(1994).
12. M. Shimokawabe, A. Ohi, and N. Takezawa, *Appl. Catal. A*, **85**, 129(1992).

Res. Chem. Intermed., Vol. 24, No. 2, pp. 123-132 (1998)

MECHANISM OF REDUCTION OF NITROGEN OXIDES WITH PROPENE IN EXCESS OXYGEN CATALYZED BY BIFUNCTIONAL CATALYSTS

MAKOTO MISONO, HIDEAKI NIIRO, and YOSHIFUMI HIRAO

Department of Applied Chemistry, Graduate School of Engineering, the University of Tokyo, Bunkyo-ku, Tokyo 113, Japan

Received 15 October 1997; accepted 22 November 1997

Abstract--HC-SCR (selective reduction of nitrogen oxide by hydrocarbon) and our view on its reaction mechanism are briefly described, and then our attempts in the last few years to design catalysts based on the concept of bifunctional catalysis for HC-SCR are reviewed. The example chosen is the NO-C_3H_6-O_2-H_2O reaction catalyzed by mechanical mixtures of transition metal oxides (e.g., Mn_2O_3) and metal-loaded ZSM-5 zeolites. The synergistic effect of two components was explained based on a reaction mechanism comprising NO oxidation, oxidative decomposition of the products of NO_2-C_3H_6 reactions and polymerization of (or carbon deposition from) C_3H_6.

INTRODUCTION

Selective Catalytic Reduction of Nitrogen Oxides with Hydrocarbons (HC-SCR)

There are two important catalytic technologies for the abatement of nitrogen oxide emissions. One is the selective reduction using ammonia as a reductant (NH_3-SCR) applied to the emission of oil- and coal-fired power plants. The other is the three way catalyst (TWC) installed in gasoline-engined automobiles. Although the two technologies were widely adopted, the concentration of NO_2 in the atmosphere does not decrease recently [1], because of the continued increase in the number of gasoline-engined cars and the emissions from diesel engines. For the latter, the above two catalytic technologies are not applicable. This is the reason why a new catalytic technology is needed for the abatement of dilute NO.

In 1986, Cu-ZSM-5 was reported to be active for the direct decomposition of NO [2]. It was subsequently disclosed in the patent literature [3] that Cu-zeolites were effective when applied to real lean-burn engines by HC-SCR (HC; hydrocarbons). In 1990-1992, a number of catalysts were found to be active for this reaction [4-6]. HC-SCR was thought very promising at first, but it soon turned out that the zeolite catalysts did not exhibit satisfactory activity and durability in the presence of water vapor.

It is now realized that HC-SCR is a very complex reaction. There are three major mechanisms proposed. (1) NO reacts directly with partially oxidized HC. (2) NO is first oxidized to NO_2 which then reacts with HC [7]. (3) The direct decomposition of NO which forms N_2 and oxidized catalyst, and the latter is reduced again by HC [8,9].

The dominant mechanism probably varies depending on the catalyst and reaction conditions. We confirmed the second NO_2 mechanism in the cases of NO_2-C_3H_6-O_2 over Ce-ZSM-5 [10,11] and NO-CH_4-O_2 over Pd/H-ZSM-5 [12,13]. We proposed that HC-SCR catalyzed by Ce-ZSM-5 proceeds in the following three-step mechanism [10,11].

$$NO \xrightarrow{O_2} NO_2 \xrightarrow{C_3H_6} [X] \xrightarrow{NO_2/O_2} N_2(+COx,\ H_2O) \qquad \textbf{(1)}$$

where X('s) is (are) organic nitro and/or nitrite compound(s). The NO oxidation to NO_2 is the initial and necessary step; Ce and/or protons catalyze this step and Ce the last steps of Eq. (1). The initial reaction between C_3H_6 and NO_2 (the second step) took place even on the surfaces of silica [15]. Oxidative decomposition of X's ends up with the formation of stable N_2 via several successive and parallel steps by the reaction with NO_2 and O_2 [16].

Bifunctional Catalysis for HC-SCR

If HC-SCR consists of several reaction steps and if is difficult to find a single catalyst component with satisfactory activity and durability, an alternative way would be to develop a composite catalyst system that is composed of multi-components each of which catalyzes only a part of the reaction but has sufficient durability. This is the idea of multifunctional catalysis. After we were convinced that the $NO_2C_3H_6$-O_2 reaction over Ce-ZSM-5 proceeds according to the reaction scheme given in Eq. (1), we attempted to apply the concept of bifunctional catalysis to HC-SCR, and demonstrated that the concept is useful and promising. The explicit and intentional applications of the concept to develop a high-performance catalyst or elucidate the reaction mechanism were attempted by our group [10,17], Kikuchi's group [18], Hamada's group [19], Resasco *et al.* [20], and Haruta and coworkers [21]. Combination of two or more catalysts having different activity patterns to improve catalytic performance was also reported [19], although this is different from the bifunctional catalysis described in this article.

MECHANICAL MIXTURES OF Mn_2O_3 WITH ZEOLITES FOR NO-C_3H_6-O_2 REACTION

Mn_2O_3 + Ce-ZSM-5

We attempted to combine Ce-ZSM-5 with a catalyst component that is more active for the first step of Eq. (1). Catalytic activities of Mn_2O_3, CeO_2, CuO and Cr_2O_3 for the oxidation of NO to NO_2 were more active than Ce-ZSM-5 [17]. Then these oxides were mechanically mixed with Ce-ZSM-5 in 1:1 ratio in weight and applied to the NO-C_3H_6-O_2 reaction. Some of the results are shown in Figure 1 [17].

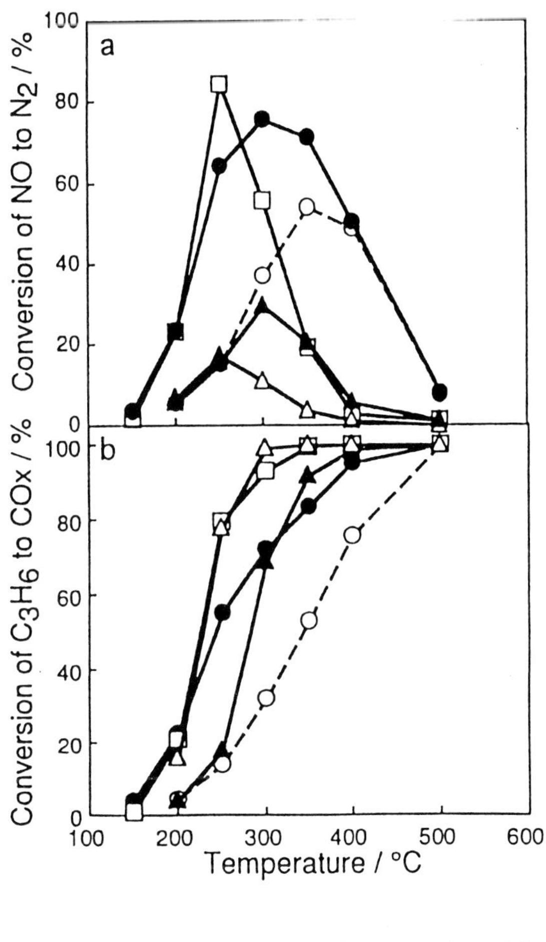

Figure 1. Temperature dependence of the conversion of NO to N_2 (a) and the conversion of C_3H_6 to COx (b) over Ce-ZSM-5 mixed with various metal oxides for the reduction of NO by C_3H_6. □, Mn_2O_3 + Ce-ZSM-5; ●, CeO_2 + Ce-ZSM-5; ▲, CuO + Ce-ZSM-5; △, Cr_2O_3 + Ce-ZSM-5; and ○, Ce-ZSM-5 alone. NO, 1000 ppm; C_3H_6, 500 ppm; O_2 2%; total flow rate, 150 cm^3 min^{-1}. Catalyst weight: Ce-ZSM-5 alone, 0.5 g; others, metal oxides (0.25 g) + Ce-ZSM-5 (0.25 g).

Ce-ZSM-5 combined with Mn_2O_3 and CeO_2 exhibited much higher activities than Ce-ZSM-5 alone, particularly in the low temperature region. On the other hand, the addition of CuO or Cr_2O_3 lowered the activity. It was noted that the addition of Mn_2O_3 or CeO_2 improved significantly the selectivity at high conversion levels of C_3H_6. The order of selectivity at 300°C agreed in general with the order of N_2 yield. There was an optimum mixing ratio of Mn_2O_3 and Ce-ZSM-5 for the reduction of NO. Mn_2O_3 itself showed much higher activity for the oxidations of NO and C_3H_6, but was not active for N_2 formation. Therefore, Ce-ZSM-5 was indispensable for the reduction of NO.

Comparison of the rates of C_3H_6-O_2, NO-C_3H_6-O_2, and NO_2-C_3H_6-O_2 provided useful information [17]. Mn_2O_3 was more active for oxidations of NO and C_3H_6 than Ce-ZSM-5. But, in the presence of NO_2, C_3H_6 was oxidized much more rapidly over Ce-ZSM-5 than over Mn_2O_3. What is most remarkable for Mn_2O_3 + Ce-ZSM-5 was the presence of (i) promotional effect of NO_2 on C_3H_6 oxidation and (ii) the interaction between NOx reduction and C_3H_6 oxidation, and this is principally due to the presence of Ce-ZSM-5.

All of these results support the bifunctional catalysis where Mn_2O_3 oxidizes NO to NO_2 and the following steps are catalyzed by Ce-ZSM-5.

The reason why the addition of Cr_2O_3 exhibited a negative influence on NO-C_3H_6-O_2 needs explanation. One factor which dominates above 300°C is the high activity of Cr_2O_3 for the oxidation of C_3H_6. Due to this, C_3H_6 required for the reduction of NO_2 over Ce-ZSM-5 is exhausted on the surface of Cr_2O_3. Actually there was little interaction between the NOx reduction and C_3H_6 oxidation for Cr_2O_3 + Ce-ZSM-5.

Below 300°C, a significant amount of C_3H_6 still remains in the gas phase, so that the above mechanism can not explain the negative effect of Cr_2O_3. We recently carried out NO-C_3H_6-O_2 reaction over Cr_2O_3 and Mn_2O_3 at 250°C. A marked difference was found in the formation of NO_2 in this reaction; NO_2 was hardly detected with Cr_2O_3, while a significant amount of NO_2 was found in the case of Mn_2O_3. So the formation of NO_2 is actually slow over Cr_2O_3, if C_3H_6 is present. There may be three possibilities for the slow formation of NO_2 over Cr_2O_3 in the presence of C_3H_6. (a) Although NO_2 was formed, it rapidly reacted with C_3H_6 to form N_2. This possibility is excluded because N_2 formation was small for Cr_2O_3 + Ce-ZSM-5. (b) Although NO_2 was formed, it rapidly returned to NO reacting with C_3H_6 as in the case of Ce-ZSM-5 [10]. This possibility is also denied, since the reaction was actually fast but the reaction rate over Cr_2O_3 was comparable with or slower than over Mn_2O_3. (c) The last possibility is that NO was oxidized very slowly by some reasons in the presence of C_3H_6. The last one was the reason and polymerization of (or carbon deposition from) C_3H_6 that took place at relatively low temperatures retarded the NO oxidation over Cr_2O_3, as described below.

When the feed was changed at 250°C from (i) NO-O_2, (ii) NO-C_3H_6-O_2 and again to (iii) NO + O_2, using Cr_2O_3 and Mn_2O_3, as shown in Figure 2, both catalysts showed similar activities for NO_2 formation in the step (i), but upon the changes of the feed from (i) to (ii) as well as (ii) to (iii), Cr_2O_3 showed very slow changes of the products and significant deficits in carbon balance during the slow changes. In the step (ii), the level of NO_2 concentration was very low for Cr_2O_3. This is in marked contrast to Mn_2O_3 which showed rapid responses, good mass balance, and considerable NO_2 formation even in the step (ii).

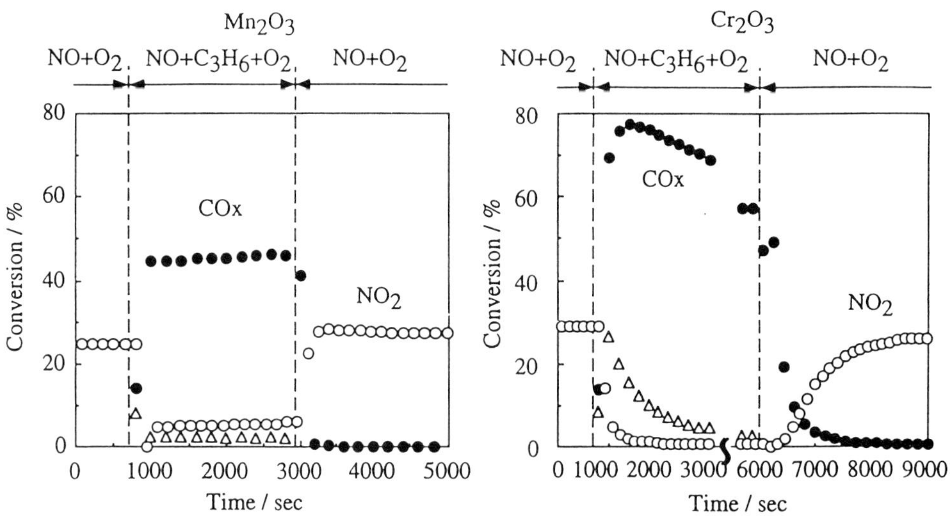

Figure 2. Effects of C_3H_6 on the oxidation of NO catalyzed by Mn_2O_3 (left) and Cr_2O_3 (right) at 250 °C. O, NO_2; ●, COx; △, carbon imbalance; NO, 1000 ppm; C_3H_6, 0 or 500 ppm; and O_2, 2%.

As described above, the concept of bifunctional catalysis much broadens the possibility of the selection for the catalysts design.

Mn_2O_3 + Sn-ZSM-5 -- A Catalyst Resistant to Water Vapor

Figure 3 compares the oxidations of NO and C_3H_6 catalyzed by Mn_2O_3 in the presence and absence of water vapor. In the case of Mn_2O_3, although the oxidation of C_3H_6 was significantly suppressed by the presence of water, the effect on NO oxidation was very small. This is in striking contrast with other catalysts. For example, the conversion of NO at 300°C decreased from 42% (under dry conditions) to 8% (in 2% water) for Cu-ZSM-5, while it is only from 40 to 38% for Mn_2O_3 under the corresponding conditions. The high resistance of Mn_2O_3 to water in the oxidation of NO is a remarkable characteristic of Mn_2O_3. The partially hydroxylated surface of Mn_2O_3 is probably quite active for the oxidation of NO.

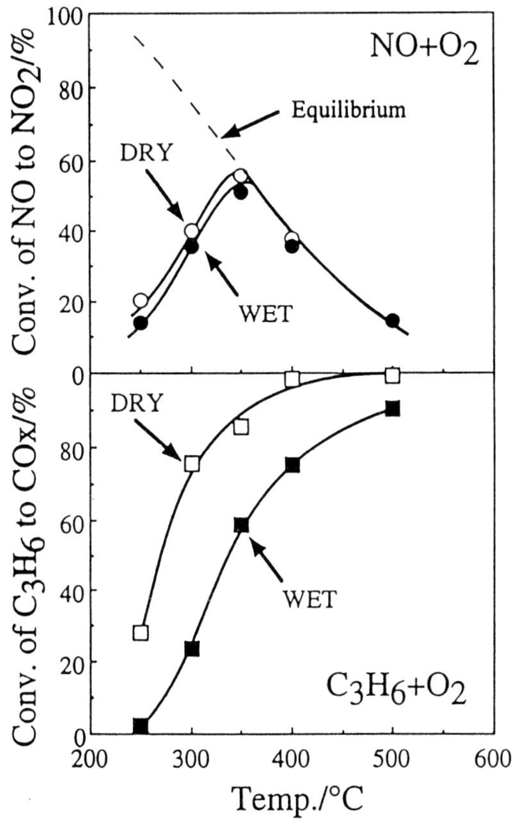

Figure 3. Oxidations of NO (top) and C_3H_6 (bottom) over Mn_2O_3 in the absence and presence of water vapor. NO, 1000 ppm (top); C_3H_6, 500 ppm (bottom); O_2, 2%; H_2O, 0 or 5% (dry or wet), (•••); % conversion at equilibrium.

Then if one could find another component which accelerates the last step (decomposition of X's) in the presence of water and combines this with Mn_2O_3, one would be able to develop a composite catalyst that catalyzes HC-SCR in the presence of water vapor. We examined various zeolite-based catalysts as a partner for Mn_2O_3. The high conversions to N_2 were observed for the combination of Mn_2O_3 with Sn- > Zn- > H- > Cu- > Co-ZSM-5 in the presence of water. In the absence of water, Ce-, In-, Pb-, Co-, and Mn-ZSM-5 mixed with Mn_2O_3 exhibited high activities. The most interesting was Sn-ZSM-5 which showed higher activity in the presence of water than in its absence. Figure 4 shows the results of NO-C_3H_6-O_2 in the presence and absence of water as a function of reaction temperature for Mn_2O_3 + Sn-ZSM-5 as well as for Sn-ZSM-5 and Mn_2O_3 alone [22].

N_2 was not formed at all over Mn_2O_3, and Sn-ZSM-5 showed only low conversions to N_2. The rate increased when Mn_2O_3 was mechanically mixed with

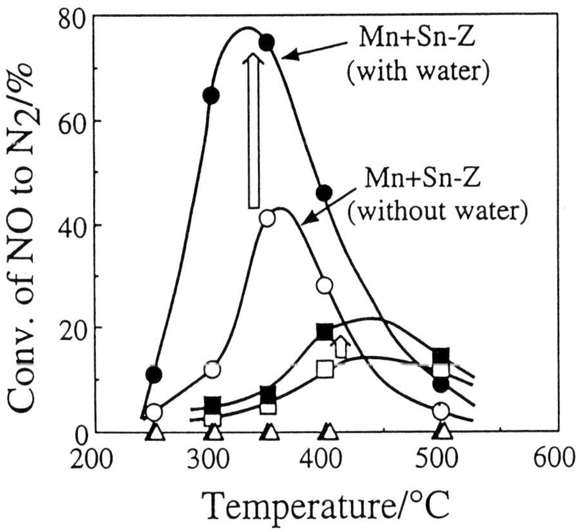

Figure 4. The influence of the presence of water vapor on the converison of NO to N_2 in the NO + C_3H_6 + O_2 reaction. ●,○; Mn_2O_3 (0.25 g) + Sn-ZSM-5 (0.25 g). ■,□ ; Sn-ZSM-5 (0.25 g). ▲,△; Mn_2O_3 (0.25 g). Solid symbols, H_2O 5.7 %; open symbols, H_2O 0%.

Sn-ZSM-5. In the presence of water vapor, the rate with Mn_2O_3 + Sn-ZSM-5 was greatly enhanced in the entire range of reaction temperature. The increase throughout the entire temperature range was unexpected and remarkable. It was noted that the conversion of C_3H_6 to COx also increased [22].

Since the effects of water vapor appeared reversibly and rapidly, the effects are not due to structural changes. As seen above in Figure 3, the rate of NO oxidation to NO_2 little changes by the presence of water, but the combustion of C_3H_6 is greatly suppressed. This is very favorable for HC-SCR, because this means that the supply of C_3H_6 to ZSM-5 increases without losing much NO_2, so that the reaction between NO_2 and C_3H_6 to form N_2 over ZSM-5 much be enhanced. As expected, with several ZSM-5 components, the N_2 yield increased in the presence of water at high temperatures where C_3H_6 was almost expired in the absence of water. The selectivity or the efficiency of utilization of C_3H_6 was enhanced by water at high conversion levels of C_3H_6 as shown in Figure 5 (250-500°C; data taken from Figure 4).

In contrast, the enhancement of N_2 formation at low temperatures (below 350°C) was found only for Sn-ZSM-5. So the low-temperature effect cannot be explained only by the function of Mn_2O_3. It was observed for Mn_2O_3 + Sn-ZSM-5 below 350°C that the activity gradually declined in the absence of water and the carbon balance of the reaction was not good in the early stage as shown in Figure 6.

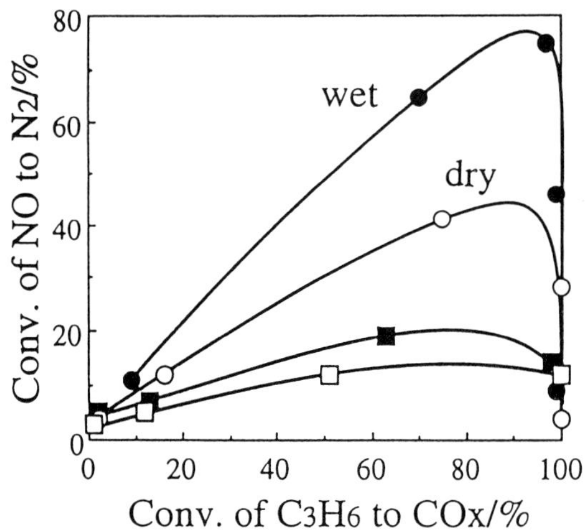

Figure 5. The selectivity or the efficiency of C_3H_6 utilization in the NO + C_3H_6 + O_2 over Mn_2O_3 + Sn-ZSM-5 and Sn-ZSM-5. Data are taken from Figure 4 (250-500 °C) and the symbols are the same as in Figure 4.

Therefore it is very likely that oligomerization or polymerization (possibly incorporating NOx in the products) leading to carbon deposits took place considerably over this catalyst and retards HC-SCR by covering the surface with less reactive species. Note that parts of the polymerized products or carbon deposits can be intermediates for N_2 formation (X's in Eq. (1)). Water probably prevents the oligomerization or polymerization to proceed too much to make the products nonreactive and cover the active sites. It is also presumable that water accelerates the decomposition of the polymerized products.

Other possible mechanisms of promotional effects of water may be the reactions such as steam reforming to form reductants like H_2 and CO, and hydration of propene to propanol. It was confirmed in the separate experiments that these two reactions did not take place to detectable extents in the absence of O_2. The other possibility may be the water gas shift reaction between CO and H_2O to produce H_2. This reaction was not checked, but as the propene did not react at least in the absence of O_2, the possibility may be small.

As described above, the low-temperature effect was observed only with Sn-ZSM-5 and therefore this must be related to unique property of Sn-ZSM-5. It is not clear yet which property is responsible for this effect, but it needs to be mentioned that the Sn-ZSM-5 used was a mixture of SnO_2 and Sn-ZSM-5; small SnO_2 particles (ca. 5 nm) covering the outersurface of ZSM-5 particles. Nominal exchange level of Sn was above 500% (ca. 36 wt%). The promotional effects at low temperatures

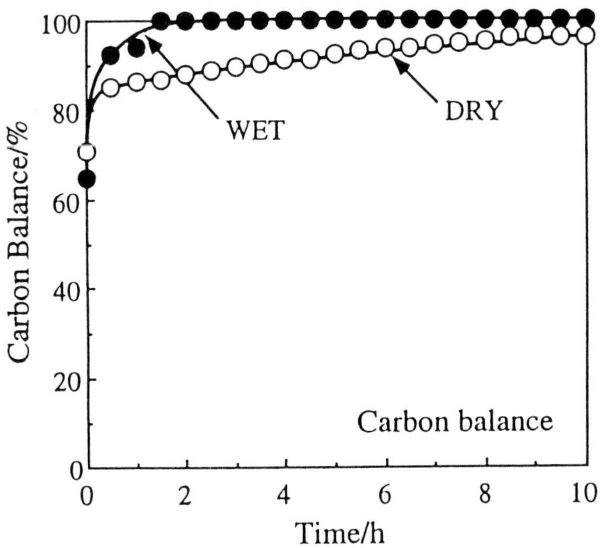

Figure 6. Conversion of NO to NO_2 and carbon balance of the NO + C_3H_6 + O_2 reaction over Mn_2O_3 + Sn-ZSM-5 at 350 °C in the absence and presence of water vapor. NO, 1000 ppm; C_3H_6, 500 ppm; O_2, 2%; H_2O, 0% (open symbols) or 5.7% (solid symbols).

were observed only when the loading level was high and a significant amount of SnO_2 was present.

Although more study is needed to elucidate the mechanism, there is certainly a hope to develop satisfactory bifunctional HC-SCR catalysts by the extension of the concept of bifunctional catalysis.

REFERENCES

1. Agency of Environment of Japan, Kankyohakusho (environment report), 1997.
2. M. Iwamoto, H. Furukawa, Y. Mine, S. Mikuriya, and S. Kagawa, *J. Chem. Soc. Chem. Comm.* 1272 (1986).
3. Ger. Offen. DE 3642018 (1987) (Volkswagen); Japan Kokai 1988-100, 919 (Toyota).
4. W. Held, A. Konig, T. Richter, and L. Ruppe, SAE paper 900496; M. Iwamoto, H. Yashiro, Y. Yu-u, S. Shundo, and N. Mizuno, *Shokubai* **32**, 430 (1990); H. Hamada, Y. Kintaichi, M. Sakai, T. Ito, and M. Tabata, *Appl. Catal.* **64**, L1 (1990); *Catal. Lett.* **6**, 239 (1990).
5. Y. Li and J. Armor, *Appl. Catal., B* **1**, 131 (1992).
6. M. Misono and K. Kondo, *Chem. Lett.* 1001 (1991).
7. H. Hamada, Y. Kintaichi, M. Sasaki, T. Ito, and M. Tabata, *Appl. Catal.* **70**, L15 (1991).
8. T. Inui et al., *Catal. Lett.* **13**, 23 87 (1992); **27**, 267 (1994); *Catal. Today* **22**, 41 (1994).
9. R. Burch et al., *J. Catal.* **139**, 435 (1993); *Appl. Catal. B* **2**, 101 (1993).

10. C. Yokoyama and M. Misono, *Shokubai* **35**, 132 (1993); *J. Catal.* **150**, 9 (1994).
11. C. Yokoyama and M. Misono, *J. Catal.* **160**, 95 (1996).
12. Y. Nishizaka and M. Misono, *Chem. Lett.* 1295 (1993); 2237 (1994).
13. M. Misono, Y. Nishizaka, M. Kawamoto, and H. Kato, 11th Intern. Zeol. Conf. Seoul, 1996.
14. H. Kato, C. Yokoyama, and M. Misono, *Catal. Lett.* **47**, 189 (1997).
15. T. Tanaka, T. Okuhara, and M. Misono, Annual Meeting of Chem. Soc. Japan, 1992; *Appl. Catal. B* **4**, L1 (1994).
16. H. Yasuda, C. Yokoyama, and M. Misono, *Shokubai* **35**, 386 (1993); *ACS Symp. Series* **587**, 110 (1995).
17. C. Yokoyama and M. Misono, *Catal. Lett.* **29**, 1 (1994).
18. M. Ogura, S. Hiromoto, and E. Kikuchi, *Chem. Lett.* 1135 (1995); M. Ogura and E. Kikuchi, *Stud. Surf. Sci. Catal.* **101**, 671 (1996).
19. H. Hamada, *Catal. Surv. Japan* **1**, 53 (1997) and references therein.
20. C.L. Loughran and D.F. Resasco, *Appl. Catal., B* **7**, 113 (1995).
21. T. Haruta, *Catal. Surv. Japan* **1**, 61 (1997).
22. Y. Hirao, C. Yokoyama, and M. Misono, *J. Chem. Soc. Chem. Comm.* 597 (1996).

Res. Chem. Intermed., Vol. 24, No. 2, pp. 133-141 (1998)

NEW TECHNOLOGY FOR SELECTIVE CATALYTIC OXIDATION OF AMMONIA TO NITROGEN

M. UESHIMA, K. SANO, M. IKEDA, K. YOSHINO and J. OKAMURA
Catalyst Research Laboratory, Nippon Shokubai Co., Ltd. 992-1 Aza Nishioki Okihama Aboshi-ku Himeji, Hyogo, 671-12, Japan

Received 12 September 1997; accepted 4 November 1997

Abstract--The selective catalytic oxidation (SCO) of ammonia to N_2 was studied by using a series of noble metal-V_2O_5-WO_3 catalysts supported on titania-silica (TS) prepared by coprecipitation method. In the V_2O_5-WO_3 catalyst system, the use of TS as a support was very effective to enhance catalytic activity compared with TiO_2 or SiO_2 alone. The addition of a slight amount of Pd and Ir to V_2O_5-WO_3/TS catalyst caused also remarkable enhancement of the catalytic activity without decreasing the selectivity to N_2. The present catalysts provide remarkably high catalytic performance for SCO of ammonia to N_2 under the practical reaction conditions for an industrial application.

INTRODUCTION

The selective catalytic reduction (SCR) of nitrogen oxides (NO_x to N_2 by ammonia on various metal oxides has been extensively investigated and high performance catalysts of V_2O_5/TiO_2 containing some metal oxides for SCR have been developed. The NO_x removal process using the catalysts are operating at power stations and chemical plants in the world. The technology development of the abatement of ammonia for waste gases in various industries is an important subject as the SCR technology for NO_x reduction. Emissions of ammonia cause environmental problems, even if ammonia is fixed in ammonium salts such as ammonium sulfate and ammonium nitrate. Many papers on the selective catalytic oxidation (SCO) reaction of ammonia have been reported in connection with the SCR process of NO_x. Most catalysts for the SCR of NO_x are also active for the SCO of ammonia. Recently gas phase SCO of ammonia to N_2 has been studied and proposed as an effective process of ammonia abatement. In an application of this NH_3 removal system, almost 100% conversion of ammonia is required in order to eliminate ammonia odor.

The catalysts for SCO of ammonia studied hitherto are Co_3O_4, MnO_2, CuO, Fe_2O_3 and V_2O_5 [1], V_2O_5 on TiO_2 [2], V_2O_5/WO_3 on TiO_2 and Fe_2O_3 on SiO_2 [3], PbO/MoO_3 on SiO_2 [4], CuO, CuO-MnO_x on TiO_2 [5] and MoO_3 on SiO_2 [6,7]. The optimum catalysts for SCO consist of an acidic metal oxide such as V_2O_5, WO_3, and MoO_3 supported on TiO_2 or SiO_2; the reaction mechanism of SCO of ammonia to N_2

over these catalysts was discussed. However, the efficiencies of these catalysts are not enough for a practical use. There is no attempt of addition of a noble metal to these metal oxide catalysts to improve the efficiency of the SCO reaction. However, application of a noble metal to the SCO catalyst seems to be interesting. In the present work, we have studied highly active and selective catalysts based on V_2O_5-WO_3 mixed oxide in the SCO reaction of ammonia to N_2 focusing on the effects of the acidic property of a support, the addition of a noble metal, and the reaction conditions from a view point of an industrial application. It has been found that titania-silica complex oxide (TS) which showed stronger acidity than TiO_2 or SiO_2 alone was an optimum support to V_2O_5-WO_3 to improve activity. Furthermore addition of a slight amount of noble metal to V_2O_5-WO_3/TS catalyst was remarkably effective to enhance the activity of the catalyst.

EXPERIMENTAL

Preparation of TiO_2-SiO_2 (TS) Support

Binary TiO_2-SiO_2 complex oxide as a catalyst support was prepared by coprecipitation of the mixture of sulfuric acid solution of titanyl sulfate (25 wt % as TiO_2) and silica sol (30 wt % as SiO_2) with 20 % aqueous ammonia solution under vigorous stirring until the pH reached 7. The atomic ratio of Ti O_2/SiO_2 in TS was fixed at 85/15. The obtained precipitate was washed with de-ionized water and dried at 200 °C for 10 h. The solid was then calcined at 500 °C in air for 3 h and crushed to obtain white powder.

Catalyst Preparation

V_2O_5-WO_3-TS catalyst was prepared by adding a 10% monoethanolamine aqueous solution of ammonium metavanadate and ammonium paratungustate to the above TS powder. The obtained mixture was kneaded and extrusion-molded into 3.3mm pitch honeycomb shape, followed by drying at 125 °C for 6 h and calcined at 450 °C for 6 h. A noble metal-V_2O_5-WO_3-TS catalyst, for example Pd-V_2O_5-WO_3-TS catalyst, was prepared by impregnation method. The above V_2O_5-WO_3-TS was impregnated with an aqueous solution of palladium nitrate for one minute, dried at 125 °C and calcined in air at 500 °C for 2h. Other TS supported noble metal catalysts were also prepared similarly by using corresponding noble metal salt solution.

Catalytic Activity Measurements

Catalytic activity measurements were carried out using a conventional continuous flow fixed bed tubular reactor. Synthetic flue gas was prepared by using a mixture of ammonia in nitrogen, air and nitrogen. Water was vaporized and mixed with the synthetic flue gas. Flow rates were controlled by mass flow controllers. Before the measurements, ammonia aging was conducted at 380 °C under a gas stream of 0.5 vol% NH_3, 30 vol% H_2O and air balance at 5000 h^{-1} of SV for 15 h. The measurement was then carried out after 2 h under appropriate experimental conditions. Inlet and outlet concentrations of ammonia were analyzed by absorbing in 1 mol / l aqueous H_3BO_3 solution, followed by NH_4^+ concentration determination using ion chromatography. Outlet concentration of NO_x was monitored by vacuum chemiluminescence method, while outlet concentration of N_2O was analyzed by thermal conductivity detector. Only NO_x and N_2O were detected as by-products except nitrogen in the measurements. Therefore, the selectivity to N_2 was calculated by subtracting the selectivities to NO_x and N_2O from 100%.

Acidic Property Measurement

The acid amount and acid strength of a support were measured by the titration with *n*-butylamine in benzene using the Hammett indicators.

RESULTS AND DISCUSSION

Selection of TS

A TiO_2 or SiO_2 is generally used as a suitable catalyst support for the SCR and SCO catalysts. A TiO_2-SiO_2 complex binary oxide (TS) prepared by coprecipitation method from titanium tetrachloride and silica sol with ammonia is also reported to be an effective catalyst support for the SCR reaction [8].

For the examination of support, the SCO reaction of ammonia was carried out over V_2O_3-WO_3 supported on TS, SiO_2 and TiO_2 (*anatase and rutile*) calcined at 500 °C. The results are shown in Table 1. Pellet type catalysts instead of the honeycomb shape were used in this case because of the poor moldability of catalysts supported on SiO_2 or TiO_2.

All catalysts showed extremely high selectivity to ammonia but low conversion of NH_3. The catalysts supported on SiO_2 and TS showed large surface areas, but the conversion of the catalyst supported on SiO_2 was very low. The order of catalytic activity in kinds of supports is TS > TiO_2 (anatase) > TiO_2 (rutile) > SiO_2.

The acidic properties (acid amount and acid strength) of these supports are shown in Figure 1. A TS showed a higher acid strength and larger acid amounts in a range of $-8.2 \leq H0 \leq +4.8$ than SiO_2 and both anatase and rutile TiO_2. The acidic properties of SiO_2 and rutile TiO_2 were very weak. The order of catalytic activity corresponds to the order of the acidic properties.

Table 1
Catalytic activity and selectivity over V_2O_5-WO_3 supported on SiO_2, TiO_2 (rutile and anatase) and TiO_2-SiO_2 at 380 °C. Gas composition (vol %): NH_3 = 1.0%, H_2O = 30%, O_2 = 14.6%/balanced N_2, SV:5000 hr^{-1}, Catalyst: 82.6 ml (pellet type).

Catalyst	Atomic ratio of V/W/Ti/Si	BET (m^2/g) Catalyst/ support	NH_3 Conv. (mol %)	Selectivity (mol %)		
				N_2	NO_x	N_2O
V_2O_5-WO_3/SiO_2	5.2/5.1/100	130/187	2.0	99.8	0.24	0
V_2O_5-WO_3/TiO_2 (rutile)	7.0/2.0/100	6.0/6.5	13.0	>99.9	0.04	0
V_2O_5-WO_3/TiO_2 (anatase)	7.0/2.0/100	17.4/19.2	18.0	99.4	0.06	0.52
V_2O_5-WO_3/TiO_2-SiO_2	7.2/2.2/100/17.6	114/154	27.0	99.5	0.01	0.46

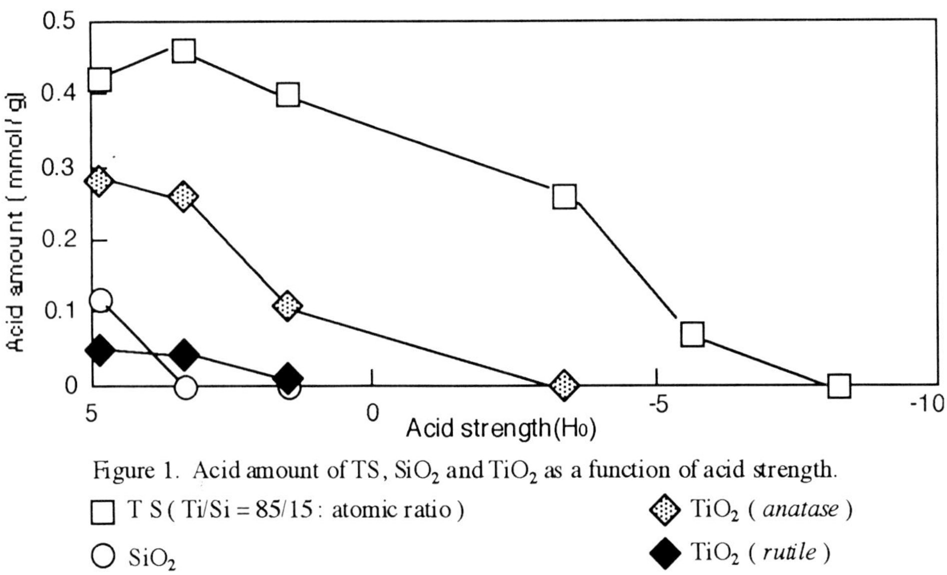

Figure 1. Acid amount of TS, SiO_2 and TiO_2 as a function of acid strength. □ T S (Ti/Si = 85/15: atomic ratio), ◈ TiO_2 (anatase), ○ SiO_2, ♦TiO_2 (rutile).

These results may suggest that catalytic activity relates to the acidic properties of support. The mechanism of the generation of acid sites by mixing of two oxides has been proposed earlier [9,10]. The generation of high acid strength of TS being considered to be due to the high electronegativity of both $Ti^{4+}$and Si^{4+}, and the excess of positive or negative charge by complexing TiO_2 and SiO_2.

Addition of Noble Metals

It has been found that TS was an effective support on enhancement of the catalytic activity. But the activity is not high enough from a viewpoint of an industrial application. It is known that noble metals are active for combustion of ammonia. However, the addition of too great an amount of a noble metal element into the catalyst results in an increase of the selectivity to NO_x and N_2O formation. Therefore the addition of a small amount of a noble metal to V_2O_5-WO_3 supported on TS catalyst was examined. The distribution of the products of ammonia oxidation over various noble metal-V_2O_5-WO_3 supported on TS is shown in Table 2. The addition of a small amount of Rh, Pd and Ir was very effective to improve the conversion of ammonia without decreasing the selectivity to N_2. Especially Pd and Ir catalysts showed a remarkable high activity and selectivity under the high reaction temperature of 380 °C.

Table 2
Catalytic activity and selectivity over noble metal-V_2O_5-WO_3/TiO_2-SiO_2 at 380 °C.

Catalyst	(metal wt%/cat.)-atomic ratio of V/W/Ti/Si	NH_3 Conv. (mol %)	Selectivity (mol %)		
			N_2	NO_x	N_2O
V_2O_5-WO_3/TiO_2-SiO_2	7.9/2.2/100/17.6	10.0	95.1	0.5	4.4
Rh-V_2O_5-WO_3/TiO_2-SiO_2	(0.68)-7.9/2.2/100/17.6	61.0	98.6	0.1	1.3
Pd-V_2O_5-WO_3/TiO_2-SiO_2	(0.68)-7.9/2.2/100/17.6	>99.9	99.0	0.3	0.7
		35.0[a]	93.3	0.02	6.7
Ir-V_2O_5-WO_3/TiO_2-SiO_2	(0.68)-7.9/2.2/100/17.6	>99.9	98.8	0.5	0.7
		76.0[a]	99.1	0.04	0.8
Pt-V_2O_5-WO_3/TiO_2-SiO_2	(0.23)-7.9/2.2/100/17.6	>99.9	80.7	1.5	17.8
		99.0[a]	68.5	0.5	31.0

[a]Results of reaction at 320 °C. Gas composition of synthetic flue gas (vol%): NH_3 = 1.0%, H_2O = 30%, O_2 = 14.6%/balanced N_2, SV: 5000 hr^{-1}, Catalyst: 82.6 ml (13.7 x 13.7 x 440 mm, honeycomb shape).

On these catalysts, the formation of NO_x (mainly NO) and N_2O was almost negligible quantity at 380 °C. On the contrary, the addition of Pt brought significant decrease of the selectivity to N_2 even at 320 °C in spite of reducing amount of Pt. This means that Pt promotes directly oxidation of ammonia to form N_2O (equation 2) in addition to the formation of N_2 (equation 1).

$$4NH_3 + 3O_2 \rightarrow 2N_2 + 6H_2O \tag{1}$$

$$2NH_3 + 2O_2 \rightarrow N_2O + 3H_2O \quad \textbf{(2)}$$

Reaction Conditions

In case of an industrial use of the SCO system, the reaction conditions are sometimes variable, because the strict control of the waste gas composition and flow rate at the reactor inlet of the SCO system is difficult . Therefore a proper SCO catalyst has to show a stable catalytic performance even under variable reaction conditions.

The SCO reaction over Pd-$V_2O_5WO_3$/TS catalyst (see Table 2) was carried out under number of different reaction conditions. Space velocity, NH_3 concentration, temperature and concentration of O_2 and H_2O were varied. The honeycomb shaped catalyst (82.6 ml; 13.7 x 13.7 x 440 mm) was used in these experiments from a viewpoint of a practical application. The results are shown in Tables 3 to 7.

These results indicate that the catalyst shows a high efficiency for wide variations of reaction conditions except for change of reaction temperature. Temperatures less than 340 °C caused drastic decreases in NH_3 conversion and an increase of N_2O as shown in Table 5. The result at 320 °C almost corresponds to that at the same temperature in Table 2 where the NH_3 concentration is 1% and the SV is 5000h^{-1} respectively. The activity of this catalyst is very sensitive to changes of the reaction temperature. The selectivity to N_2 increased with increasing temperature. It seems very likely that at lower temperatures the side reaction of N_2O formation (equation 2) proceeds in parallel with the main reaction to form N_2 (equation 1). Upon increasing temperature the reaction of equation 1 is preferentially promoted to form N_2. Therefore N_2O is most likely an intermediate in the SCO reaction of ammonia.

The optimum space velocity is a range of 5000 to 10000 h^{-1}. Lower space

Table 3
Effect of space velocity at 340 °C. Gas composition: NH_3 = 0.5%, H_2O = 30%, O_2 = 14.6%/balanced N_2.

SV (hr^{-1})	NH_3 Conv. (mol %)	Selectivity (mol %)		
		N_2	NO_x	N_2O
2500	>99.9	95.8	0.02	4.14
5000	99.6	98.2	0.05	1.80
10000	97.8	98.1	0.06	1.85
15000	76.0	98.1	0.08	1.81

Table 4
Effect of NH_3 concentration at 340 °C. Gas composition: H_2O = 30%, O_2 = 14.6%/balanced N_2, SV = 10000 hr^{-1}.

NH_3 Conc. (vol %)	NH_3 Conv. (mol %)	Selectivity (mol %)		
		N_2	NO_x	N_2O
0.25	97.2	98.0	0.09	1.90
0.50	97.8	98.1	0.06	1.85
0.75	98.1	98.1	0.06	1.88
1.00	99.2	98.3	0.08	1.65

Table 5
Effect of reaction temperature. Gas composition: NH_3 = 0.5%, H_2O = 30%, O_2 = 14.6%/balanced N_2, SV = 10000 hr^{-1}.

Temp (°C)	NH_3 Conv. (mol %)	Selectivity (mol %)		
		N_2	NO_x	N_2O
300	24.0	88.3	0.01	11.7
320	34.0	93.3	0.02	6.49
340	97.8	98.1	0.06	1.85
400	99.8	98.6	0.37	1.07

Table 6
Effect of O_2 concentration at 340 °C. Gas composition: NH_3 = 0.5%, H_2O = 30%/balanced N_2, SV = 10000 hr^{-1}.

O_2 Conc. (%)	NH_3 Conv. (mol %)	Selectivity (mol %)		
		N_2	NO_x	N_2O
0.5	95.8	98.8	0.42	0.80
2.5	97.6	98.9	0.24	0.90
5.0	97.8	98.7	0.16	1.12
14.6	97.8	98.1	0.06	1.85

velocity causes increase of N_2O, while high space velocity results in low conversion. Upon varying NH_3 and O_2 concentration (Tables 4, 6) the catalyst provides a stable high activity and selectivity to N_2. Especially in the presence of excess oxygen of 14.6% (Table 6) both conversion and selectivity to N_2 are held at the same high level as in the case of 0.5% oxygen concentration which is close to the theoretical amount to form N_2. This behavior seems to be a special feature of this catalyst, because Pd

Table 7
Effect of H_2O concentration at 340 °C. Gas composition: NH_3 = 0.5%, O_2 = 14.6%/balanced N_2, SV = 10000 hr^{-1}.

H_2O Conc. (%)	NH_3 Conv. (mol %)	Selectivity (mol %)		
		N_2	NO_x	N_2O
0	98.8	96.5	0.09	3.36
5	98.2	97.9	0.08	2.02
10	97.4	98.2	0.08	1.75
30	97.8	98.1	0.06	1.85

which generally promotes the combustion of NH_3 to NO_x does not form NO_x and N_2O upon varying the oxygen concentration. The impact of varying H_2O concentration is shown in Table 7. The presence of water promotes enhancement of the selectivity to N_2 without decreasing the catalytic activity. From these results, it is concluded that Pd-V_2O_5-WO_3 supported on TiO_2-SiO_2 catalyst is very effective for the SCO reaction of NH_3 to N_2 under the variable reaction conditions assumed for an industrial application.

CONCLUSIONS

A titania-silica complex oxide (TS) prepared by coprecipitation method showed a stronger acidity than TiO_2 or SiO_2. TS was the optimum support to enhance the catalytic activity of the SCO catalyst based on V_2O_5-WO_3 binary oxides. The addition of a small amount of noble metal such as Pd or Ir to V_2O_5-WO_3/TS catalyst was extremely effective for enhancement of the catalytic activity without decreasing the selectivity to N_2. Especially Pd-V_2O_5-WO_3/TS catalyst showed a high catalytic performance under a wide range of reaction conditions assumed for an industrial use.

REFERENCES

1. N.I. Il'echenko and G.I. Golodets, *J. Catal.* **39**, 57 (1975).
2. F. Cavani and F. Trifiro, *Catal.Today* **4**, 253 (1989).
3. F.J.J.G. Janssen and F.M.G. van den Kerkhof, *KEMA Sci. & Techn. Report* **3**, 71 (1985).
4. J.J.P. Biermann and F.J.J.G. Janssen, *J. Mol. Catal.* **60**, 229 (1990).
5. A. Wöllner, F. Lange, H. Schmelz, and H. Knözinger, *Appl. Catal. A* , **94**, 181 (1993).
6. M. de. Boer, H.M. Huisman, R.J.M. Mos, R.G. Leliveld, A.J. van Dillen, and J.W. Geus, *Catal. Today* **17**,189 (1993).
7. J.J.P. Biermann and F.J.J.G. Janssen, *Appl. Catal. A* **86**, 165 (1992).

8. M. Kobayashi, *Analysis and Design of Functional Surface*, edited by The Society of Chemical Engineers, Japan, 1991, p. 174.
9. M. Ito, H. Hattori, and K. Tanabe, *J. Catal.* **35**, 225 (1974).
10. K. Tanabe, T. Sumiyoshi, K. Shibata, T. Kiyoura, and J. Kitagawa, *Bull. Chem. Soc. Jap.* **47**, 1064 (1974).

Res. Chem. Intermed., Vol. 24, No. 2, pp. 143-149 (1998)

DESIGN OF UNIQUE TITANIUM OXIDE PHOTOCATALYSTS BY AN ADVANCED METAL ION-IMPLANTATION METHOD AND PHOTOCATALYTIC REACTIONS UNDER VISIBLE LIGHT IRRADIATION

MASAKAZU ANPO*, YUICHI ICHIHASHI, MASATO TAKEUCHI, and HIROMI YAMASHITA

Department of Applied Chemistry, Osaka Prefecture University, Sakai, Osaka 599, Japan

Received 29 September 1997; accepted 22 December 1997

Abstract--The metal ion-implantation of titanium oxide with various transition metal ions was carried out by applying by high voltage acceleration. The subsequent calcination of the implanted TiO_2 in oxygen at around 723 K resulted in a large shift in the absorption spectra of the TiO_2 toward visible light regions, its extent being dependent on the amount and the kind of metal ions implanted. Such metal ion-implanted titanium oxide catalysts were active in carrying out various photocatalytic reactions such as the decomposition of NO into N_2, O_2 and N_2O at 275 K under irradiation with visible light longer than 450 nm. The application of this advanced, high energy metal ion-implantation method enables the novel design of titanium oxide photocatalysts which can absorb and initiate vital reactions under visible light and will contribute to the development of catalytic systems utilizing solar energy.

INTRODUCTION

Environmental pollution and destruction on a global scale have drawn attention to the vital need for totally new environmentally-friendly, clean chemical technologies and processes. This can be considered one of the most important challenges facing chemical scientists for the 21st century. Strong contenders as environmentally-harmonious catalysts are photocatalysts which are able to operate at room temperature and in a clean manner. The application of such safe photocatalytic systems are urgently desired for the purification of polluted water, the decomposition of offensive atmospheric odors and toxins, the fixation of CO_2 and the decomposition of NOx as well as chlorofluorocarbons, on a huge and global scale [1-3].

Recently, investigations to address such concerns using various kinds of titanium oxide photocatalysts have been actively carried out and to this end we have been moving in a very positive direction with various practical applications already at hand. However, conventional titanium oxides such as TiO_2 powdered semiconducting catalysts cannot utilize visible light and make use of only the 3-4% of the solar beams that reach the earth, necessitating the use of an ultraviolet light source such as a mercury lamp. Investigations into the chemical doping of metal ions into the TiO_2 catalyst to explore the possibility of visible light absorption have also been intensively carried out. However, in these systems, it was found that the metal ions doped were present only as impurities enhancing the recombination of the photo-formed electrons and holes, and did not enhance reactions with any noticeable effect neither under visible light nor UV light.

The present study deals with detailed investigations into designing chemically stable, non-toxic and efficient titanium oxide photocatalysts which are able to absorb and thus operate under irradiation with solar or visible light [4-7].

EXPERIMENTAL

The TiO_2 powder was supplied by the Catalysis Society of Japan as one of the standard reference catalysts (JRC-TIO-4: surface area ~ 50 m^2/g, bandgap ~ 3.5 eV, purity > 99.5 %). The transition metal ion-implantation was carried out by using an ion-implanter consisting of a metal ion source, mass analyzer, high voltage ion acceleration (150 keV), and a high vacuum pump. The catalysts were calcined at 725 K for 5 h, degassed at 575 K for 2 h, heated in O_2 at the same temperature for 2 h and finally outgassed at 475 K to 10^{-6} Torr before use in the photoreactions. Irradiation of the catalysts in the presence of NO (12.8 μmol) was carried out using a high pressure Hg lamp (Toshiba SHL-100UV) ($\lambda > 450$ nm as visible light) and $\lambda \sim 380$ nm as UV light) at 275 K. The reaction products were analyzed by gas chromatography. The UV-VIS spectra (diffuse reflectance) of the catalysts were measured using a Shimadzu UV-2200A spectrophoto-meter at 295 K. ESR spectra were recorded at 77 K with a Bruker ESP300E spectrometer (X-band) using a 100 kHz field modulation and a 10-20 Gauss standard modulation. The 3D images of the metal ion-implanted catalysts were obtained by detailed SIMS analysis.

RESULTS AND DISCUSSION

As shown in Figure 1-A, the absorption band of the original pure TiO_2 photocatalysts does not overlap with the solar spectrum, indicating that these catalysts do not react under visible light. However, as can be seen in Figure 1-A, the absorption band of the metal ion-implanted TiO_2 photocatalyst is found to shift smoothly to visible light regions, the extent of the red shift being dependent on the amount and type of metal ions implanted. The absorption spectra of the Cr ion-implanted catalysts without calcination showed no difference with that of the unimplanted original catalysts. Therefore, calcination in conjunction with ion-implantation was found to be essential in the shift of the spectrum to visible light regions. The calcination of the ion-implanted TiO_2 catalysts at around 725 - 875 K was found to show the longest and most remarkable shift to visible light regions. As can be seen in Figure 1-A, the absorption band of the TiO_2 photocatalysts implanted with Cr ions shifts smoothly to visible light regions, the extent depending on the amount of Cr ions implanted, while the absorption maximum and minimum

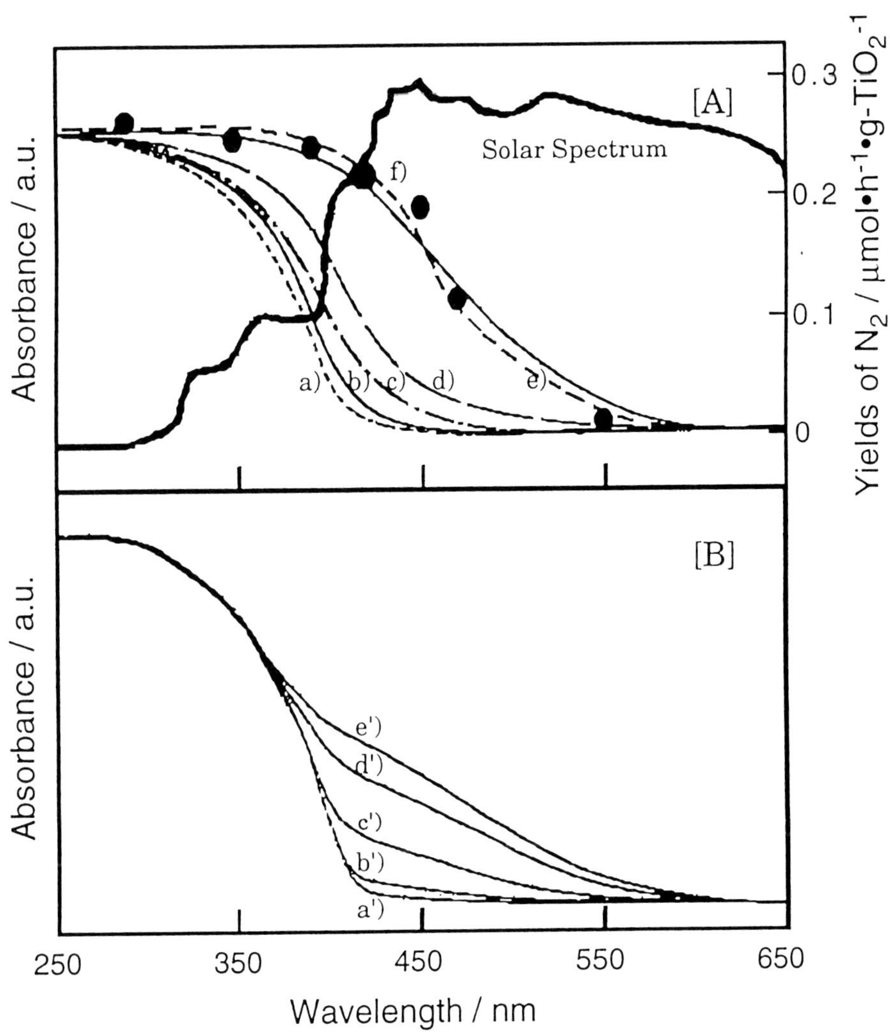

Figure 1. [A] UV-VIS spectra (diffuse reflectance) of the un-implanted original TiO_2 (a), the TiO_2 implanted with Cr ions (b-e) with different amounts, the action spectrum of the photocatalytic decomposition of NO (circle points) (f), and solar beam spectrum on the ground. [B] UV-VIS spectra (diffuse reflectance) of the original TiO_2 (a') and the TiO_2 chemically impregnated with Cr ions (b'-e') with different amounts. (Amounts of Cr ion impregnated: (in 10^{-7} mole/g-TiO_2) b: 2.2, c:6.6, d: 13.0, e: 26.0, and amounts of Cr ion impregnated: (in 10^{-6} mole/g-TiO_2) b′: 1.9, c′: 19.0, d′: 97, e′: 190)

values remained constant, allowing the Cr ion-implanted TiO_2 to use solar beams more effectively and efficiently up to 20-30% [4-5].

On the other hand, Figure 1-B shows the absorption bands of the TiO_2 photocatalysts impregnated with Cr ions by a conventional chemical process. The Cr ion-impregnated TiO_2 shows no shift in the absorption band, but a new absorption band at around 420 nm due to the appearance of the Cr ions (and/or oxide forms) as impurities can be seen, its intensity increasing with the amount of Cr ions chemically doped.

As shown in Figure 2, visible light irradiation ($\lambda > 450$ nm) of the Cr ion-implanted TiO_2 in the presence of NO at 275 K leads to the direct decomposition of NO into N_2, O_2, and N_2O with a good linearity against the irradiation time. As can also be seen in Figure 2, under the same conditions of visible light irradiation, these reactions did not proceed on the original non Cr ion-implanted TiO_2 nor on the TiO_2 chemically impregnated with Cr ions. Moreover, it has been well established that when metal ions were chemically doped into the catalysts, the reaction efficiency

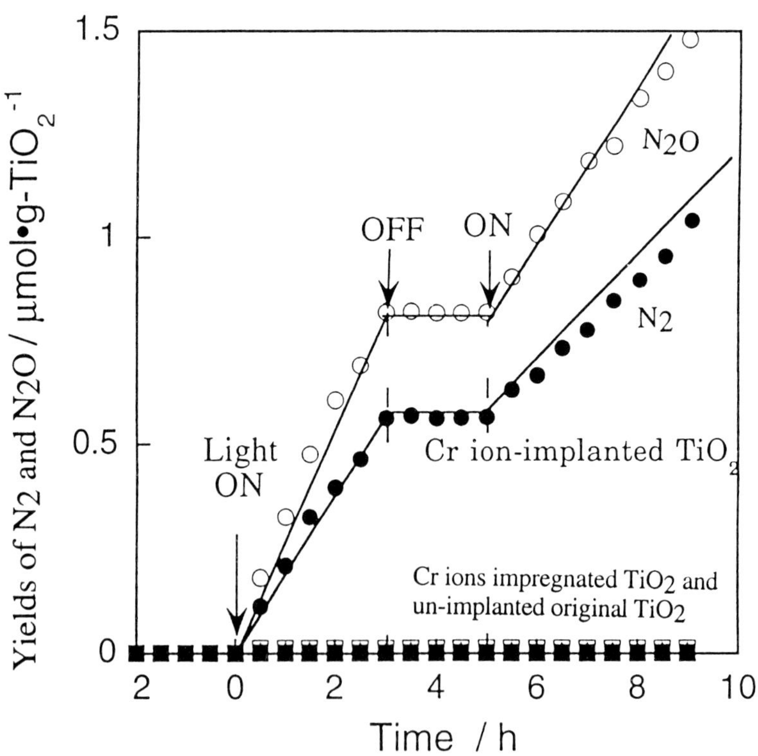

Figure 2. Reaction time profiles of the photocatalytic decomposition of NO into N_2, O_2 and N_2O on the Cr ion-implanted TiO_2 under visible light irradiation ($\lambda > 450$ nm). The reaction does not proceed on the TiO_2 nor on the TiO_2 chemically doped with Cr ions.

decreased drastically under UV light irradiation (or bandgap irradiation) because metal ions doped were present as impurities enhancing the recombination of the photo-generated electrons and holes.

However, as can be seen in Figure 3, the Cr ion-implanted TiO_2 catalyst shows exactly the same photocatalytic efficiency as the unimplanted TiO_2 catalyst under bandgap UV irradiation. Furthermore, as can also be seen in Figure 1-A (circle points), the action spectrum for the photocatalytic reaction on the Cr ion-implanted TiO_2 catalyst is in good agreement with the absorption spectrum of the catalyst, indicating that only Cr ion-implanted photocatalysts were effective for the photocatalytic decomposition of NO. Thus, it is confirmed that the metal ion-implanted TiO_2 catalysts enabled the absorption of visible light up to a wavelength of 400-600 nm and were able to operate effectively as photocatalysts under visible light irradiation. From this point of view, these metal ion-implanted TiO_2 can be named "second-generation titanium dioxide photocatalysts" [4-5].

We have carried out various other significant reactions not only with Cr ions but also with other various transition metal ions such as V ion-implanted TiO_2 under

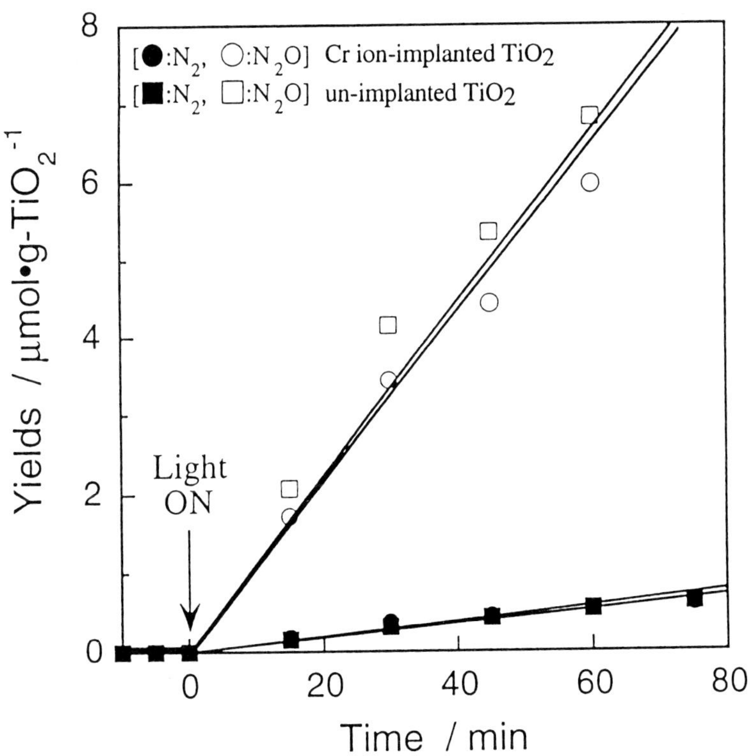

Figure 3. Reaction time profiles of the photocatalytic decomposition of NO into N_2, O_2, and N_2O under UV light irradiation on the unimplanted original TiO_2 and on the TiO_2 implanted with Cr ions. x 10^{-4}).

visible light irradiation (λ > 450 nm) at 275 K, such as the photocatalytic isomerization of cis-2-butene to produce trans-2-butene and 1-butene as well as the photocatalytic hydrogenolysis of unsaturated hydrocarbons such as $CH_3C{\equiv}CH$ with H_2O [4-6]. Furthermore, our investigations with various other semiconducting catalysts such as ZnO as well as non-semiconducting catalysts such as Si-Ti binary oxide catalysts implanted with Cr or V ions by high energy implantation also exhibit large shifts in their absorption bands toward visible light regions, the extent depending on the amount of metal ions implanted. These results clearly indicate that the modification of the TiO_2 photocatalyst by metal ion-implantation is closely associated with the electronic interaction which arises between the host oxide and guest metal ions implanted and not by the formation of the impurity energy levels within the bandgap of the TiO_2 catalysts which can often be observed in the chemical doping of the metal ions into semiconducting materials [4,7].

The photocatalyst developed may take any form, be it powder, granules, pellets, or thin film, and may be used as prepared or mixed with a binder to form a mixture that can be applied or supported on any substance or materials, demonstrating the many applications that are possible. In fact, under outdoor solar light irradiation at ordinary temperatures, the metal ion-implanted TiO_2 catalyst exhibited photocatalytic reactivity several times higher than the original non ion-implanted TiO_2 catalyst, showing the potential of utilizing solar energy with greater effectiveness and efficiency by applying second-generation titanium dioxide photocatalysts [4,6].

In summary, our studies clearly show that, after calcination in oxygen at around 723 K, TiO_2 catalysts incorporated with transition metal ions by a high energy metal ion-implantation method, specifically Cr or V ion-implanted TiO_2, allow the efficient absorption of visible light, thus enabling the photocatalysts to operate effectively under visible or solar light irradiation. The present research has opened the way to many innovative possibilities, significantly to address urgent environmental concerns, and can also be considered an important breakthrough in the utilization of solar energy

Acknowledgments

This work has been supported by the Petroleum Energy Center (PEC) under the sponsorship of the New Energy and Industrial Technology Development Organization (NEDO), the Ministry of International Trade and Industry (MITI) of Japan and the Research Institute of Innovative Technology for the Earth (RITE).

REFERENCES

1. M. Anpo, H. Yamashita, and S.G. Zhang, *Current Opinion. Solid State Mater. Sci.*, **1**, 630 (1996).
2. M. Anpo, H. Yamashita, and Y. Ichihashi, Proc. *Res.Opportunity Photochem. Sci.*, (Colorado), 219 (1996).
3. M. Anpo, *Catal. Surveys Jpn.*, **1**, 169 (1997).
4. M. Anpo, Y. Ichihashi, H. Yamashita, Y. Tamada, P. Heide, R.R. Martin and T. Yoshinari, *Optronics*, **186**, 161 (1998).
5. M. Anpo, Y.Ichihashi, Y. Tamada, H. Yamashita, T. Yoshinari, and Y. Suzuki. In: *Photoelectrochemistry*, K. Rajeshwar, and A. Fujishima., (Eds.), The Electrochemical Society, INC, 1997, p. 331.
6. M. Anpo, H. Yamashita, Y. Ichihashi, and M. Takeuchi, unpublished data.
7. B. Morin, A. Davidson, M. Che, Y. Ichihashi, H. Yamashita, and M. Anpo, *J. Phys. Chem.*, (1998) (to be submitted).

Res. Chem. Intermed., Vol. 24, No. 2, pp. 151-168 (1998)

STRUCTURES AND DYNAMIC BEHAVIOR OF CATALYST MODEL SURFACES CHARACTERIZED BY MODERN PHYSICAL TECHNIQUES

Y. YAMAGUCHI[1], W.-J. CHUN[1], S. SUZUKI[1], H. ONISHI[1], K. ASAKURA[2], and Y. IWASAWA[1]*

[1]*Department of Chemistry, Graduate School of Science, the University of Tokyo, Hongo, Bunkyo-ku, Tokyo 113, Japan*

[2]*Research Center of Spectrochemistry, Faculty of Science, the University of Tokyo, Hongo, Bunkyo-ku, Tokyo 113, Japan*

Received 19 September, 1997; accepted 7 November, 1997.

Abstract-- This paper reports the visualization of mobile pyridine on the terraces and site-specific adsorption of pyridine on the particular step sites on a TiO_2(110)-(1x1) surface by scanning tunneling microscopy (STM), the anisotropic structure and reactivity of molybdenum oxides dispersed on TiO_2(110)-(1x1) characterized by polarization-dependent total-reflection fluorescence x-ray absorption fine structure (PTRF-EXAFS), and the energy dispersive and real-time images of Au mesh on Si(111) recorded by a new x-ray photoemission electron miscroscopy (XPEEM).

INTRODUCTION

Metal oxides and supported metals find application in a variety of technologies where surface science is critical to success, including catalysis, gas sensing, photo electrolysis, ceramic electronics, semiconductor device processing, coatings, lubrication, etc [1]. Understanding and controlling oxide surfaces and metal-oxide interfaces are the key issues for the development of industrial catalysts, but those surfaces are in general heterogeneous and complicated, and hence have been little studied so as to put them on a scientific basis by traditional approaches. Recent progress in in-situ characterization of catalysts and catalytic phenomena by modern physical techniques may contribute to a complete understanding of the origin of heterogeneous catalysis and therefore new strategies for the development of new types of promising catalysts [1,2]. In-situ observation of atoms and adsorbed molecules on surfaces is a crucial issue in molecular-scale understanding of many surface technological applications [1-3]. Practically the inherent compositional and structural inhomogeneity of catalyst surfaces makes the problem of identifying the essential issues for their catalytic performance extremely difficult. In consequence, the development and application of in-situ surface analysis techniques for non-ordered surfaces with multi-sites are expected.

Scanning tunneling microscopy (STM) has particularly great potential for in situ studies [4-11]. While our present knowledge of the atomic structure of catalyst surfaces is largely limited to those structures which are stable in ultra-high vacuum before and after reaction, STM may provide an insight into both adsorbate and catalyst surface structure in situ during the reaction. Pyridine is a typical Lewis base molecule and often employed as a probe molecule to examine the acidic property of oxide catalyst surfaces, which is relevant to performance of acid-oxide catalysis and surface chemical processes. There is, however, little information on atomic or molecular structures and behaviors of surface acid sites and adsorbed probe molecules.

In-situ polarization-dependent total-reflection fluorescence (PTRF) x-ray absorption fine structure (XAFS) is a suitable tool to identify the quasi-three dimensional structure around X-ray absorbing atoms like dispersed metal atoms [12, 13]. The technique also allows the determination of their local structures under the presence of a gas phase [14]. Catalysts are often used as metals or metal oxides dispersed on robust oxide supports. Anisotropic/asymmetric bonding and structural features at metal-support interfaces are the key issues relevant to the catalytic properties of supported catalysts such as activity, selectivity, deactivation, etc. We have studied three dimensional structures of metals and metal oxides on an α-Al_2O_3(0001) single crystal plane; eg., CoO_x/α-Al_2O_3(0001) and $[Pt]_x/\alpha$-Al_2O_3(0001) [15,16].

Recently, X-ray photoemission electron microscopy (XPEEM) has been demonstrated to be available for surface imaging [17,18]. Independently, XPEEM with a conventional X-ray tube has been newly developed by us [19], which can provide chemical mapping of elements at surfaces in situ under surface processes such as diffusion, phase transformation, adsorption, etc.

In this study the following examples of in-situ surface characterization by means of STM, PTRF-EXAFS and XPEEM are reported; (1) site-specific adsorption and dynamic behavior of pyridine on a TiO_2(110) single crystal surface, (2) anisotropic structure of Mo oxides dispersed on a TiO_2(110) surface before and after the reaction with ethanol, and (3) development of a new XPEEM system and its application to imaging of Au mesh on Si(111).

METHODOLOGY AND EXPERIMENTAL SECTION

Scanning tunneling microscopy (STM)

The experiments were performed in an UHV microscope (JSTM-4500VT, JEOL) equipped with an Ar^+ gun and LEED-AES optics. A polished $TiO_2(110)$ wafer of 6.5x1x0.25 mm^3 (Earth Chemicals) was annealed in air at 1100 K for 1 h. Nickel film was then deposited on the back side of the wafer for resistive heating on the microscope stage. The temperature of the surface was monitored with an IR radiation thermometer. After cycles of Ar^+ sputtering (3 keV, 0.3 μA) and vacuum annealing at 900 K, a sharp (1x1) LEED pattern was yielded on the surface. The (1x1) surface was cooled at room temperature and then exposed to pyridine (research grade, Wako Pure Chemicals) vapor. Constant current topography of the pyridine-exposed surface was continuously determined with an electrochemically etched W tip, and recorded in video. A positive sample bias voltage (Vs) of +2.5 V and a tunneling current (It) of 0.05 nA were employed.

Polarization-dependent Total-reflection Fuorescence Extended X-ray Absorption Fine Structure (PTRF- EXAFS)

Extended x-ray absorption fine structure (EXAFS) appearing at ca.50-1000 eV above the x-ray absorption edge is explained by the single scattering theory and provides information about the interatomic distance and coordination number around the absorbing atom [14,20]. EXAFS oscillation $\chi(k)$ also depends on the angle θ_i between the polarization direction of the incident x-ray and the i-th bond direction as shown in eq.(1);

$$\chi(k) = \sum_i 3\cos^2\theta_i \cdot \chi_i(k) \tag{1}$$

where k and $\chi_i(k)$ are the wavenumber of photoelectron and the EXAFS oscillation accompanying the i-th bonding. The local structure of x-ray absorbing atoms dispersed on a flat oxide substrate can be determined in three different directions, i.e. three dimensionally, by changing the θ_i by rotating the substrate plane against the polarization vector of x-rays [13]. We have developed an in-situ total-reflection fluorescence EXAFS cell which can be applied to catalytically interesting systems as described hereinafter.

A polished rutile $TiO_2(110)$ single crystal was annealed at 823 K for 2 h in air and impregnated with an ultra-pure aqueous solution of $(NH_4)_6Mo_7O_{24}\cdot 4H_2O$, followed by calcination at 773 K for 3 h. The PTRF-EXAFS spectra were measured in the three directions; the polarization vectors of the incident x-rays were parallel to [110], [001] and [1$\bar{1}$0] directions of the $TiO_2(110)$ surface. EXAFS measurements were carried out at BL 14A wiggler beam line in the Photon Factory (KEK-PF) with a positron energy of 2.5 GeV and a beam current of 300 mA. A Si(311) double-crystal monochromator was used to monochromatize x-rays stemming from synchrotron radiation.

EXAFS analysis was performed by subtracting background using a cubic smoothing method and normalized by the edge height. Backscattering amplitude, phase shift function and mean free path of electron were calculated using FEFF ver.6.0 [21].

X-ray Photoemission Electron Microscopy (XPEEM)

Recently photoemission electron microscopy (PEEM) has been developed and demonstrated to be available for surface imaging during catalytic reaction conditions, e.g. CO oxidation on Pt(100) [22]. However, this method is not element specific, which may be disadvantageous when it is applied to catalytically interesting systems. If the excitation source of PEEM which is usually a deuterium lamp or mercury lamp is replaced by an X-ray source such as Al K_α and Mg K_α, and an electron energy analyzer is installed in the optics of the microscope, a new technique, XPEEM, can be developed to obtain chemical mapping of surface elements in a mesoscopic scale in situ under reaction conditions. We succeeded in designing and developing a new PEEM system as described in Results and Discussion.

RESULTS AND DISCUSSION

STM Study on Site-specific Adsorption and Dynamic Behavior of Pyridine on $TiO_2(110)$

Rutile TiO_2 is a typical transition metal oxide used for catalysts, gas sensors, electrodes, cosmetics, etc. The (110) surface of rutile TiO_2 is regarded as being a stable truncation [23]. An atomically-resolved topography of a TiO_2(110)-(1x1) surface was observed by STM with a small sample bias voltage of +1.0 V (tunneling current: 0.3 nA), as shown in Fig.1(a) [24-26]. There were spots ordered in a rectangular periodicity of 0.65x0.30 nm^2. This was the size of the (1x1) unit cell, 0.649x0.296 nm^2 in Fig.1 (b). The exposed 5-fold coordinated Ti^{4+} ion is a responsible source of the bright contrasts. Fig.1(b) illustrates a model proposed for the stoichiometric TiO_2(110)-(1x1) truncation [27]. The alternative alignment of the exposed, 5-fold coordinated Ti-row, and the bridging O-ridge results in the characteristic anisotropy on this surface. Surface x-ray diffraction [28], medium-energy backscattered electron diffraction (MEED) [29], and first-principle ab-initio calculations [23, 30] support the model as essence. The surface x-ray diffraction also demonstrated the structural relaxation of TiO_2(110)-(1x1) with the magnitudes from 0 to 0.027 nm, leading to rumpling of the titanium planes [28].

In-situ observation of adsorbed organic molecules is indispensable to control many technological applications precisely, e.g. catalysis, gas sensing, corrosion,

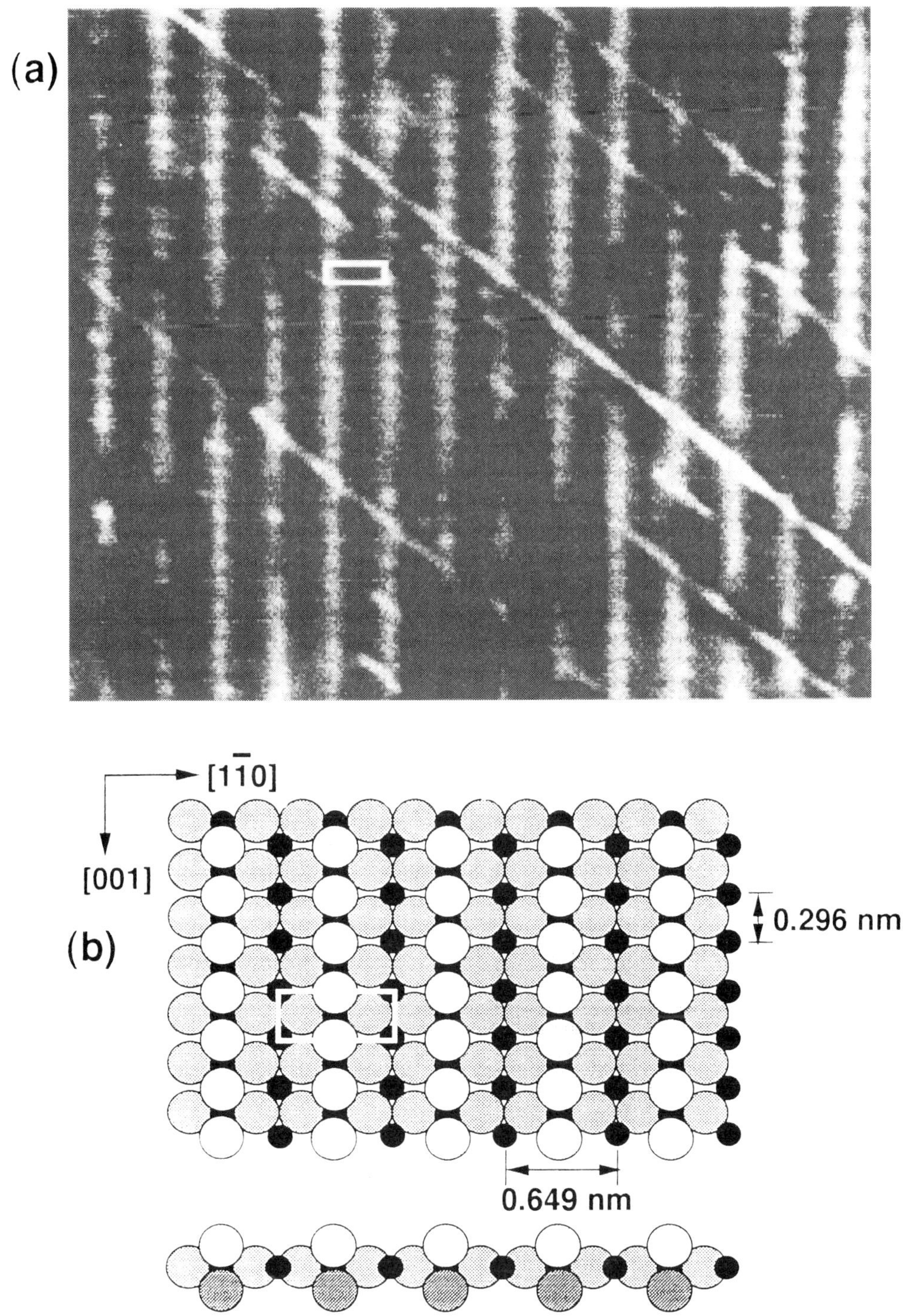

Figure 1. (a) STM image (10x10 nm^2) of TiO_2(110)-(1x1), (b) a stoichiometric truncation of TiO_2(110)-(1x1). The (1x1) unit is also shown in (a) and (b).

lubrication, device processing, etc. Development of STM opens a novel way to observe the dynamic behavior of individual molecules at solid surfaces [1-3,10,31].

The feature and strength of the interaction between probe pyridine molecules and surface acidic sites reflect the character and reactivity of the acidic sites of oxide surfaces. The relationship between atomic-scale structure of acidic sites and their reactivity is, however, not straightforward. For example, the highly ionized Ti atoms exposed to the (110)-(1x1) surface of rutile TiO_2 are assumed as Lewis acid sites favorable for pyridine chemisorption in the acid-base points of view. However, our previous study [32] revealed that pyridine was only weakly adsorbed on the surface without definite N-Ti bonding. There is a clear need to give light into the atomic-scale structural feature of surface acidic sites and the behavior of individual adsorbed probe molecules.

Adsorbed pyridine molecules on $TiO_2(110)$ were traced as many bright spots in constant current STM images [32]. Fragmented shapes of the molecular image suggest vital migration of adsorbed pyridine molecules. When an adsorbed molecule migrates quickly, its whole shape cannot be imaged with scanning microscopy. The number of pyridine images (bright spots) reduced with time when the surface was maintained at room temperature on the microscope stage. The observed depopulation is ascribed to the desorption, since the desorption of pyridine occurs at room temperature as shown in the TPD spectra. The average height of the pyridine image was 0.26 nm with cross-section analysis. The observed topographies are regarded to represent the shape of unoccupied states localized on adsorbed pyridine molecules. The lowest unoccupied molecular orbitals (LUMOs) of a free pyridine molecule are nearly degenerated π^* states of the aromatic ring, 3b1 and 2a2 at 3.0 eV above the vacuum level [33]. In addition to the pyridine images, the admolecules mainly stayed on the Ti atoms, though they were mobile at this temperature.

Figure 2 shows a typical STM image of pyridine molecules adsorbed on a $TiO_2(110)$-(1x1) surface with terraces and steps. The annealed (1x1) surface was exposed to pyridine vapor of 0.01 L at room temperature in the treatment chamber and then transferred to the microscope stage. An upper (1x1) terrace, a lower (1x1) terrace, and steps of single atom height were imaged in Fig.2. White particles located on several sites along the steps were assigned to individual pyridine molecules. These step-bonded molecules were imaged completely, which indicates that they are tightly bound to the adsorption sites along the steps, whereas many images of fragmented shapes due to migration were observed over the upper terrace.

The rows of the five-fold coordinated Ti^{4+} atoms were resolved as white lines of slight contrast in the topography of the upper terrace in Fig.2. The step-bonded pyridine molecules were always adsorbed on the Ti-rows. Four-fold coordinated Ti atoms should be exposed at a step along the $[1\bar{1}2]$ direction, where each Ti atom originally five-fold coordinated loses one oxygen ligand. Pyridine molecules

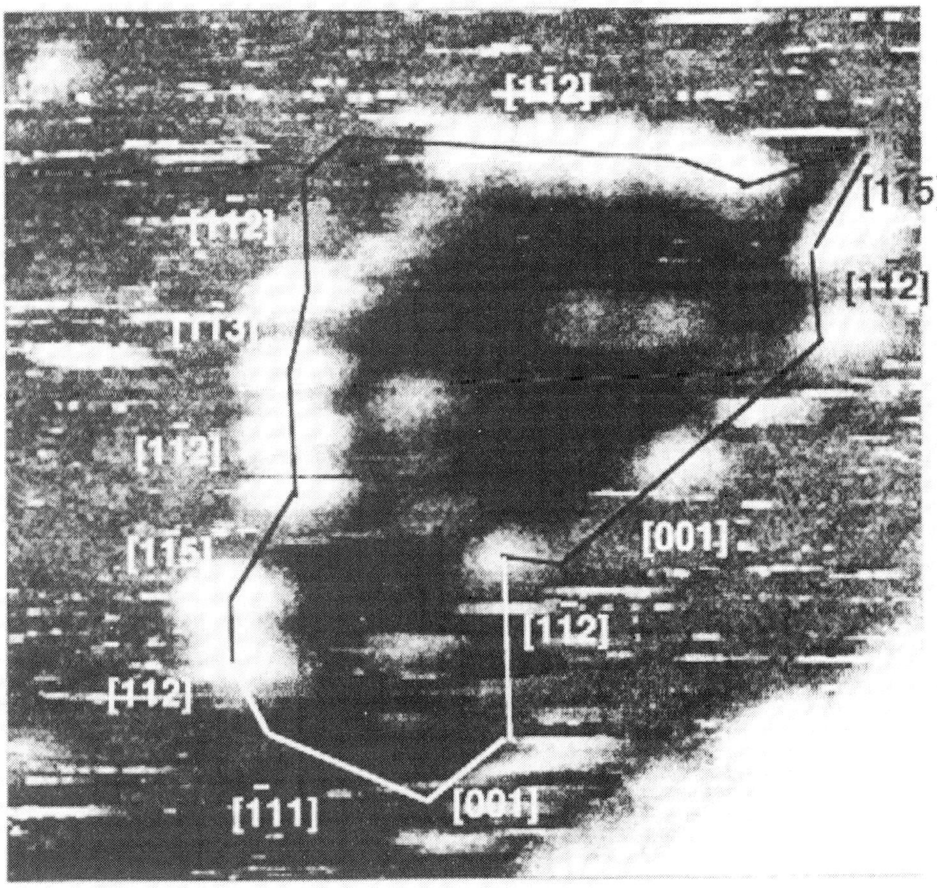

Figure 2. An STM image of the pyridine-exposed $TiO_2(110)$-(1x1) surface and a schematic drawing of the single-atom height steps; V_s= +2.5 V, I_t=0.05 nA.

adsorbed more strongly on the coordinatively unsaturated (four-fold coordinated) Ti atoms exposed on the $[1\bar{1}2]$ steps, than on the five-fold coordinated Ti atoms over the (110)-(1x1) terrace [34].

The molecular topographies of step-bonded and terrace-adsorbed pyridine resembled each other. The image height of step-bonded species was 0.25 nm. A similar image height of terrace-adsorbed molecules (0.26 nm) was determined on the surface cooled at about 100 K to freeze the migration of adsorbed pyridine molecules. The molecular dynamics calculation of a single pyridine molecule adsorbed on a perfect (110) terrace revealed that the adsorbed pyridine was most stable when the aromatic ring is parallel to the surface [32]. If the nitrogen atom of pyridine bonds directly with a Ti atom at a step, the aromatic ring should stand up. Pyridine adsorbed in this manner was found on a $ZnO(10\bar{1}0)$ surface in an X-ray absorption study [35]. The identical topography does not suggest the upright geometry of the step-bonded species.

It has been pointed out that the multiplicity of coordination vacancy is a key factor controlling the reactivity of the metal centers of metal oxides [36]. The present study reports the first evidence of coordination-controlled adsorption on a metal oxide surface. Furthermore, the affinity of the four-fold coordinated Ti atom at a step also depended on the direction of the step in azimuth as shown in Fig.2. The steps in Fig.2 on which pyridine molecules were adsorbed run parallel to the $[\bar{1}12]$ direction. Similar activity was observed on the steps parallel to $[\bar{1}13]$, $[\bar{1}14]$, and $[\bar{1}15]$ directions. In contrast, steps of $[\bar{1}11]$, $[\bar{1}10]$ and [001] directions were not active for the adsorption. The missing Ti atom may perturb the electronic states of the adsorption site through the bridging O atom in between [34]. The step-direction dependent adsorption may be relevant to the key issue of the acidic property of coordinatively unsaturated Ti sites.

Sequential imaging of the pyridine-exposed surface revealed that terrace-adsorbed molecules were transformed to step-bonded species and vice versa. Figure 3 shows sequential three frames recorded on the surface of Fig. 2 at an interval of 18.3 s/frame. The number and position of step-bonded pyridine molecules fluctuated through Fig. 3 (a) to (c). The fluctuation suggests that the exchange between step-bonded and terrace-adsorbed species occur in the time scale of image acquisition (order of 10 s). No pyridine vapor was in the microscope chamber during the imaging. The observed exchange allows us to build a picture that the terrace-adsorbed state is an intermediate to produce the step-bonded species [34].

PTRF-XAFS study on structure and reactivity of Mo oxides dispersed on $TiO_2(110)$

If the incidence angle of the X-rays is below the critical angle(θ_c)(eq.(2), total reflection occurs.

$$\theta_c = N(Z/A)\rho e^2\lambda^2/\pi m_c c_o^2 \qquad \textbf{(2)}$$

N, Z, A, ρ, e, λ, m_c, and c_0 represent Avogadro's number, the atomic number of the absorber atom, the atomic mass of the absorber atom, the density of the support, the elementary charge, the wavelength of the incident X-ray, the electron rest mass and the speed of light in vacuum. The critical angle of x-ray in the energy region for the PTRF-EXAFS of Mo samples is as small as 1.7 mrad. To achieve the adjustment of the sample position to the total reflection condition, a high precision four-circle goniometer was employed with a minimum step of 0.17 mrad [37].

Mo oxides are regarded as active species or precursors for important

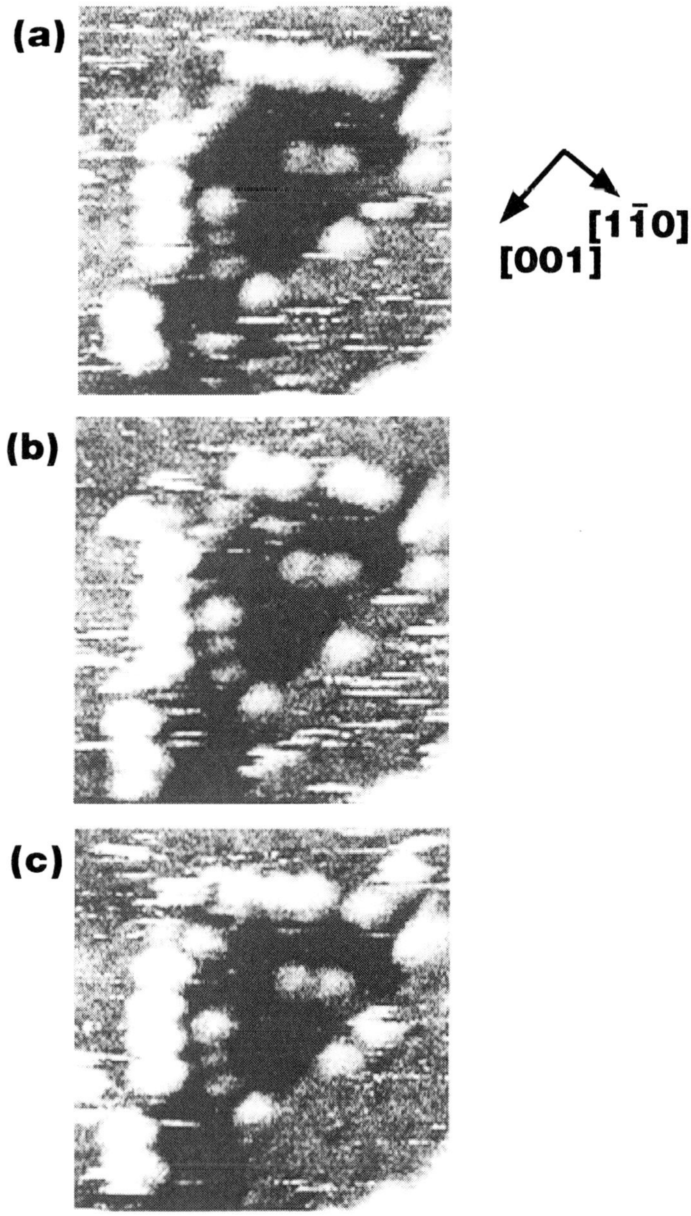

Figure 3. Serial STM images (11.5x10.0 nm^2) of pyridine at steps at an interval of 18.3 s around 12 h after exposure to 3 L pyridine; V_s= +2.5 V, I_t=0.05 nA.

catalysis such as partial oxidation, metathesis, hydrodesulfurization and so on. Ethanol oxidation is often used as a test reaction to examine the catalytic property of Mo oxides. In the present study Mo oxides (0.2 ML) dispersed on TiO_2(110)-(1x1) were characterized by using the PTRF-EXAFS cell which we designed to measure the PTRF-EXAFS spectra in three different [$\bar{1}$10], [001] and [110] directions.

We carried out the reaction of EtOH and Mo/TiO_2(110). The reaction temperature was 473 K and the pressure of ethanol was 13.3 kPa. We measured the total reflection fluorescence EXAFS using a home-made in-situ cell before and after the reaction.

EXAFS analysis was carried out by subtracting background using a cubic smoothing method and normalized by the edge height [38]. Extracted EXAFS oscillations in the three directions are shown in Fig.4 (a). Assuming an appropriate model structure and comparing the calculated oscillations with the observed ones in the three directions, we determined a possible three-dimensional structure around Mo atoms. In fact, twenty-nine atoms were included in the simulation of model structures. The residual factor ε^2 was calculated according to the following eq.(3);

$$\varepsilon^2 = \sum ((\chi_{obs}(k_i) - \chi_{cal}(k_i))/error(k_i))^2 \qquad (3)$$

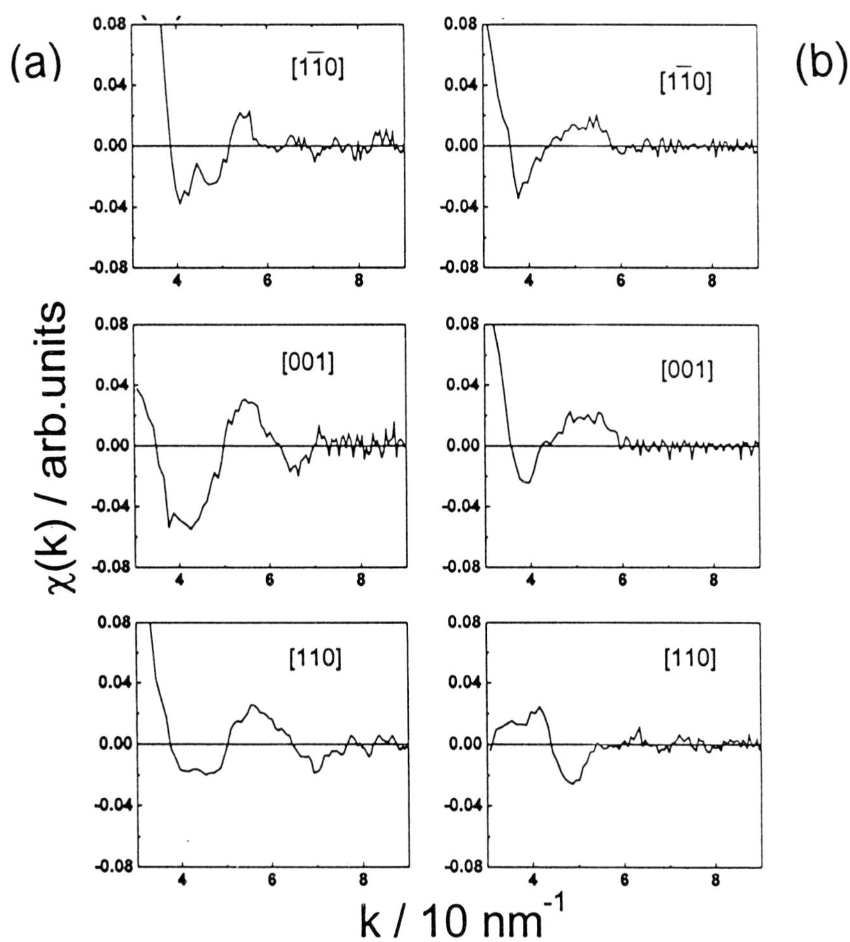

Figure 4. PTRF-EXAFS oscillations for Mo/TiO_2(110) before (a) and after (b) reaction with C_2H_5OH at 473 K.

where only static errors were taken into account [37]. The evaluation of ε^2 was carried out in the range of 40-75 nm^{-1}. Models satisfying ε^2 less than unity for all three directions were adopted as plausible ones.

The structural information on bonding with surface normal is selectively obtained from the PTRF-EXAFS of **E**//[110], and the structure along the bridging oxygen rows is obtained from the spectra of **E**//[001]. The bonding feature perpendicular to those two directions is analyzed by the data for **E**//[$\bar{1}$10]. The PTRF-EXAFS spectra along **E**//[$\bar{1}$10] showed the short periodic oscillation, indicating the presence of Mo-Mo bonding in this direction. No short-periodic oscillation was observed in the high k-region of the spectra for **E**//[001] and **E**//[110] in Fig. 4 (a). Thus it seems that the Mo species on TiO_2 (110) has an anisotropic structure.

We have calculated the EXAFS oscillations for several plausible models derived from various literature [39-42]. Before calculation, the symmetric structures such as tetrahedral and octahedral structures were excluded. The FEFF simulations were begun with (100), (010) and (001) faces of MoO_3 crystal as potential surface structure. Although many publications suggested the growth of MoO_3(100) and (010) planes on TiO_2, the theoretical simulation did not reproduce the PTRF-EXAFS data [37,43]. Polymolybdates typically like $Mo_7O_{24}^{6-}$ never reproduced the experimental spectra in the three polarization directions. Oscillations calculated for monomeric structures were entirely different from the observed ones though isolated oxo-molybdenum species were reported in the literature [44, 45]. As for Mo dimer structures, we calculated the PTRF-EXAFS oscillations for two kinds of dimers, corner-shared and edge-shared dimer models. The corner-shared Mo dimer did not fit the observed data, while edge-shared Mo dimer model was the first to give ε^2 value less than unity for the [$\bar{1}$10] and [001] directions. Eventually we have calculated several hundred dimer models to optimize the fitting and succeeded in finding the best-fit result on the Mo-oxide structure dispersed on TiO_2(110). Figure 5 shows the calculated oscillations based on the optimized Mo-dimer structure together with the experimentally observed data in the three directions The Mo-Mo distance was found to be 0.335 nm along **E**//[$\bar{1}$10] and the Mo-Ti distance was obtained at 0.296 nm. The separation between the two Mo-atoms level and the five-coordinated Ti-atoms level was 0.220 nm. This vertical separation was most sensitive to the simulation, which gave a validity to the FEFF simulation based on the dimer-model structure. The isolated anisotropic Mo-dimer structure on TiO_2(110) is illustrated in Fig. 6.

When ethanol was introduced to the Mo/TiO_2(110) system, surface reaction took place as proved by change of the Mo structure in Fig. 4 (b). Upon exposure to ethanol of 13.3 kPa at 473 K for 2 h the short period oscillation in the **E**//[$\bar{1}$10] direction disappeared. The disappearance of the oscillation can be explained by breaking of the Mo-Mo bonding. In the other directions the oscillation

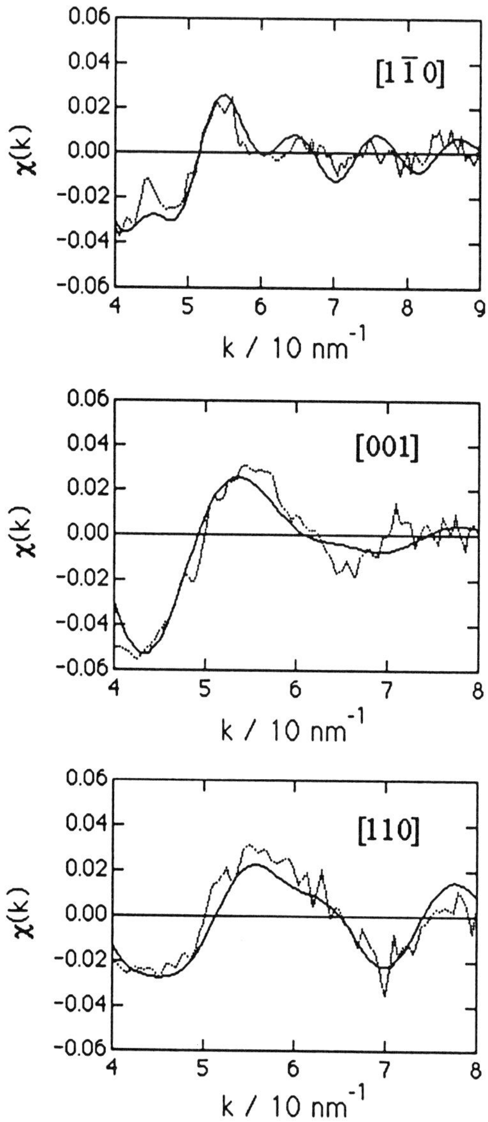

Figure 5. Comparison between the observed and calculated PTRF-EXAFS data based on the model shown in Fig. 8 ; Dotted line : observed; solid line : calculated.

corresponding to the Mo-Mo were not observed. Thus it is likely that the Mo dimer structure was destroyed to monomeric species by reaction with ethanol vapor at 473 K. An alternative explanation is due to the formation of Mo ensembles which have largely distorted and disordered structures, where the EXAFS oscillations may be canceled to diminish to a negligible level. Although the data are noisy, it can be safely suggested that the PTRF-EXAFS oscillations in three polarization directions are not the same as each other, indicating an asymmetric structure of Mo species produced by the reaction with ethanol.

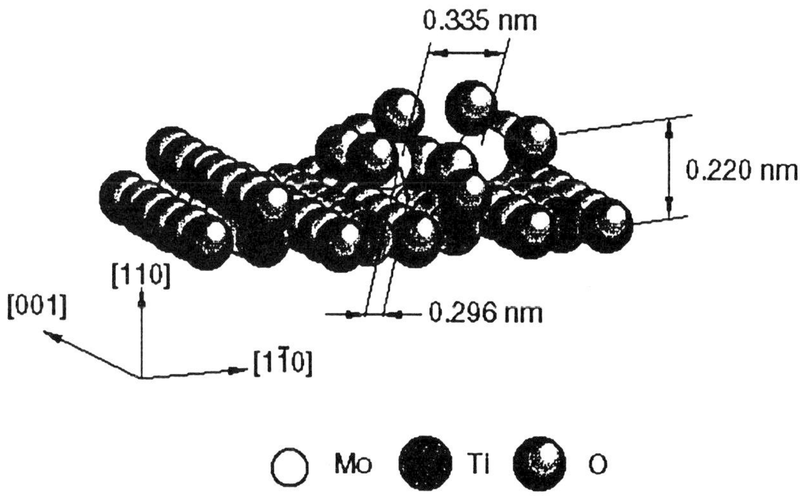

Figure 6. A proposed dimer structure for Mo oxides on $TiO_2(110)$.

We carried out one-term curve fitting analyses for the three directions, **E**//[$\bar{1}$10], **E**//[110] and **E**//[001], and found that Mo-O distance was about 0.18 nm. The value indicates double bond character. The presence of double bond oxygen was also confirmed by PTRF x-ray absorption near-edge structure (XANES) spectra at Mo K-edge. The 1s → 4d forbidden transition peak intensity has been related to the number of double bond oxygens around the Mo atom because the peak intensity increases by the rehybridization of 4d with 2p of the double bond oxygen [46,47]. The intensities of the pre-edge peaks in the three directions were similar before and after the ethanol reaction at 473 K. Under the present reaction conditions the number of double bond oxygens coordinated to the Mo atom did not significantly change by the reaction.

XPEEM study on imaging Au mesh on Si(111)

Direct observation of the distribution and migration of elements at surfaces is one of the interesting issues in the field of surface science and catalytic chemistry. For example, aggregation and redispersion of metal particles or metal oxides are related to deactivation and regeneration of catalysts, and diffusion and phase transformation of multi-elements are related to the origin and mechanism of surface catalysis. However, it is not easy to achieve the element-distribution analysis under reaction conditions. We have developed a new real-time imaging technique, XPEEM with a spatial resolution of 0.1-1.0 μm to obtain the information on mesoscopic surface chemical processes [19].

Figure 7 shows the schematic drawing of a new XPEEM (X-ray photoelectron emission microscopy) instrument. Al Kα radiation was used as an X-

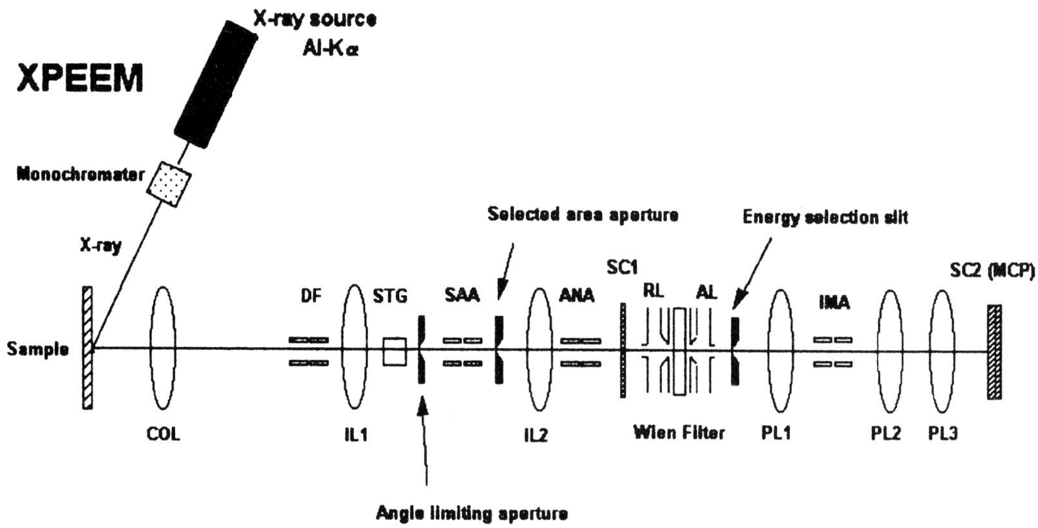

Figure 7. A schematic drawing of XPEEM; Col: Objective lens; DF, SAA, ANA, and IMA : Double stage diffrection system; STG: Quadropole stigmator, IL1 and IL2: intermediate lens; RL and AL: Retarding and accelerating lens; SC1: First imaging point; PL1 and PL2: Projective lens; SC2: Final imaging point.

ray source. As -8.5 kV ~ -10 kV is charged on the sample and objective lenses are grounded in the system, photoemission electrons with about 10 keV or less go into the electron microscopic system. The X-ray source is separated from a main chamber by an Al window and evacuated using an ion pump. The open hole of the objective lens is a few millimeters and the whole microscope is also differentially pumped. The alignment and pumping system in Fig.7 allows us to measure XPEEM images in situ under the ambient of 10^{-3} Pa pressure. The photoemission electron image is magnified and goes into an energy analyzer. We used a Wien filter as energy analyzer because the optical axis is not changed in the Wien filter. In the Wien filter the electric field and the magnetic field are provided at the same time with the perpendicular directions to each other and to the electron propagation direction. Thus the electrons feel the electric field as well as the Lorenz field as shown by eq. (4);

$$\boldsymbol{F} = e\boldsymbol{E} - e\boldsymbol{B}\bullet v \qquad (4)$$

where $\boldsymbol{E}$ and $\boldsymbol{B}$ are the electric field and the magnetic field, respectively, and v is the velocity of photoemission electrons.

When $\boldsymbol{E}$ and $\boldsymbol{B}$ are properly chosen, the electrons with a certain energy can go straight in the Wien filter and the electrons with other energies will be bent. If the energy selection slit is put after the Wien filter, one can choose the photoemission electrons with a selected kinetic energy. The electrons are then projected on MCP fluorescence screen to produce imaging of the selected electrons stemming from the specified surface elements. When the energy selection slit is removed and the lens conditions are changed, an energy dispersion image can be obtained.

Figure 8 is an energy dispersion image of Au on Si(111), where Au 4f photoelectron spectrum is imaged as lines. Since Au $4f_{7/2}$ (84.0 eV) and Au $4f_{5/2}$ (87.7 eV) signals were separated, the energy resolution is estimated to be better than 2 eV.

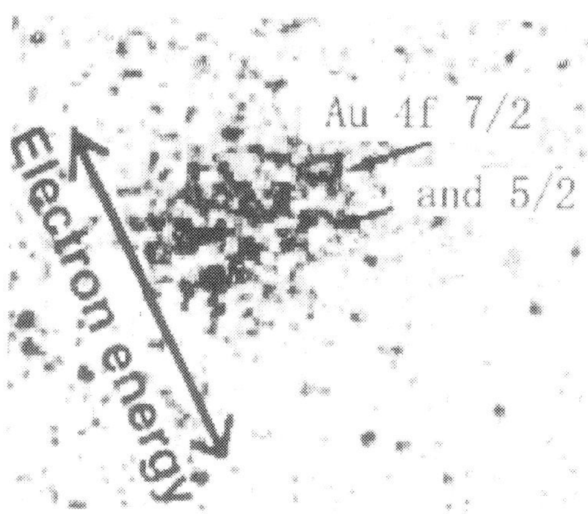

Figure 8. Energy dispersive image for Au $4f_{7/2}$ and $4f_{5/2}$ levels (marked by arrows).

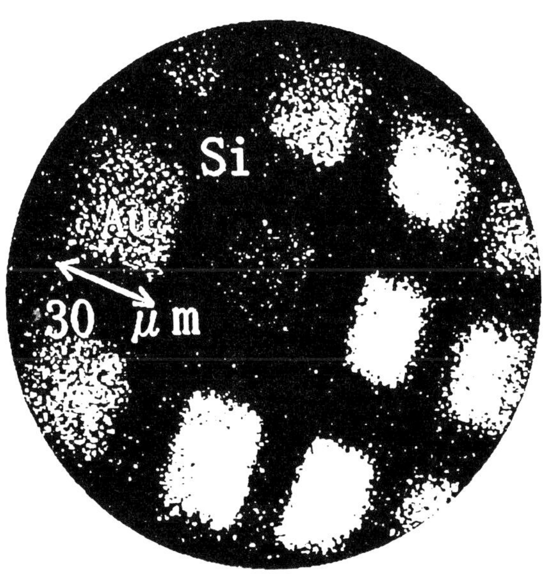

Figure 9. Real time video image of non-energy selected XPEEM for Au mesh on Si(111).

Figure 9 shows real-time imaging of X-ray photoemission electron for Au mesh on Si(111) without energy selection. Because a large amount of electrons was emitted from the Au surface as compared to the Si surface, the brighter imaging area is referred to the Au area. We are now attempting to visualize element distribution of oxide surfaces under reaction conditions by this new XPEEM.

CONCLUSIONS

No strong chemical bonding between the nitrogen atom of adsorbed pyridine and the Ti^{4+} atom at the terrace surface of $TiO_2(110)$ was suggested by STM, despite the presence of enough geometric space for the bonding on the Ti^{4+} atoms at the surface. Furthermore, the fact that adsorbed pyridine probe molecules migrate dynamically at the surface before and during the desorption may give some caution and implication to explanations for static acid-base interaction and temperature programmed desorption in a traditional manner. The origin of strong acid-base interaction at oxide catalyst surfaces may involve more complicated structural factors around acidic sites in addition to a simple coulombic interaction.

PTRF-EXAFS is a unique tool for three-dimensional surface structure analysis of dispersed species. It is applicable under any condition, e.g, from high pressure to UHV conditions. Mo oxides dispersed on a $TiO_2(110)$-(1x1) surface were determined to have an anisotropic Mo-dimer structure with Mo-Mo separation of 0.335 nm and Mo-Ti separation of 0.296 nm. Upon reaction with ethanol at 473 K for 2 h, the Mo-dimer structure was transformed to Mo species which showed no Mo-Mo bonding.

XPEEM is a new technique for surface imaging of elements or chemical states for model surfaces as well as real catalyst surfaces. This technique has the potential to provide information on dynamic surface processes such as diffusion, adsorption and phase transformation in situ under the working conditions. As the first examples recorded by the XPEEM, energy dispersive Au $4f_{7/2}$ (84.0 eV) and Au $4f_{5/2}$ (87.7 eV) signals and real time image of non-energy selected XPEEM for Au mesh on Si(111) were reported.

These three techniques are complimentary with each other. PTRF-EXAFS is element-specific and provides the information on three-dimensional local structures, but has no spatial resolution and is insensitive to chemical states. STM has an atomic resolution but not element-specific. XPEEM has a low spatial resolution as compared to STM but can image the mesoscopic distribution of elements and chemical states. Combinatory work of these three in-situ techniques may provide new picture and principle of surface dynamic processes like catalysis and new implication for development of efficient catalytic materials.

Acknowledgment

The authors would like to express their thanks to Dr. Kishimoto for his technical help in measurements of TPRF-EXAFS in the BL14A station (Proposal No. 95G001). They also thank Dr. Sakai and Dr. Kato of JEOL Co. for their technical supports in constructing the XPEEM system. This work was supported by CREST(Core Research for Evolutional Science and Technology) of Japan Science and Technology Corporation (JST). A part of XPEEM study was supported by Grant-in-Aid on Priority-Area-Research: Functionally Graded Materials (No. 09229218) from the Ministry of Education, Science, Sports and Culture, Japan.

REFERENCES

1. Y. Iwasawa, *Stud. Surf. Sci. Catal.*(Proc.11th Int. Congr. Catal., Baltimore), **101**, 21 (1996).
2. Y. Iwasawa, *Catal. Survey Jpn.*, Baltzer, **1**, 3 (1997).
3. H. Onishi and Y. Iwasawa, in: *Monograph on Chemistry for the 21st Century*: Interface Science, M.W. Roberts (Ed.), Blackwell Science, Oxford, 1997, in press.
4. G. Binnig, H. Rohrer, Ch. Gerber, and E. Weibel, Physica (Utrecht), **107 B+C**, 1335, (1981); **109 & 110 B**, 2075 (1982).
5. G. Binnig, H. Rohrer, Ch. Gerber, and E. Weibel, *Phys. Rev. Lett.*, **49**, 57 (1982).
6. R. Wisesendanger, *Scanning Probe Microscopy and Spectcroscopy*, Cambridge Univ. Press, Cambridge, 1994.
7. B.J. McIntyre, M.B. Salmeron, and G.A. Somorjai, *Catal. Lett.*, **14**, 263 (1992); *Rev. Sci. Instrum.*, **64**, 687 (1993).
8. H. Onishi, Y. Yamaguchi, K. Fukui, and Y. Iwasawa, *J. Phys. Chem.*, **100**, 9582 (1996).
9. H. Onishi and Y. Iwasawa, *Surf. Sci.*, **357/358**, 773 (1996).
10. H. Onishi, K. Fukui, and Y. Iwasawa, *Colloids Surf. A: Physicochem. Eng. Aspects*, **109**, 335 (1996).
11. H. Onishi and Y. Iwasawa, *Phys. Rev. Lett.*, **76**, 791 (1996).
12. M. Heald, E. Keller, and E.A. Stern, *Phys. Lett.*, **103A**, 155 (1984).
13. M. Shirai and Y. Iwasawa, in: *X-ray Absorption Fine Structure for Catalysts and Surfaces*, Y. Iwasawa, (Ed.), World Scientific, Singapore, 1996, p.332.
14. D.C. Koningsberger and R. Prins, *X-ray Absorption, Principles, Applications, Techniques of EXAFS, SEXAFS, and XANES*, John Wiley, New York, 1988, p.92.
15. M. Shirai, K. Asakura, and Y. Iwasawa, *Catal. Lett.*, **15**, 247 (1992).
16. M. Shirai, K. Asakura, and Y. Iwasawa, *Chem. Lett.*, 1037 (1992).
17. A. Komakov, J. Kovac, S. Günther, L. Casalis, L. Gregoratti, M. Marsi, and M. Kiskinova, *Phys. Rev. B*, **55**, 4101 (1997).
18. S. Günther, A. Komakov, J. Kovac, L. Casalis, L. Gregoratti, M. Marsi, and M. Kiskinova, *Surf. Sci.*, **377/379**, 145 (1997).
19. Y. Yamaguchi, K. Asakura, and Y. Iwasawa, *Hyomen (Surface)*, in press.
20. Y. Iwasawa (Ed.), *X-ray Absorption Fine Structure for Catalysts and Surfaces*, World Scientific, Singapore, 1996.
21. J.J. Rehr, J .Mustre de Leon, S.I. Zabinsky, and R.C. Albers, *J. Am. Chem. Soc.*, **113**, 5135 (1991).
22. H.H. Rotermund, W. Engel, M. Kordesch, and G. Ertl, *Nature*, **343**, 355 (1990).

23. M. Ramamoorthy, D. Vanderbilt, and R.D. King-Smith, *Phys. Rev. B*, **49**, 16721 (1994).
24. H. Onishi and Y. Iwasawa, *Chem. Phys. Lett.*, **226**, 111 (1994).
25. H. Onishi and Y. Iwasawa, *Surf. Sci. Lett.*, **313**, L783 (1994).
26. H. Onishi, K. Fukui, and Y. Iwasawa, *Bull. Chem. Soc. Jpn.*, **68**, 2447 (1995).
27. V.E. Henrich and P.A. Cox, *The Surface Science of Metal Oxides*, Cambridge Univ. Press, Cambridge, 1994.
28. G. Charlton, P.B. Howes, C.L. Nicklin, P. Steadman, J.S.G. Taylor, C.A. Muryn, S.P. Harte, J. Mercer, R. McGrath, D. Norma, T.S. Turner, and G. Thornton, *Phys. Rev. Lett.*, **78**, 495 (1997).
29. B.L. Maschhoff, J.-M. Pan, and T.E. Madey, *Surf. Sci.*, **259**, 190 (1991).
30. U. Diebold, J.F. Anderson, K.-O. Ng, and D. Vanderbilt, *Phys. Rev. Lett.*, **77**, 1322 (1996).
31. H. Onishi and Y. Iwasawa, *Langmuir*, **10**, 4414 (1994).
32. S. Suzuki, Y. Yamaguchi, H. Onishi, T. Sasaki, K. Fukui, and Y. Iwasawa, *J. Chem. Soc., Faraday Trans.*, in press.
33. W.L. Jorgensen and L. Salem, in: *The Organic Chemist's Book of Orbitals*, Academic Press, 1973, p.263.
34. S. Suzuki, Y. Yamaguchi, H. Onishi, K. Fukui, T. Sasaki, and Y. Iwasawa, to be published.
35. G. Thornton, in H.-J. Freund and E. Umbach (Eds.), *Adsorption on Ordered Surfaces of Ionic Solids and Thin Films*, Vol.33, Springer-Verlag, Berlin, 1993, p.115.
36. M.A. Barteau, *Chem. Rev.*, **96**, 1413 (1996).
37. W.J. Chun, Thesis, The University of Tokyo, 1997.
38. K. Asakura, in: *X-ray Absorption Fine Structure for Catalysts and Surfaces* Y. Iwasawa, (Ed.), World Scientific, Singapore, 1996, p.33.
39. T.Machej, B. Doumain, B. Yasse, and B. Delmon, *J. Chem. Soc., Faraday Trans.1*, **84**, 3905 (1988).
40. J.C. Volta and J.L. Portefaix, *Appl. Catal.*, **18**, 32 (1985).
41. H. Shimada, N. Matsubayashi, T. Sato, T. Yoshimura, and A. Nishijima, *J. Catal.*, **138**, 746 (1992).
42. K.Y.S. Ng and E. Gulari, *J. Catal.*, **92**, 340 (1985).
43. Y. Iwasawa, *J. Phys. IV France*, **7**, C2-67 (1997).
44. C.C. Williams, J.G. Ekerdt, J.-M. Jehng, F.D. Hardcastle, A.M. Turek, and I.E. Wachs, *J. Phys. Chem.*, **95**, 8781 (1991).
45. F.H. Jones, R. Dixon, J.S. Foord, R.G. Egdell, and J.B. Pethica, *Surf. Sci.*, **376**, 367 (1997).
46. F.W. Kutzler, C.R. Natoli, D.K. Misemer, S. Doniach, and K.O. Hodgson, *J. Chem. Phys.*, **73**, 3274 (1980).
47. Y. Satow, K. Asakura, and H. Kuroda, *J. Phys. Chem.*, **20**, 5027 (1987).

Res. Chem. Intermed., Vol. 24, No. 2, pp. 169-181 (1998)

INTEGRATED COMPUTATIONAL CHEMISTRY STUDY FOR ZEOLITE MICROPOROUS MATERIALS

A. MIYAMOTO*, A. CHATTERJEE, M. KUBO, H. TAKABA and Y. OUMI
Department of Materials Chemistry, Graduate School of Engineering, Tohoku University, Aoba-ku, Aramaki, Sendai 980-77, Japan

Received 16 September 1997; accepted 4 November 1997

INTRODUCTION

The knowledge of the structure, dynamics and energies of solid porous catalysts under the reaction conditions are fundamentally important information to develop efficient catalysts. Computational chemistry studies can contribute significantly in achieving an understanding of structure - property relationships by the synthesis of current understanding and data, and by their perspicacity in revealing critical conceptual issues whose resolution demands additional experimentation. In relevance with the importance of the zeolite catalysts in various petroleum, petrochemical and related processes [1], as well as due to its growing importance as environmental catalysts [2], as catalysts for the synthesis of fine chemicals [3], as advanced materials in separation technology [4] and as electronic device materials [5], the number of reports of computational studies of zeolites are in proportion. From our earlier experience of applying the molecular dynamics (MD) and computer graphics (CG) techniques to derive valuable insight on the structural aspects of various materials such as supported metal catalysts [6], superconducting materials [7], epitaxial heteroconjunctions [8] etc. we extended the application of these techniques to study zeolite catalysts [9]. The structure and dynamics of the framework atoms, extra framework atoms and guest molecules have been widely studied by various research groups [10-13]. Recently, there have been considerable improvements in the methodology and the cluster models adopted and sophisticated problems such as the adsorption configuration of various molecules and the mechanisms of various reactions have been tackled by QC techniques [14-15].

In this present communication we will give an overview of the computational chemistry methodology used and developed by our recent investigation, followed by the application of the computational chemistry study on different aspects of zeolites such as structure - property correlation, adsorption phenomenon and ion-exchange behavior. This will be followed by the examples of novel application of computational chemistry study in the area of zeozymes, de-NOx processes and in the exciting area of permeation.

COMPUTATIONAL METHODOLOGY

MD Methodology

The MD calculations were carried out with the MXDORTO program developed by Kawamura [16]. The Verlet algorithm was used for the calculation of atomic motions, while the Ewald method was applied for the calculation of electrostatic interactions. Temperature and pressure were controlled by means of scaling of atomic velocities and unit cell parameters under 3-dimensional periodic conditions. The two body central force interaction potential was used for all the calculations. Calculations were performed for 200,000 to 500,000 steps, each step corresponds to 1.5×10^{-15} sec. We modified the program time to time in order to keep the temperature values of the solid and gas phases at desired levels during the simulation.

MC Methodology

In conventional Grand Canonical MC (GCMC) simulation, the simulations were performed at a certain temperature under the three dimensional periodic boundary condition, e.g. by using Sorption module of BIOSYM/MSI [17]. This kind of simulation is performed in cycles, and each cycle consists of randomly selected attempts either to displace, to create or to destroy a molecule. We developed a new GCMC simulation code where along with the conventional procedure it is possible to displace extra framework cations for the metal-exchanged zeolite [18], which is a very useful tool for analyzing metallosilicates. Our GCMC has the following features : (i) either GCMC or NVT simulations can be performed with a movable zeolite framework, (ii) calculation and visualization of the adsorbate-zeolite potential energy map in 3-D space is possible and finally (iii) the interacting system can be visualized during the simulation process as an interface.

Semiempirical (AMPAC) and Accurate (DFT) Quantum Chemical (QC) Calculations

The semiempirical quantum chemical method using AM1 hamiltonian is adopted to perform the cluster calculations for large tetrameric cluster (four SiO_4 terahedra linked by bridging oxygen) models [19]. The calculations were carried out using the AMPAC code (QCPE Program No. 539) [20]. Density Functional Theory (DFT) was applied for studying structure property relationship for zeolites along with adsorption phenomenon. The calculations were carried out using the self consistent

Kohn Sham formalism [21] as implemented in the DMOL program of Biosym/MSI based on the work of von Barth and Hedin [22]. The geometry optimization calculations were carried out using Double Numerical with Polarization basis set (DNP) [23]. Local density approximation was performed using JMW (Janak-Moruzzi-Williams) functional as well as using VWN (Vasko-Wilk-Nusair) functional. The non-local gradient corrections were used for getting accurate energy values. We have used non local gradient corrections for exchange Becke [24] and correlation Lee-Young-Parr [25] to calculate self consistently the final energy.

Static Visualization

The static visualization by computer graphics (CG) of molecular configurations as snapshots during MD simulations as well as the plots of optimized molecular geometries from DFT calculations were made with the INSIGHTII code of BIOSYM/MSI technologies Inc. on a Silicon Graphics IRIS-4D/25TG engineering workstation. The dynamic visualization for final checking the MD derived equilibrium was made with MOMOVIE and RYUGA [26] programs developed in our laboratory on OMRON-LUNA-88K and HP9000 Model 715/33 workstations, respectively.

Molecular Electrostatic Potential (MEP) Map from the Quantum Chemical Calculations

We have devised a methodology [27], where the electrostatic potential can be quantified to understand the spatial volume and the intensity of interaction. The MEP at a certain `r` in space represents a first order approximation to the interaction energy of the molecular charge distribution with the probe of unit charge at that point. The molecular electrostatic potential between the molecular electronic and nuclear charge distribution as well as an external proton at all the points formed by a grid 25 x 25 x 25 points are calculated by a method proposed by Tomasi [28]. The isopotential points are connected to form a contour.

RESULTS AND DISCUSSION

A computational chemistry study which has been performed on zeolites in our laboratory can be divided into two major areas namely: (1) structure and property correlation in zeolites including various phenomenon like adsorption, ion-exchange etc. and (2) novel application areas of computer simulation in the novel fields in zeolite catalysis. We have used MD, MC, Quantum chemical calculations on zeolites to rationalize a better understanding in this area of research.

Structure and Property Correlation in Zeolites

Zeolite structures. The selective substitution of a silicon by titanium in MFI lattice was investigated by MD methods and CG [29]. It was demonstrated that using a two body interatomic potential of MD is effective to reproduce the framework structure. All eight silicons in the T12 site were replaced by Ti leading to an Si/Ti ratio of 11. The anisotropic lattice expansion due to Ti substitution in various crystallographic sites of MFI lattice was studied. The favorable site for Ti substitution was discussed by comparing the expansions of lattice parameters reported by X-ray diffraction (XRD) studies. It is observed that the T8 site is the most probable site for isomorphous substitution of Ti.

A further study on structure and dynamic behavior of copper, gallium and cerium cations in cation-exchanged ZSM-5 framework was performed using MD and CG [30]. Various oxidation states of these metals were simulated and their relevance in the catalytic activity is also discussed. Depending on the starting position of Cu^+, either the migration readily occurred even at 300 K when it is closer to T12 site, or a higher temperature, namely 600 K was needed when it is further from T12 site. At lower temperatures, Cu^{2+} ion in Cu-ZSM-5 is considered to be a hydrated state as $[Cu(OH)]^+$. Under the reaction conditions, the copper is expected to be in a dehydrated state and hence we have simulated both Cu^{2+} and [CuO] inside ZSM-5 channels. It was observed that both these species also prefer a location closer to the $[AlO_4]^-$ tetrahedron. The variation of the coulomb energy of the lattice with the distance between the positive and negative charges indicates that the coulomb energy is less favorable, when two aluminium atoms are present at a distance 11Å (at two T12 sites) with a Cu^{2+} ion. Thus, Cu^{2+} is not stable and it prefers to be Cu^+ even under reaction conditions. Various Gallium species such as Ga^{3+}, $[GaO]^+$, $[Ga(OH)]^{2+}$ and $[Ga(OH)_2]^+$ were simulated inside ZSM-5. Gallium also migrates similarly towards T12 site where Al is substituted. Cerium also follows the same trend as that of gallium. The reversible formation of oxidized and reduced states of cations particularly that of copper are the reason for catalytic activity. Whereas the increased dynamic freedom of the oxygen attached to gallium and cerium and the low coordination of these cations in ZSM-5 may be the cause of their improved catalytic activity.

MD and QC (DFT) calculations have been performed in a recent study [31] to understand the structure and electronic properties of metallosilicates. For the first time, the local structure of the metal substituted zeolite-framework is derived from MD calculations and its electronic structure is determined by DFT calculations for different metal substituted clusters. Our calculations showed that the net electron density on the MO_4 unit, where M is Ti^{4+}, Al^{3+}, Ga^{3+} or Fe^{3+}, is a reliable descriptor for the acidity of metallosilicates. Furthermore the analysis of MEP maps for the cluster models predict the polarization resultant from various substitutions which further justifies the acidity trend as observed by experiment.

Ion-exchange properties. The applicability of MD and CG to investigate the structure and function of ion-exchanged zeolites was demonstrated [32-33] for (i) reproducing the known structures of various zeolites, (ii) determining the positions of Al and exchanged cations in the framework and (iii) understanding the dynamic mechanism in the molecular sieving process of O_2 and N_2 in A-type zeolites. It is observed that Na ion cannot reach the vicinity of Al cation at 300k, while it can migrate easily at higher temperature. H and Li having smaller ionic radii are located near the vicinity of oxygen anions whereas Na, K or Cs due to their larger ionic radii spread out in the micropores of ZSM-5. The dynamic visualization of the diffusion process of O_2 and N_2 at different temperatures suggested that the Na ion at the window plays the crucial role for determining the effective pore radius of NaA ion.

The effectiveness and applicability of MD, NMR simulation and CG to the investigation of the sites and distribution of framework aluminums in faujasites, is shown [34]. By using MD calculations, we successfully predicted the adequate aluminum distribution in the NaY zeolite (Si:Al = 2.43), as it reproduces the sites and occupancies of Na^+ cations which have been further supported by neutron diffraction techniques. The simulated neutron diffraction patterns show the differences in the distribution of cations among the I` sites in the different patterns at low angle region. The validity of our aluminum distribution model in Na-Y has been tested with simulated ^{29}Si MAS NMR spectra, which shows agreement with both in terms of chemical shifts and intensities.

Adsorption and diffusion behavior. MD and CG were applied to study the dynamic behavior of adsorbed molecules with different diameters and weights in ZSM-5 and mordenite [35]. The medium size monatomic molecules diffused 3-dimensionally in the micropores of ZSM-5. The mobility of the adsorbed molecule decreases with increase in its size. The higher diffusivity is observed in the case of ZSM-5 in comparison to mordenite. On further study it is revealed that the flexibility of the framework is present in the case of ZSM-5 or mordenite to allow the diffusion. The MD simulation results also show that the collision between adsorbed molecules greatly decreased the diffusion rate for mordenite, while the dynamics and rapid diffusion were not much retarded by collision for ZSM-5. This is consistent with the experimental results.

In another study with a combination of GCMC, DFT and CG the adsorption mechanism of CFCs (chlorofluorocarbons) in zeolite CsNaY are successfully predicted [36]. CFCs are widely used as solvents, refrigerants and propellants. However, the emission of CFCs led to stratospheric ozone depletion and greenhouse effects. The accurate distribution of Al^{3+}, Cs^+ and Na^+ in the CsNaY (Si/Al = 2.43) was obtained from MD calculations and NMR simulation, while the potential parameters for CF_2Cl_2 were determined by DFT calculations. The experimental adsorption isotherm of the CF_2Cl_2 in CsNaY was fully reproduced by GCMC simulation as shown in Figure 1. Furthermore, CG visualization of the calculated results revealed the presence of two different mechanisms of CF_2Cl_2 adsorption in CsNaY: (i) at low pressure, the CF_2Cl_2 selectively and drastically adsorbed on the Cs cation of CsNaY and (ii) at high pressure, the aggregation of CF_2Cl_2 in the supercages of CsNaY occurred.

In an earlier review, computational methodology was used to design synthetic sorbents for selective adsorption of molecules [37]. The attempts made to design synthetic sorbents at a molecular level are thoroughly discussed.

Novel Application Areas of Computer Simulation in the Novel Fields in Zeolite Catalysis

Zeozymes. Zeozymes are a type of materials which can act as enzymes after encapsulating microporous crystals inside zeolites and can act as selective catalysts.

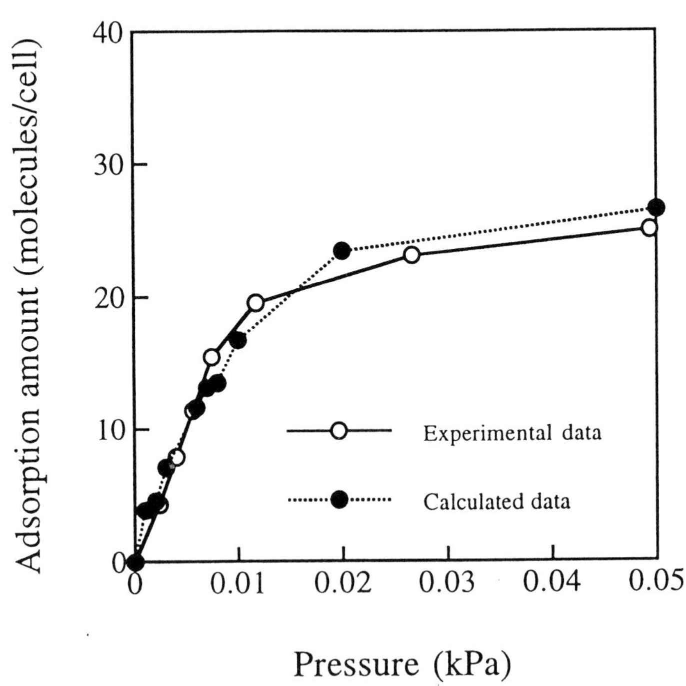

Figure 1. Calculated adsorption isotherm of CF_2Cl_2 in CsNaY (Si/Al = 2.43) with the experimental result.

Our aim is to design these catalysts by carrying out electronic structure calculations and to study the encapsulation of enzyme-like active sites into porous mantles. Porphyrins have a significant role in biological actions such as photosynthesis, as oxygen carriers in the decomposition of water. Although porphyrin based catalysts undergo oxidative dimerization as well as degradation, their encapsulation inside microporous materials extends their catalytic life, activity and selectivity. In our recent study [38], we studied the configuration of metal (Mg, Co, Fe) porphyrin complexes inside Y-type zeolites using MD simulations. Figure 2 shows the orientation of the Mg-porphyrin complex in absence of Na cations; the Mg-porphyrin is located above the six membered ring composed of zeolite oxygens with the metal center oriented towards the oxygen atoms. Whereas in the presence of Na the metal center (Mg) of the porphyrin is oriented towards the center of the supercage as shown in Figure 3, which is attributed to the electrostatic repulsion with sodium cations.

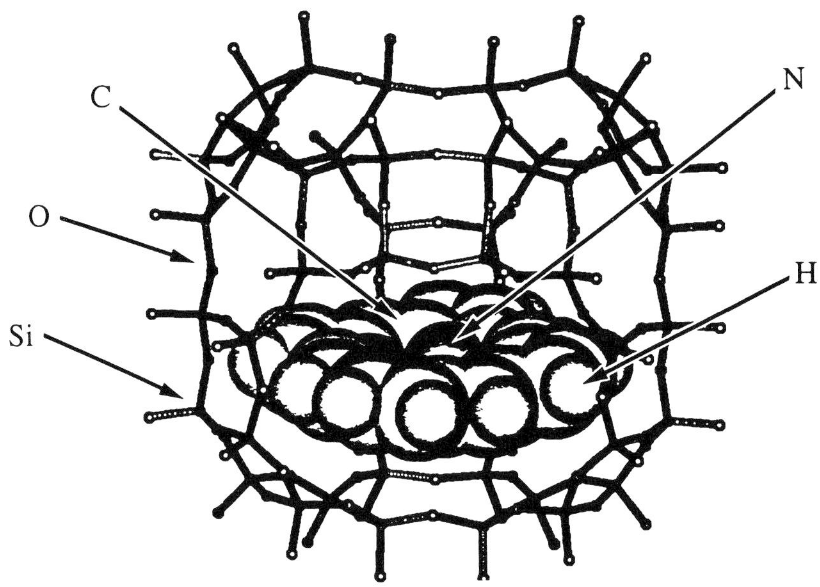

Figure 2. Mg-porphyrin complex encapsulated in USY zeolite at 300K. The metal center of the porphyrin interacts with zeolite oxygen anions.

In a recent study we have used QC and CG methods to study the docking of organic molecules pyrrole and acetaldehyde inside faujasite zeolites [39]. The low-energy binding sites for the organics inside the supercage of the faujasite framework were derived from the energy minimization procedure. The exact orientation of these molecules are derived from QC DFT cluster model calculations.

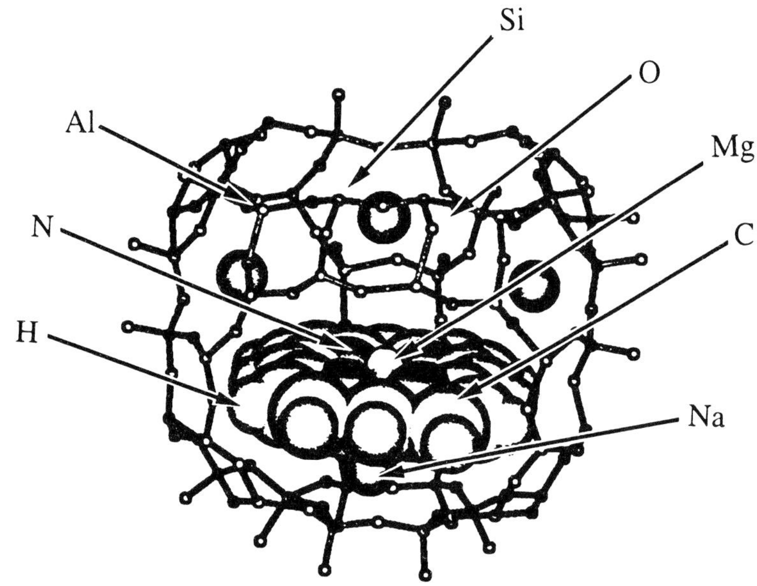

Figure 3. Mg-porphyrin complex encapsulated in NaY zeolite at 300 K. The Na cation is located above zeolite oxygens and the Mg atom is oriented toward the supercage center.

The electronic structure of pyrrole and acetaldehyde as well as their adsorption complexes over a suitable cluster model of the faujasite framework are reported. The configuration of tetramethylporphyrin formed by the condensation of pyrrole and acetaldehyde is observed to be inclined inside the supercage.

It is difficult experimentally to obtain information about the configuration of the porphyrin inside the zeolite, the effect of exchanged cations, or the size limitations due to the presence of organic substituents on the porphyrin. We therefore performed MD simulation to study the configuration of metal (Mg, Co, Fe) porphyrin complexes encapsulated in Y-type zeolites [40]. It is observed that as the zeolite supercage can hold only one porphyrin the oxidative dimerization which occurs in homogeneous systems can be avoided by the encapsulation. Further, when the zeolite does not contain exchanged cations, the metal center of the porphyrin goes towards the center of the six member ring formed by framework oxygens. In this orientation, the metal interacts with the zeolites; therefore, a low activity of the porphyrin-zeolite catalyst is expected. Now in the situation with exchanged metal ion it is oriented towards the supercage center, so the active site is available to react and a better porphyrin-zeolite catalyst is expected.

De-NOx process. The utilization of zeolite catalysts as prospective environmental catalysts has been explored for the past few years. Metal exchanged zeolites have emerged as efficient and economical catalysts for the removal of nitric oxide from combustion exhaust and automobile exhaust gases. Despite the extremely challanging character of the topic, the mechanism of the process is basically unknown; for example, little is known about the mechanism of water poisoning or the specific role of the reductant and the catalyst in the reaction; also the identification of intermediates has not yet been accomplished. We have performed a combination of MD and QC DFT calculations [41] to form an efficient route towards understanding a complex reaction sequence in catalytic reduction of NOx. It is shown on the basis of our result that the large nucleophilic space (Figure 4) around GaO is capable of activating even non-polar hydrocarbons. This nucleophilicity arises from the coordination unsaturation around Ga^{3+}. The proposed mechanism of the interaction of methane with the Ga site in Ga-ZSM-5 agrees well with recent experimental results on efficient catalytic performance of the catalyst.

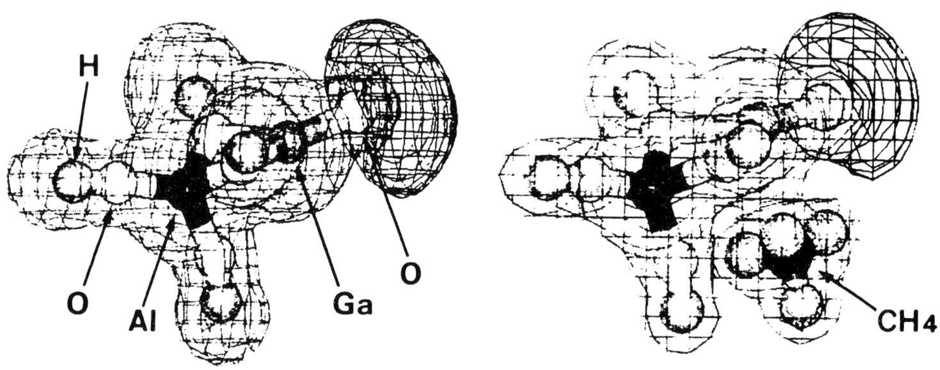

Figure 4. The MESP map around $[H_4AlO_4][GaO]$ cluster. Dark contour corresponds to nucleophilic region (-0.05 to -1.00 a.u.) and light contour corresponds to electrophilic region (+0.05 to +1.00 a.u.).

A MD, QC and CG combination methodology has been further applied to rationalize the multistep removal of NOx by methane activated Ga-ZSM-5 [42]. It is obseved that the methane molecule may undergo weak adsorption with polarization of an intramolecular C-H bond prior to the dissociation of the bond. Two types of methane dissociation are proposed, namely, one with the methyl group attached to the exchange Ga ion and the dissociated H atom connected to extraframework oxygen and the other one with the methyl group bonded to the extraframework oxygen and the hydrogen atom attached to the exchanged Ga ion. The former one is the preferred geometry in view of energy stabilization, while both forms highly stabilize the system in comparison to physisorbed methane molecule.

It is also observed that the water molecule may be easily dissociated without physisorption on the Ga site, and the dissociated form of water strongly stabilizes the system energetically. This greatly decreases the adsorption energy of the methane molecule on the Ga site and, as a consequence, prevents the activation process of methane.

We performed few exhaustive studies in rationalizing the understanding of the adsorption and activation mechanisms of methane and NOx molecules on Ga-ZSM-5, reaction profile and transition state calculations using a combination of MD. QC and CG methodologies [43-45].

Zeolite membranes. Much attention has been given recently to the design and use of inorganic membranes in separation processes. The use of zeolites as inorganic membranes will allow the combination of their catalytic activity and separation ability, which make them act as bifunctional systems. For the first time, we used computational chemistry methodology based on interatomic interactions, such as MD and MC techniques for understanding the dynamic behavior and transport mechanism of butane through the zeolite membrane [46]. The permeation mechanism of n-butane is primarily dominated by surface diffusion or capillary condensation at 373K as shown in Figure 5, since the aggregation of molecules inside the silicalite pore was observed. At higher temperature i.e. 773 K the adsorption becomes smaller and the mechanism can be described as an activated diffusion. In the gas mixture, the permeation of n-butane through the membrane was observed at 373 K, where as iso-butane did not permeate. However diffusion of both isomers individually into the zeolite pores was observed. Our MD results show an enhanced adsorption of iso-butane and a lower permeance of n-butane compared with the behavior of single gases. Since the diffusion of iso-butane is significantly lower than that of n-butane, the few diffused iso-butane molecules reduce the diffusion and permeability of n-butane.

CONCLUSION

The content of this review could be summarized to conclude that computational chemistry studies are having a substantial impact on our understanding of structural properties, adsorption, ion-exchange, encapsulation, de-NOx removal and membranes at present and will have enormous predicting power in the near future. The tactics needed to attack the problems lie in a judicious use of different combinations of computational techniques. In this review, it has only been possible to highlight the outline of our studies in the field of zeolite catalysis. All these methodologies are essentially applicable to any other unsolved challenging areas of

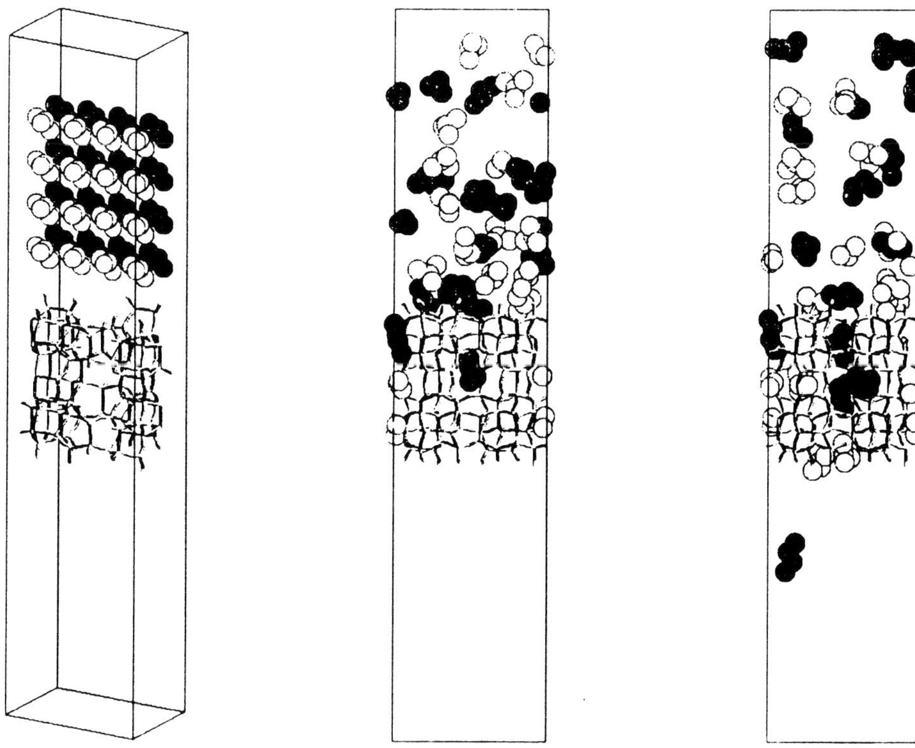

Figure 5. Permeation of n-butane gas through a silicalite membrane at 373 K.

interest in zeolites and other correlated fields and essentially reminds us that a great deal remains to be done. Future explorations should deal with the elimination of the approximations that are often being made, such as QC AMPAC methods. Particularly with the advent of limitless computing power, more accurate QC and MD calculations will become feasible in the foreseeable future. As our future task, we are engaging ourselves in deriving new softwares for applying in this challenging area. So the computational chemistry study has now established its role in exploring the limitaions of experiment to propose plausible explanation for the critical situations.

REFERENCES

1. N.Y. Chen and T.F. Degnan, *Chem. Eng. Prog.* Feb., 32 (1988).
2. H. Hamada, Y. Kintaichi, T. Yoshinari, M. Tabata, M. Sasaki, and T.Ita, *Catal. Today* **17**,

111 (1993).
3. W. Holderich, M. Hesse, and F. Naumann, Angew. *Chem. Int. Ed. Engl.* **27**, 226 (1988).
4. R.V. Jasra, N.V. Choudary, and S.G.T. Bhat, *Sep. Sci. Technol.* **26**, 885 (1991).
5. G.A. Stucky, *Science* **247**, 669 (1990).
6. A. Miyamoto, T. Hattori, and T. Inui, *Appl. Surf. Sci.* **60/61** 660 (1992); A. Miyamoto, R. Yamauchi, and M.Kubo, *Appl. Surf. Sci.* **75**, 51 81994).
7. A. Miyamoto, T. Hattori, and T.Inui, *Physica C* **190**, 93 (1991).
8. A. Miyamoto, K. Takeichi, T. Hattori, M.Kubo, and T. Inui, *Jpn. J. Appl. Phys.* **31**, 4463 (1992).
9. A. Miyamoto, K. Matsuba, M.Kubo, K. Kawamura, and T. Inui, *Chemistry Lett.* 2055 (1991).
10. P. Demontis, G.B. Suffritti, S. Quarteri, E.S. Fois, and A. Gamba, *Zeolites* **7**, 522 (1987).
11. P. Santikary, S. Yashonath, and G. Anathakrishna, *J. Phys. Chem.* **96**, 10469 (1992).
12. J.O. Titiloye, S.C. Parker, F.S. Stone, and C.R.A. Catlow, *J.Phys. Chem.* **95**, 4038 (1991).
13. J.B. Nicholas, F.R. Trouw, J.E. Mertz, L.E. Iton, and A.J. Hopfinger, *J. Phys. Chem.* **97**, 4149 (1993).
14. J. Sauer, *Chem. Rev.* **89**, 199 (1989).
15. S. Beran. In: *Theoretical Aspects of Heterogeneous Catalysis*, J.B. Moffat (Ed.), van Nostrand Reinhold, New York, (1990), p. 160, .
16. K. Kawamura. In: *Introduction to Molecular Simulations*, I.Okada and E. Osawa (Eds.), Kaibun-do, Tokyo, 1989, Chapters 6 and 7.
17. *Cerius2 Computational and Analytical Instrumentation User's Guide*, Release 1.0, Molecular Simulation Inc., Massachusette, USA, 1994.
18. A. Hirotani, K. Mizukami, R. Miura, H. Takaba, T. Miya, A. Fahmi, A. Stirling, M. Kubo, and A. Miyamoto, *Appl. Surf. Sci.*, in press.
19. S. Prasad and R. Vetrivel, *J. Mol. Catal.* **84**, 299 (1993).
20. M.J.S. Dewar, E.G. Zoebisch, E.F. Healy, and J.J.P. Stewart, *J. Am. Chem. Soc.* **107**, 3902 (1985).
21. W. Kohn and L.J. Sham, *Phys. Rev. A* **140**, 1133 (1965).
22. U. von Barth and L. Hedin, *J. Phys. C* **5**, 1629 (1972).
23. B. Delley, *J. Chem. Phys.* **92**, 508 (1990).
24. A. Becke, *J. Chem. Phys.* **88**, 2547 (1988).
25. C. Lee, W. Yang, and R.G. Parr, *Phys. Rev. B* **37**, 786 (1988).
26. R. Miura, H. Yamano, R. Yamauchi, M. Katagiri, M. Kubo, R. Vetrivel, and A. Miyamoto, *Catal. Today* **23**, 409 (1995).
27. R. Vetrivel. R.C. Deka, A. Chatterjee, M. Kubo, E. Broclawik, and A. Miyamoto, *Theor. Comp. Chem.* **3**, 509 (1996).
28. J. Tomasi. In: *Chemical Application of Atomic and Molecular Electrostaic Potentials*, P. Politzer and D.G. Truhler (Eds.), Plenum, New York, 151, 1981.
29. Y. Oumi, K. Matsuba, M. Kubo, T. Inui, and A. Miyamoto, *Microporous Mater.* **4**, 53 (1995).
30. A. Miyamoto, H. Himei, Y. Oka, E. Maruya, M. Katagiri, R. Vetrivel, and M. Kubo, *Catal. Today* **22**, 87 (1994).
31. Y. Oumi, M. Yamadaya, T. Kanougi, M. Kubo, A. Stirling, R. Vetrivel, E. Broclawik, and A. Miyamoto, *Catal. Lett.* **45**, 21 (1997).
32. A. Miyamoto and M. Kubo, *Stud. Surf. Sci. Catal.* **83**, 117 (1994).
33. A. Miyamoto, H. Himei, E. Maruya, M. Katagiri, R. Vetrivel, and M. Kubo, *Stud. Surf. Sci. Catal.* **90**, 217 (1994).
34. H. Himei , M. Yamadeya, Y. Oumi, M. Kubo, A. Stirling, R. Vetrivel, E. Broclawik, and A. Miyamoto, *Microporous Mater.* **7**, 235 (1996).

35. M. Kubo, A. Miyamoto, and T. Inui, *Stud. Surf. Sci. Catal.* **80**, 333 (1993).
36. K. Mizukami, H. Takaba, Y. Oumi, M. katagiri, M. Kubo, A. Stirling, and A. Miyamoto, *Stud. Surf. Sci. Catal.* **105** , 1811 (1997).
37. R. Vetrivel, H. Takaba, M. Katagiri, M. Kubo, and A. Miyamoto, *Stud. Surf. Sci. Catal.* **99**,3 (1996).
38. T. Sato, K. Sugao, Y. Oumi, R. Vetrivel, M. Chatterjee, A. Chatterjee, M. Kubo, A. Stirling, A. Fahmi, and A. Miyamoto, *Appl. Surf. Sci.*, in press.
39. A. Chatterjee, R. Vetrivel, M. Kubo, and A. Miyamoto, *Stud. Surf. Sci. Catal.* **105**, 2339 (1997).
40. T. Sato, K. Sugao, Y. Oumi, R. Vetrivel, M. Chatterjee, A. Chatterjee, M. Kubo, A. Stirling, A. Fahmi, and A. Miyamoto, *Appl. Surf. Sci.*, in press.
41. R. Vetrivel, M. Kubo, H. Himei, E. Maruya, M. Katagiri, E. Broclawik, and A. Miyamoto, *Stud. Surf. Sci. Catal* **92** , 233 81995).
42. H. Himei, M. Yamadaya, M. Kubo, R. Vetrivel, E. Broclawik, and A. Miyamoto, *J. Phys. Chem.* **99**, 12461 (1995).
43. E. Broclawik, H. Himei, M. Yamadaya, M. Kubo, A. Miyamoto, and R. Vetrivel, *J. Chem. Phys.* **103**, 2102 (1995).
44. M. Yamadaya, H. Himei, T. Kanougi, Y. Oumi, M. Kubo, A. Stirling, R. Vetrivel, E. Broclawik, and A. Miyamoto, *Stud. Surf. Sci. Catal.* **105**, 1485 (1997).
45. T. Kanougi, K. Furukawa, M. Yamadaya, Y. Oumi, M. Kubo, A. Stirling, A. Fahmi, and A. Miyamoto, *Appl. Surf. Sci.*, in press.
46. H. Takaba, R. Koshita, K. Mizukami, Y. Oumi, N. Ito, M. Kubo, A. Fahmi, and A. Miyamoto, *J. Membrane Sci.*, in press.

Res. Chem. Intermed., Vol. 24, No. 2, pp. 183-196 (1998)

HYDRAZINE SYNTHESIS: COMMERCIAL ROUTES, CATALYSIS AND INTERMEDIATES

HIROMU HAYASHI
Department of Chemical Science and Technology, University of Tokushima, Minamijosanjima, Tokushima 770 JAPAN

Received 3 September 1997; accepted 18 December 1997

CONTENTS

I. INTRODUCTION

Hydrazine is one of the most versatile chemicals, finding use in a wide spectrum of applications such as an energy source in fuel cells and rocket propulsion, synthesis of organic nitrogen compounds for medical and agricultural use, a blowing agent for polymers, and an oxygen scavenger for boiler water [1]. Comparing aquo- (H-O-H) and ammono- (H-NH-H) systems, hydrazine (H-NH-NH-H) can be regarded as the ammono analogue of hydrogen peroxide (H-O-O-H). Hydrazine and its commercial production are well documented elsewhere [1-5].

Interest in our laboratory has centered on a cyclic catalytic oxidation process for hydrazine synthesis from benzophenone, ammonia and molecular oxygen via the azine with recycling of benzophenone. A summary of the results is given in the present review along with a brief survey on competing hydrazine technologies, and the possible intermediates in titanium silicalite (TS-1) catalyzed hydrogen peroxide oxidation of ketone-ammonia system are also discussed.

II. FORMATION OF HYDRAZINE: PUBLICATION STATISTICS AND INSTALLED CAPACITY

Formation of hydrazine has been reported in a variety of fields of science and technology, and the publication statistics during 1945-1988 given in Ref. [1] shows four major areas in research of: (1) Raschig synthesis, (2) irradiation/radiolysis, (3) discharge, and (4) nitrogen fixation. The chemistry of hypochlorite oxidation of ammonia was established by Raschig [6] as early as in 1907. The reaction still provides the scientific basis for the commercial production of hydrazine at present, and the *Raschig process* was an exclusive technology until the early 1970s as shown in Figure 1. Attempts were made over 30 years to obtain hydrazine from ammonia by *glow discharge*. *Radiolysis/irradiation* and *nitrogen fixation* showed boom-and-bust cycles in the mid.1960s and early 1980s, respectively.

The problems encountered in hydrazine synthesis directly from ammonia would be the difficulties, not in the formation of the N-N bond but in the recovery of hydrazine without decomposition. Thus the commercial route to hydrazine has been directed toward the cyclic processes for protecting the N-N bond, tentatively in the structure of organic nitrogen compounds. Hydrazine installed capacity given in Figure1 evidently shows that Raschig synthesis has been replaced by the Bayer process and/or the modified chlorine oxidation technology based on diaziridine and (ket)azine as intermediates since 1973.

New processes by-passing chlorine have received increasing attention in recent decades, due to the current tight chlorine situations with a large amount of by-product chlorides. Attempts to avoid chlorine date back to 1959. Meyer *et al.* [7] claimed copper-catalyzed air oxidation of benzophenone imine to the azine. Another approach based on hydrogen peroxide oxidation was disclosed by Schirmann *et al.* [8] in 1971 with a series of patents.

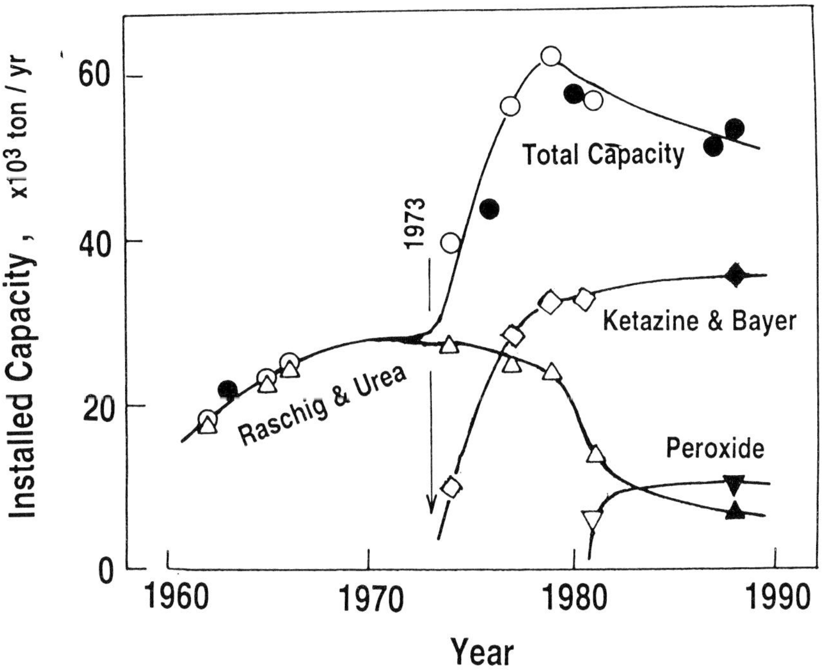

Figure 1. Hydrazine installed capacity ([1], revised). Circles: Total capacity; triangles: Raschig and Urea; squares: Ketazine and Bayer. White symbols: Schmidt [3], Black symbols: Schirmann [2].

III. COMMERCIAL ROUTES TO HYDRAZINE: CHEMISTRY AND INTERMEDIATES

A brief survey on the commercial routes to hydrazine is given in this chapter to elucidate both advantages and drawbacks in the current technology.

Raschig Synthesis ($NaOH$-Cl_2-NH_3)

The reaction of ammonia with sodium hypochlorite yields chloramine as an intermediate (Eq. 2), which reacts with additional ammonia to form hydrazine (Eq. 3).

$$2\,NaOH + Cl_2 \longrightarrow NaOCl + NaCl + H_2O \quad (1)$$

$$NaOCl + NH_3 \longrightarrow NH_2Cl + NaOH \quad (2)$$

$$2\,NH_3 + NH_2Cl \longrightarrow N_2H_4 + NH_4Cl \quad (3)$$

Unfortunately, this process entails a large quantity of by-product chlorides, 6 pounds per every pound of hydrazine [9]. Moreover, both chloramine and hydrazine are unstable at higher concentrations, decomposing into nitrogen and ammonium chloride. Thus the concentration of hydrazine in the reaction liquor is usually maintained at less than 1%, leading to a sizable consumption of primary energy for distillation.

Schestakov Synthesis (NaOH-Cl_2-Urea)

The Schestakov synthesis, in which the ammonia of the Raschig synthesis is replaced by urea, also includes similar drawbacks.

$$(NH_2)_2CO + NaOCl + 2\,NaOH \longrightarrow N_2H_4 + NaCl + Na_2CO_3 + H_2O \quad (4)$$

The reaction bears a formal analogy to the Hofmann degradation of amides to primary amines, but it is not certain whether hydrazine is formed by a Hofmann rearrangement of the N-halouride anion [5].

Bayer Process (Ketone-NH_3-Cl_2)

An aliphatic ketone forms *imine* by dehydrative condensation with ammonia. The imine thus formed reacts with chloramine in the presence of ammonia to afford a three-membered ring compound, *diaziridine*, and ammonium chloride [10] (Scheme 1). The subsequent reaction of diaziridine with an additional ketone gives *azine*, which is hydrolyzed quantitatively to form hydrazine with the regeneration of ketone.

Although the drawbacks inherent in the chlorine oxidation still remain, the improvements in the Bayer Process might be evaluated as in the following features: The reaction of chloramine with aliphatic ketone is fast enough even at room temperature, while the reaction with ammonia requires a higher temperature of 125-

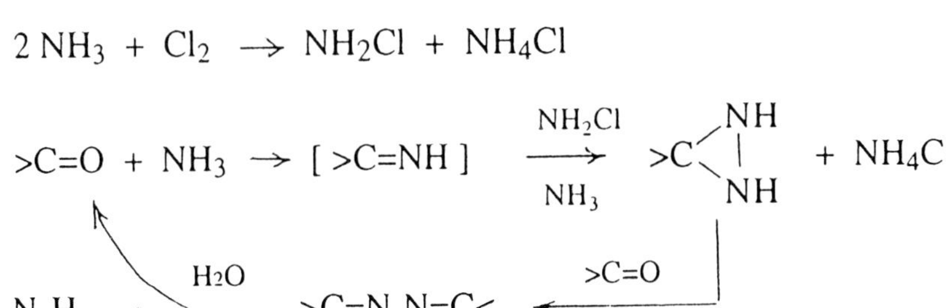

Scheme 1

155 °C. Thus a ketone is used as a nitrogen carrier to activate ammoniacal nitrogen into imine. Another improvement is the stabilization of N-N bond of unstable hydrazine in the structure of azine.

Ugine Kuhkmann (Atochem) Process (Ketone-NH_3-H_2O_2)

Oxidation of Schiff bases by hydrogen peroxide in the presence of a nitrile gives *oxaziridines* in high yields (Eq. 5) [11]. The reaction is reminiscent of the epoxidation of olefins [12].

(5)

Schirmann and Weiss [11] extended the reaction to the *in situ* oxidation of intermediate imine. They obtained *azine* in high yields via the hydrogen peroxide oxidation of ketone-ammonia system in the presence of a nitrile. The peroxide route, which is similar to the Bayer process (Scheme 1), bypasses chlorine, leading to a breakthrough of hydrazine technology. Presumably, azine was formed through oxaziridine intermediate unsubstituted at the N-site, followed by the reactions with ammonia and ketone (Scheme 2) [2].

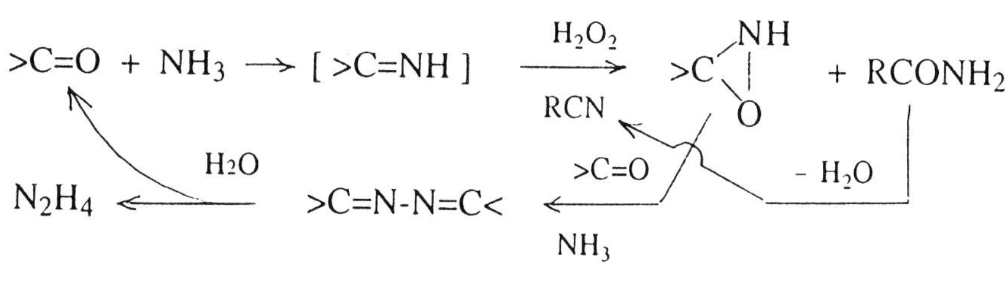

Scheme 2

IV. CATALYTIC OXIDATION WITH MOLECULAR OXYGEN

Benzophenone azine, which gives hydrazine quantitatively on acid hydrolysis [13,14], is prepared by the oxidative coupling of diphenylmethanimine in the presence of copper catalysts [15,16]. Benzophenone reacts selectively with ammonia to afford diphenylmethanimine [17]. Thus a cyclic catalytic process for hydrazine synthesis from benzophenone, ammonia, and molecular oxygen via the azine intermediate with recycling of benzophenone is possible (Eqs. 6-8).

$$Ph_2C{=}O + \boxed{NH_3} \longrightarrow Ph_2C{=}NH + H_2O \quad (6)$$

$$2\ Ph_2C{=}NH + 1/2\ O_2 \longrightarrow Ph_2C{=}N{-}N{=}CPh_2 + H_2O \quad (7)$$

$$Ph_2C{=}N{-}N{=}CPh_2 + 2\ H_2O \longrightarrow 2\ Ph_2C{=}O + \boxed{N_2H_4} \quad (8)$$

Chemistry of Imines

The first step in the cyclic process is the dehydrative condensation of a ketone with ammonia to form imine, which is implicitly anticipated as the intermediate precursor for the second step and should be unsubstituted at the N-site. The chemistry of imines has been reviewed extensively [18,19]. It is well known that substituted imines (Schiff base) could be obtained in high yield by dehydrative condensation of a ketone with an amine, preferably employing azeotropic removal of water. However, elaborate *preparations of monomeric imines unsubstituted at the N-site* were difficult except for benzophenone. Condensation of cyclohexanone with ammonia resulted in the formation of N-cyclohexenylcyclohexanone imine, due to the reaction of tautomeric enamine with an additional ketone.

Copper-Catalyzed Oxidative Coupling of Imine: Azine Synthesis

a) Copper complexes. Benzophenone azine $Ph_2C{=}N{-}N{=}CPh_2$ was obtained in high yield by the oxidative coupling of diphenylmethanimine $Ph_2C{=}NH$ in the presence of cuprous chloride. Oxygen is necessary to form azine, and a complex of cuprous chloride with the imine $CuCl{\bullet}Ph_2C{=}NH$ was isolated under nitrogen. The complex absorbs 1/4 mol of oxygen in solution to form $Cu(OH)Cl{\bullet}Ph_2C{=}NH$:

$$2\ Cu^{I}Cl{\cdot}Ph_2C{=}NH + 1/2\ O_2 + H_2O \longrightarrow 2\ Cu^{II}(OH)Cl{\cdot}Ph_2C{=}NH, \quad (9)$$

which contains the N-H bond of imine coordinated on divalent copper (3280 cm^{-1}) in the IR spectra, implying that dehydrogenation at the N-site of the imine has not yet occurred. The decomposition in solution above 120°C resulted in the formation of azine with reduction of Cu(II) to Cu(I).

b) Monodentate pyridines. A temperature above 120°C is required for the copper-catalyzed oxidative coupling of diphenylmethanimine to benzophenone azine, while in the presence of an excess monodentate pyridines, such as pyridine, γ-picoline and isoquinoline, the azine was obtained in high yield even at room temperature [15].

c) Active copper-species. It was expected that azine would be formed at a lower temperature in the presence of cupric chloride with excess pyridine, as in the case of cuprous chloride, but no azine was obtained. Cupric hydroxide and cupric nitrate were also inactive as catalysts as shown in Table 1. However, it is of interest that cupric chloride and cupric hydroxide were active in the presence of an equimolar amount of potassium hydroxide and hydrochloric acid, respectively. Basic copper chloride Cu(OH)Cl is suggested as an active intermediate:

$$4\,CuCl + O_2 + 2\,H_2O \longrightarrow 4\,Cu(OH)Cl \tag{10}$$

$$CuCl_2 + KOH \longrightarrow Cu(OH)Cl + KCl \tag{11}$$

$$Cu(OH)_2 + HCl \longrightarrow Cu(OH)Cl + H_2O \tag{12}$$

An analogous complex $Cu(OCH_3)Cl$ obtained by the oxidation of cuprous chloride in methanol revealed excellent activity [16].

Table 1
Active copper species in oxidative coupling of diphenylmethanimine[a] [15].

Catalyst	Benzophenone azine yield (%)
Oxygen preabsorbed CuCl	84.7
$CuCl_2$	0
$CuCl_2$ + KOH	73.1
$CuCl_2$ + 2 KOH	1.1
$Cu(OH)_2$	0
$Cu(OH)_2$ + HCl	52.9
$Cu(NO_3)_2 \cdot 3H_2O$	0

[a] $Ph_2C{=}NH$, 3.2g; Cu / imine = 1/7; pyridine, 50 cm^3. Oxygen was absorbed for 3h at 50°C.

d) Structure of copper complexes. Far-infrared spectra of $CuCl_2{\bullet}2Ph_2C{=}NH$ were quite similar to that of copper(II) bipyridine bichloride $Cupy_2Cl_2$, of which polymeric octahedral structure was determined by X-ray crystallography [20], and the magnetic susceptibility of 1.90BM suggests a structure with negligible Cu-Cu interaction. The polymeric octahedral structure bridged through the long Cu•••Cl linkage, similar to $Cupy_2Cl_2$ (1.8BM) and $CuCl_2{\bullet}2H_2O$ (1.87BM), would be reasonable.

Pyridine copper(II) methoxide chloride $pyCu(OCH_3)Cl$ is mononuclear at infinite dilution and binuclear at higher concentrations in pyridine [16]:

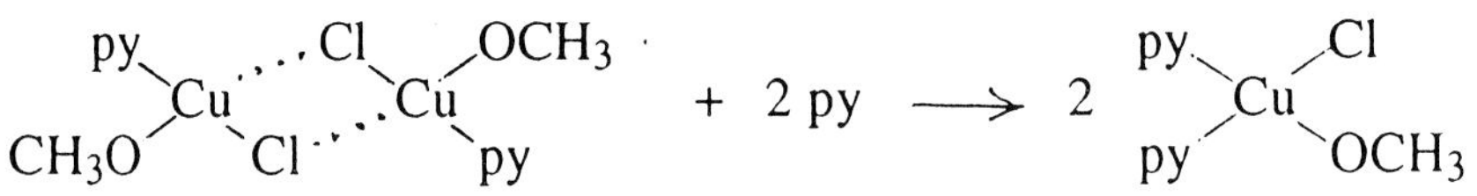

(13)

e) Copper-loaded polymer catalyst. Attempts were made to prepare copper-loaded polymer catalysts insoluble in the reaction media in oxidative coupling of diphenylmethanimine to azine [16]. Copolymers of 4-vinylpyridine and divinylbenzene were impregnated with copper methoxide chloride. The colorless or white copolymer suspended in methanol became green in color, and about 92-95% copper was found in the resultant catalysts. The highest activity was found in the macroreticular region of the copolymers. The catalyst was recovered by filtration and recycled four times without any chemical regenerations, to give a constant yield of azine of 91±2% [16].

Recovery of Hydrazine

Hydrazine was obtained in high yields in homogeneous hydrolysis of bezophenone azine in the presence of strong acids such as sulfuric acid and *p*-toluenesulfonic acid. The acids, however, formed inactive salts with the basic product hydrazine and should be used at least stoichiometric amounts. Lipophilic sulfonic acids showed high activity for conversion of benzophenone azine into hydrazine under phase-transfer conditions [13]. Well-designed sulfonated resins with moderate acidity and with uniform acid-site distribution prepared by sodium *p*-styrenesulfonate as the starting monomer revealed excellent activity as the immobilized phase-transfer catalyst [14]. A reactor device capable of continuous flow operations for the triphase system without the running-out of catalyst particle has been proposed by the authors and demonstrated over 50 h-on stream with 90% conversion of azine [14].

V. ATTEMPTED CATALYTIC HYDROGEN PEROXIDE OXIDATION

Activation of Hydrogen Peroxide : Epoxide and Oxaziridine

In the presence of a stoichiometric amount of a nitrile, hydrogen peroxide is activated to form an intermediate iminohydroperoxide (peroxycarboimidic acid), which is capable of epoxidizing olefins under mild conditions [12].

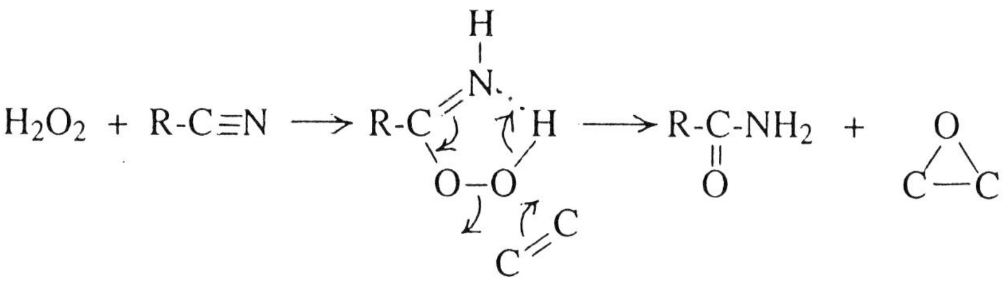

(14)

A commercially available 30% aqueous solution of hydrogen peroxide is enough to prepare the active intermediate, which behaves similarly as peracids. The aqueous solution can be handled more safely and is effective even under basic conditions. Thus the oxidation of C=N gives oxaziridines (Eq. 5) [11], and the reaction was extended to the >C=O / NH_3 system to afford azine [8] (Scheme 2), leading to a new route in hydrazine technology, and the peroxide process is now on-stream in a commercial scale [2].

Titanium Silicalite-1

Titanium silicalites (TS-1) with *MFI* topology [21] provided new opportunities during the last decade in catalytic oxidation [22-26], among which epoxidation of olefins [22-24] is of interest for reference analogy in the catalytic hydrogen peroxide oxidation of C=N bond to oxaziridines. Applications of TS-1 catalyzed hydrogen peroxide oxidation to ketone-ammonia system are of more recent trends [26-32] in selective oxidation chemistry.

Oxiaziridine Intermediate

On the analogy of oxidation with H_2O_2-RCN; C=C to epoxide and C=N to oxaziridine, it is of interest to apply the epoxidation catalyst, TS-1, in the oxidation of ketone-NH_3 system. *If* the three-membered ring compound, *oxaziridine* unsubstituted at the N-site >C-NH (with O bridging C and N), was the predominant intermediate, subsequent reaction of the unstable oxaziridine with ammonia through *cleavage at the N-O bond* gives hydrazone followed by the condensation with an additional ketone to afford *azine* as the final product (Eq. 15), while *cleavage at the C-O bond* through hydration gives *oxime*, an isomer of the oxaziridine intermediate (Eq. 16).

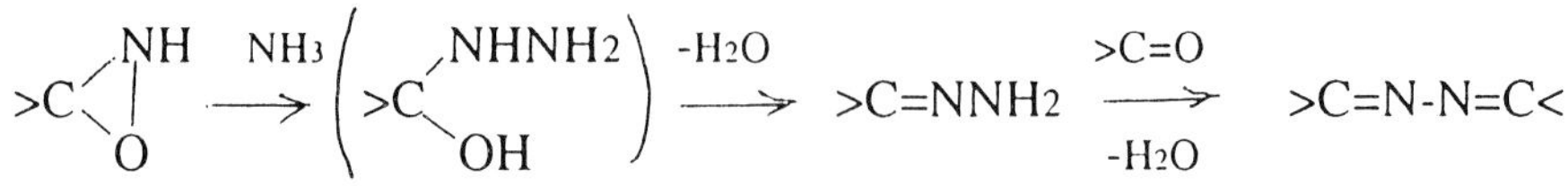 **(15)**

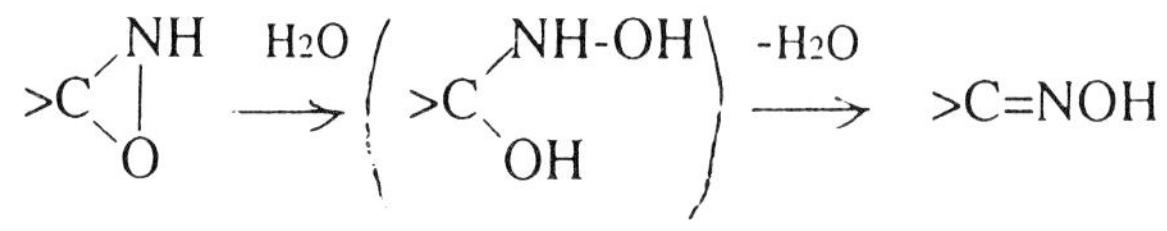 **(16)**

Hydroxylamine

A highly selective formation of cyclohexanoxime by direct ammonia oxidation with hydrogen peroxide in the presence of cyclohexanone, using titanium silicalite (1.5% Ti) as catalyst, was reported in 1989 [26]. The reaction in water / t-butanol at 80°C for NH_3:H_2O_2:cyclohexanone = 2.1:1.05:1 (mol) with 2 wt% catalyst gave 99.9% conversion of cyclohexanone to yield 93.2% oxime based on H_2O_2. The results of a breakthrough in caprolactam chemistry by the TS-1 catalyzed *ammoximation* were reproduced in various laboratories [27-31] including many other ketones. The *imine* (>C=NH; a precursor of oxaziridine) has been postulated as intermediate in the ammoximation [27,28]. However, hydroxylamine was obtained in high yields of 50-70% based on H_2O_2 in TS-1 catalyzed ammonia oxidation in the *absence* of ketone [31], and ketones with large molecular size, unable to diffuse into the zeolite pore, give good yields in ammoximation [31].

Hydrazine

Another application of TS-1 catalyzed hyrogen peroxide oxidation of ketone-NH_3 system is to hydrazine synthesis via azine. The reaction of heptan-4-one, NH_3 (28% aq. solution) and H_2O_2 (30% aq.solution) with an ingredient ratio of 1 : 17.09 : 0.51 (mol) at 55°C in the presence of 0.2 g of TS-1 (Ti/Si=0.026) and related catalysts has been reported [32] and a part of the results are given in Table 2. The major product in the earlier stage of reaction (less than 5 h) was oxime as anicipated, but hydrazine was obtained with a yield as high as 55.6% after 24 h with a recovery of heptanone, namely, ketone conversion decreased at a longer time of reaction. The results are interpreted [32] similarly as in the copper-catalyzed cyclic process given in Eqs. 6-8 that the reaction involves the condensation of heptan-4-one with ammonia to the corresponding imine in the presence of TS-1, followed by oxidative coupling of imine to azine, and hydrolysis of azine into hydrazine with recovery of heptan-4-one. The authors note that heptan-4-one azine was not detected either and may react very quickly to yield hydrazine [32], but ketazines are generally stable for hydrolysis and should be detected, if really formed, as a stable precursor of

Table 2
Hydrazine synthesis by catalytic oxidation over Ti-silicalite of *MFI* structure[a] [32].

Time (h)	Ketone conv. (mol%)	Product distribution (mol%)			H_2O_2 decomp. (mol%)
		Imine	Oxime	Hydrazine	
1	20.9	11.0	75.2	13.8	6.5
3	38.2	9.8	71.0	19.2	9.7
5	51.0	4.5	73.2	22.3	12.7
24	22.3	5.2	39.2	55.6	---

[a]Conditions: catalyst TS-1 (Ti/Si=0.026); temperature, 328K; molar ratio, heptan-4-one:NH_3:H_2O_2 = 1:17.09:0.51.

hydrazine. Methyl ethyl ketazine is hydrolyzed by the reactive distillation under pressure (0.8-10 MPa) in the commercial peroxide process to give concentrated aqueous hydrazine as bottom and overhead ketone, which is recycled [2].

It appears generally accepted in literatures [26-31] that the major product in TS-1 catalyzed oxidation of ketone-ammonia system is *oxime*, and a detailed byproduct study in ammoximation of cyclohexanone on TS-1[30] reported the formation of nitrocyclohexane, cyclohexenyl cyclohexanone, cyclohexenone oxime and cyclohexanone azine, among which azine detected was at most trace amounts of 50 ppm. Experimental detail especially for analysis in the reaction in Table 2 is not given in the ref. [32], and it is difficult to follow the procedure exactly, but hydrogen peroxide oxidations of heptan-4-one-NH_3 system catalyzed by TS-1 *with and without added azine* would provide interesting information with regard to hydrazine synthesis.

Prior to the reaction studies, authentic samples of heptan-4-one oxime (M=129; bp.190-195°C) [33] and heptane-4-one azine (M=224; bp.188°C)[33] were prepared in our laboratory by the reactions with hydroxylamine hydrochloride and hydrazine hydrate, respectively. The products were identified by a quadrapole mass spectrometer (direct injection mode, GCMS-QP1000, Shimadzu, Kyoto) as the parent peaks of M^+ = 129 and 224 for oxime and azine, respectively, and used for calibration in GC analysis by 15% PEG 4000 / Uniport B, 3^{Φ}-x 2m, 150°C, 263-FID, Hitachi, Tokyo. Hydrazine in aqueous phase was analyzed by the spectrophotometry (UV-2500, Shimadzu) with *p*-dimethylaminobenzaldehyde in acidic ethanol. Trace amounts of hydrazine as low as 1-20 μmol/*l* could be detected at 458 nm due to the intense color of *p*-dimethylaminobenzaldehyde azine [34]. Preparation and characterization of titanium silicalite-1 were described in the previous paper [25].

The TS-1 catalyzed hydrogen peroxide oxidation was carried out in the following manner: To a 50-*ml* round bottom flask equipped with a reflux condenser was added 1.68 *ml* (1.37g; 12.0 mmol) of heptan-4-one, 17.2 *ml* (22 wt%, 15.8 g;

206 mmol NH_3) of aqueous ammonia, 0.55 *ml* (33.3 wt%, 0.616 g; 6.03 mmol H_2O_2) of aqueous hydrogen peroxide, and 0.2g of TS-1. After magnetically stirred for a given time at 55 °C, the milky white heterogenous liquid was centrifuged for 30 min to separate the catalyst. The liquid was extracted with ether, which was then evaporated off, and the colorless oily residue was diluted with ethanol exactly to 10 ml to analyze for unreacted ketone, product oxime and azine (if present) by gas chromatography. An aliquot of the aqueous phase remained in ether extraction was taken for iodometric titration of unreacted hydrogen peroxide. Table 3 summarizes

Table 3

Hydrogen peroxide oxidation of heptan-4-one-NH_3 system catalyzed by titanium silicalite-1 with and without added azine[a].

Catalyst	Time	Ketone[b]		H_2O_2		Oxime		Azine
(mol%Ti)	h	mmol	conv.%	mmol	conv.%	mmol	sel.%[c]	mmol
TS-1	0	12.0	0	6.03	0	0	0	0
(3.43%Ti)	5	3.26	72.8	4.25	29.5	1.30	73.1	0
	24	4.08	66.0	0.22	96.4	3.92	67.5	0
TS-1	0	12.0	0	6.03	0	0	0	4.05
(4.27%Ti)	5	5.93	50.5	4.28	29.0	1.22	69.7	4.04

[a]Conditions: same as in Table 2.
[b]Heptan-4-one.
[c]Based on H_2O_2 consumed.

the results for hydrogen peroxide oxidation of heptan-4-one-NH_3 system catalyzed by TS-1 with and without added azine. Thus heptan-4-one oxime was obtained as the major product in 67-73% selectivity based on H_2O_2 consumed, in rather good agreement with the results given in Table 2. We could not detect azine either in organic phase by gas chromatography. Hydrazine detected in aqueous phase by the spectrophotometry [34] with *p*-dimethylaminobenzaldehyde was a trace amount of 0.17 μmol at 24 h, where 3.92 *m*mol of oxime was obtained, and the molar ratio of hydrazine was negligibly small as low as 1/23,060 as compared with oxime. The reaction repeated under the same conditions except for the *added* azine in the initial ingredients yielded *the azine remained intact* under the reaction conditions for 5 h.

VI. CONCLUSION

A cyclic catalytic oxidation process for hydrazine synthesis from benzophenone,

significant effect if the reaction condition is chosen such that all palladium salt was dissolved in the reaction mixture. Employment of excess aniline and small concentration of Pd in our case should have satisfied this condition.

One interesting finding in the present carbonylation system was that much more aniline was consumed than nitrobenzene, especially when reaction mixtures of high aniline/nitrobenzene ratios were charged (Table 1). The possibility of a

Table 1
Effect of different initial charge of aniline (AN) and nitrobenzene (NB) in N,N'-diphenylurea synthesis at 120 °C and 40 bar.

Initial Charge[a]/mmol			Amount Consumed/mmol	
AN	NB	Reaction Time/h	AN	NB
50	50	4.0	49	43
100	50	6.0	98	49
150	50	4.5	121	50
300	50	2.5	144	50

[a]$Pd(CH_3COO)_2$, 0.67 mmol; PPh_3, 3.8 mmol; NEt_4Cl, 2g; xylene, 60 g. The typical isolated yield of the urea product was 98% of the value calculated from reacted aniline and nitrobenzene.

reaction between aniline itself is excluded because no reaction takes place without nitrobenzene. There must be a reaction that consumes more aniline than nitrobenzene. A possible candidate is reaction 2, which consumes 5 mol of aniline per mol of nitrobenzene.

$$PhNO_2 + 5PhNH_2 + 3CO \rightarrow 3PhNHCONHPh + H_2O \qquad \textbf{(2)}$$

The formation of water was also confirmed. Tsuji *et al.* [23] proposed that both reactions 1 and 2 might take place with a catalyst system of $PtCl_2(PPh_3)_2$-$SnCl_4$-NEt_3. However, no experimental results were provided indicating the importance of reaction 2.

In order to obtain more information about the reaction, we studied the reaction between fully deuterated nitrobenzene-d_5 and the ordinary aniline and CO under otherwise same conditions. The main product DPU and an impurity (less than 1 wt%) product N,N'-diphenyloxamide (DPOA) were subjected to mass spectroscopic analysis to determine the deuterium distribution. Three forms each of DPU (PhNHCOHNPh) and DPOA (PhNHCOCOHNPh) were found depending on the number of deuterated phenyl rings (Ph). We denote these as x, y and z forms when, respectively, 0, 1 and 2 deuterated phenyl rings are present. Obviously these species have originated from two molecules of aniline (x), one molecule each of nitrobenzene and aniline (y), and two molecules of nitrobenzene (z). The abundance of x, y and z species in DPU and DPOA determined from the relative height of molecular peaks in mass spectra are summarized in Table 2. Also shown are ratios

Table 2
Mass spectroscopic analysis of solid products formed in catalytic reaction of nitrobenzene-d_5, aniline, and carbon moxoxide[a].

Initial Charge Ratio AN/NB	Product[b]	Relative Abundance x	y	z	C_6H_5/C_6D_5 in the Product
6.0	DPU	3.7	1.0	<0.1	7.0
	DPOA	17.6	1.0	-	34.8
2.0	DPU	1.6	1.0	0.2	2.8
1.0	DPU	0.9	1.0	0.4	1.5
	DPOA	4.5	1.0	0.3	6.0
0.5	DPU	0.7	1.0	0.5	1.2

[a]Nitrobenzene-d_5 50 mmol; $Pd(CH_3COO)_2$, 0.67 mmol; PPh_3, 3.8 mmol; NEt_4Cl, 2g; toluene, 60g; 120 °C; CO, 40 bar.
[b]N,N'-diphenylurea (DPU) or N.N'-diphenyloxamide (DPOA) in solid products.

of total undeuterated to deuterated phenyl groups contained in DPU or DPOA. This ratio represents the relative amounts of aniline and nitrobenzene-d_5 consumed to form that product. The ratios for DPU are always greater than 1 even when initial charge ratios are equal or less than 1. The results unequivocally show that the stoichiometry of the reaction 1, which consumes nitrobenzene and aniline on an equimolar basis cannot account for the palladium-catalyzed DPU synthesis. Reaction 2 becomes particularly important when excess aniline is employed relative to nitrobenzene.

When the filtrate was analyzed by GC-MS, two interesting observations were made: (i) deuterated aniline, $C_6D_5NH_2$ was found, and (ii) the amount of aniline remaining in the filtrate was always significantly greater than had been calculated from the balance between initial charge and the amount used for DPU formation. These indicate that some nitrobenzene (d_5) has been converted to aniline during DPU synthesis. This explains why C_6H_5/C_6D_5 ratios in DPU (Table 2), could exceed aniline/nitrobenzene ratios consumed (Table 1) and even the value predicted by reaction 2, namely 5.0.

The presence of DPOA in the product with C_6H_5/C_6D_5 ratios greatly different from those for DPU could provide a clue for the reaction intermediate. Let us consider possible reaction paths leading to the formation of DPU as shown in Figure 1. Path **a** assumes initial coordination of nitrobenzene to a Pd complex [Pd] and the formation of a nitrene intermediate PhN[Pd] by deoxygenation with CO. Carbonylation of nitrene forms a phenylisocyanate complex PhNCO[Pd], which subsequently reacts with aniline to give the desired DPU. This path requires equal moles of nitrobenzene and aniline to form DPU and satisfies reaction 1. It has been proposed previously as a mechanism for DPU synthesis from nitrobenzene, aniline, and CO in the presence of Pd [21], Ru [28], and Pt [23] complexes. In the case of Ru complex catalysts, the nitrene intermediate was isolated and its reactivity was studied under the conditions of alkyl phenylcarbamate synthesis (another important intermediate in phosgene-free MDI synthesis) without aniline [28,29].

When aniline is initially coordinated to the metal, two paths are possible depending on whether the removal of hydrogen by nitrobenzene from the active catalytic complex comes first (path **b**), or CO insertion does (path **c**). Path **b** involves the nitrene intermediate, like path **a**, while path **c** involves a carbamoyl intermediate PhNHCO[Pd]. Steps following the formation of phenylisocyanate complex are the same as path **a**. Both paths **b** and **c** satisfy reaction 2, because a mole of nitrobenzene-d_5 oxidizes 6 moles of aniline and itself is converted to aniline-

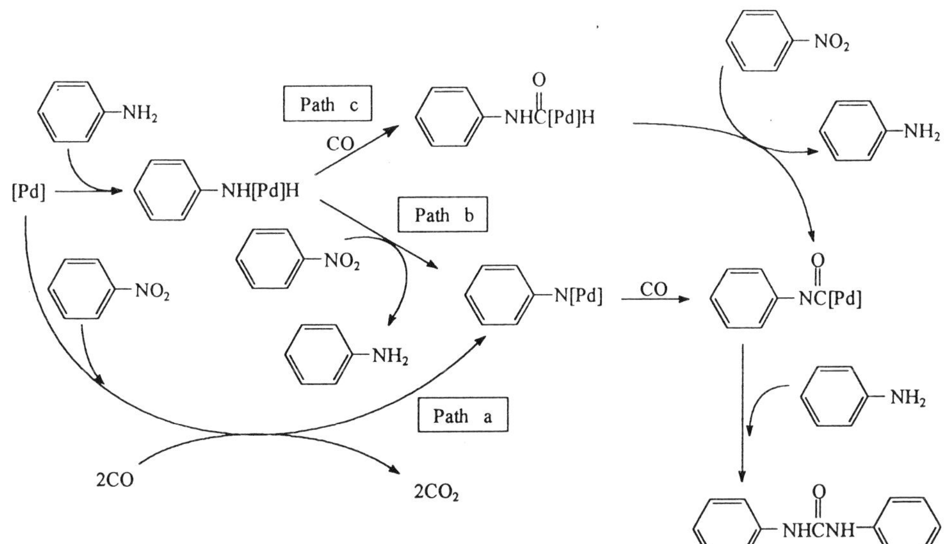

Figure 1. Possible reaction pathways of Pd-catalyzed synthesis of N,N'-diphenylurea from nitrobenzene, aniline and CO.

d_5. The carbamoyl metal species have been thought to play important roles in carbonylation of amines [23,30] although there has never been positive identification of the complexes under conditions of DPU or alkyl phenylcarbamate synthesis.

Among the possible intermediates in DPU synthesis shown in Figure 1, only the carbamoyl intermediate or the phenylisocyanate complex could form DPOA by a coupling reaction. However, the much greater values of C_6H_5/C_6D_5 in DPOA than that in DPU suggest that its source is most probably the carbamoyl intermediate, which has only a small probability of containing the deuterated phenyl group, namely, the case when deuterated aniline formed from nitrobenzene-d_5 starts the catalytic cycle through path **c**.

The carbonylation reaction was tested in a continuous mini-pilot plant together with other subsequent steps to produce MDI. All results obtained in the batch reactor were reproduced and long-term stability of the catalyst was also confirmed. The new process was found to be economically competitive compared with the current phosgene process in an economic feasibility study [31].

Carbonylation of Formaldehyde to Methyl Glycolate over Cationic Ion Exchange Resins

Present commercial production of ethylene glycol is based on the partial oxidation of ethylene to ethylene oxide which is then hydrated to ethylene glycol [32]. In spite of more than 50 years of commercial operation, the overall yield of the process is limited to the vicinity of 70%. A number of routes have been explored to synthesize ethylene glycol by C_1 chemistry: (i) direct production from syngas [33], (ii) hydroformylation of HCHO to glycolaldehyde [34,35], (iii) carbonylation of HCHO to glycolic acid, (iv) oxidative coupling of CO to oxalate esters [36], and (v) redox reaction of CH_3OH and HCHO [37].

We explored acid-catalyzed carbonylation of HCHO to glycolic acid and its esterification to methylglycolate (MG).

$$HCHO + CO + H_2O \rightarrow HOCH_2COOH \qquad \textbf{(3)}$$

$$HOCH_2COOH + CH_3OH \rightarrow HOCH_2COOCH_3 + H_2O \qquad \textbf{(4)}$$

The MG is catalytically hydrogenated to ethylene glycol. This route was once practiced commercially by the Du Pont company. However, the process required 130-200 °C and 900 bar with H_2SO_4 as a catalyst. This condition gives high corrosion rates with conventional materials of construction, and thus the process was discontinued in 1968. More recent studies employing strong acid catalysts such as HF [38] or concentrated H_2SO_4 and copper carbonyls [39] demonstrated that the carbonylation could proceed under extremely mild conditions. However, the use of strong liquid acids has been hampered by stringent environmental regulations. The problem could be overcome by employing solid acid catalysts. We found that strongly acidic ion exchange resins showed excellent activity and methyl glycolate (MG) selectivity [16]. They exhibited significantly improved performance over the previously reported similar solid acid catalyst systems [40,41]. Such solid acid catalysts are expected to provide decreased corrosion, facile separation of product from catalyst, and safer handling compared to liquid acid catalysts.

The carbonylation of HCHO produces a complicated spectrum of products. In addition to the desired MG, the reaction product contained methyl formate (MF), methyl methoxy acetate (MMAc, $CH_3OCH_2COOCH_3$), and some unidentified by-products of high molecular weights denoted as "unknowns". The MMAc, if hydrogenated, produces glycol ethers instead of the desired ethylene glycol. Inclusion of these by-products seen by GC, however, could not account for all the HCHO that had converted. There are obviously other by-products not detected by GC. A possible candidate is the acid-catalyzed condensation of HCHO to polyhydroxy aldehyde as proposed by Hendriksen [40].

$$(n + 1)\ HCHO \rightarrow H[CH(OH)]_nCHO \qquad \textbf{(5)}$$

The aldehyde could be subsequently carbonized. Indeed, some catalysts turned brown after reaction when the selectivity to MG was low.

The rate of HCHO conversion and selectivities to MG after 1 h of the reaction with various catalysts are compared in Table 3. Per proton basis, Nafion showed the highest activity. This reflects the stronger acidity of Nafion compared with polystyrenesulfonic acid resins due to the electron-withdrawing effect of its perfluorocarbon backbones. Yet, the concentration of acid sites in Nafion is much less, and thus its specific activity of HCHO conversion is lower. All the polystyrenesulfonic acid resins showed similar performance. The negligible difference between macroreticular type (Amberlyst and Dowex) and gel type (Amberlite) indicates that the reaction takes place in the swelled bulk phase of the resins. Furthermore, the effects of surface area (Amberlyst 15 *versus* XN-1010), degree of cross-linking (Dowex 50′8-100 *versus* 50′2-100), and bead size (Dowex 50′8-400 *vs.* 50′8-100) were not evident.

The HCHO conversion increased with temperature, yet the MG yield showed a maximum near 135 °C. MG yield also showed a maximum at a certain

Table 3
Carbonylation of HCHO over ion exchange resin catalysts[a].

Catalyst	HCHO Reacted/H^+ (mol/equiv)	HCHO Reacted/gcat (mmol/g)	MG Selectivity[b] (%)
Nafion NR-50	15.7	12.72	81.6
Amberlyst 36	5.4	25.38	87.9
Amberlyst 15	5.2	24.44	79.6
Amberlyst XN-1010	4.5	14.85	58.9
Amberlite IR 120	4.4	21.15	61.9
Dowex 50x8-100	4.5	22.95	88.9
Dowex 50x8-400	4.4	22.71	85.1
Dowex 50x 2-100	3.0	15.32	84.7

[a]1 h reaction with 4 g of catalyst at 135 °C, 240 bar CO, water/HCHO = 1 mol/mol, 80 cm^3 of 1,4-dioxane.
[b]Selectivity = 100 x mol of MG produced/mol of HCHO reacted.

reaction time. Hence, the formed glycolic acid appears to undergo a secondary reaction to form glycolic acid oligomers such as reaction 6.

$$nHOCH_2COOH \rightarrow H(OCH_2CO)_nOH + (n-1)H_2O \qquad \textbf{(6)}$$

Methyl formate may originate from dimerization of formaldehyde and its formation showed a slight decrease with temperature. Above 150 °C, it was found by GC/MS analysis that some of the 1,4-dioxane solvent decomposed. The effect of CO pressure was tested. After 2 h at 135 °C, almost complete conversion of HCHO was obtained at CO pressure above 50 bar. Yet, MG yields greater than 80% were obtained only above 200 bar. At low pressures, HCHO appears to be consumed by side reactions such as reaction 5 instead of being carbonylated.

The effect of solvent is summarized in Table 4 in the absence of and presence of water. Solvents showed different behaviors depending on the presence or absence of water in the reaction mixture. Without water, acetic acid gave the highest yield of MG. Acetic acid is the solvent recommended by Sugita [41] for this reaction over sulfonic acid resin catalysts. With the acetic acid solvent, the carbonylation product is not glycolic acid, but acetylglycolic acid (CH_3COOCH_2COOH). The latter could be converted to MG and methyl acetate in the subsequent esterification step. However, methyl acetate should be hydrolyzed to acetic acid for solvent recycle. 1,4-Dioxane is inert under the

Table 4
Effect of solvent for HCHO carbonylation over Amberlyst 15 with and without water[a].

Water	Solvent	HCHO conv. (%)	Yield (%)				MG Selectivity (%)
			MG	MMAc	MF	Unknowns	
without	1,4-dioxane	99.9	53.31	1.84	0.03	9.72	53.4
	acetic acid	99.6	91.23	-	2.59	6.97	91.6
	methyl benzoate	97.7	83.14	5.94	1.38	8.37	83.0
	methyl formate	99.7	43.42	62.37	-	38.58	44.3
	methyl acetate	89.3	51.25	20.39	0.97	37.70	57.4
with	1,4-dioxane	93.8	78.52	1.53	4.57	3.62	83.6
	acetic acid	95.7	58.61	-	3.84	37.09	61.2
	formic acid	92.9	80.63	-	-	12.43	86.7
	methyl benzoate	97.7	83.81	6.16	5.32	4.25	85.8
	methyl formate	99.7	55.82	47.53	-	42.43	57.1
	methyl acetate	95.8	66.57	26.20	4.31	73.64	69.4
	methanol	-	2.93	0.03	-	-	-
	CH_2Cl_2	83.6	61.24	0.32	4.72	24.21	73.1

[a]$HCHO/H_2O/H^+$ = 10/10/1 for 0.12 mol of HCHO, 2 h, 135 °C, 240 bar, solvent 80 cm^3.

present condition. Hence, the carbonylation product cannot be stabilized as glycolic acid in the absence of sufficient water. Instead, glycolide or polyglycolides may be formed by oligomerization of the intermediate in a similar way as in reaction 6 [38,40]. These materials could have been depolymerized in the subsequent esterification step. The fact that the MG yield in 1,4-dioxane is substantially lower in the absence of water indicates, however, that the depolymerization is not complete.

Some esters have been tested as solvents. They gave good yields of MG, yet produced large amounts of by-products such as MMAc and unknowns. Methyl benzoate, however, showed excellent selectivities. Ester solvents are reactive under the present reaction conditions and participate in carbonylation reactions. They are also converted to unknown products as evidenced by the fact that the sum of product yields exceeds 100%. The good MG selectivity with benzoate may arise from the stability of the intermediate compound $C_6H_5COOCH_2COOCH_3$ or its precursor complex due to an electron-withdrawing effect of the benzene ring. The reaction did not proceed appreciably with methanol as a solvent. Dichloromethane also produced a large amount of unknowns. Except for acid solvents, all solvents showed better MG selectivities in the presence of water. The effects of various reaction variables reported above could be understood on the basis of a reaction network shown in Figure 2.

The present work proposes an efficient catalytic system consisting of a sulfonic acid resin catalyst and 1,4-dioxane solvent, and employing aqueous HCHO in order to manufacture methyl glycolate as a precursor to ethylene glycol. The system showed a good stability in a life test performed in a continuous reaction system. Although the reaction conditions are severer than the efficient liquid acid catalyst systems [38,39], the present solid catalyst system offers facile separation of product from catalyst, decreased corrosion, and safer handling. It is environmentally more sound. Compared with other solid acid systems previously reported, polystyrene sulfonic acid resin catalysts show higher specific activities than more expensive Nafion [40]. The present system has advantages over the same sulfonic acid resin catalyst system of Sugita [41] in that it employs unreactive 1,4-dioxane solvent and aqueous HCHO compared to the anhydrous acetic acid system of Sugita. Both the use of anhydrous HCHO and employment of one more hydrolysis step for acetic acid recovery could give a significant economic penalty.

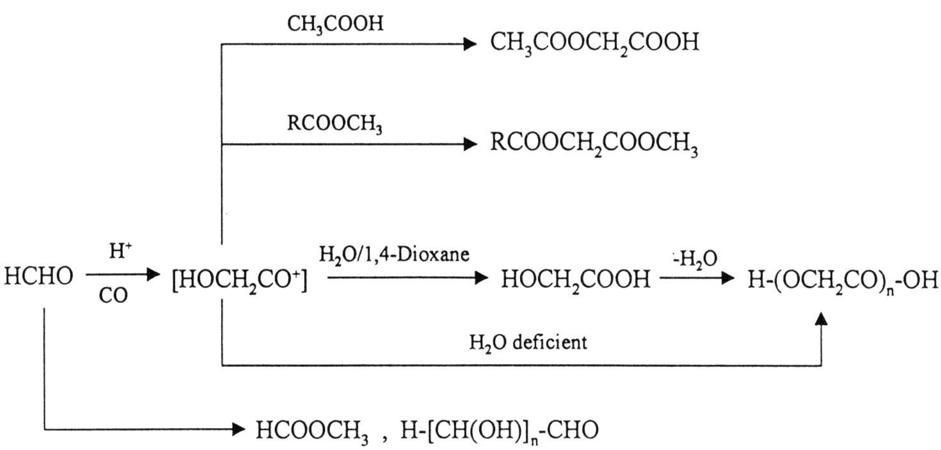

Figure 2. Reaction network of acid-catalyzed HCHO carbonylation.

Regioselective Synthesis of Ibuprofen via the Palladium Complex-Catalyzed Hydrocarboxylation of 1-(4-isobutylphenyl) Ethanol

There have been numerous publications related to the hydrocarboxylation of alkenes and alcohols to carboxylic acids with carbon monoxide and water in the presence of palladium complex catalysts [42-47]. The reaction is related to the last step in a new process for the production of ibuprofen, a large-volume nonsteroidal antiinflammatory drug. Among various processes developed for the manufacture of ibuprofen, the new process of Hoechst-Celanese involves three elegant steps, starting with 4-isobutylbenzene; catalytic acylation, hydrogenation, and carbonylation. More recently much attention has been paid to synthesizing selectively the (S)-(+)-ibuprofen enantiomer which has the active effect [48-51].

Our previous studies [19-20] demonstrated that the highly regioselective carbonylation of 4-methylstyrene to α-(4-methylphenyl) propionic acid and its ester form could be achieved in the presence of $PdCl_2$-$CuCl_2$-PPh_3 catalysts. We applied a similar catalyst system to the preparation of α-(4-isobutylphenyl) propionic acid, ibuprofen, from 1-(4-isobutylphenyl) ethanol (IBPE).

A typical change with reaction time in the concentrations of the reactant and products are shown in Figure 3. The reaction appears to proceed through a sequence of dehydration to 4-isobutylstyrene (IBS) and its subsequent carbonylation reactions. Hence, the balance in catalyst between dehydration and carbonylation functions should be important. Under some conditions, 1-(4-isobutylphenyl) ethylchloride (IBPCl) or 1-(4-isobutylphenyl) ethane is also formed.

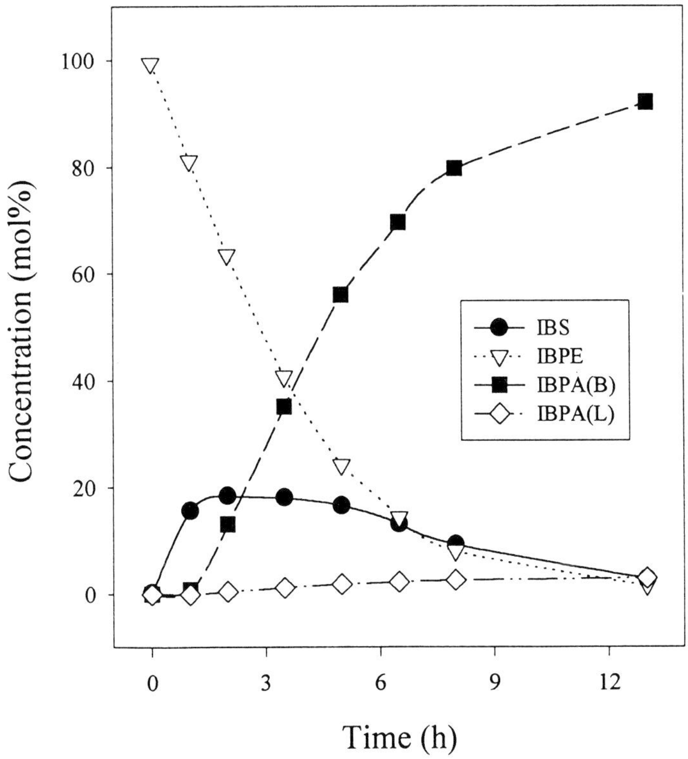

Figure 3. The hydrocarboxylation of 1-(4-isobutylphenyl) ethanol at 40 bar of CO and 125 °C. IBPE = 5 g, PPh_3/Pd = 2.2, 10% HCl = 10 g, 3-pentanone = 81 cm^3, 13 h.

Table 5 shows the effect of mole ratio of IBPE to palladium complex catalyst while the ratio of phosphine ligand to palladium was fixed at 2.2. When the IBPE/Pd mole ratio changed from 511 to 64 by increasing Pd concentration, the conversion of IBPE increased only slightly. The most conspicuous effect was the shift of selectivity from the dehydration product (IBS) to carbonylated products (IBPA). Furthermore, the selectivity to branched acid over linear isomer decreased. Because of the last effect, to achieve more than 98% purity of ibuprofen, the preferred IBPE/Pd mole ratio had to be more than 128.

Table 5
Hydrocarboxylation of 1-(4-isobutylphenyl) ethanol (IBPE) with different IBPE/Pd ratios[a]

IBPE/Pd (mol/mol)	Conversion (%)	Selectivity (%)				B/L Ratio
		IBS	IBPCl	IBPA(B)	IBPA(L)	
511	97.6	27.4	5.1	66.5	0	∞
255	99.6	0.4	0	98.9	0.6	138.2
128	100	0	0	98.0	2.0	47.8
64	100	0	0	93.3	6.7	14.0

[a]IBPE = 5 g, PPh3/Pd = 2.2, 10% HCl = 10 g, 3-pentanone = 81 cm^3, 125 °C, 50 bar CO, 13 h.

The effects of PPh_3 are summarized in Table 6. Without PPh_3 added, only the dehydration reaction occurred, which revealed that the PPh_3 was the essential component for the palladium complex to be effective for the carbonylation step. The increase of PPh_3/Pd mole ratio improved the reaction rate but reduced the selectivity to branched acid. There have been many reports in which excess

Table 6
Effects of PPh_3 on hydrocarboylation of 1-(4-isobutylphenyl) ethanol[a].

Pph_3/Pd	Conversion (%)	Selectivity (%)			B/L Ratio
		IBS	IBPA(B)	IBPA(L)	
0	97.1	94.2	5.8	0	∞
2.2	100	0	98.0	2.0	47.8
4.4	100	0	95.4	4.6	20.6

[a]IBPE = 5 g, $PdCl_2$ = 0.039 g, 10% HCl = 10 g, H_2O = 17 g, 3-pentanone = 81 cm^3, 125 °C, 50 bar CO, 13 h.

phosphine was used in order to prevent the decomposition of palladium-phosphine ligand complexes.

The rate of carbonylation reaction increased slightly as the pressure increased. A more significant effect was observed on the ratio of branched to linear IBPA (B/L ratio). Higher CO pressures gave higher B/L ratios. When total reaction pressure was maintained at 40 bar with 10 bar of CO balanced with nitrogen, the results were similar to those for 10 bar of pure CO, indicating that partial pressure of CO is important. This implies that abundant CO which dissolved in the liquid phase is very important for the carbonylation rate and selectivity [19,52].

The carbonylation of IBPE hardly occurred in the absence of HCl. When Br or I was employed, the reaction rate increased slightly whereas the B/L ratio decreased significantly. In the case of HF, the reaction was so slow and the carbonylation step did not occur at all. Interestingly, when HI was added in place of generally used HCl or HBr, the hydrogenation of IBPE or IBS occurred and gave fully hydrogenated product, 1-(4-isobutylphenyl) ethane.

In order to separate the effects of proton and chloride ion, sulfuric acid and metal halides were used as proton and chloride ion sources respectively. When only sulfuric acid was employed, IBS was formed predominantly. When only KCl was provided carbonylation did not take place for 13 hours. When H_2SO_4 and KCl were supplied together, then, ibuprofen was prepared in good yields. This suggests that the carbonylation of IBPE requires strongly acidic condition as well as a halide ion added.

The hydrocarboxylation of IBPE to ibuprofen was achieved in excellent regiospecificity in the presence of $PdCl_2$-PPh_3-HCl. Important reaction variables that affect the reaction rate and selectivity to branched IBPA were the amount of catalytic complex, solvent, the ratio of phosphine ligand to palladium, pressure, temperature, and the nature of halide ion used. HCl was an essential co-catalyst which provided both proton and chloride ion. The proton is necessary as a dehydration catalyst and chloride ion is one of the main factors determining the selectivity to branched acid in the carbonylation step.

REFERENCES

1. W. Keim (Ed.), *Catalysis in C_1 Chemistry*, D. Reidel, Dordrecht, 1983.
2. D.R. Fahey (Ed.), *Industrial Chemical via C_1 Processes*, American Chemical Society, Washington D.C., 1987.
3. J.S. Lee, J.C. Kim, and Y.G. Kim, *Appl. Catal.* **57**, 1 (1990).
4. M. Cheong, S.H. Lee, J.C. Kim, J.S. Lee, and Y.G. Kim, *J. Chem. Soc., Chem. Commun.* 661 (1990).
5. H.C. Woo, K.Y. Park, Y.G. Kim, I. Nam, J.S. Chung and J.S. Lee, Appl. Catal. **75**, 267

(1991).
6. M. Cheong, S.H. Lee, Y.S. Sa, J.C. Kim, J.S. Lee, and Y.G. Kim, *J. Mol. Catal.* **68**, 277 (1991).
7. K.Y. Park, W.K Seo, and J.S. Lee, *Catal. Lett.* **11**, 349 (1991).
8. S.J. Choi, J.S. Lee, and Y.G. Kim, *J. Mol. Catal.* **85**, 109 (1993).
9. K.M. Kim, M. Cheong, J.C. Kim, J.S. Lee, and Y.G. Kim, *Korean J. Chem. Eng.* **7**, 259 (1990).
10. E.J. Kim, S.H. Lee, K.H. Park, and J.S. Lee, *Korean J. Chem. Eng.* **8**, 53 (1991).
11. J.S. Lee, J.W. Woo, C.W. Lee, K.S. Hong, and J.K. Yeo, *Korean J. Chem. Eng.* **7**, 145 (1990).
12. J.S. Lee, C.W. Lee, S.M. Lee, and K.H. Park, *J. Mol. Catal.***61**, L15 (1990).
13. J.S. Lee, S.M. Lee, J.K. Yeo, C.W. Lee, and J.S. Lee, *Ind. Eng. Chem. Res.***30**, 1456 (1991).
14. S.M. Lee, N.S. Cho, K.D. Kim, J.S. Oh, C.W. Lee, and J.S. Lee, *J. Mol. Catal.* **73**, 43 (1992).
15. K.D. Kim, S.M. Lee, N.S. Cho, J.S. Oh, C.W. Lee, and J.S. Lee, *J. Mol. Catal.* **75**, L1 (1992).
16. S.Y. Lee, J.C. Kim, J.S. Lee, and Y.G. Kim, *Ind. Eng. Chem. Res.* **32**, 253 (1993).
17. C.W. Lee, J.S. Lee, S.M. Lee, N.S. Cho, K.D. Kim, and J.S. Oh, *J. Mol. Catal.* **81**, 17 (1993).
18. C.W. Lee, S.M. Lee, J.S. Oh, and J.S. Lee, *Catal. Lett.* **19**, 217 (1993).
19. H.S. Yun, K.H. Lee, and J.S. Lee, *J. Mol. Catal. A*, **95**, 11 (1995).
20. J.-Y. Yoon, E.J. Jang, K.H. Lee, and J.S. Lee, *J. Mol. Catal. A*, **118**, 181 (1997).
21. H.A. Dieck, R.M. Laine, and R.F. Heck, *J. Org. Chem.* **40**, 2819 (1975).
22. P. Giannoccaro, *J. Organometall. Chem.* **336,** 271 (1987).
23. Y. Tsuji, Y. Takeuchi, and Y. Watanabe, *J. Organometall. Chem.* **290**, 249 (1985).
24. 24. B.K. Nefedov, N.S. Lergeeva, and Ya. T. Eidus, *Izv. Akad. Nauk SSSR, Ser. Khim.* 349 (1976).
25. B.K. Nefedov, N.S. Sergeeva, and Ya. T. Eidus, *Izv. Akad. Nauk SSR, Ser. Khim.* 807 (1973).
26. V. Macho, J. Hudec, M. Polievka, and M. Filadelfyova, *Chem. Prum.* **25**, 140 (1975).
27. K. Kondo, N. Sonoda, and S. Tsutsumi, *J. Chem. Soc., Chem. Commun.* 307 (1972).
28. S. Senti, C. Crotti, M. Pizzotti, and F. Porta, *J. Org. Chem.* **53**, 1243 (1988).
29. S.H. Han, J.S. Song, P.D. Macklin, S.T. Nguyen, and G.L. Geoffroy, *Organometallics* **8,** 2127 (1989).
30. S. Fukuoka, M. Chono, and M. Kohno, *J. Org. Chem.* **49**, 1458 (1984).
31. R.M. Smith, *PEP Review 94-1-2*, SRI International, Menro Park, 1995.
32. J.M. Berty. In: *Applied Industrial Catalysis*, B.L. Leach (Ed.), vol. 1, Academic Press, New York, 1983, p.207.
33. R.L. Pruett and W.E. Walker, *U.S. Patent 3 957 857*, 1976.
34. S.E. Jacobson, *J. Mol. Catal.* **41**, 163 (1987).
35. A. Spencer, *J. Organomet. Chem.* **194**, 113 (1980).
36. K. Nishimura, S. Uchinmi, K, Fujii, and H. Itattani, *Prepr.-Am. Chem. Soc., Div. Pet. Chem.* **24**, 355 (1979).
37. J. Kollar, *Chemtech* 504 (1984).
38. S. Suzuki, J.B. Wikes, R.G. Wall, and S.J. Lapporte. In: *Catalytic Conversion of Synthesis Gas and Alcohols to Chemicals*, R.G. Herman (Ed.), Plenum, New York, 1984, p.221.
39. Y. Souma and H. Sano, *Chem. Soc. Jpn.* **2**, 263 (1982).
40. D.E. Hendriksen, *Prepr.-Am. Chem. Soc., Div. Fuel Chem.* **28**, 176 (1983).

41. N. Sugita, *Sekiyu Gakkaishi,* **34**, 13 (1991).
42. G. Cavinato, and L. Toniolo, *J. Organomet. Chem.* **398**, 187 (1990).
43. C. Pisano, A. Mezzetti, and G. Consiglio, *Organometallics* **11,** 20 (1992).
44. G. Consiglio, L. Kollar, and R. Koller, *J. Organomet. Chem.* **396**, 375 (1990).
45. G. Consiglio, S.C. A. Nefken, C. Pisano, and F. Wenzinger, *Helvetica Chimmica Acta* **74**, 323 (1991).
46. J.C. Barborak, and K. Cann, *Organometallics* **1**, 1726 (1982).
47. R. Bertani, G. Cavinato, L. Toniolo, and G. Vasapollo, *J. Mol. Catal.* **84,** 165 (1993).
48. G. Consiglio, *Catalytic Asymmetric Synthesis* 273 (1982).
49. H. Alper and N. Hamel, *J. Am. Chem. Soc.* **112**, 2803 (1990).
50. H.J. Federsel, *Chemtech,* 24 (1993).
51. A.S.C. Chen, *Chemtech,* 76 (1993).
52. H. Alper, J.B. Woell, B. Despeyroux, and D.J.H. Smith, *J. Chem. Soc. Chem. Commun.* 1270 (1983).

Res. Chem. Intermed., Vol. 24, No. 2, pp. 213-225 (1998)

A NEW OXIDATION SYSTEM USING NITRITE OXIDANTS

ASUMARU NAKAMURA and TOKUO MATSUZAKI
Ube Laboratory, Ube Industries, Ltd., Kogushi, Ube 755, Japan

Received 19 September 1997; accepted 9 November 1997

Abstract--The oxidative reactions using alkyl nitrites (RONO) as oxidant have been developed by Ube Industries, Ltd. The substrates such as CO and/or unsaturated- and/or carbonyl-compounds are oxidized by RONO over the palladium catalyst with no contact of molecular oxygen. Dialkyl carbonates, dialkyl oxalates and other useful chemicals are synthesized efficiently under moderate conditions by this alkyl nitrite technology.

INTRODUCTION

Many oxidative reactions with Pd compound catalysts have been studied for the past few decades, and some are of great importance in the chemical industry as well as the chemistry of oxidation. The Wacker reaction [1] is an interesting oxidative reaction in the catalytic cycle, in which Pd(II) chlorides oxidize the substrates such as olefins and the resulting Pd(0) species are reoxidized by copper chlorides and oxygen.

The Pd(II) compounds are the active species for oxidizing substrates, so the reoxidation of Pd(0) to the Pd(II) compounds is of critical importance in the catalytic reactions. Gaseous oxygen and high concentrations of $CuCl_2$ in the reaction solution are employed for the reoxidation of Pd(0) in the Wacker reaction.

In most oxidative reactions including the Wacker reaction molecular oxygen is used as the most convenient oxidant and introduced directly into the reaction system, but the oxidation by O_2 brings two major problems: (i) the deactivation of catalyst due to the H_2O generated during the oxidation by O_2, and (ii) the increase of CO_2 and other oxidative by-products due to the co-presence of oxygen with substrates and/or products over the catalyst. Moreover the catalyst efficiency, which is commonly expressed as the turnover frequency of catalyst redox cycle, TOF (h^{-1}) for short, is not sufficiently high and a considerable amount of chloride compound is used to promote the redox cycle of the catalyst. For instance, in the case of the oxidative carbonylation of methanol by O_2 over $PdCl_2$-$CuCl_2$ catalyst [2,3], in which H_2O as well as the carbonate and/or oxalate compound was formed, it was difficult to obtain high catalytic activity and selectivity. Generally it is known that H_2O causes the reduction of the metal chloride by the reaction with CO, which often results in the deactivation of catalyst.

$$MCl_2 + CO + H_2O \rightarrow M(0) + CO_2 + 2HCl \quad (1)$$

It is suggested that the oxidative reaction with no contact of molecular oxygen would reduce such disadvantages or problems as mentioned above. The new oxidation system, in which the substrates were oxidized with alkyl nitrite (RONO) over the catalyst in the absence of any O_2, was discovered at Ube Industries, Ltd. (UBE, for short). Actually in several processes based on this technology an extremely high catalytic activity has been demonstrated.

The physical properties of some alkyl nitrites are listed in Table 1. Generally they are sensitive and unstable to light or heating [4], and they are also

Table 1
Physical properties of alkyl nitrites.

	MW	b.p./°C	m.p./°C	d.p./°C
Methyl nitrite CH_3ONO	61	-12	-17	141
Ethyl nitrite CH_3CH_2ONO	75	17	-35	116
n-Butyl nitrite $CH_3(CH_2)_3ONO$	103	78	no data	96

catalyzed to dissociate to alkoxyl radicals and nitric oxide on the palladium. The high performance of this new oxidation system seems to be brought about by those properties of alkyl nitrites.

$$RONO \rightarrow RO\bullet + NO \quad (2)$$

This new technology with alkyl nitrites will be applied to the wide-ranging oxidative reactions as a promising approach to improve the catalyst efficiency.

DIALKYL CARBONATE (DRC) SYNTHESIS

Conventional Methods of DRC Synthesis

Dimethyl carbonate (DMC) which has been used for the specialty chemicals syntheses and it has been conventionally prepared by the reaction of alcohol and phosgene, but this preparation method has been often undesirable because of the use of toxic phosgene. A lot of research for new DMC preparation methods have been carried out worldwide for over twenty years. Several routes to prepare the carbonate without the use of phosgene or toxic chemicals have been reported [5-8]. One of the favorable methods of DMC preparation might be the oxidative carbonylation of alcohol in the presence of the metal complex catalyst.

I. L. Mador *et al.* [2] disclosed the synthesis of carbonate using $PdCl_2$ as oxidant to the substrates and $CuCl_2$ as a reoxidation agent for the reduced Pd(0).

$$PdCl_2 + CO + 2CH_3OH \rightarrow Pd(0) + CO(OCH_3)_2 + 2HCl \quad (3)$$

$$Pd(0) + CuCl_2 \rightarrow PdCl_2 + 2CuCl \quad (4)$$

This reaction, however, failed to proceed in catalytic cycle.

U. Romano *et al.* [5] succeeded in catalytic oxidative carbonylation by using CuCl in the slurry state.

$$CO + 2CH_3OH + 1/2\,O_2 \xrightarrow[CuCl]{} CO(OCH_3)_2 + H_2O \quad (5)$$

This direct oxidative carbonylation method of DMC preparation was commercialized by EniChem Synthesis S.p.A. (Italy) in 1983. In the process the catalytic reaction is carried out under the conditions of 120 to 140 °C and 15 to 40 kg/cm^2 in the presence of CuCl catalyst.

In this liquid phase direct oxidative carbonylation, high pressure conditions and high catalyst concentration are employed and some amount of CO is consumed through CO_2 by-production because of the co-presence of O_2 and CO over the metal complex catalyst.

New Oxidation Method

Dimethyl carbonate (DMC) synthesis using methol nitrite. The DMC synthesis from CO and methyl nitrite was discovered in the course of the study on the alkyl nitrite reactions at UBE [9]. This reaction is represented by the following equation (6), in which palladium compounds are employed as the effective catalysts [10].

$$CO + 2CH_3ONO \xrightarrow[\text{Pd cat.}]{} CO(OCH_3)_2 + 2NO \qquad (6)$$

The various dialkyl carbonates could be also synthesized by using the corresponding alkyl nitrites in the same reaction. The DMC synthesis is carried out by the gas phase reaction, as the reaction temperature is about 70 to 140 °C and the boiling point of methyl nitrite is as low as -12 °C. The alkyl nitrite is not only a substrate for the dialkyl carbonate formation, but also it works as an oxidant to maintain the palladium catalyst in the active state [11]. Direct introduction of O_2 into the reaction mixture can be avoided in this system.

In the direct oxidative carbonylation of MeOH the introduction of O_2 into the reaction system brings some formation of H_2O. In the alkyl nitrite reaction, however, the formation of DMC is not accompanied by the generation of water as shown in equation (6). This non-hydrous condition makes the high catalytic activity and long catalyst life.

The methyl nitrite (MN, for short) used in the DMC synthesis is synthesized by O_2 oxidation of NO and methanol [12]. The methyl nitrite formation takes place at a sufficient rate even at room temperature with no catalyst.

$$2CH_3OH + 2NO + 1/2O_2 \rightarrow \underset{\text{methyl nitrite(MN)}}{2CH_3ONO} + H_2O \qquad (7)$$

The methyl nitrite from reaction (7) is supplied to reaction (6), and the NO produced in the DMC synthesis reaction is re-usable in the reaction (7). Therefore the DMC preparation method using alkyl nitrite consists of the two reactions, those are, DMC synthesis reaction (6) and MN synthesis reaction (7).

Palladium species for DMC synthesis. Some examples of DMC synthesis reaction according to the equation (6) over the various kinds of supported catalysts are shown in Table 2. These results indicate that DMC formation is catalyzed efficiently and selectively under the mild conditions over the Pd(II) halides on the supports.

Generally speaking, the Pd(II) species show high catalytic activities for the DMC formation in the reaction of CO and alkyl nitrite. The Pd(0) species without

Table 2
Catalyst activity.

Catalysts	Space Time Yield (g/l-cat.,h)	
	DMC	DMO
Pd^0/AC	28	162
Pd-Cl/AC	183	12
Pd-Cu-Cl/AC	553	29
$Pd\text{-}Cu\text{-}Cl/SiO_2$	58	tr
$Pd\text{-}Cu\text{-}Cl/Al_2O_3$	246	7

React. cond.: temp. 120 °C, P. 3 Kg/cm^2, SV 4000 h^{-1}. Catalysts meatal loading: Pd 1 wt%, Cu 1.2 wt%.

halogen anions, on the other hand, show lower activities for DMC formation, though they show high catalytic activities for dimethyl oxalate (DMO) formation under the similar reaction conditions as referred to later.

The Pd(0) species are, however, reoxidized by MN in the presence of HCl and they change into the active catalyst species for the DMC formation [10]. This means that MN oxidizes the Pd(0) species to the Pd(II) species in the presence of Cl^- anions easily.

According to the experimental results on the catalyst activity of the Pd(II) compounds with various ligands, it is concluded that halogen anions are the most suitable counter anions. Especially the Pd(II) compounds with Cl^- anions or Br^- anions showed the prominent catalytic activity for the DMC synthesis.

Reaction mechanism. Halogen anions are involved in the formation of methyl chloroformate, probably through the formation of Pd methoxycarbonyl chloride as the reaction intermediate. So the catalytic mechanism of dialkyl carbonate formation from CO and alkyl nitrite are supposed as shown in Figure 1. According to this mechanism the complex of Pd methoxycarbonyl chloride is formed at first in the reaction of CO, alkyl nitrite and palladium chloride, and then this intermediate complex is attacked by the alkyl nitrite to from dialkyl carbonate and NO is released. The catalyst activity is maintained by the presence of small amount of HCl or methyl chloroformate with alkyl nitrite according to the following reactions (8) or (9), even though less active Pd(0) species are formed.

$$Pd(0) + 2ClCOOCH_3 + 2CH_3ONO \rightarrow PdCl_2 + 2CO(OCH_3)_2 + 2NO \quad (8)$$

or

$$Pd(0) + 2HCl + 2CH_3ONO \rightarrow PdCl_2 + 2CH_3OH + 2NO \quad (9)$$

As methyl chloroformate is synthesized from methyl nitrite, CO and HCl over the Pd catalyst, the addition of HCl into the reaction gas brings the same effect as the addition of methyl chloroformate. Small amount of methyl chloroformate brings the high activity of the catalyst via its oxidative addition to the Pd(0) species.

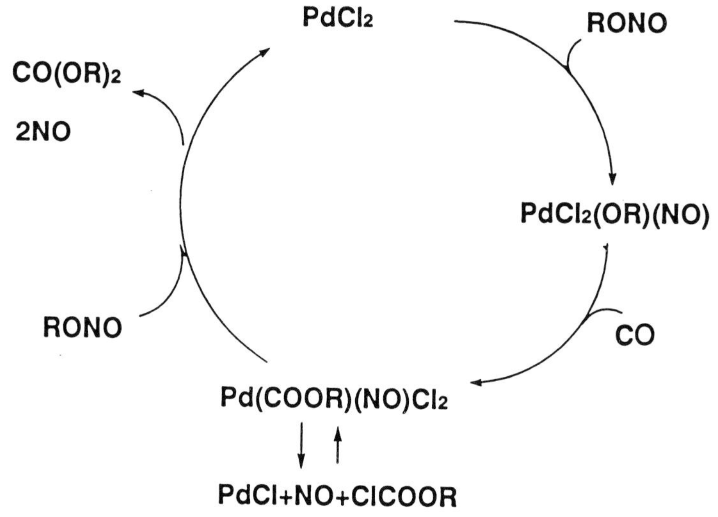

Figure 1. Catalytic cycle for dialkyl carbonate formation using alkyl nitrite.

In this alkyl nitrite reaction under the presence of a small amount of halogen anions, the $PdCl_2$ supported catalyst even in the absence of $CuCl_2$ hardly deactivates, whereas it is known that in the Wacker type reaction the $PdCl_2$ catalyst without co-presence of $CuCl_2$ deactivates quickly.

The kinetics of this DMC synthesis reaction was also studied. The rate does not depend on the partial pressure of MN, but it depends on about first order of the partial pressure of CO, from which the rate determining step is considered to be CO concerning step, that is, the formation step of the Pd methoxycarbonyl chloride complex from CO and MN in Figure 1.

$$PdCl_2 + CO + CH_3ONO \rightarrow PdCl_2(NO)(COOCH_3) \quad (10)$$

The rate of the subsequent step to form the DMC from that Pd complex and MN may be fast. The apparent activation energy of this DMC synthesis reaction, which was calculated by the measurement of the temperature dependency of the DMC formation rate, was about 14 kcal/mole. In connection with the reaction mechanism, it seems that this apparent activation energy is mainly related to the formation of the complex of the Pd methoxycarbonyl chloride.

Though the palladium chloride itself shows the high catalytic activity as mentioned above, the binary catalyst systems such as $PdCl_2$-$CuCl_2$, $PdCl_2$-$FeCl_3$, $PdCl_2$-$BiCl_3$ and so on [10] also show high catalytic activity. One of possible explanations for the role of those co-catalysts is that they prevent the aggregation of Pd species of low oxidation state and promote the reoxidation of them.

"UBE" gas phase DMC process. The DMC process based on the above mentioned gas phase DMC synthesis reaction using MN was developed by UBE [13]. This so called "UBE gas phase DMC process" is illustrated in Figure 2, and it is composed of the reaction part, the DMC separation part, and the DMC purification part.

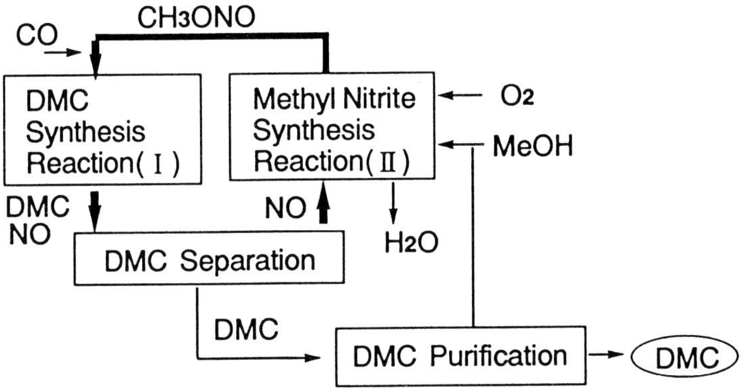

Figure 2. Block flow diagram of "UBE" gas phase DMC process. Reaction (I): $2CH_3ONO + CO \rightarrow (CH_3O)_2CO + H_2O$. Reaction (II): $2CH_3OH + 2NO + 1/2\ O_2 \rightarrow 2CH_3ONO + H_2O$.

The reaction parts are two reactors, one for the DMC synthesis and the other for the MN synthesis. In the separation part the effluent gas from the DMC synthesis reactor is separated into the gas fraction containing NO and the condensate containing DMC. In the purification part the DMC and other by-products are separated and the DMC is purified from methanol and other by-products through distillation columns. The detailed conditions at each part are as follows:

A gas mixture containing CO and MN is introduced into the DMC synthesis reactor filled with the Pd and co-catalyst components supported solid catalyst and the DMC synthesis reaction takes place according to the equation (6). The effluent gas from the DMC synthesis reactor contains DMC, NO, unreacted CO and MN, and other constituents including the inert ones and it is led to an absorption column, where DMC is absorbed into the liquid phase. The non-condensed gas fraction containing NO is led to the MN synthesis reactor, so called MN regeneration column, in which NO is converted to the MN by the reaction with oxygen and methanol according to equation (7). This MN synthesis reaction is carried out in liquid phase at temperatures lower than 60 °C. The water generated by this reaction is taken out in methanol from the bottom of this column. The gas containing regenerated MN mixed with make-up CO is introduced to the DMC synthesis reactor.

In this gas recycle system consisting of the two reactors neither O_2 nor H_2O is introduced into the catalyst bed, which is one of the reasons why high activity and selectivity of the catalyst is maintained for a long time. Furthermore any pollutants or toxic materials are not exhausted from the process because of the completely closed gas recycle system.

UBE constructed the DMC 3,000 t/y plant of the gas phase process applying the methyl nitrite reaction and started operation of the plant from 1993. Good performance without any trouble has been testified and a full confidence on this process has been obtained. Recently the operation rate of DMC production was raised to 4,500 t/y, based on improvement of the catalyst.

DIALKYL OXALATE (DRO) SYNTHESIS

Oxidative Coupling

D.M. Fenton *et al.* [3] reported the oxalate synthesis by the direct oxidative coupling of carbon monoxide before our finding of the novel method of oxalate preparation. According to their method the oxalate is synthesized from CO, alcohol and O_2 in the presence of Pd chloride catalyst and dehydrant under a relatively high pressure by the reaction equation (11):

$$2CO + 2ROH + 1/2\,O_2 \xrightarrow[PdCl_2]{} (COOR)_2 + H_2O \qquad \textbf{(11)}$$

This method was, however, not industrialized because of a couple of disadvantages. First of all the stoichiometric amount of alkyl orthoformate is required as dehydrant to keep the reaction system water-free, otherwise this reaction does not proceed catalytically.

$$CH(OR)_3 + H_2O \rightarrow HCOOR + 2ROH \qquad \textbf{(12)}$$

Moreover the selectivity to the oxalate is not so high because a fair amount of carbonate, RCl and ROR are formed as by-products.

New Oxidation Methods Using Alkyl Nitrites

Liquid phase process of oxalate synthesis. We found the oxalic acid diester synthesis by using alkyl nitrite for the first time [14]. In this oxalate synthesis, a dehydrant is not essential in the reaction system and the oxalate is synthesized with high catalyst efficiency and high selectivity.

The details of this reaction are described in the literature [14], in which dibutyl oxalate (DBO) is synthesized oxidatively from CO and butyl nitrite in the presence of pallaidum catalyst suspended in the liquid phase according to the reaction equation (13). In this reaction butyl nitrite is formed *in situ* as the intermediate from butanol, O_2 and NO.

$$2CO + 2ROH + 1/2\,O_2 \xrightarrow[\text{Pd/C-RONO}]{} (COOR)_2 + H_2O \tag{13}$$

The reaction stages of the oxalate formation are considered as such that CO reacts with alkyl nitrite over the Pd/C catalyst to form the oxalate with NO release,

$$2CO + 2RONO \xrightarrow[\text{Pd/C}]{} (COOR)_2 + 2NO \tag{14}$$

and then the generated NO reacts with O_2 and alcohol to form alkyl nitrite and H_2O in the same reaction system.

Although the water generated by the formation of the oxalate is present in the reaction system, the catalyst is hardly deactivated by the water existing in the azeotropic mixture with butanol.

The critical difference between the reactivities in the alkyl nitrite system and the Fenton's system under the presence of water is due to the differences of the catalysts and oxidants. Chlorine free Pd(0) catalyst supported on carrier is used in the alkyl nitrite system, whereas the $PdCl_2$-$CuCl_2$ catalyst is used in the Fenton's system. The substrates are oxidized through the redox cycle with alkyl nitrite over chlorine free Pd(0) catalyst, that is relatively stable even in the presence of water in general.

Gas phase process of DMO synthesis. An alternate alkyl nitrite reaction for the oxalate synthesis, that is, gas phase dimethyl oxalate (DMO) synthesis has been also developed [15]. The gas phase oxalate synthesis has been derived from the study of the liquid phase DBO synthesis. In this gas phase process DMO is synthesized by the oxidative coupling reaction of CO using the methyl nitrite over the supported Pd catalyst:

$$2CO + 2CH_3ONO \xrightarrow{\text{Pd cat.}} (COOCH_3)_2 + 2NO \tag{15}$$

This reaction is carried out under moderate conditions as at the temperature of 80 to 140 °C and the pressure of 1 to 5 kg/cm^2.

OTHER OXIDATIVE REACTIONS USING ALKYL NITRITES

Alkyl nitrite reaction was applied to the oxidations of olefins [16]. By the oxidations of olefins with alkyl nitrites over the Pd catalyst the acetals or ketals are obtained with high yields and selectivities [17]:

$$CH_2{=}CHX + 2RONO \xrightarrow{\text{Pd cat.}} (RO)_2CHCH_2X + 2NO \tag{16}$$

Under the reaction conditions similar to those of the Wacker reaction, the oxidation rates of the olefins substituted with an electron withdrawing group such as -CN or -COOR are very slow. On the other hand in the alkyl nitrite reaction, no oxygen is introduced into the reaction of substrates, which brings extremely high selectivity to form acetal or ketal in high reaction rate. Even the olefins substituted with an electron withdrawing group are oxidized to acetals or ketals efficiently by the alkyl nitrite reactions. Examples of the olefin oxidations with alkyl nitrites are shown in the following reaction equations:

$$CH_2{=}CH_2 + 2CH_3ONO \rightarrow (CH_3O)_2CHCH_3 + 2NO \tag{17}$$

$$CH_2{=}CHCOOCH_3 + 2CH_3ONO \rightarrow (CH_3O)_2CHCH_2COOCH_3 + 2NO \tag{18}$$

$$CH_2{=}CHCN + 2CH_3ONO \rightarrow (CH_3O)_2CHCH_2CN + 2NO \tag{19}$$

These oxidations are conveniently carried out in the liquid phase in the presence of a small amount of palladium chloride catalyst, though the gas phase reactions are possible by using the solid catalyst. The oxidation of ethylene with methyl nitrite give high yield of 1,1-dimethoxyethane. Methyl acrylate is also oxidized with methyl nitrite to methyl 3,3-dimethoxypropionate, which is used as a reagent for fine chemicals syntheses. The 3,3-dimethoxypropanenitrile is synthesized by the reaction of acrylonitrile and methyl nitrite. The catalytic cycle for the oxidation of olefins with alkyl nitrites is proposed as shown in Figure 3.

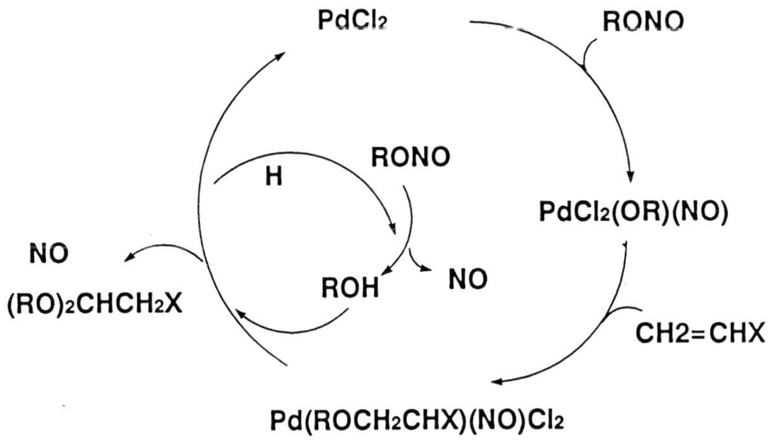

Figure 3. Catalytic cycle for acetalization reaction using dialkyl nitrite.

The process flow of the olefin oxidation with alkyl nitrite is shown in Figure 4. In this process gaseous methyl nitrite is fed into the olefin-methanol solution in the presence of $PdCl_2$ catalyst. The NO gas produced by the acetalization or ketalization is converted to the methyl nitrite in another reactor and the regenerated methyl nitrite is reused for the olefin oxidation. The water produced in the methyl nitrite synthesis is removed out of the MN regeneration reactor and no oxygen is introduced into the olefin oxidation reactor, so that high catalyst efficiency with high selectivity to form acetal or ketal is obtained.

Moreover, alkyl nitrite reactions have been extensively studied and the novel

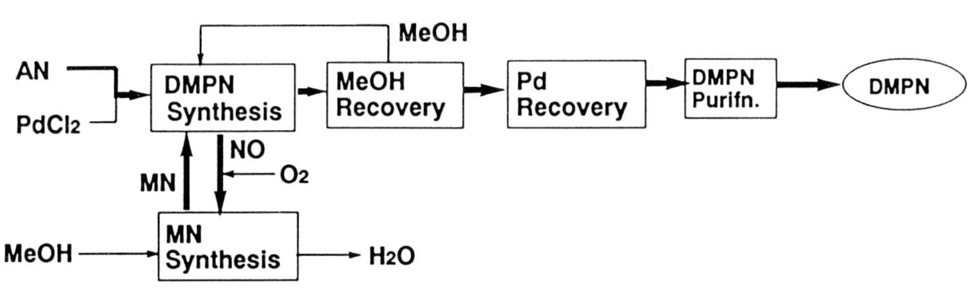

Figure 4. Process flow of Dimethoxy propanenitorile (DMPN) synthesis as example of acetalization reaction CH_2=CH-CN + 2MeONO → $(CH_3O)_2CHCH_2CN$ + 2NO.
AN DMPN

synthetic methods [18,19] of malonic acid diester, succinic acid diester and so on were revealed as the following reaction equations:

$$CH_2{=}C{=}O + CO + 2RONO \rightarrow CH_2(COOR)_2 + 2NO \qquad \textbf{(20)}$$

$$CH_2{=}CH_2 + 2CO + 2RONO \rightarrow (CH_2COOR)_2 + 2NO \qquad \textbf{(21)}$$

$$\begin{array}{l} CH_2{=}C-O \\ \quad\; | \qquad | \\ \quad CH_2\text{-}C{=}\,O \end{array} + CO + 2RONO \rightarrow O{=}C(CH_2COOR)_2 + 2NO \qquad \textbf{(22)}$$

In the reactions with the alkyl nitrites over Pd compounds, according to the reaction mechanism, the oxidation products are defined as the results of addition of two alkoxy groups, or two alkoxycarbonyl groups, or one alkoxy group and one alkoxycarbonyl group. The derivatives of the primary products might be formed under some varying conditions.

CONCLUSIONS

Ube Industries, Ltd. has developed the new category of oxidative reaction, and the concepts or the features of this reaction are summarized as follows;

(i) The substrates are not oxidized directly by molecular oxygen, but oxidized through the alkyl nitrite with high reactivity and selectivity.

(ii) This indirect oxidation makes it possible for the substrates such as CO or olefinic compounds not to contact with molecular oxygen over the palladium catalysts.

(iii) The palladium (II) catalysts, which are generally known to be deactivated by H_2O, are maintained to be active for a long time in this system because of no presence of H_2O over the catalysts. Neither H_2O nor CO_2 generates over the catalysts due to the absence of O_2.

(iv) The alkyl nitrite is very effective for Pd species to be activated from low oxidation state to high in the redox cycle.

(v) The oxidation products are defined as the results of addition of two alkoxy groups, or two alkoxycarbonyl groups, or one alkoxy group and one alkoxycarbonyl group.

This alkyl nitrite reaction has been applied to the dialkyl carbonate synthesis, dialkyl oxalate synthesis, acetal or ketal synthesis, etc. Some of them already have been in commercial operation with good performances. This alkyl nitrite technology will open further new chemical synthesis fields.

REFERENCES

1. J. Smidt, W. Hafner, R. Jira, R. Sieber, J. Sedlmeier, and A. Sabel, *Angew. Chem.* **74**, 93 (1962).
2. National Distillers and Chemical Corp., U.S.P. 3,114,762.
3. D.M. Fenton and P.J. Steinwand, *J. Org. Chem.* **39**, 701 (1974).
4. L. Batt, *Interntnl. J. Chem. Kinetics* **11** 977 (1979); L. Batt, T.S.A. Islam, and H. Scott, *Interntnl. J. Chem. Kinetics* **10**, 1195 (1978).
5. U. Romano, R. Tesei, M.M. Mauri, and P. Rebora, *Ind. Eng. Chem. Res. Dev.* **19**, 396 (1980); U. Romano, *Chim. Ind.* **75**, 303 (1993); EniChem Synthesis, U.S.P. 4,318,862.
6. G.L. Curnutt and A.D. Harley. In: Oxygen Complexes and Oxygen Activation by Transition Metals, A.E. Martell and D.T. Sawyer (Eds.), Plenum, New York, 1988, 21; Dow Chem. Co., U.S.P. 5,387,708; Mitsubishi Chemical Corp., Jpn. Kokai Tokkyo Koho JP 05,255,199; Daicel Chemical Ind., Jpn. Kokai Tokkyo Koho JP 06,238,165.
7. J.F. Knifton and R.G. Duranleau, *J. Mol. Catal.* **67**, 389 (1991); M. Kirishiki, Y. Onda, and H. Tsuneki, *Shokubai* **38**, 92 (1996); A.G. Bayer, EP 0,530,15 A2; Asahi Chemical Ind., Jpn. Kokai Tokkyo Koho JP 05,078,284; Nippon Shokubai Co., Jpn. Kokai Tokkyo Koho JP 06,238,165.
8. J. Kizlink and I. Pastucha, *Collect. Czech. Chem. Comm.* **60**, 687 (1995); Q. Jiang, Q. Lin, and Z. Huang, *Cuihuia Xuebao* **17**, 91 (1996).
9. Ube Ind., U.S.P. 4,384,133.
10. K. Nishihira, T. Matsuzaki, and S. Tanaka, *Shokubai* **37**, 68 (1995); Ube Ind., U.S.P. 5,162,563; Ube Ind., U.S.P. 5,292,916; Ube Ind., U.S.P. 5,214,184; Ube Ind., Jpn. Kokai Tokkyo Koho, JP 05,043,517.
11. T. Matsuzaki, Y. Yamamoto, K. Abe, K. Ohdan, S. Tanaka, K. Nishihira, and A. Nakamura, *Shokubai* **36**, 127 (1994).
12. Union Carbide Comp., *Jpn. Kokai Tokkyo Koho JP* 57,139,041.
13. Ube Ind., U.S.P. 5,214,184; Ube Ind., Jpn. Kokai Tokkyo Koho, JP 06,025,104; Ube Ind., Jpn. Kokai Tokkyo Koho, JP 06,295,151; Ube Ind., Jpn Kokai Tokkyo Koho, JP 06,299,679.
14. S. Uchiumi and M. Yamashita, *J. Jpn. Petrol. Inst.* **25**, 197 (1982); K. Nishihira and S. Uchiumi, *Kagaku to kogyo* **34**, 243 (1982); A. Yasui, *Shokubai* **23**, 23 (1981); S. Uchiumi, *Shokubai* **23**, 477 (1981), Ube Ind., Jpn Kokai Tokkyo Koho JP 53,015,313.
15. Ube Ind., U.S.P. 4,874,888; Ube Ind., U.S.P. 4,410,722; Ube Ind., Eur. Patent Appl. 108,359; Ube Ind., Jpn. Kokai Tokkyo Koho JP 57,013,904; Ube Ind., Jpn. Kokai Tokkyo Koho JP 57,142,942.
16. M. Yamashita, *Petrotech* **15**, 564 (1992).
17. Ube Ind., U.S.P. 4,501,705; Ube Ind., U.S.P. 4,504,421; Ube Ind., U.S.P. 4,504,422.
18. Ube Ind., U.S.P. 4,518,793; Ube Ind., Jpn. Kokoku Tokkyo Koho JP 82,035,897; Ube Ind., Jpn. Kokoku Tokkyo Koho JP 81,035,658.
19. K. Ataka and S. Uchiumi, *Shokubai* **34**, 130 (1992).

Res. Chem. Intermed., Vol. 24, No. 3, pp. 227-258 (1998)

USE OF SHORT CHAIN ALKYLAMINES AS COMPLEXING-MOBILIZING AGENTS: AN ALTERNATIVE TO THE "FLUORIDE ROUTE" FOR THE SYNTHESIS OF ZEOLITE METALLOSILICATE CATALYSTS

Z. GABELICA* and S. VALANGE
ENSC Mulhouse, Laboratoire des Matériaux Minéraux, 3, Rue A. Werner, F-68093 Mulhouse Cedex, France

Received 9 October 1997; accepted 21 December 1997

Abstract --A wide series of zeolite metallosilicates was successfully synthesized using a series of short alkyl chain amines as new potential mineralizing agents. In the particular case of ZSM-5 containing various framework heteroatoms, the mobilizing-complexing role of methylamine, as compared to that of fluoride ions, can be very different and basically depends on the nature of the metallic ion. The relative strength of the various fluoro or amino complexes could be exploited to monitor and control the selective (simultaneous or successive) incorporation of various metallic species in the MFI framework. The incorporation, occurring in variable amounts, can be arbitrary considered as total (i.e. with more than 2 ionic species incorporated per MFI unit cell) for Zn^{2+}, Al^{3+}, Ga^{3+}, Fe^{3+}, B^{3+} and Ti^{4+}. Cd^{2+} and Cr^{3+} are only partly incorporated (M/u.c.< 2) while Cu^{2+}, Co^{2+}, Mn^{2+} and Mo^{5+} never incorporate the MFI lattice. Besides the fact that these ions do not readily achieve a tetrahedral coordination, the other possible reason could be that their polycondensation, if any possible, does not occur when these species are involved in too strong complexes with methylamine. In some cases (Cu^{2+}, Fe^{2+}, Ag^{+}), methylamine behaves as a reducing agent in the synthesis medium and the corresponding metallic species are found admixed with silicalite. At similar complexing power, the (alkali-free) "amine route" is preferred to the now more conventional "fluoride route", that involves lower pH and less supersaturated solutions, requiring very long crystallization times. These latter can also be drastically reduced by using fluoride-amine admixtures.

INTRODUCTION

Molecular sieves involving heteroatoms are conventionally called "metallosilicates" when they involve isomorphously substituted key framework Si atoms by other elements involving various oxidation states.

The replacement of Si and/or Al atoms in the T (tetrahedral) positions of a crystalline zeolite lattice by other divalent (Zn, Be...), trivalent (Ga, Fe, B, Cr...) or tetravalent (Ge, Ti...) elements [1] with ionic radii and coordination requirements compatible with the characteristics inherent to the T site (see below), has been reported and regularly reviewed in the literature. The resulting materials, referred to as metallosilicate analogs of zeolites, are often responsible for unique catalytic performances [2]. Most of these new materials and their properties have been specifically described in the open or patent literature and critically evaluated in a series of recent review papers [3-12].

To undergo a possible substitution, the heteroelements (T) often obey a certain number of theoretical rules such as the Pauling's first rule requiring a specific range for the ratio of the ionic radii between M^{n+} and O^{2-}, although exceptions are observed for frameworks that tolerate mixed iono-covalent bondings. Other requirements for a true substitution imply some flexibility of the heteroelements to accept the tetrahedral coordination and an oxidation state compatible with the overall (stable) framework charge.

Some elements that have, for example, the appropriate size (ionic radius) required for a stable tetrahedral coordination, are potential candidates for an extended incorporation. This is obviously the case of Al but also Ge and, to a lesser extent, Be, Ga, Fe and B.

Other more bulky elements like Zr, V, Cr, Mo, Sn, In, Zn or even Ti are less readily accommodated in tetrahedral sites. At the best (Cr, V, Ti, Sn, Zn, In...), their actual amount in metallosilicates is small [13-15]. In some cases (Zr, Mo, Sb, Co, Mn, Cu...), the unambiguous proof of their true lattice incorporation is still questioned. In such a case, the final phase could possibly consist of admixtures of zeolite and some extra framework, probably well dispersed, metallic species, thus potentially conferring to the material multifunctional catalytic sites.

Chemical factors also do play a predominant role. For example, an ionic species should be able to undergo an easy mobilization in the gel phase that is currently achieved through solubilization in basic or fluoride (neutral or acidic) media. They also must be ready to further condense under hydrothermal conditions at relatively low temperatures. Framework substitutions are nevertheless difficult to generalize and seem to be both metal and structure specific.

In this review, we will more specifically evaluate a few new synthesis pathways leading to an efficient formation and stabilization of a series of industrially important framework metal substituted molecular sieves with the emphasis on recent research dealing with the solubilization and mobilization of the various metallic species in the hydrogel phase.

During the last years, we have extensively explored the possibility of using other types of molecules than the commonly used OH^- or fluoride ions to behave as new potential complexing-mineralizing agents for a variety of metallic ions in zeolite syntheses. After a thorough screening of a wide series of candidates, we came to the conclusion that <u>short chain alkylamines</u> and possibly their protonated forms were in all aspects the most promising mineralizers.

Some of the most prominent results dealing with <u>methylamine</u> complexing power towards metallic ions aimed to be incorporated in the framework of ZSM-5 are summarized below. All the data are based on author's recent home work. In most cases, the previous (published) results have been very recently completed by further observations and new experimental data, neither published nor discussed so far. In some cases, a critical comparison between the mineralizing efficiency of methylamine and fluoride ions is presented and discussed.

USE OF SHORT CHAIN ALKYLAMINES AS COMPLEXING-MOBILIZING AGENTS

Short chain alkylamines present several advantages with respect to classical preparations achieved in alkali basic media, and exhibit interesting properties, comparable and often complementary to those shown by fluoride ions.

- As bases, amines first bring and stabilize the synthesis pH range to basic values (10-12), needed for a rapid zeolite nucleation and growth. This is essentially valid in the case of silicate solution or gel species that are readily solubilized in these conditions. There is little evidence that amines further form stable or soluble complexes with Si compounds.

- As small entities, amines show a poor structure-directing (templating) ability, especially when used along with strong templating agents, such as Pr_4N^+ (TPA) ions. Upon screening the wide literature dealing with synthesis of various molecular sieves using organic templates, only Barrer and Denny [16] reported the partial formation of ill defined methylammonium-harmotome type phases at high temperature and of phillipsite type frameworks that crystallized in the presence of mixed K^+-methylamine media [17].

- Used free from traces of (often not desired) alkali cations, even if incorporated during growth along with other true templating molecules, for example as pore fillers (bases) or framework counterions (protonated forms), short chain alkylamines are readily eliminated from the as-synthesized metallosilicate zeolites by direct calcination, yielding in one step the acidic form of the zeolite, the protons being generated through the Hoffman type thermal degradation of the template. This latter property was advantageously exploited in large scale (industrial) syntheses of metallosilicates [18-23].

- The very low amount of alkali cations present in commercial (aqueous solutions or pure liquid) alkylamines along with their basic character can be advantageously exploited to avoid the use of costly TPAOH solutions as source of TPA (template for ZSM-5) and replace it by crystalline alkali free and inexpensive TPABr. This was successfully applied for the preparation of ZSM-5 metallosilicates that can not nucleate and grow in the presence of alkali cations, such as Ge-MFI [24-

27] or Ti-MFI [14,28].

- The most remarkable property of small amine molecules however consists in their excellent ability to form amino complexes of variable strength in aqueous or gel solutions, with a variety of metallic ions, thereby increasing the general "mobility" of the M^{n+} ions or posibly of the $(Si\text{-}O\text{-}M)^{n+}$ oligomers, desired for an efficient crystallization of the zeolite from the gel phase, through a dissolution-condensation process.

We summarize below the main features observed when a series of selected metallic species are complexed and mobilized with methylamine in order to be incorporated into the framework of ZSM-5 zeolites and underline the most consequent findings.

EXPERIMENTAL

All MFI metallosilicates were prepared by heating (typically at 185°C, in Teflon autoclaves, static conditions, 3-5 days) hydrogels containing the metallic salt pre-dissolved in a TPA-methylamine admixture, and a reactive silica source. Various additives such as the fluoride source were optionally admixed during the various synthesis steps, as specifically indicated. The as-synthesized materials were characterized by a combination of techniques, involving chemical analysis, SEM-EDX, XRD, thermoanalytical methods (TG-DTA-DTG-DSC), NH_3-TPD, n-hexane sorption and high resolution solid state multinuclear (^{11}B, ^{27}Al, ^{29}Si, ^{71}Ga...) NMR [27,29,30], Ga and Zn depth profiling by Rutherford Back Scattering (RBS) [31], XPS and FTIR [14,29].

RESULTS AND DISCUSSION

The Case of Germanium

A wide series of crystalline (Si, Ge)-MFI zeolites with variable amounts of framework Ge concentrations have been synthesized in hydrothermal conditions over the whole pH range, in the presence of TPA^+ ions optionally admixed with fluoride anions, $GeCl_4$ and Aerosil silica [24-27].

The fact that many Ge compounds very readily precipitate under these conditions in the presence of even traces of alkali and ammonium cations, forced us to search conditions to achieve syntheses in an appropriate (basic) pH range, avoiding mineral bases, and without necessarily using expensive ingredients such as TPAOH.

Short chain alkylamines were soon found to be the ideal molecules achieving the required conditions for the preparation to yield highly crystalline pure

(Si, Ge)-MFI, in good yield and over a wide (Si-Ge) initial compositional range, obviously for a series of reasons listed above. A further adding of fluoride mineralizers was shown to considerably reduce the crystallization time. Another side role of fluoride ions was to favor the formation of TPA-F ionic pairs at the intersection of the zeolite channels at the expense of TPA^+ that lead to the building of highly defected (Si, Ge)-MFI frameworks.

The Ge-MFI crystals obtained for the first time by this method showed interesting properties, never suspected *a priori.*

The template free crystals, monoclinic at ambient temperature, became orthorhombic at higher temperatures, the monoclinic-orthorhombic transition temperature showing a perfect regular increase as a function of unit cell Ge content. This increase, along with the increase of the unit cell volume, also linear with Ge loading (Table 1), were the first and very simple probing tests demonstrating the true Ge lattice incorporation.

Table 1
Linear variation of unit cell volume and of monoclinic-orthorhombic transition temperature of some $Si_{96-x}Ge_xO_{192}$ calcined MFI phases, as a function of Ge loading.

x (chem. anal)	Unit cell volume ($Å^3$)	T trans. (°C)
0.0 (silicalite)	5345	80
17.1	5391	170
32.8	5428	250

The true framework Ge incorporation was further substantiated by the presence of Si(nGe) lattice configurations, as seen on ^{29}Si-NMR spectra, or sorption capacities. A further examination of our germano-zeolites by EXAFS and XANES definitely confirmed the Ge lattice insertion [32].

The second surprise was that the incorporation of (bulky) Ge^{4+} cations could be extended up to a Ge/(Si + Ge) ratio of 0.33 (Si/Ge = 2), this corresponding to the presence of about 33 Ge atoms per 96 T atoms constituting one MFI unit cell, a degree of incorporation never approached so far. Conventional degrees of Si substitutions by any other element, including Al, were, for the first time, far largely overstepped.

A pure Ge end member ("germanolite") of MFI topology could never be prepared, probably because this would have required more than 2 Ge atoms in one (5-1) Secondary Building Unit (SBU) from which the MFI structure is supposed to build up during crystallization. This situation would have supposed the formation of Ge-O-Ge bonds, probably very unstable.

^{29}Si-NMR along with TG-DTA data further complemented by ^{19}F- and ^{13}C-NMR spectroscopic evidences, soon showed that, because of its large size and its lower ability to polycondense into regular Ge-O-Si bridges, most of the germanium ions were probably not fully linked to the oxygens of the MFI structure. This latter was indeed shown to be highly defected, involving many terminal Ge-OH internal bonds. The MFI topology was however integrally maintained (XRD data). The fact that the unit cell of Ge-rich zeolites showed a marked expansion suggested a true Ge participation to the network, contrary to the case of indium that definitely undergoes a very different structural behavior within the MFI lattice (see below).

This first attempt to use small amines to replace undesirable alkali cations proved advantageous in many other aspects than simply avoiding the use of alkali bases. The most important (and preliminary unsuspected) property was that these amines actually proved to be very efficient mobilizing agents towards Ge, most probably by solubilizing substantially the Ge- or the (Si, Ge)-oligomers in the hydrogel phase.

This soon suggested us to compare the advantageous use of the smallest amine of the series, namely methylamine (MA), as the main mobilizing agent, in the case of syntheses of other MFI - type metallosilicates with the idea that methylamine would rather solubilize most of the metallic ions by forming with them water soluble complexes like ammonia.

Pure Silica MFI (Silicalite)

Although methylamine (MA) does not seem to form soluble complexes with silica, silicalite (large, elongated crystals along the c axis, with well defined faces) readily crystallizes (12-24 h heating at 180°C) from TPABr-MA-SiO_2 solutions, due to the intrinsic solubility of silica at these pH values (12-12.5). When prepared in the presence of methylamine, silicalite crystals exhibit a homogeneous particle size and involve a low defected framework, as evidenced by high resolution solid state ^{29}Si-NMR [33]. Other silicalite characteristics already partly discussed elsewhere [26-27] will be compared with those of other metallosilicates synthesized using the same method.

The Case of Aluminum

The ideal candidates for a ready MFI framework incorporation in methylamine medium are the series of trivalent ions Al, B and Ga. The strength of the corresponding amino complexes is intermediate, so as to favor a sufficient mobility of either the metallic ions or possibly of the Si-O-M^{3+} small oligomers and to allow their further polycondensation during crystallization.

Indeed, MFI alumino-, boro- and gallosilicates, as well as the various bimetallic variants are very easily and rapidly prepared by using methylamine as mobilizer. While the extent of the incorporation of each element is about the same as in the case of syntheses achieved in conventional alkali basic media, generally larger, more uniformly sized and less defected crystals are obtained. These properties, often desired for various applications, probably stem from the particularly low supersaturation conditions achieved in the methylamine bearing gel phase prior to crystallization.

Aluminum is readily complexed by methylamine, forming water soluble Al-MA associations. Al-MFI could be easily crystallized in pure form from methylamine bearing Si-Al hydrogels, after 3 to 7 days heating at 185° for initial Si/Al ratios between 25 and 100 . The crystallization is slower if compared to pure silicalite (less than one day heating under the same conditions), in line with what is observed for Al-MFI synthesized in alkali media [34], thus confirming that Al insertion in the MFI framework is a difficult process irrespectively to the nature of the mineralizing agent [35]. As in the case of Ga, Al was found to be quasi uniformly partitioned throughout the crystal particles (see the gallium case, discussed below).

Because the rather strong complexing power of methylamine towards Al, it was necessary to control very carefully the actual amount of MA in the synthesis gel. The optimum MA/Si needed to prepare pure Al-MFI with a 100% synthesis efficiency, was about 2 for a Si/Al ratio of 24.

More methylamine (e.g. MA/Si $\geq$ 4) resulted in the formation of stronger MA-Al complexes at the expense of the usual weaker (Si-Al-MA) entities. While the latter are probably those species that easily undergo polycondensation in the building zeolite nuclei and lead to a true Al framework incorporation, the former stay stable and thus retain part of the Al^{3+} in solution so that the final zeolite would contain less Al than expected from the ingredients.

We have prepared such a sample from MA rich gels and checked the Al structural state in the final MFI phase by NMR. The amount of tetrahedrally coordinated Al in the as synthesized sample, as evaluated quantitatively by ^{27}Al-NMR [30], actually well corresponds to the amount expected from the ingredients. Upon heating the material at 550°C in nitrogen and air flows so as to decompose the TPA templates, some methylamine was evolved around 250-300°C (automatic base titration during heating), thus before the TPA started to decompose. The ^{27}Al- MAS NMR spectrum of this calcined phase showed a decrease of the intensity of the central line at c.a. 54 ppm corresponding to tetrahedrally coordinated Al, while another smaller line appeared at - 1.4 ppm. It corresponds to a new Al species formed after the calcination procedure (Figure 1). The decrease of the 54 ppm line does not correspond to any dealumination process, that never occurred under these "mild" conditions, at least for samples synthesized in the presence of methylamine

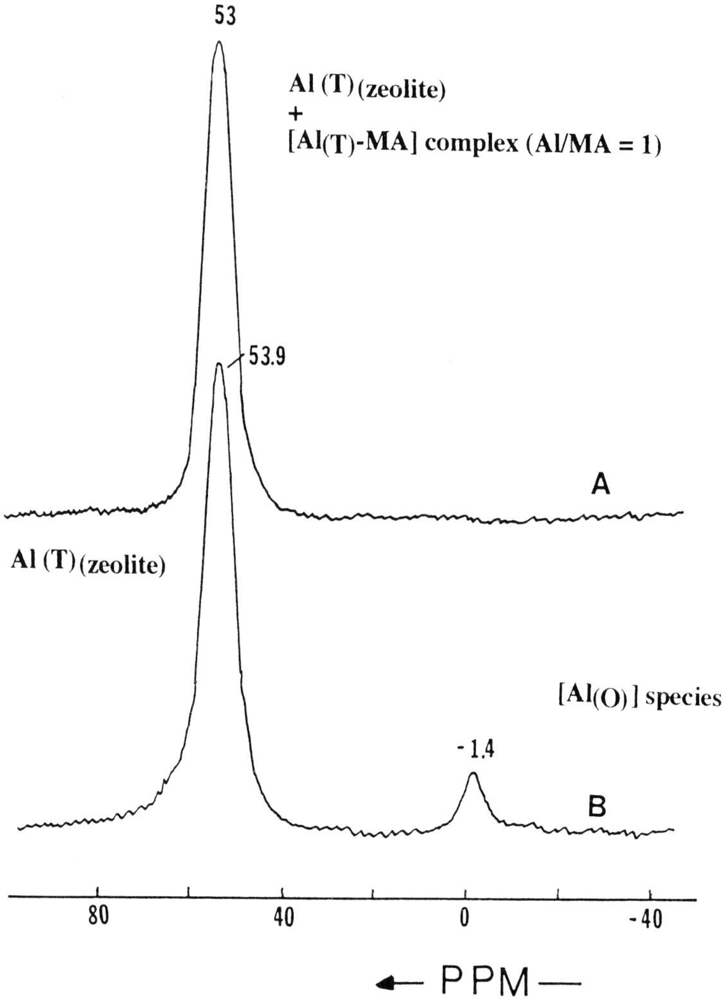

Figure 1. ^{27}Al-MAS NMR spectra of (A): as-synthesized Al-MFI synthesized in the presence of an excess of methylamine (Si/Al = 24, MA/Si = 5); (B): sample A, calcined at 550°C (nitrogen, then air flow), equilibrated in ambient air.

[30]. Indeed, far more drastic heating and the presence of steam were required for dealumination to start [21-23].

The decrease of the NMR line, parallel to the evolution of gaseous methylamine, was attributed to the low temperature decomposition of methylamine-Al complexes that remained stable during synthesis and overcoated the zeolite crystals at the end of the crystallization process. They proved stable upon washing and drying but started to decompose at about 250°C. The direct analysis by combined TG-DTA coupled with NH_3-TPD of the amount of methylamine involved in such associations and released at about 250°C, combined with the indirect determination of the remaining tetrahedrally coordinated Al by ^{27}Al-MAS NMR, suggested that the Al-MA complex involved equivalent amounts of both species (Al/MA = 1). This stoichiometry might appear surprising for a true coordination complex but less surprising if we consider that ill defined, amorphous and rather labile associations are readily formed in MA rich solutions.

The NMR line at -1.4 ppm was used to characterize the new octahedrally coordinated Al species that were formed upon the decomposition of the Al-MA complex entity. The amount of these Al species, as evaluated by NMR, assuming that the relaxation times of all the Al species were similar, fairly well corresponded to the amount of tetrahedrally coordinated Al that had disappeared upon the complex destruction.

Another side experiment consisted in ageing at 50°C for 2 hours in <u>open</u> systems, gels involving the correct MA/Si ratio of 2 and Si/Al of 24, with the aim to partly remove some methylamine from the system upon progressive evaporation. The pH, that is directly fixed by the amount of methylamine in the gel, appeared as the main variable to control in this case. For pH values $\geq$ 12 i.e. for MA/Si $\geq$ 2, pure Al-MFI always crystallized. As soon as the pH dropped beyond 9-9.5 upon evaporation of part of methylamine, Al-MFI involving less than 4 framework Al per unit cell readily formed but some Al hydroxy-oxy species also co-crystallized as side phase. However, one should remember that for these pH values one can very easily crystallize pure Al-MFI with a 100% incorporation efficiency using classical procedures involving Na bearing basic media such as Na silicate [34,36] or NaOH [37]. The above results therefore suggest that upon a partial removal of MA during the ageing process, some "useful" (Si-Al-MA) entities are destroyed as well, releasing some Al that would more readily undergo some hydrolytic reaction and precipitate as hydroxy-oxidic species. For some reason, either the kinetics of their dissolution into the corresponding amphoteric form is slow, or possibly they do not dissolve totally due to the (uncontrolled) continuation of the pH decrease. These extra framework Al species could be detected by DTA (dehydration endotherm at 255°C, Figure 2, A) and by ^{27}Al-NMR that detected some octahedral Al species (spectra not shown). The fact that TPA decomposed in 2 steps corresponding to the presence of both Si-O-TPA (weaker) and Al-O-TPA (stronger) associations (endotherms occurring respectively at 415 and 480°C, Figure 2, A) [38,39], confirms that the amount of framework Al in the final zeolite is lower than 4 per unit cell. Actually, in this case, only 0.7 Al per unit cell did incorporate the MFI framework (^{27}Al-NMR of the final zeolite from which the Al extra phase was removed by mild washings with HCl). The weak DTA peak at 480°C and the particular elongated morphology of the MFI crystals along the c axis (Figure 2, A) actually suggest [40] that the zeolite composition is quite close to that of pure silicalite.

In contrast, when the same gel was aged at 50°C for 2 h in a <u>covered</u> beaker, the final pH remained stabilized at 12.5 and the final zeolite contained about the

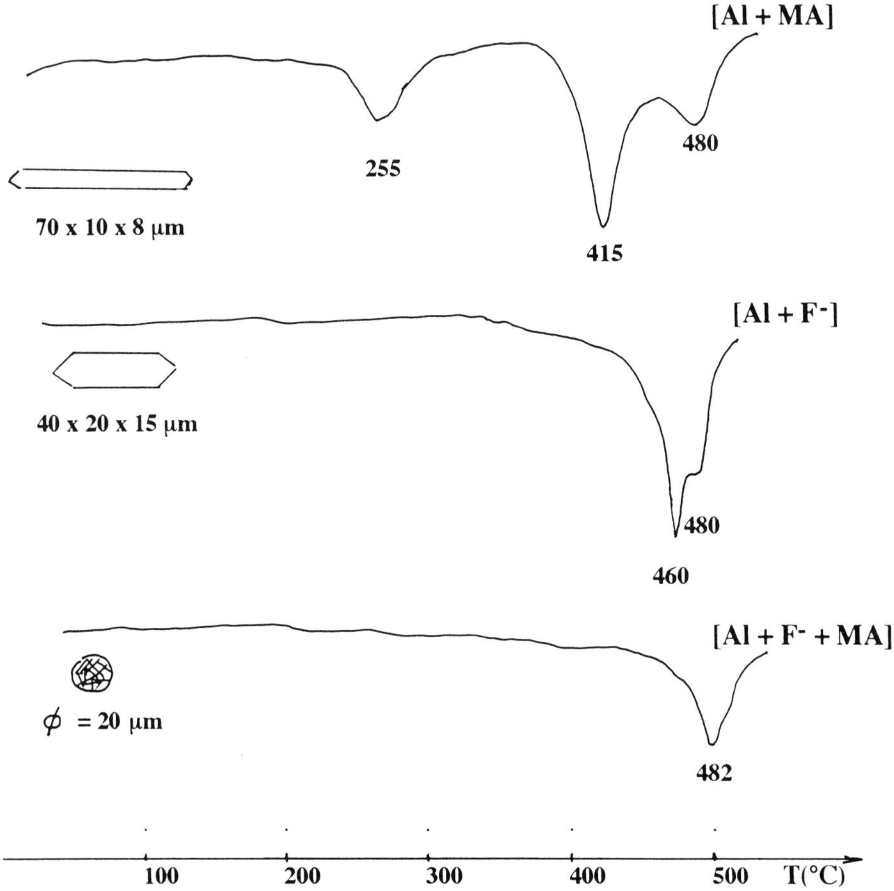

Figure 2. DTA trace of Al-MFI zeolites synthesized in the presence of various mobilizing agents (as indicated on the figure) with the schematic outline of the corresponding crystal morphologies and sizes.

same amount of Al (4 per unit cell) as introduced in the gel (Si/Al = 24). No extra framework Al oxy-hydroxy species were detected by DTA and only one endotherm appeared around 485 °C, characterizing only Al-O-TPA associations (diagrams not shown).

When fluoride ions were used instead of methylamine, all other experimental conditions unchanged, a pure Al-MFI (Si/Al = 24.5) was formed (Figure 2, B). The morphology of the large, well outlined crystals, similar to that of the same zeolite crystallized in the presence of methylamine (MA/Si = 2) (not shown on the figure) suggests that both mobilizing species form rather soluble complexes of similar strength and that the supersaturation of the solution (gel) is still low.

Surprisingly, when both F^- and MA mineralizing agents were admixed, although pure Al-MFI (Si/Al = 25) was still rapidly formed as expected (only one

DTA peak at 482°C, Figure 2, C), its morphology was very different: the uniform spherical polycrystalline aggregates suggested that a rather high supersaturation was achieved and was rapidly "broken", possibly through the salting-out effect due to methylammonium fluoride that readily formed. Such effects are also observed when NaCl breaks the supersaturation in Na^+ bearing gels precursors to (Na, Al)-MFI [41].

The Case of Boron

MFI type borosilicates, currently prepared through hydrothermal synthesis in basic (alkali) [42-45] or fluoride [46] media could also be readily synthesized in the presence of methylamine.

When boric acid was used as boron source, pure rounded B-MFI aggregates were formed after less than 1 day heating at 185°C, (Figure 3). They contained about 3.5 B/unit cell (^{11}B-NMR data [43,45,47]), for gels involving Si/B ratios between 10 and 25 (B/96T atoms, with T = Si + B, being between 4 and 9). After very long crystallization times (up to 30 days) all the B-MFI phases remained perfectly stable in methylamine medium at 185°C and no other crystalline phases were detected.

The case of boron is *a priori* more simple than the Al case in the sense that B^{3+} ions are probably rapidly and totally solubilized as $B(OH)_4^-$ species in the pH range 12-12.5 and thereby do not form any complex or association with

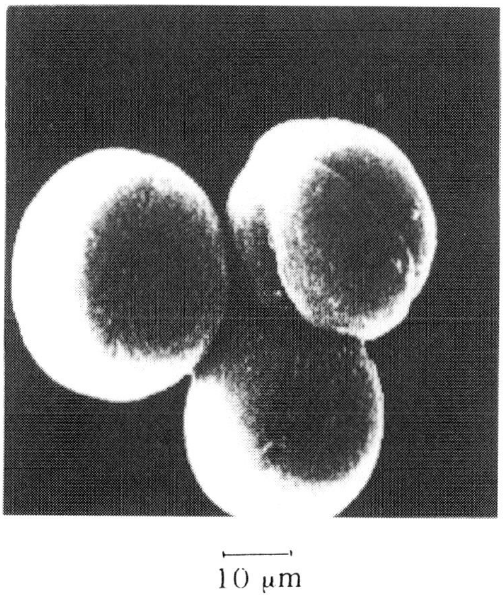

Figure 3. SEM micrograph of B-MFI synthesized in the presence of an excess of methylamine (MA/Si = 8).

methylamine. Its behavior is therefore expected to be similar to silica species that also show a (far more moderate) solubility in alkaline media.

An important question, related to potential catalytic properties of borosilicates rapidly arose: would the methylamine route lead not only to a homogeneous crystal size distribution but also to a uniform B concentration gradient, like in the case of Al and Ga-MFI synthesized in the presence of methylamine? Answering such a question required an in depth investigation of the formation of borosilicates in methylamine media involving two different silica sources that would undergo hydrolysis at very different rates and possibly form the appropriate borosilicate oligomers to variable extents prior to condensation. The sources selected were SiO_2 (Aerosil) and $SiCl_4$. The results, detailed in a recent publication [48], can be summarized as follows:

For syntheses involving SiO_2 and H_3BO_3, the "induction" time (time before the first borosilicate crystals start to be detected by XRD) was about 6 h and 100% crystallinity was attained after 24 h heating at 180°C, just as in the case of silicalite.

For syntheses involving only $SiCl_4$, the induction time (of silicalite) was of about 4 days and its degree of crystallization was only about 60 % after 10 days heating. Addition of hydrogen fluoride, which is known to increase the solubility of silica, markedly increased the crystallization kinetics: 100% crystalline silicalite was obtained after 6 days heating. These results suggest that silicon oxidic- or mixed $SiO_xCl_y(OH)_z$ species were formed upon hydrolysis of $SiCl_4$. They are probably difficult to rearrange to polysilicate oligomeric species required for a further polycondensation so as to yield the first viable nuclei, in the sole methylamine medium. It is also possible that methylamine, by forming some specific associations with these partly hydrolysed Si species, hinders their further transformation into more soluble species. Such a solubilization obviously readily occurs through the use of fluoride ions.

Interestingly, the addition of H_3BO_3 to the system involving $SiCl_4$ (4B/96T) dramatically increased the rate of crystallization of the corresponding MFI borosilicate. The zeolite started to grow after an induction period of less than 20 hours and the crystallization was completely achieved after 3 days heating. Contrary to the case of Al, the presence of boron dramatically accelerated the rearrangement of the silicate species into "more useful" borosilicate entities, for a ready crystallization. Such a quite surprising behavior was interpreted by considering more in depth kinetic effects [48].

Regarding the variation of the average crystallite size, there was a tendency that the low B-concentration gels yielded large needle-like crystals, while the borosilicate phases grown from gels involving larger B amounts exhibited smaller crystal sizes. A remarkable characteristic of the synthesis from $SiCl_4$ bearing gels was the formation of large crystals, which may be due to a restricted nucleation, this latter being itself governed by the slow $SiCl_4$ hydrolysis. For comparison, the $SiCl_4$

bearing reaction gel involving 8B/96T yielded hexagonal prisms of about 43 μm length, while gels involving Aerosil silica produced, in similar conditions, small prismatic crystals of about 8 μm length.

The repartition of B throughout the zeolite crystallites could be evaluated by monitoring (^{11}B-NMR) the evolution of differently coordinated boron species during the two types of crystallization processes.

Aerosil silica bearing gels suggested that SiO_2 was readily solubilized in the basic methylamine medium so as to generate "useful" oligomeric species that were incorporated, either along with $B(OH)_4^-$ soluble species or as mixed Si-O-B soluble oligomers, through a classical solution transportation mechanism. Both Si and B species were probably incorporated at a similar rate, as suggested by the quasi constant tetrahedral B concentration in the crystals during growth [48], with, as a result, the probable quasi homogeneous boron lattice distribution in the resulting B-MFI crystals.

In contrast, in $SiCl_4$ bearing gels, intermediate phases (amorphous gel and zeolite admixtures) already contained tetrahedrally coordinated boron before heating (^{11}B-NMR data). This confirmed that $SiCl_4$ started to hydrolyze and to form an amorphous phase that also rapidly incorporated boron species probably at a concentration roughly corresponding to 4B/96T. This phase was formed before any viable zeolite crystallite did appear. The fact that the total boron concentration did not change along the crystallization process, clearly suggested that we may have been dealing with, at least partly, a gel restructuring type mechanism. Indeed, while the total boron was always constant, the lattice tetrahedral boron increased only when the zeolite crystals started to be detected by XRD. Although the hydrogel restructuration mechanism (also considered partly as a surface nucleation), as early proposed by Breck and Flanigen [49] and McNicol et al. [50] was not very often evidenced in zeolite synthesis, it has been recently proposed to explain the formation of (Na, Al)-MFI [34] and Ga-faujasite [51] zeolites under particular synthesis conditions.

The fact that boron was incorporated through two different mechanisms could potentially subtly influence the catalytic properties of these materials.

The cases of <u>boron</u> and <u>aluminum</u>, not worth more comments as such, will be more thoroughly debated when these ions are used along with <u>titanium</u> or <u>gallium</u> to make mixed (bimetallic) MFI frameworks (see below).

The Case of Gallium

With the aim to create novel bifunctional Ga bearing type zeolite catalysts desired for the very important aromatization processes, we have explored a series of new synthesis routes leading to alkali-free MFI-gallo- and gallo-alumino silicates that readily yield acidic catalysts upon straightforward calcination. Again short chain alkylamines were the best potential candidates as complexing-mobilizing alkali-free species to assist successfully these types of syntheses.

In contrast to the fluoride route, Ga^{3+} could be easily, very rapidly (3 days heating) and quasi homogeneously (RBS data, see below) incorporated in the MFI lattice, in variable amounts (up to a maximum of about 5 Ga/u.c.), with a 100% efficiency (^{71}Ga-NMR), by using the methylamine route [18-20,30,31]. Under similar heating conditions but in the presence of sole fluoride mineralizing agents, Ga-MFI takes 21 days to crystallize [52]. The more detailed operatory conditions emphasising the useful role played by methylamine were the following:

A series of (Al)-, (Ga)- and mixed (Ga, Al)-MFI zeolites were synthesized from hydrogels containing reactive silica (Aerosil 200, 970 $m^2\ g^{-1}$) and Al^{3+} and/or Ga^{3+} salts pre-dissolved in variable amounts of an aqueous TPABr solution. The basic pH (about 12) was adjusted by adding appropriate amounts of methylamine, this avoiding the use of alkali cationic bases. Unless specifically mentioned, the optimum MA/T ratio (T = Si + Al + Ga...) of 2 was selected. Some other synthesis parameters were adequately adjusted so as to yield zeolites with variable but quasi homogeneous Al-Ga distribution desired for their use as bifunctional catalysts in aromatization reactions [31].

Although 100% crystalline [Al_x, Ga_y]-MFI zeolites were formed after already 20 h heating (crystallinities evaluated by isothermal n-hexane adsorption on in-situ calcined samples), the gels were typically heated in Teflon-coated stainless steel autoclaves for 3 days at 185 °C. Both Ga and Al easily incorporated the MFI framework to quite appreciable extents, namely to about 4Al/96T and 5Ga/96T. A silicalite (x=0, y=0) was synthesized under identical conditions, as reference.

The progressive substitution of Ga for Si in the silicalite framework was marked by a parallel decrease of the elongated morphology of the crystallites along the c axis (silicalite, Figure 4 A) yielding more isometric prismatic crystals for higher Ga loadings (Figures 4 B and 4 C) and finally rounded polycrystalline aggregates for MFI involving 3.5 Ga per unit cell (Figure 4 D).

The order of addition of the reactants, the stirring conditions, the silica source and the Ga and/or Al concentration [53,54] were among the (minor) parameters to control for achieving a nearly simultaneous and therefore a quasi homogeneous Ga and Al incorporation. The relative M^{3+} (M standing here for Al and/or Ga only) site homogeneity of the samples, as suggested by XPS, was confirmed by the presence of a single DSC (TPA^+ decomposition) and TPD (NH_3 release from $NH4^+$ exchanged samples) peak for each [Ga, Al] sample, as well as the smooth variation of these peaks as a function of the framework Ga/Al ratio.

The single TPD peak varied regularly from 385 °C (x = 0, y = 2.45) to 450 °C (x = 4, y = 0), indicating that the corresponding acid sites were substantially stronger in our catalysts than in zeolites synthesized from conventional NaOH media that exhibit the corresponding TPD peaks at 335 °C and 385 °C respectively [55].

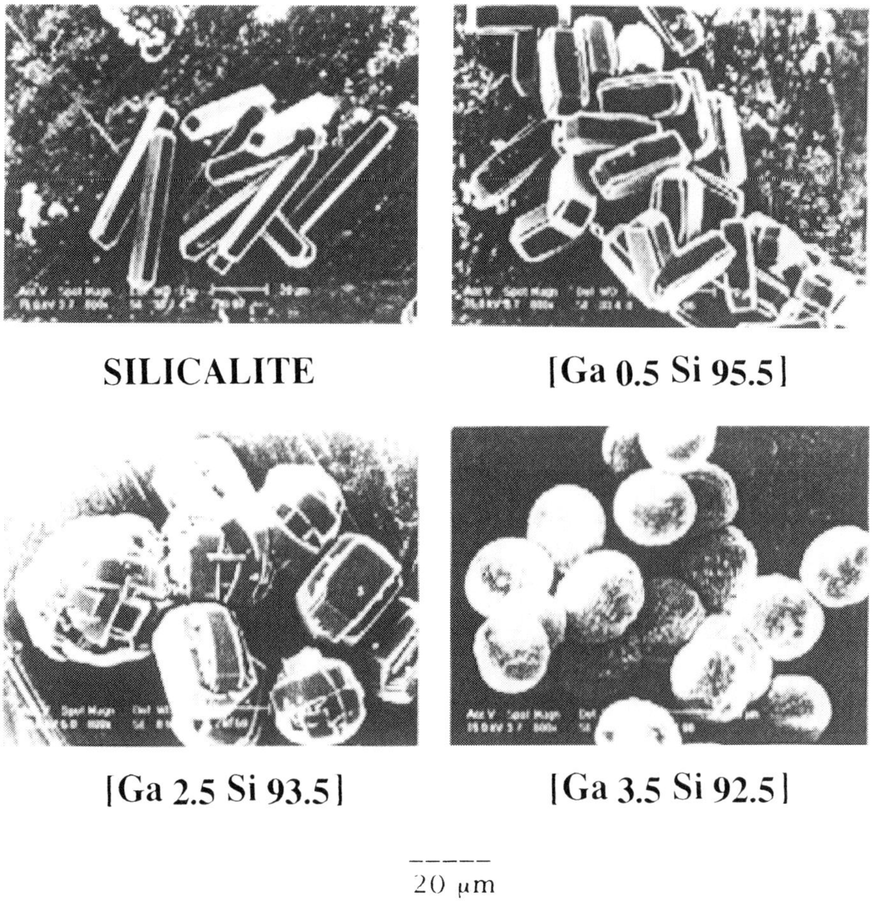

Figure 4. SEM micrographs showing the morphology evolution of Ga-MFI zeolites synthesized in the presence of methylamine, as a function of Ga framework incorporation. Sample compositions, as indicated.

Moreover, in this latter case, two distinct DSC peaks indicative of the presence of two different TPA^+ species were observed, thereby suggesting the presence of more pronounced M^{3+} site heterogeneities.

The selective incorporation of Ga and/or Al in the as synthesized zeolitic lattice, and their further stability upon various post synthesis (hydro)thermal treatments could be accurately determined by combining ^{27}Al- and ^{71}Ga-NMR along with the Ga (wet) chemical analysis and TPD of ammonia. ^{71}Ga-NMR was previously demonstrated to be of quantitative use under carefully defined conditions to measure the amount of framework Ga concentrations in MFI zeolites [30,56]. By evaluating the intensity of the central NMR line at about +150 ppm with the help of adequate calibration curves of NMR line intensities versus the Ga elemental composition established for each type of synthesis (NaOH, methylamine or fluoride routes), we were able to determine quantitatively the actual amount of framework Ga in each case [30]. This technique indicated that the lattice incorporation of gallium from gels was quantitative in methylamine medium, in the explored concentration range of 0.6 to 5 Ga per unit cell of zeolite.

Gallosilicates and silicalite also involved traces of amine in their pore volume, that seemed not to be in strong interaction with Ga. This apparent greater stability of the Al-amine association with respect to Ga, may cause a slightly preferential Ga incorporation in the beginning of the crystallization process, resulting in a hardly, but definitely detectable Ga-Al zoning in the final crystallites (XPS) [30]. These trends were very recently confirmed by probing Ga depth profiling using RBS technique specifically adapted to the case of quasi spherical Ga-MFI particles [31].

Ga-Al zoning and Ga homogeneous distribution: an RBS evidence. Four different Ga and (Al, Ga)- MFI samples, 100% crystalline, have been investigated by a combination of techniques, including Rutherford Back Scattering (RBS) [31].

Sample coded ZN-1 exhibited a very homogeneous Ga concentration at least through a depth of 400 nm. The Ga concentration could not be evaluated at higher depths and we therefore arbitrary considered that the amount of Ga calculated from the RBS plateau ending at a depth of about 400 nm, further extrapolates through the bulk of the crystallites. The actual amount of Ga, as calculated for a matrix supposed to involve theoretically 192 oxygens, 92 Si (and 4 Ga) atoms per unit cell of MFI zeolite, the target density being estimated to 2, thus between the compact atomic arrangement of quartz (2.2) and that of porous pure silica ZSM-5 (1.76), was perfectly constant along the first 400 nanometers from the surface and was equal to 3.63 ± 0.2 atoms per unit cell (Table 2).

This value fairly well fitted the one obtained by both chemical analysis and NMR, this latter technique only detecting framework Ga. RBS therefore brought for the first time a major proof that the methylamine route leads to a homogeneous Ga insertion during the crystallization stage of Al free Ga-ZSM-5.

Sample ZN-2 showed a gradient in Ga concentration, namely a slight Ga enriched outer surface (2.4 Ga / u.c. within the first 85 nm) and a more Ga depleted inner part of the crystal (a fairly constant amount of 1.3 Ga/u.c. found in between 85 and 400 nm from the outer rim - spectra not shown). One can speculate that both Ga and Al undergo mixed complexation with methylamine and that Al, more readily decomplexed than Ga, incorporates the zeolite lattice at a reasonable rate. Al to some extent prevents the final Ga insertion (only 1.3 Ga per unit cell were detected in the last crystal layers). At the very end of the growth process, RBS suggests that, as already observed in the case of Ti-ZSM-5 [57] and Cr-ZSM-5 (see below), some

Table 2
Ga- and (Al, Ga)-MFI samples and their physico chemical characteristics.

Sample	M^{3+}/96T in gel	As-synthesized zeolite						
		Nature (XRD)	% Cryst. (n C6)[a]	Crystal. size (SEM)	Al/96T (NMR)[b]	Ga/96T		
						Chem.	NMR	RBS
ZN-1	4Ga	MFI	96	4 μm	-	3.71	3.37	3.63±0.2
ZN-2	2 Ga, 2 Al	MFI	99	10 μm	2.01	1.78	1.72	gradient
ZN-3	4 Ga (NH_4F)	MFI	(103)	2 μm	-	4.0	3.77	gradient
ZN-4	3.5 Ga, 0.5 Al	MFI	96	20 μm	0.6	3.36	2.92	3.39

[a]Crystallinities evaluated by n-hexane (n-C6) adsorption, as described elsewhere [38].
[b]Exclusively framework Ga evaluated from appropriate calibration curves using in each case standard gallosilicate samples prepared under identical conditions [29].

unreacted Ga-amino complex or possibly some other soluble gallate type phase are indeed overcoating the very outer rim of the crystals.

The fact that large amounts of Al prevent any complete Ga incorporation by possibly forming mixed complexes is substantiated by the behavior of sample ZN-4 that involves four times less Al and in which nearly all Ga was found homogeneously distributed throughout at least the last 400 nm of the crystallites (Table 2). NMR data might however suggest that only about 90 % of these Ga species are truly framework Ga^{3+} ions.

Finally sample ZN-3, synthesized in the presence of fluoride mobilizing agent, does show a Ga gradient: 5.8, 5.0, 4.2 and 3.4 Ga/96 T were detected respectively within "slices" of 25 nm depth starting from the surface towards the core. This indicates that residual Ga fluoro complexes may overcoat the outer surface, as suggested by the differences between chemical and NMR analyses (Table 2). The "bulk" value (3.4 Ga/u.c.) suggests a fairly homogeneous incorporation, that was shown to occur only when initial gel Si/Ga ratios are smaller than 25 [52]. Larger ratios yield crystallites with strong Ga gradients [52,58].

The above findings do complement our previous results [29,30] and shed some new light on the actual competing of Al^{3+} and Ga^{3+} ions for methylamine during synthesis of zeolites in these media. They provide new possibilities to monitor the relative concentration of acid sites (framework Al or Ga) and of extra lattice Ga species (both involved in the mechanism that governs olefin aromatization reactions), by an adequate adjustment of the synthesis conditions and post synthesis treatments [19,20,54].

The Case of Chromium

Conventional synthesis procedures for Cr-MFI conducted in alkaline media lead to chromosilicate zeolites in which Cr is essentially outside the framework, probably weakly linked to oxygen atoms so as to form oxides or oxidic clusters similar to Cr_2O_3 [59,60,61]. Such a behavior is in line with the prediction that Cr^{3+} ions are relatively unstable in tetrahedral coordination [5]. Moreover, at high pH values, hydrated Cr species readily undergo dimerization via hydroxyl groups so that the bulky Cr dimers can not easily incorporate the silicalite framework.

Among other mobilizing agents in aqueous solution, ammonia [62,63] and fluoride ions [63] were tested in the case of synthesis of Cr-MFI. Although ammonia was used to stabilize monomeric $[Cr(H_2O)_6]^{3+}$ species, only loosely bound Cr aggregates were formed. They were readily leached from the zeolite during liquid phase oxidations with *tert* - butyl hydroperoxide [93].

We have attempted to prepare a wide series of Cr-MFI zeolites in methylamine or methylamine-fluoride mixed systems. The general trends observed can be summarized as follows:

Methylamine appears to behave as an excellent mobilizing agent for Cr, at least in terms of solubilizing abilities. A sample (coded Z-1) synthesized from a gel involving the following typical molar composition:

$$SiO_2\text{-}0.01\ Cr_2O_3\text{-}4\ MA\text{-}0.25\ TPABr\text{-}40\ H_2O$$

was prepared as follows: the clear dark green solution containing Cr nitrate dissolved in H_2O + TPABr was added dropwise to an aqueous solution of methylamine. The order and the rate of addition were important. Indeed, the Cr^{3+} ions in each droplet of the first solution must rapidly undergo a temporary precipitation and a ready redissolution either as amphoteric CrO_2^- species or most probably as soluble methylamine complexes. After mixing, the resulting solution exhibited a red-pink color, indicating that a Cr-MA complex was formed (such complexes easily form with Cr^{3+} and various amines or ammonia). After the final (rapid) addition of SiO_2, a pinkish-red fluid gel was obtained that is further stirred for 2 h until complete homogeneization and heated at 180°C, in static conditions, for 5 days.

To compare the mobilizing-complexing power of methylamine and fluoride ions, a similar preparation (code Z-2) was achieved from a gel in which 0.5 mole of HF have been added in the methylamine solution. The final pH was still of 12.5 because of the excess of methylamine. Crystals of sample Z-1 showed an elongated morphology typical of silicalite and were admixed with some amorphous phase (Figure 5 A) that could be readily removed upon ultrasonic separation (Figure 5 B). Crystals of sample Z-2, free from any amorphous phase, were somewhat smaller but showed a similar outline (Figure 5 C). A closer view revealed that these crystals

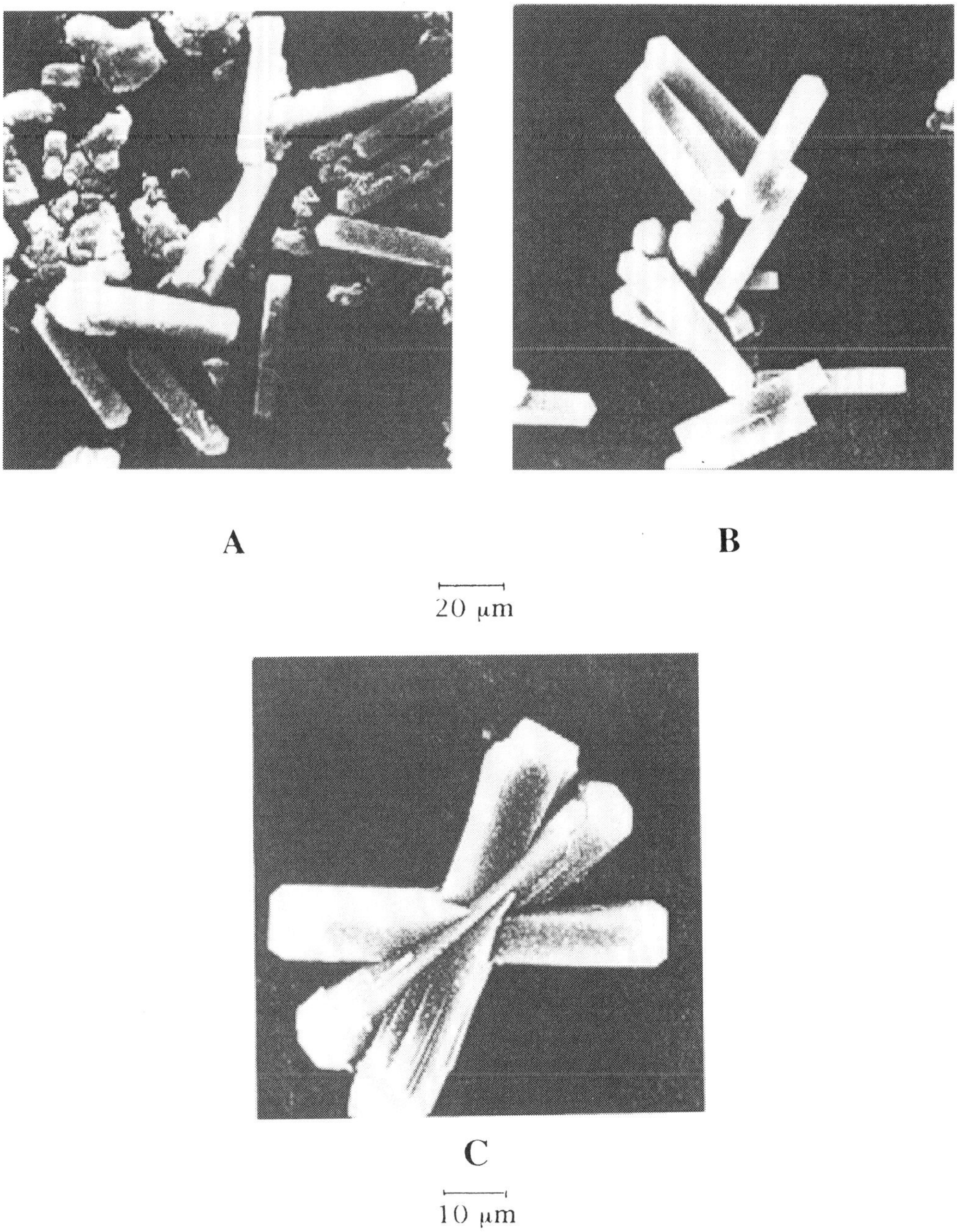

Figure 5. SEM micrographs showing the morphology of (A): Cr-MFI (sample Z-1) synthesized in the presence of methylamine, the as made sample, (B): sample A, from which the residual amorphous phase was removed by ultrasonic cleaning; (C): Cr-MFI (sample Z-2) synthesized in the presence of HF; notice a thin film (Cr-bearing phase) overcoating the crystals.

were coated with a thin film possibly composed of a multitude of very small spherical particles.

Bearing in mind that Cr^{3+} forms very strong fluoride complexes (e.g. CrF_6^{3-}) involving quite strong Cr-F bonds not readily subject to hydrolytic cleavages even in basic media, one may suppose that these Cr species will not readily undergo "decomplexation" as to be able to further condense with the gel Si oligomeric species and form the first crystallites of Cr-MFI in the (heated) gel. The crystals of sample Z-2 indeed proved to be quasi pure silica MFI, although XPS revealed traces of a Cr bearing species at the very outer surface of the crystallites. We believe that most of the Cr species were eliminated through filtration as strong water soluble fluoro complex. It is also possible that part of this complex did overcoat the silicalite crystals during the last stages of crystallization, possibly impeding to some extent their further growth. Indeed, pure silicalite synthesized in same conditions (fluoride ions as mineralizing agent, absence of any Cr salt), yielded larger crystals.

Crystals of Z-1, separated from the residual amorphous phase, proved to be pure silicalite in which no traces of Cr^{3+} could be detected. As no Cr was ever found in the liquid phase separated by filtration, it was supposed that most of the Cr was present in the residual unreacted gel. This could be confirmed through EDX (bulk) and XPS (surface) analysis of the amorphous phases removed by ultrasonic cleaning, that revealed the presence of Si and Cr. In addition, Cr was probably associated with some methylamine, as suggested by automatic titration of NH_3 evolved during a further heating of this amorphous phase at 500°C under nitrogen flow. Upon reoxidation in air of this residue, Cr^{3+} could never be oxidized into Cr^{6+} (TG/DTA experiments), in contrast to the Cr^{3+} "admixed" with MFI zeolite synthesized from classical alkali media. This suggests that the Cr^{3+} ions readily agglomerate so as to form a bulk type oxide upon heating.

The fact that most of the chromium was found in the amorphous phase rather than within the zeolite framework suggests that Cr^{3+} was present as an (hydroxy)-oxide type phase in the final admixture, as a result of the partial (hydro)thermal degradation of the MA-Cr^{3+} complex. In contrast, Cr-F complexes are probably more stable, do not undergo decomposition at 180° and are eventually recovered as such in the liquid phase.

Our investigations demonstrate that Cr^{3+} ions are never incorporated in the MFI lattice when methylamine or fluoride ions are used as mobilizing agents. Rather strong (soluble) Cr complexes are immediately formed in the gel precursor. They have variable stabilities at higher (crystallization) temperatures. The strong Cr-F associations never release the appropriate Cr species for a possible condensation with silica so as to form a Cr bearing MFI framework. At the best, the silicalite crystals formed in these conditions are partly overcoated by traces of these complexes. These weaker Cr-MA associations essentially decompose into an amorphous Cr bearing phase that is admixed with well formed silicalite crystals. A further calcination of both systems would yield an MFI phase admixed (or coated) with tiny particles of various Cr oxides that may play an interesting catalytic role in selected systems.

The Case of Indium

Little has been published about MFI zeolites containing indium. Vorbeck *et al.* [64] claimed having substituted isomorphously indium ions for silicon in the MFI framework to a limited extent (0.6 to 1.2 In^{3+}/unit cell, corresponding to only 30-40% of the initial amount of indium used as starting ingredient). Rather weak Brönsted acid sites were generated. The thermal stability of these indosilicates proved quite low, indium showing the neat tendency to leave the framework upon prolonged heat treatments.

Our own experience showed that trivalent indium ions could neither be incorporated into the MFI lattice in the presence of fluoride ions in neutral or acidic media, nor by using alkali free short chain alkylamines as sole mineralizing species. Methylamine brings pH values of about 12 that were not high enough to readily solubilize indium hydroxide. Mixtures of silicalite and indium hydroxide are the only crystalline phases present. Figure 6 A shows the presence of small $In(OH)_3$ crystallites of typical cubic morphology admixed with silicalite that readily crystallizes in the presence of methylamine as large elongated crystals. After calcination, such admixtures may nevertheless show potential dehydrogenating catalytic properties, as do their Ga or Zn analogs.

Basic media were more favorable to In^{3+} incorporation, particularly pH values higher than 13, when the most abundant anionic species are $In(OH)_4^-$. As alkali cations do not prevent the formation of these anionic species, NaOH could be used. ZSM-5 crystals (XRD) free from any amorphous phase (Figure 6 B) were obtained. Indium could be detected by spot EDX chemical analysis in individual clean crystallites with no amorphous phase around. The presence of acid sites (NH_3-TPD data, see below) was not in complete contradiction with an In^{3+} framework siting, although their strength was quite larger than that reported for MFI indosilicates [64].

Unexpectedly, XRD did not show any unit cell expansion as a function of In content. We have concluded that indium was incorporated in the silicalite lattice defects (T missing sites, various macropores...) readily generated at high pH values [33], as voluminous $In_xO_y(OH)_z$ entities, not being really linked to the framework and thereby not affecting the unit cell dimensions. Actually, Vorbeck *et al.* [64] did record a slight unit cell increase as a function on In loading but their syntheses were achieved under milder conditions (pH about 10-11 obtained upon using Na silicate instead of NaOH) that favor the (slower) formation of less defected MFI crystals [33].

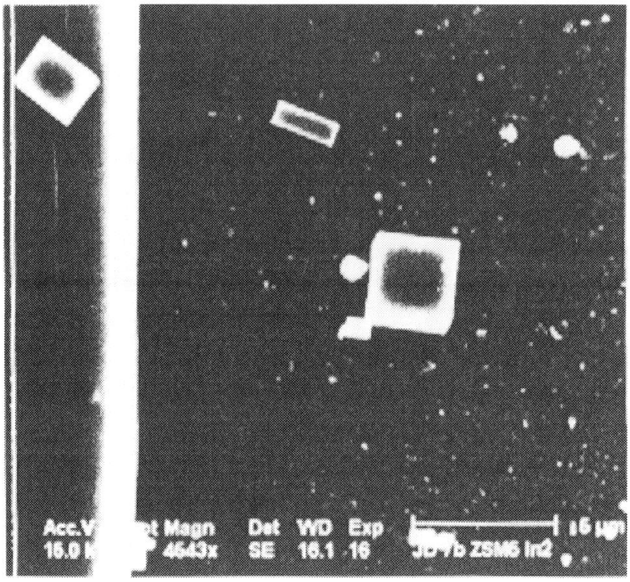

A

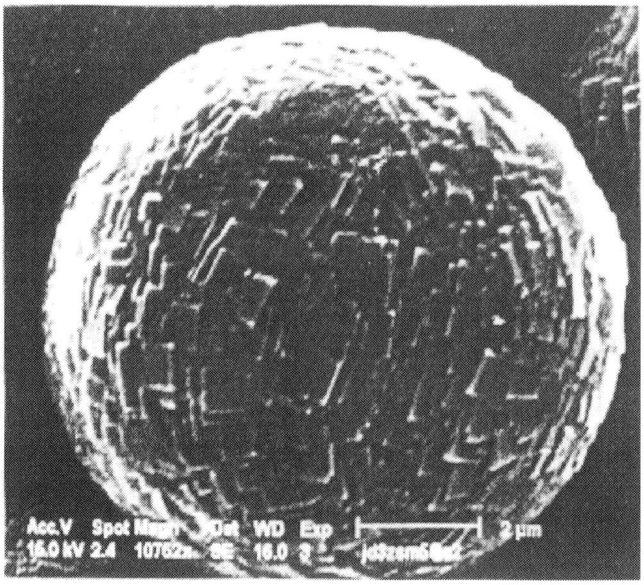

B

Figure 6. SEM micrographs of (A): $In(OH)_3$ cubic crystallites next to a large silicalite crystal, both generated by heating Si and In ingredients in the presence of methylamine: (B): In-MFI rounded polycrystalline aggregates, synthesized in the presence of NaOH.

Upon calcination, the In species basically migrate towards the surface of the crystallites, probably as tiny and well dispersed extra framework particles that involve quite strong acid sites. The surface migration could be visualized by RBS [65].

The Case of Titanium and Boron-titanium Admixtures

Incorporation of Ti in the MFI framework requires drastic preparation conditions such as, for example, the absence of any trace of alkali cations. Consequently, the various routes leading to Ti-MFI or (Ti, B)-MFI necessarily involve elaborated and costly reactants such as e.g. various Ti-alkoxides and alkali free TPAOH [66].

We have recently explored a series of more readily accessible, less time consuming and inexpensive alternative routes, yielding both (Ti)- and (Ti, B)- MFI zeolites, based on the use of methylamine [14,28,57,67].

Crystallization of titanosilicates from methylamine-fluoride admixtures. A wide series of preparation procedures leading to Ti-MFI (TS-1) has been tested [14,28,57]. All were based on the use of alkali free complexing agents also able to solubilize silica, thus exhibiting basic properties. Those were a series of short chain amines, such as diethanolamine (DEA) and triethanolamine (TEA) and mostly methylamine (MA), that are not supposed to compete with TPA for the templating role. Their basic properties further allowed the use of alkali free inexpensive TPABr.

The Si source was either SiO_2 (Aerosil-200) or $SiCl_4$, this latter being easy to co-hydrolyze along with $TiCl_4$, the main Ti source used.

Ingredients were mixed in various ways, under variable pH ranges, of which a close control was often required. Various additives, among which HF or HCl, used as efficient complexing agents for Ti to prevent its ready hydrolysis in acidic media, but also other Ti co-complexing agents such as citrates, oxalates or H_2O_2, were also occasionally added.

Gels involving 4Ti/96T ratios were typically heated in Teflon coated autoclaves at 185 °C for periods of time extending from 2 to 15 days.

While 100% crystalline silicalite was formed after 2 days heating in the presence of methylamine, no crystalline phase was detected after heating the same gel for 14 days in the presence of $TiCl_4$ (Si/Ti = 25) [14]. The absence of crystalline TiO_2 indicated that the amorphous phase was still composed of stable Ti-O-Si oligomers; indeed, $TiCl_4$ admixed with MA and heated under similar conditions, yielded pure anatase. Replacement of $TiCl_4$ by Ti acetylacetonate and adding HF to the $TiCl_4$ bearing mixture slightly improve the crystallinity of the final Ti-MFI (30 % after 10 days heating and 75 % after 6 days heating respectively). The crystallization rate was also shown to increase with decreasing amounts of MA, despite of a pH drop from 12.5 to 11.

Materials involving up to 1.7 Ti per MFI unit cell could be readily prepared in pure form. The framework Ti content increased with the increasing initial Ti/96T ratio in the gel up to 4Ti/96T (Table 3). Larger ratios yielded TS-1 crystals contaminated by anatase (for example, the total Ti/96T in sample 3 was 4.9, as measured by ICP, of which only 1.7 Ti were true framework atoms, as measured from unit cell expansion)

Table 3
Properties of some MFI zeolites prepared from dual methylamine + HF media (adapted from ref. [28]).

Sample N°	Ti/96T in gel	F/MA in gel	Unit cell volume (Å^3)	Framework Ti (from u.c. increase)
1 (SiO_2)	2.0	3	5362	1.3
2 (SiO_2)	4.0	3	5368	1.6
3 (SiO_2)	6.0	3	5370	1.7
4 (SiO_2)	4.0	1	5340	0.1
5 ($SiCl_4$)	4.0	12	5342	0.2
silicalite-1	0.0	3	5339	-

Although the presence of Ti ions in the reaction gel slowed down the crystallization of MFI zeolites, this latter could be dramatically accelerated by further adding fluoride ions, that act as co-mineralizing agent along with methylamine (compare samples 2 and 4, Table 3).

In the case of samples synthesized either in the presence of sole methylamine and in dual HF-methylamine media (fluoride ions do not favor a better Ti incorporation but just accelerate the reaction kinetics), IR and UV-vis, along with the lattice constant expansion (Table 3), confirmed the true incorporation of titanium into the MFI framework.

The use of $SiCl_4$ that efficiently forms Si-O-B gel complexes in MFI borosilicate syntheses (see above), lead to a slow crystallization and a poor framework Ti incorporation, suggesting that $SiCl_4$ is not an appropriate Si source for TS-1 synthesis.

Decreasing the quantity of methylamine in the gel improved the apparent crystallization rate. However, the amount of lattice titanium in the crystals dramatically decreased. These findings indicated that a larger quantity of methylamine is required to generate the appropriate Ti-O-Si oligomers that would potentially form the TS-1 framework. When the amount of methylamine is not sufficient, these oligomers are not readily assembled. The Ti would form a separate phase and only silicalite-1 would crystallize (samples 4 and 5, Table 3).

Crystallization of boro-titanosilicates from methylamine media. MFI type boro-titanosilicates could also be very efficiently synthesized in methylamine medium [67]. The crystallization rate of the boro-titanosilicates was slightly improved with respect to that of titanosilicates. Indeed, as a gel involving 4Ti/96T was still amorphous after 14 days heating in the presence of methylamine as the only mobilizing agent, by simultaneous adding of the Ti source to a borosilicate gel involving 4B/96T, 34% crystalline (B, Ti)-MFI was obtained after the same heating time. Again, in the presence of dual HF-MA mineralizing agents, the overall crystallization time was shortened; 85 % crystalline boro-titanosilicate was formed after 10 days heating.

^{11}B-NMR, IR, XRD and UV-vis confirmed that Ti was readily incorporated into the MFI framework along with boron and showed that both ions were homogeneously distributed throughout the crystallites. The amount of boron in the boro-titanosilicate decreased with respect to that of pure borosilicate. The titanium amount in the crystals also decreased along with the simultaneous incorporation of boron, suggesting that the stabilities of both B-MA and Ti-MA complexes were mutually affected [104].

The incorporation of boron in the zeolite framework leads to a contraction of the MFI unit cell because the atomic radius of boron is smaller than that of silicon (1.61 and 1.39Å for Si-O and B-O distances, respectively). Indeed, while the unit cell volume of borosilicates synthesized in methylamine media varied linearly as a function of framework boron quantities determined by ^{11}B NMR [71], the values of the unit cell volume of the boro-titanosilicates were slightly but definitely larger than those characterizing borosilicates. This relative expansion of the unit cell volume was caused by the presence of Ti-O bonds (1.79Å), longer than the Si-O bonds. This constitutes an additional proof of the incorporation of titanium ions in the zeolite framework.

The Case of Copper and Zinc

Except the work of Eränen *et al.*, who used copper impregnated silica fibre as the silica source in the gel precursor to ZSM-5 [68], to our knowledge, the preparation of Cu-MFI through direct synthesis has not been reported so far in the literature. On the other hand, Hagen *et al.* [15] seem to be the only team having attempted to prepare Zn-MFI through direct synthesis by using the rapid crystallization method of Inui [69]. XANES and EXAFS studies suggested that the lattice Zn^{2+} ions were surrounded by 4 framework oxygens and 2 other counter ions, either Na^+ in the as synthesized Zn-MFI or protons in the acid exchanged form, so as to achieve a sixfold coordination.

As part of an extensive research aiming to prepare Cu-bearing MFI, Zn-bearing MFI and their bimetallic (Cu, Zn)-bearing MFI counterparts through another route than the conventional impregnation that often yields non homogeneous copper-zeolite admixtures, namely by direct synthesis [70], we have tried to increase and to monitor the overall Cu and Zn concentration in the zeolite by introducing directly various Cu and/or Zn salts in the gel precursor prior to hydrothermal crystallization.

Direct synthesis of Cu-MFI was expected to yield as-synthesized materials with variable Cu^{2+} structural states that would generate upon further controlled calcination in various atmospheres, either the active Cu species, for example those extra- or intra-framework hydroxy-oxy polynuclear complexes that were shown to be good precursors to Cu active species in NO_x reduction [71], or dispersed metallic copper, the main active phase (especially when in synergetic interaction with ZnO), for CO_x hydrogenation or water gas shift conversion [72]. In such a case, both Cu and Zn species, either entrapped in the zeolite pores or intimately admixed with the zeolite crystals, were supposed to undergo modification (possibly through solid state reactions) upon zeolite calcination/reduction and interact so as to generate mixed oxide phases with enhanced activity in hydrogenation. The zeolite acidity, due either to residual framework Zn^{2+} ions left after calcination or to Al^{3+} ions introduced by purpose during zeolite synthesis, was expected to further convert the methanol formed into hydrocarbons.

Pure Zn-MFI was prepared as reference to provide information on the behavior of Zn in the MFI lattice and its eventual changes and structural rearrangements upon calcination. All the syntheses were achieved using

Synthesis of Cu-MFI, Zn-MFI and mixed (Cu, Zn)-bearing MFI phases [70]. A wide series of Cu- and Zn-containing ZSM-5 materials were prepared by heating appropriate (Si)- or (Si, Al)-hydrogel precursors involving variable amounts of Cu^{2+} or Zn^{2+} nitrates (or their admixtures), TPABr as template for MFI, and ammonia, methylamine or diethylamine as mineralizing or complexing agents towards both metallic ions, optionally admixed with fluoride ions.

The short chain alkylamine route presented at least the following advantages in the case of the two transition metal ions:

i) The good complexing power of amines towards Cu^{2+} and Zn^{2+} ions would favor a ready mobilization of these ions in solution.

ii) The absence of any traces of alkali cations would be even a more serious advantage because these cations, even in minor amounts, were shown to strongly decrease the catalytic activity of Cu-MFI catalysts in NO_x conversions [71].

The nature of the base and its concentration were among the critical variables that could be controlled to monitor an optimized incorporation (or admixing) of the metallic species in (or with) the final zeolite. The amount of Zn and Cu present in the final ZSM-5 markedly depended on the actual complexation of these ions by ammonia or various amines during the gel preparation or upon its hydrothermal heating. Methylamine, diethylamine and ammonia proved the most efficient complexing agents to bring Zn^{2+} ions in the MFI framework, the incorporation efficiency increasing in the following sequence: methylamine < diethylamine < ammonia = (ammonia + NH_4F). EDX spot analysis of single (Cu, Zn)-MFI crystals grown from hydrogels involving 2 Zn and 2 Cu per 96 T atoms (T = Si + Cu + Zn), from which any residual amorphous phase was removed by ultrasonic cleaning, confirmed the presence of respectively 0.7, 1.2, 2.1 and 2 Zn per MFI unit cell, in that sequence. Fluoride ions added along with ammonia did not improve the Zn incorporation but resulted in an increased MFI crystallinity. Conversely, NH_4F alone did yield silicalite but did not bring any Zn^{2+} in its framework, even when the pH was raised to basic values by e.g. NaOH. This suggests that the Zn species are definitely more efficiently mobilized in the gel phase by forming soluble amino complexes with ammonia or, to a lesser extent, with the short chain amines.

Increasing the initial amount of Zn in the pure silica gel yielded a zincosilicate MFI phase with no more than about 2.5 Zn per MFI unit cell, as ascertained by chemical analysis combined with ^{29}Si-NMR data showing Si(1Zn) configurations. FTIR was used to characterize framework Zn^{2+} ions through a new OH vibration at 3640 cm^{-1} and, indirectly, through CO or NH_3 interaction with the Zn species dispersed in the MFI lattice [73].

Despites the high solubility of Cu^{2+} ions in different amines and ammonia, these ions were never found incorporated within the MFI crystals grown in the presence of both Cu and Zn salts. Selected EDX probing of both the pure crystals and the residual amorphous phase separated from the latter, confirmed that Cu was involved as a hydroxy-oxidic type species admixed with silica. After separating the zeolite from the mother liquor, the blue color of the filtrate suggested that some Cu was also removed from the system as soluble complex. These results indicate that even if a metallic species is adequately mobilized through complexation and further released from the complex for a possible condensation with silica so as to build the first zeolite units, this condensation will not occur if the ions do not readily achieve the appropriate (tetrahedral) coordination.

The Case of Some Other Transition Metal Ions

<u>Sn</u>$^{4+}$ and <u>Th</u>$^{4+}$ ions could be incorporated in the MFI framework under very drastic synthesis conditions, in very small amounts, as suggested by crystal morphology and

TPA decomposition pattern changes (with respect to silicalite), but their presence dramatically slowed down the zeolite crystallization.

- Vanadium (VO^{2+} ions) seems to form rather strong complexes with methylamine so that their lattice incorporation would be limited, although probably reel (EPR, DTA); vanadium ions possibly also partly stay admixed with the final zeolite crystals, as (MA-VO^{2+}) complexes. By carefully monitoring various synthesis parameters, large elongated crystals (45 x 5 x 5 μm) or extremely small rounded crystallites (about 0.1 μm in diameter) could be favored. Vanadium could be detected in all crystallites as well as in some residual amorphous phases contaminating some of the syntheses.

- The same was probably also true with molybdenum (introduced in the gel as NH_4^+ molybdate), although EPR did not bring definite evidence for its partial framework incorporation (in a not defined oxidation state). The final MFI crystals, elongated along the c axis, show a morphology common to silicalite synthesized under the same conditions. DTA suggested that both silicalite and a (a small amount of) Mo-bearing MFI zeolite could be present. This latter was more abundant when $MoCl_3$ is used as starting material (DTA trace corresponding to TPA release) although only one type of morphology was detected by SEM. Spot EDX analysis (Si/Mo ratios) detected from 0.1 to 0.2 wt. % Mo on clean individual crystallites but such small amounts have to be handled with caution on gold coated samples, due to the close energies and possible interferences between Au and Mo X-ray lines.

Incidently, Mo-MFI could be more successfully crystallized from ammonium molybdate in strong basic NaOH media. From 1.9 to 3.4 wt % Mo could be detected by EDX in individual crystallites, free from any amorphous phase. The structural state of these Mo species could not be defined by this technique and is currently under investigation.

- Nickel, manganese and cobalt divalent cationic species most probably do behave similarly to chromium, namely they are quite strongly complexed by methylamine and yield some metal rich residual amorphous phase at the end of crystallization. Methylamine and NH_4F admixture yielded larger and better defined Co-MFI crystallites than the sole methylamine medium. As in the case of chromium, these crystallites also appeared coated by a thin film of some methylamine-cobalt complex, suggesting that mixed F-MA-Co or sole F-Co strong associations would definitely impede the final Co incorporation in the MFI lattice. Ammonia, ammonia admixed with urea and various alkanolamines admixed with HF were among the other mineralizing agents yielding highly crystalline Co-MFI. The cobalt structural state in these phases is currently studied.

- Cadmium proved similar to zinc, by forming complexes with methylamine so that their incorporation in the MFI lattice is *a priori* easy. Preliminary investigations indicated that while zinc incorporated in appreciable amounts (up to about 2 - 2.5 Zn^{2+} ions per MFI unit cell, see above), the more bulky Cd^{2+} ions were also possibly incorporated in the MFI lattice but to more limited extents. In both cases, ammonia proved to be at least an as efficient mobilizing-complexing agent as short chain alkylamines.

EDX probing of selected Cd-MFI crystallite areas allowed us to detect less than 1 Cd per unit cell. ^{29}Si-NMR data indeed suggested the possible presence of Si(nCd) configurations. However, none of these techniques should be considered as 100% reliable in this case. EDX could also probe Cd stemming from residual amorphous Cd-rich phase located at the vicinity of the zeolite crystallites, while the NMR line at about - 103 ppm characterizing the Si(nCd) configurations can also be attributed to defect silanol groups of the MFI structure [74]. The true incorporation of Cd in the MFI lattice is still being questioned.

- Finally, methylamine can also behave, in particular conditions, as a reducing agent, towards metallic ions such as <u>Cu</u>$^{2+}$ (in the absence of Zn^{2+}), <u>Ag</u>$^{+}$ or (partly) <u>Fe</u>$^{2+}$, with, as a result, the easy preparation of silicalite (or Al-MFI) crystallites, intimately admixed with the corresponding (very small) metallic particles, that potentially can constitute further interesting multifunctional catalytic systems.

Acknowledgements

This review paper is dedicated to Prof. Tomoyuki Inui, close friend and skilful scientific guide of one of us (Z.G.) during his stay in Kyoto university in 1986. This first involvement into the synthesis of metallosilicate zeolites, the real outset to a large part of the present work, is basically due to Prof. Inui's bright vision of the future needs in metal modified zeolite synthesis and catalysis. This first research could also be successfully achieved thanks to the diligence of Prof. Inui's former collaborators in Kyoto, Dr. Akira Miyamoto and Hirokazu Matsuda.

The authors are much grateful to their colleagues, professors and researchers, J.L. Guth (Mulhouse) and R. Galiasso, G. Giannetto and R. Monque (Caracas), for their fruitful collaboration in the early research dealing with the first uses of alkylamines as mobilizing agents, to L. Gilbert and E. Fache (Aubervilliers), M.F. Ribeiro (Lisbon), J.L.G Fierro (Madrid), I. Ivanova (Moscow), G. Demortier (Namur), J. Barrault (Poitiers), S. Kaliaguine (Quebec), T. Ito (Sapporo) and E. Garrone (Torino) for having characterized some of the metallosilicate samples prepared by the methylamine route by various complementary techniques, to M. Shibata (KAO Corp. Tokyo) for his extensive contribution to the synthesis and characterization of B- and Ti-MFI zeolites during his stay in Namur university, to students, C. Torre-Abreu (Lisbon), A. Smirnov and S. Nikon (Moscow), J.C. Bertrand, J. Duchenne, J. Gérard, G. Jodocy and D. Solbreux, (Namur), J.M. Clacens (Poitiers) and B. Onida and C. De Feo (Torino), for their help in the final presentation of their most recent results, and, finally, to M. Wiame, C. Mayenez and

G. Daelen (Namur), for their invaluable technical help.

The EEC is particularly acknowledged for the Human Capital and Mobility contract and grant to one of us (S.V.).

REFERENCES

1. Neither the isomorphous substitution of Si by pentavalent elements like P, As...thus leading to the well known large family of microporous (silico)aluminophpsphates (arsenates...), nor the new microporous Ge and Sn sulfides based on tetrahedral sulfide frameworks, nor a series of newly reported miscellaneous frameworks briefly mentioned in ref. [9], are discussed here.
2. As a quite complete and recent review on the use of heteroatom molecular sieves as shape selective catalysts, see S.M. Csicsery, *Stud. Surf. Sci. Catal.*, **94**, 1 (1995), and references therein.
3. R.M. Barrer, *Hydrotharmal Chemistry of Zeolites*, Academic Press, London 1982, 543 pp.
4. M. Tielen, M. Geelen, and P.A. Jacobs, *Acta Phys. Chem.* **31**, 1 (1985).
5. K.G. Ione and L.A. Vostrikova, *Russ. J. Chem. Rev.* **56**, 231 (1987).
6. S. Kaliaguine, J. B.Nagy, and Z. Gabelica, *Stud. Surf. Sci. Catal.* **35**, 381 (1988).
7. P. Ratnasamy and R. Kumar, *Catalysis Today* **9**, 329 (1991).
8. H. Kessler. In: *Proc. 9th Intern. Zeolite Conf. Montreal, 1992*, R. Von Ballmoos et al. (Eds.), Butterworths-Heinemann, Boston, 1993, p. 73.
9. H. Kessler. In: *Comprehensive Supramolecular Chemistry*, J.L. Atwood et al. (Eds), Pergamon Press, Oxford, Vol 7, 1996, p. 425.
10. G. Bellussi and M.S. Rigutto, *Stud. Surf. Sci. Catal.* **85**, 177 (1994).
11. S.L. Burkett and M.E. Davis. In: *Comprehensive Supramolecular Chemistry*, J.L. Atwood et al. (Eds.), Pergamon Press, Oxford, Vol 7, 1996, p. 465.
12. T. Inui, *Stud. Surf. Sci. Catal.* **105**, 1441 (1997).
13. T. Inui, D. Medhanavyn, and A. Miyamoto, *React. Kin. Catal. Lett.*. **34**, 69 (1987).
14. M. Shibata, J. Gerard, and Z. Gabelica, *Stud. Surf. Sci. Catal.* **105A**, 245 (1997).
15. A. Hagen, K.H. Hallmeier, C. Hennig, R. Szargan, T. Inui, and F. Roessner, *Stud. Surf. Sci. Catal.* **94**, 195 (1995).
16. R.M. Barrer and P.J. Denny, *J. Chem. Soc.* 971 (1961).
17. Ref. [4], p. 159.
18. Z. Gabelica, R.J. Monque, and G. Giannetto, *U.S .Patent* 5,354,719 (1994).
19. G. Giannetto, G. Leon, J. Papa, R. Monque, R. Galiasso, and Z. Gabelica, *Catal. Letters* **22**, 981 (1993).
20. G. Giannetto, R. Monque, R. Galiasso, J. Papa, and Z. Gabelica, *Stud. Surf. Sci. Catal.* **84B**, 381 (1994).
21. G. Giannetto, R. Monque, A. Montes, and Z. Gabelica. In: *Proc. XV Iberoamerican Symposium on Catalysis, Cordoba, Argentina, 1996*; E. Herrero et al., (Eds.), Unica Edicion, Buenos Aires, Vol. I, 1996, p. 83.
22. A. Montes, Z. Gabelica, A. Rodriguez, and G. Giannetto. In: *Proc. XV Iberoamerican Symposium on Catalysis, Cordoba, Argentina, 1996*; E. Herrero et al., (Eds.), Unica Edicion, Buenos Aires, Vol. I, 1996, p. 87.
23. G. Giannetto, G. Leon, J. Papa, R. Monque, R. Galiasso, and Z. Gabelica, *Catalysis Today* **31**, 317 (1996).
24. Z. Gabelica and J.L. Guth, *Eur. Patent* 337,385 (1993).
25. J.L. Guth and Z. Gabelica, *U.S. Patent* 5,371,307 (1994).

26. Z. Gabelica and J.L. Guth, *Angew. Chem. Int. Ed. Engl.* **28**, 81 (1989).
27. Z. Gabelica and J.L. Guth, *Stud. Surf. Sci. Catal.* **49A**, 421 (1989).
28. M. Shibata and Z. Gabelica, *Zeolites* **19**, 246 (1997).
29. Z. Gabelica, C. Mayenez, R. Monque, R. Galiasso, and G. Giannetto. In:: *Synthesis of Microporous Materials: Vol I: Molecular Sieves*, M.L. Occelli and H. Robson (Eds.), Van Nostrand Reinhold, New York, 1992, p. 190.
30. Z. Gabelica, G. Giannetto, F. Dos Santos, R. Monque, and R. Galiasso. In:: *Proc. 9th Intern. Zeolite Conf. Montreal*, R. Von Ballmoos et al. (Eds.), Butterworths-Heinemann, Boston, 1993, p. 231.
31. Z. Gabelica and G. Demortier, *Nucl. Instr. & Methods in Phys. Res. B.*, in the press (1997).
32. A. Lopez, *PhD Thesis*, University of Mulhouse, France, (1990).
33. J.M. Chezeau, L. Delmotte, J.L. Guth, and Z. Gabelica, *Zeolites* **11**, 598 (1991).
34. E.G. Derouane, S. Detremmerie, Z. Gabelica, and N. Blom, *Appl. Catal.* **1**, 201 (1981).
35. L.D. Rollmann and E.W. Valyocsik, *Eur. Patents,* 21,674 and 21,675 (1981).
36. Z. Gabelica, N. Blom, and E.G. Derouane, *Appl. Catal.* **5**, 227 (1983).
37. Z. Gabelica, E.G. Derouane, and N. Blom, *ACS Symp. Series* **248**, 219 (1984).
38. Z. Gabelica, J.B.Nagy, E.G. Derouane, and J.P. Gilson, *Clay Miner.* **19**, 803 (1984).
39. Z. Gabelica, J.B.Nagy, P. Bodart, and A. Nastro, *Thermochim. Acta* **93**, 749 (1985).
40. J.C. Jansen, C.W.R. Engelen, and H. van Bekkum. In: *Zeolite Synthesis*, M.L. Occelli and H.E. Robson (Eds.), ACS, Washington, DC, 1989, p. 257, (ACS Symposium Series **398**).
41. E.G. Derouane and Z. Gabelica, *J. Solid State Chem.* **64**, 296 (1986).
42. M. Taramasso, G. Manara, V. Fattore, and B. Notari, *U. S. Patent* 4,656,016 (1987).
43. Z. Gabelica, J. B.Nagy, P.Bodart, and G. Debras, *Chem. Letters* 1059 (1984).
44. R. de Ruiter, J. C. Jansen, and H. van Bekkum, *Zeolites* **12**, 56 (1992).
45. Z. Gabelica, G. Debras, and J. B.Nagy, *Stud. Surf. Sci. Catal.* **19**, 113 (1984).
46. H. Kessler, J. M. Chezeau, J.L. Guth, H. Strub, and G. Coudurier, *Zeolites* **7**, 360 (1987).
47. D. Trong On, M.P. Kapoor, L. Bonneviot, S. Kaliaguine, and Z. Gabelica, *J. Chem. Soc. Faraday Trans.* **92**, 1031 (1996).
48. M. Shibata and Z. Gabelica, *Appl. Catal. A*, (1997), in press.
49. D.W. Breck and E.M. Flanigen. In: *Molecular Sieves,* Soc. Chem. Ind. London, 1968, p. 47.
50. B.D. McNicol, G.T. Pott, R.K. Loos, and N. Mulder, *Adv. Chem. Ser.* **121**, 152 (1973).
51. Z. Gabelica, V. Norberg, and T. Ito, *Stud. Surf. Sci. Catal.* **105**, 285 (1997).
52. A. Seive, *PhD Thesis*, University of Mulhouse, France, (1989).
53. G. Giannetto, F. Dos Santos, R. Monque, R. Galiasso, and Z. Gabelica, *Zeolites* **15**, 719 (1995).
54. G. Giannetto, J. Papa, J. Perez, L. Garcia, R. Monque, and Z. Gabelica, *Zeolites* **14**, 549 (1994).
55. R. Challoner, R.K. Harris, S.A. Barri, and M.J. Taylor, *Zeolites* **11**, 827 (1991).
56. A.P.M. Kentgens, C.R. Bayense, J.H.C. Van Hooff, J.W. De Haan, and L.J.M. Van de Ven, *Chem. Phys. Lett.* **176**, 399 (1991).
57. M. Shibata, J. Gerard, and Z. Gabelica, *Microporous Mater.* **12**, 141 (1997).
58. J.L. Guth, H. Kessler, J.M. Higel, J.M. Lamblin, J. Patarin, A. Seive, J.M. Chezeau, and R. Wey. In: *Zeolite Synthesis*, M.L. Occelli and H.E. Robson (Eds.), ACS, Washington, DC, 1989, p. 176, (ACS Symposium Series **398**).
59. P. Tsai and J.R. Anderson, *J. Catal.* **80**, 207 (1983).
60. A. Miyamoto, D. Medhanavyn, and T. Inui, *Chem. Express* **1**, 73 (1986).
61. A.V. Kucherov, A.A. Slinkin, G.K. Beyer, and G. Borbely, *Zeolites* **15**, 431 (1995).
62. J.S.T. Mambrin, E.J.S. Vicki, H.O. Pastore, C.U. Davanzo, H. Vargas, E. Silva, and O. Nakamura, *J. Chem. Soc., Chem. Commun.* 923 (1991).

63. H.E.B. Lempers and R.A. Sheldon, *Stud. Surf. Sci. Catal.* **105**, 1061 (1997).
64. G. Vorbeck, J. Jänchen, B. Parlitz, M. Schneider, and R. Fricke, *J. Chem. Soc., Chem. Commun.* 123, (1994).
65. S. Valange, Z. Gabelica, and G. Demortier, in preparation.
66. D. Trong On, S. Kaliaguine, and L.J. Bonneviot, *J. Catal.* **157**, 235 (1995)
67. M. Shibata and Z. Gabelica, *Microporous Mater.* **11**, 237 (1997).
68. K. Eränen, N. Kumar, and L.E. Lindfors, *Appl. Catal. B* **4**, 213 (1994).
69. T. Inui, *ACS Symp. Series*, **398**, 479 (1989).
70. S. Valange, Z. Gabelica, M. Lopez Granados, S. Rojas, and J.L.G. Fierro. In: *Discussions on Zeolites and Microporous Materials,* H. Chon and Y.S. Uh (Eds.), Hanrimwon Publ. Co., Seoul, Korea, 1997, p. 203.
71. M. Iwamoto, *Stud. Surf. Sci. Catal.* **84**, 1395 (1994) and references of the same author therein.
72. J.L.G. Fierro. In: *Proc. 8° Seminario Brasileiro de Catalise* Eds: IBP., Anais, Rio de Janeiro, Brazil, 1995, vol. II, p. 544.
73. S. Valange, Z. Gabelica, B. Onida, and E. Garrone, submitted for publication.
74. J. B.Nagy, Z. Gabelica, and E.G. Derouane, *Chem. Letters* 1105 (1982).

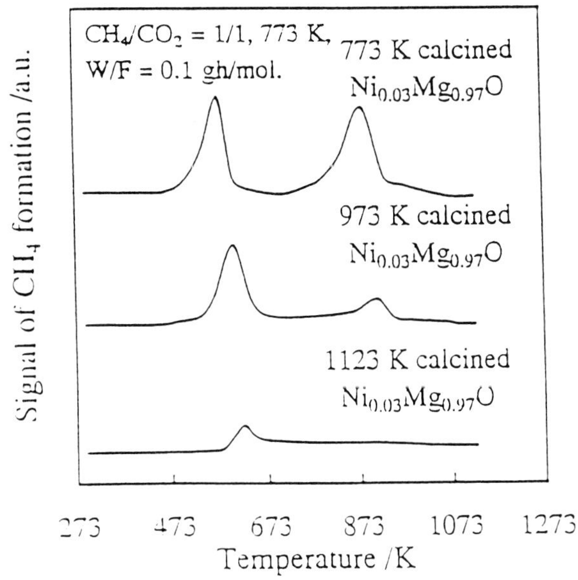

Figure 2. Temperature programmed hydrogenation profiles of $Ni_{0.03}Mg_{0.97}O$ calcined at various calcination temperatures after CH_4-CO_2 reaction at 773 K for 1 h. Catalytic reaction conditions: 773 K, W/F = 0.1 gh/mol, CH_4/CO_2 = 1/1, total pressure 0.1 Mpa. TPH conditions: heating rate 20 K/min, 298-1123 K, H_2 flow 50 ml/min.

Carbon deposition is known to be favorable at lower reaction temperature [1], therefore the evaluation of resistance to carbon deposition was carried out at low temperature. In TPH profiles, two peaks were observed. The lower temperature peak was mainly attributed to adsorbed CO_2 on the support surface, and higher temperature peak was deposited carbon as reported in detail in another article [20]. On the sample at lower calcination temperature, a larger signal due to carbon deposition was observed, but on $Ni_{0.03}Mg_{0.97}O$ solid solution catalyst, almost no carbon was deposited. Solid solution formation is favorable for the inhibition of carbon formation.

Effect of Reduction Pretreatment Temperature

Figure 3 shows the catalytic stability of $Ni_{0.03}Mg_{0.97}O$ and 3 mol% Ni/MgO reduced at various temperatures. By reducing the sample at 973 K, catalytic activity began to appear, but at this reduction temperature the catalyst stability was very low. At higher reduction temperature than 1023 K, the catalytic activity became high and stable under W/F = 1.2 gh/mol. After the catalytic reaction under W/F = 1.2 gh/mol, the color of the sample was not black but green. This means that the deactivation

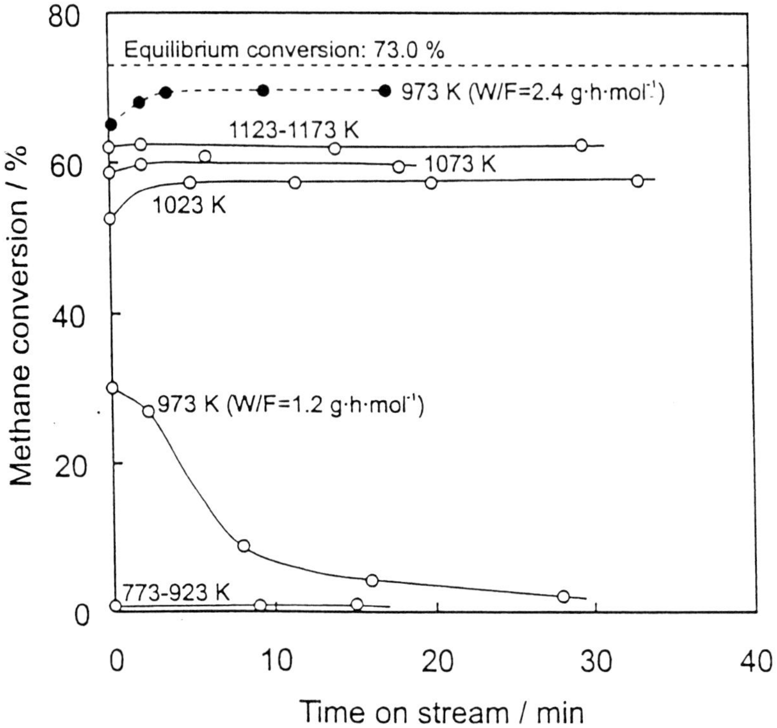

Figure 3. Effect of H_2 reduction temperature on CH_4-CO_2 reaction activity on $Ni_{0.03}Mg_{0.97}O$ solid solution catalyst. Reaction conditions: 973 K, W/F = 1.2 gh/mol, CH_4/CO_2 = 1/1, total pressure 0.1 MPa.

was caused by not carbon formation but the oxidation of reduced Ni which was formed during the pretreatment procedure with CO_2 and/or H_2O, which is from reverse water gas shift reaction ($H_2 + CO_2 \rightarrow H_2O + CO$). But under W/F = 2.4 gh/mol condition, the catalyst reduced at 973 K was found to have high and stable activity. This indicated that the partial pressure of the products (H_2 and CO) is related to the catalyst stability.

Effect of the Addition of Hydrogen to the Reactant Gas

Figure 4 shows the catalyst stability for CH_4-CO_2 and CH_4-CO_2-H_2 reactions on $Ni_{0.03}Mg_{0.97}O$ solid solution and 3 mol% Ni/MgO catalysts under low W/F = 0.1 gh/mol. As shown in previous chapter, $Ni_{0.03}Mg_{0.97}O$ solid solution catalyst deactivated with time on stream rapidly under low W/F. This deactivation is due to the oxidation of reduced Ni by CO_2 and/or H_2O which is formed via reverse water gas shift reaction. Under low W/F, the partial pressure of CO and H_2 is not high

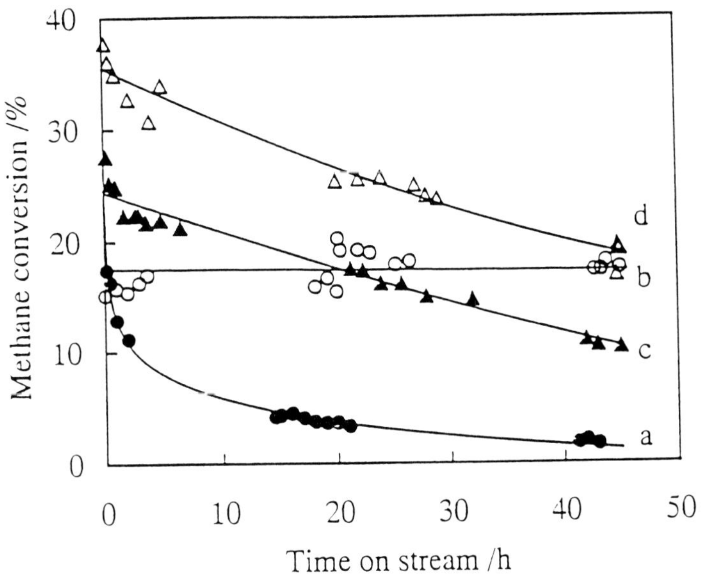

Figure 4. Catalytic stability of $Ni_{0.03}Mg_{0.97}O$ (a, b) and 3 mol% Ni/MgO (c,d) under 1023 K and the atmosphere of $CH_4/CO_2/N_2$ = 50/25/25 (a,c) and $CH_4/CO_2/H_2/N_2$ = 50/25/5/20 (b,d). W/F = 0.5 gh/mol for $Ni_{0.03}Mg_{0.97}O$ and W/F = 0.2 gh/mol for 3 mol% Ni/MgO, total pressure 0.1 MPa.

enough to maintain Ni in reduced state. In contrast, when methane conversion is high under high reaction temperature and high W/F, the partial pressure of CO and H_2 is very high enough to maintain Ni in reduced state. In this condition, $Ni_{0.03}Mg_{0.97}O$ exhibited excellent stability shown in Figure 1. The deactivation was also observed on 3 mol% Ni/MgO and this seems to be due to aggregation of nickel and carbon deposition. When hydrogen was fed with the reactant gas ($CH_4/CO_2/H_2/N_2$ = 50/25/5/20), the activity decreased and the deactivation could not be inhibited on 3 mol% Ni/MgO. In contrast to this result, the activity was maintained and at the same time the deactivation could be completely inhibited on $Ni_{0.03}Mg_{0.97}O$. This indicated that the hydrogen addition is very effective for avoiding the deactivation due to oxidation of nickel.

Effect of Calcium Addition by Coprecipitating Preparation Method

Figure 5 shows the catalytic activity of CO_2 reforming of methane on $Ni_{0.03}Mg_{0.97}O$ and $Ni_{0.03}Ca_{0.03}Mg_{0.94}O$ catalysts under severer conditions (CH_4/CO_2 = 2/1, total pressure 0.32 MPa). Under these reaction conditions, the deactivation due to aggregation of nickel and carbon deposition was observed even on $Ni_{0.03}Mg_{0.97}O$. $Ni_{0.03}Ca_{0.03}Mg_{0.94}O$ had more stable activity, though lower initial activity than

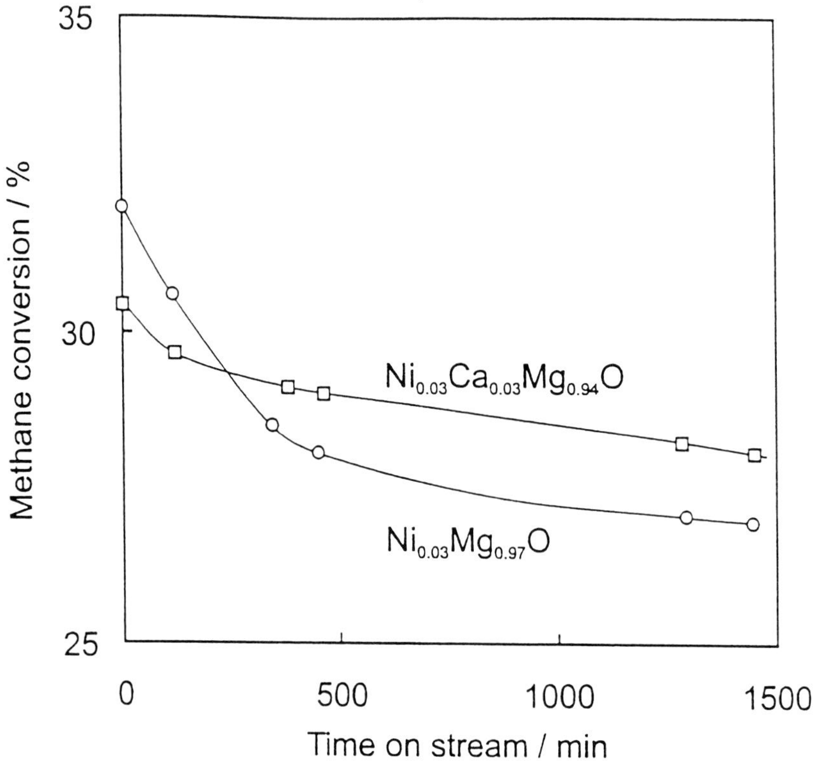

Figure 5. Reaction time dependence of methane conversion in CO_2 reforming of methane on $Ni_{0.03}Mg_{0.97}O$ and $Ni_{0.03}Ca_{0.03}Mg_{0.94}O$ catalysts. Reaction conditions: 1123 K, W/F = 0.6 gh/mol, CH_4/CO_2 = 2/1, total pressure 0.32 MPa.

$Ni_{0.03}Mg_{0.97}O$. The addition of calcium promoted the catalyst stability through the inhibition of carbon deposition.

Figure 6 shows the relation between added calcium amount and catalytic activity. There is an optimum amount of calcium; small amounts of calcium promoted catalytic activity, but larger amounts decreased it. The results of BET surface area on these samples were listed in Table 1. The BET surface area was maintained on the catalyst with small amount of calcium addition, but BET decreased with larger amounts of addition. This seems to be related to the composition limit of solid solution formation between CaO and MgO. Figure 7 shows the catalytic activity normalized by BET surface area. The addition of calcium apparently promoted the catalytic activity. Figure 8 shows the dependence of catalytic activity on CO_2 partial pressure over $Ni_{0.03}Mg_{0.97}O$ and $Ni_{0.03}Ca_{0.13}Mg_{0.84}O$. The activity increased with the CO_2 partial pressure, and it was saturated under high CO_2 pressure. The methane conversion rate on

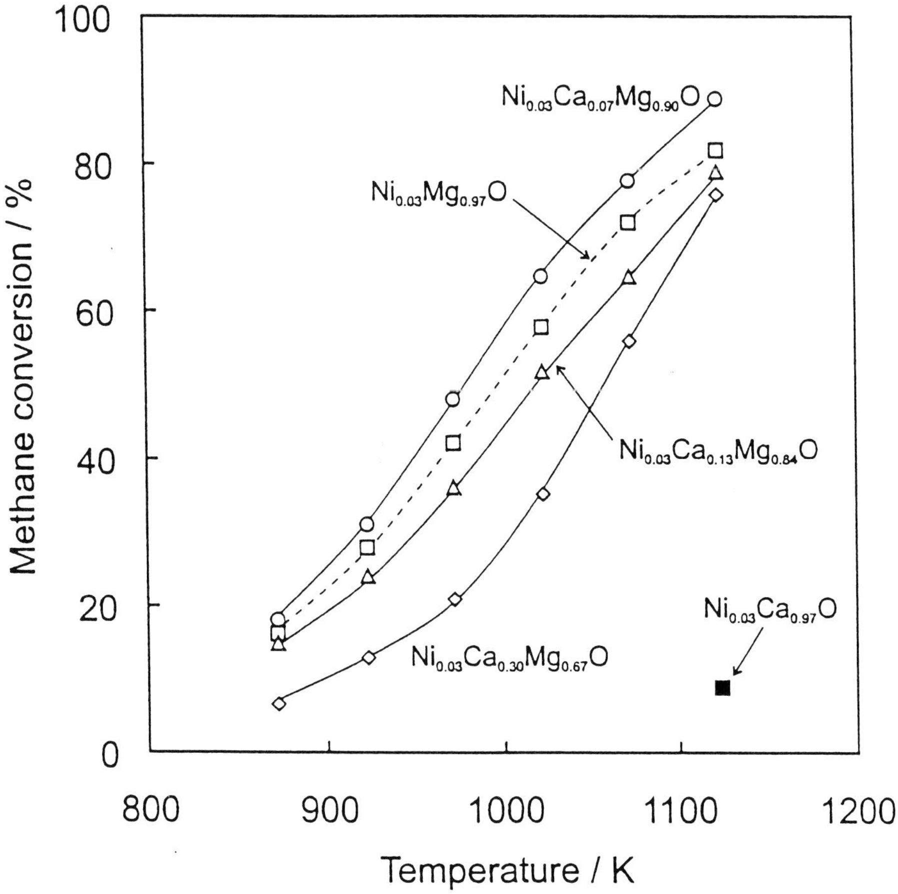

Figure 6. Effect of the addition of calcium ions to $Ni_{0.03}Mg_{0.97}O$ catalyst on initial catalytic activity. Reaction conditions: W/F = 0.6 gh/mol, CH_4/CO_2 = 1/1, total pressure 0.1 MPa.

Table 1
BET surface area of $Ni_xCa_yMg_{1-x-y}$ catalysts.

Catalyst	BET surface area / m^2g^{-1}
$Ni_{0.03}Mg_{0.97}O$	21.6
$Ni_{0.03}Ca_{0.07}Mg_{0.90}O$	23.1
$Ni_{0.03}Ca_{0.13}Mg_{0.84}O$	17.4
$Ni_{0.03}Ca_{0.30}Mg_{0.67}O$	10.3

$Ni_{0.03}Ca_{0.13}Mg_{0.84}O$ reached the saturation level under lower CO_2 partial pressure than that on $Ni_{0.03}Ca_{0.13}Mg_{0.84}O$. This means that $Ni_{0.03}Ca_{0.13}Mg_{0.84}O$ has stronger

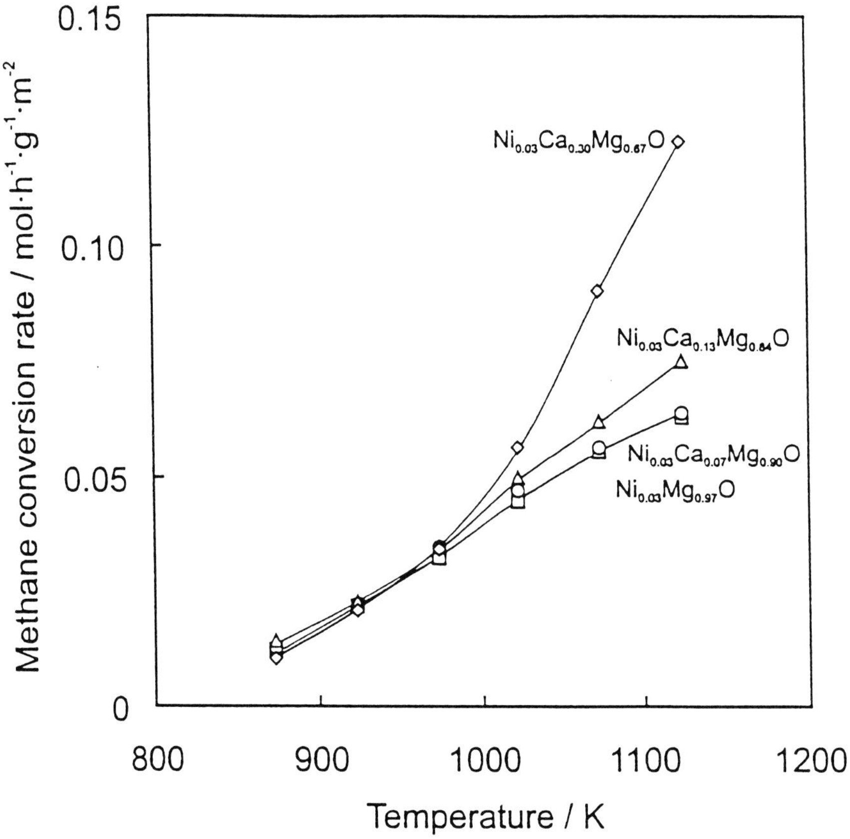

Figure 7. Effect of the addition of calcium ions to $Ni_{0.03}Mg_{0.97}O$ catalyst on initial catalytic activity normalized by BET surface area. Reaction conditions: W/F = 0.6 gh/mol, CH_4/CO_2 = 1/1, total pressure 0.1 MPa.

interaction with CO_2 and higher basicity than $Ni_{0.03}Mg_{0.97}O$. It has been reported previously that small doses of CaO create population sites with enhanced Lewis basicity on the MgO surface [21]. Current observations correspond well with this previous report. The promotion of activity and stability by the addition of calcium is caused by enhancement of surface basicity and probably the activation of adsorbed CO_2.

Catalytic Behavior Under High Pressure

The carbon dioxide reforming reaction in the previous section was performed under lower total pressure conditions than practical methane reforming reaction conditions (2.0-4.0 Mpa). Figure 9 shows the conversion of methane and carbon dioxide

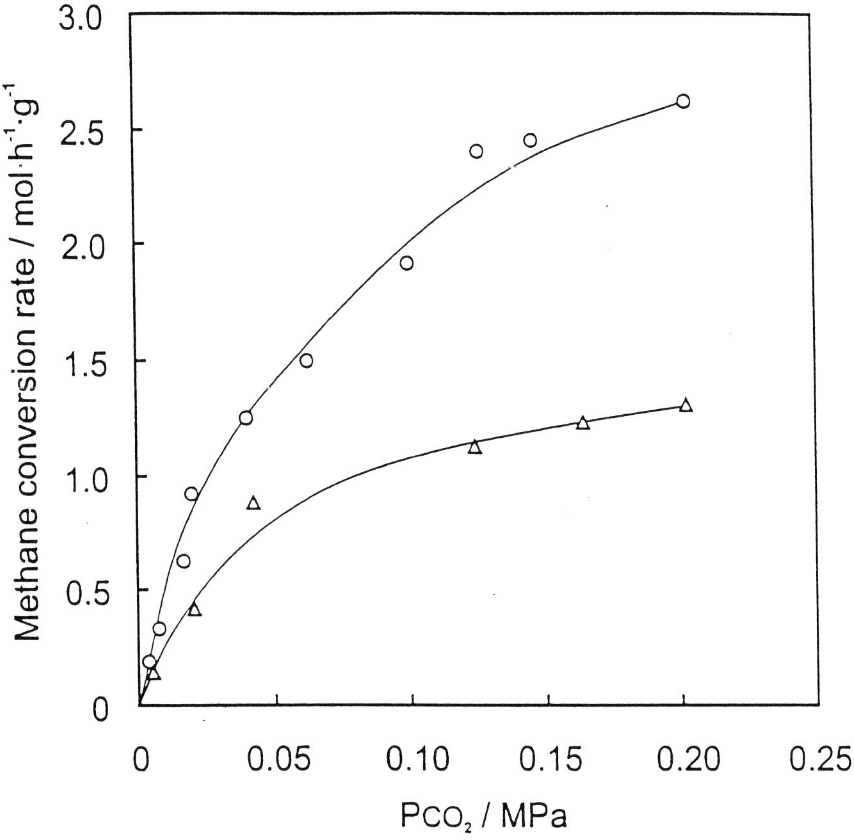

Figure 8. Dependence of methane conversion rate on CO_2 partial pressure over $Ni_{0.03}Mg_{0.97}O$ (O) and $Ni_{0.03}Ca_{0.13}Mg_{0.84}O$ (△). Reaction conditions: 1123 K, W/F = 0.3 gh/mol, P_{CH4} = 0.16 MPa, total pressure = 0.32 MPa, N_2 balance.

in CO_2 reforming of methane on Ca (1 wt%)/$Ni_{0.03}Mg_{0.97}O$ catalyst. Under this reaction condition (W/F = 1.44 gh/mol), the conversion reached the equilibrium level. In this case CO_2 conversion was much higher than CH_4 conversion. The H_2/CO ratio (1.5) under this reaction condition was higher than for the stoichiometric reaction (1.0). This should be due to additional conversion of CO_2 by reverse water gas shift reaction ($H_2 + CO_2 \rightarrow H_2O + CO$), which seems to be equilibrium at the temperature of the reactor exit. On $Ni_{0.03}Mg_{0.97}O$ catalyst the very similar result to Ca (1 wt%)/$Ni_{0.03}Mg_{0.97}O$ was observed apparently. But the behavior of carbon deposition on each catalyst was different. According to the TG measurements, on $Ni_{0.03}Mg_{0.97}O$ catalyst after 22-h reaction, 20.8 wt% carbon was deposited, while only 2.4 wt% carbon was deposited on Ca (1 wt%) /$Ni_{0.03}Mg_{0.97}O$ catalyst after 42-h reaction. These results indicated that the addition of calcium is

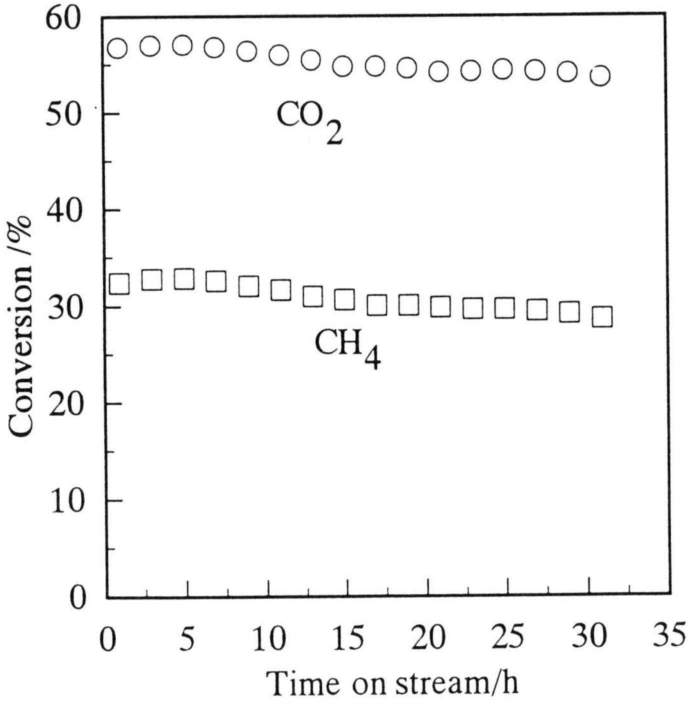

Figure 9. Conversion of methane and carbon dioxide in CO_2 reforming of methane on Ca (1 wt%)/$Ni_{0.03}Mg_{0.97}O$ catalyst under high total pressure. Reaction conditions: 1023 K, W/F = 1.44 gh/mol, $CH_4/CO_2/H_2$ = 47/47/6, total pressure 2.0 MPa.

very effective for the inhibition of carbon formation. This is due to the higher basicity of calcium oxide, which causes the strong interaction between catalyst surface and CO_2. As reported previously, it is suggested that the rapid oxidation by oxygen is supplied from the dissociation of adsorbed CO_2 at the metal-support interface [22, 23]. The effect of calcium addition can be interpreted by this suggestion.

CONCLUSION

We have developed a catalyst with high resistance to carbon formation during CO_2 reforming of methane. The $Ni_{0.03}Mg_{0.97}O$ solid solution catalyst showed much greater resistance to carbon formation than other reforming catalysts. The enhancement of surface basicity by the addition of calcium oxide was found to inhibit the carbon deposition. The CO_2 reforming of methane is essential for utilizing remote natural gas and CO_2 in it. It may become the key technology for energy development in the 21st century.

REFERENCE

1. J.R. Rostrup-Nielsen. In: *Catalysis Science and Technology*, J.R. Anderson and M. Boudart (Eds.), Springer, Berlin, Germany, **5**, 3 (1984).
2. J.P. Van Hook, *Catal. Rev. Sci. Eng.* **21** , 1 (1980) .
3. C.H. Bartholomew, *Catal. Rev. Sci. Eng.* **24**, 67 (1982).
4. I. Puskas, *CHEMTECH*, December, 43 (1995).
5. J.R. Rostrup-Nielsen, *Stud. Surf. Sci. Catal.* **81**, 25 (1994) .
6. S. Wang and G.Q. Lu, *Energy and Fuel* **10** , 896 (1996).
7. For example, *Oil and Gas Journal*, Nov. 21, 30 (1994).
8. A.T. Aschcroft, A.K. Cheethan, M.L.H. Green, and P.D.F. Vermon, *Science* **352**, 225 (1991).
9. J.R. Rostrup-Nielsen and J.H.B. Hansen, *J. Catal.* **144** , 38 (1993) .
10. D. Qin and J. Lapszewics, *Catal. Today* **21** , 551 (1994).
11. J.T. Richardson and S.A. Paripatyadar, *Appl. Catal.* **61**, 293 (1990) .
12. Z. Zhang and X.E. Verykios, *Appl. Catal. A:General* **138**, 109 (1995) .
13. Z. Zhang, X.E. Verykios, S.M. MacDonald, S. Affrossman, *J. Phys. Chem.* **100**, 744 (1996).
14. O. Yamazaki, T. Nozaki, K. Omata, and K. Fujimoto, *Chem. Lett.* 1953 (1992).
15. K. Fujimoto, K. Omata, T. Nozaki, O. Yamazaki, and Y. Han, *Energy Conversion and Management* **33**, 529 (1992).
16. O. Yamazaki, K. Tomishige, and K. Fujimoto, *Appl. Catal. A:General* **136**, 49 (1996) .
17. E. Ruckenstein and Y.H. Hu, *Appl. Catal. A:General* **149**, 149 (1995).
18. Y.H. Hu and E. Ruckenstein, *Catal. Lett.* **36**, 145 (1996).
19. Y.H. Hu and E. Ruckenstein, *J. Catal.* **163**, 306 (1996).
20. Y. Chen, K. Tomishige, and K. Fujimoto, *Appl. Catal. A:General*, in press.
21. R. Philipp and K. Fujimoto, *J. Phys. Chem.* **96**, 9035 (1992).
22. K. Tomishige, Y. Chen, K. Yokoyama, Y. Sone, and K. Fujimoto, *Shokubai* **39**, 70 (1997).
23. Y. Chen, K. Tomishige, K. Yokoyama, and K. Fujimoto, *Appl. Catal. A:General*, in press.

Res. Chem. Intermed., Vol. 24, No. 3, pp. 273-289 (1998)

ZEOLITE CONTAINING CATALYTIC MEMBRANES AS INTERPHASE CONTACTORS

SHANQIANG WU[a], CHRISTIAN BOUCHARD[b], and SERGE KALIAGUINE[a,*]
[a]*Department of Chemical Engineering and* [b]*Department of Civil Engineering, Université Laval, Sainte-Foy, Québec, Canada G1K 7P4*

Received 29 September 1997; accepted 16 December 1998

Abstract -- In this study, a catalytic membrane reactor was developed. A titanium silicalite (TS-1) containing polydimethylsiloxane (PDMS) catalytic composite membrane was placed at the interface between the two immiscible phases containing respectively n-hexane (organic phase) and a solution of hydrogen peroxide (aqueous phase). This allowed adequate transport of both reactants to the catalyst surface, without using a co-solvent. This concept of zeolite containing catalytic membrane as interphase contactor, which may be applicable to numerous multiphase reactions, has been tested for the oxyfunctionalization of n-hexane to a mixture of hexanols and hexanones using H_2O_2 as the oxidant. It was shown that the oxyfunctionalization products are formed in and separated by the catalytic membrane. The experimental results illustrated the technical advantages of such a catalytic membrane reactor since the observed conversion and selectivity are similar to the ones obtained with the same catalyst in a conventional reactor. The various factors (membrane thickness, catalyst loading and membrane modifications) which may affect the membrane catalytic and permeation performances were investigated.

1. INTRODUCTION

In the last decade, there has been an explosion of interest in the development of catalytic membrane reactors [1]. The membrane reactors take advantage of the separation function of a membrane to affect selectively the concentrations of reactants and/or products at specific locations within the reaction medium. Such functions of membranes are successfully applied in some biological and chemical reactions [2,3]. In most of these applications the reactants and products are miscible. We recently demonstrated that there is a great potential interest to running a multiphase reaction in a catalytic membrane reactor [4]. In the type of membrane reactor used, a catalytic membrane is placed at the interface between two immiscible reacting phases and allows adequate transport of both reactants towards the catalyst surface. In other words, in this membrane reactor, the membrane should play a role similar to the one of the solvent in a conventional multiphase reactor by affecting the reactants concentrations at the catalyst surface. Such a catalytic membrane reactor using a TS-1/PDMS composite membrane could indeed be run with the catalytic reaction rate of n-hexane oxyfunctionalization comparable to the rates observed with the suspended catalyst [5]. For most organic / aqueous / solid catalyst triphasic reactions when a conventional batch slurry reactor is used, the reaction rates are quite low because the immiscibility of the two liquid phases often leads to a very low concentration of the organic reactants in the aqueous phase where the reactions take place. In such conventional reactors, the use of a co-solvent is required to increase the organic reactants concentration at the catalyst surface and therefore increase the reaction rates [6]. The potential advantage of this membrane interphase contactor is to avoid the use of co-solvents at industrial scale.

The objective of this work was to demonstrate the feasibility of the concept of zeolite containing catalytic membranes as interphase contactors, by performing the n-hexane oxyfunctionalization with hydrogen peroxide in the presence of a TS-1/PDMS catalytic membrane.

2. CHEMICAL REACTIONS

The oxyfunctionalization of n-hexane (n-C_6H_{14}) is a typical example of the partial oxidation reactions of organic compounds that the titanium silicalites allow to perform [6,7,8,9] with dilute aqueous (30 wt %) solutions of hydrogen peroxide (H_2O_2). In these conditions H_2O_2 is considered as a clean and safe oxidant. In the oxyfunctionalization of n-hexane with H_2O_2 several simultaneous reactions are actually taking place (Eqs. 1 to 3). Primary reactions produce hexanols, 2- or 3-($C_6H_{13}OH$). Each of these reactions is followed by a secondary reaction which produces hexanones, 2- or 3-($C_6H_{12}O$). Simultaneously hydrogen peroxide decomposes into water and oxygen (Eq. 3).

$$n\text{-}C_6H_{14} \xrightarrow{+H_2O_2} 2\text{-}C_6H_{13}OH \xrightarrow{+H_2O_2} 2\text{-}C_6H_{12}O \quad (1)$$

$$n\text{-}C_6H_{14} \xrightarrow{+H_2O_2} 3\text{-}C_6H_{13}OH \xrightarrow{+H_2O_2} 3\text{-}C_6H_{12}O \quad (2)$$

$$H_2O_2 \rightarrow H_2O + 1/2\,O_2 \quad (3)$$

Previous studies [6] indicated that these oxyfunctionalization reactions can be performed under very mild conditions in a conventional reactor if an appropriate catalyst is used. When performed with a solid catalyst, the oxyfunctionalization process is triphasic: there are two liquid phases which are the organic (n-hexane) and the aqueous phase (H_2O_2 solution), and a solid phase which is the catalyst. As mentioned above, a co-solvent, such as acetone, methanol or acetonitrile, must be used to increase the reaction rates and selectivities. A kinetic investigation was performed in our group [10,11]. It was demonstrated that the effect of the co-solvent is to increase the solubility of n-hexane in the aqueous phase, and to decrease the hydrogen peroxide concentration in this phase.

3. CATALYTIC MEMBRANES AND CONCEPT OF INTERPHASE (GAS-LIQUID) CONTACTOR

The catalytic membranes consist of the polymeric matrix filled with catalyst particles. The polydimethylsiloxane (PDMS) was chosen in this study because of its CH_3 groups compatible with n-hexane [12,13] and its high resistance to oxidation due to the O-Si-O bridges involved. The zeolitic titanium silicalite-1 (TS-1) was chosen as catalyst since it is efficient for organic compounds direct oxidation by hydrogen peroxide dilute solution. The zeolite particles were embedded in the polymeric matrix and thought to provide the hydrophilicity necessary to allow a proper transport of H_2O_2 to the catalytic surface. In several tests, some additional hydrophilicity was also given to the polymeric matrix by grafting polar functional groups to the PDMS matrix using a dimethylsiloxane-ethyleneoxide-propyleneoxide copolymer modifier.

In the present work the oxyfunctionalization of n-hexane was performed by contacting the TS-1/PDMS composite membranes, on one side with the saturated n-hexane vapor and on the other side with the H_2O_2 aqueous solution. Thus, the reactants reach the catalysts surface by diffusion through the membrane from each side. As the catalyst is embedded in the polymeric matrix, the concentration of n-hexane at the catalyst surface is higher than the one of in a conventional reactor due to its high solubility in the PDMS matrix. The polymeric matrix plays therefore a role similar to the one of a co-solvent in the case of a conventional reactor. The reactions take place within the membrane medium. When the oxygenate products and unreacted n-hexane permeate through the catalytic membrane they are all quickly removed from the membrane top side surface, since this organic compounds are essentially immiscible in the H_2O_2 aqueous phase and of lower density. Using this innovative vapor-liquid membrane interphase contactor it was found possible to get high reaction conversions and selectivity by choosing proper reaction conditions.

4. EXPERIMENTAL PROCEDURES

4.1 Synthesis of Titanium Silicalite Zeolite Particles

The titanium silicalite zeolite (TS-1) was synthesized according to a procedure proposed by Thangaraj *et al.* [14]. The key of the zeolite synthesis was to obtain sufficiently small and monodispersed size zeolite particles to allow easy incorporation of these particles into the polymeric matrix. Scanning electron micrographs (SEM) showed that the synthesis conditions used in this study allowed to obtain uniform and cubic shape zeolite crystals with a 0.2μm average size (Figure 1a). Prior to the incorporation in the membrane, the prepared TS-1 zeolites (Ti/Si+Ti =0.019) were tested in a conventional slurry reactor for their activity and selectivity in n-hexane oxyfunctionalization.

4.2 Composite Membrane Preparation

The catalytic composite membranes were prepared by embedding the zeolite particles into a polymeric matrix using the casting technique.

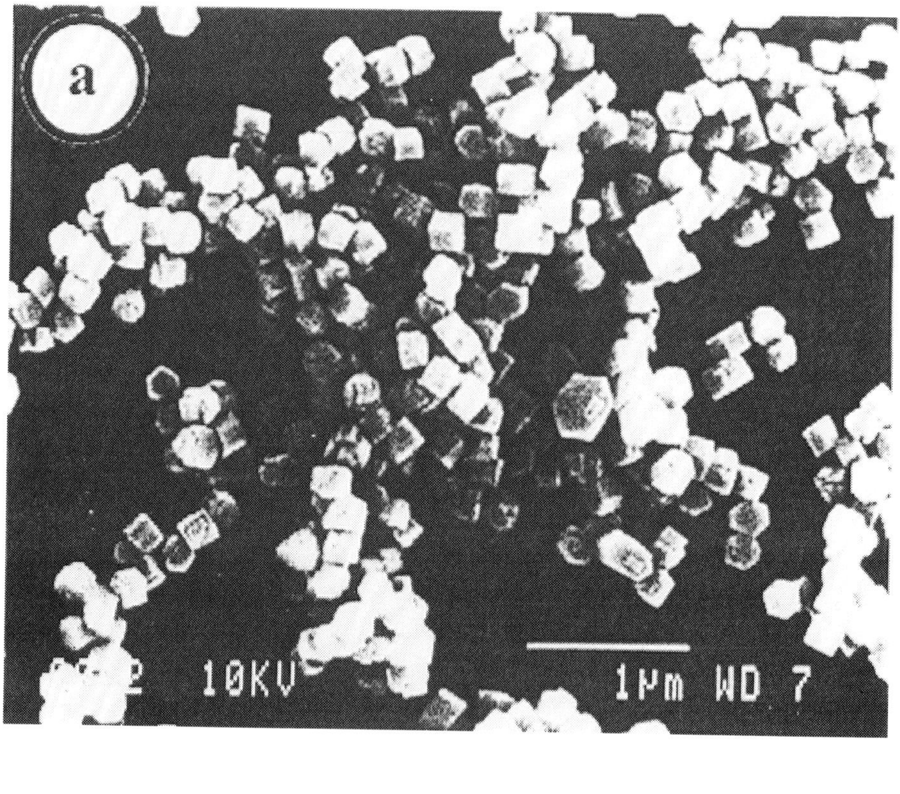

Figure 1. Scanning Electron Micrograph of a) TS-1 particles, b) and c) TS-1/PDMS Composite Membranes. b) Cross section of membrane; c) overview of membrane.

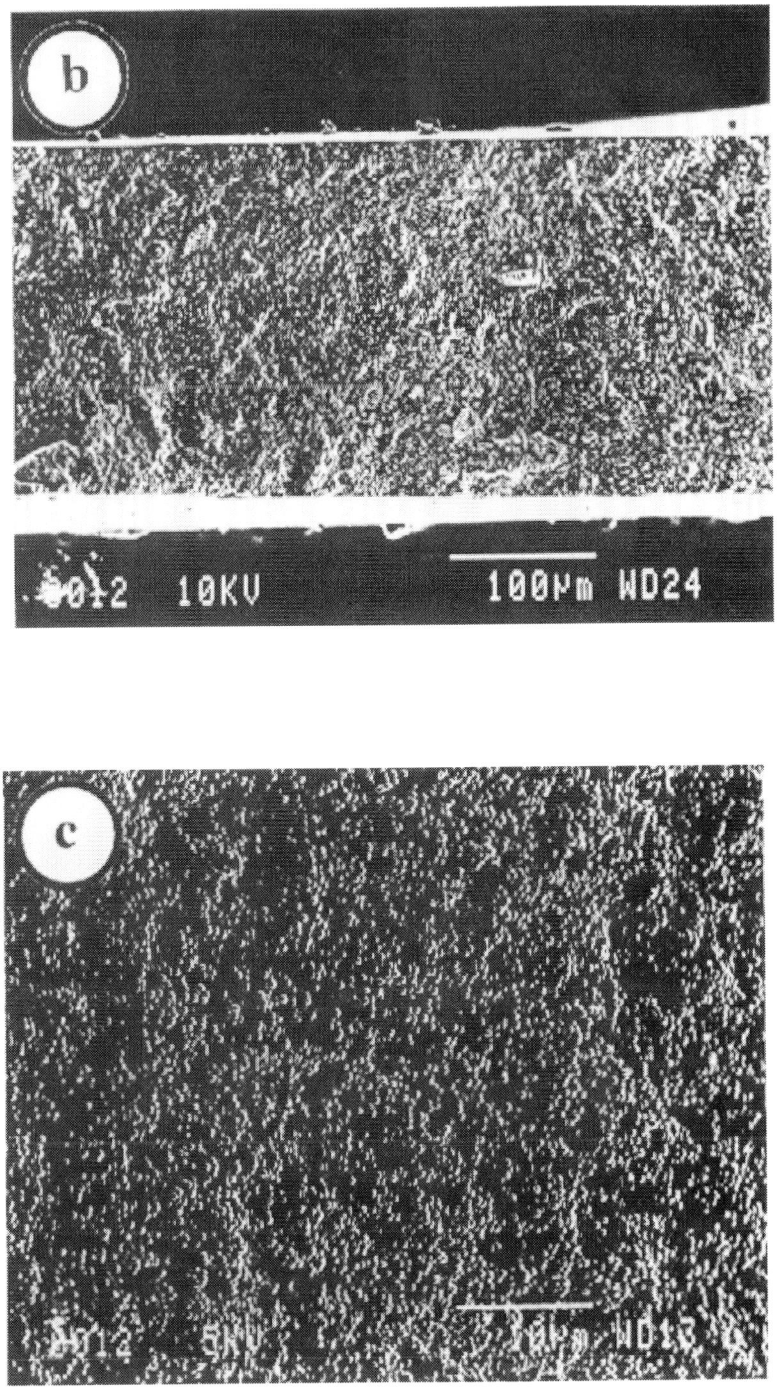

Figure 1 (continued). Scanning Electron Micrograph of a) TS-1 particles, b) and c) TS-1/PDMS Composite Membranes. b) Cross section of membrane; c) overview of membrane.

The membrane preparation is described below. The PDMS matrix was prepared from commercial base components RTV-615 A (vinyl methylpolysiloxane) and RTV-615 B (polydimethyl hydrogen siloxane) supplied by GE Co. These two components were dissolved in toluene controlling the concentration as 30% and mixed with the TS-1 powder. When necessary, a 10 wt% co-polymer modifier polydimethysiloxane-ethyleneoxide-propyleneoxide (United Chemical Technologies Inc., PS702) was also added to this blend. The mass ratio of components A and B was kept constant as 9:1 whereas the zeolite content was varied between 40-65 wt%. After vigorously stirring this mixture for 4 hours, a stable suspension was formed. This viscous suspension was then cast on a Teflon flat mold and left to dry at room temperature for 10 hours to allow solvent evaporation. The membrane was then kept at 358 K in a vacuum oven overnight to ensure complete curing. The membrane was cut into disks of 60 and 50 mm diameter for permeation and reaction tests respectively. The membrane thickness was measured both with a Palmer and from SEM pictures of cross sections of the membranes. Figures 1 b and c show the SEM pictures of overview and cross section of the TS-1/PDMS catalytic membrane respectively.

4.3 Membrane and Conventional Reactors

Membrane reactor: An atmospheric pressure water jacketed glass reactor with two compartments separated by the catalytic composite membrane was used in this study (Figure 2). The catalytic membrane was contacted with the H_2O_2 aqueous phase (on the upper side) and saturated n-hexane vapor (on the lower side). The reactor is temperature controlled by a constant temperature circulating bath and equipped with a reflux condenser. The temperature of the condenser can be adjusted down to 269K using an external refrigerated bath, in order to condense efficiently all of the organic chemicals. A gas collector is provided for oxygen measurement. The n-hexane is introduced batchwise in the lower compartment of the reactor and it takes about 1/4 volume of the lower compartment. Openings in the upper compartment allow the introduction of a batch feed of H_2O_2 solution. It allows also the sampling of unconverted n-hexane and oxygenates using a syringe and to check the temperature inside the reactor. The upper and lower compartment volumes are respectively 100 ml and 80 ml.

Conventional reactor: A 180 ml cylindrical Pyrex reactor is used in this study. It is equipped with a sampling port and a water jacket connected to a constant temperature circulating bath and also connected to a reflux condenser and an oxygen collection device . A magnetic stirrer ensures the mixing inside the reactor.

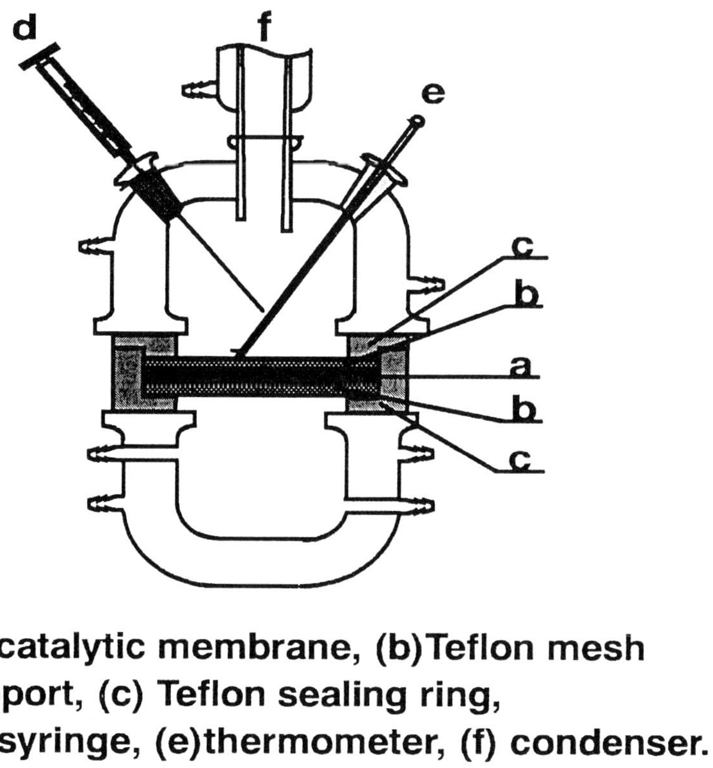

Figure 2. Membrane reactor.

4.4 Catalytic Reaction Performance

Most of the catalytic reaction tests, with the conventional or membrane reactor, were performed at 328K. The experiment length was standardized at 5h.

For the tests performed in the conventional reactor, 250 mg of TS-1 catalyst powder, 10g of n-hexane, 5 g of methanol (only when using a co-solvent) and 5 g of hydrogen peroxide solution (30% wt%) were used for each experiment. The reaction rates and product distributions were measured by collecting aliquots of both the aqueous and organic phase (0.1ml) at the end of each run. The n-hexane and oxygenates were analyzed using a gas chromatograph (HP 5890) equipped with a capillary column (DB wax) and a flame ionization detector. The volume of oxygen produced was measured as function of reaction time. At the end of each reaction experiment, the two liquid phases were separated, filtrated and analyzed. The concentration of unconverted H_2O_2 in the aqueous phase was determined by iodometric titration.

For the tests performed in the membrane reactor, 10 g of n-hexane and 20 g of H_2O_2 solution (30 % wt%) were used for each experiment. The membrane area was 10.75 cm^2. The mass of TS-1 catalyst embedded in the membrane was varied from 120 to 380 mg. The membrane thickness was varied between 250-600μm. Gas chromatographic analysis of organic samples taken from both the lower and the upper compartment allowed to follow the evolution of the primary and secondary products concentrations in the feed and organic permeation phase. The H_2O_2 concentration in the aqueous phase was measured by using the iodometric titration. The volume of oxygen generated was measured as a function of time using the gas collector. At the end of each run, the amounts of organic reactant and products, dissolved in the aqueous phase were also measured. The average reaction rate is expressed in μmol oxygenates per gram TS-1 catalyst per hour. The product distributions are represented as the molar ratios :

$$alcohol/ketone = \frac{2\text{-}hexanol + 3\text{-}hexanol}{2\text{-}hexanone + 3\text{-}hexanone} \quad \textbf{(4)}$$

$$2\text{-}/3\text{-} = \frac{2\text{-}hexanol + 2\text{-}hexanone}{3\text{-}hexanol + 3\text{-}hexanone} \quad \textbf{(5)}$$

4.5 Pervaporation Measurements.

It was expected that permeation data may provid a qualitative insight in the transport phenomena which take place in the zeolite filled PDMS membrane during the reaction process. Investigations of permeation behaviors of catalytic membranes on pure water and n-hexane were performed using the pervaporation process. For a better understanding of the transport behavior of aqueous compound under the swollen membrane conditions, pervaporation tests of a mixture emulsion of water and n-hexane were also carried out in this study. A magnetic stirrer ensures a good mixing inside the solution feed tank. A high circulation flow rate ensures the good emulsion inside the system. These pervaporation experiments were performed, at room temperature, for each of those membranes used in the catalytic reaction tests, in a standard tangential flow permeation cell. This cell allows testing of flat circular coupons of the membrane (15.90 cm^2). The feed side was under atmospheric pressure while the pressure on the downstream side was kept at 0.6 Torr. The permeate was collected in two successive liquid nitrogen cold traps to ensure complete collection. The permeation flux was expressed in gram per square meter per hour.

5. RESULTS AND DISCUSSION

5.1 Comparison of the Catalytic Results in Conventional and Membrane Reactor

In order to compare the n-hexane oxyfunctionalization rates observed in both types of reactor, the reaction was first run in a conventional slurry reactor using TS-1 from the same catalyst batch as the one introduced in the composite membranes. Some typical results are reported in Table 1 as the average oxyfunctionalization rate and n-hexane conversion over a five hour reaction time. The reaction rate is expressed in μmol per gram of TS-1 per hour. The test indicated as TS-1, is in the standard conditions (see section 4.4) in absence of a co-solvent, whereas the one indicated TS-1 + MeOH was made in the presence of the optimum amount of methanol (maximum reaction rate as demonstrated in reference [6]). The third test reported in Table 1 was performed in the membrane reactor with a 250 μm thickness and 50 wt% catalyst loading TS-1/PDMS membrane, whereas the fourth one, indicated as TS-1/PDMS/Modifier, was done with a membrane which was modified by copolymerization with a dimethylsiloxane - ethylene oxide - propylene oxide oligomer (UCT, PS702). In this last case the membrane which contained 10 wt% of PDMS polymer matrix modifier, had the same thickness and particle loading as the TS-1/PDMS membrane.

Table 1
Comparison of typical reaction rate results in a conventional and a membrane reactor.

Catalysts	TS-1 + (MeOH)*	TS-1	TS-1/PDMS	TS-1/PDMS/Modifier
n-hexane conversion, %/5h	2.8	1.1	1.7	5.7
reaction rate, $\mu mol/g_{cat} \cdot h$	2600	1000	800	1400
alcohol / ketone	0.33	0.35	0.59	0.38

* Experiment run in conditions (5g CH_3OH) where the reaction rate was shown to be maximum [6].

These results indicate that the experiment performed with the TS-1/PDMS membrane yields a rate value comparable to the one of the standard test in the conventional reactor. This means the use of the composite membrane does not involve a significant decrease in rate per unit mass of catalyst. Moreover using the modified TS-1/PDMS composite membrane yields an increase in reaction rate. This increase in reaction rate and n-hexane conversion is probably due to the interaction between the polar functional groups of the modifier and the H_2O_2 molecular.

Comparing oxygenates distribution in a conventional and membrane reactor, indicates that the use of the membrane reactor affects significantly the alcohol/ketone ratio. This ratio reflects the relative rate of the secondary oxidation of the alcohols to ketones compared to the rate of the primary oxidation. The rapid transport of the alcohols out of the catalytic membrane may decrease the reaction contact time with the catalyst surface. The results in the Table 1 demonstrate that not only the reaction rates but also the selectivities may be controlled by proper design of the catalytic membranes.

5.2 *Effects of Membrane Thickness on the Catalytic and Permeation Performance*

To study the effects of membrane thickness on the membrane catalytic and permeation performance, three different TS-1/PDMS catalytic membranes with thicknesses 250, 400 and 600 μm having the same particle size and catalyst loading were selected. The results obtained are summarized in Table 2a and b. The n-hexane permeation flux during the reaction test (organic permeate in Table 2a) decreases as the membrane thickness increases. It should be noted that the thicker one has a much higher n-hexane conversion (about 4 times the one of the thinner) and a significantly lower organic permeate flux. The higher conversion means a higher oxygenates concentration in the permeate. Moreover a decrease in average reaction rate as the membrane thickness increases was observed. This implies that the catalyst efficiency

Table 2a

Effects of Membrane Thickness on Catalytic Performance

TS-1:PDMS, w/w	50:50	50:50	50:50
membrane thickness, μm	250	400	600
organic permeate, g/5h	2.9	1.92	1.01
n-hexane conversion, %/5h	1.7	3.2	7.6
oxygenates in permeate phase, μmol/5h	450	560	700
oxygenates in feed phase, μmol/5h	100	85	62
oxygenates in aqueous phase, μmol/h	60	65	60
total oxygenates, μmol/5h	610	710	822
O_2 gas collected, ml/5h	135	150	160
ol/one	0.59	0.55	0.52
-2/-3	1.67	1.54	1.67
average reaction rate, μmol /g_{cat}. h	800	617	484

Table 2b
Effects of Membrane Thickness on Permeation Performance

Membrane	TS-1PDMS	PDMS	TS-1/PDMS	PDMS	TS-1/PDMS	PDMS
TS-1:PDMS, w/w	50:50	0:100	50:50	0:100	50:50	0:100
membrane thickness,μm	250	250	400	400	600	600
H_2O permeation flux ,g/h m^2	47.9	3.4	32.2	2.1	17.1	1.1
n-hexane permeation flux, g/h m^2	2652	5332	1756	3107	924	1593

of each membrane is different. The thinner one shows a higher efficiency compared to the other two.

The results of the investigation of membrane thickness on permeation performance are given in Table 2b. Three pure PDMS membranes with thicknesses corresponding to those of the three composite membranes were also tested for comparison. The results show that adding zeolite TS-1 to the polymer matrix improves the water permeation flux but decreases the n-hexane permeation flux.

From the data reported in Table 2a, the qualitative description of the various concentration profiles in the depth of the membrane (Figure 3) may be proposed. As a net upward flux of n-hexane is obtained (organic permeate) the n-hexane concentration profile must show a decreasing gradient in the upward direction. This is indeed understandable since the lower face of the membrane is in contact with n-hexane vapor, whereas the upper face is contacted with the aqueous phase at very low n-hexane concentration. Since no water or H_2O_2 is found in the lower compartment of the reactor, the H_2O_2 concentration gradient must be zero at the lower face. Thus the H_2O_2 concentration profile must be as shown on Figure 3, with a non-zero value only in a limited depth below the upper face. This depth defines the reaction zone since the two reactants are only present simultaneously above this plane. The oxygenates concentration profile has non-zero gradient at both membrane faces since oxygenates were found both in the permeated organic phase and in the feed phase in the lower compartment. Oxygenates being generated in the reaction zone, their concentration must show a maximum at a given depth within this zone.

The performance of the catalytic membrane is therefore a complex interplay of reaction rates and transport phenomena.

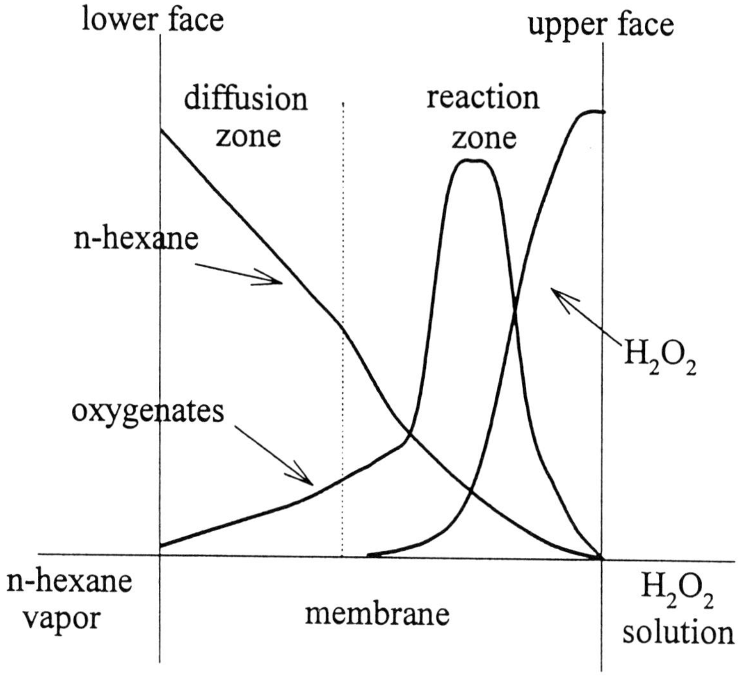

Figure 3. Schematic description of concentration profiles in the catalytic memebrane.

The data in Table 2b suggest that the n-hexane transport is through the solubilization-diffusion mechanism in the polymer matrix. Simultaneously the H_2O/H_2O_2 mass transfer must be through a pathway associated with the catalyst particles. At this point it is not yet possible to discriminate between two possibilities of a diffusion of the aqueous phase through the micropores of TS-1 or by a diffusion of the aqueous phase through the voidages or slits surrounding the solid particles in the polymer matrix.

5.3 Effects of the Catalyst Loading on the Catalytic and Permeation Performance

The investigation of the effects of the catalysts loading on the catalytic and permeation performance was carried out with TS-1 catalyst at 40, 50 and 65 wt% in the composite membranes. Typical results are given in Table 3a and 3b. The pervaporation fluxes are also reported in Figure 4. These data indicate that a higher catalyst loading yields a significant raise of n-hexane conversion and an increase on the average reaction rate. These changes are associated to both a significant decrease in organic permeate flux and an increased reaction rate due to the increased density of active sites. The results of permeation tests (Figure 4) provide additional information on the mass transfer mechanisms over these membranes. The experiments performed with pure n-hexane show clearly that the dominant n-hexane

Table 3a
Effects of Catalyst Particle Loading on Catalytic Performance

Membrane	TS-1/PDMS	TS-1/PDMS	TS-1/PDMS
TS-1:PDMS, w/w	40:60	50:50	65:35
membrane thickness, μm	600	600	600
organic permeate, g/5h	1.24	1.01	0.65
n-hexane conversion, %/5h	3.8	7.6	15.6
oxygenates in permeate phase, μmol/5h	410	700	1000
oxygenates in feed phase, μmol/5h	50	62	58
oxygenates in aqueous phase, μmol/h	62	60	60
total oxygenates, μmol/5h	522	822	1118
O_2 gas collected, ml/5h	140	160	190
ol/one	0.60	0.52	0.39
-2/-3	1.45	1.67	2.02
average reaction rate, μmol /g_{cat}. h	372	484	502

Table 3b
Effects of Catalyst Particle Loading on Permeation Performance

Membrane	PDMS	TS-1/PDMS	TS-1/PDMS	TS-1/PDMS
TS-1:PDMS, w/w	0:10	40:60	50:50	65:35
membrane thickness,μm	250	250	250	250
H_2O permeation flux, g/h m^2	3.4	18.8	47.9	76.5
n-hexane permeation flux, g/h m^2	5332	3120	2652	1976
H_2O/n-hexane emulsion permeation flux, g/h m^2	12/3980	136/1588	320/1383	442/1060

permeation process is through the PDMS matrix as the pervaporation flux is essentially proportional to the PDMS content. The experiment conducted with pure water shows the reverse trend, confirming that the water permeation follows a pathway associated with the solid particles.

From the experiments conducted with the n-hexane-water emulsion, it is seen that the n-hexane fluxes are lower in presence of water. As these fluxes do not tend to zero at zero PDMS content, it is suggested that a pure TS-1 layer would allow some n-hexane permeation. This result is not unexpected because n-hexane must diffuse in the TS-1 pores for the oxyfunctionalization reaction. It is also clearly established that the presence of n-hexane in the emulsion increases drastically the water permeation flux. This is fortunate because this effect will improve the H_2O_2 diffusion and thus enlarge the reaction zone allowing the catalytic membrane oxyfunctionalization tests. It is likely associated with the swelling of the PDMS matrix which enlarges the slits around the catalysts particles. This interpretation would imply that diffusion within these slits is a major process of H_2O_2 mass transfer in the catalytic membranes.

Regarding the oxygenates distribution in the oxyfunctionalization tests, a high alcohol/ketone ratio is found at lower catalyst loading. This is likely due to the hexanols following the same diffusion pathway as n-hexane, their permeation rate out of the membrane is faster, remaining in a lower contact time with the catalyst in the reaction zone. The selectivity to hexanones in therefore lowered.

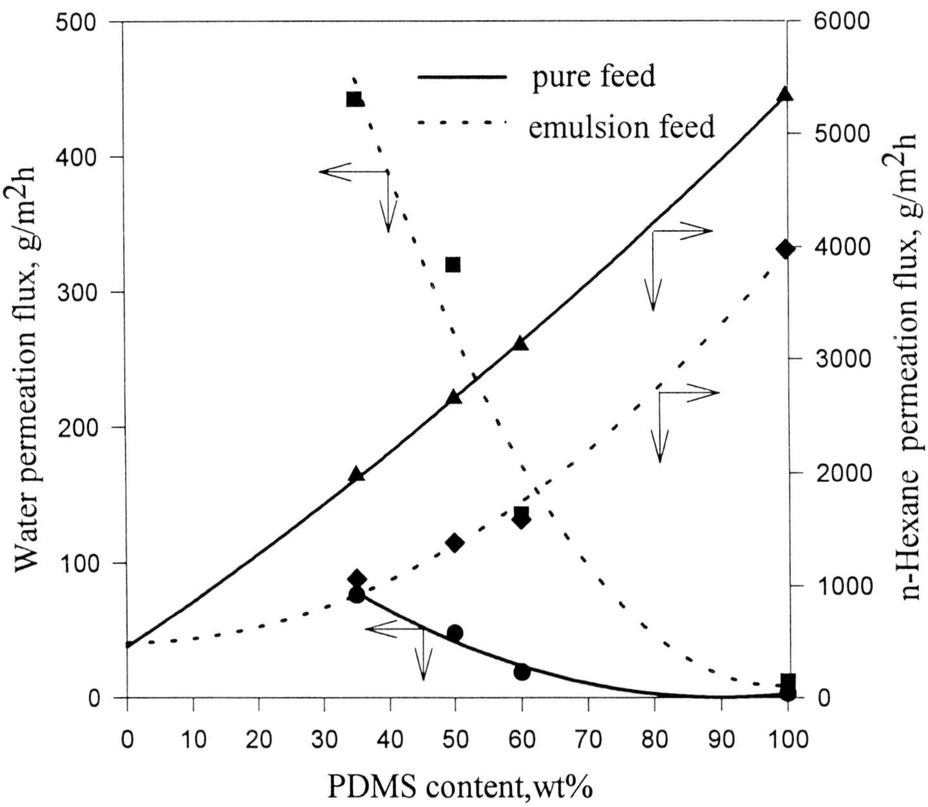

Figure 4. Effects of membrane composition on permeation fluxes.

5.4 Effect of Membrane Modifications on Catalytic Performance

Table 4 shows the results of reaction tests performed with modified membranes. In some experiments (column 2 in Table 4) a mixed nitrate and acetate cellulose ester membrane (Millipore Co., RAWP 04700) was used as a support (NC). This means that the TS-1/PDMS membrane was cast on a 100 μm thickness and 1.2μm pore size commercial film of nitrate and acetate cellulose ester. This film was firmly sticking to the PDMS membrane and acted as a polar support. During the catalytic tests it was in contact with the aqueous phase. It is believed that this polar material with higher H_2O_2 adsorption capacity, should yield higher H_2O_2 concentration at the upper face of the TS-1/PDMS membrane. It should thus increase the inlet flux of H_2O_2 during catalytic tests.

The results in Table 4 (column 2) both show the corresponding increase in average catalytic rate in spite of a decreased organic permeate flux compared to the non-supported membrane of same thickness. The decreased alcohol/ketone ratio is correlated with the decreased hexanols permeation flux.

Table 4
Effects of Porous Support Membrane and Chemical Modifier on Catalytic Performance

Membrane	TS-1/PDMS	TS-1/PDMS	TS-1/PDMS	TS-1/PDMS
TS-1:PDMS:modifier ,w/w	50:50	50:50	50:40:10	50:40:10
membrane thickness, μm	600	600	600	600
porous membrane support		NC		NC
functional group containing modifier*			PDMS -EO-PO	PDMS -EO-PO
organic permeate, g/5h	1.01	0.65	0.91	0.82
n-hexane conversion, %/5h	7.6	14.9	10.8	23.3
oxygenates in permeate phase, μmol	700	850	922	1100
oxygenates in feed phase, μmol	62	200	100	170
oxygenates in aqueous phase, μmol	60	80	60	80
total oxygenates, μmol/5h	822	1130	1080	1350
O_2 gas collected, ml/5h	160	150	155	140
ol/one	0.52	0.32	0.42	0.27
-2/-3	1.67	1.60	1.52	1.59
average reaction rate, μmol /g_{cat}· h	484	706	635	794

* Dimethylsiloxane-Ethylene oxide-Propylene oxide Copolymer

In another set of experiments (column 3 in Table 4) the PDMS matrix was functionalized by introducing a commercial copolymer polydimethysiloxane-ethyleneoxide-propyleneoxide. This should make the PDMS matrix more polar and thus reduce the diffusion coefficient of n-hexane and hexanols. Comparing with the non-modified membrane (column 1) the n-hexane conversion is higher due to both lower permeation fluxes and higher average reaction rates. In this case the alcohol/ketone ratio of the oxygenate products is also decreased.

The last set of experiments in Table 4 (column 4) is for membranes combining the two modifications; a nitrate and acetate cellulose ester membrane support was added to a chemically modified PDMS matrix. The results show that the two effects are adding. Surprisingly high conversions (close to 24%) were reached due to lower permeation fluxes and higher catalytic rates.

These last results are specially interesting because they show that the catalytic membrane can not only reach high reaction rates, but also that the membrane can perform simultaneously some separation of the reaction products. In the mean time the reactant phases are kept separated.

6. CONCLUSIONS

In this exploratory work, it was possible to demonstrate that the principle of using a catalytic membrane as interphase contactor in a biphasic reaction is feasible. In the case of the oxyfunctionalization of n-hexane it was found that the TS-1/PDMS membrane can not only work as a catalyst to produce hexanols and hexanones but also essentially separate these products form the organic feed.

A qualitative description of the pseudo-steady state concentration profiles within the membrane was proposed. The effects of the various factors affecting both the reaction performance and permeation fluxes were examined experimentally and discussed in the light of the above description.

It was found that the mass transport mechanism for n-hexane and H_2O_2 reactants were very different. n-Hexane is dissolved in the PDMS matrix whereas the aqueous solution diffuses in the voidages which surround the TS-1 particles.

REFERENCES

1. H.P. Hsieh, *Catal. Rev.- Sci. Eng.*, **33**, 1-70 (1991).
2. A.M. Vaidya, G. Bell and P.J. Halling, *J. Membrane Sci.*, **71**, 139 (1992).
3. J.N. Armor, *Applied Catalysis*, **49**, 1 (1989).
4. S. Kaliaguine, C. Bouchard, S.Q. Wu and J. Shu, *Canadian Patent*, Submitted, 1997.
5. S.Q. Wu, C. Bouchard and S. Kaliaguine, *Proceeding of Annual Meeting of North American Membrane Society*, Ottawa, May 1996.

6. H. Fu and S. Kaliaguine, *J. Catal.*, **148**, 540 (1994).
7. M. Taramasso, G. Perego and B. Notari, *US Patent* 4,410,501 (1983).
8. D.R.C. Huybrechts, L. De Bruycker and P. A. Jacobs, *Nature*, **345**, 240 (1990).
9. T. Tatsumi, M. Nakamura, S. Negeshi and H. Tominaga, *J. Chem. Soc. Chem. Commun.*, 476 (1990).
10. J.E. Gallot, H. Fu, M.P. Kapoor and S. Kaliaguine, *J. Catal.*, **161**, 789 (1996).
11. J.E. Gallot, Do Trong On, M.P. Kapoor and S. Kaliaguine, *IEC J.*, **36** (9) (1997b).
12. R.F. Parton, I.F.J. Vankelecom, M.J.A. Casselman, C.P. Bezoukhanova, J.-B. Uytterhoven and P.A. Jacobs, *Nature*, **370**, 541 (1994).
13. C. Dotremont, B. Brabants, K. Geeroms, J. Mewis, C. Vandecasteele, *J. Membrane Sci.*, **104**, 109 (1995).
14. A. Thangaraj, M. J. Eapen, S. Sivasanker and P. Ratnasamy, *Zeolites*, **12**, 943 (1992).

Res. Chem. Intermed., Vol. 24, No. 3, pp. 291-307 (1998)

CHEMISTRY OF μ-η^2:η^2-PEROXO DIMETAL COMPLEXES: A BIOINORGANIC MODEL TRIGGERS A NEW TREND OF TRANSITION METAL PEROXO SPECIES

MUNETAKA AKITA, KIYOSHI FUJISAWA, SHIRO HIKICHI AND YOSHIHIKO MORO-OKA*

Research Laboratory of Resources Utilization, Tokyo Institute of Technology, 4259 Nagatsuta, Midori-ku, Yokohama 226, Japan

Received 16 October 1997; accepted 25 November 1997

CONTENTS

I. INTRODUCTION

The dioxygen molecule is the most abundant and cheapest oxidant, and many oxidative transformations of organic substrates are catalyzed by a transition metal species including both homogeneous and heterogeneous catalysts [1]. However, oxidation with dioxygen molecule is not always feasible, and much effort has been paid toward development of new (and, hopefully, catalytic) organic transformations. Needless to say, dioxygen molecule plays very important roles in the living systems as well, and many enzymatic processes (e. g. energy production (respiration) and metabolism) are catalyzed by metalloproteins which contain transition metal component(s) in their active sites [2]. In these chemical and biological systems, dioxygen molecule is activated in a coordination sphere of the transition metal component, and substrates are oxygenated through the action of activated oxygen species such as dioxygen [M(O_2)], hydroperoxo [M-OOH] and oxo species [M=O]. Related species such as alkyl- [M-OOR] and acyl-peroxo species [M-OOC(=O)R] are also involved as intermediates of oxidative transformations.

In order to understand the reaction mechanisms of such oxidative transformations and develop new oxidative processes, it is essential to clarify the interaction mode of dioxygen molecule with the transition metal center at a molecular level. In the case of metalloproteins, however, active sites are surrounded by proteins, which make it difficult to handle them and study inorganic chemistry of the active metal sites. One effective solution is synthesis and characterization of artificial discrete transition metal complexes which can reproduce the coordination environment around the metal component and the functions of the catalysis, and the bio-related studies are called bioinorganic chemistry [3]. For successful modeling it is essential to design ligand systems which can mimic the coordination environment of the metal component in the catalyst system to be studied.

Transition metal dioxygen complexes have long been a research subject of inorganic chemistry, and they are obtained mainly by oxidative addition of dioxygen molecule. As a result, various types of coordination modes of the dioxygen ligand have been revealed as summarized in Figure 1 [3]. The coordinated dioxygen ligand can be classified according to the extent of electron transfer from the metal center. 2e-reduction gives a peroxide ligand (O_2^{2-}), whereas single electron transfer produces a superoxide ligand (O_2^-). Until recently mononuclear species (types **A** and **B**) and type **C** dinuclear species had been the major research targets of inorganic chemistry, and much knowledge about the structure and reactivity has been accumulated for such types of complexes. The most prominent result may be the picket-fence porphyrin-O_2 adduct which solved the problem of oxygen transport with *hemoglobin* [4]. However, compared to heme chemistry based on the

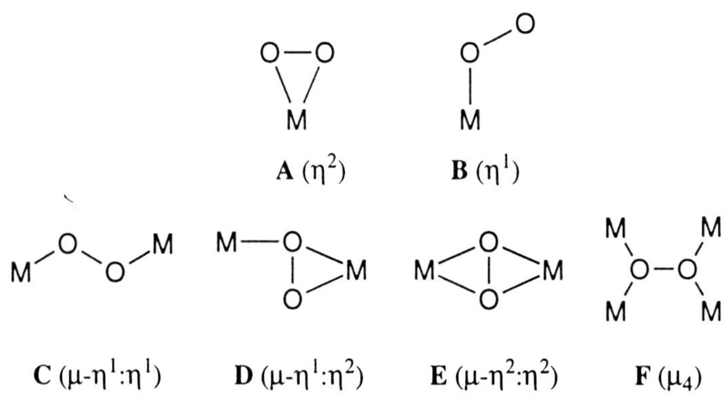

Figure 1.

porphyrin ligand which is rigid and shows characteristic spectroscopic features, the chemistry of non-heme proteins remains far less developed mainly due to the non-rigid structure supported by the flexible protein backbone and lack of spectroscopic features to be generally observed.

Little attention had been paid to polynuclear dioxygen complexes, although dinuclear structures of types **C** and **D** were known. About twenty years ago we reported the first synthesis of the μ-η^2:η^2-O_2 of Pd (**1**) [5] and Rh (**2**) [6] (type **E**) which were obtained by redox metathesis reaction with the anionic dioxygen species (KO_2: potassium superoxide) (Scheme 1). It is notable that they also carried organic substrates to be oxidized (e. g. olefin and alkyl), but their reactivity turned out to be sluggish. Under forced conditions less selective oxygen transfer to the cyclooctadiene ligand of the Rh complex was observed, and instead the μ-O_2 part in the complexes was found to be so basic as to abstract proton from acidic organic compounds such as alcohol, cyclopentadiene and acetylacetone [7].

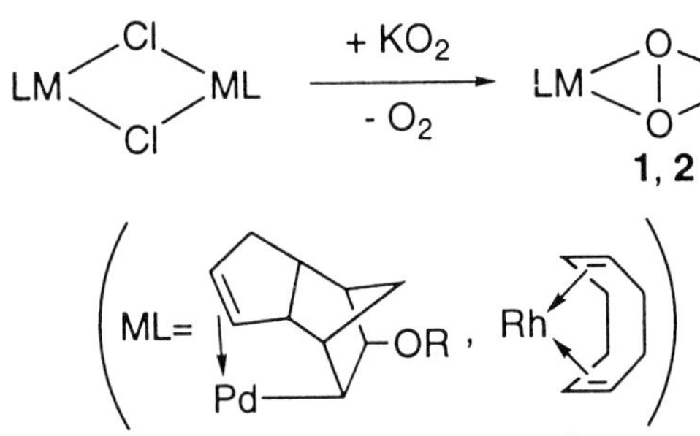

1 2

Scheme 1

Thus the type **E** complexes of Pd and Rh did not show remarkable reactivities, but attention has been focused on such structures again from bioinorganic viewpoints as will be described below.

II. SYNTHESIS OF μ-η^2:η^2-PEROXO DICOPPER COMPLEX AS A MODEL FOR THE OXIDIZED FORM OF HEMOCYANIN

It had been known that, in *hemocyanin* (the oxygen transport proteins of invertebrates such as squid and horseshoe crab), reversible dioxygen binding took place at a dicopper center separated by the distance of ca. 3.6 - 4.6 Å and each copper center was coordinated by three imidazolyl groups from the histidyl residues of the protein chain [8]. However, the coordination mode of dioxygen had long been a matter of controversy. Many coordination structures had been proposed for the dicopper-O_2 adduct and model compounds including type **C** complex had been also prepared [9]. But none of them could accurately reproduce the spectroscopic features (the UV and Raman spectra and the unique diamagnetism) of the dicopper center of the oxidized form of *hemocyanin* (*oxyhemocyanin*) partly due to preconceptions such as the presence of another bridging ligand in addition to O_2 [10].

We have employed tripodal hydrotris(pyrazolyl)borates (Tp^R, in particular, the diisopropyl derivative (Tp^{iPr}); Figure 2 [11]) as a supporting ligand, which can mimic the coordination environment created by a set of facially located three imidazolyl groups of histidine residues frequently found in metalloproteins. During the course of our study we succeeded in synthesis of the dicopper-O_2 adduct **3** via oxidative addition of dioxygen molecule to the reduced copper (I) center **4** (Scheme 2) [12]. Later we developed a new synthetic method for the Cu-O_2 adduct **3**, i.e. dehydrative condensation of the hydroxo complex **5** with hydrogen peroxide, and this method turned out to be effective for synthesis of dioxygen complexes of other transition metals (see below). The adduct **3** was concluded to be a dinuclear species mainly on the basis of its diamagnetism resulting from antiferromagnetic coupling of the two copper centers, and *oxyhemocyanin* was also found to be diamagnetic. In addition to the magnetic property, complex **3** resembled *oxyhemocyanin* and its analogue (*oxytyrosinase*) in its UV-vis spectrum and the O-O stretching vibration as summarized in Table 1. None of model complexes reported previously could reproduce the requisite physicochemical properties as satisfactorily as **3** did. Therefore we proposed complex **3** as an accurate model for the dicopper center in *oxyhemocyanin*.

As for the coordination mode of dioxygen, resonance Raman study of **3** obtained from an isotopic mixture of O_2 ($^{16}O_2$, $^{16}O^{18}O$ and $^{18}O_2$) indicated a symmetrical structure. Combined with the quite low $\nu_{O\text{-}O}$ value (741 cm^{-1}), the coordination mode was assigned to the μ-η^2:η^2 structure (type **E**), which was later

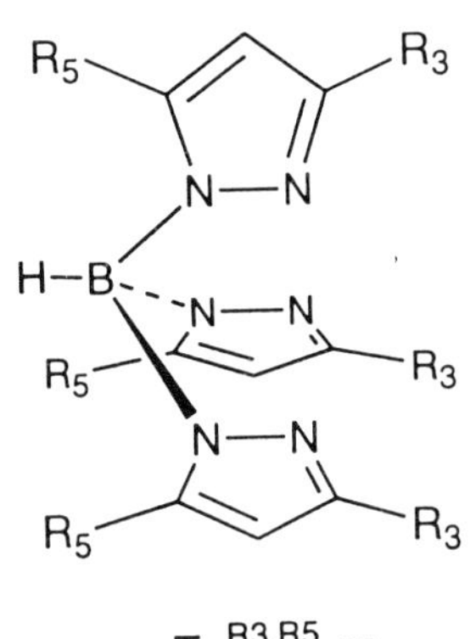

$Tp^{R3,R5}$ or

Tp^R (when R_3= R_5= R)

Figure 2.

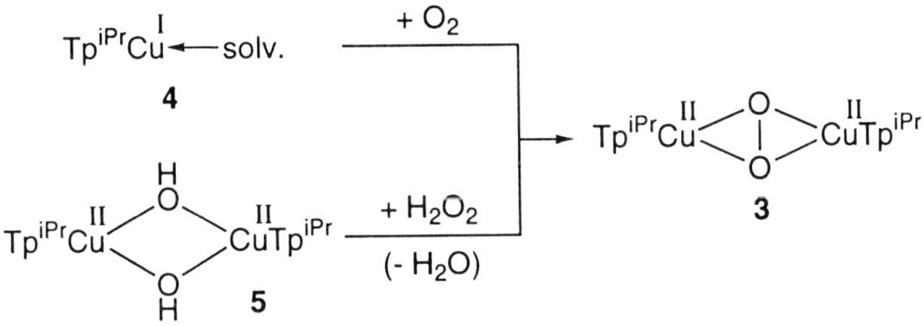

Scheme 2

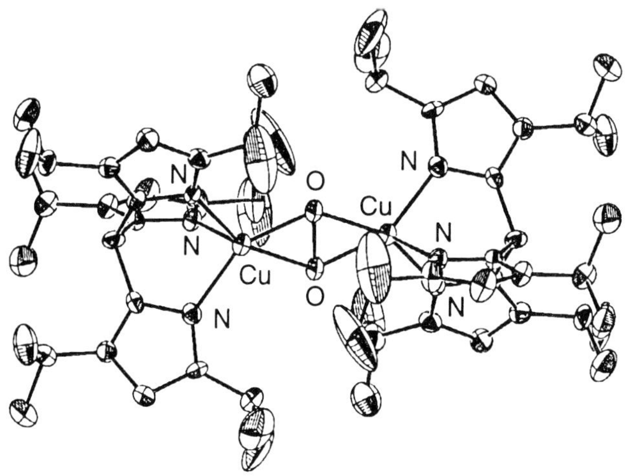

Figure 3.

verified by the single crystal X-ray diffraction study (Figure 3). The structural analysis also revealed (1) the Cu---Cu separation (3.56Å) comparable to those in the copper enzymes (Table 1), (2) the elongated O-O distance (1.41Å), and (3) the distorted square-pyramidal geometry of the copper centers. The unusual Raman and structural features of the O_2 moiety were ascribed to the back-donation from the cupric centers to the vacant, antibonding σ* orbitals of the O-O bond.

On the basis of these results, we predicted that the coordination mode of dioxygen in *oxyhemocyanin* should be a μ-η^2:η^2- structure (type **E**). Several years later our prediction was verified by the structure determination of *oxyhemocyanin* of *Limulus polyphemus* [13]. Thus our study has been recognized as a successful synthetic study of a bioinorganic model compound for *oxyhemocyanin*.

Although a number of transition metal dioxygen complexes had been reported as mentioned in the introductory part, the wavenumbers of their O-O stretching vibration were larger than 800 cm^{-1} and, prior to our studies, there was no example which exhibited the υ_{O-O} band below 800 cm^{-1}. The red shift of the υ_{O-O}

Table 1
Comparison of selected spectroscopic and structural parameters for μ-η^2:η^2-O_2 dimetal species and related species.

compound	λ_{max} / nm (ε / $M^{-1}cm^{-1}$)	$\nu_{O\text{-}O}$ (cm^{-1})	O-O (Å)	Cu-Cu (Å)
(μ-η^2:η^2-O_2)($CuTp^{iPr}$)$_2$ (**3**)	349 (21000), 551 (790)	741	1.41	3.56
(μ-η^2:η^2-O_2)($CuTp^{Me}$)$_2$	338 (20800), 530 (840)	731		
(μ-η^2:η^2-O_2)($CuTp^{Ph}$)$_2$	355 (18000), 542 (1040)	759		
oxyhemocyanin	340 (20000), 580 (100)	744 - 752		3.5 - 3.7
oxytyrosinase	345 (18000), 600 (1200)	755		~3.6
(μ-η^2:η^2-O_2)($CoTp^{iPr}$)$_2$ (7)	350 (8900), 493 (2900)	651		
(μ-O_2)($NiTp^{iPr}$)$_2$	304 (12000), 404 (10500)	*a*		
(μ-η^2:η^2-O_2)($CuTACN^{iPr3}$)$_2$ (**15**)	366 (22500), 510 (1300)	722		
(μ-O)$_2$($CuTACN^{iPr3}$)$_2$ (**16**)	324 (11000), 448 (13000)	*a*	2.97	2.79

[a] not located.

band suggested weakening of the O-O bond, in other words, a higher reactivity was anticipated. Complex **3** showed basic character as observed for the μ-η^2:η^2-O_2 complexes of Pd and Rh (see above), and could oxidize phenolic substrates as discussed below. However, the finding of the unique coordination structure of the *bioinorganic* dioxygen complex **3** is now recognized as a clue to expansion to a new trend of transition metal dioxygen species.

III. EXPANSION OF CHEMISTRY OF μ-η^2:η^2-PEROXO DIMETAL SPECIES

For elucidation of the principles of transition metal-dioxygen interaction, systematic synthesis and comparative study of LM-O_2 complexes supported by the same ligand (L) is crucial but such chemistry has not appeared so far. Following the copper chemistry, we have extended our study to other metal complexes supported by a series of Tp^R ligands. The metals involve bio-related first row metals (Fe and Mn), non-bio-related first row metals (Co and Ni), and second row metals. We have found that, of the metal complexes studied, the Co and Ni systems also involve the chemistry of μ-η^2:η^2-O_2 species, which can be generated successfully by dehydrative condensation method developed for the dicopper complex (Scheme 2). Their reactivity, however, is considerably different from that of the Cu complex **3**.

1. Cobalt μ-η^2:η^2-Peroxo Complex and Related Species [14]

Treatment of the bis(μ-hydroxo) complex **6** with an equimolar amount of H_2O_2 at -50°C afforded a dark brown compound **7** (Scheme 3). Although the reaction product **7** could not be fully characterized due to its thermal instability, it has been identified as a dinuclear Co(II) μ-η^2:η^2-peroxo complex, $[Tp^{iPr}Co]_2(\mu\text{-}O_2)$, by comparison of its spectral features with those of the copper complex **3** as summarized in Table 1. The notable red shift of the O-O stretching vibration is presumably attributed to the increased back-donation from the metal center to the antibonding σ^* O-O orbital, because the energy levels of the Co d-orbital are higher than those of the Cu d-orbitals. This indicates the O-O bond weakening in the Co complex **7** compared to the O-O bond in the Cu complex **3**, and a new reactivity of **7** has been expected.

The Co(II)-peroxo complex **7** decomposed spontaneously upon warming to ambient temperature to give three different products depending on the reaction conditions (Scheme 3). Decomposition of **7** *in the absence of H_2O_2* produced *mono-oxygenated dinuclear* compound **8**, whereas decomposition *in the presence of an excess amount of H_2O_2* resulted in the formation of a mixture of two products, deep

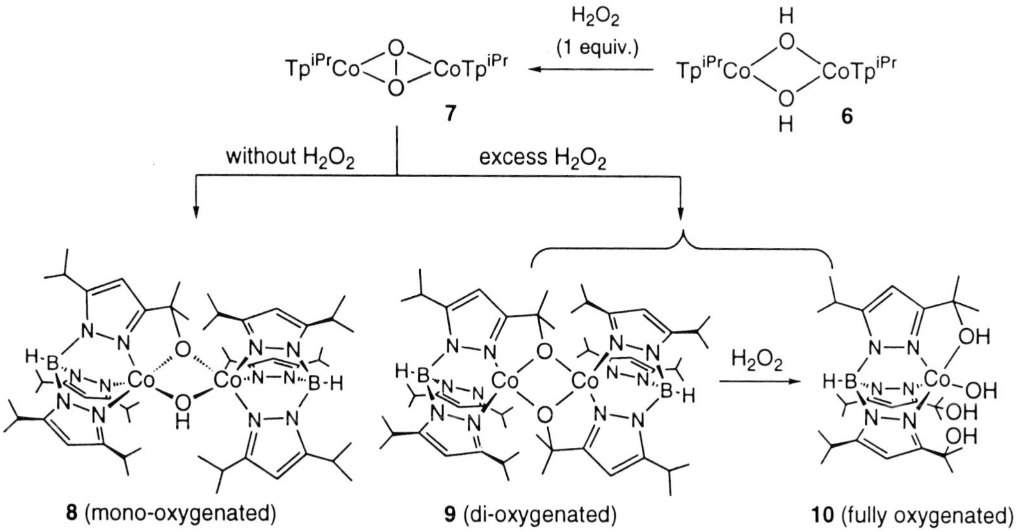

Scheme 3

blue-green *di-oxygenated dinuclear* complex **9** and pale brown *fully oxygenated mononuclear* complex **10**. To be noted is that, in every product, the isopropyl methine parts proximal to the metal center are oxygenated as confirmed by X-ray crystallography, and the number of the oxygenated sites increases from one to two and then to six per the Co_2 unit as going from **8** to **10**. Furthermore treatment of an isolated sample of the *di-oxygenated* product **9** with H_2O_2 caused conversion into the *fully oxygenated* complex **10**.

The formation mechanism of the mono-oxygenated product **8** from **7** (Scheme 4) should involve a reaction sequence of (i) homolysis of the O-O bond, (ii) abstraction of a proximal isopropyl methine hydrogen atom by one of the two resulting oxyl radicals (**I**), and (iii) coupling between the formed tertiary alkyl radical and the remaining oxyl radical (**II**). In the case of the decomposition of **7** in the presence of excess H_2O_2, protonolysis of the Co-O part with H_2O_2 forms two equivalents of the mononuclear hydroperoxo species **III**. Subsequent radical processes analogous to those mentioned above give the chelating alkoxo intermediate **VI**, dimerization of which gives the *di-oxygenated* product **9**. But a part of **VI** may further react with excess H_2O_2 to produce another hydroperoxo intermediate **VII**. Repetition of the oxygenation processes analogous to **III**(**VII**) → **IV** → **V** → **VI** → **VII** finally leads to the formation of the *fully oxygenated* product **10**.

A closely related system was reported by Theopold and his coworkers (Scheme 5). The first example of a symmetrical side-on superoxo complex, $Tp^{t\text{-}Bu,Me}Co(O_2)$ (**11′**: t-Bu derivative of **11**), was prepared by oxidative addition of O_2 to a Co(I) precursor [15]. Although H-abstraction from the t-Bu group was

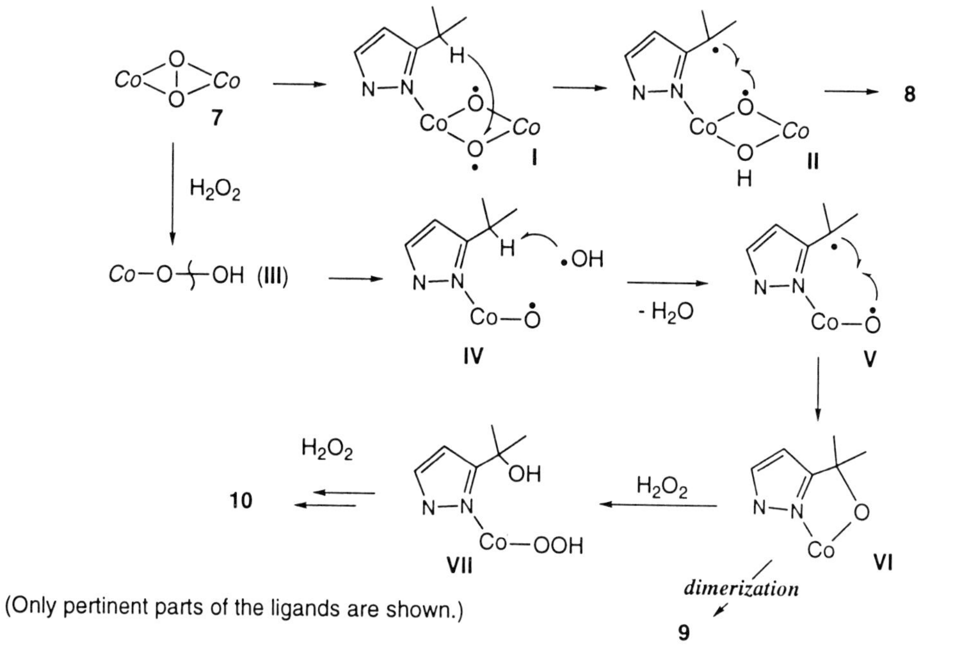

Scheme 4

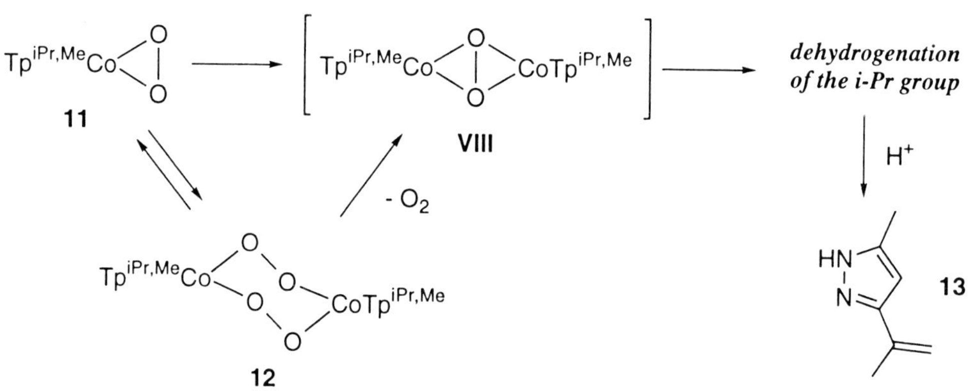

Scheme 5

observed no oxygenated product could be detected in this system. However, introduction of the less hindered isopropyl ligand, $Tp^{iPr,Me}$, resulted in dehydrogenation of the proximal isopropyl group giving isopropenylpyrazole **13** after acidic workup [16]. It was also noted that, in a solution, the mononuclear superoxo species, $Tp^{iPr,Me}Co(O_2)$ **11**, was in equilibrium with the structurally characterized, dimeric structure **12** with a chair-like six-membered cyclic Co-O_2-Co-O_2 linkage [17], deoxygenation from which would lead to the μ-η^2:η^2- O_2

intermediate **VIII**. The kinetic analysis of the decomposition process also supports the dinuclear μ-η^2:η^2-O_2 intermediate **VIII**, and O-O bond cleavage and H-abstraction analogous to our system would cause isopropyl dehydrogenation.

Thus the μ-η^2:η^2-peroxo cobalt species **7** has been successfully formed and characterized. Instead of its thermal instability, **7** shows a unique reactivity via radical intermediates resulting from O-O bond homolysis. Analogous reactions are also observed for the related Co-OOH (**III**) and Co-OOR species as well as for the nickel system [18].

2. Bis(μ-oxo)dimetal Intermediates

The stability and reactivity of the (μ-η^2:η^2-O_2)M_2 species are dependent on the central metals as described above. Similar aliphatic C-H bond activation on the analogous μ-η^2:η^2-O_2 dicopper complexes has been reported recently [19, 20]. The results presented in the previous sections indicate the importance of contribution of the bis(μ-oxyl) (**IX′**) or bis(μ-oxo) species (**IX″**) arising from O-O bond cleavage for the functionalization of aliphatic substrates (Scheme 6).

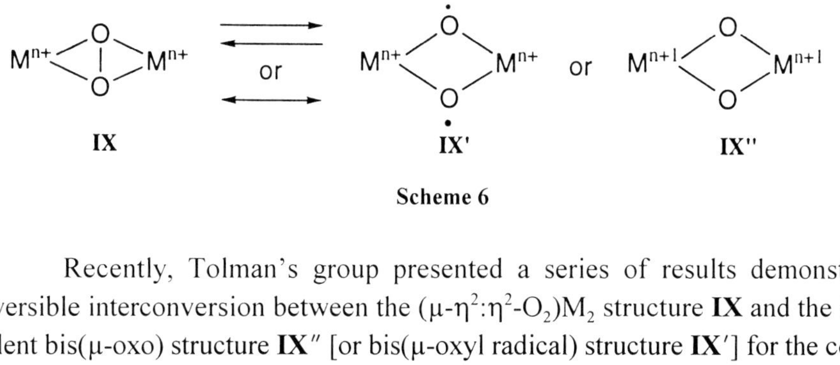

Scheme 6

Recently, Tolman's group presented a series of results demonstrating reversible interconversion between the (μ-η^2:η^2-O_2)M_2 structure **IX** and the higher valent bis(μ-oxo) structure **IX″** [or bis(μ-oxyl radical) structure **IX′**] for the copper-dioxygen complex supported by substituted 1,4,7-triazacyclononane (TACN) ligands, N_3-ligands analogous to Tp^R (Scheme 7) [21]. Oxidative addition of O_2 to a Cu(I) precursor, $[TACN^{iPr3}Cu(NCMe)]^+$ **14**, in CH_2Cl_2 at a low temperature gave the (μ-η^2:η^2-O_2)$[CuTACN^{iPr3}]_2^{2+}$ adduct **15** which showed spectroscopic features quite similar to those of (μ-η^2:η^2-O_2)$(CuTp^{iPr})_2$ (cf. Table 1). However, oxygenation in THF gave a different species **16** (Table 1). The Raman band observed at 600 cm^{-1} was not assigned to an O-O stretching vibration but to one associated with Cu-O vibration, and the O-O bond [O---O: 2.974(2) Å] was found to be cleaved as confirmed definitely by X-ray crystallography of the tribenzyl derivative. Thus the isomer **16** has been characterized as a bis(μ-oxo)Cu(III)$_2$ species [$(\mu$-$O)_2Cu_2$]. Similar O-O bond cleavage on a copper complex leading to the bis(μ-oxo)Cu(III)$_2$ structure was reported by Stack [22]. Remarkably, the $(TACN^{iPr3}Cu)_2$-O_2 adduct

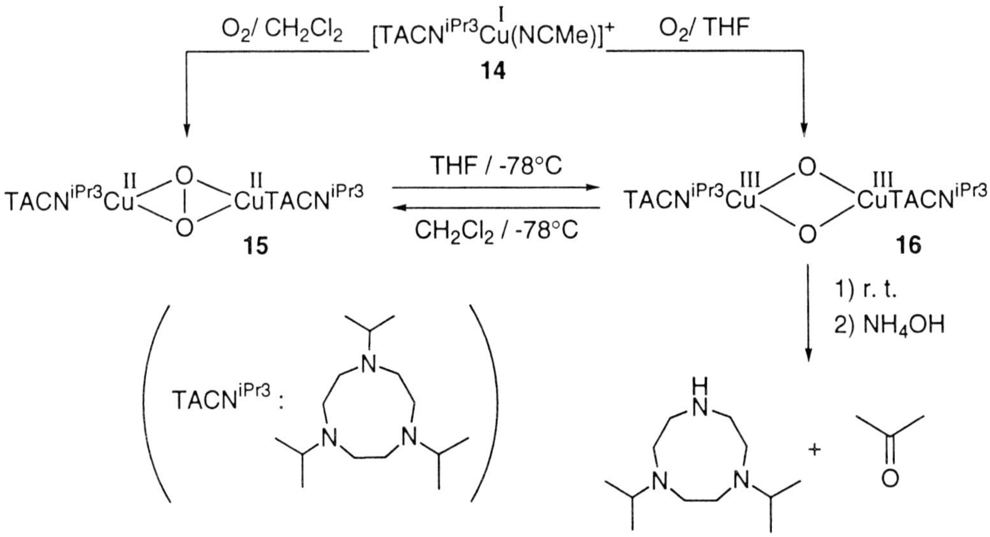

Scheme 7

showed solvatochromism. Upon dissolution of the μ-η^2:η^2-structure **15** (obtained in CH_2Cl_2) in THF, the UV-vis spectrum showed the feature of the $(\mu\text{-}O)_2Cu_2$ form **16**, which could be converted back to the $(\mu\text{-}O_2)Cu_2$ species **15** by re-dissolution in CH_2Cl_2.

Thus reversible cleavage and formation of the O-O bond was observed on a dicopper center. In particular, the formation of O-O bond from the $(\mu\text{-}O)_2M_2$ form is relevant to the pivotal process in oxygen evolution in photosynthesis. As for the oxidizing ability, N-dealkylation was observed for the $(\mu\text{-}O_2)Cu_2$ form upon warming to room temperature [20]. This process involves C-H bond activation of tertiary alkyl (i-Pr) and benzylic moieties effected by the μ-oxo species as confirmed by labeling experiments.

Combined with our result, these observations provide us with a clue to a novel activation mode of the dioxygen molecule by transition metal species: the μ-η^2:η^2- peroxo intermediate has the potential to exhibit high oxidizing power via the μ-oxo intermediate resulting from O-O bond homolysis. The high reactivity of the $M(\mu\text{-}O)_2M$ core should originate from its electrophilic nature typically expressed by the higher oxidation state form **IX**″, and the $[M\text{-}O]_n$ system of high oxidation state would work as an efficient oxidizing agent like mononuclear electrophilic oxene species (M=O).

IV. APPLICATIONS TO CATALYTIC REACTIONS

In this section, catalytic aspects of $(\mu\text{-}\eta^2\text{:}\eta^2\text{-}O_2)M_2$ intermediates will be discussed, though only a few systems involving such intermediates are known.

1. Reaction Mechanism of Tyrosinase

Tyrosinase containing a dicopper center in its active site is a monooxygenase which catalyzes conversion of phenols into *o*-benzoquinones **17** in the presence of dioxygen (Scheme 8) [23]. Because its O_2-adduct (*oxytyrosinase*) shows spectroscopic features similar to those of *oxyhemocyanin* (Table 1), the coordination mode of peroxide in the *oxytyrosinase* is assumed to be the $(\mu\text{-}\eta^2\text{:}\eta^2\text{-}O_2)Cu_2$ form. In the case of sterically hindered phenol, oxidative dimerization leading to diphenoquinone **18** is also catalyzed by *tyrosinase*.

Treatment of the dicopper complex **3** with substituted phenols resulted in the formation of diphenoquinone **18** and, in the presence of O_2, a small amount of *o*-benzoquinone **17** was formed. The formation of **18** has been explained in terms of the following reaction sequence (Scheme 9) [24]: (i) hydrolysis of the $(\mu\text{-}\eta^2\text{:}\eta^2\text{-}O_2)Cu_2$ core with phenol giving phenoxo intermediate **X** or **XI**, (ii) Cu-O bond homolysis giving phenoxo radical **XI**, (iii) carbon-centered dimerization of two phenoxo radicals, and (iv) dehydrogenation. As for the formation mechanism of **17**, a conclusive result could not be deduced from our results. However, the pair of phenoxo and hydroperoxo intermediates **X** should be responsible for it, because the

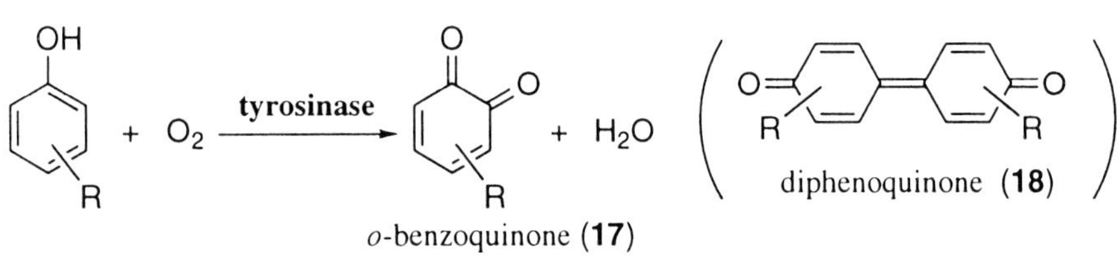

Scheme 8

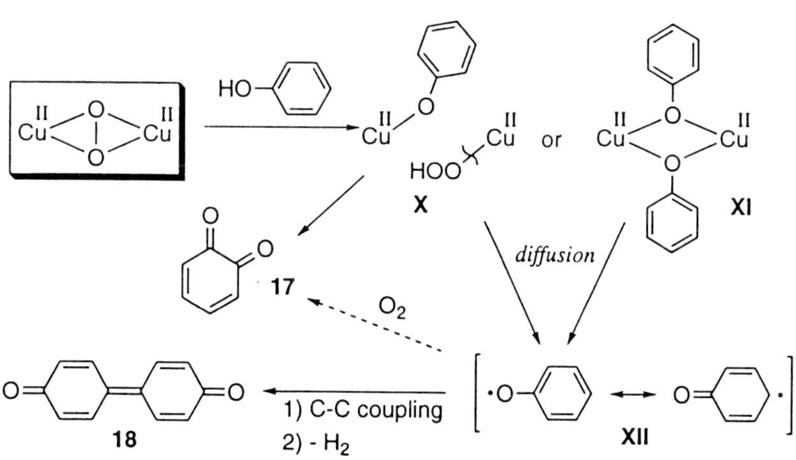

Scheme 9

incorporation of two molecules of phenol (like **XI**) may not be possible owing to the limited size of the cavity of the active site in *tyrosinase*. Although the reactivity of hydroperoxo species is not well understood, a radical mechanism via Cu-O homolysis is plausible as suggested by the reactivity of alkylperoxo complexes, $Tp^{iPr}Cu$-OOR, which have been structurally characterized [25]. The formation of **17** from **3** in the presence of O_2 may involve oxygenation of the phenoxo radical **XII**. Although a couple of mechanisms such as electrophilic addition of the μ-O_2 moiety toward the arene ring and radical reaction of the $(\mu\text{-}O)_2Cu_2$ intermediate have been proposed [26], no conclusive results have previously been obtained.

2. Regioselective Dehydrogenative Polymerization of Phenol Leading to Poly(p-phenylene Oxide) [27]

Poly(*p*-phenylene oxide) has attracted much attention due to its properties applicable to engineering plastics. Although it is available via multi-step stoichiometric reactions, catalytic dehydrogenative coupling of phenol is highly desirable because of the cheapness of the starting material and the simplicity and easiness of the catalytic process. Cu(I) ion has been known to be active for such transformation, but the problem of selectivity concerning the linked position remains to be solved [28]. Contamination with a small amount of the other linkages (*o*- and *m*-conjunction) causes significant deterioration of its performance. Recently, the μ-η^2:η^2-O_2 dicopper(I) complexes supported by Tp^R or $TACN^{R3}$ ligands have been found to be more effective for the selective *p*-coupling of phenol (Scheme 10). The steric protection of the copper center by the pyrazolyl substituents at the 3-position (sites proximal to the metal center) may hinder formation of undesirable *o*- and *m*-linkage.

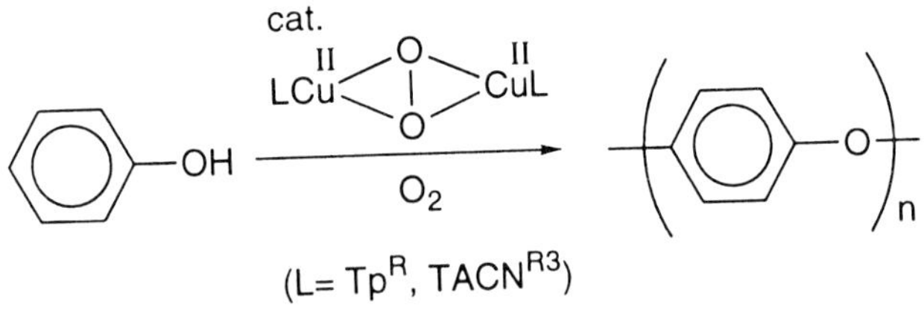

Scheme 10

3. Catalytic Oxidation of Hydrocarbons Based on the Tp^RFe System

Before closing this section, some more bioinorganic studies based on the Tp^RFe system will be discussed, though a $(\mu\text{-}\eta^2\text{:}\eta^2\text{-}O_2)M_2$ species is not necessarily involved.

Methanemonooxygenase (MMO) [29] is an enzyme which hydroxylates

alkanes including methane under very mild conditions (r. t., normal pressure) to give corresponding alcohols (Scheme 11), and it contains a diiron center in the active site. Because such transformation can not be realized by conventional catalytic systems, much attention has been focused on its diiron core structure and reaction mechanism. In this case, too, a number of model studies have appeared. In 1993, Lippard et al. succeeded in structural analysis of *Methylococcus capsulatus* (Bath), which contained the diiron structure bridged by hydroxo, acetato and glutamato ligands [30]. A similar core structure is found for *hemerythrin* [31], an O_2-carrier of some kind of invertebrate phyla: sipunculans, brachiopods, priapulids, and annelids. O-O bond homolysis leading to higher valent oxo species is now getting much attention [32].

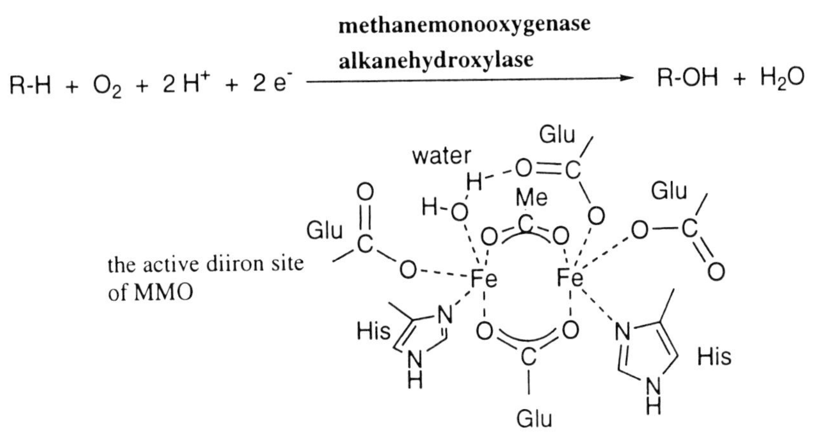

Scheme 11

Before the report by Lippard, we had been studying model systems for MMO. By mimicking the diiron core by the μ-oxo diiron complex **19**, we succeeded in aerobic oxygenation of alkanes and arenes at room temperature (Scheme 12) [33]. Stoichiometric amounts of proton (CF_3COOH) and electron sources (Zn) are consumed in the present catalytic reaction. Alkanes are oxidized to corresponding alcohol and ketone, and epoxidation of olefin and hydroxylation of arenes are also observed with high efficiency.

IV. CONCLUDING REMARKS

Let us point out the relationship between the reactivity of dioxygen complexes and O-O stretching vibration. This is a historical view of the electronic state of the O-O

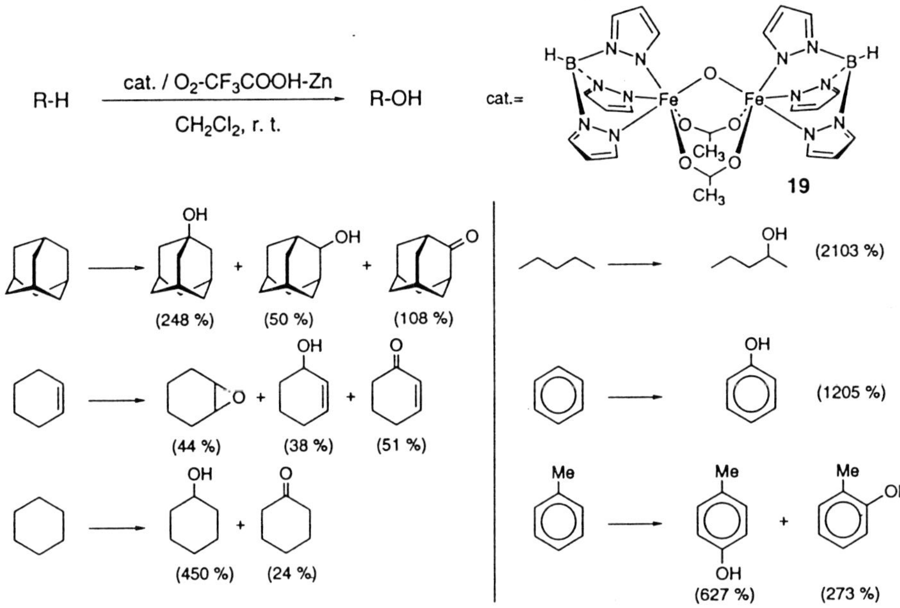

Scheme 12

moiety. ν(O-O) values of mononuclear complexes **A** and **B** (>800 cm^{-1}), the previous major research target, indicate multiple bond character and, actually, they show nucleophilic reactivity. No dioxygen complex showing a ν(O-O) band lower than 800 cm^{-1} was known before our report on the type **E** dinuclear complexes [5, 6,12]. Upon complexation with the second metal center, the ν(O-O) band shifts to lower energies (<750 cm^{-1}), indicating that the O-O bond order is considerably reduced as compared with organic peroxide [R-O-O-R: (890 - 830 cm^{-1})] but the electron transfer to the antibonding O-O orbital, which weakens the O-O bond, is still insufficient for O-O bond cleavage. The introduction of more electron-donating elements (Co, Ni) or more electron-donating ligands such as TACN causes homolytic scission of the O-O bond and, as a result, aliphatic C-H hydroxylation is observed.

The μ-η^2:η^2-coordination mode of dioxygen ligand was initially established for the organometallic (**1** and **2**) and bioinorganic complexes (**3**). Although their unique structures attracted much attention, in particular, from relevance to the O_2-binding mode of *hemocyanin*, their reactions were not selective as was expected. However, through the studies of other transition metal dioxygen complexes, the μ-η^2:η^2-binding mode has been accepted as not only a binding mode but also a general leading concept which triggers development of novel types of reactivity. Extension to other metal systems and (catalytic) oxidation of external substrates are now under study in our laboratory.

VI. REFERENCES

1. (a) R.A. Sheldon and J.K. Kochi, *Metal-Catalyzed Oxidation of Organic Compounds*, Academic Press, New York (1981). (b) T. Mukaiyama and T. Yamada, *Bull. Chem. Soc. Jpn.*, **68**, 17 (1995).
2. (a) the thematic issue of *Chem. Rev.* (Metal Diooxygen Complexes), *Chem. Rev.*, **94**, 567 (1994). (b) the thematic issue of *Chem. Rev.* (Bioinorganic Enzymology), *Chem. Rev.*, **96**, 2239 (1996).
3. (a) J.A. Lyons, *Aspects Homog. Catal.*, **3**, 1 (1977). (b) M.H. Gubelmann and A.F. Williams, *Struct. Bonding*, **55**, 1 (1983). (c) N. Kitajima, M. Akita and Y. Moro-oka, In *Organic Peroxides*, W. Ando (Ed.), Wiley, New York (1992).
4. (a) J.P. Collman, R.R. Gagne, C.A. Reed, T.R. Hulbert, G. Lang and W.T. Robinson, *J. Am. Chem. Soc.*, **97**, 1427 (1975). (b) J.P. Collman, R.R. Gagne, C.A. Reed, W.T. Robinson and C.A. Rodley, *Proc. Natl. Acad. Sci. U.S.A.*, **71**, 1326 (1974).
5. (a) H. Suzuki, K. Mizutani, Y. Moro-oka and T. Ikawa, *J. Am. Chem. Soc.*, **101**, 748 (1979). (b) P.J. Chung, H. Suzuki, Y. Moro-oka and T. Ikawa, *Chem. Lett.*, 63 (1980).
6. F. Sakurai, H. Suzuki, Y. Moro-oka and T. Ikawa, *J. Am. Chem. Soc.*, **102**, 1749 (1980).
7. R. Sugimoto, H. Suzuki, Y. Moro-oka and T. Ikawa, *Chem. Lett.*, 1863 (1982)
8. (a) *Structure and Function of Hemocyanin*, J.V. Bannister (Ed.), Springer-Verleg, Berlin (1977). (b) E.I. Solomon, K.W. Penfield and D.E. Wilcox, *Struct. Bonding*, **53**, 1 (1983). (c) E.I. Solomon, M.J. Baldwin and M. Lowery, *Chem. Rev.*, **92**, 521 (1992).
9. (a) T.R. Jacobson, Z. Tyeklar, A. Farooq, K.D. Karlin, S. Liu and Z. Zubieta, *J. Am. Chem. Soc.*, **110**, 3690 (1988). (b) K.D. Karlin, S. Kaderli and A. D. Zuberbühler, *Acc. Chem. Res.*, **30**, 139 (1997).
10. (a) K.D. Karlin, R.W. Cruse, Y. Gultneh, J.C. Hayes and J. Zubieta, *J. Am. Chem. Soc.*, **106**, 3372 (1984). (b) K.D. Karlin, R.W. Cruse, Y. Gultneh, A. Farooq, J.C. Hayes and J. Zubieta, *J. Am. Chem. Soc.*, **109**, 2668 (1987).
11. (a) N. Kitajima and W. B. Tolman, *Prog. Inorg. Chem.*, **43**, 419 (1995). (b) S. Trofimenko, *Chem. Rev.*, **93**, 943 (1993)
12. (a) N. Kitajima, K. Fujisawa, C. Fujimoto, Y. Moro-oka, S. Hashimoto, T. Kitagawa, K. Toriumi, K. Tatsumi and A. Nakamura, *J. Am. Chem. Soc.*, **114**, 1277 (1992). (b) N. Kitajima and Y. Moro-oka, *Chem. Rev.*, **94**, 737 (1994). (c) N. Kitajima and Y. Moro-oka, *J. Chem. Soc., Dalton Trans.*, 2665 (1993). (d) N. Kitajima, *Adv. Inorg. Chem.*, **39**, 1 (1992).
13. (a) K.A. Magnus, B. Hazes, H. Ton-That, C. Bonaveture, J. Bonaveture and W.G.J. Hol, *Proteins*, **19**, 302 (1994). (b) K.A. Magnus, H. Ton-That and J.E. Carpenter, *Chem. Rev.* **94**, 727 (1994).
14. (a) S. Hikichi, H. Komatsuzaki, N. Kitajima, M. Akita, M. Mukai, T. Kitagawa and Y. Moro-oka, *Inorg. Chem.*, **36**, 266 (1997). (b) S. Hikichi, H. Komatsuzaki, M. Akita and Y. Moro-oka, submitted for publication.
15. J.W. Egan, Jr., B.S. Haggerty, A.L. Rheingold, S.C. Sendlinger and K.H. Theopold, *J. Am. Chem. Soc.*, **112**, 2445 (1990).
16. O.M. Reinaud and K.H. Theopold, *J. Am. Chem. Soc.*, **116**, 6979 (1994)
17. O.M. Reinaud, G.P.A. Yap, A.L. Rheingold and K.H. Theopold, *Angew. Chem. Int. Ed. Engl.*, **34**, 2051 (1995).
18. S. Hikichi, M. Akita and Y. Moro-oka, to be submitted.
19. W.E. Allen and T.N. Sorrell, *Inorg. Chem.*, **36**, 1732 (1997).
20. S. Mahapatra, J.A. Halfen and W.B. Tolman, *J. Am. Chem. Soc.*, **118**, 11575 (1996).
21. (a) J.A. Halfen, A. Mahapatra, E.C. Wilkinson, S. Kaderli, V.G. Young, Jr., L. Que, Jr., A.

D. Zuberbühler and W.B. Tolman, *Science*, **271**, 1397 (1996). (b) A. Mahapatra, J.A. Halfen, E.C. Wilkinson, G. Pan, X. Wang, V.G. Young, Jr., C.J. Cramer, L. Que, Jr., and W.B. Tolman, *J. Am. Chem. Soc.*, **118**, 11555 (1996). (c) W.B. Tolman, *Acc. Chem. Res.*, **30**, 227 (1997).

22. A.P. Cole, D.E. Root, Mukherjee, E.I. Solomon and T.D.P. Stack, *Science*, **273**, 1848 (1996).
23. D.E. Wilcox, A.G. Porras, Y.T. Hwang, K. Lerch, M.E. Winkler and E.I. Solomon, *J. Am. Chem. Soc.*, **107**, 4015 (1985).
24. N. Kitajima, T. Koda, Y. Iwata and Y. Moro-oka, *J. Am. Chem. Soc.*, **112**, 8833 (1990).
25. N. Kitajima, T. Katayama, K. Fujisawa, Y. Iwata and Y. Moro-oka, *J. Am. Chem. Soc.*, **115**, 7872 (1993)
26. (a) P.K. Ross and E.I. Solomn, *J. Am. Chem. Soc.*, **112**, 5871 (1992). (b) P.K. Ross and E. I. Solomon, *J. Am. Chem. Soc.* **113**, 3246 (1993).
27. H. Higashimura, K. Fujisawa, Y. Moro-oka, M. Kubota, A. Terahara, A. Siga, H. Uyama and S. Kobayashi, to be reported.
28. H.L. Finkbeiner, A.S. Hay and D.M. White, *Polymerization Processes*, C.E. Schildknecht and I. Skeist (Eds.), Wiley-Interscience, New York, pp. 537 (1977).
29. (a) J.D. Lipscomb, *Annu. Rev. Microbiol.* **48**, 371 (1994). (b) K.E. Liu and S.J. Lippard, *Adv. Inorg. Chem.*, **42**, 263 (1995).
30. (a) A.C. Rosenzweig, C.A. Frederick, S.J. Lippard and P. Nordlund, *Nature*, **366**, 537 (1993). (b) A.C. Rosenzweig, P. Nordlund, P.M. Takahara, C.A. Frederick and S.J. Lippard, *Chem. Biol.*, **2**, 409 (1995).
31. (a) R.E. Stenkamp, *Chem. Rev.*, **94**, 715 (1994). (b) R.E. Stenkamp, L.C. Sieker and L.H. Jensen, *J. Am. Chem. Soc.*, **106**, 618 (1984). (c) S. Sheriff, W.A. Hendrickson and J.L. Smith, *J. Mol. Biol.*, **197**, 273 (1987). (d) M.A. Holms and R.E. Stenkamp, *J. Mol. Biol.*, **220**, 723 (1991).
32. (a) L. Shu, J.C. Nesheim, K. Kauffmann, E. Münck, J.D. Lipscomb and L. Que, Jr., *Science*, **275**, 515 (1997). (b) L. Que, Jr. and Y. Dong, *Acc. Chem. Res.*, **29**, 190 (1996).
33. (a) N. Kitajima, H. Fukui and Y. Moro-oka, *J. Chem. Soc., Chem. Commun.*, 485 (1988). (b) N. Kitajima, M. Ito, H. Fukui and Y. Moro-oka, *J. Chem. Soc., Chem. Commun.*, 102 (1991).

Res. Chem. Intermed., Vol. 24, No. 3, pp. 309-327 (1998)

PHOTOOXIDATION OF LIGHT ALKANES OVER ALKALI-ION-MODIFIED VANADIUM CATALYSTS

S. YOSHIDA*, S. TAKENAKA and T. TANAKA
Department of Molecular Engineering, Graduate School of Engineering, Kyoto University, Sakyo-ku, Kyoto, 606-01, Japan

Received 11 September 1997; accepted 4 November 1997

CONTENTS

I. INTRODUCTION

The selective conversion of light alkanes is of current interest from both the scientific and industrial viewpoints; the challenges include difficult activation of saturated C-H bonds and utilization of the abundant resource of liquefied petroleum gas (LPG) containing mainly propane and butanes. With this regard, oxidative dehydrogenation of propane to produce propene [1] and aromatization of propane [2] have been investigated widely. It has been attempted to synthesize butenes and butadiene from *n* - butane over V-Mg-O catalysts [3] and maleic anhydride from *n*-butane over V-P-O catalysts [4]. In these reactions, high temperatures are very often required even at a low conversion level. On the other hand, the reactions using photocatalysts take place under mild conditions and can be controlled easily [5]. However, there have been few papers on the photooxidation [6] and most of them report low activity and/or low selectivity to partial oxidation.

Among the photocatalysts, V_2O_5/SiO_2 is known to be effective for selective photooxidation of olefins and alcohols [7]. In a previous work on the photooxidation of propene over V_2O_5/SiO_2 and Na^+-modified V_2O_5/SiO_2 [8], we found that oxidative cleavage of a C=C bond is dominant over irradiated V_2O_5/SiO_2 to form ethanal while over Na^+-modified V_2O_5/SiO_2, the formation of acraldehyde is preferred. The addition of sodium ions to V_2O_5/SiO_2 brings about the reduction of acidity of the catalyst as well as the formation of different surface vanadate species, which actively attack C-H bonds. The results motivated us to explore partial oxidation of light alkanes by alkali-ion-modified V_2O_5/SiO_2, and we have found that propane and 2-methylpropane are selectively converted to propanone over the photoirradiated catalysts. In this article, we summarize our recent studies [9-15] on the activity and selectivity, characterization of the active species and electronic states of a series of V_2O_5/SiO_2 catalysts modified by Na^+, K^+, and Rb^+.

II. EXPERIMENTAL PROCEDURES

Catalysts

V_2O_5/SiO_2 (VS; V_2O_5 2.5 wt%) was prepared by impregnating SiO_2 (BET surface = 588 m^2 g^{-1}) with an aqueous solution of ammonium metavanadate in the same manner in ref. [16]. The BET surface area was 465 m^2 g^{-1}. Alkali-ion-modified V_2O_5/SiO_2 catalysts (Na-VS, K-VS and Rb-VS) were prepared by impregnating VS with aqueous solutions of NaOH, KOH and RbOH containing the desired amounts of Na^+, K^+ and Rb^+, respectively, followed by calcination in a dry air stream at 773 K for 5 h. The contents of alkali ions are 1, 2 and 4 wt% as Na_2O, K_2O and Rb_2O, respectively, corresponding to almost equimolar amounts of vanadium in VS. The BET surface areas of Na-VS, K-VS and Rb-VS were 261, 165 and 170 m^2 g^{-1}, respectively.

Reactions

The reactions were performed in a conventional closed circulating system (dead space: 280 cm^3) described elsewhere [17]. The powder catalyst (500 mg) was spread on the flat-bottom of a quartz reactor and irradiated at room temperature by UV light from 250 W ultrahigh pressure Hg lamp through glass filters, UV-31 or UV-39, which permits light of wavelength, l > 310 nm or λ > 390 nm, respectively. In the dark, the reactions did not take place. After each run, the catalyst bed was heated to 573 K for collecting products which were tightly adsorbed on the catalyst at room temperature. The desorbed products were frozen out in a trap cooled by liquid nitrogen, and analyzed by GLC and GLC-mass spectrometry.

Characterization

Characterization of catalysts was done by means of diffuse reflectance UV-vis, photoluminescence, Raman, and X-ray absorption spectroscopies. For assignment of intermediates, FT-IR spectra were recorded *in situ*.

III. EFFECT OF ALKALI IONS ON THE PHOTOADSORPTION AND EMISSION

Figure 1 shows diffuse reflectance UV-vis spectra of VS and alkali-ion-modified VS [13]. VS exhibited the charge-transfer bands at 250 nm and 290 nm. These bands were assigned to VO_4 tetrahedral species on silica [18]. This result indicates

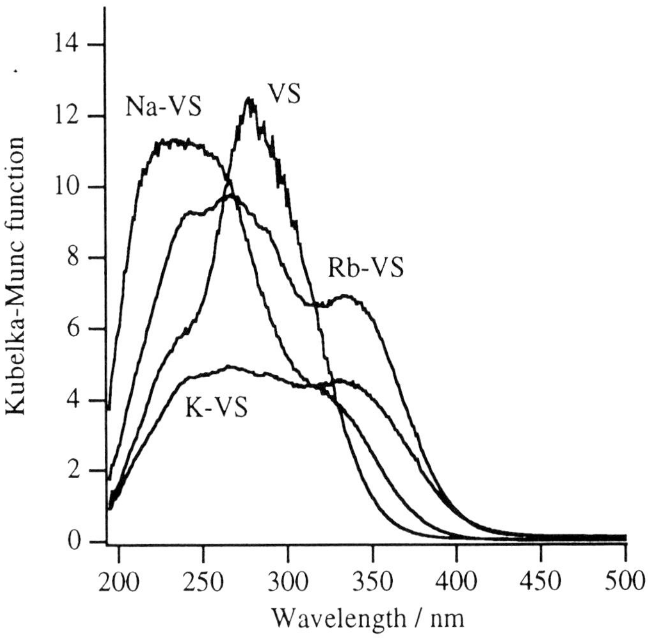

Figure 1. Diffuse reflectance UV-vis spectra of V_2O_5 (2.5 wt%)/SiO_2 (VS) and alkali-ion-modified VS.

that the structure of vanadium species in VS is a VO_4 tetrahedron mainly. For alkali-ion-modified VS samples, the charge-transfer bands were observed around 240 nm and 280 nm and the positions were different from those of VS, suggesting that alkali ion interacted with the vanadium species on silica. In addition, the new charge-transfer bands were found at around 350 nm and enhanced with the larger atomic numbers of added alkali ions.

Phosphorescence of highly dispersed vanadium oxide on silica is well known and assigned to the triplet state of $(V=O)O_3$ cluster excited by the light of

λ = 310 - 370 nm [19]. The triplet state results from the charge transfer localized in the V=O bond and the emission spectrum shows fine structure associated with the vibration of V=O, as seen in Figure2 (a) [10]. By addition of alkali ions, the intensity of the vibrational fine structure decreased and a shoulder peak was seen in the vicinity of 550 nm, indicating that phosphorescent emission spectra of the modified catalysts excited by 310 nm consist of two kinds of emission bands. This is confirmed by excitation spectra monitored at 530 nm, in which only a tail of absorbance of the band centered at 290 nm was observed for VS, but a new band in the region of 360 - 420 nm was observed for modified catalysts, and correspondingly, emission spectra excited by 400 nm light were recorded as shown in Figure 2.

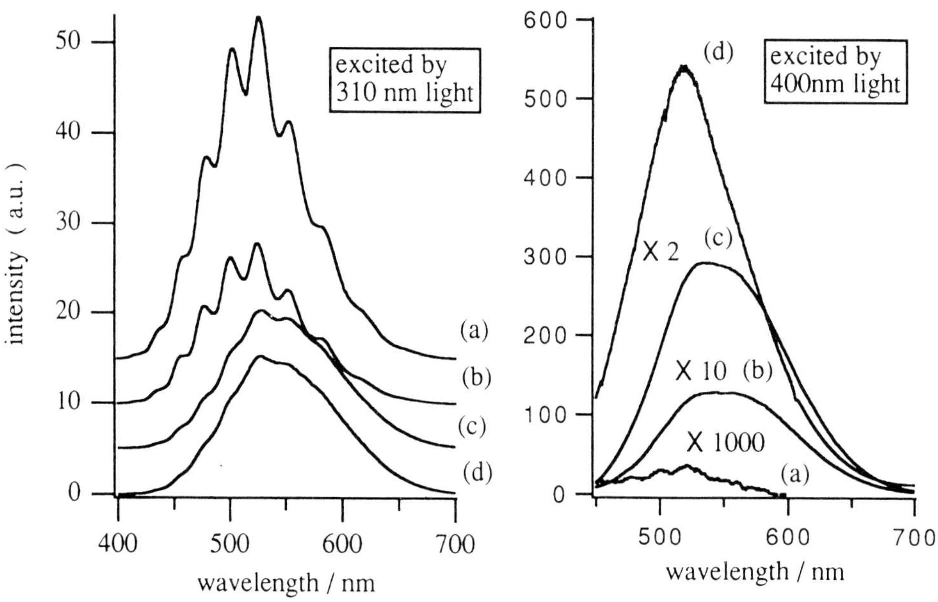

Figure 2. Photoluminescent emission spectra excited by 310 nm and 400 nm light at 77 K. (a) VS, (b) Na-VS, (c) K-VS, and (d) Rb-VS. X n stands for n-fold amplification in recording.

In a previous report, we tentatively concluded the formation of two kinds of emission centers based on the report by Kazansky *et al.* [20]. However, later studies by Raman and X-ray absorption spectroscopies indicated a single surface species in the alkali-ion-modified catalysts as described in later sections. Thus, we conclude that two kinds of emission bands are associated with the two different triplet states of the same species. The important conclusion deduced from the results mentioned above is that the modification of VS by alkali ions brought about the formation of a new species which is excited by both light of wavelengths shorter than 320 nm and longer than 350 nm. Especially, the effectiveness of 400 nm (visible light) for emission should be noted.

IV. PHOTOOXIDATION OF PROPANE AND 2-METHYLPROPANE

Photooxidation of propane [9,10] and 2-methylpropane [11,12] was carried out over VS and the alkali-ion-modified catalysts. Tables 1 and 2 show the activity and selectivity in the photo-oxidation of propane and 2-methylpropane, respectively, under irradiation by the light of $\lambda > 310$ nm or $\lambda > 390$ nm, where the selection of light components was done by UV-31 and UV-39 filters. Generally speaking, the activity was enhanced by the addition of alkali ions and the effect was more remarkable for a heavier alkali ion. Furthermore, the addition of alkali ions caused a high selectivity to propanone.

The most striking effect of the modification was observed in the case of irradiation by light of $\lambda > 390$ nm. The activity of VS was very low, as expected from the phosphorescent excitation and emission spectra shown in Figure 2. On the other hand, the alkali-ion-modified catalysts showed considerably high activity. Furthermore, total oxidation was significantly depressed compared with that in the

Table 1
Photooxidation of propane by oxygen gas over VS and alkali-ion-modified VS at room temperature[a].

λ[b] /nm	Catal.	Conv.[c] /%	Selectivity[d] / % C_3H_6	C_2H_5CHO	$(CH_3)_2CO$	C_2H_4	CH_3CHO	COx
> 310	VS	63.4	10	12	16	2	17	42
	Na-VS	83.8	4	1	33	4	2	56
	K-VS	85.2	2	1	37	1	1	57
	Rb-VS	88.5	1	0	33	1	1	63
> 390	VS	5.0	43	19	15	tr	13	9
	Na-VS	60.3	8	7	52	1	7	25
	K-VS	67.0	4	2	51	1	3	39
	Rb-VS	67.1	3	1	60	1	1	34

[a]Catalysts 0.5 g, closed circulating system (280 cm^3), propane 70 μmol, oxygen 140 μmol, and irradiation time 1 h.
[b]Irradiated through UV-31 filter for $\lambda > 310$ nm and UV-39 filter for $\lambda > 390$ nm.
[c]Based on introduced propane.
[d]Based on converted propane.

Table 2
Photooxidation of 2-methylpropane by oxygen gas over VS and alkali-ion-modified VS at room temperature[a].

λ[b] /nm	Catal.	Conv.[c] /%	IB	PR	AA	AC	TB	BA	ML	COx
			Selectivity[d] / %							
> 310	VS	58.4	8	6	10	36	0	9	1	29
	Na-VS	84.6	7	1	2	41	1	3	1	39
	K-VS	90.7	6	1	2	51	5	2	1	31
	Rb-VS	94.3	6	1	1	51	5	3	1	31
> 390	VS	5.4	36	8	5	21	0	20	7	0
	Na-VS	58.3	18	2	1	39	5	11	5	16
	K-VS	77.6	6	1	1	48	16	7	2	15
	Rb-VS	82.2	7	1	0	54	22	7	2	1

[a]Catalysts 0.5 g, 2-methylpropane 70 μmol, oxygen 140 μmol, irradiation time 1 h.
[b]Irradiated through UV-31 filter for λ > 310 nm and UV-39 filter for λ > 390 nm.
[c]Based on introduced 2-methylpropane.
[d]Based on converted 2-methylpropane. IB: 2-methylpropene, PR: propene, AA: ethanal, AC: propanone, TB: 2-methylpropan-2-ol, BA: 2-methylpropanal, ML: metacrolein, and COx: CO and CO_2.

reactions under irradiation by light of λ > 310 nm. It is noteworthy that the selectivity to propanone and 2-methylpropan-2-ol increased remarkably in the photooxidation of 2-methylpropane, indicating preferential activation of the tertiary C-H bond. The effect was most pronounced for Rb-VS. This suggests that the alkali ions are incorporated in the active species and the basicity of the active species is an important factor for activation of tertiary C-H bonds, as the base strength of alkali ions is higher with a larger atomic number. Other basic elements such as Mg, Eu, Pr and Ce were not found to be effective in formation of an active center for enhancement of activity for photooxidation of propane [12].

V. STRUCTURE OF ACTIVE SPECIES IN THE MODIFIED CATALYSTS

As described previously, the active center of VS is ascertained to be a highly dispersed tetrahedral VO_4 species stabilized on a silica surface [21]. The results mentioned in previous sections indicate the formation of a new active species in alkali-ion-modified catalysts. We investigated the structure of the new species to elucidate the effect of the alkali ions at a molecular level by means of Raman and

X-ray absorption spectroscopies [13]. Raman spectra of VS and alkali-ion-modified VS were recorded at room temperature, and are presented in Figure 3. A band at 1035 cm^{-1} is seen in the spectrum of VS. The band was assigned to the V=O stretching vibration of monomeric vanadium species, (V=O) O_3, bound directly to silica [22,23]. The result is consistent with the conclusion deduced from the UV-vis spectrum, described in section III.

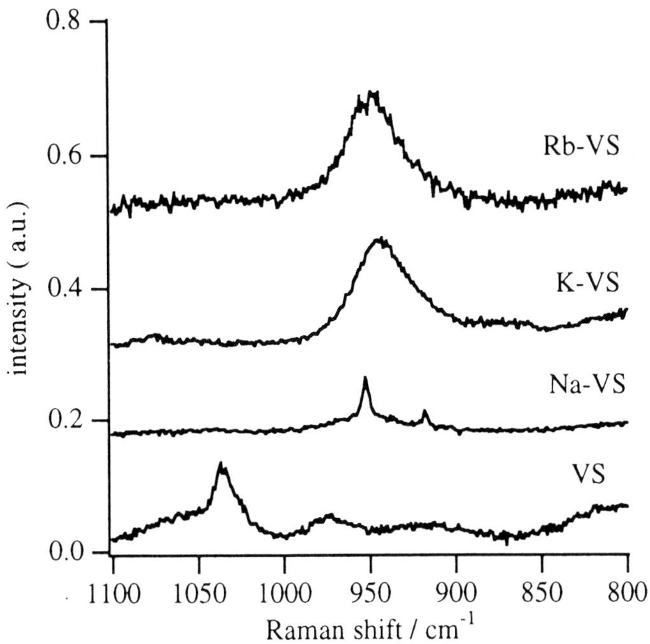

Figure 3. Raman spectra of VS and alkali-ion-modified VS.

In the spectra of alkali-ion-modified VS, the lack of the band at 1035 cm^{-1} and the appearance of the bands at *ca.* 940 cm^{-1} are noticed; addition of alkali ions to VS resulted in the shift from 1035 cm^{-1} to 940 cm^{-1}. The shift means that the bond length of V=O in a VO_4 cluster on silica is elongated by the addition of alkali ions, and oxygen atoms of V=O interact with some alkali ions. In addition, a complete disappearance of the band at 1035 cm^{-1} in alkali-ion-modified VS indicates that all the surface vanadium oxide moieties in alkali-ion-modified VS interact with alkali ions. Wachs *et al.* examined the effect of potassium ion on the structure of surface vanadium oxide in V_2O_5/TiO_2 by Raman spectroscopy [24]. The presence of potassium ion on V_2O_5/TiO_2 shifted the Raman band (1027 cm^{-1}) assigned to the V=O bond stretching in a VO_4 tetrahedron on a titania support to lower wavenumbers (880-910 cm^{-1}). They concluded that potassium ion interacted with the terminal V=O bond and/or the vanadium-oxygen-support bridging bond.

Our results of Raman spectra are compatible with that in V_2O_5/TiO_2. Raman spectra of $RbVO_3$, Rb_3VO_4, and $Rb_4V_2O_7$ were recorded and each spectrum was completely different from that of Rb-VS. Therefore, we can exclude formation of such alkali vanadates in Rb-VS.

More detailed information about the local structure of vanadium species can be obtained by analysis of X-ray absorption spectra. Vanadium K-edge XANES spectra are sensitive to the symmetry of the vanadium oxide moiety, and useful to understand the local structure of vanadium oxide on silica. Figure 4 shows XANES spectra of VS, alkali-ion-modified VS samples and reference compounds. All the spectra have a sharp pre-edge peak. The peak is attributed to a 1s-3d transition, which is mainly caused by mixing of 4p orbitals of vanadium atoms and 2p orbitals of the oxygen atoms with the 3dπ orbitals of vanadium atoms. The coordination of vanadium ions in K_3VO_4 and KVO_3 is tetrahedral, and their pre-edge peaks are intense. It was suggested that when the vanadium ion was in a tetrahedron coordination, a pre-edge peak was more intense than that of a square-pyramidal one [24]. XANES spectrum of VS had a strong pre-edge peak, such as those of K_3VO_4 and KVO_3. This result suggests that the structure of a vanadium cluster in VS is of a tetrahedron. XANES spectra of alkali-ion-modified VS also had almost the same characteristic peaks as the spectrum of VS. From these results, the presence of alkali ions did not change the structure of a surface vanadium species greatly, and the tetrahedral structure was retained. XANES spectrum of K-VS is different from those of K_3VO_4, KVO_3, and $K_4V_2O_7$, indicating that alkali ion

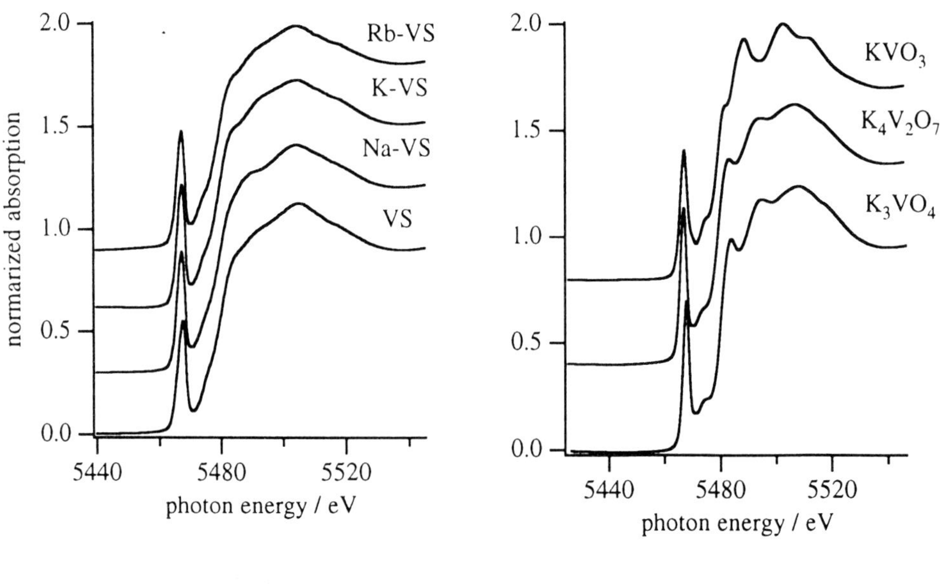

Figure 4. V K-edge XANES spectra of VS, alkali-ion-modified VS samples, and reference samples.

addition to VS does not cause the formation of such alkali-vanadates.

The structural information beyond the nearest coordination sphere as well as structural parameters can be obtained by analysis of EXAFS. At first, a Fourier-transform was performed on EXAFS in the 3.5-12.5 $Å^{-1}$ region, to obtain the radial structure function (RSF). The RSFs are shown in Figure 5. In RSF of VS, two peaks appearing at 0.8-2 Å are due to V-O bonds and a small peak is found at 3.1 Å. The latter peak may show the presence of neighboring vanadium atoms. In alkali-ion-modified samples, the two peaks appearing at 1-2 Å are attributed to V-O bonds and are broader than those of VS. This shows that the V-O bond lengths of VO_4 in alkali-ion-modified VS contain a variety of V-O bondlengths, compared with those in VS. The profiles in alkali-ion-modified VS are similar to each other, suggesting that the structure of vanadium species in the modified catalysts is almost the same, despite the different kinds of added alkali ions. In the case of Rb-VS, a small peak at around 3.2 Å is found as in the case of VS. The peak height changes with the loading of rubidium ions and the envelope function of the inversely transformed spectrum of the peak has the maximum at higher k value than that for VS. This implies that the small peak is mainly due to the presence of rubidium atoms adjacent to vanadium atoms. In the case of Na-VS, we can not confirm the presence of sodium ions adjacent to vanadium atoms because of considerable noise in the RSF. However, the presence of the adjacent sodium ions is strongly suggested by the similarity of nature of V-O bonds in Na-VS and Rb-VS, as described below.

We carried out further analysis of the EXAFS oscillations employing non linear curve-fitting against the Fourier-filtered EXAFS spectra of VS and alkali-ion-modified VS. The results are summarized in Table 3. In VS, there are two

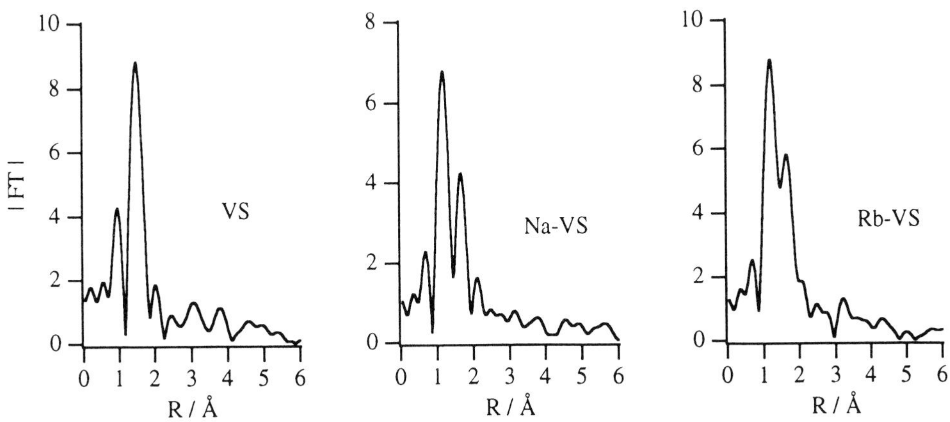

Figure 5. Fourier-transforms of k^3-weighted V K-edge EXAFS spectra of VS and alkali-ion-modified VS samples.

Table 3
The stractural parameters estimated by the curve-fitting of EXAFS of VS and alkali-ion-modified VS.

Catalysts	Shell	C.N.[a]	R /Å [b]	σ^2 /Å^2 [c]
VS	V-O	1.0	1.60	-0.00501
	V-O	3.0	1.79	-0.00363
	V-V	0.3	3.38	0.00027
Na-VS	V-O	2.1	1.65	-0.00363
	V-O	2.0	1.84	-0.00333
Rb-VS	V-O	2.0	1.65	-0.00522
	V-O	2.0	1.86	-0.00459
	V-Rb	0.8	3.30	0.00937

[a]C.N.: coordination numbers.
[b]R: interatomic distances.
[c]σ^2: Debye-Waller factors.

kinds of V-O bonds different in bond lengths, 1.60 and 1.79 Å. The shorter one possibly has a double bond character. The structural parameters indicate that the structure of vanadium species in VS is of a distorted tetrahedron, as suggested from the results of UV-vis spectrum, Raman spectrum, and XANES. The presence of neighboring vanadium atoms at 3.35 Å are also detected, and the coordination number is estimated to be 0.3. Although this value indicated that some vanadium species in VS were aggregated, Raman spectrum and UV-vis spectrum showed that the structure of them are mainly tetrahedral. Taking account of the results mentioned above, we can conclude that vanadium species in VS exists as a mononuclear species mainly.

As for Rb-VS, the curve-fitting analysis showed that the V-O bonds were two short bonds (1.65 Å) and two long bonds (1.85 Å), and rubidium atoms were present at 3.30 Å from vanadium atoms. The interatomic distance between a vanadium atom and a rubidium atom shows that the rubidium atom is present adjacent to vanadium atoms, and V-O-Rb bonds are possibly formed by addition of rubidium ions. The shorter bonds in alkali-ion-modified VS are longer than the V=O bond in VS, suggesting that the interaction between the vanadium oxide species and rubidium ion weakens the double bond character of V=O. The longer ones in alkali-ion-modified VS are also elongated compared with those of VS. The structural parameters of V-O bond lengths for Na-VS are almost the same as those for Rb-VS, indicating that the structure of the vanadium species in alkali-ion-modified VS samples is similar to each other despite the kinds of added alkali ions. On the basis of these results, we assume that the structure of the vanadium species in alkali-ion-modified VS is of a VO_4 tetrahedron as depicted in Figure 6. Malet *et al.* [25] and Courcot *et al.* [26] reported the effect of alkali ion addition to V_2O_5/TiO_2. In these reports, they concluded that the presence of alkali ion did not change the type of VO_x at a low alkali ion loading, and VO_x located beside added alkali ions. Irusta *et al.* [27] carried out characterization of surface species on V_2O_5/SiO_2 and Na^+-added V_2O_5/SiO_2 by Raman spectroscopy and they concluded that sodium ion addition to V_2O_5/SiO_2 caused the formation of non-distorted orthovanadate-like tetrahedra with very low or no vanadyl character and this vanadate species possessed four almost equivalent V-O bonds. Our conclusion is consistent with these reports.

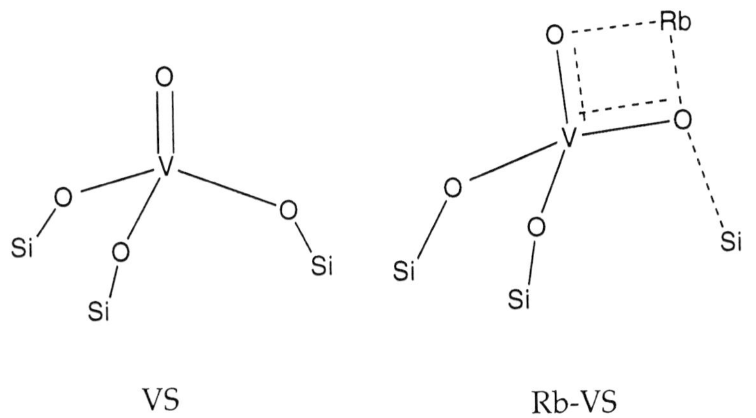

Figure 6. Proposed structures of vanadium species in VS and alkali-ion-modified VS samples.

The proposed structures depicted in Figure 6 is supported by *ab initio* molecular orbital calculations at Hartree-Fock level with GAUSSIAN 94 program [13]. The model clusters used in the calculations and the optimized structural parameters are shown in Figure 7. Terminal Hs are located to represent Si atoms adjacent to the cluster. The calculated values of V-O and V=O bond lengths in the $VO(OH)_3$ model for VS are in good agreement with the values estimated by EXAFS analysis mentioned above. The optimized structural parameters are compatible with our previous reports [28]. This result supports that the proposed structure of vanadium species in VS is reasonable.

As for the model for Rb-VS, the structure was optimized as shown in Figure 7. The calculated values of the model for Rb-VS also coincided with the values estimated by EXAFS analysis mentioned above. A rubidium atom was stabilized at the position shown in Figure 7, and interacted with two oxygen atoms of V-O directly. By the interaction of a rubidium atom, V-O_1 and V-O_2 bonds of the

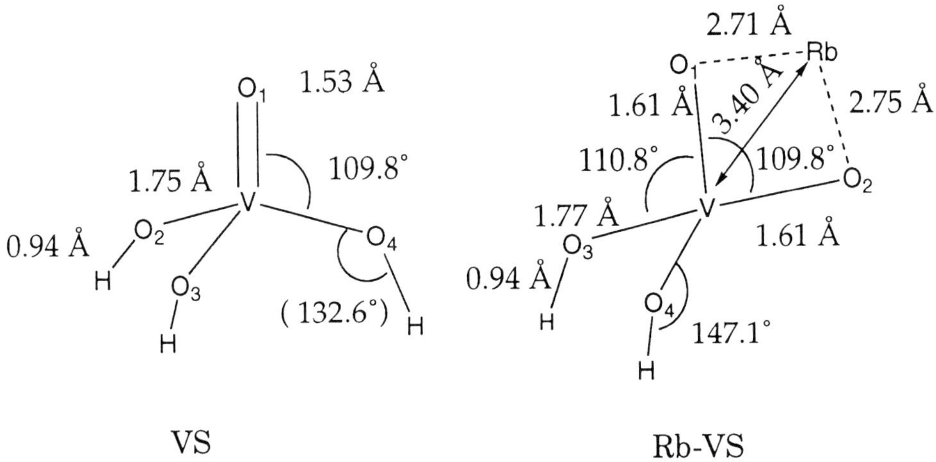

Figure 7. Model clusters for vanadium species in VS and Rb-VS. Optimized bond lengths and bond angles are also shown.

model for Rb-VS become nearly equivalent, and the bond lengths of V-O_3 and V-O_4 are also the same.

VI. ELECTRONIC STATES OF ACTIVE SPECIES

The most interesting point of the effects of modification by alkali ions is that the modification caused the vanadium species to be active for photooxidation of light alkanes under irradiation by visible light. This should be due to a change in the electronic state. To elucidate the electronic states, we carried out a theoretical study by a density functional theory (DFT) method with the program GAUSSIAN-94 BP86 option. It was found that the calculation for the model cluster of VS by the DFT method gave optimized structures more consistent with the experimental results than the Hartree-Fock calculation described in section V [15]. In the following calculations, a potassium-ion-modified cluster model (model K-VS) is adopted by replacing the rubidium atom to a potassium atom for saving computing time.

Figure 8 shows the energy levels as well as the main bonding components of important MOs for model VS and model K-VS in the ground state. In the case of model VS, the highest occupied MOs (HOMO, ϕ28 and ϕ29) are degenerate and composed of O 2p non-bonding orbitals in V=O mainly. The lowest unoccupied MOs (LUMO, ϕ30 and ϕ31) are constituted of the V 3d orbital parallel to the basal plane of the model VS and ϕ30 and ϕ31 are degenerate MOs. ϕ32 and ϕ33, which

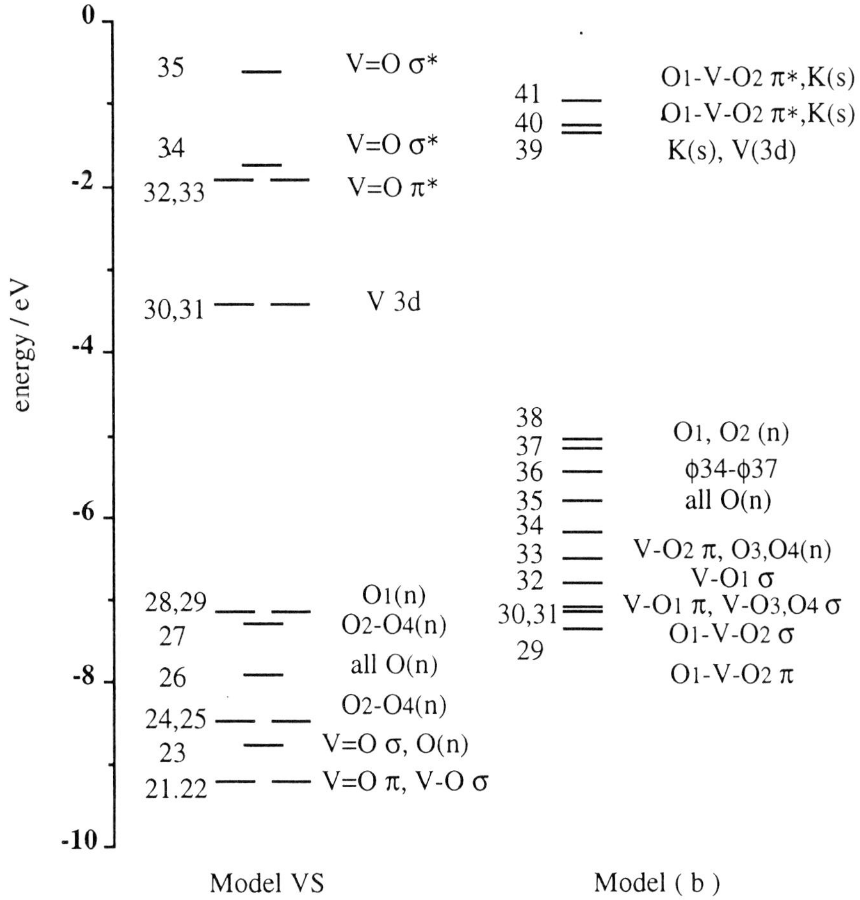

Figure 8. The energy and bonding nature of the higher-lying MOs for the model VS and K-VS in the singlet state (the ground states).

are also degenerate MOs, are composed of V=O π*-like orbitals. The V-O_1 bond, which is the shortest of four V-O bonds in lengths, is described by the σ-like bonding MO (ϕ23) and π-like bonding MOs (ϕ21 and ϕ22), while the other V-O bonds are done by only the σ-like bonding MOs (ϕ19-ϕ22). This result indicates that the V-O_1 bond has double bond character.

On the other hand, in the case of model K-VS, HOMO (ϕ38) is made up of non-bonding orbitals of O_1 and O_2 and is destabilized compared with that of VS. LUMO (ϕ39) is constituted of 4s AO of the potassium atom and 3d AO of the vanadium atom. ϕ40 and ϕ41 consist of π*-like MOs of O_1-V-O_2 mainly, and contain the small fraction of 4s AO of the potassium atom.

The energy levels of important MOs for both models in the excited state (the lowest triplet state) are shown in Figure 9. In both models, the order of MOs in the

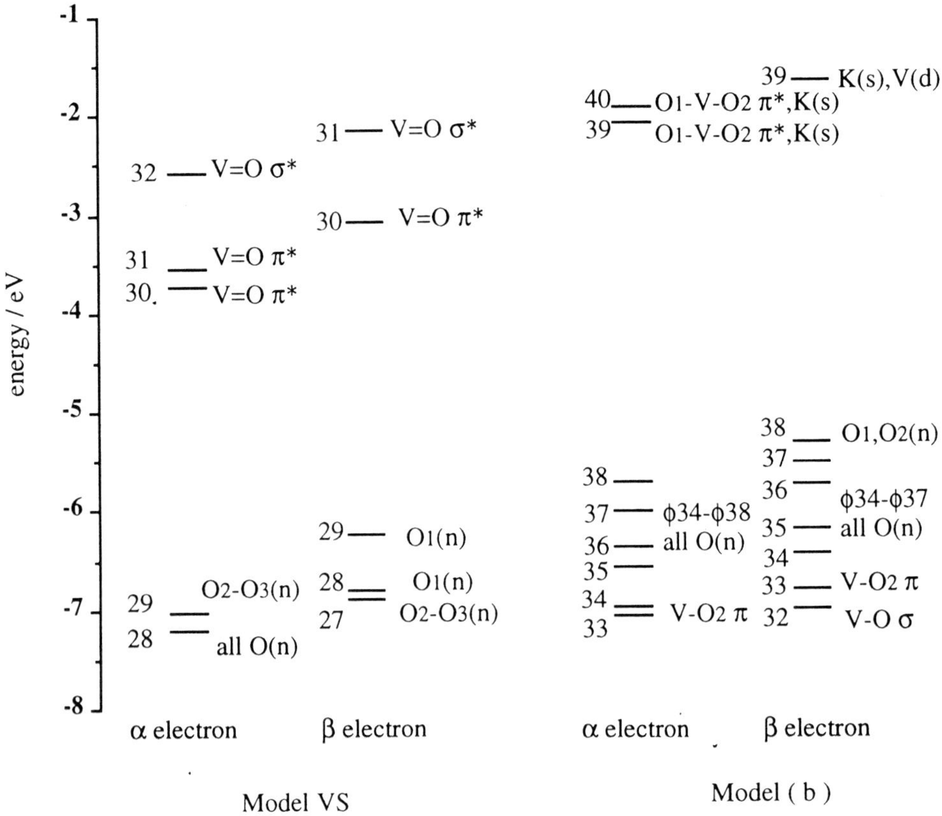

Figure 9. The energy and bonding nature of the higher-lying MOs for the model VS and K-VS in the triplet state (the excited states).

triplet states is not very much different from that of MOs in the ground states. Because the transition from the ground state to the triplet state is assignable to the one-electron excitation, in the triplet state an occupied orbital and an unoccupied orbital are formed in the orbitals of α-electrons and of β-electrons, respectively. In the model VS, the occupied and unoccupied orbitals formed in the triplet state are the anti-bonding π-like orbital of $V{=}O_1$ (α electron ϕ30) and the non-bonding orbital of oxygen in $V{=}O_1$ (β electron ϕ29), respectively, indicating that the excitation from the ground state to the triplet state is assigned to one-electron transition from ϕ28-29 to ϕ32-33 in Figure 8. On the other hand, in the model K-VS, the occupied and unoccupied orbitals formed in the triplet state are the anti-bonding π-like orbital of O_1-V-O_2 (α electron ϕ39) and the non-bonding orbitals of O_1 and O_2 atoms (β electron ϕ38), respectively. The excitation is assignable to the transition from ϕ38 to ϕ40 and/or ϕ41 in Figure 8. It is clear that the transition energy for model K-VS is lower than that for model VS. This explains the excitation of K-VS by the light of a long wavelength qualitatively.

VII. INTERMEDIATES

The study of the reaction mechanism is not yet completed but some information about the intermediates is already available [14]. In this section, we will describe the intermediates.

No adsorption of propane and oxygen was observed in the dark but photoadsorption of propane occurred on the irradiated catalysts. The photoadsorption was confirmed by quenching of photoluminescent emission from the catalysts in contact with propane. The propane was not desorbed by evacuation at room temperature. Oxygen was not photoadsorbed on the irradiated bare catalysts. However, the photoadsorption of oxygen was observed on the catalysts pre-adsorbing propane and the adsorption rate is expressed by the equation;

$$rate = k[C_3H_8(ads)]p$$

where $[C_3H_8(ads)]$ is the concentration of adsorbed propane and p is the pressure of oxygen. This is consistent with the formation of a complex comprised of one propane molecule and one oxygen molecule as an intermediate. However, this does not imply that the oxygen species incorporated in oxygenated products is the adsorbed oxygen. Thus, photooxidation of propane by $^{18}O_2$ resulted in a gradual increase in ^{18}O concentration in formed propanone and propanal with reaction time. No oxygen exchange reaction with the lattice oxygen was confirmed, so that the result indicates the lattice oxygen being incorporated at first. The same conclusion was deduced for photooxidation of 2-methylpropane to propanone, however, the ^{18}O content in 2-methylpropan-2-ol was already high (85 %) in the first stage of the reaction and kept almost constant, indicating direct incorporation of adsorbed oxygen to the alcohol. The result shows that formation of alcohol and aldehydes proceeds *via* different paths.

In situ FT-IR spectroscopy was applied to detect intermediates formed in the photooxidation of propane over Rb-VS irradiated by the light of $\lambda > 390$ nm. The spectra shown in Figure 10 were recorded with the following procedure. At first, Rb-VS was irradiated by the light of $\lambda > 390$ nm in the presence of gaseous propane for 30 min at room temperature, and evacuated at room temperature in the dark for 30 min (Figure 10 (a)). Subsequently, gaseous oxygen was introduced into the *in situ* cell at room temperature and the Rb-VS was irradiated by the light of $\lambda > 390$ nm at room temperature. During the irradiation, the IR spectrum was recorded at 3, 6, 13 and 30 min irradiation (Figure 10 (b) - (e)). Finally, after 40

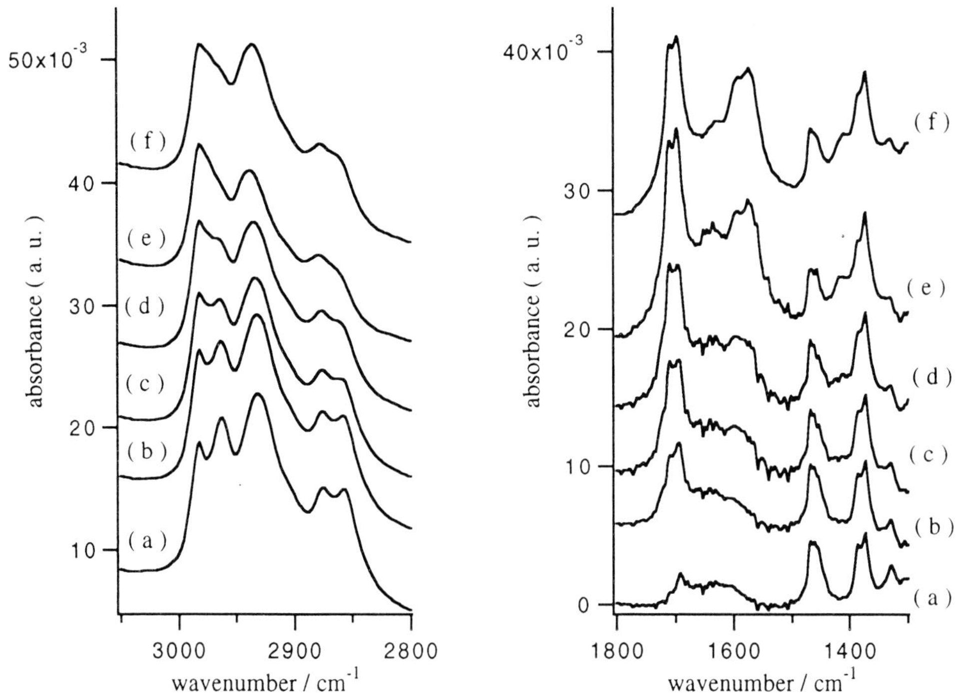

Figure 10. In situ FT-IR spectra of the adsorbed species arising from propane photoadsorption (a), and subsequent oxygen photoadsorption on Rb-VS (b) - (f). Photoirradiation time; (b) 3 min, (c) 6 min, (d) 13 min, (e) 30 min, (f) 40 min and evacuation in the dark.

min irradiation, the cell was evacuated in the dark at room temperature for 10 min and the spectrum was recorded (Figure 10 (f)). In spectrum (a), five bands due to C-H stretching vibrations were observed at 2980, 2965, 2936, 2876, and 2855 cm^{-1}. Other bands in the region of C-H and C-C deformations were also found at 1460, 1387, 1372 and 1328 cm^{-1}. The spectrum indicates that propane molecules are photoadsorbed on Rb-VS. The adsorbed species are expected to have strong chemical interaction with Rb-VS, since the adsorbed species remain on Rb-VS after evacuation treatment for 30 min at room temperature. It is likely that propane is attached to the vanadium species in Rb-VS with the cleavage of the C-H bond of propane. By contact with oxygen, new bands appeared at 1700, 1712, 1570 and 1410-20 cm^{-1}, and the intensity of the bands at 1460, and 1328 cm^{-1} due to the adsorbed species arising from propane photoadsorption diminished slightly. In the region of ν(C-H), the intensities of the bands except for that at 2980 cm^{-1} became low.

The bands 2965, 2936, 2876, 1460, 1387, 1372 and 1328 cm^{-1} in spectrum (a) are assignable to an isopropoxide group attached to vanadium atom by comparison with the IR spectrum of $VO(O\text{-}CH(CH_3)_2)_3$ [29]. As for newly appearing bands by contact with oxygen, the bands at 1712, 1700 and 1420 cm^{-1} are assignable to adsorbed propanone, while those at 1570 and 1410 cm^{-1} are assignable to acetate by comparison with the spectra of adsorbed propanone and acetic acid on Rb-VS. The bands due to adsorbed propanone appeared immediately by contact with gaseous oxygen and the bands became stronger with irradiation time. On the other hand, the bands due to acetate could not be seen at the initial stage of photoirradiation and grew with the irradiation time. The results mentioned above indicate that an isopropoxide-type intermediate is, at first, formed by photoadsorption of propane on Rb-VS and then transformed to propanone in the presence of gaseous oxygen under irradiation by visible light, while a part of propanone is converted to acetate by further oxidation.

VIII. CONCLUSION

Photooxidation of hydrocarbons has been investigated intensively from a view point of preservation of the environment, where the high activity to total oxidation is required. However, from the view point of resource and energy problems, selective photooxidation of light alkanes will be a very important target in the next century. So far, the effective photocatalysts had not been reported for partial oxidation of light alkanes. In this review paper, we described our recent results on the photooxidation of propane and 2-methylpropane over alkali-ion-modified vanadium oxide supported on silica. The most important result is that the modification by alkali ions generates a photoactive species under irradiation by visible light. The selectivity to partial oxidation is high. The high selectivity could be associated with the basicity of the active species. The alkali ion interacts directly with distorted tetrahedral VO_4 species. The interaction does not change the tetrahedral structure but deeply affects the bonding nature of V-O bonds and reduces the excitation energy for transition from the ground state to the active triplet state. The reaction is initiated by photoadsorption of alkanes, forming an isopropoxide-type intermediate, followed by photoadsorption of oxygen to form a 1:1 complex which is then oxidized to propanone.

Acknowledgements

The studies presented in this article are partially supported by Grant-in-Aids' from the Ministry of Education, Science, Sports and Culture of Japan (No. 06239237 and 08405052). Computer time was provided by the Supercomputer Laboratory, Institute for Chemical Research, Kyoto University. The authors thank Professor H. Kobayashi of Kurashiki university of science and the arts for his valuable discussion. The X-ray absorption experiment was performed under the approval of the Photon Factory Program Advisory Committee (Proposal No. 93G-170). S. T. acknowledges support by the Fellowship of JSPS for Japanese Junior Scientists.

REFERENCES

1. J.R.H. Ross, R.H.H. Smits, and K. Seshan, *Catal. Today*. **16**, 503 (1993); M.C. Kung and H.H. Kung, *J. Catal.,* **134**, 668 (1992).
2. G.L. Price and V. Kanazirev, *J. Catal.,* **126**, 267 (1990); G. Giannetto, A. Montes, N.S. Gnep, A. Florentino, P. Cartraud, and M. Guisnet, *J. Catal.,* **145**, 86 (1993).
3. D. Patel, P.J. Andersen, and H.H. Kung, *J. Catal.* **125**, 132 (1990) and references thererin.
4. N.H. Batis, H. Batis, A. Ghorbel, J.C. Vedrine, and J.C. Volta, *J. Catal.,* **128**, 248 (1991); H. Igarashi, K. Tsuji, T. Okuhara, and M. Misono, *J. Phys. Chem.,* **97**, 7065 (1993).
5. R.I. Bickley. In: *Catalysis, Specialist Periodical Report*, G.C. Bond, and G. Webb (Eds.), The Chemical Society, London, Vol. 5, 1982, p. 308; S. Yoshida, Y. Nishimura, T. Tanaka, H. Kanai, and T. Funabiki, *Catal. Today.* **8**, 67 (1990).
6. N. Djeghri and S.J. Teichner, *J. Catal.* **62**, 99 (1980); K. Marcinkowska, S. L. Kaliaguine, and P.C. Roberge, *J. Catal.* **90**, 49 (1984).
7. T. Tanaka, M. Ooe, T. Funabiki, and S. Yoshida, *J. Chem. Soc., Faraday Trans. 1,* **82**, 35 (1986); M. Anpo and Y. Kubokawa, *Rev. Chem. Interm.*, **8**, 105 (1987) and references therein.
8. T. Tanaka, Y. Nishimura, S. Kawasaki, M. Ooe, T. Funabiki, and S. Yoshida, *J. Catal.,* **118**, 327 (1989).
9. T. Tanaka, S. Takenaka, T. Funabiki, and S. Yoshida, *Chem. Lett.*, 1585 (1994).
10. S. Takenaka, T. Kuriyama, T. Tanaka, T. Funabiki, and S. Yoshida, *J. Catal.*, **155**, 196 (1995).
11. T. Tanaka, S. Takenaka, T. Funabiki, and S. Yoshida, *J. Chem. Soc., Faraday Trans.*, **92**, 1975 (1996).
12. S. Takenaka, T. Tanaka, T. Funabiki, and S. Yoshida, *Catal. Lett.*, **44**, 67 (1997).
13. S. Takenaka, T. Tanaka, T. Yamazaki, T. Funabiki, and S. Yoshida, *J. Phys. Chem. B, in press.*
14. S. Takenaka, T. Tanaka, T. Funabiki, and S. Yoshida, *J. Chem. Soc., Faraday Trans.*, *in press.*
15. S. Takenaka, T. Tanaka, T. Funabiki, and S. Yoshida, *J. Phys. Chem. B, submitted.*
16. S. Yoshida, T. Matsuzaki, T. Kashiwazaki, K. Mori, and K. Tarama, *Bull. Chem. Soc. Jpn.,* **47**, 1564 (1974).
17. S. Yoshida, Y. Matsumura, S. Noda, and T. Funabiki, *J. Chem. Soc., Faraday Trans. 1,* **77**, 2237 (1981).
18. U. Scharf, M. Schrami-Marth, A. Wokaun, A. Baiker, *J. Chem. Soc., Faraday Trans.*, **87**, 3299 (1991).
19. A.M. Gritscov, V.A. Shvets, V.B. Kazansky, *Chem. Phys. Lett.,* **35**, 511 (1975).
20. A. M. Gritscov, V. A. Shvets, and V. B. Kazansky, *Dokl. Akad. Nauk SSSR*, **221**, 868 (1975).
21. S. Yoshida, T. Tanaka, Y. Nishimura, H. Mizutani, and T. Funabiki, *Proc. 9th Intern. Congr. Catal.*, M.J. Phillips and M. Ternan (Eds.), The Chemical Institute of Canada, 1988, p. 1473.
22. S.T. Oyama, G.T. Went, K.B. Lewis, A.T. Bell, and G.A. Somorjai, *J. Phys. Chem.*, **93**, 6786 (1989); N. Das, H. Eckert, H. Hu, I.E. Wachs, J.F. Walzer, and F.J. Feher, *J. Phys. Chem.*, **97**, 8240 (1993).
23. G. Deo and I. E. Wachs, *J. Catal.,* **146**, 335 (1994).
24. T. Tanaka, H. Yamashita, R. Tsuchitani, T. Funabiki, and S. Yoshida, *J. Chem. Soc., Faraday Trans. 1*, **84**, 2987 (1988).
25. P. Malet, A. Muñoz-Páez, C. Martíne, and V. Rives, *J. Catal.*, **134**, 47 (1992).

26. D. Courcot, B. Grzybowska, Y. Barbaux, M. Rigole, A. Ponchel, and M. Guelton, *J. Chem. Soc., Faraday Trans.*, **92**, 1609 (1996).
27. S. Irusta, A.J. Marchi, E.A. Lombardo, and E.E. Miró, *Catal. Lett.*, **40**, 9 (1996).
28. H. Kobayashi, M. Yamaguchi, T. Tanaka, Y. Nishimura, H. Kawakami, and S. Yoshida, *J. Phys. Chem.*, **92**, 2516 (1988).
29. M. Schraml-Marth, A. Wokaun, and A. Baiker, *J. Catal.*, **124**, 86 (1990).

Res. Chem. Intermed., Vol. 24, No. 3, pp. 329-336 (1998)

SELECTIVE OXIDATION OF PROPYLENE OVER GOLD DEPOSITED ON TITANIUM-BASED OXIDES

M. HARUTA[1], B.S. UPHADE[1], AND S. TSUBOTA[1], and A. MIYAMOTO[2]
[1]*Osaka National Research Institute, AIST, Midorigaoka, Ikeda 563, Osaka, Japan*
[2]*Department of Materials Chemistry, Graduate School of Engineering, Tohoku University, Aoba-ku, Aramaki, Sendai 980-77, Japan*

Received 17 October 1997; accepted 24 December 1997

Abstract--Gold supported on titanium-based metal oxides can assist the selective partial oxidation of propylene at temperatures from 313 K to 573 K in a gas containing both H_2 and O_2. The preparation method was found to be crucial in controlling the selectivities. In general, impregnation and chemical vapor deposition methods do not produce selective catalysts. Only the deposition-precipitation method makes gold selective to propylene oxide or propanal, suggesting that a strong contact between the gold particles and the titanium ion sites on the support is important. The effect of changing the support was also dramatic; the use of the anatase form of TiO_2 and Ti-MCM-41 results in propylene oxide production, while the rutile structure of TiO_2 caused complete oxidation to CO_2. Microporous crystalline titanium silicates such as TS-1, TS-2, and Ti-β zeolite make gold relatively selective to propanal and of the three TS-1 gives the highest selectivity. These results indicate that the oxidation of propylene in the copresence of H_2 must involve the surface of the supports and that the reaction takes place at the interface perimeter around the gold particles.

INTRODUCTION

The selective oxidation of hydrocarbons is an important chemical technology for the conversion of oil- and natural gas-based feedstocks to high value products such as oxygenates [1], which are used as key intermediates for polymer and organic syntheses. For economic and environmental reasons, most of the current processes for oxygenates production require further improvements. A mainstream strategy currently being considered is to develop a direct gas-phase one-pass reaction using molecular oxygen as the primary oxidant. In the case of gas-phase propylene oxidation, acrolein is presently the only product that can be industrially produced by gas phase oxidation with O_2 and occurs over the Bi_2MoO_6 catalysts [1]. Other products, such as propylene oxide [2], propionaldehyde (propanal), and acetone as yet cannot be selectively produced. We have however recently reported that gold supported on TiO_2 and Ti-MCM-41 can selectively produce propylene oxide in the copresence of oxygen and hydrogen [3,4]. Although the conversion of propylene remains very low, 1-2%, and hydrogen consumption exceeds propylene conversion by three times to stoichiometric values, the selectivity exceeds 90%. In an attempt to overcome the problems of the low yield and high H_2 consumption we therefore have examined the influence of the preparation methods for the gold catalysts, and also the support effect on the product selectivities. This paper is a continuation of our work on gold catalyst to search for new routes to produce oxygenates.

EXPERIMENTAL

Catalyst Preparation

Supported gold catalysts were prepared by deposition-precipitation (DP) [5], impregnation (IMP), and CVD [6] methods. The support materials used were TiO_2 (Nippon Aerosil, P-25, specific surface area; 45 m^2/g, mainly composed of anatase), two types of anatase TiO_2 (specific surface area; 279 m^2/g and 57 m^2/g), two types of rutile TiO_2 (specific surface area; 39 m^2/g and 40 m^2/g), amorphous TiO_2 (specific surface area; 110 m^2/g), Ti-MCM-41 (Ti/Si = 3/100), titanium silicalites (TS-1 and TS-2), and Ti-β zeolite. The precursors were calcined in air at 673 K for 3 hours to deposit fine gold metal particles. In the case of DP methods, actual gold loadings of Au/TiO_2 samples obtained were smaller than the gold contents in the starting solutions and are expressed as wt% relative to the support.

The particle size of gold was determined by TEM (Hitachi H-9000NA). At least 200 particles were chosen to determine the mean diameter by using a computerized image analyzer (Nippon Avionics C. Ltd., EXCEL).

Catalytic Tests

Catalytic performance tests were made under atmospheric pressure in a quartz reactor using a continuous-flow system at a temperature range of between 273 K and 623 K. A mixture of propylene, oxygen, hydrogen, and argon as diluent was passed through a fixed bed containing the catalyst samples. Samples were sieved between 125 and 212 μm of particles size and in most cases weighed to 0.5 g. The temperature of the catalyst bed was measured by using a thermocouple placed in a glass sleeve in such a way that the tip of the thermocouple just touches the catalyst bed. The flow rate was 4,000 ml/hr unless otherwise stated. All reaction products were identified by standard GC-Mass and NMR spectroscopic techniques and analyzed by two on-line gas chromatograms, using three types of columns for thermal conductivity detectors (TCD). Porapak Q (3 m) was used for CO_2, propylene, and propane detection, molecular sieve 5A (3 m) for CO, and activated carbon (3 m) for H_2 and O_2. A HR-20M (0.53 mm x 50 m, Shinwa Chemical Industries, Co Ltd.) capillary column was used for organic compounds with a flame ionization detector (FID).

RESULTS AND DISCUSSION

Table 1 shows that among the three preparation methods only the DP method produces gold which is selective for the production of propylene oxide or propanal. In the case of Au/TiO_2, the DP catalyst gives propylene oxide with a selectivity above 99% while impregnation gives mainly CO_2 and as a major byproduct propane. This difference in product selectivity can be related to the state of dispersion and the contact structure of gold particles on the surface of TiO_2. On the DP catalyst gold particles of diameters from 3~5 nm are homogeneously dispersed with a standard deviation of around 30%. They are hemispherical in shape and are attached to the support at their flat planes, usually at the [111] plane [7]. This contact structure provides the longest distance around the perimeter interface of gold particles. In the IMP catalyst, spherical gold particles are simply loaded on the TiO_2 support resulting in a very short perimeter distance. Their size distribution is also very wide.

Although the DP catalyst has a lower Au loading (actually around 0.3 wt% with respect to a nominal loading of 1.0 wt%) and is used at a lower temperature than the IMP catalyst, the conversion of C_3H_6 is much higher over the DP catalyst. On the other hand, the conversion of hydrogen is higher over the IMP sample, indicating that hydrogen consumption to form water takes place more preferentially. Since hydrogen oxidation takes place over the gold surface and the rate of reaction is almost independent of the type of metal oxide supports and of the metal-support interaction [8], the above results suggest that the perimeter sites around gold particles at the TiO_2 surface play an important role for selective oxidation and that the high selectivity over the DP catalyst arises from the homogeneity in size of the gold particles. The formation of propane over the IMP catalyst indicates that small gold clusters of less than 2.0 nm in size [9] are also present in addition to large gold particles with diameters above 2.0 nm.

In the case of TS-1, a DP catalyst gives propanal as a main product with a selectivity above 70% and CO_2 as the major byproduct. The CVD catalyst produces CO_2 as a main product and propanal and acetone as byproducts. The DP method is usually selective in depositing gold on the metal oxide supports and because of this, on TS-1, gold is deposited only on the Ti cation sites. $Au(OH)_3$ precipitate cannot be deposited on SiO_2 surface because the point of zero charge (PZC) is at about pH 2. In contrast, the CVD method is not selective in the deposition of gold and may deposit gold nanoparticles even on the surface of SiO_2 [10]. Taking into consideration the similarities in product distribution between the two CVD samples, Au/TS-1 and Au/SiO_2, it is likely that the higher selectivity to propanal over Au/TS-1 prepared by DP can therefore be ascribed to gold particles deposited on the Ti cation of TS-1.

It should be noted that the Au/TS-1 catalyst needs a higher temperature for the oxidation of propylene and produces propanal at temperatures above 573 K. The

Table 1
Influence of preparation method on the product selectivity of gold catalysts in the reaction of propylene with oxygen and hydrogen.

Au loading wt%	Support	Prepar. Method	React. Temp. K	Conversion %		Product Selectivity %						
				C_3H_6	H_2	PO	AL	PA	AT	AA	CO_2	C_3H_8
1[a]	TiO_2[c]	DP	323	1.1	3.2	>99	0	0	0	0	0	0
1	TiO_2[c]	IMP	353	0.2	8.9	0	-	-	0	-	>70	<10
8[b]	TS-1	DP	473	0.6	21	4	-	74	0	4	12	0
1	TS-1	CVD	473	0.9	96	0	0	31	15	7	40	0
1	SiO_2	CVD	523	0.2	98	0	28	5	17	0	50	0

[a and b]Au loading in starting solution, actual loading is about 0.3 wt% and 1.0 wt%, respectively.
[b]TiO_2, P-25, mainly composed of anatase. PO: propylene oxide, Al: acrolein, PA: propanal, AT: acetone, AA: acetaldehyde.
[c]Ref. [3].

conversion to propanal may be either caused by a reaction involving propylene oxide as an intermediate, or the hydrogenation of acrolein.

Table 2 shows how the product selectivity of gold catalysts prepared by DP changes depending on the types of support. Both anatase samples with different surface areas (279 m^2/g and 57 m^2/g) provide excellent selectivities to propylene oxide at above 95%, while the two rutile samples are not selective at all and cause complete oxidation to CO_2. This surprising difference can be correlated to the atomic configuration of TiO_2 crystals: anatase has Ti cations relatively isolated from each other whereas rutile has Ti cations more densely packed. Figure 1a and 1b show two typical pictures for a gold cluster deposited on the [100] plane of anatase and the [110] plane of rutile, on which gold can be deposited with epitaxial contact.

Amorphous TiO_2 normally produces CO_2, acetone, and propylene oxide. Since amorphous TiO_2 is partly transformed into anatase and rutile when it is calcined in air at 400 °C, after gold hydroxide is deposited by DP method, the product selectivities appear to result from a combination of anatase and rutile. Probably because the structure of rutile is not fully crystallized, the oxidation of acetone to CO_2 is however more depressed than on well crystallized rutile.

The group of titanium silicates also provides an interesting series of supports. As previously reported by us [4], Ti-MCM-41 gives propylene oxide as the main product and that, TS-1, TS-2, and Ti-β produce, at much higher temperatures, propanal, acetic acid and in addition CO_2. TS-1 was the most selective support for propanal.

Ti-MCM-41 consists of amorphous silica with regular pore structure in which Ti^{4+} is present as isolated tetrahedrally coordinated ions, partly at the surface of the wall, and partly embedded in the walls by combining with four O-Si units [10]. The Ti^{4+} species can be regarded as equivalent in nature to those contained in TiO_2 on SiO_2 gel. Thus, since TiO_2 deposited on SiO_2 gel by using titanylacetylacetonate works as an effective support for gold to produce propylene oxide [3], it is reasonable that Ti-MCM-41 support also leads to the production of propylene oxide.

Microporous crystalline titanium silicalites, TS-1 and TS-2, have small pores with a diameter of 5.5 Å and tetrahedrally coordinated Ti^{4+} cations occupying framework positions at random [10]. The Ti-β zeolite is similar to TS-1 and TS-2 except for the larger pore diameter (7 Å) and the presence of aluminium which imparts strong protonic acidity [10] and therefore shows similar catalytic behaviour. However, the use of supports with small pore sizes is not favourable for depositing gold hydroxide precipitate from neutralized aqueous $HAuCl_4$ solutions. The resulting very low loadings of Au in TS-1 and TS-2 was confirmed by TEM observation, and explains why the reaction apparently takes place at much higher temperatures than TiO_2 and Ti-MCM-41, producing mainly propanal. The lower conversion of propylene over Au/Ti-β zeolite can be explained by the lower actual

Table 2
Support effect on the product selectivity of gold catalysts in the reaction of propylene with oxygen and hydrogen.

Au loading wt%	Support	Surface Area m^2/g	Reaction Temp. K	Conversion %		Selectivity %						
				C_3H_6	H_2	PO	AL	PA	AT	AA	CO_2	C_3H_8
1.0	anatase	279	313	0.17	6.2	99	0	0	1	0	0	0
11.4	anatase	279	313	1.4	13	97	0	0	1.5	0	1.5	0
1.1	anatase	57	313	0.26	3.0	98	0	0	2.0	0	0	0
1.0	rutile	39	353	0.095	16	0	0	0	3.0	0	97	0
1.0	rutile	40	423	1.3	30	0	0	0	2.3	1.6	96	0
10.3	rutile	40	353	0.13	47	0	0.5	0	7.5	0	92	0
10.3	amorphous	113	313	0.57	21	23	0	0	38.5	0	38.5	0
8.0	Ti-MCM[a]	1096	373	1.8	36	95	0	0	0	0	5	0
8.0	TS-1	469	523	1.8	38	1	0	69	0	8	13	0
8.0	TS-2	333	523	1.6	43	0	0	31	1	13	44	0
8.0	Ti-β	518	523	0.67	17	0	0	54	0	12	25	0

[a]Ref. [4].

(a)

(b)

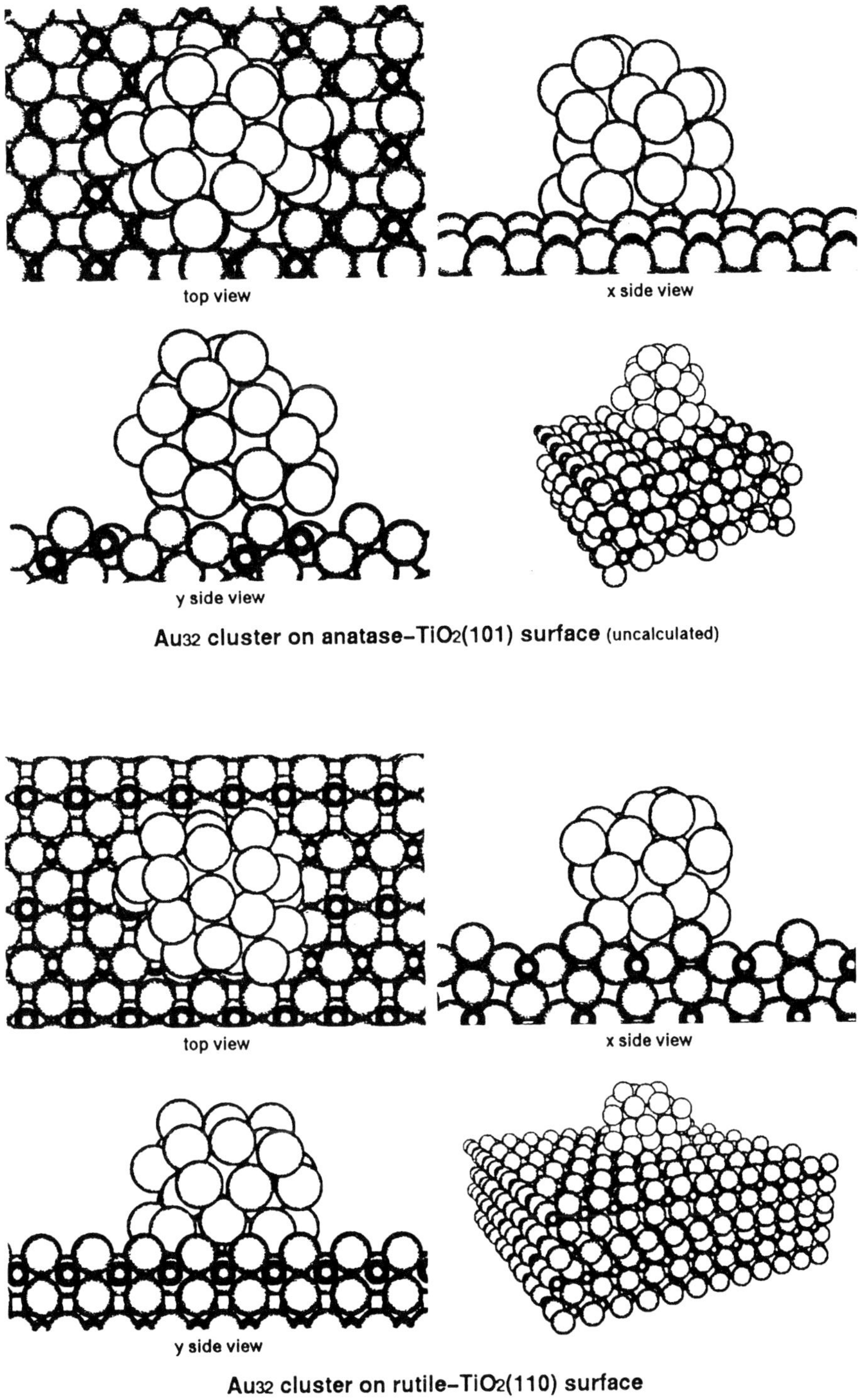

Figure 1. Computer graphics for a Au cluster of 32 atoms supported on the surface of TiO_2. a: [100] plane of anatase, b: [110] plane of rutile.

gold loading because of the strong acidity of the support.

CONCLUSION

In the gas-phase reaction of propylene with O_2 and H_2 over supported gold catalysts at temperatures of 313 K to 573 K, the product selectivity markedly depends on the catalyst preparation and the types of titanium dioxide and titanium-containing microporous and mesoporous silicates used.

1) Only the deposition-precipitation method brings about selective gold catalysts and impregnation and CVD methods result in poorly selective catalysts which produce CO_2 as a main product.

2) Propylene oxide is produced with selectivities higher than 90% when anatase TiO_2 and Ti-MCM are used as the support.

3) Propanal is produced with selectivities from 30 to 70% when TS-1, TS-2, and Ti-β zeolite are used as the support.

Acknowledgements
The authors thank Mr. H. Tsuchiya for his contribution to part of the experimental work. Thanks are also extended to Mr. T. Hayashi of Nippon Shokubai Co. Ltd. for supplying TS-1, TS-2, and Ti-β materials.

REFERENCES

1. R.A. Sheldon. In: *Catalytic Oxidation: Principles and Applications*, R.A. Seheldon and R.W. van Santen (Eds.), World Sci., Singapore, 1995, p. 1.
2. U.S. Pat. 5,241,088 (1993) and Eur. Pat. 0640598A1 (1994).
3. T. Hayashi, K. Tanaka, and M. Haruta, *J. Catal.*, accepted for publication.
4. Y.A. Kalvachev, T. Hayashi, S. Tsubota, and M. Haruta. In: *Stud. Surf. Sci. Catal.*, R.K. Grasselli, S.T. Oyama, A.M. Gaffney, and L.E. Lyons (Eds.), Vol. 110, Elsevier, Amsterdam, 1997, p. 965.
5. S. Tsubota, D.A.H. Cunningham, Y. Bando, and M. Haruta. In: *Preparation of Catalysts VI*, G. Poncelet et al. (Eds.), Elsevier, Amsterdam, 1995, p. 229.
6. M. Okumura, K. Tanaka, A. Ueda, and M. Haruta, *Solid State Ionics* **95** 143 (1997).
7. M. Haruta, S. Tsubota, T. Kobayashi, H. Kageyama, M.J. Genet, and B. Delmon, *J. Catal.* **144**, 175 (1993).
8. M. Haruta, N. Yamada, T. Kobayashi, and S. Iijima, *J. Catal.* **115**, 301 (1989).
9. K. Tanaka, T. Hayashi, and M. Haruta, *J. Japan Inst. Metals* **60**, 693 (1996).
10. M. Okumura, S. Tsubota, T. Nakamura, and M. Haruta, *Catal. Lett.*, accepted for publication.
11. B. Natari. In: *Advances in Catalysis*, D.D. Eley, H. Pines, and P.B. Weisz (Eds.), Vol 41, Academic Press, New York, 1996, p. 253.

oxide is obtained with 93 % selectivity at 94 % H_2O_2-conversion. This synthetic route is environmentally friendly and offers a genuine alternative to the technical epoxidation process presently in operation such as the chlorohydrin (about 2 tons $CaCl_2$ per ton PO and chlorinated hydrocarbons are inevitable by-products) and hydroperoxide methods via oxygen. Enichem is developing right now this process in a pilot plant.

These examples demonstrate that oxidation reactions in the presence of hydrogen peroxide as oxidant are possible over zeolitic materials in the liquid phase in a batch reactor. Thus, in general zeolites can also be used efficiently in liquid phase using slurry technology. This is particularly advantageous if equilibrium reactions should be shifted to the desired products favoured at low temperatures.

Zeolites for Removal of Nitrogen Oxides

The role of catalysis in the so called "end of pipe"-strategy to avoid environmental pollution is well known. One of the most significant and severe problems is the removal of NO_x, the air contaminant of automotive lean burn and diesel engines, of stationary diesel engines and industrial boilers accompanied by a large excess of oxygen. The decomposition or reduction of NO is a major target to be achieved. M. Iwamoto *et al.* [38] found first that Cu-ZSM-5 is active for the decomposition of NO into N_2 and O_2. Among various metal ion-exchanged zeolites only the Cu- and Co-ZSM-5 exhibited stable activities. However this activity is strongly diminished in the presence of O_2 and SO_2.

A breakthrough in the NO-removal was then the selective reduction of NO with hydrocarbons in oxidizing atmosphere [39]. The efficiency of the reduction is remarkably enhanced by the presence of O_2. ZSM-5 based catalysts are influenced by SO_2. A research boom in the development of catalysts and various reduction systems for the so called SCR-HC technology could be observed, particularly in Japan. Among the Japanese groups T. Inui *et al.* have studied various metallosilicates prepared by the rapid crystallization [21] for the NO reduction in the presence of hydrocarbons and oxygen [21,40]. They proposed the microscopic sequential mechanism (MSR mechanism) based on the effect of the carbon numbers of the hydrocarbons on the SCR-HC method [41].

III. ZEOLITE CATALYSIS IN THE 90S

High research activities in the removal of NO_x maintain in the present decade. Also, there is still a strong continuation in the use of zeolite catalysts for the organic syntheses of intermediates. Thereby base zeolite catalysts find a revival. Furthermore, this field of zeolite catalysts gets more and more expanded into the area of fine chemicals. A key target is the enantioselective synthesis using zeolites as carrier for chiral homogeneous catalysts via various immobilization methods. Research activities in the application of multifunctional zeolitic catalysts for highly attractive but risky synthetic routes. Such catalysts enable several reaction steps to be combined in mutually optimized functions to a direct one step procedure with the idea to develop new, environmentally benign and highly economic processes. With respect to process technology, liquid phase reactions in slurry reactions gain more in importance. Those conversions can be carried out even at temperatures $\leq 0\,°C$.

Zeolites for the Synthesis of Intermediate and Fine Chemicals

As a starting material for the nylon-6 synthesis, ϵ-caprolactam is of high industrial importance. The classical route starts from the hydrogenation of benzene to cyclohexane which is subsequently oxidized to form a cyclohexanone / cyclohexanol mixture. This mixture is oxidized to cyclohexanone in the next step. After the reaction of anone with hydroxylamine in the presence of sulfuric acid, cyclohexanone oxime is formed and rearranged to ϵ-caprolactam. This Beckmann rearrangement is also carried out in sulfuric acid. Although very high yields are obtained, the large amount of ammonium sulfate due to the neutralization of the oleum used in the oximation as well as in the rearrangement is considered to be an environmentally unacceptable drawback. Recently a new route for the production of ϵ-caprolactam and adipic acid has been introduced. This route is also based on benzene as starting material but on the use of heterogeneous zeolite catalysts. Thereby, the selective partial hydrogenation of benzene over a Ru-catalyst to form cyclohexene and the succeeding highly selective hydration in the presence of a special H-ZSM-5 zeolite at 120 °C in the liquid phase (Asahi-Process, 60,000 t/a since 1990) are the first steps towards a new adipic acid and caprolactam process. In the case of caprolactam, after the oxidation of the cyclohexanol to cyclohexanone, the oximation to cyclohexanone oxime is carried out in the presence of ammonia and hydrogen peroxide as oxidant over TS-1 (Enichem-Process, 12,000 t/a demonstration plant). There are several studies on the rearrangement of the cyclohexanone oxime to ϵ-caprolactam over zeolitic catalysts. This new route provides a number of advantages compared to the conventional method and is an energetically and economically favorable as well as an environmentally friendly alternative [30,42]. However, the heterogeneously catalyzed rearrangement step is not commercialized so far, because of problems such as catalyst lifetime. The Beckmann rearrangement of cyclohexanone oxime to ϵ-caprolactam has been studied using a B-MFI zeolite. Optimized process parameters were found, i.e., at reduced overall pressure of 0.1 bar in addition to a temperature of 300 °C the highest selectivity of about 95 % was achieved at approx. 100 % conversion [30,42]. The insufficient catalyst service time necessitates regeneration of the B-MFI zeolite. It was shown that regenerating with air at an elevated temperature of about 500 °C leads to complete reactivation of the catalyst. Regeneration in a fixed bed, though, is not very ingenious. A suitable process of simultaneous reaction and regeneration, both inside a fluidized bed, should provide technical and economical feasibility for industrial use. In the case of using a fluidized bed of laboratory scale, selectivity and starting conversion showed good correspondence with the results obtained in the fixed bed reactor. α-Pinene oxide is a very reactive, sensible substrate which rearranges easily under the influence of acid catalysts [30,43]. Thereby many products such as isomeric campholenic aldehyde, trans-carveol, trans-sobrerol, isopinocamphone and *p*-cymene are observed. For example, if the isomerization is carried out at temperatures higher than 100 °C, more than 200 compounds are found [30,43]. The industrially most desired compound is campholenic aldehyde, an important intermediate for the synthesis of sandalwood-like fragrance chemicals like santalol. The zinc halogenides $ZnBr_2$ and $ZnCl_2$ are known as effective homogeneous catalysts having known drawbacks. A new catalytic process was found, combining good selectivities with environmental compatibility and reusability [43]. A strongly dealuminated H-US-Y zeolite was used for the isomerization of α-pinene oxide to campholenic aldehyde. By applying low temperatures, e.g. 0 °C in a slurry reactor, high selectivities of 75 % campholenic aldehyde can be achieved within 24 h reaction time. This is surprising because diffusion limitations would be expected to prevent reactions in zeolites at such low temperatures. This unique catalyst shows highly dispersed Lewis acidic species immobilized in the zeolite framework in which activity and access are not hindered by amorphous silica materials. To remove amorphous silica species, the catalyst has to be pretreated with an excess of 0.01 molar HCl.

Zeolites as Base Catalysts

As in the conventional acid catalyzed processes in homogeneous liquid-phase, similar problems exist for the use of typical basic catalysts such as NaOH and KOH. Therefore, there is an ambitious interest for finding and developing new and various solid base catalysts for the organic synthesis of fine and intermediate chemicals. Furthermore by using solid bases as catalysts it is also possible to run the reactions in vapor phase using fixed bed reactor technology. With regard to this general trend basic zeolites found renewed interest in organic syntheses in recent years even though it was known for a long time that Rb and Cs modified X- and Y-zeolites can be applied for side chain alkylation of alkylaromatics and double bond isomerisation in olefins [28,29,44]. An outstanding example is the synthesis of 4-methyl-thiazole (4-MT) which is needed for manufacturing thiobendazole, a systematic fungicide used for spoilage control of citrus fruit and for preventing Dutch elm disease in trees and fungal disease in seed potatoes [23,44]. Surprisingly, 4-MT can be synthesized over a basic zeolite in a single reaction step from acetone imine and SO_2. In the presence of Cs-doped ZSM-5 (2 wt-% Cs) this cyclization reaction occurs at 470°C yielding 70 % of 4-MT. In contrast, basic CaO or MgO are less effective. Concerning the alkali modification, it was found that potassium does not work at all, caesium is most effective, and rubidium is less efficient. This new process is extremely environmentally benign compared to the conventional five step route using several hazardous chemicals such as CS_2, NH_3, chloroacetone etc. and resulting in inevitable salt formation.

Zeolites as Carriers for Homogeneous Catalysts

Enantiomers show different physiological effects. Because of that impact, a wide spread development in the area of asymmetric reactions is going on. The attractive application of solid chiral catalysts for the enantioselective synthesis is disclosed in the excellent review articles of H.U. Blaser [45,46]. Zeolites became involved in the discussion because of the fact that the Sharpless-epoxidation can be improved by the addition of molecular sieves such as 3 Å or 4 Å [47] and when the material scientists realized that zeolite beta should have chiral centers at the interconnections of the channels. Up to date, it was not possible to synthesize a zeolite beta with an excess of one site to enable us to carry out enantioselective acid catalyzed reactions.

Recently, A. Corma *et al.* [48] published very stimulating results of enantioselective hydrogenations of prochiral alkenes over new Rh-complexes containing N-based chiral ligands and anchored on modified USY-zeolite. Using such catalysts, the hydrogenation of N-acylphenylalanine derivatives at 65°C under 5 atm of hydrogen results in the corresponding hydrogenated compounds with quantitative conversion and with high enantioselectivity (97,9 % ee at 100 % conversion). The "ship-in-the-bottle"-concept [49,50] is also applied. It was recognized [51] that Pt-salen complexes could be occluded in the cavities of USY-zeolite using N-N-bis(salicylidene)-diaminocyclohexane i.e. the Jacobsen-ligand without tert-butyl groups. In the hydrogenation of pyruvate derivatives having different molecular bulkiness to the corresponding lactates over these Pt-catalysts reactant shape selectivity was demonstrated. A few months later, A. Corma *et al.* were successful to use a similar Mn-salen complex (at 5°C) for the enantioselective epoxidation of olefins achieving ee-values up to 60 % at 76 % epoxide selectivity [52]. In the transhydrogenation of acetophenone over a Co-salen "ship-in-the-bottle"-complex at 25°C an ee-value of 8 % was obtained at 70 % conversion and 98 % selectivity [53]. In contrast such a result was found for the homogeneous Co-salen complex only at -10°C. There must be a zeolitic effect involved to explain these findings. Assumably, the steric hindrance of the zeolite cavity causes a conformation which is needed for the enantioselective transhydrogenation at 25°C. In the homogeneous case it is necessary to lower the temperature to -10°C to freeze such a conformation of the Co-salen complex.

Zeolites as Multifunctional Catalysts

In the field of organic synthesis increasing attention is being paid to multifunctional catalysis in which chemical conversion consisting of a number of individual reactions are brought about in the minimum number of steps [24]. Increasing interest is also in simple "one pot"-reactions for the synthesis of important chemicals which have been produced hitherto in more complicated and very often environmentally unfriendly operation sequences. Such reactions are for example the direct amination of benzene to form aniline [54] or the oxidation of propylene with oxygen in the presence of hydrogen [55,56]. In the latter case the "one-pot"-reaction over a Pd/Pt-TS-1 is favoured to the two step reaction path i.e. in the first step conversion of H_2 and O_2, in the consecutive step the reaction of the prepared highly diluted H_2O_2 with propylene to form propylene oxide (PO). The formation of PO was found to be favoured by a high fraction of Pd (II) species and by small Pd clusters whereas fully reduced Pd and large clusters favoured the propane formation. The part of Pd (II) was increased by autoreduction and adding minor amounts of platinum at a temperature above 150°C. Since platinum effects the Pd cluster size and distribution and catalyzes also the hydrogenation of propylene to propane there is an optimum Pt loading to achieve maximum PO yields and selectivities. Using a Pd/Pt-TS-1 (0,02 wt-% Pt; 1,0 wt-% Pd) at 175 °C in an autoclave the best result obtained so far is around 40 % PO yield.

IV. CONCLUSION AND FORTHCOMING TRENDS

This review has given only a selected view of topics. For very good reasons other examples could have been equally discussed, certainly with almost similar results. The example chosen should illustrate the numerous applications of zeolite catalysts to indicate the opportunities offered by these shape selective acidic or basic catalysts and to underscore the future importance of such systems. The main application is still in the field of acid catalyzed reactions via a carbenium ion mechanism and industrial processes are well established, although recently the advent of basic and multifunctional zeolites has opened up new perspectives and opportunities. Basic sites in hydrophobic micropores environment should be more water and carbon dioxide resistant than other solid bases. Acid-base bifunctional zeolites should be explored in the future in order to use the synergism of both sites. Zeolitic materials having redox-properties such as Ti-containing zeolites will gain even more in importance than in the past. It is likely that in the future a continuation and expansion can be expected in the application of zeolites in organic catalysis of intermediates, fine chemicals and pharmaceuticals. There is no doubt about a great need and an extremely ambitious and exciting challenge to find guidelines in the development of chiral heterogeneous catalysts to carry out enantioselective reactions. Zeolites as carriers for chiral homogeneous catalysts (including the "ship-in-the-bottle" concept) can help in this respect. An even more challenging and intrigueing task would be the development of a pure chiral zeolite. In general, the use of zeolites as carriers of catalytically active compounds and homogeneous catalysts or even enzymes has a promising future.

That is also true for developing multifunctional catalysts in order to shorten the reaction paths. Then in addition to the catalytic behaviour of the modifying components the intrinsic acidic or basic or redox properties of zeolites combined with shape selective features are still present. Such multifunctional catalysts can help to combine several catalytic steps and tailor them optimally to one another. And this will certainly make chemical processes most cost-efficient and will help to minimize the formation of by-products which are produced in the conventional multi-step synthesis of the downstream chemicals. Short synthesis routes of attractive, highly valuable chemicals are future targets. With respect to the starting materials the C_2-C_4 olefins and butadiene as well as the BTX-aromatics are still the original base for the chemical industry. However, the catalyst scientist should take a look at the renewable raw materials more than they have done in the past. Nature should be the architect of the necessary chemical building units, compounds having preformed the desired C-framework. Attention is already paid to the broader use of alkanes as feedstock but that challenge should be deepened e.g. the oxidation of alkanes to linear alcohols. Concerning social political aspects heterogeneous catalysis has to play an increasing role in environmental protection because the public is more and more concerned with chemicals effecting pollution and the governmental regulations become more and more stringent. Therefore, in catalytic technology the targets are replacing the homogeneous Lewis acids and bases as well as Brönsted acids and bases by heterogeneous systems. The development of highly selective catalysts and the achievement to carry out reactions with no or only a little formation of undesirable and sometimes toxic by-products should have the highest priority, even in spite of eventually very low conversion of around 10 %. This production integrated with environmental protection is one of the most challenging research targets in the future. This is in addition to further progress in the field of "end-of-pipe"-environmental protection. With respect to reactor design and process engineering it was demonstrated that zeolite catalysts can be applied in gas-, liquid- as well as in supercritical phase. The latter one should be more investigated in the future. Membrane reactors using zeolites simultaneously as membrane as well as catalysts seem to be worthwhile for future exploration.

Although catalysis research is still dominated by experimental and empirical studies, theoretical efforts and the use of computers are becoming increasingly important. Computer aided catalyst design opens new vistas. The computer images allow us to model complicated processes and to visualize them in graphic form to extend our range of ideas in the synthesis of fine and intermediate chemicals.

In conclusion zeolite catalysis is rich in promise for the future and will help in protection of the environment.

V. Acknowledgement

The author W.H. would like to express his sincere gratitude to Prof. Tomoyuki Inui for his kind invitation in taking part and celebrating his retirement from Kyoto University as well as presenting this review dedicated to him. This paper describes part of our research activities which have been of our mutual interest and gave us a lot to discuss since we met the first time in Karlsruhe, Germany 1982.

VI. REFERENCES

1. P.B. Weisz, V.J. Frilette, R.W. Maatman and E.B. Mower, *J. Catal.* **1**, 307 (1962).
2. B.P. Venuto and E.Th. Habib, *Fluid Catalytic Cracking with Zeolite Catalysts*, Chemical Industries Vol. 1, Marcel Dekker, Inc., New York, 1979.
3. N.Y. Chen, W.E. Garwood and F.G. Dwyer, *Shape Selective Catalysis in Industrial Applications*, Marcel Dekker, New York, 1989, pp. 1-303.
4. A.P. Bolton, *ACS Monograph* **171**, 714 (1976).
5. W.F. Hölderich and E. Gallei, *Ger. Chem. Eng.* **8**, 337 (1985).
6. K. Weissermel and H.J. Arpe, *Industrielle Organische Chemie*, Verlag Chemie, Weinheim, Germany, 1978.
7. G.T. Kokotailo and W.M. Meier, *Spec. Publ. Chem. Soc.* **33**, 133 (1980)
8. P.B. Weisz, *Pure Appl. Chem.* **52**, 2091 (1980).
9. N.Y. Chen, J. Maziak, A.B. Schwartz and P.B. Weisz, *Oil Gas J.* **66**, 154 (1976).
10. W.E. Garwood, *ACS Symp. Ser.* **218**, 383 (1983).
11. K.W. Smith, W.C. Starr and N.Y. Chen, *Oil Gas J.* **70**, 75 (1980).
12. S.C. Meisel, J.P. McCullough, C.J. Lechthaler and P.B. Weisz, *Chemtech* **6**, 86, (1976).
13. W.O. Haag in: *Proceedings of the 6th Int. Zeol. Conf. Reno, NV (USA) 1983*, D.H. Olson et al. (Eds.), Butterworth, Guildford, 1984, p.466.
14. C.D. Chang, *Hydrocarbons from Methanol*, Marcel Dekker, New York, 1983, pp. 1-129.
15. F.G. Dwyer, *Catalysis of Organic Reactions* in: *Chemical Industries*, W.R. Moser (Ed.), Vol. 5, Marcel Dekker, New York, 1981, p. 39.
16. W.O. Haag and F.G. Dwyer, *AIChe*, 8th Meeting, Boston, 1979.
17. W.W. Reading, C. Chu, L.B. Young and S.A. Butter, *J.Catal.* **69**, 392 (1981).
18. D.G. Parker, *Appl. Catal.* **9**, 53 (1984).
19. W.W. Keading and S.A. Butter, *J. Catal.* **61**, 15 (1980).
20. T. Inui, T. Ishihara, N. Morinaga, G. Takeuchi, H. Matsuda and Y. Takegami, *Ind. Eng. Chem. Prod. Res. Dev.* **22**, 26 (1983).
21. T. Inui, *Stud. Surf. Sci. Catal.* **105** Part B, 1441 (1996).
22. W. Hölderich, H. Eichhorn, R. Lehnert, L. Marosi, W.D. Mroos, R. Reinke, R. Ruppel and W. Schlimper, in: A *Proceedings of the 6th Int. Zeol. Conf., Reno, NV, USA 1983*, Bisio et al. (Eds.), Butterworth, London 1983, p.545.
23. W.F. Hölderich in: *Comprehensive Supramolecular Chemistry*, T. Bein, J.H. Lehn (Eds.), Vol. 7 Part II, Elsevier, Oxford, 1996, pp. 671.

24. W.F. Hölderich, *Stud. Surf. Sci. Catal.* **49A**, 69 (1989).
25. W.F. Hölderich, M. Hesse and F. Näumann, *Angew. Chem.* **27**, 226 (1988).
26. W.F. Hölderich and H. van Bekkum, *Stud. Surf. Sci. Catal.*, **58**, 631 (1991).
27. W.F. Hölderich, *Stud. Surf. Sci. Catal.*, **67**, 257 (1991).
28. P.B. Venuto, *Microporous Mater.*, **2**, 297 (1994).
29. W.F. Hölderich, *Stud. Surf. Sci. Catal.*, 75, 127 (1993)
30. W.F. Hölderich, J. Röseler, G. Heitmann and A.T. Liebens, *Cat. Today*, **37**, 353 (1997).
31. W.F. Hölderich and G. Heitmann, *Cat. Today*, **38**, 227 (1997).
32. A. Corma and H. Garcia, *Cat. Today*, **38**, 257 (1997).
33. M. Misono and N. Nojiri, *Appl. Catal.*, **64**, 1 (1990).
34. W.S. Fong, *SRI-International*, PEP Review No. 89-3-4, 1991.
35. I. Mochida, A. Yasutake and H. Fujitsu, *J. Catal.*, **82**, 313 (1983).
36. A. Chauvel, B. Delmon and W.F. Hölderich, *Appl. Catal. A: General*, **115**, 173 (1994).
37. B. Notari, *Stud. Surf. Sci. Catal.*, **37**, 413 (1987).
38. M. Iwamoto in: *Future Opportunities in Catalytic and Separation Technology*, M. Misono et al. (Eds.), Elsevier, Amsterdam, 1990, pp. 121.
39. M. Iwamoto, *Proceedings of the Meeting on Catalytic Technology for Removal of Nitrogen Monoxide*, Tokyo, Japan, 1990, pp.17.
40. T. Inui, *Catal. Lett.*, **13**, 87 (1992).
41. T. Inui, S. Iwamoto, S. Kojo, S. Shimizu and T. Hirabayashi, *Cat. Today* **22**, 41 (1994).
42. J. Röseler, G. Heitmann and W.F. Hölderich, *Appl. Catal. A: General* **144**, 319 (1996).
43. A.T. Liebens, G. Mahaim and W.F. Hölderich, *Stud. Surf. Sci. Catal.* **108**, 587 (1997).
44. Y. Ono and T. Baba, *Cat. Today*, **38**, 321 (1997).
45. H.U. Blaser and M. Müller, *Stud. Surf. Sci. Catal.* **59**, 737 (1991).
46. R.R. Bader and H.U. Blaser, *Stud. Surf. Sci. Catal.* **108**, 17 (1997).
47. R.M. Hanson, K.B. Sharpless, *J. Org. Chem.* **51**, 1922 (1986).
48. A. Corma, M. Iglesias, C. del Pino and F. Sanchez, *J. Chem. Soc. Chem. Commun.* 1253 (1991).
49. T. Bein in: *Comprehensive Supramolecular Chemistry, Vol. 7 Part II*, T. Bein and J.H. Lehn (Eds.), Elsevier, Oxford, 1996, pp. 579.
50. D.E. de Vos, P.L. Buskens, D.L. Vanoppen, P.-P.-Knops-Gerrito, and P.A. Jacobs in: *Comprehensive Supramolecular Chemistry, Vol. 7 Part II*, T. Bein and J.H. Lehn (Eds.), Elsevier, Oxford, 1996, pp.647.
51. W. Kahlen, A. Janssen and W.F. Hölderich, *Stud. Surf. Sci. Catal.* **108**, 469 (1997).
52. M.J. Sabatier, A. Corma, A. Domensch, V. Forres and H. Garcia, *Chem. Commun.* 1285 (1997).
53. W. Kahlen and W.F. Hölderich, *Catal. Lett.* submitted.
54. J. Becker and W.F. Hölderich, *Catal. Lett.* submitted.
55. R. Meiers, U. Dingerdissen and W.F. Hölderich, *J. Catal.* accepted.
56. R. Meiers, R.P. Schulz and W.F. Hölderich, *Chem. Comm.* submitted.

Res. Chem. Intermed., Vol. 24, No. 3, pp. 349-357 (1998)

GLOBAL COMPETITION: TOWARDS CONTINUAL IMPROVEMENT ON CATALYSIS IN CHEMICAL INTERMEDIATES PRODUCTION

T. ONODA
Mitsubishi Chemical Corporation, Chiyoda-ku, 1000 Tokyo, JAPAN

Received 24 September 1997; accepted 22 December 1997

Abstract--The outlook for the future growth of Japanese petrochemical business can be expressed as "distressful" especially when considering business characteristics and global competition. For Japanese companies, the most critical issue is whether petrochemistry can support industrial production of a wide range of chemical products. Business, production, and technology development must all be executed with exceptionally qualitative strategies and rationalization. Technology development of catalysis, which is the key technology for petrochemistry, must also be correlated with such strategies.

PETROCHEMICAL BUSINESS: CHARACTERISTICS AND PROSPECTS

It goes without saying that the most rational petrochemical business was spun out of a huge petrochemical complex of naphtha based ethylene cracking plants and a wide range of derivative production plants. This leading position is most likely to last for some time more. As a result, huge petrochemical complexes need an enormous amount of investment including land, harbors, roads, energy centers in order to provide the utilities, and infrastructure required by the employees.

It is no exaggeration to say that already most of the essential technologies are now perfected and exist at a matured level. In the extreme, anyone with enough capital can purchase these essential technologies and participate in the business. Another advantage is that international standards are almost firmly established for the manufactured petrochemical products. For these reasons, the technology barriers for business entry is very low. It can be stated with confidence that a large increase in supply and demand for petrochemicals can not be expected in industrially developed countries. The growing market for petrochemicals is mainly in the economically developing countries around Asia. Therefore, the keys for success of the future petrochemical business are extremely simple; the cost competitiveness and price domination (supply and demand domination ability). In another words, the advantages of large production scale shall become more and more conspicuous.

Considering the above mentioned characteristics and the environment surrounding petrochemical business, it becomes distinct who will be the leading actors in the future. First, they are the oil producing countries with their overwhelming dominance over raw naphtha and low energy costs. Second, they are the chemical companies under oil majors with their abundant capital along with the stable supply routes for raw naphtha. Third, they are companies with commercial capital, especially the ones at new location sites. Their business entry is most likely to become active owing to the strength of their market penetration and financial advantage.

Hereafter, the leading actors are only the actors who offer foundation capital as coagents of the businesses. They will need to either purchase winning technologies or to win the participation of such technology owners, in order to possess the diversified derivative products that construct the modern petrochemical businesses. An assumption to be drawn from this is that many business consortiums will emerge in oil producing countries or near the growing markets.

For economic and chemicals production expansion, the existence of fundamental and global restriction such as resource and environmental issues are acknowledged facts. However, the surge of developing countries, which trace recklessly after the paths of economically developed countries, that enjoyed the benefits of the rapid consumption civilization, can not be now slowed down. Nor can the severe economic competition among the developed countries, that expands through detecting business opportunities in emerging developing countries, be held back immediately. In reality, we must all make a progress through this vortex by finding a better if not the best way.

JAPANESE PETROCHEMICAL BUSINESS: FUTURE PROSPECTS

Despite the fact that the Japanese petrochemical industry is the second largest to that of the USA in scale, the core of its businesses is weak when focused on the domestic production facilities. The production scale of world's new petrochemical complexes is at least in the order of one million tons for ethylene cracking plants. Even considering the small advantage of low fixed cost for old domestic petrochemical complexes, with production in the 0.5 million ton order, only a grim prospect can be drawn. Therefore it appears indispensable to reorganize to a more rationalized production and business, with a one million ton scale production center (complex) by M&A, S&B, or by any other methods. Mitsubishi Chemical Corporation (MCC) has the largest petrochemical business domestically, however, the three ethylene centers at Kashima, Mizushima, and Yokkaichi need to be drastically reorganized with future business strategy.

Similarly, for the petrochemical derivative businesses, elevation of cost competitiveness must be accomplished through decreasing fixed costs and expanding business scope and by rationalizing production by means of M&A or S&B. A general criticism is that Japan has to also resolve the fact that it has a low aptitude as the production region for the future petrochemical businesses.

What one must not forget, however, is the fact that manufacturing (production) of a variety of Japanese chemical products depends heavily on the infrastructure of the petrochemical industrial complex. Chemical companies usually acquire beneficial environment from the huge infrastructure of the petrochemical complex, either directly or indirectly such as variegated raw chemicals, utility, wastes and waste water countermeasures, plant operation, and management etc. Taking into account the current situation, Japanese chemical industry must set its major premise to continue and maintain domestic petrochemical competitiveness.

On the other hand, in the focus on petrochemistry as business, Japanese chemical companies have sufficient potential of global leadership position as either a member of the allied companies, or a competitive and independent derivative business player, even if not as a leading independent major as previously mentioned. To realize such a case, preconditions would be to set either technology or market dominancy to be highly superior over global majors at least, and for the less superior (technology or market dominancy) to be of a certain level as well. In fact, technological superiority can be said to be an essential condition as fierce cost competition is too easily assumed.

Globally competitive technology can not be realized by only being at an extremely high level in the domestic market. It must be adaptable to all sorts of regional conditions as well as construction and operation conditions. Of all, what is most important is the sufficient development potential for technological improvements. It is often the case for petrochemical goods to be produced via many derivatives. In order to increase profitability and to stabilize businesses, construction of a downstream product structure like a derivatives tree is desirable. Therefore, to design qualitative strategy, to select and prioritize business scopes, and to intensify technology an ability to adapt is essential.

Petrochemical products of MCC's Group can be categorized to groups and the following business candidates can be listed. Those listed below are the main businesses and technology development targets. Not surprisingly, most development work is centered around catalyst technology development.

Polyester Business: Terephthalic acid, ethylene glycol, and polyethylene terephthalate
Oxo Business: 2-Ethyl hexanol, various alcohols, and plasticizers
Acrylic acid Business: Acrylic acid, acrylic esters, and derivatives
Styrene Business: Styrene, polystyrene, and ABS resins
Vinyl chloride Business: Vinyl chloride and polyvinyl chlorides
Polypropylene Business: Polypropylene
C_4 derivatives Business: Maleic anhydride, 1,4-butanediol,

tetrahydrofuran, gammabutyrolactone, N-methyl-2-Pyrrolidone, polybutadiene terephthalates, polytetramethylene ether glycol, and isononyl alcohol

MCC'S C_4 DERIVATIVES BUSINESSES: CATALYSIS TECHNOLOGY

In this section, details of technology development of C_4-derivatives is given, as an example of MCC's core business, relating to its business expansion strategy and catalysis technology. A systematic flow diagram of the C_4-derivatives is shown in Figure 1. Shown here, 1,4-butanediol (1,4BD) and maleic anhydride (MAA) are the central source of business.

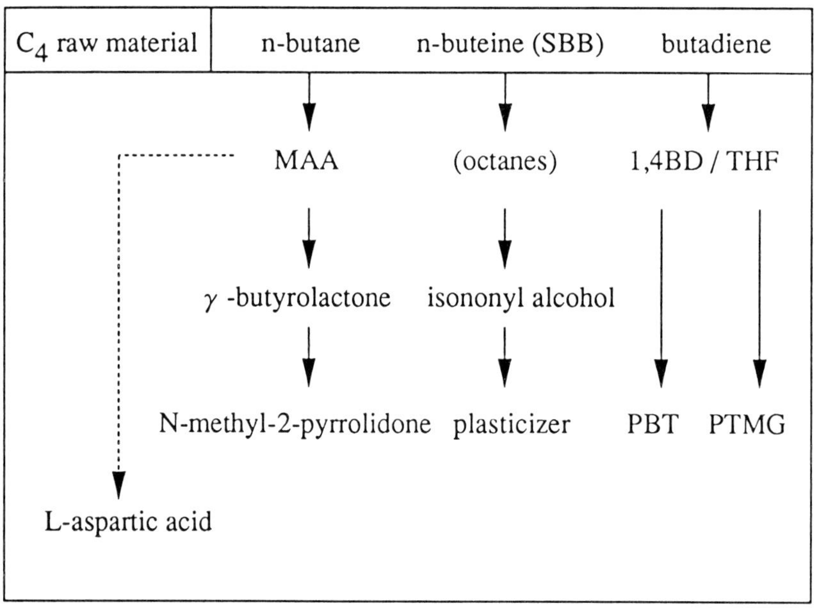

Figure 1. Schematic business flow of C_4 derivatives at MCC.

MAA has a long history as the raw material for unsaturated polyester resins. At MCC, maleic anhydride was at first produced by gas phase oxidation of benzene, using a fixed bed catalyst. In 1970, however, MCC succeeded in the industrial application of the new process of fluidized bed catalytic gas phase oxidation, a first in the world. This new process utilizes spent C_4 fraction (SBB) as the raw material. SBB is obtained after extracting butadiene from the C_4 fraction of naphtha crackers.

The fluidized bed process was then subsequently converted to butane process in order to further improve economical competitiveness, and even at present, the process of optimization and catalyst improvement is continuously developing. Therefore, process of MAA manufacturing at MCC can be said a history of raw material conversion; benzene process → SBB process → butane process.

Gammmabutyrolactone (GBL) is one of the most promising products as a derivative of MAA. The market for GBL is expanding, as GBL can be used as a solvent or as a composition element for electrolytes for batteries and condensers. GBL was first co-produced with tetrahydrofuran (THF) by liquid phase high pressure hydrogenation by using suspended catalysts. The production of GBL by this process, however, was terminated after THF was produced by the new butadiene process. Thus, for the manufacture of GBL, a new gas phase low pressure hydrogenation process had to be developed. To further rationalize and increase production, however, the new process was replaced by a low pressure liquid phase hydrogenation process, using a homogeneous Ru-complex catalyst. The newest process was industrially applied with great success earlier this year with the capacity of 10,000T/Y. Simultaneously, production capacity for N-methyl-2-pyrrolydone (NMP), a derivative of GBL, was increased to 8.000T/Y. Again, processes of GBL manufacturing at MCC followed a series of technology improvements which followed the series; high pressure/suspended catalyst/liquid phase → low pressure/fixed bed catalyst/gas phase → low pressure / homogeneous catalyst / liquid phase.

At present, the technology development of a new process to produce L-aspartic acid (L-ASP), which is a new derivative of MAA, is underway and a pilot plant test is in progress. L-ASP and its polymer have good biodegradability and thus are attracting attention as coagulation material or as chelating agent that is environmentally friendly. It is noteworthy that the process technology now under development is a hybrid process of chemical reaction and bio reaction [1].

For a long time, manufacturing of 1,4BD was monopolized by the Reppe reaction using acetylene as the raw material. In 1982, a new process to manufacture 1,4BD by oxidative acetoxylation process using butadiene as a raw material was applied industrially at MCC. The process is the world's first fixed bed liquid phase oxidation process by using a unique catalyst, that includes Pd-Te intermetallic compounds [2]. At first, regeneration of catalyst was needed after 6 months. Repeated improvements have however increased catalyst life to more than 2 years with increased selectivity and the production cost has decreased remarkably. Currently, production capacity is 50,000 T/Y as the result of repeated rationalization, which is a sharp rise from the 12,000 T/Y capacity obtained at the very beginning. Plans for production in Korea and Taiwan are now under development. Although industrial application of a new hydrogenation process of MAA has become active world wide, the butadiene process at MCC, which can simultaneously control production amount of THF, is acknowledged as being the most cost competitive. Even taking into account the price difference of butane and butadiene as raw materials, the rationalness of butadiene process can be easily agreed by immediate intuition in reference to the overall reaction equation below. (Note the difference in the consumption amount of oxygen and hydrogen, and the amount of waste water) .

Butane process: $C_4H_{10} + 3O_2 + 4H_2 \rightarrow$ 1,4BD + $4H_2O$
Butadiene process: $C_4H_6 + 1/2O_2 + H_2 + H_2O \rightarrow$ 1,4BD

The market for 1,4BD is expanding, for example, in the engineering plastics business areas, as the usage of polybutylene terephthalate (PBT) is expanding in electronic instruments business. THF is mainly used as a solvent or for polytetramethylene ether glycol (PTMG) to make elastic fibers. All of these derivatives are also manufactured at MCC.

A raw material for the manufacturing of isononyl alcohol (INA) is octenes that are produced by dimerization of n-butenes derived from SBB. At present, MCC relys on outside manufacturers for the production of octenes. INA is produced by oxo (hydroformylation) process, reacting carbon monoxide and hydrogen to octenes. Producers other than MCC in the world use the oxo process with cobalt carbonyl as the catalyst, but MCC obtains high selectivity by using rhodium-phosphine-phosphine oxide catalyst which cleverly uses the weak ligand effect of phosphorus compounds [3].

INA business also relates to oxo business, which represents one of MCC's core businesses. Ceaseless research to further rationalize, in other words, lower reaction pressure and improve quality is in progress.

The above is a summary of the history of the progress of C_4 derivatives business and catalyst technology at MCC. The presented series of C_4 derivatives consist of moderate scale catalyst reaction processes in which most of the derivatives have strong business opportunities. All derivatives are closely related to highly functional products that are expected to have high future growth. Thus, business globalization is now underway based on the thorough observation and analysis of global market trends in which companies take into account the issues of raw material cost, location, competition and alliances. To maintain globally competitive technologies, ceaseless technology development centered around catalyst development is indispensable. It is worth mentioning that, in general research results have a high probability of business success, once research objectives are correctly framed.

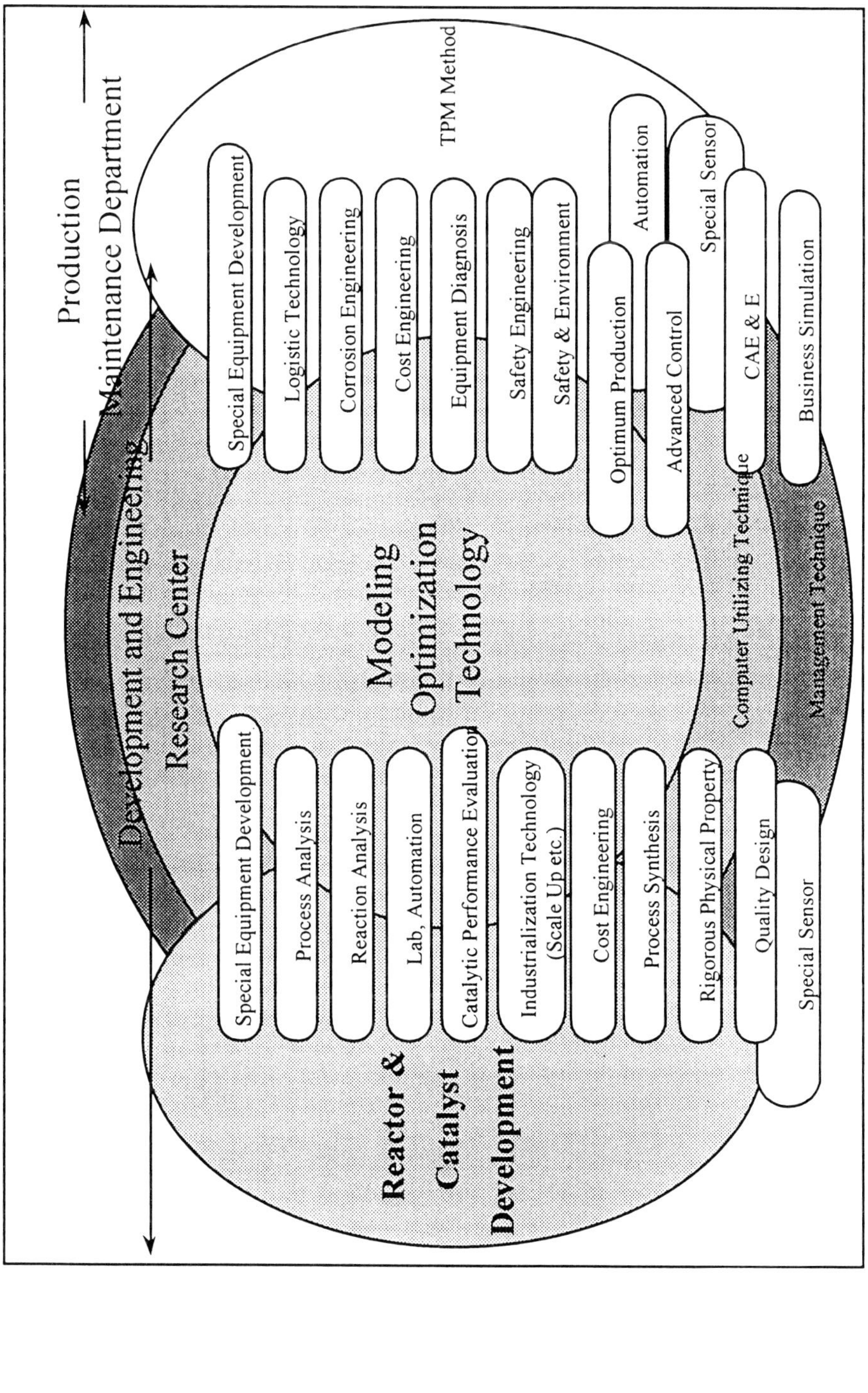

Figure 2. Image of technology fusion.

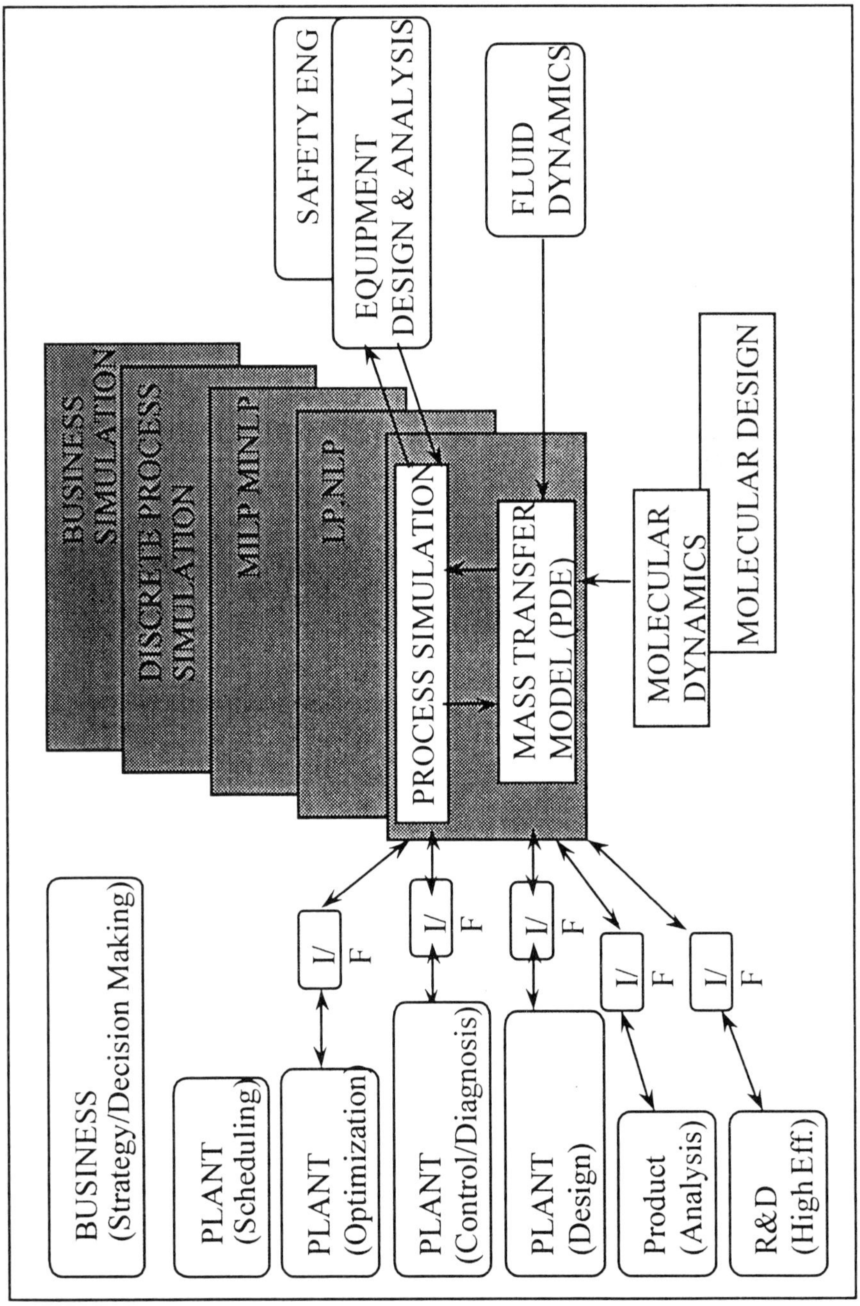

Figure 3. Scope of modeling technology.

TECHNOLOGY FUSION [4]

For the Chemical Industry to be the "Bright Chemical Industry of 21st Century", it has to be globally competitive. The major premise is to acquire both qualitative strategy and for rationalize all functions of businesses, production and technology development. Focusing on technology aspects, there are many scientific and engineering disciplines, between which there is a strong need to tighten the mutual interaction. Thus, the inevitability of technology fusion arises, and it is apparent that what will contribute most to such technology fusion will be advanced computational technology.

At the center of the technical fusion concept is modeling technology. It is essential that modeling technology must be further optimized. As shown in Figure 2, numerous scientific and engineering disciplines are included in the technology fusion model. Figure 3 shows an example of a schematic flow procedure constructed using modeling technology.

Although process simulation technology is already well established and there are powerful engineering skills, modeling technology is still missing from its vision. Now, the success of applying modeling technology is near to being realized due to the advancement of both computational technology and validation technology.

In general, there have been two large separate functions; research and development (R&D) groups and production & maintenance departments in Japanese Chemical Industry. The mutual link between the two has not been at all strong. In order to close the gap between the two functions, as a first step at MCC, both production technology and engineering technology have been reinforced and Production Technology and Engineering Centers (PTEC) have been established at main plants. At the end of last year, as a second step, R&D function and PTEC function have been organizationally combined. Fruits of integrated technology are to be awaited with expectation. At MCC, the catalyst development is conducted with the position demonstrated as in Figure 2 and Figure 3. It is expected that such positioning will accelerate the effectiveness and efficiency of technology development, by catalyst scientists who are given a wider scope of self identity.

REFERENCES

1. M. Terasawa and H. Yukawa. In: *Industrial Application of Immobilized Biocatalysts*, A. Tanaka et al. (Eds.), Marcel Dekker, Inc., New York, 1993, p. 37.
2. T. Onoda, *PETROTECH* **10**, 810 (1987).
3. T. Onoda, *CHEMTECH*, September **1993**, 34 (1993).
4. Y. Natori and L.O'Young, *1st PTEC Symposium (Mitsubishi Chem. Corp.)*, September 1996.

Res. Chem. Intermed., Vol. 24, No. 4, pp. 359-372 (1998)

A KEY TO PROGRESS: IDENTIFICATION OF THE STATE OF CATALYSTS DURING THE CATALYTIC REACTION

B. DELMON*, S. ZEYSS, E. GAIGNEAUZ** and P. RUIZ
Université catholique de Louvain, Unité de catalyse et chimie des matériaux divisés, Place Croix du Sud, 2/17, B-1348 Louvain-la-Neuve (Belgium)
*** Aspirant, Fonds National de la Recherche Scientifique (Belgium)*

Received 16 September , 1997; accepted 25 November 1997

Abstract -- We stress the importance of identifying the state in which catalysts are during the catalytic reaction. Catalysts as prepared do not correspond to the real working catalyst. Several examples will be recalled, and a new one (Mo-Te-O catalyst) will be mentioned. In a more detailed way, it will be shown that the working state of MoO_3 is Mo_xO_{3x-2} with $8 \leq x \leq 18$. The pitfalls encountered when focusing on catalysts as prepared (as opposed to catalysts "as they work") is illustrated in the case of mixed $Sb_xMo_yO_z$ oxides which decompose spontaneously during the reaction.

INTRODUCTION

Although all top scientists working in catalysis are convinced that catalysts undergo changes when they are in the conditions of the catalytic reaction, the number of studies taking this fact into consideration is still disappointingly small. The objective of the present contribution is to show that considerable progress can be achieved by understanding the real state of catalysts when they are catalyzing a reaction. We shall illustrate the argument by selective oxidation, which is coming to the forefront of catalysis. The real working catalyst usually escapes physico-chemical characterization. The behavior of MoO_3 constitutes a relatively simple example. A comprehensive approach can lead to a correct understanding of the problems and correlatively helps find practical solutions.

SOLID STATE PROCESSES IN CATALYSTS DURING THE CATALYTIC REACTION

General

High activity and selectivity demand that the active atoms be in a very precise environment, namely a precise coordination. In spite of this, the common approach to heterogeneous catalysis remains almost unchanged. It still rests heavily (often exclusively) on correlations between catalytic properties on the one hand and the properties of the catalysts *as prepared* by the experimenter on the other hand. When the coordination of surface atoms in the metal, oxide or sulfide phases is considered, this is that of the *precursor* (the solid "*as prepared*") and not of the "working" catalyst. Actually, at any given point of the surface, this coordination varies cyclically under the influence of the dynamics of the catalytic reaction, during adsorption or coadsorption of the reactants, breaking or formation of new chemical bonds, and desorption of the products. This may bring about extensive changes in the course of time. Very few papers allude to methods which could be envisaged to control coordination at a stage of catalyst life posterior to preparation, namely when the catalyst is in a state (e.g. the activated form) closer to the functioning form.

Catalysis is a dissipative phenomenon (in the sense of so-called "irreversible thermodynamics"). Energy is continuously released at the catalyst surface. This facilitates the movement of atoms. This is the reason why loose or flexible structures form during the catalytic process. They possess an excess of free energy. The corresponding high reactivity unfortunately makes it such that these structures may tend to form more stable, catalytically less active phases. The consequence is that the changes occurring continuously in the dynamic state trigger a part, or possibly all, of the solid state reactions that lead to deactivation. Particularly in the field of oxidation, the dynamic reduction-oxidation process to which the working catalysts is subjected can easily trigger such solid state reactions.

Examples of Solid-Sate Transformation in Catalysts During Catalysis

A first example of misinterpretation of the origin of catalytic activity in oxidation reactions is very old. The activity of $V_xMo_yO_z$ catalysts in the selective oxidation of naphthalene to phthalic anhydride had been attributed in the 60's to a mixed phase ($y/x \sim 0.3$) which actually constitutes the major phase in the *precursor* of the most active catalysts. But the elements in this phase segregate irreversibly when the catalysts work ($V_xMo_yO_z \rightarrow V_wO_u + MoO_3$). A new explanation of the activity of this family of catalysts has still to be found.

A very recent example concerns mixed oxides containing tellurium. Tellurium is present in selective oxidation or ammoxidation catalysts and acts synergistically with them (e.g. in propylene reaction to acrylonitrile, Te is used in one of the two most widespread commercial catalysts) [1-3]. When used in conjunction with molybdenum, the effect is believed to be due to the formation of tellurium molybdates [1-6]. We tested Te_2MoO_7 in the oxidation of iso-butene [3]. The conversion increases with increasing reaction temperature. But if the temperature is decreased to 420°C, the conversion does not return to the original value. Carrying out the catalytic test a second time, the same conversions as for the first test could be obtained up to 400°C. But for higher temperatures, a significant decrease of the iso-butene conversion is again observed, indicating some additional deactivation. XRD and IR after the first test indicate that a reaction in the catalysts bulk occurs (Figure 1) [3]. Besides TeO_2 (which is also present in small amounts in the fresh catalyst because of a non-completed solid state reaction) new telluromolybdate mixed phases $TeMo_4O_{13}$ and $TeMo_5O_{16}$ [7] could be identified, and some metallic Te was found. Some reduction of the catalyst occurs, explaining a partial volatilization of tellurium during the catalytic test [2] and the formation of telluoromolybdate mixed phases ($TeMo_4O_{13}$ and $TeMo_5O_{16}$) with low tellurium content. But the $TeMo_5O_{16}$ phase is poorly active. The conclusion is that, although Te_2MoO_7 is more active than $TeMo_5O_{16}$ in the selective oxidation of propene [8] it decomposes to the less active $TeMo_5O_{16}$ [3]. The key to stable catalytic activity is hidden in the various interactions which can take place during the catalytic reaction above 420 °C between the decomposition products of Te_2MoO_7, namely Te, TeO_2, Te_2MoO_7, $TeMo_4O_{13}$ and $TeMo_5O_{16}$ (figure 1). These interactions have still to be identified.

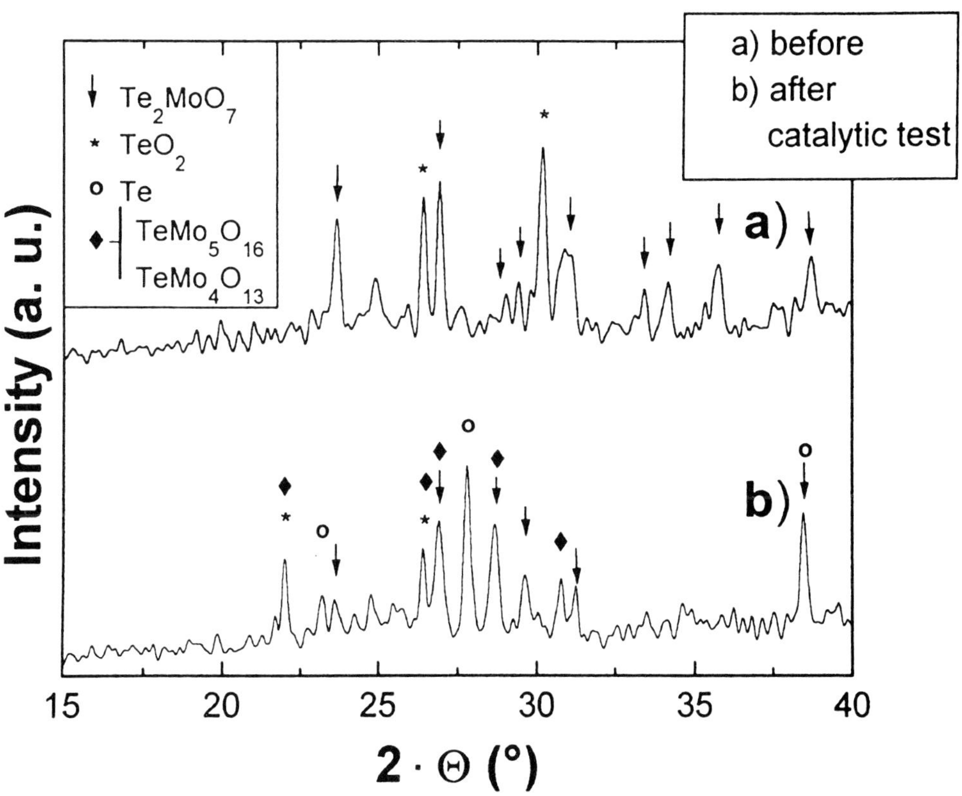

Figure 1. X-ray diffraction of Te_2MoO_7 before and after the catalytic test in the presence of isobutene and oxygen.

Still another example dealt with the synergy between separate phases in a mechanical mixture of $Fe_2(MoO_4)_3$ and α-Sb_2O_4 [9]. These reacted to give a mixture of non-transformed $Fe_2(MoO_4)_3$ and α-Sb_2O_4, with $FeSbO_4$ and MoO_3 as new phases. The catalysts were tested in the selective oxidation of isobutene to methacrolein [10]. Pure α-SbO_4 is inert, $Fe_2(MoO_4)_3$ is active, but its selectivity is very poor. The fresh (non-calcined) mechanical mixture exhibits a high synergy in the yield and selectivity to methacrolein. The synergy increases significantly in calcined mixtures, as shown in Table 1. The same changes seem to take place when the initial mixture is contacted with isobutene and air in the catalytic reaction conditions. None of the isolated phases (i.e. α-SbO_4, $Fe_2(MoO_4)_3$, $FeSbO_4$ or MoO_3) alone has specially high activities or selectivities [10]. The higher performances are due to synergetic effects that the real catalytic system develops after some time. This demonstrates that explanations of catalytic activity should not rest on the existence of these starting phases, but on the understanding of the cooperation taking place in the catalyst having undergone solid-state transformations during a catalytic reaction conducted in normal conditions for catalysis.

Table 1
Catalytic activity of mechanical mixtures of α-Sb_2O_4 and $Fe_2(MoO_4)_3$ [10]. The values in parenthesis mentioned for the non-calcined samples are those that correspond to the addition of the values corresponding to the pure catalytic phases.

Sample	Conversion (%)	Yield (%)	Selectivity
$Fe_2(MoO_4)_3$ + α-Sb_2O_4 (non calcined)	19.6 (18.7)	5.7 (3.1)	31.0 (16.3)
$Fe_2(MoO_4)_3$	31.2	5.1	16.3
$Fe_2(MoO_4)_3$ + α-Sb_2O_4 (calcined 6 days)	36.0	9.4	51.5

CONTROL OF STRUCTURE AND TEXTURE OF THE WORKING FORM OF CATALYSTS

The traditional approach to catalyst improvement is based on modifications of the precursor form, disregarding the real working catalysts. Remarks concerning this approach [11,12] are summarized here.

Doping with small amounts of elements supposed to get inserted in an oxide lattice or spread atomically on the surface is in principal a valuable approach for modifying catalysts. However, it has seldom been checked that the doping elements were truly incorporated in the host oxide, and if so, did not spontaneously segregate out. In almost all the cases where the check has been made, a segregation was shown to occur as in the cases of the solid solution of molybdenum mentioned above or antimony in respectively vanadium or tin oxide. In theory, the addition of dopants can allow a control of valency of active elements in the mixed oxide, a change in the rate of the reduction-oxidation process at the surface or changes of "acidity". But the correlations with the effects observed are not valid if the concepts do not apply to the real catalytic form. The basic ideas are correct in principle, but may lead only to uncertain conclusions or advances for lack of adequately planned experiments. Nowadays, microanalysis makes possible the determination of composition at the nanometer scale. A larger number of accurate analyses of phases in working catalysts (in the state they are during use) or quenched to room temperature after use, correlated with activity and selectivity measurements, will certainly lead to substantial progress in the future, when the doping action will be exerted on the catalytically useful phases.

Supports are traditionally presumed to be inert (e.g. SiO_2) or to permit atomic dispersion or formation of a monolayer (the most famous example is that of anatase when V_2O_5 is the active phase). But this does not imply that epitaxial layers detected in the precursor form of catalysts are necessarily the active and selective species. The existence of an atomic dispersion or monolayer structure in working catalysts has not often been verified. For example, strong doubts are raised concerning the atomic dispersion (monolayer) of V_2O_5 on anatase. The most likely structure is that of a several-layer thick suboxide of vanadium (V_6O_{13}) [13,14]. The corresponding islets are probably stabilized by epitaxy, thus permitting the V_6O_{13} to smoothly lose and reincorporate oxygen during the catalytic cycle without sustaining an irreversible transformation to a different structure: V_6O_{13} in epitaxy with anatase could be the working form of the catalyst. Systematic examination of used catalysts could avoid misleading interpretations. There are cases where epitaxy is doubtful. It has been claimed that Sb_2O_4 crystallites decorated epitaxially $FeSb_2O_4$. But this cannot be detected. If the picture of oppositely oriented flat vanadium phosphate pyramids as active sites for butane oxidation is correct (and this is presently gaining credibility) the possible role of an epitaxy between various vanadium phosphates, which has not been proven, may become irrelevant. A detailed work carried out recently on the activation of these catalysts [15,16] suggests the picture of a catalyst with a very flexible ("wobbly") structure and gives no indication of epitaxy.

The parameters that literature considered in the last 25 years are relevant in principle. However, a fundamental precaution is to check whether the feature mentioned (e.g. doping or epitaxy) can still be observed after the catalyst has worked in realistic conditions. Ideally, this should be done *in situ, in operandi* (i.e. during the reaction), but observations on quenched catalysts usually give extremely valuable information, account taken of the relatively low rate of structural and textural modifications in solids. When looking at the literature on selective catalytic oxidation (and other fields as well) in a critical way, it is striking that the most active catalysts, especially those used in industry, contain several phases. This suggests that many catalysts (mixed oxides, doped oxides) which were thought to be constituted of a single phase, actually contained several distinct phases in their "working state". To illustrate the consequences of this fact, the following section focuses on mixtures of α-Sb_2O_4 with MoO_3.

AN EXAMPLE: STARTING MoO_3 IN PROMOTED OXIDATION CATALYSTS, AND THE WORKING STATE OF THE CATALYST

This part suggests the kind of developments which could be expected following the lines suggested above. A major component in selective oxidation catalysts is frequently MoO_3. This is usually associated (pure or in the form of molybdates) with various metals like iron, cobalt or bismuth. Taking into account the fact that a suboxide of MoO_3 is the active state in the selective oxidation of olefins to unsaturated aldehydes, it is possible to expect new phenomena to take place. This will be shortly outlined. It will also be shown how it is possible to discard the view that a mixed oxide (here: $Sb_xMo_yO_z$) could be the active phase.

Structural Aspects

The selective oxidation actually consists in a catalytic cycle involving reduction of the catalyst by the hydrocarbon and subsequent reoxidation by molecular oxygen. Let us recall that, in the case of the so-called allylic oxidation, e.g. propene to acrolein or isobutene to methacrolein, two atoms of oxygen extracted from the catalyst are used. One of them removes two hydrogen atoms from the olefin to form H_2O. The other one gets inserted into the hydrocarbon molecule to form the unsaturated aldehyde. The structure of fully oxidized MoO_3 consists of distorted corner-sharing octahedra of MoO_6 conventionally represented as viewed from one corner. During reaction, the MoO_6 octahedra can also form edge-sharing pairs if one oxygen is removed. The cyclic reduction-oxidation of MoO_3 corresponds to a cyclic transformation of the surface in which twin adjacent pairs of corner sharing octahedra lose two oxygen atoms to form two adjacent pairs of edge-sharing octahedra. This is represented in Figure 2.

In the present case, as in many selective oxidation reactions, the extraction of oxygen from the catalyst surface (reduction) is more rapid than the reoxidation (this is indicated by the thickness of the arrows in fig. 2). The imbalance in the

evolution to preferred *textures*. Our preliminary results suggest that various Mo_xO_{3x-2} phases form at places where the dynamics of gas phase diffusion in the porosity of the MoO_x particles, and the flux of spillover oxygen create the dynamic reduction-oxidation conditions adequate for each of these phases.

PERSPECTIVES

Identifying the real catalytically active state of catalysts certainly constitutes one of the most important, perhaps the most important, advance in the development of catalytic science and technology. Knowing the structure and texture of this state would permit, in particular:

- to improve fabrication and activation, namely to devise methods which would produce directly the wanted "wobbly" material with maximum efficiency;
- to optimize the operation conditions of the catalytic reactor. For doing so presently, the guide is essentially to follow conversion and selectivity. If the real working state is known, an additional, very reliable guideline would be to search for ways to maintain the corresponding structure and texture;
- to identify the source of the mechanisms leading to deactivation, namely to identify the parameters which lead to impairing the ideal structure and texture, e.g. through nucleation;
- and last but not least, to provide the most reliable basis for interpretation of results, modeling, and further improvements of activity, selectivity or resistance to deactivation, e.g. through adequate doping.

One can easily imagine the enormous advantage to using samples of the real working catalyst for *in situ* spectroscopies, pulse experiments or isotopic exchange. The quality of the insight into intermediates, mechanisms or side reactions would then benefit from a quantum jump. The real structure and texture of the catalytic material being known, it would also be possible to use rationally all the concepts of solid state chemistry, like doping, control of vacancies and other defects, or epitaxy. To take specifically the case of two-phase catalysts where spillover oxygen is necessary (remote control), it would be possible to adjust the nature, dispersion and amount of donor of spillover oxygen as a function of the demands.

In the "riser" reactor of the DuPont-Monsanto process for oxidizing butane to maleic anhydride using VPO catalysts, the reduction and oxidation steps occur in separate parts of the reactor. We recently made the remark [26] that the catalyst used in the conventional fixed bed seems to go through an activity maximum (with, presumably, the largest amount of the "real active phase") in the course of months. We speculated, by analogy with MoO_3, that increasing the number of defects is first beneficial (as we observed when going from MoO_3 to $Mo_{18}O_{52}$ in the previous section) then harmful (as on going from Mo_8O_{23} to Mo_3O_{11}). It is therefore logical to try and maintain the VPO catalyst in the "safe range" (with the analogy of molybdenum oxide: between $Mo_{18}O_{52}$ and Mo_8O_{23}). This can be done by forcing it to go from one end of the "safe range" to the other and back, respectively in the separate reduction and oxidation steps of the riser reactor.

A still more exciting line of research has opened recently. It is based also on modifications of feed composition. This is the addition of *hydrogen* in catalytic oxidation. Corresponding results report the unexpected direct synthesis of phenol by reaction with oxygen and hydrogen [27,28], the transformation of propene to propene oxide on Au/TiO_2 [29,30] or the selective oxidation of methane or ethane on iron phosphate [31,32], both in the presence of hydrogen. In a recent contribution Y. Moro-Oka reviews his own work and these more important results concerning what he calls the reductive activation of dioxygen [33]. The role of hydrogen is certainly not to react with oxygen to produce water, because the amount is too small to affect the reaction. It could possibly produce heat locally on the surface of the catalyst, or perhaps create new surface intermediates like peroxo or hydroperoxo species as proposed by the authors [30,32,33]. But, if we accept that the spillover species can play a role (remote control), we could also speculate in a direction which comes in complement to that of the authors. The surface must be kept relatively reduced in alkane oxidation, in order to initiate the first activation steps (this is the case with VPO in butane oxidation). We could therefore speculate that hydrogen (spillover hydrogen?) could impose the adequate reduced state to the catalysts. A recent review paper stresses in a more general way the importance of the reactive atmosphere, in particular steam, in selective oxidation [34].

The search for the real active state of catalysts is difficult. Investigations *in situ* on the catalyst while it works (*in operandi*) are ideally suited, but very difficult. X-ray or neutron diffraction, EXAFS or similar methods and sometimes spectroscopy (Raman) techniques can be used. Nearly *in operandi* observations can now be made with electron microscopy techniques [17,35,36]. Surface science techniques applied to catalyst transferred from the reaction to the measurement chamber can also bring valuable information. The examples we have shown with MoO_3, comparing samples subjected or not to spillover oxygen, show that more indirect methods can bring extremely precise information. It is certain, therefore, that investigation centred on the real state of the working catalyst will develop strongly in the next years.

Acknowledgements
The authors thank the Deutscher Akademischer Austauschdienst (postdoctoral fellowship to SZ) and the Fonds National de la Recherche Scientifique (Belgium) (fellowship to EG).

REFERENCES

1. P. Forzatti, P.L. Villa, D. Ercoll, G. Ercoll, F. Gasparini and F. Trifirò, *Ind. Eng. Chem., Prod. Res. Dev.*, **16**, 26 (1977).
2. R.K. Grasselli, G. Centi and F. Trifirò, *Appl. Catal.* **57**, 149 (1990).
3. S. Zeyß, P. Ruiz and B. Delmon, *submitted.*
4. J.C.J. Bart and N. Giordano, *J. Catal.*, **64**, 356 (1980).
5. B. Gryzbowka, A. Mazurkiewicz and J. Slocynski, *Appl. Catal.*, **13,** 223 (1985).
6. H. Hayashi, S. Suigiyama, N. Masaokaa and N. Shigemoto, *Ind. Eng. Chem. Res.*, **34**, 135 (1995).
7. J.C.J. Bart, G. Petrini and N. Giordano, *Z. Anorg. Allg. Chem.*, **413**, 180 (1975).
8. T.V. Andrushkevich, G.K. Borskov, L.L. Kuznetsova, L.M. Plyasova, Y.N. Tyurin and Y.M. Shchekochikhin, *Kinet. Katal.*, **15**, 424 (1974).
9. L.E. Cadus, F.J. Gotor, D. Acosta, J. Naud, P. Ruiz and B. delmon, *Solid State Ionics*, **63-65**, 743 (1993).
10. L.E. Cadus, Y.L. Xiong, F.J. Gotor, D. Acosta, J. Naud, P. Ruiz and B. Delmon, in: *New Developments in Selective Oxidation II*, V. Cortés Coberan and S.V. Bellon (Eds.), Elsevier, Amsterdam, 1994, p. 41.
11. B. Delmon, P. Ruiz, S.R.G. Carrazán, S. Korili, M.A. Vicente Rodriguez and Z. Sobalik, in: *Catalysts in Petroleum Refining and Petrochemical Industries 1995*, M. Absi-Halabi, J. Beshara, H. Qabazard and A. Stanislaus (Eds.), Elsevier, Amsterdam, 1996, pp. 1-25.
12. B. Delmon, Lectrue, 3rd World Congress on Oxidation Catalysis, San Diego (USA), Sept. 21-26,1997, *accepted for publication.*
13. T. Machej, M. Remy, P. Ruiz and B. Delmon, *J. Chem. Soc., Faraday Trans.*, **86**, 715 (1990).
14. T. Machej, M. Remy, P. Ruiz and B. Delmon, *J. Chem. Soc., Faraday Trans.*, **86**, 723 (1990).
15. C.J. Kiely, A. Burrows, S. Salip, G.J. Hutchings, M.-T. Sananes, A. Tuel and J.C. Volta, *J. Catal.*, **162**, 31 (1996).
16. J.C. Volta, *Catal. Today,* **32**, 29 (1996).
17. P.L. Gai-Boyes, *Catal. Rev. - Sci. Eng.*, **34**, 1 (1992).
18. E.M. Gaigneaux, P. Ruiz and B. Delmon, *Catal. Today*, **32**, 37 (1996).
19. E. Gaigneaux, E.E. Wolf, P. Ruiz and B. Delmon, (6th Iketani Conf., Tokyo, Nov. 25-27, 1996) , *Appl. Surf. Science, in press.*
20. R.G. Teller, M.R. Antonio, J.F. Brazdil, M. Mehicic and R.K. Grasselli, *Inorg. Chem.***24**, 3370 (1985).
21. M. Parmentier, Ch. Gleitzer and R.J.D. Tilley, *J. Solid State Chem.*, **31**, 305 (1980).
22. E. Gaugneaux, *to be published.*
23. L.T. Weng and B. Delmon, *Appl. Catal., A,* **81**, 141 (1992).
24. B. Zhou, E. Sham, R. Machej, P. Bertrand, P. Ruiz and B. Delmon, *J. Catal.*, **132**, 157 (1991).
25. E. Godard, E. Gaigneauz, P. Ruiz and B. Delmon, *to be published.*
26. B. Delmon, Lecture, 7th Intern. Symp. On catalyst deactivation, Cancún, Mexico, Oct. 5-8, 1997, *accepted for publication.*
27. M. Hamada, H. Niwa, Y. Sasaki, M. Oguri, US Patent: 5,426,243 (to Tosoh Co.) (June 20, 1995).
28. T. Miyake, M. Hamada, Y. Sasaki, M. Oguri, *Appl. Catal., A,* **131**, 33 (1995).
29. T. Hayashi and M. Haruta, *Shokubai*, **37**, 72 (1995).
30. M. Haruta, *Catal. Today,* **36**, 153 (1997).

31. Y. Wang and K, Otsuka, *J. Catal.,* **155**, 256 (1995).
32. Y. Wang and K. Otsuka, *J. Catal.,* **171**, 106 (1997).
33. Y. Moro-Oka, invited commun., OA 21, First Asia-Pacific Corgr., Kyongju (Korea), Nov. 2-5, 1997.
34. E.A. Mamedov, *Appl. Catal,* **116**, 49 (1994).
35. P.J.R. Uwins, G.J. Millar and M.L. Nelson, *Microscopy Res. And Techn.,***36**, 382 (1997).
36. G.J. Millar, M.L. Nelson and P.J.R. Uwins, *J. Catal.,* **169**, 143 (1997).

Res. Chem. Intermed., Vol. 24, No. 4, pp. 373-390 (1998)

NON-LINEAR ASPECTS IN CATALYSIS AND IMPORTANCE OF THEIR CONTROL TO SOLVE ENERGY AND ENVIRONMENTAL PROBLEMS

TOMOYUKI INUI
Department of Energy and Hydrocarbon Chemistry, Graduate School of Engineering, Kyoto University, Sakyo-ku, Kyoto 606-01, Japan

Received 13 October 1997; accepted 18 November 1997

CONTENTS

I. INTRODUCTION

The relationship between adsorption and catalysis had attracted particular attention from the last century to the beginning of this century as the most essential aspect to understand the catalytic phenomena [1]. In the early part of this century, the adsorption phenomena could be systematized at the beginning by the mono-layer adsorption theory proposed by Langmuir [2]. Since Hinshelwood [3] applied the

Langmuir theory successfully to explain various vapor phase catalytic reactions occurring on solid catalysts, the reaction mechanism, which enables an application to the practical reactions, has been called the Langmuir-Hinshelwood mechanism. This reaction mechanism then played a central role for the understanding and analysis of catalytic reaction kinetics. However, this theory is applicable under conditions which lead to linear phenomena, in which adsorption rates of reactants are essentially higher than surface reaction rates or consumption rate of those reactants.

In recent years, problems to be solved by the applications of catalysts have appeared, which had never been faced previously. This concerns the elimination of toxic materials, which are contained in the exhaust gases as extraordinary low concentrations in the mixtures, which consequently, are far deviating from the reaction stoichiometry. Moreover, the coexistence of highly concentrated undesired components for catalysis such as excess oxygen and steam strongly interfere with the smooth progress of the catalytic reaction. Diminution of carbon dioxide emission from fossil fuel combustion, new routes for the synthesis of fuels and energy carriers, and the use of raw materials from alternative resources for the chemical industries are also unavoidable tasks for the catalyst chemists.

However, these tasks need extremely rapid conversion rates in comparison to the conventional conversion rates for the production of compounds in usual industries, and the selective elimination of very dilute amounts of undesired compounds in the presence of high concentrations of retarding compounds, especially in the exhaust gas from diesel engine and lean-burn facilities of large factories.

Approaches based on conventional linear phenomena in catalysis cannot alone respond to these difficult demands. In order to realize novel development in catalytic reactions, it should be necessary to understand and control non-linear phenomena which can be observed in catalytic reactions under special circumstances in both the catalyst structure and the reaction conditions. In this review article, recent advances in catalysis for solving the energy and environmental problems are presented. Typical rapid and effective catalytic reactions will be cited as examples. In addition, if the oxidation state of the catalyst surface changes very easily, it is difficult to maintain stable and intermediate conversion levels, and this problem must be examined also.

II. DYNAMIC BEHAVIORS OCCURRING IN HETEROGENEOUS CATALYSIS

The oxidation-reduction state of metal and/or metal oxide catalysts is, in practice, sensitively changed by the reaction environment, such as reaction-gas composition, temperature, pressure, retardation by the reactants and products, and even by the degree of conversion of the reactants. However, in the conventional analysis on catalytic surface reactions, these practical factors; which often correspond to non-linear phenomena, are usually ignored or neglected for the sake of simplicity [4]. The state of catalyst surface during a reaction at almost zero-conversion level or in differential reaction conditions, and too low temperature conditions as in a circulating reactor system, do not reflect directly the real state of the catalyst surface under the practical reaction conditions in which the desired reaction progresses rapidly. In catalytic combustion of fuels and/or pollutants exhaust gas purification by catalytic processes [5], catalytic reforming of methane [6], catalytic decomposition of methanol to syngas [7], and methanation of carbon dioxides [8], for example, only the total conversion of the target reactants is practically significant, and intermediate conversions have almost no meaning.

Self-oscillations of Heterogeneous Catalytic Reaction Rates

Self-oscillating reactions, in which reaction rates vary periodically, closely relate to the biochemical reactions such as pulsation of heart. Homogeneous and enzyme-containing systems exhibiting oscillations have been studied extensively [9] from the beginning of this century [10,11]. In the early 1970's, such reaction behaviors in catalytic reactions over metallic surface, mainly the oxidation of CO and H_2 on Pt, and Ni catalysts, have been reported [12,13]. In general, however, the kinds of reaction with such behaviors and methods for detection were limited, and the reaction rates could be measured using simple detection methods only in case of using a catalyst having a very low heat capacity, such as catalysts supported on an unwoven sheet of ceramic fiber [14] or the high surface area of small particle silica [15]. The self-oscillation in reaction rate observed in the catalytic combustion of gaseous fuels is essentially a non-linear phenomenon. However, by applying a consistent analysis on the basis of the surface reaction mechanism, an approximate solution reflecting the data can be obtained [14]. From these analyses, a reason for the self-oscillating behavior of the reaction is proved as the following sequential changes in the reaction circumstance. As a consequence of the rapid reaction on a catalyst having a low heat capacity, the reaction heat accumulates in the catalyst, and consequently, the reaction rate and the catalyst temperature rise in an accelerating way. As a consequence of the temperature rise, catalyst coverage by the fuel molecule decreases, while the relative oxygen coverage increases and becomes predominant. Dissociative oxygen is adsorbed irreversibly and partially forms an oxide layer on the surface of metallic catalyst. The concentration of CO_2 increases with the temperature rise, and is strongly adsorbed on the oxygen-adsorbed surface. This adsorption exerts a strongly inhibiting action, and causes the rapid decrease in the reaction rate. Consequently, the catalyst temperature falls and then coverage of fuels recovers. The oxidized surface is reduced and the oxidation activity increases again. Thus, the reaction cycle is repeated automatically.

Forced-oscillating Reactions - Observation of the Change in Oxidative State of Catalyst Surface during the Reaction

From the viewpoint of microscopic sequential reaction mechanism, the redox cycles in the reaction described above occurs in every catalyst surface during the combustion reaction. However, on the ordinary catalysts the temperature generated on the active sites is diffused to the vicinity and averaged, and, in practice, the rise of temperature to an apparent steady state can only be observed. However, when one of the reaction parameters, involved in the oscillating reactions, such as temperature is varied systematically, the features of oscillating reaction appear clearly. The author proposed this method as "the forced oscillating reaction method" [16].

As a typical example, combustion of methane on a supported Pt catalyst was investigated. A ceramic foam prepared by Bridgestone Co. Ltd. was used as the catalyst support. The dimension of foam was about 0.25 mm. Since the surface area of the ceramic foam was very small (0.10 m^2/g), this was coated with an alumina sol before supporting the catalyst material. The increase in BET surface area was 12 m^2/g after the alumina loading. Platinum load was 0.33 wt%. A continuous flow reaction apparatus was used under atmospheric pressure. The catalyst, cut to 10 mm diameter and 50 mm length, was packed in a quartz tube of 11.0 mm inner diameter. The reaction gas mixture composed of methane, oxygen and nitrogen, was allowed to flow with a space velocity of 1,400 h^{-1}.

The composition of the reaction gas mixture was varied as follows; (a) excess oxygen condition (CH_4 6.5%, O_2 19.6 %, N_2 balance; oxidative catalytic condition), (b) stoichiometric condition (CH_4 6.5%, O_2 13.0 %, N_2 balance), and (c) excess methane condition (CH_4 38%, O_2 13.0 %, N_2 balance; reductive catalytic condition). The hysteresis of conversion was measured, which is caused by the elevation of the reaction temperature with a constant heating rate and followed by descending the reaction temperature in the same way. The temperature was raised at a constant rate of 10°C/min until methane or oxygen conversion reached 100%, and then the temperature was lowered. The temperature was measured by thermocouples placed at the centers of both inlet and outlet of the catalyst. The reaction gas was analyzed by gas chromatographs, and/or a infra-red analyzer for continuous analysis of CO_2. Under the oxygen excess condition a minor hysteresis was observed as shown in Fig. 1 (a). At the stoichiometric condition, the descending line was considerably shifted to the lower temperature side (Fig. 1 (b)), and this tendency increased with increasing methane in the feed. Under a CH_4 excess condition, the catalytic combustion of methane maintained at a temperature as low as 150~180°C with total consumption of oxygen (Fig. 1 (c)), even when the heat supply of the electric furnace was cut off. This kind of hysteresis reflects the oxidation-reduction states of the catalyst. On the reduced surface, methane can be combusted even at a very low temperature. As for a supported Pt catalyst with a very low loading such as 0.33 wt%, the change in weight between the oxidized and reduced states is very small and cannot be detected unambiguously even by a sensitive microthermogravimertric analyzer. Different from such precious metal catalysts, a Cu catalyst with higher Cu loading exhibits both oscillating reaction and weight change caused by the change in oxidation state of copper during the reaction.

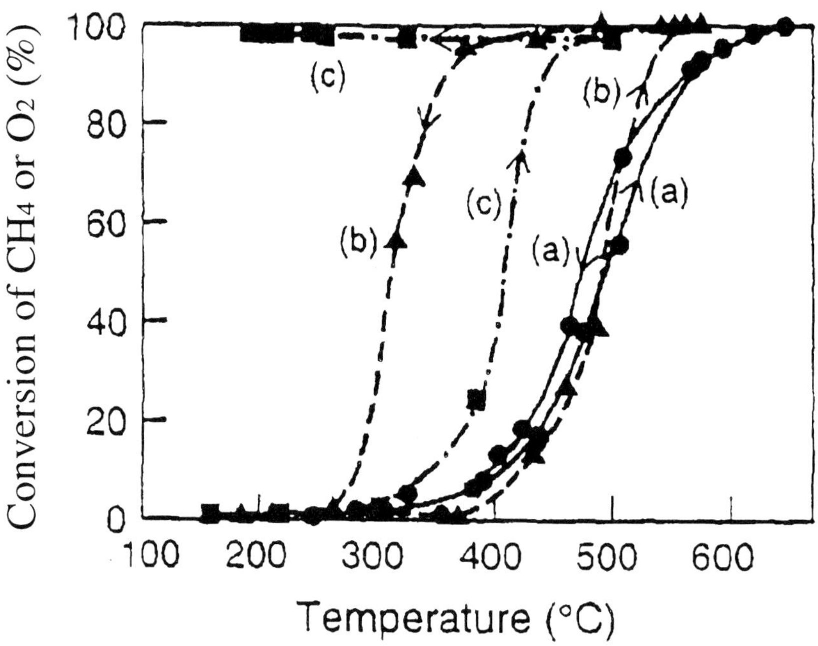

Figure 1. Hysteresis curves for catalytic activity observed using the forced-oscillating reaction method. Catalyst: 0.33 wt% Pt supported on a wash-coated ceramic form carrier; SV: 1,400 h^{-1}. (a) Oxidative catalytic condition; (b) Stoichiometric condition for complete combustion; (c) Reductive catalytic condition. In case of (a) and (b), the vertical axis expresses the experimental measurement where conversion of methane was actually observed. In case of (c), the axis expresses the conversion of oxygen.

Almost the same situations concerning the relation between the gas composition and the reactivity in CO oxidation were verified to occur on a supported Cu catalyst and Cu-containing A type zeolites [17]. Figure 2 shows a typical example, which indicates the result of simultaneous measurement of the activity expressed by O_2 conversion and the weight change of the copper in the catalyst, under reductive catalytic conditions. The figure clearly proves the cause of the activity change and hysteresis in the measurement of the forced-oscillating reaction.

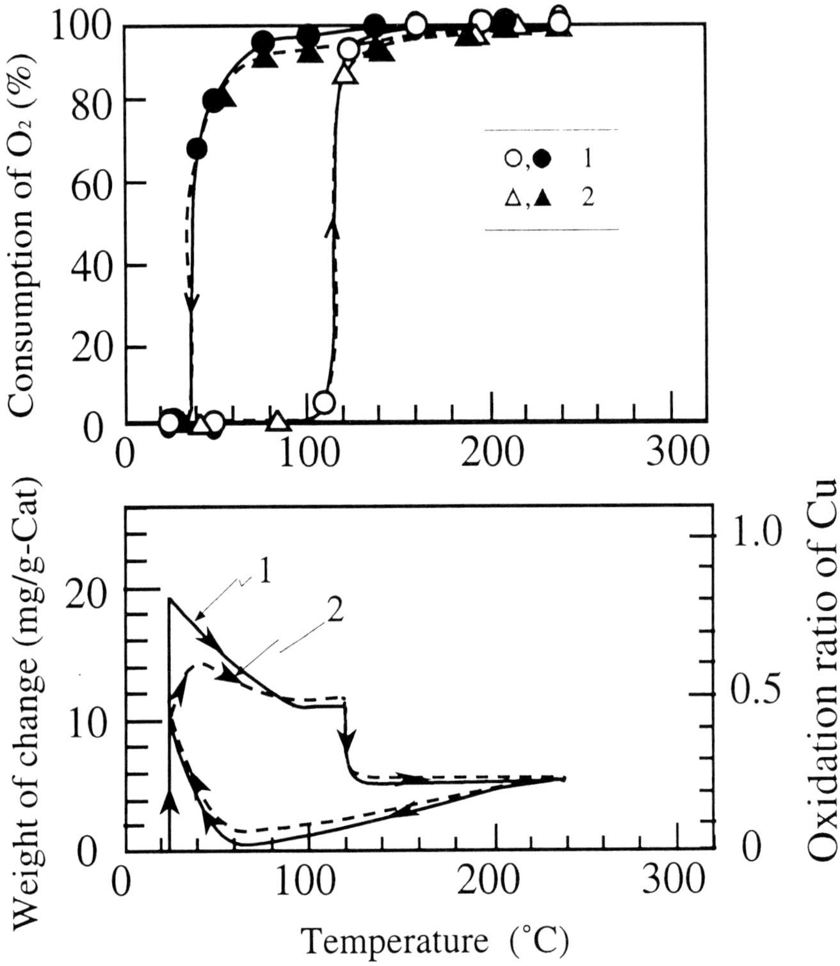

Figure 2. Relation between the activity in CO oxidation and the oxidation state of copper in a catalyst as observed by the forced-oscillating reaction method. Catalyst: 8.6 wt% Cu supported on silica gel calcined at 500°C; reaction gas: 5.0% O_2, 76.2% CO, 18.8% N_2; 350 ml/h; 50 mg catalyst; a gas chromatograph Shimadzu GC8A was used for the measurement.

Autocatalysis and Abrupt Change in Activity and Selectivity

Catalytic reaction processes such as the methanol to hydrocarbon conversion on H-ZSM-5 catalyst [18-20], and the oligomerization of light olefins [21], involve autocatalysis. The co-existence of a very low concentration of alcohols [22, 23] and a concentration of light olefins higher than a certain critical concentration [21] markedly enhances the reactions, respectively. By taking advantage of this phenomenon, an extraordinary high reaction rate was achieved using a high concentration of the reactant [24]. For example, on an H-Fe-silicate catalyst non-diluted propylene fed with a gaseous space-velocity of 4,500 h^{-1} almost completely converted to gasoline-range liquid hydrocarbons, which consisted mainly of internal iso-mono olefins, and the space-time yield amounted to a surprising value as high as 8.09 kg/l·h.

Retardation Caused by Reactants and Products

Different from complete conversion reactions such as combustion of hydrocarbons as an example of complete oxidation, or methanation of carbon oxides as a typical complete hydrogenation, the reactions, in which intermediate products are desired products, occur on the catalyst surface of intermediate oxidation states. For example, the oxidation of propylene on CuO_y supported on a ceramic carrier, the optimum performance for acrolein formation was achieved with y value of 0.41 [25]. The results are summarized in Table 1. As shown in this table, the y value at the steady state strongly depends on the initial oxidation or reduction state of the catalyst metal. The proportion of the adsorbates on the surface also determines the value of y as mentioned hereafter. The strong retardative action of small amounts of adsorbate was investigated by varying the reaction-gas composition and temperature over a wide range after previously setting y at a certain level as shown

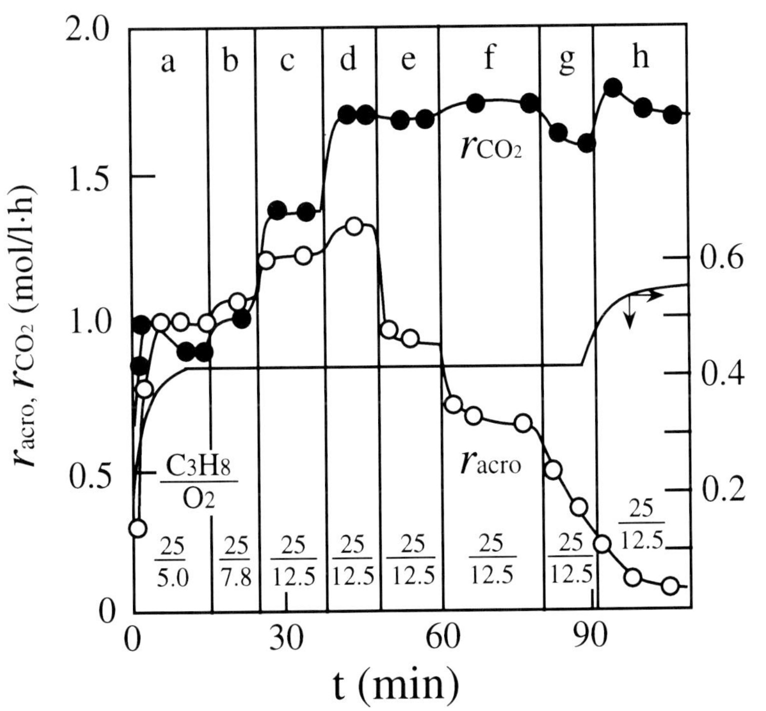

Figure 3. Responses of the oxidation state of copper (y in CuO_y) and rates (r) to changes in gas composition.

Table 1

Performance parameters for propylene oxidation starting from various oxidation states of the catalyst as measured by using a microgravimeter.

Exp. No.	Initial state of the catalyst	Oxidation degree		Rate and selectivity at the steady state							
		y_0	y	r_{acro} (mol/l·h)	S_{acro} (%)	r_{ach} (mol/l·h)	S_{ach} (%)	r_{CO2} (mol/l·h)	S_{CO2} (%)	r_{C3H8} [a] (mol/l·h)	r_{O2} [a] (mol/l·h)
(1) [b]	R	0.00	0.41	0.96	51.6	0.065	3.5	0.82	44.1	1.86	4.69
(2)	R → O	0.38	0.50	0.61	39.2	0.047	3.0	0.90	57.8	1.56	4.68
(3)	O → R	0.38	0.59	0.74	39.5	0.067	3.6	1.07	60.5	1.88	5.90
(4)	O → R	0.69	0.78	0.35	20.7	0.040	2.4	1.30	76.9	1.69	6.22
(5)	O	1.00	0.98	0.12	7.3	0.26	1.7	1.51	91.0	1.66	6.93

Note: Catalyst: 9.7 wt% Cu supported on a porous SiO_2-Al_2O_3 calcined at 1500°C, 20-30 mesh in size, 19 mg; Test conditions: 25% C_3H_8, 5% H_2O and 68% N_2, flow rate 33 ml/min, 300°C. R: completely reduced state (Cu); O: completely oxidized state (CuO); R → O: reduced catalyst was oxidzed by O_2 until the y_0 listed; O → R: oxidized catalyst was reduced with H_2 until the y_0 listed; y_0: initial y. Acro = acrolein; ach = acetaldehyde.

[a] Consumption rate.

[b] Small amount of propionaldehyde ($S = 0.6\%$) and acetone ($S = 0.2\%$) formed in this case.

in Fig. 3. After taking a constant y (0.41) with the reductive gas as well as the conditions in Table 1, the inlet gas composition was varied stepwise into an oxidative one, with increasing oxygen or decreasing propylene. Rate values (r) immediately showed a variation corresponding to the gas composition; however, y remained constant even with a very oxidative gas, such as the composition indicated as f in the figure.

Taking into account both the time for the initial change of y and for the change in the oxygen treatment, the interval between each conditions in Fig. 3 was sufficient for the change to occur if retardation did not exist. But in another experiment, even when the reaction time for condition f was extended by 120 min, no change occurred in the effluent gas composition or in y.

In the first instance, an abrupt increase of y occurred 10 min after the gas changed into an extremely oxidative composition, with propylene concentration decreasing to a very low level, as shown in g. It was considered that at this time the adsorbates were burned off, and their retardative effect for the oxidation of catalyst metal ceased. After that (period h), the gas composition was restored to that used in e. Then y attained a new level; namely 0.55, which was constant. However, r_{acro} was quite different from that previously in e, and had a very small value. The temperature was then raised at a constant rate to 400°C for 100 min; nevertheless, y did not change within the accuracy of the apparatus. Thus, once y had been fixed, it remained constant over a wide range of reaction conditions. On the basis of these facts the action of the adsorbates was deduced to be as follows;

(1) The adsorbates not only strongly retarded the oxidation and reduction of the catalyst metal, but also played an important role in controlling the selectivity to the partial oxidation product (acrolein) by retarding the rate of deep oxidation.

(2) Active sites, which are regenerated repeatedly by redox cycles, occupy only an extremely small portion of the catalyst surface because rates changed rapidly with quite constant y. This further suggests that the adsorbed oxygen and not the lattice oxygen is of significance.

(3) In particular, the complete-combustion activity is determined directly by the equilibrium bulk composition of copper and copper oxide mixtures.

III. INTRODUCTION OF THE APPROPRIATE STRUCTURAL DEVICE IN CATALYSTS TO CONTROL THE CATALYSIS

The high activity level of reduced Cu catalyst observed in the forced-oscillating reaction method can be induced also by creating an adequate catalyst structure. Fig. 4 shows a typical example which could be realized by combination of a very diluted Rh catalyst with a Cu catalyst supported on an alumina wash-coated ceramic fiber shaped in plate shape of 1 mm thickness [26]. A single ceramic fiber has a diameter of about 5~10 μm and void space of the support namely between fibers amounted to 88%.

An evidence of periodic self-oscillation in the reaction rate was observed in the rapid catalytic combustion of city gas (Composition of city gas: 46% H_2, 22% CH_4, 5% CO, 5% $C_2 \sim C_4$ hydrocarbon, 12% CO_2, 2% O_2, 8% N_2; city gas/air = 1/1.56) on a copper catalyst loaded in a proportion of 7 wt% on the ceramic fiber support. A typical example of the temperature change in a single cycle is shown in Fig. 4. The oscillations disappeared by the addition of small amount of platinum group metals, and a stationary oxidation occurred. The magnitude of the effect differed with the metal; the order was Rh > Pt > Pd. As shown in Fig. 4 the average space-time conversion (STC) of O_2 consumed by the city gas combustion on the 7% Cu catalyst during the cycle, was 173 mol/l·h. On the other hand, the catalyst containing 0.001% Rh on the same support exhibited an O_2 STC 45 mol/l·h without oscillations. When a very thin (80 μm) platelet of the 0.001% Rh catalyst was attached to the Cu catalyst of 1 mm thickness, the oscillation on the Cu catalyst stopped, and a surprisingly high stationary reaction level, i.e. 940 mol/l·h expressed

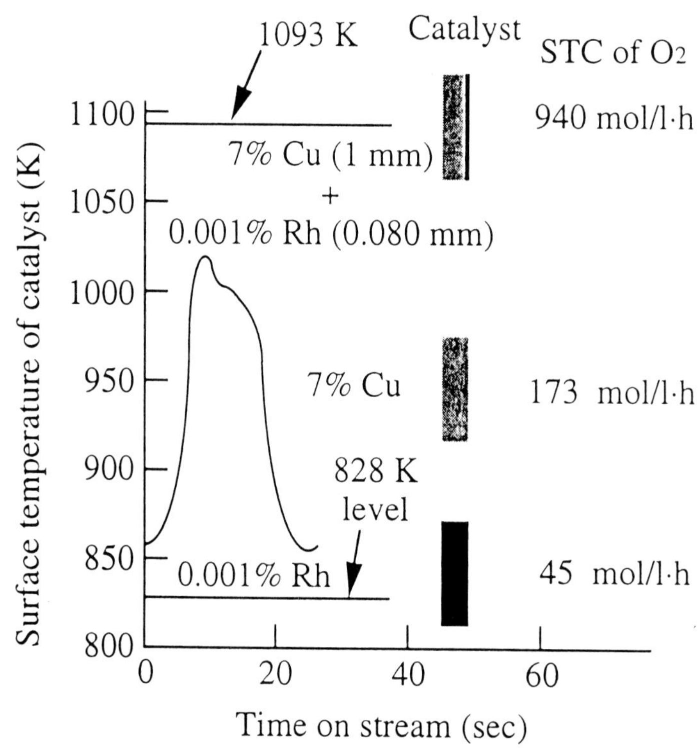

Figure 4. A prominent enhancing effect of the contact of Rh catalyst with a Cu catalyst upon the combustion of city gas. Catalyst volume: 0.028 ml (including void space); Composition of city gas: 46% H_2, 22% CH_4, 5% CO, 5% C_2~C_4 hydrocarbon, 12% CO_2, 2% O_2, 8% N_2; City gas/air = 1/1.56; Gas flow rate 340 l/h.

obtained on H-ZSM-5, but light paraffins also formed owing to the strong hydrogen-shift function of the Al in H-ZSM-5. When H-Ga-silicate was used in the second reactor, the selectivity and the space-time yield of gasoline increased to 53.6% and 328 g/l·h, respectively. The reason for the increase is ascribed to the fact that hydrogen is removed from the catalyst, because the Ga part in H-Ga-silicate plays a role of the porthole for inverse hydrogen spillover.

V. CONCLUSION

Non-linear phenomena in catalysis have been often overlooked or neglected in the analysis of the catalytic reaction kinetics and constructive application to improve the performance of catalyst. However, as overviewed in the present review article, it has a high potential for application especially in the energy and environmental problems in 21st century.

Acknowledgments

The present review reported results which are connected with deals with concerns one of the core parts of the author's researches on catalysis for the past 42 years, from 1956 to 1998, in Kyoto University. The author expresses his sincere thanks to Emeritus Professors H. Shingu, Y. Takegami, and all of the collaborators who contributed so much to conduct the studies and contributed greatly to the progress of research. Our work has been achieved thanks to the invaluable financial supports from Japanese Government, the Petroleum Institute of Japan, and many Companies acting in the chemical industries.

REFERENCES

1. H.S. Taylor, In: *A History of Industrial Chemistry*, Heineman Ltd., p. 411 (1957).
2. I. Langmuir, *J. Amer. Chem. Soc.*, **40**, 1391 (1918).
3. C.N. Hishelwood and C.R. Pritchard, *J. Chem. Soc.*, **127**, 1552 (1925).
4. O. Levenspiel, In: *Chemical Reaction Engineering*, John Wiley and Sons INC, p. 450 (1962).
5. T. Inui, Y. Miyamoto and Y. Takegami, *Stud. Surf. Sci. Catal.*, **17**, 181 (1983).
6. T. Inui, *Appl. Surf. Sci.*, in press.
7. Y. Nakazaki and T. Inui, *IEC Res.*, **28**, 1285 (1989).
8. T. Inui, M. Funabiki and Y. Takegami, *IEC Prod. Res. Dev.*, **19**, 385 (1980).
9. H. Higgins, *Ind. Eng. Chem.*, **59**, 19 (1976).
10. A.J. Lotka, *J. Phys. Chem.*, **14**, 27 (1910).
11. A.J. Lotka, *J. Amer. Chem. Soc.*, **42**, 1595 (1920).
12. P. Hugo and M. Jakubith, *Chem. End. Tech.*, **44**, 383 (1972).
13. M.G. Slinko and M.M. Slinko, *Catal. Rev. Sci. Eng.*, **17**, 119 (1978).

14. T. Inui and T. Iwana, *Stud. Surf. Sci. Catal.*, **19**, 205 (1984).
15. M. Kobayashi, T. Kanno, H. Takeda and S. Fujisaki, *Stud. Surf. Sci. Catal.*, **101** Part B, 1059 (1996).
16. T. Inui, H. Wakita and H. Fukuzawa, *MRS Intern. Mtg. on Advan. Mats.*, **2**, 273 (1989).
17. Y. Ono, M. Shibata and T. Inui, *J. Catal.*, submitted.
18. N.Y. Chen and W.J. Regan, *J. Catal.*, **59**, 123 (1979).
19. Y. Ono, Z. Imai and T. Mori, *J. Phys. Chem.*, **115**, 99 (1979).
20. Y. Ono and T. Mori, *J. Chem. Soc. Faraday I*, **77**, 2209 (1981).
21. T. Inui, *React. Kinet. Catal. Lett.*, **35**, 227 (1987).
22. B.E. Langner, *Appl. Catal.*, **2**, 289 (1982).
23. T. Inui, T. Takaki, T. Hagiwara and M. Yoshikawa, *Stud. Surf. Sci. Catal.*, **34**, 639 (1987).
24. T. Inui, F. Okazumi, J. Tarumoto, O. Yamase, H. Matsuda, H. Nagata, N. Daito and A. Miyamoto, *J. Jpn. Petrol. Inst.*, **30**, 249 (1987).
25. T. Inui, T. Ueda and M. Suehiro, *J. Catal.*, **65**, 166 (1980).
26. T. Inui, *Stud. Surf. Sci. Catal.*, **77**, 17 (1993).
27. T. Inui, M. Funabiki and Y. Takegami, *J. Chem. Soc. Faraday I*, **76**, 2237 (1980).
28. T. Inui, K. Ueno, M. Funabiki M. Suehiro, T. Sezume and Y. Takegami, *J. Chem. Soc. Faraday I*, **75**, 1495 (1979).
29. T. Inui, H. Hara, T. Takeguchi and J. Kim, *Catal. Today*, **36**, 25 (1996).
30. T. Yamamoto and T. Inui, *Proc. 4th Intern. Congr. on Carbon Dioxide Utilization*, Elsevier, 1998, in press.
31. E. Kikuchi, *CATTECH*, **1**, 67 (1997).
32. T. Inui, *Sekiyu Gakkashi,* **40**, 243 (1997).
33. H. Hara, T. Takeguchi, T. Inui, *Proc. 4th Intern. Congr. on Carbon Dioxide Utilization*, Elsevier, 1998, in press.

Res. Chem. Intermed., Vol. 24, No. 4, pp. 391-399 (1998)

SELECTIVE REDUCTION OF NO_X WITH C_3H_6 OVER CU INCORPORATED INTO SILICOALUMINO-PHOSPHATE (SAPO)

HIROYASU NISHIGUCHI*, SHUJI KIMURA, TATSUMI ISHIHARA AND YUSAKU TAKITA
Department of Applied Chemistry, Faculty of Engineering, Oita University, Dannoharu 700, Oita 870-11, Japan

Received 22 November 1997; accepted 23 December 1997

Abstract--Cu ion-exchanged SAPO-34 which is a highly active catalyst for selective reduction of NO with C_3H_6, was synthesized and the Cu ion was characterized in the SAPO-34 by ESR, XPS, IR, and XAFS. ESR study indicated that two kinds of Cu(II) with different environments exist in SAPO-34. The Cu(II) species disappeared immediately by C_3H_6 treatment and recovered by oxygen treatment. XAFS study indicated that most of the Cu(II) was reduced to Cu(I) under the presence of C_3H_6.

These studies revealed that a selective NO reduction by C_3H_6 over SAPO-34 should proceed by the redox reaction of a copper ion between Cu(I) and Cu(II).

INTRODUCTION

Aluminophosphate (AlPO) is a new family of molecular sieves and possesses the characteristics of a molecular sized pore structure and acidity. In particular, AlPO exhibits extremely high thermal stability compared to that of synthesized silica-alumina zeolite. It was reported that the crystal structure of AlPO-5 was sustained up to 1473 K [1]. Similarly, the crystal structure of SAPO-34 is also sustained up to 1273 K even in the presence of humidity [2]. Silico-aluminophosphates (SAPO) have a cation exchange property induced by isomorphous substitution of P in AlPO by Si. Therefore, AlPO and SAPO have a great possibility as thermostable catalysts. However, the advantages in thermal stability of SAPO or AlPO have not been thoroughly utilized in any conventional application.

Selective reduction of NOx under an oxidizing atmosphere with hydrocarbons has attracted attention as a new process for the catalytic removal of NOx in exhaust gases of diesel or lean-burn engines [3-6]. It is reported that Cu-ion exchanged ZSM-5 is highly active for the selective reduction of NO [7]. However, the high activity of Cu-ZSM-5 for NO reduction decreased gradually with the reaction time at temperatures higher than 973 K. One reason for this decreased activity seems to result from the poor thermal stability of ZSM-5 [8]. In addition, dealumination easily proceeds on ZSM-5 under a humidified atmosphere resulting in the deactivation for NOx reduction. We have reported that, the Cu-SAPO-34 has high activity for NOx reduction with C_3H_6 under an oxidizing atmosphere with high thermal stability [9,10].

In this study, in situ oxidation states of copper ion in SAPO-34 were investigated to clarify the mechanisms of selective reduction of NO with C_3H_6.

EXPERIMENTAL

SAPO-34, (Si content; 1.77 mmol•g^{-1}) was synthesized according to the Patent [11]. Synthesized SAPO was then ion exchanged with Cu(II) in Cu(II) acetate aqueous solution. Exchanged amounts of Cu(II) for SAPO-34 were estimated to be about 3 wt% from ICP analysis. Before measuring the activity for NO reduction with C_3H_6, the catalysts were calcined at 773 K for 4 h in a He stream.

The catalytic activities for NO reduction were measured by using a fixed-bed micro flow reactor. A mixed gas consisting of NO (1000 ppm), C_3H_6 (1000 ppm), O_2 (5%), and He (balance) was fed to the reactor at W/F=0.3 g-cat•s•cm^{-3} (S.V.= c.a. 8500 h^{-1}), where W and F stand for the catalyst weight and flow rate, respectively. Conversion into N_2 was determined from the analysis of N_2 concentration in effluent gas by gas chromatography.

Valence of Cu ions in SAPO-34 was determined by ESR and XPS (Shimadzu ESCA-850) and XAFS (EXAFS and XANES) spectra. The ESR measurement was performed at room temperature using ESP380 (Brüker) and diphenylpicrylhydrazyl (DPPH) was used as an external standard for the calibration of g value. The XPS measurement of the ion-exchanged Cu in SAPO-34 was performed after the pretreatment in oxygen (1 atm) at 623 K for 3 h in the pretreatment chamber.

The XAFS spectra were recorded at room temperature using EXAC820 (Technos) equipped with an in situ cell. (EXAFS: 20 KV 100 mA Ge(220), XANES: 25 KV 200 mA Ge(220)).

Before the ESR and XAFS measurement, the catalyst was evacuated at 773 K for 5 h and calcined in O_2 (100 torr) at 673 K, for 4 h and then evacuated at 573 K, for another 4 h.

RESULTS AND DISCUSSION

It was reported that the activity of ZSM-5 was strongly affected by an ion-exchanged metal cation [12]. Reduction of NO with C_3H_6 proceeded on all the SAPO-34 catalysts exchanged by metal cations with almost the same maximum conversion of

NO into N_2. However, temperatures for the maximum NO conversion strongly depend on the metal cations by which SAPO-34 is exchanged. Fe-SAPO-34 exhibited a high activity in the low temperature range from 473 to 873 K. On the other hand, high activity was attained at temperatures higher than 873 K on Ag-SAPO-34. Among the metal cations examined in this study, copper is found as the most suitable ion-exchanged cation for SAPO-34 considering the high catalytic activity and the wide temperature range showing high NO conversion [9].

Oxidation State of Cu in SAPO-34 Under NO Reduction

It is reported that the valence of Cu in ZSM-5 changes between monovalent and divalent under the direct decomposition of NO and the adsorbed NO is reduced to N_2 by the redox of Cu ion [13]. Therefore, it is also expected that the redox of Cu ion has an important role for the selective reduction of NO. The ESR spectra of Cu give useful information on the valence of copper [14]. Figure 1 shows the ESR spectra of Cu in SAPO-34 after heating at 623 K for 1 h under C_3H_6, NO, and O_2. The ESR signals which could be assigned to Cu(II) were observed on Cu-SAPO-34

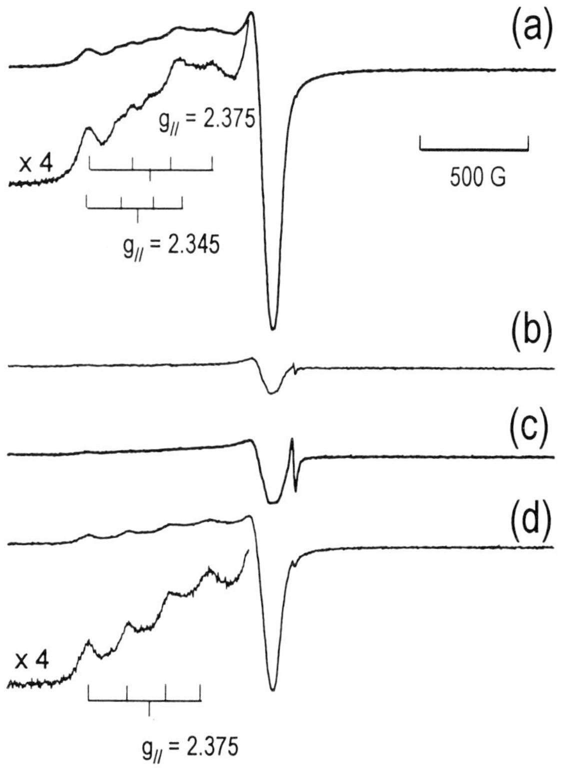

Figure 1. ESR spectra of Cu-SAPO-34 in various treatments. (a) pretreatment, (b) C_3H_6 100 torr at 623 K 1 h, (c) C_3H_6, NO, O_2 at 623 K 1 h, (d) re-oxidized (623 K, O_2 200 torr).

after evacuation at 523 K as shown in Figure 1(a). Furthermore, two kinds of hyperfine structure possessing different g values were observed in the ESR spectra of Cu-SAPO-34. This suggests that two kinds of Cu(II) with different environments were highly dispersed in SAPO-34. These two kinds of Cu(II) in SAPO-34 were also reported by Zamadics and assigned to Cu(II) existing near the plane of the six-membered ring and those in the hexagonal prism [15]. On the other hand, the ESR signals assigned to Cu(II) have almost disappeared by heating in C_3H_6 at 623 K for 1 h as shown in Figure 1(b). Although it is reported that the ESR spectrum of Cu(II) ions appear when NO was exposed over the Cu-Y zeolite reduced with CO, introduction of NO at 623 K gave no change in the ESR spectrum of Cu-SAPO-34. This may suggest that reduced state of Cu in SAPO-34 is easy to form by C_3H_6 treatment compared with Y-zeolite. However, the ESR signals assigned to Cu(II) as well as the hyperfine structure were restored by following calcination at 623 K for 5 h in O_2. Since the ESR signal of Cu(II) was restored by oxidizing treatment, the disappearance of ESR signal of Cu(II) by C_3H_6 treatment can not be explained by the aggregation of Cu ions but the reduction of Cu(II). The similar changes in ESR spectra with the treatment of C_3H_6 are reported for Cu ion exchanged KY [16] or ZSM-5 zeolite [17].

The spectrum (c) in Figure 1 shows the Cu-SAPO-34 catalyst which is quenched to ambient temperature during the NO decomposition reaction at 623 K. The signal of Cu(II) is somewhat larger than that of spectrum (b) suggesting that the Cu-SAPO-34 sample during the catalytic reaction possesses a significant amount of Cu(II). Figure 1(d) is the spectrum of the sample after the treatment at 100 torr for 30 min. The signal of Cu(II) with hyperfine structure again appeared. It can be concluded from these results that the selective reduction can be effected by the redox reactions between Cu(II) and Cu(I) and that most of the Cu exists as Cu(I) during the reaction.

The change in the valence of Cu ions on the catalyst surface was also studied by means of XPS [9]. After heating in O_2 at 723 K followed by the evacuation at room temperature peaks were produced which have binding energies of $Cu2p_{3/2}$ and Cu LMM consistent with those of CuO and the shakeup satellite peaks at 940 to 950 eV suggesting that the oxidation state of Cu in SAPO-34 seems to be Cu(II). The XPS results are fully consistent with the results of ESR measurements. On the other hand, the calcination in C_3H_6 at 723 K caused the disappearance of satellite peaks, the shift of $Cu2p_{3/2}$ peak to lower binding energy, and a small shift of Cu LMM peak to higher binding energy. The XPS spectrum of Cu in SAPO-34 after calcination in C_3H_6 was similar to that of Cu_2O, but dissimilar to metallic Cu. This clearly shows that Cu(II) is reduced to Cu(I) by the calcination of Cu-SAPO-34 in C_3H_6 at 723 K.

To estimate valence of Cu at the working state in the catalytic reaction, XPS measurement of the Cu-SAPO-34 catalyst quenched to ambient temperature during

NO reduction at 623 K was carried out. The spectrum obtained closely resembled that of Cu-SAPO-34 after calcination in C_3H_6.

Figure 2 shows the XANES(I) and EXAFS(II) spectra at the Cu K-edge of Cu-SAPO-34 and reference compounds of Cu. At the K edge, the absorption event involves the promotion of a 1s core electron to some bound states near the continuum. The energy of the final state and of the corresponding resonance will therefore depend upon the chemical bonding of absorbing element. As can be seen from Figure 2, the spectrum (d) of as synthesized Cu-SAPO-34 is almost identical to that of CuO (spectrum c). On the other hand, The XAFS spectra of Cu_2O (b), Cu-SAPO-34 treated in C_3H_6 (e) and Cu-SAPO-34 treated in the reaction gas mixture (f) exhibited sharp pre-edge peak at around 8985 eV due to the 1s - 4pπ* transition. This sharp pre-edge peak is assigned to the peak of Cu(I) species.

Figure 3 shows the normalized FT-EXAFS spectra of Cu-SAPO-34 derived

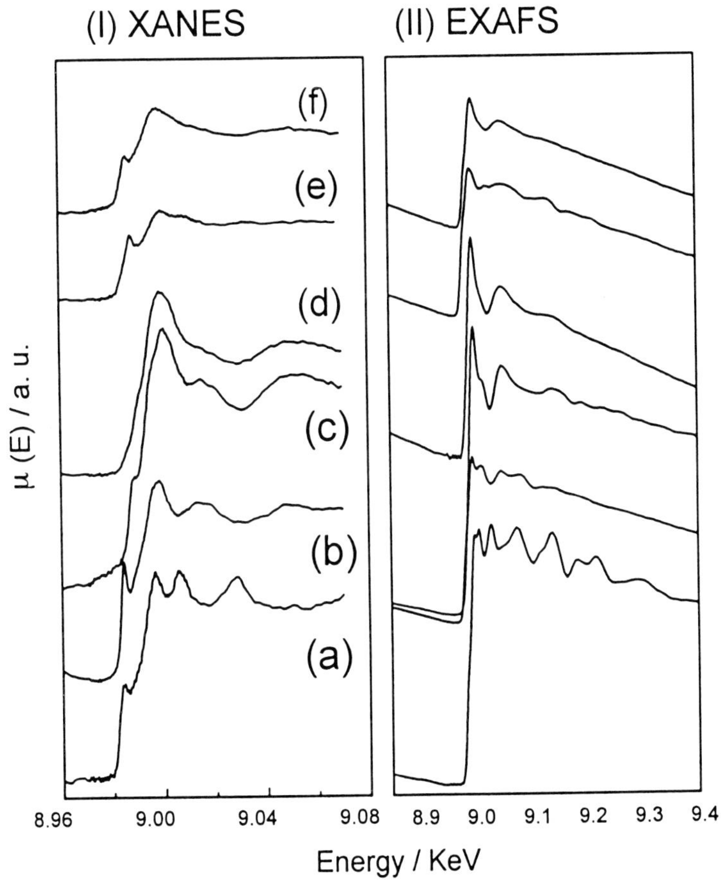

Figure 2. In-situ XANES (I) and EXAFS (II) spectra at the Cu K-edge of Cu-SAPO-34 and Cu related samples. (a) Cu metal, (b) Cu_2O, (c) CuO, (d) Cu-SAPO-34 as synthesized, (e) 623 K in C_3H_6 (100 torr), (f) 623 K in NO (100 torr), C_3H_6 (100 torr), and O_2 (100 torr).

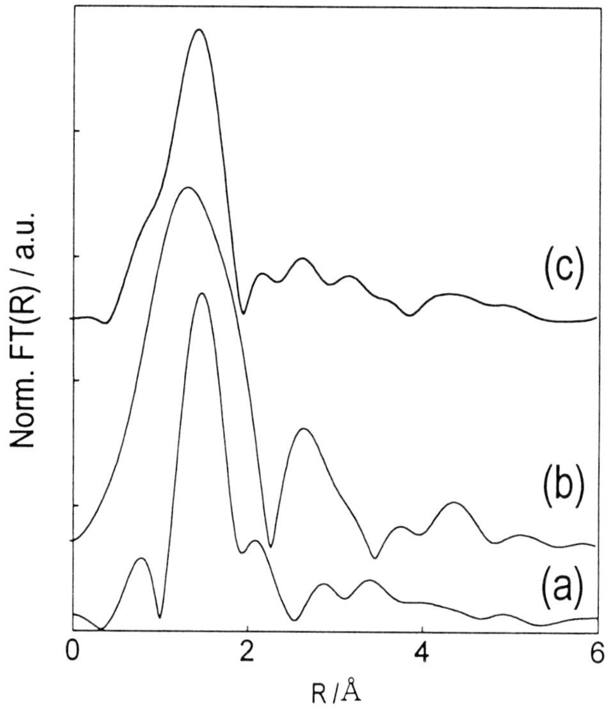

Figure 3. FT-EXAFS spectra of Cu-SAPO-34. (a) as synthesized (b) 623 K in NO, C_3H_6, and O_2 (c) after reaction.

from EXAFS spectra shown in Figure 1. All data are presented without any correction for phase shifts.

All Cu-SAPO-34 samples exhibit a large peak at around 1.8 Å which is attributed to the neighboring O atoms (Cu-O) of the zeolite framework and surface OH groups or water and did not show the peak of neighbouring copper atom (Cu-Cu).

Kevan *et al.* have reported that the ESR parameters of the hydrated Cu-SAPO-34 are characteristic of octahedrally coordinated Cu(II) and simulation of the ESEM (Electron Spin Echo Modulation) signal indicated that Cu(II) ion is directly coordinated to three water molecules and three framework oxygens [21]. The distance of Cu-O is about 2.1 Å.

Table 1 shows the results of the curve fitting of Cu K-edge EXAFS data for Cu-SAPO-34. The coordination number (CN) of as synthesized sample was 4.8 suggesting that the mixture of Cu(II) with 4 coordination and 6 coordination, perhaps $Cu(O_F)_3(OH)$ and $Cu(O_F)_3(H_2O)_3$, where O_F refers to the framework oxygen of SAPO-34. By the introduction of reaction gas mixture at 623 K, CN of Cu-SAPO-34 decreased to 2.7. This may be the state of $Cu(O_F)_3$. Cooling to ambient

Effect of Hydrocarbon Injection

A separate study was also conducted whereby a reactive hydrocarbon, C_4H_8 was injected by a pulse during CO oxidation at 600 °C to deposit carbon intentionally on the Al_2O_3. The injection was made intermittently after steady condition was reached. Fig. 6 shows the results. During the first 120 minutes, no hydrocarbon was injected and the CO conversion reaches a steady value of 26%. 4 ml C_4H_8 was injected after 137 minutes and the CO conversion goes up to 37%. After 160 minutes, an additional 3 ml of the hydrocarbon was injected and the maximum CO conversion was attained at 45%. No more hydrocarbon was injected after this and the activity declines to an almost constant value of 41%. This is indicative of carbon formation from injected C_4H_8 and its positive role for CO oxidation.

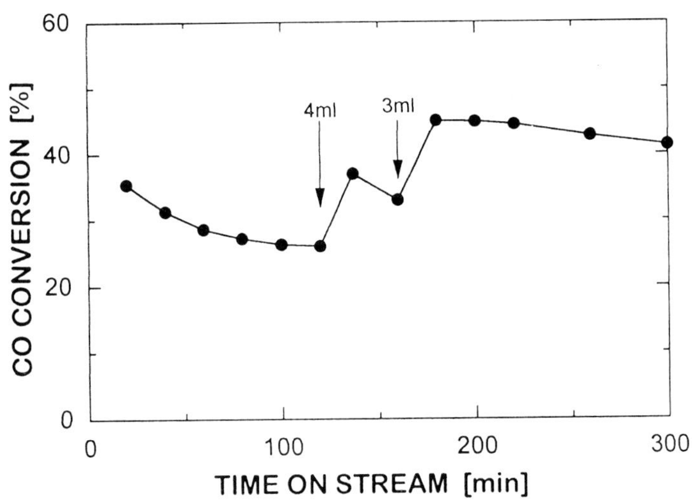

Figure 6. Time course activity test during CO oxidation on alumina when C_4H_8 is injected by pulse. Arrows and values beside the arrows express the injection times and the injection volumes.

The surface area of the used catalysts from the CO oxidation experiments with and without hydrocarbon injection was measured: the fresh one being 142 m^2/g, the used one without hydrocarbon injection being 154 m^2/g and used one with hydrocarbon injection being 158 m^2/g. The area of the used catalysts is larger than that of the fresh one.

Thermogravimetric Analysis

A separate study on determining the optimal temperature where maximum carbon is deposited during methanol oxidation on Al_2O_3 was also done using thermogravimetric analysis. The catalyst was placed in TG apparatus and subjected to the same reaction atmosphere used in the flow reactor system. A rate of temperature change of 5°C/minute was programmed during the heating from 400°C to 700°C and the cooling from 700°C to 400°C. Fig. 7 shows the result. The carbon formation, as evidenced from the weight increase, occurs from 400°C to 550°C while its combustion starts to occur at 550°C. The maximum amount of carbon deposition occurs at a temperature of 550°C. These results clearly demonstrate that the carbons on the surface are in a dynamic balance of deposition and combustion: at the lower temperature the carbons are formed, and at the higher temperature they are combusted.

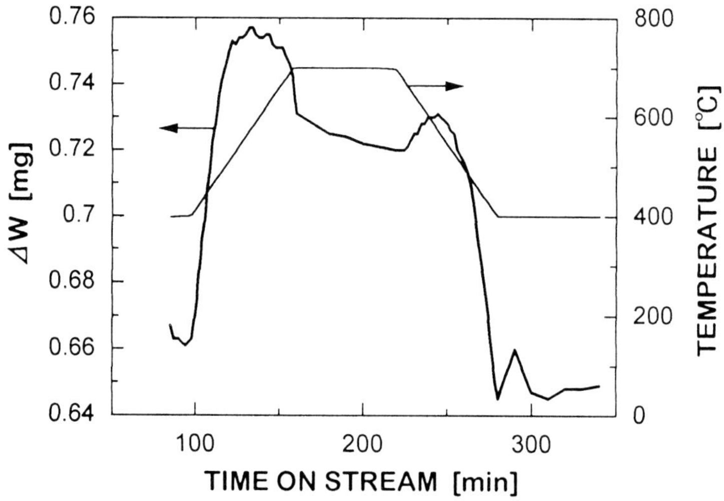

Figure 7. Thermogravimetric analysis to determine optimal temperature for carbon formation during methanol oxidation using 5°C/min.

Nature of Active Sites: In-situ formation of carbon

Several results discussed above gives the concept of an in-situ prepared catalyst on Al_2O_3: that is, carbon deposit produced during the catalysis becomes an active site. Here we discuss on what kind of site the carbon is produced. It seems reasonable to consider that the active sites for CO oxidation are the same as that for methanol oxidation.

In order to understand the rather complicated results given in Fig. 3, it might be good to first review the probable reaction paths with participating active sites.

It can be seen in the Fig. 3 that, at the high temperature range (600 - 700°C), the CO_2 yield is lower while CO and CH_4 yields are higher after experiencing highest temperature operation, where carbon is combusted. It is clearly interpreted based on, (1) oxidation is accelerated by carbon deposition, (2) carbon formation occurs

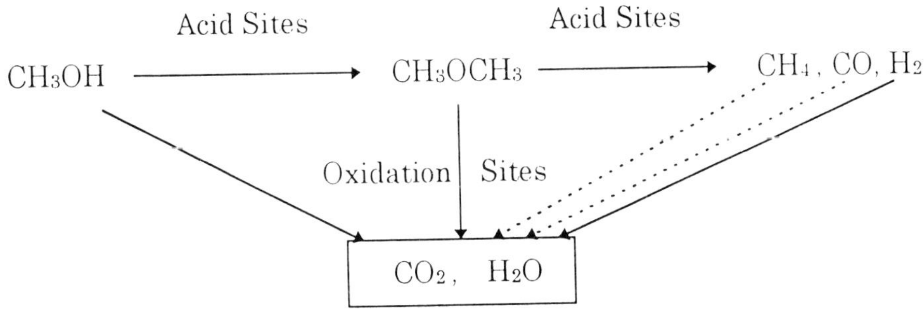

Reaction Mechanism of Methanol Oxidation

on acid sites, and (3) carbon formation occurs at 450 - 500°C.

The comparative study of methanol oxidation on Al_2O_3 and the alkaline treated one gives insight into the nature of active sites upon which carbon is formed and its relation to hysteresis. The lessening of surface acidity of Al_2O_3 may have led to the suppression of carbon formation and leading to the disappearance of the hysteresis for any one of the products, as revealed in Fig. 4. Another comparative study is shown in Fig. 5. As discussed earlier, the ratio of CO_2/CO increases with time until a steady activity is attained. This ratio becomes smaller with time when the oxidation of methanol takes place on CeO_2-Al_2O_3 where no carbon is formed. This gives us an idea that carbon formation increases the selectivity to CO_2 because carbon reduces the acidity of the Al_2O_3 surface.

The small temperature dependency of CO oxidation on Al_2O_3 in Fig. 1 may be attributed to two reaction mechanisms that are taking place. At lower temperature, CO oxidation takes place on the carbon sites and at higher temperature, oxidation takes place on carbon-free Al_2O_3. Hence the carbon deposit decreases with temperature, it partially compensates the increase of reaction rate with temperature. A clearer result was shown earlier in methanol oxidation where carbon formation is expected to occur more easily than in CO oxidation.

Finally, the surface areas of the used catalysts were larger than the fresh alumina, indicating the formation of new surface being developed by the reaction deposition.

All these results show that carbon formation plays a vital role in CO oxidation on Al_2O_3. At the same time, it is confirmed that methanol oxidation also proceeds on Al_2O_3. Active sites formed by carbon is very interesting to study further in relation to auto emission control, because unburned hydrocarbons are present in the emission which may become carbon sources under practical conditions.

Acknowlegement
The authors wish to thank the Japan Society for the Promotion of Science and the Department of Science & Technology of the Philippines for the support given.

REFERENCES

1. J. B. Claridge, M. L. H. Green., *Catalysis Letter* **22**, 299(1993).
2. A. Takano, T. Tagawa and S. Goto, *Sekiyu Gakkaishi* **39**, 144 (1996).
3. S.B. Tang, *Catalysis Today* **24**, 253 (1995).
4. T. Osaki, T. Horiuchi, K. Suzuki and T. Mori, *Catalysis Letter* **35**, 39 (1995).
5. Y.Q. Deng, T. G. Nevell and M. G. Jones, *Catalysis Letter* **34**, 313 (1995).
6. Y. Kintaichi, H. Hamada, M. Tabata M. Sasaki, T. Ito, *Catal. Lett.*, **6**, 239 (1990).
7. T. Aida, R. Kawakami, H. Hayashi and H. Niiyama, *Proceedings of the Regional Symposium on Chem. Eng.*, Oct. 1994, Manila, Philippines.
8. S. Gallardo, T. Aida and H. Niiyama, *Proceedings of the Regional Symposium on Chem. Eng.*, Oct. 1995, Bangkok, Thailand.
9. S. Gallardo and H. Niiyama, *DLSU Eng. Journal* XI. 57 (1997).

Res. Chem. Intermed., Vol. 24, No. 4, pp. 411-423 (1998)

IR OBSERVATION OF ADSORPTION AND INITIAL REACTIONS OF OLEFINS ON BRØNSTED ACID SITES OF A DEUTERATED ZSM-5

KAZUNARI DOMEN[1]*, JUNKO N. KONDO[1] and FUMITAKA WAKABAYASHI[2]

[1]*Research Laboratory of Resources Utilization, Tokyo Institute of Technology, 4259 Nagatsuta, Midori-ku, Yokohama 226, Japan*

[2]*Department of Science and Engineering, National Science Museum, 3-23-1 Hyakunin-cho, Shinjuku-ku, Tokyo 169, Japan*

Received 10 September 1997; accepted 4 November 1997

Abstract -- Adsorption of linear olefins (C_2-C_4) on a deuterated H-ZSM-5 (D-ZSM-5) was studied by infrared (IR) spectroscopy. The initial interaction of the olefins with Brønsted acidic OD groups was hydrogen-bonding to form π-complexes at low temperatures. The adsorbed ethene and propene desorbed by heating under evacuation, while various reactions took place for adsorbed 1-butene; double bond migration (DBM) to 2-butene below 230 K followed by dimerization at room temperature. An unusual reaction path was deduced for DBM of 1-butene, where proton transfer from Brønsted acid sites (BAS) to the adsorbed 1-butene was not essential.

INTRODUCTION

Zeolites are representative materials of solid acids, which possess ordered microprobes structure and show strong acidity. The emergence of synthetic zeolites has enabled their use as practical catalysts in many industrial plants [1-5]. The primary studies on the acidic sites on zeolites were carried out by IR observation of Brønsted acidic OH groups. On the other hand, Lewis acidic sites (LAS) generated by dehydration of zeolites were probed by IR studies on pyridine adsorption as well as ammonia temperature programmed desorption (TPD). Both of the techniques have been used to characterize the acid strength and amount of BAS and LAS independently [1,5,6]. Recent IR studies have employed small molecules with weak basicity as probe molecules to detect strong acid sites even in small pores [7,8]. The interaction and reaction of BAS with the reactants of practical systems such as alcohols [9-16] and olefins [14-24] have been recently examined by NMR [9,10,16-18] and IR [11-15, 19-24]. Various types of cyclo- or π-conjugated carbocations have been detected by NMR very recently during acid catalyzed reactions [9,10,16]. The existence of such carbocations supports that the same reaction mechanisms as those in acid solutions are applicable on solid acids. However, the results from theoretical calculations point out the difference of reaction intermediates on solid acids from acidic solutions; the stable intermediate is not ionic on solid acids but covalent bonding is rather likely [25].

In this study, initial interaction of BAS of D-ZSM-5 with some linear olefins was observed at low temperatures by IR methods. Various reactions of the adsorbed 1-butene found below room temperature are shown.

EXPERIMENTAL

A thin disk of a H-ZSM-5 zeolite (Si/Al = 50) was placed in an IR cell connected to a closed gas-circulation system. The sample disk was treated under the condition that no LAS were produced [8]. Deuteration of H-ZSM-5 to form D-ZSM-5 was conducted by exposing the H-ZSM-5 disk to D_2 at 673 K. A small amount of butenes (less than 3 μmol) was introduced below 150 K and was immediately evacuated. Less than 80% of BAS are occupied by adsorbate, and the intermolecular interaction was thus avoided. The IR cell was then gradually warmed (*ca.* 6 $K{\bullet}min^{-1}$) while evacuating. If the reactant remains in gas phase at higher temperatures, protonation and reaction of the protonated species with molecules supplied from gas phase occur simultaneously. On the other hand, this method enables the observation of the initial interaction of BAS and adsorbed olefins, and the following stepwise thermal reactions.

RESULTS AND DISCUSSION

Adsorption of Ethene and Propene

Ethene and propene adsorbed on D-ZSM-5 show IR spectra shown in Figure 1. Since a background spectrum of D-ZSM-5 measured in evacuation was subtracted, a negative band at 2671 cm^{-1} appeared due to the decrease of isolated OD band. The isolated OD band shifted to the hydrogen-bonded one appearing at 2380 and 2315 cm^{-1} for ethene and propene adsorption system, respectively. The structures of hydrogen-bonded 1:1 complex of the acidic OD groups with C=C bonds of ethene and propene are illustrated as follows.

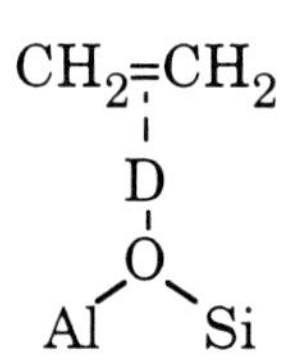

and

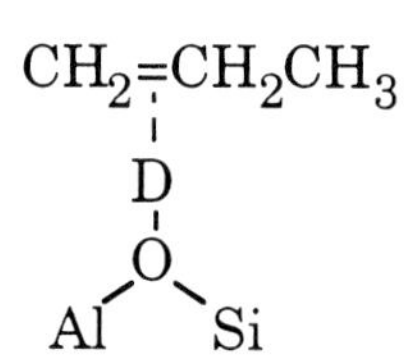

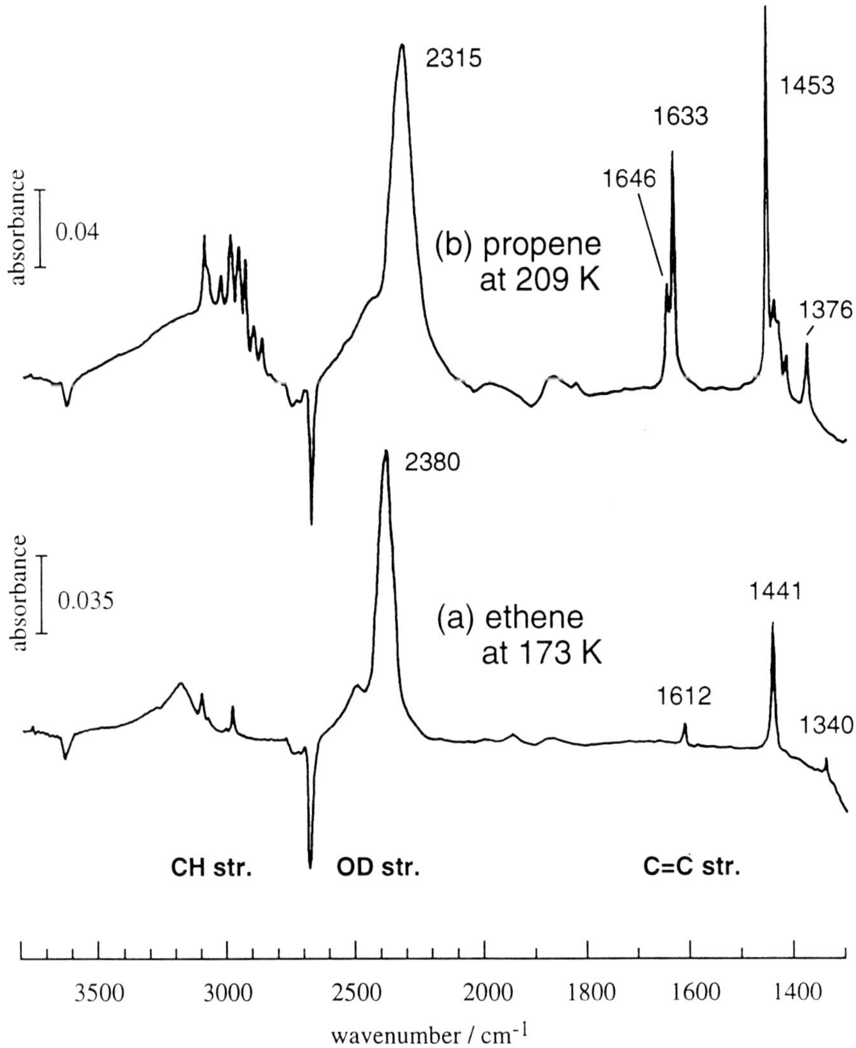

Figure 1. IR spectra of ethene and propene adsorbed on D-ZSM-5 in evacuation at 173 and 209 K, respectively.

The C=C stretching bands of ethene and propene in the π-complex appeared at 1612 and 1633 cm^{-1}, respectively, while those of free molecules are observed at 1623 (Raman) and 1647 cm^{-1}, respectively. The appearance of an originally IR inactive mode of ethene and the low frequency shift of the band of propene are consequences of the participation of the C=C bond in the interaction. Another C=C band of propene at 1646 cm^{-1} is due to the weakly adsorbed propene in the same analogy of the weakly adsorbed 1-butene as mentioned below in detail.

The spectra of π-bonded (molecularly adsorbed) ethene and propene were observed only below room temperature in evacuation. Complete disappearance of all the IR bands occurred at 273 K in the case of ethene, which indicates a simple molecular adsorption-desorption of ethene in the present experimental condition. In the case of propene, gradual desorption was observed by increasing temperature in evacuation to 273 K followed by oligomerization of the remaining species (less than 20%). The detailed reaction mechanisms of oligomerization and polymerization have been reported by Zecchina's group [20, 21], which take place in the presence of gaseous molecules.

Adsorption of 1-Butene on D-ZSM-5 below 210 K

IR spectra of 1-butene on D-ZSM-5 at low temperatures are shown in Figure 2. The spectrum of D-ZSM-5 measured in the absence of adsorbates was subtracted. 1-Butene was introduced at *ca.* 140 K followed by evacuation and gradual heating to 206 K. Either decrease of the isolated OD band (2671 cm^{-1}) or increase of the hydrogen-bonded OD band (2000-2500 cm^{-1}) was not clearly observed at 153 K, while bands due to the adsorbed 1-butene were evident. The IR spectra of adsorbed 1-butene are characterized by bands of olefinic CH stretching at 3078 cm^{-1}, C=C stretching at 1627 cm^{-1} and skeletal vibrations between 1500 and 1300 cm^{-1}. The adsorbed 1-butene at 153 K showed its C=C stretching band at the same frequency as that in Ar matrix [26]. Therefore, the C=C bond of the adsorbed 1-butene at 153 K is considered to be free from any interaction with the acidic OD groups. By gradual heating, conversion of the peak position of C=C stretching band was observed from 1641 to 1627 cm^{-1}, while maintaining the same intensity of the other bands due to 1-butene. The transformation of C=C stretching band was accompanied by the alteration of the OD band from the isolated to the hydrogen-bonded; decrease of the isolated OD band and increase of the hydrogen bonded one (2302 cm^{-1}). The linear correspondence of the C=C stretching band at 1627 cm^{-1} and the decrease of the isolated OD bands clearly indicate the 1:1 interaction of the OD groups of BAS with the adsorbed 1-butene showing its C=C stretching band at 1627

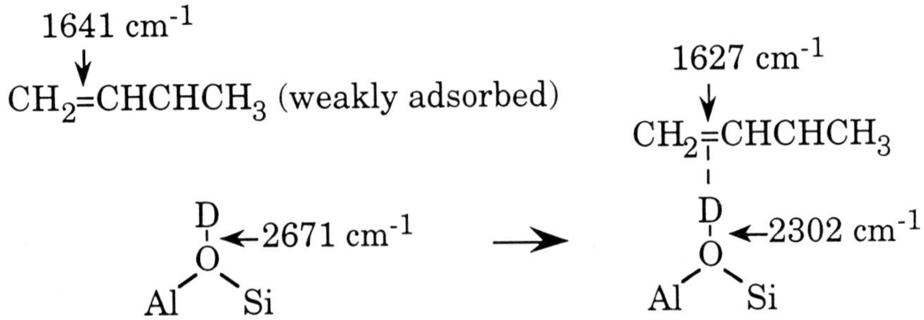

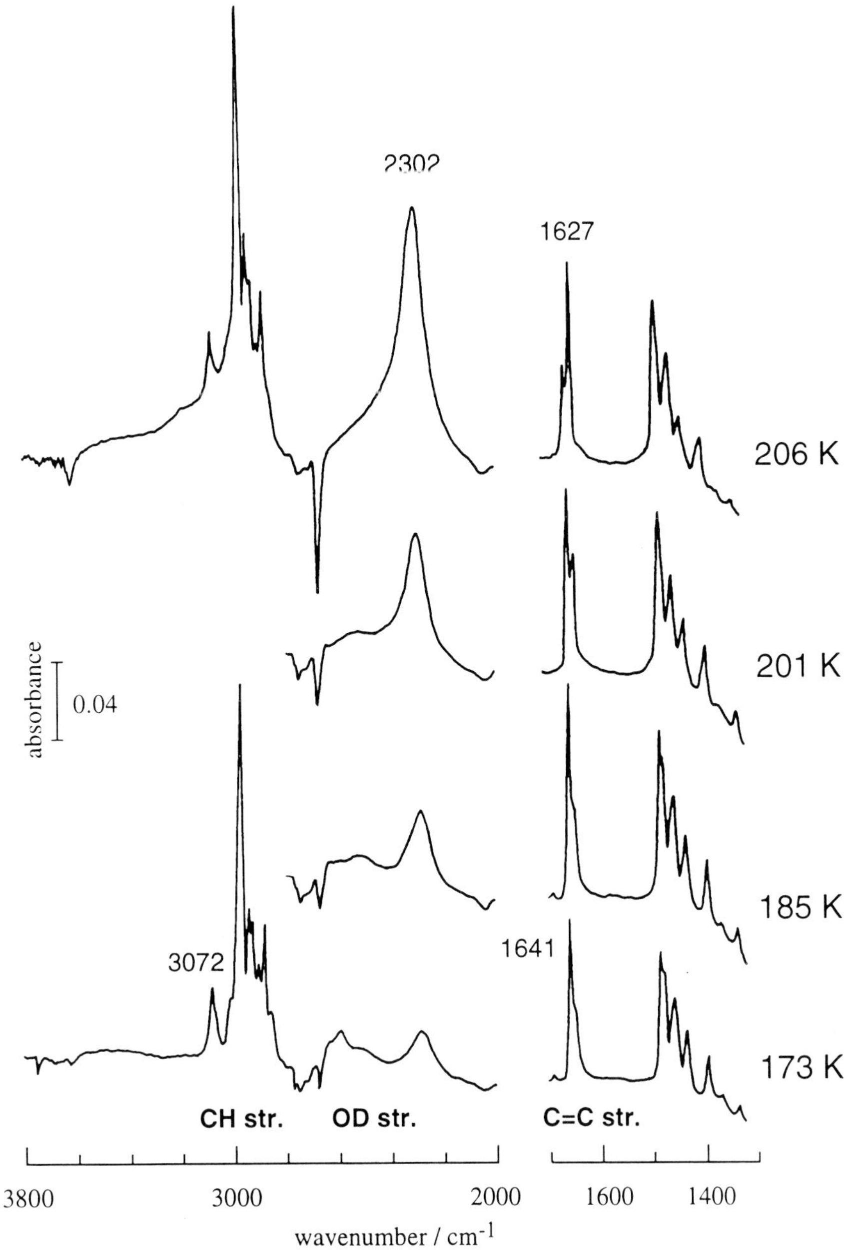

Figure 2. IR spectra of 1-butene on D-ZSM-5 at low temperatures.

cm^{-1}. The lower C=C stretching frequency of 1-butene in 1:1 π-complex than that of the weak adsorbates is a result of the direct interaction of π-electron of 1-butene with BAS.

Reaction of the Adsorbed 1-Butene on BAS

A reaction of the π-bonded 1-butene was noticed by further heating from 200 K. Increasing the temperature, IR spectrum started to change at 222 K from that measured at 204 K (Figure 3). The increase of the peak intensity of the reverse band of the isolated OD band (BAS) as well as those of the bands at 2302 and 1627 cm^{-1} at 222 K is attributed to the increase of the amount of the 1:1 π-complex of 1-butene with BAS as mentioned above. At 230 K IR spectrum completely changed. All the

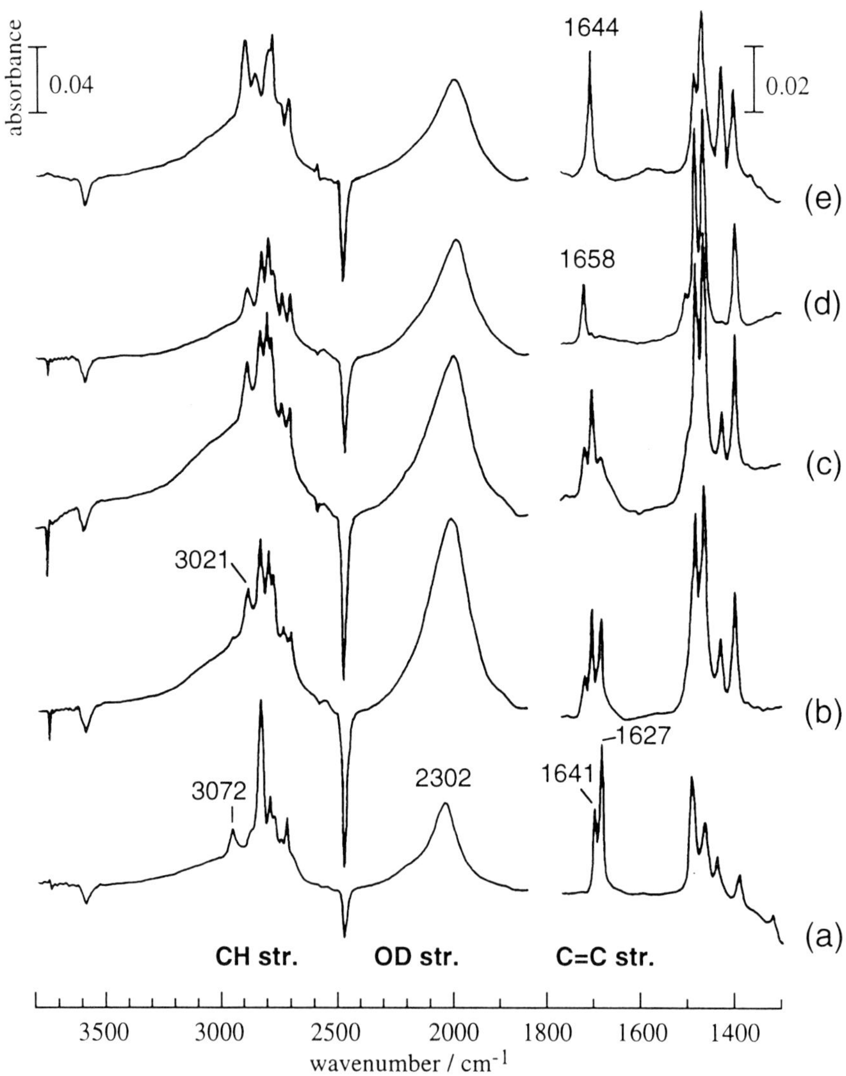

Figure 3. IR spectra of adsorbed 1-butene heated to 204 K (a), 222 K (b) and 230 K (c) and those of trans- (d) and cis- (e) 2-butenes.

characteristic bands of the adsorbed 1-butene almost disappeared. Alternatively, new bands appeared and increased in intensity; an olefinic CH stretching band at 3021 cm^{-1}, two bands in the C=C stretching region at 1658 and 1644 cm^{-1} and those in the skeletal vibration region. These results evidence the occurrence of the reaction of most of the 1-butene in a 1:1 π-complex with BAS below 230 K. It should be noted that the hydrogen-bonded OD band only shifted from 2302 to 2258 cm^{-1} but still remained after the reaction. The absence of any evident hydrogen-bonded OH band was noticed in Figure 3(c). Therefore, it is concluded that the reaction of 1-butene proceeds without H/D isotope exchange of the OD groups of BAS. At higher temperatures than 230 K, the intensity of the hydrogen-bonded OD band decreased accompanied by the growth of the hydrogen-bonded OH band around 3100 cm^{-1}, indicating the occurrence of the H/D isotope exchange reaction (not shown).

The spectrum of the reaction product (Figure 3(c)) was compared in with those of cis- (Figure 3(e)) and trans- (Figure 3(d)) 2-butenes adsorbed on D-ZSM-5. Cis- and trans-2-butenes show their C=C stretching bands at 1644 and 1658 cm^{-1}, respectively, and the hydrogen-bonded OD bands appeared at almost the same frequency. The C=C stretching of 1- and cis-2-butenes appear at 1641 and 1668 cm^{-1} in Ar matrix, while that of trans-2-butene is IR inactive (1682 cm^{-1} by Raman). For the adsorbed butenes they were observed at 1627, 1645 and 1657 cm^{-1}, shifting to lower frequencies and becoming IR active. These are the evidence of the existence of direct interactions of C=C bonds with OD groups of BAS. By comparison of the IR spectrum of the reaction product with those of 2-butenes it becomes clear that the adsorbed species produced from 1-butene are the mixture of cis- and trans-2-butenes. The shift of the hydrogen-bonded OD stretching from 2302 (Figure 3(a)) to 2258 cm^{-1} (Figure 3(c)), which was observed when the reaction of the adsorbed 1-butene occurred, further confirms the production of 2-butene. Note that only *ca.* 80 % of the isolated OD groups were altered to the hydrogen bonded ones in Figure 2. Therefore, all the 1-butene molecules are stabilized on BAS, and intermolecular interaction is disregarded. It is also recalled that the reaction proceeds on D-ZSM-5 where no LAS exist. Therefore, one comes to the conclusion that adsorbed 1-butene was converted to either cis- or trans-2-butene on Brønsted acidic OD groups below 230 K in the absence of proton transfer from D-ZSM-5.

The activation energy for the DBM of 1-butene on D-ZSM-5 was estimated

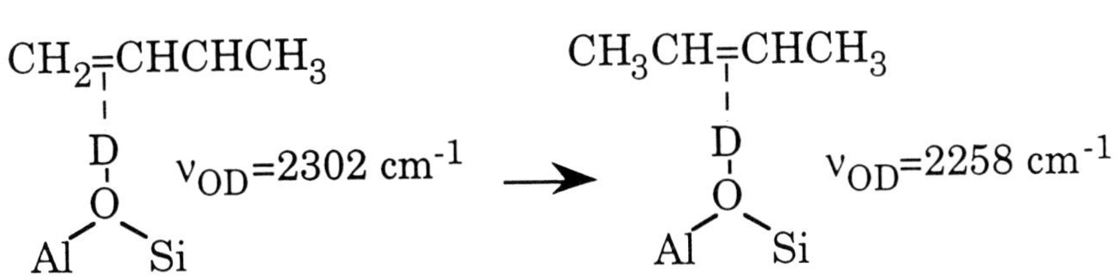

between 218 and 242 K as 49 ± 4 kJ•mol^{-1} using the decrease of the integrated intensity of the C=C stretching band at 1627 cm^{-1} of 1-butene. The slow H/D isotope exchange reaction of BAS (31 ± 8 kJ•mol^{-1}) was concurrently observed although the rate of the reaction was much smaller than that of the DBM. This suggests the slight contribution of DBM *via* the protonated intermediate.

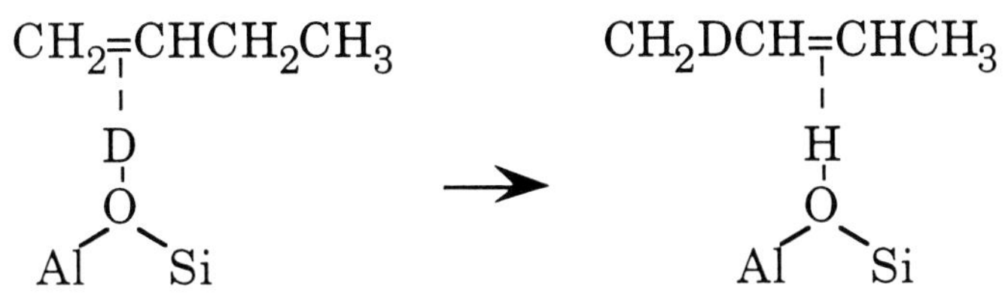

However, the pre-exponential factors of DBM and H/D isotope exchange reactions are 5.7 x 10^9 and 8.0 x 10^4, respectively. The large difference between them confirms the difference of their mechanisms. Moreover, as mentioned above and noted in Figure 3, 1-butene has already transformed to 2-butene before H/D isotope exchange reaction of BAS. Therefore, contribution of the DBM accompanied by the protonation from the BAS can be disregarded. The H/D isotope exchange reaction of BAS would occur rather with the formed 2-butene.

Adsorption of 1-Butene at Room Temperature

The adsorption and the reaction of 1-butene following to the double bond migration was studied next at room temperature. The H-ZSM-5 sample was exposed to 1.6 μmol of 1-butene and the system was immediately evacuated. The IR spectrum of the adsorbed species formed by 1-butene adsorption at room temperature is shown in Figure 4(a), which are assigned to the 2-butenes. The IR band of the hydrogen-bonded OH groups appeared at *ca.* 3100 cm^{-1}. Then, the system was maintained in evacuation for 120 min at the same temperature (Figure 4(b)). The hydrogen-bonded OH groups observed at *ca.* 3100 cm^{-1} disappeared after 120 min, and a small band at *ca.* 3500 cm^{-1} appeared. Concurrently, the negative band due to the isolated OH groups decreased. The bands of the adsorbed species also changed in the CH stretching (2800-3100 cm^{-1}), C=C stretching (1600-1700 cm^{-1}) and deformation (1200-1500 cm^{-1}) regions. The subtracted spectrum of (a) from (b) shown in Figure 4(c) clearly demonstrates the above-mentioned changes in spectra.

The change of the integrated intensity of the isolated OH band of BAS in the time course of the reaction observed in Figure 5 is quantitatively described in Figure 5. The amount of the isolated OH groups recovered as the reaction proceeded after it had sharply decreased by butene adsorption. The extent of the recovery was estimated by an extrapolated value of the time 0 as about 40-50 % of the initial

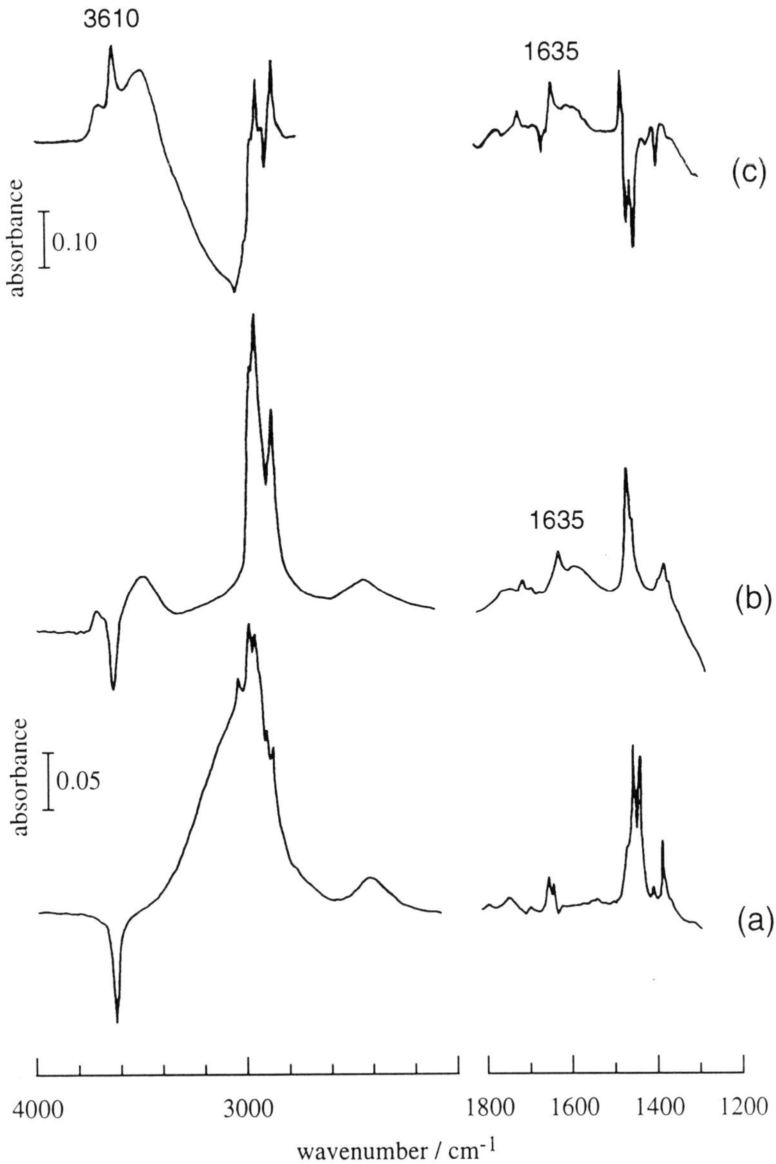

Figure 4. Reaction of dosed 1-butene at room temperatures; soon after adsorption (a), 120 min after adsorption in evacuation (b) and a subtracted spectrum of (a) from (b).

decrease. It is noted that further recovery was not observed after the reaction completed. Therefore, the reaction of butenes on H-ZSM-5 at room temperature is deduced as dimerization assuming the homogeneous reactivity of adsorbed butenes. In the case where oligomerization or polymerization proceeds, a greater amount of the isolated OH groups than that seen in the figure may be recovered from hydrogen-

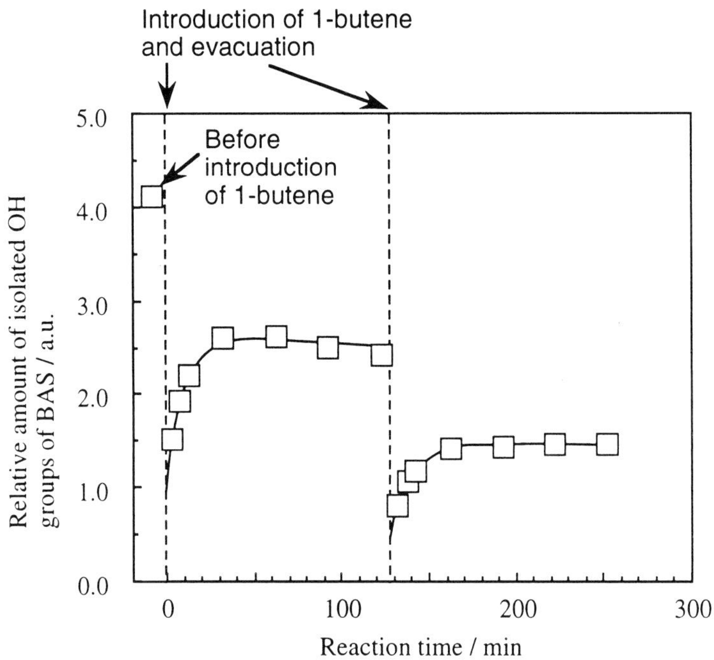

Figure 5. Time course of the isolated OH groups of H-ZSM-5 during the reaction in Figure 5.

bonded ones as a result of the decrease in the amount of the adsorbed butene.

In order to confirm the reaction to be dimerization, additional adsorption of 1-butene was carried out to the H-ZSM-5 observed in Figure 4 (b). The change of the spectra (subtracted spectra), where the spectrum of Figure 4 (b) was used as a background, during the reaction for 120 min was almost the same as those in Figure 4(c) except the overall intensity of the absorbance. This indicates that the same reaction proceeded on the vacant BAS, and the product of the first reaction independently existed on the H-ZSM-5 surface. The time course of the reaction was also the same in the first and the second runs as plotted in Figure 5; about half of the isolated OH groups which was decreased by butene adsorption again recovered from the hydrogen-bonded OH groups. If oligomerization or polymerization is assumed to proceed, it would extensively dominate the reaction more in the second run than in the first run. Therefore, it is concluded that the reaction of the adsorbed butenes terminates at dimerization and that oligomerization or polymerization does not proceed.

The geometrical and adsorption structures of the produced dimer compound are then considered. The characteristics of the IR spectrum (Figure 4(c)) of the adsorbed dimer are as follows: (1) the hydrogen-bonded OH groups with the dimer was observed at around 3500 cm^{-1}, (2) a weak C=C stretching band was observed at

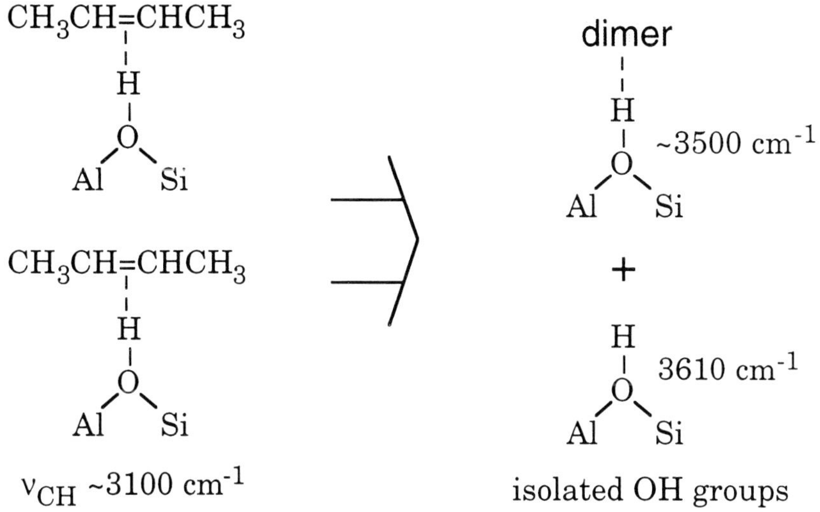

1635 cm^{-1}, and (3) any olefinic CH stretching band was not observed. The shift of *ca.* -100 cm^{-1} of the OH band to low frequency region by hydrogen-bonding was observed when a saturated hydrocarbon (n-heptane) adsorbed on H-ZSM-5, which indicated a weak hydrogen-bonding of OH groups of BAS with an alkyl group [21]. The alkyl chains of the oligomer or polymer produced by reaction of ethene and propene on H-ZSM-5 were found to have such a weak hydrogen-bonding with the OH groups of BAS, by comparison with that of n-heptane [21]. In the present study the IR band of the hydrogen-bonded OH groups with the dimer compound was observed in the same manner, and adsorption of the dimer is regarded as a weak hydrogen-bonding between OH groups and alkyl chains of the dimer. The product of the dimerization of 2-butene is regarded as being branched, and the weak C=C stretching band of it indicates its highly symmetric structure. If one notices that the hydrogen-bonding of OH groups is not what is observed with the C=C bond but that with the alkyl chains of the dimer, the C=C bond is considered to be sterically hindered to approach the OH groups. Combining the fact that the dimer did not show any olefinic CH stretching band above 3000 cm^{-1} with the above mentioned consideration, the structure of the dimer is expected to be as follows.

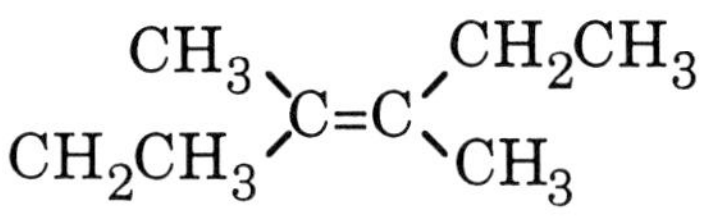

3, 4-dimethyl-3-hexene

Only the trans- form is shown for simplicity. This compound would be produced by a reaction of carbenium ion and 2-butene. There are other possible structures for the product of dimerization of 2-butene such as 3,4-dimethyl-2-hexene. However, they would be isomerized to a stable tetra-substituted ethene, 3,4-dimethyl-3-hexene, as a result of double bond migration on BAS.

SUMMARY

IR spectra of adsorbed olefins were observed from 150 K to room temperature in the absence of any interaction of the adsorbed molecules with gaseous species. The linear C_2-C_4 olefins were found to adsorb on Brønsted acidic OD groups forming 1:1 π-complexes as an initial interaction. Ethene simply desorbed while oligomerization of a small part of propene remaining on BAS proceeded. The adsorbed 1-butene reacted to form a mixture of cis- and trans-2-butene (DBM) in the absence of the proton transfer from the acidic OH (OD) groups of zeolites below 230 K. The reaction was clearly distinguished from the H/D isotope exchange reaction of the OH (OD) groups of BAS, which was expected to occur if the alkoxy or carbenium intermediate was assumed. The observed reaction path was not interpreted by the conventional acid-catalyzed mechanism, and could be attributed to a unique property of solid surfaces. The reaction of the hydrogen-bonded 2-butene at room temperature was found to be dimerization, and that of oligomerization did not proceed.

REFERENCES

1. D.W. Breck, *Zeolite Molecular Sieves*, Wiley, New York, 1974.
2. J.M. Thomas, *Scientific American* **266**, (1992).
3. S.L. Masel, J.P. McVullough, C.H. Lechthaler, and P.B. Wisz, *Chemtech* **6**, 86 (1976).
4. D.C. Chang, *Catal. Rev.-Sci. Eng.* 25, **1** (1983).
5. J.W. Ward. In: *Zeolite Chemistry and Catalysis*, J.A. Rabo (Ed.), Am. Chem. Soc., Washington D.C., 1976.
6. W.E. Farneth and R.J. Gorte, *Chem. Rev.* **95**, 615 (1995).
7. J.A. Lercher, C. Gründling, and G. Eder-Mirth, *Catal. Today* **27**, 353 (1996).
8. F. Wakabayashi, J. N. Kondo, K. Domen, and C. Hirose, *J. Phys. Chem.* **100**, 4154 (1996), and the references there in.
9. E.A. Lombardo, J.M. Dereppe, G. Marcelin, and W.K. Hall, *J. Catal.* **114**, 167 (1988).
10. M.T. Aronson, R.J. Gorte, W.E. Farneth, and D. White, *J. Am. Chem. Soc.* **111**, 840 (1989).
11. F. Wakabayashi, M. Kashitani, T. Fujino, J.N. Kondo, K. Domen, and C. Hirose, *Stud. Surf. Sci. Catal.*, **105** 1739 (1997).
12. M.T. Aronson, R.J. Gorte, and W.E. Farneth, *J. Catal.*, **105** 455 (1987).
13. T.R. Forester and R.E. Howe, *J. Am. Chem. Soc.* **109**, 5076 (1987).
14. A. Kogelbauer, J.G. Goodwin, Jr. and J.A. Lercher, *J. Phys. Chem.*, **99**, 8777 (1995).
15. M.C. Grandy and R.J. Gorte, *J. Phys. Chem.* **89**, 1305 (1985).
16. J.F. Haw, J.B. Nicholas, T. Xu, L.W. Beck, and D.B. Ferguson, *Acc. Chem. Res.* **29**, 259 (1996), and the references there in.

17. A.G. Stepanov, M.V. Luzgin, V.N. Romannikov, and K.I. Zamaraev, *Catal. Lett.* **24**, 271 (1994), and the references there in.
18. R. Robert, P.R. Rajamohanan, S.G. Hogde, A.J. Chandwadkar, and P. Ratnasamy, *J. Catal.* **155**, 345 (1995).
19. A.K. Ghosh and R.A. Kydd, *J. Catal.*, **100**, 185 (1986).
20. F. Geobaldo, G. Spoto, S. Bordiga, C. Lamberti, and A. Zecchina, *J. Chem. Soc. Faraday Trans.*, **93**, 1243 (1997), and the references there in.
21. G. Spoto, S. Bordiga, G. Ricchiardi, D. Scarano, L. Carnelli, A. Zecchina, and C.O. Areán, *J. Chem. Soc., Faraday Trans.* **90**, 2827 (1994).
22. J.N. Kondo, K. Domen, and F. Wakabayashi, *J. Phys. Chem.* **101**, 5477 (1997).
23. J.N. Kondo, L. Shao, F. Wakabayashi, and K. Domen, *J. Phys. Chem.*, in press.
24. J.N. Kondo, L. Shao, F. Wakabayashi, and K. Domen, *Catal. Lett.*, accepted.
25. V.B. Kazansky, *Acc. Chem. Rev.* **24**, 379 (1991).
26. A.J. Barnes and J.D.R. Howells, *J. Chem. Soc., Faraday Trans. 2.* **69**, 532 (1973).

Res. Chem. Intermed., Vol. 24, No. 4, pp. 425-437 (1998)

DETERMINATION OF SURFACE V=O SPECIES OF V_2O_5 CATALYSTS DURING REACTION CONDITION BY USING *IN-SITU* NARP TECHNIQUE

ATSUSHI SATSUMA[1*], JUN-ICHI OKUBO[1], KEIKO MATSUHISA[1], TADASHI HATTORI[1], and YUICHI MURAKAMI[2]

[1]*Department of Applied Chemistry, Graduate School of Engineering, Nagoya University, Furo-cho, Chikusa-ku, Nagoya 464, Japan*

[2]*Present address: Department of Industrial Engineering, Aichi Institute of Technology, Yagusa 470-03, Japan*

Received 29 September 1997; accepted 25 December 1997

Abstract-- NARP (Nitric oxide-Ammonia Rectangular Pulse) technique was applied for the *in-situ* measurement of the number of surface V=O species of V_2O_5 catalysts. When V_2O_5 catalyst was exposed to flowing benzene/O_2/N_2 mixture gas at 573-673 K, the surface concentration of V=O species decreased and finally reached a steady state which was dependent on the $[O_2]$/[Benzene] ratio in the feed gas. It was confirmed by XPS spectra that the decrease in the surface V=O species is attributed to the reduction of the V_2O_5 surface. Although calcination of V_2O_5 at above 873 K reduced the surface concentration of V=O species, it recovered after a cyclic reduction-oxidation treatment at 673 K, which modifies the near-surface layer of V_2O_5. The dependence of the surface V=O species on calcination temperature was correlated with the flatness of the surface and the diffusion of lattice oxygen in near-surface layer.

INTRODUCTION

V_2O_5 is one of the most extensively studied catalysts, because of its wide application in the chemical industry and high activity and selectivity in various chemical processes, such as the selective reduction of NO with NH_3, oxidation of *o*-xylene to phthalic anhydride, ammoxidation of toluene, and so on. For the good understanding of the catalysis over V_2O_5, the determination of the active surface area is essential. Oxygen chemisorption is the most widely applied, as reported by Fierro *et al.* [1], Nag *et al.* [2] and Oyama *et al.* [3]. Tanaka and coworkers [4] applied isotopic oxygen exchange between ^{18}O-labeled acetone for the measurement of the surface V=O species which play important role in catalytic reactions [5-9]. NARP (Nitric oxide-Ammonia Rectangular Pulse) technique was developed by our group [10] and has been applied to various supported and promoted V_2O_5 catalysts [11,12].

In these methods, V_2O_5 catalysts are first pre-treated under the appropriate conditions before the measurements. However, for the discussion on the catalytic performance, the number and the property of the active sites during reaction conditions are more important. Taking structural changes of catalysts during reaction conditions into account, "in-situ" measurement is more essential for better understanding of catalysis. Raman spectroscopy can be applied to "in-situ" measurement of supported and promoted V_2O_5 catalysts, as reported by Wachs *et al.* [13] and Went *et al.* [14]. However, the "in-situ" method for the determination of the surface active species has not been well developed and the changes in the surface active species during the reaction conditions have not been well examined.

In this study, we applied NARP technique for "in-situ" measurement of the surface V=O species during the oxidation of benzene which is a typical oxidation reaction over bulk V_2O_5 [15,16], and discussed the structure sensitivity of the number of surface V=O species by using model V_2O_5 catalysts prepared by various methods.

EXPERIMENTAL

Preparation of Model V_2O_5 Catalysts.

Three types of model V_2O_5 catalysts named V_2O_5-U, V_2O_5-F and V_2O_5-RO were prepared [17]. V_2O_5-U (Unfused) catalyst was prepared by a thermal decomposition of NH_4VO_3 in flowing O_2 at 773 K for 3h. V_2O_5-F (Fused) catalyst was prepared by fusing V_2O_5-U catalyst at 1073 K for 18 h in flowing O_2, followed by gradual cooling to room temperature. V_2O_5-RO (Redox treated) catalyst was obtained from the V_2O_5-F catalyst by a reduction-oxidation treatment as follows; the reduction in flowing H_2 at 673 K for 1 h followed by re-oxidation in flowing O_2/N_2 (O_2/N_2 = 20/80) at 673 K for 1h and this cycle was repeated 5 times. All of the diffraction lines in XRD diagrams of these V_2O_5 were assigned to the V_2O_5 phase.

Characterization.

X-ray photoelectron spectra (XPS) were measured with KRATOS ZSAM800 with Mg-K(α) (1253.6 eV). The binding energy was corrected with C_{1s}=285.1 eV as a reference. The weight change in the RO treatment and temperature programmed reduction (TPR) were measured with SEIKO TG-20 thermal gravimeter. For the measurement of weight change during the RO treatment, the samples were treated in flowing O_2/N_2 mixture (O_2/N_2 = 20/80) or in the flowing H_2 at 673 K. In the TPR measurement, the samples were reduced in flowing H_2 from 673 K to 773 K at a rate of 0.2 K min^{-1}.

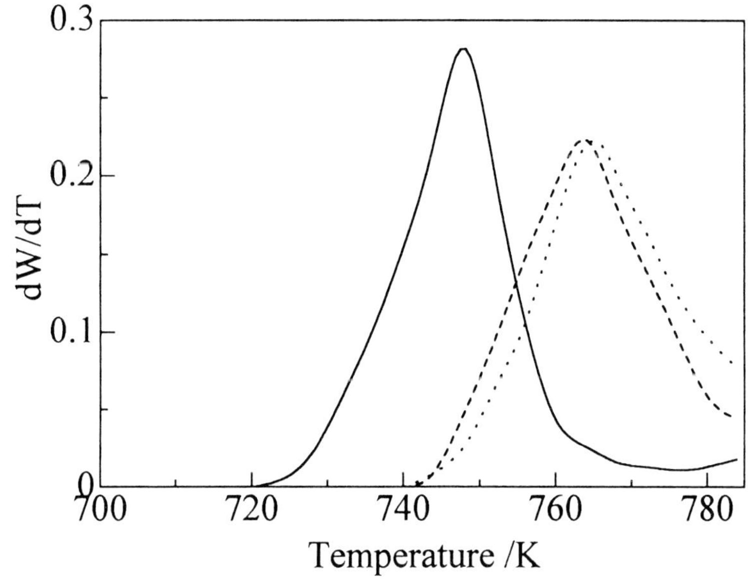

Figure 6. TPR spectra of the V_2O_5-U (——), the V_2O_5-F (-------), and the V_2O_5-RO (.........).

products, e.g., maleic anhydride or any carbonaceous materials. If so, the surface concentration of V=O species should depend on the conversion of benzene. However, that was independent of the reaction temperature, while the conversion of benzene varied from 0.2 to 11.8%. Moreover, the intensity and the profile of C_{1s} signal in XPS spectra were invariant before and after the oxidation of benzene, suggesting any carbonaceous materials were not deposited on the surface.

As shown in Figure 2, the ratio of $[V{=}O]_{IS}/[V{=}O]_{FR}$ was dependent on the $[O_2]/[\text{Benzene}]$ ratio in the feed gas: the ratio of $[V{=}O]_{IS}/[V{=}O]_{FR}$ became lower as the ratio decreased. The $[O_2]/[\text{Benzene}]$ ratio also modified XPS spectra: the low ratio resulted in the shift of the $V_{2p3/2}$ signal to the lower binding energy, which clearly indicates the reduction of the catalyst surface. The relative amount of V^{5+} ions estimated form $V_{2p3/2}$ signal were 93%, 76% and 53% at the spectrum *b*, *c* and *d*, respectively. The trends qualitatively correspond with the $[V{=}O]_{IS}/[V{=}O]_{FR}$ ratios, which were 79%, 60% and 41% at the same samples. Therefore, the decrease in the surface V=O species should be caused by the reduction of the surface oxidation state, i.e., the transformation of the surface V=O species into the surface V^{4+} ions. The regeneration of the surface V=O species by heating in gaseous oxygen also supports this conclusion, and that by heating in the absence of oxygen suggests the re-oxidation process with lattice oxygen.

Dependence of V_2O_5 Structure on Calcination Temperature and RO Treatment. The $[V{=}O]_{IS}/[V{=}O]_{FR}$ ratio depends on the calcination temperature of V_2O_5 catalysts and on the RO treatment (Figure 4). The surface area of the V_2O_5-F was extremely smaller than the V_2O_5-U, suggesting the significant sintering by fusing. As for morphology of the crystallites, Mori *et al.* reported that the V_2O_5-U has a very porous and rugged surface, while the surface of the V_2O_5-F and V_2O_5-RO is much smoother and flatter [17]. This is reasonable, since fusion of a solid would generally lead to a smooth surface. Mori *et al.* [17] also confirmed using UV-visible and XPS spectroscopy that there is no significant change in the electric state and no impurities on the surface. In this study, any additional signals were not observed in XPS spectra and only the diffraction lines assignable to V_2O_5 were observed in XRD patterns. These results suggest that the modification of V_2O_5 by the calcination was significant at the surface.

In the case of the V_2O_5-RO which was obtained by the RO treatment of the V_2O_5-F, there was no significant difference with the V_2O_5-F in TPR spectra, BET surface area, and SEM images. These results indicate that the RO treatment did not affect the bulk structure of the V_2O_5-F. A small weight change as small as 0.1% was observed during the RO treatment, suggesting that the modification was limited to small part of V_2O_5. The weight change of 0.1% corresponds to the reduction of V_2O_5 to V_2O_3 in less than 10 layers from the external surface, which was calculated from the surface area (0.44 m^2g^{-1}) and cross section area of V_2O_5 layer (1.27 x10^5 m^2 mol^{-1}). Therefore, the RO treatment is considered to result only in the modification of surface layers of V_2O_5 crystal.

It is generally known that a severe reduction and oxidation treatment would lead to a rugged surface. V_2O_5 transforms into V_2O_4 through Wadsly phase with edge-sharing of each octahedra with oxygen vacancy, and then reduced to V_2O_3 through the Magli phase [21-23]. During this process, the formation of defects and cracks leads to a destruction of vanadium oxide crystal. Actually, in the separate experiment, the reduction of the V_2O_5-F catalyst in hydrogen at 773 K for 1h resulted in the increase in the surface area from 0.44 to 11.0 m^2 g^{-1}. The defect and cracks are also expected to be formed through the RO treatment, but they should be limited in the surface layers because of little effect the RO treatment on the bulk.

Dependence of V=O Species on Near-Surface Structure of V_2O_5.
The surface concentrations of V=O species on fresh catalyst, $[V{=}O]_{FR}$, were almost equal to each other, while the V_2O_5-F exhibited the lower ratio of $[V{=}O]_{IS}/[V{=}O]_{FR}$ than the V_2O_5-U (Table 3). The result indicates that the surface concentration of V=O species on fresh catalyst is structure insensitive, but that during the reaction conditions is structure sensitive. On the other hand, the ratio of $[V{=}O]_{IS}/[V{=}O]_{FR}$ on the V_2O_5-RO was equal to that of the V_2O_5-U. As discussed above, the RO treatment

dose not affect the bulk structure of V_2O_5-F, but it modifies only near-surface layers and leads to the reduction of the surface flatness. From these results, it can be concluded that the ratio of $[V{=}O]_{IS}/[V{=}O]_{FR}$ of V_2O_5 catalysts depends on the surface flatness of V_2O_5.

As mentioned above, the ratio of $[V{=}O]_{IS}/[V{=}O]_{FR}$ is also related to the reduction-oxidation properties of V_2O_5, i.e., the decrease in the surface V=O species is due to the formation of the surface V^{4+} ions. The regeneration of V=O species in the absence of gaseous oxygen suggests a contribution of the lattice oxygen. Actually, the catalytic oxidation of benzene over V_2O_5 catalysts is suggested to proceed with the reduction-oxidation mechanism (or Mars-van Krevelen mechanism) [15,24]. Thus, it can be expected that the dependence of the surface V=O species on the surface flatness is due to the difference in the regeneration rate of the surface V=O species with lattice oxygen.

Figure 7 shows a schematic model of V_2O_5 surface during the reaction. Since the decrease in the surface V=O species on a rugged surface (V_2O_5-RO) is lower than on a flat surface (V_2O_5-F), the surface V^{4+} ions, probably V^{4+}-OH [25-28], near side faces can be expected to be reoxidized more easily than those on the flat surfaces. This expectation can be rationalized by assuming that gaseous oxygen is adsorbed and reduced on the dangling bond exposed to the side face, i.e., on the kink and step sites, and oxygen anions thus formed diffuse to surface V^{4+} ions to oxidize them into the surface V=O species.

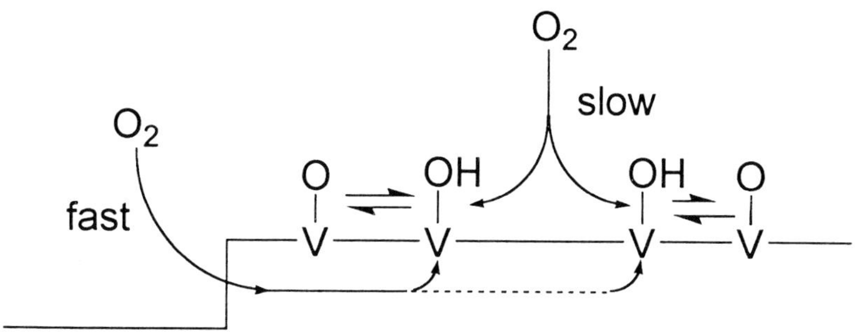

Figure 7. Schematic model of the redox process in near-surface layers of V_2O_5 catalysts.

CONCLUSION

NARP technique was applied for the determination of the surface concentration of V=O species of V_2O_5 catalysts during the reaction conditions. It was found that the surface V=O species significantly decreased during benzene oxidation,. The decrease in the surface V=O species was attributed to the reduction of V_2O_5 surface caused by the smaller rate in regeneration of surface V=O species. The examination using model V_2O_5 catalysts clarified the strong dependence of the surface concentration of V=O species on the surface flatness of V_2O_5.

Acknowledgment

We thank Dr. Katsumi Kuwabara (Nagoya University) for his contributions throughout the measurement of SEM. And we also thank Mr. Masatoshi Kato and Mr. Akitaka Hattori (Tosoh Co-oporation) for valuable discussions and contributions throughout the measurement of XPS spectra. This work was partially supported by a Grant-in-Aid from the Ministry of Education, Science and Culture, Japan (No. 03750601).

REFERENCES

1. J.L.G. Fierro, L.A. Gambaro, T.A. Cooper and G. Kremen, *Appl. Catal.*, **6**, 363 (1983).
2. N.K. Nag, K.V.R. Chary, B.M. Reddy, B.R. Rao and V S. Subrahmanyam, *Appl. Catal.*, **9**, 225 (1984); N.K. Nag, K.V.R. Chary, B.R. Rao and V.S. Subrahmanyam, *Appl. Catal.*, **31**, 73 (1987).
3. S.T. Oyama, G.T. Went, K.B. Lewis, A.T. Bell and G.A. Somorjai, *J. Phys. Chem.*, **93**, 6786 (1989).
4. T. Tanaka, R. Tsuchitani, M. Ooe, F. Funabiki and S. Yoshida, *J. Phys. Chem.*, **90**, 4905 (1986).
5. K. Tarama, S. Teranishi, S. Yoshida and N. Tamura, *Proc. Int. Congr. Catal.* **3**, 282 (1965); K. Tarama, S. Yoshida, S. Ishida and H. Kakioka, *Bull. Chem. Soc. Jpn.*, **41**, 2840 (1968).
6. M. Ai, *J. Catal.*, **54**, 426 (1978).
7. H. Miyata, Y. Nakagawa, T. Ono and Y. Kubokawa, *J. Chem. Soc. Faraday Trans. 1*, **79**, 2343 (1983).
8. P.J. Gellings, in: *Catalysis*, vol.7, G. C. Bond and G. Webb, (Eds.), Royal Society of Chemistry, Burlington House, London, p.105, 1985 and references cited therein.
9. G.C. Bond and S.F. Tahir, *Appl. Catal.*, **71**, 1 (1991).
10. A. Miyamoto, Y. Yamazaki, M. Inomata and Y. Murakami, *J. Phys. Chem,*. **85**, 2366 (1981); M. Inomata, A. Miyamoto and Y. Murakami, *J. Phys. Chem.*, **85**, 2372 (1981).
11. M. Inomata, K. Mori, A. Miyamoto, T. Ui and Y. Murakami, *J. Phys. Chem.*, **87**, 754 (1983); M. Inomata, K. Mori, A. Miyamoto and Y. Murakami, *J. Phys. Chem.*, **87**, 761 (1983).
12. A. Satsuma, A. Hattori, K. Mizutani, A. Furuta, A. Miyamoto, T. Hattori and Y. Murakami, *J. Phys. Chem.*, **92**, 6052 (1988); *J. Phys. Chem.*, **93**, 1484 (1989).
13. T. Machej, J. Haber, A. M. Turek and I. E. Wachs, *Appl. Catal.*, **70,** 115 (1991); J. M. Jehng, and I. E. Wachs, *Catal. Lett.*, **13**, 9 (1992); G. Deo and I. E. Wachs, *J. Catal.*, **146**, 323 (1994).
14. T. Went, L. J. Leu and A.T. Bell, *J. Catal.*, **134**, 479 (1992); G. T. Went, L. J. Leu, R. R. Rosin and A. T. Bell, *J. Catal.*, **134**, 492 (1992); G. T. Went, L. J. Leu, S. J. Lomardo and A. T. Bell, *J. Phys. Chem.*, **96**, 2235 (1992).
15. K. Mori, M. Inomata, A. Miyamoto and Y. Murakami, *J. Chem. Soc. Faraday Trans. 1* **80**, 2655 (1984).
16. A. Satsuma, F. Okada, A. Hattori, A. Miyamoto, T. Hattori and Y. Murakami, *Appl. Catal.*, **72**, 295 (1991).
17. K. Mori, A. Miyamoto and Y. Murakami, *J. Phys. Chem.*, **88**, 2735 (1984).

18. P. Ciambelli, G. Bagnasco, L. Lisi, M. Turco, G. Chiarello, M. Musci, M. Notaro, D. Robba and P. Ghetti, *Appl. Catal. B*, **1**, 61 (1992).
19. H. Morishige, J. Tamaki, Y. Teraoka, N. Miura and N. Yamazoe, *Nippon Kagaku Kaishi*, 1983 (1989).
20. R.C. Weast, M.J. Astle, W.H. Bayer (Eds.), *CRC Handbook of Chemistry and Physics, 66th ed.*, B-144, CRC Press, Boca Raton, Florida (1985).
21. A.D. Wadsley, *Acta Cryst.*, **10**, 261 (1957).
22.. K. Nagasawa, *Mat. Res. Bull.*, **6**, 853 (1971).
23. A. F. Wells, Structural Inorganic Chemistry, Fourth Edition, p. 467, Clarendon Press, Oxford, 1975.
24. P. Mars and D.W. van Klevelen, *Chem. Eng. Sci.*, **3**, 41 (1951).
25. G. Ramis, G. Busca, F. Bregani and P. Forzatti, *Appl. Catal.* **64**, 259 (1990).
26. W.M. Riggs and M.J. Parker, in: *Methods of Surface Analysis*, A.W. Czanderna, (Ed.), p. 103, Elsevier, Amsterdam, 1975.
27. Z. Sobalik, D. Kozlowski, and J. Haber, *J. Catal.*, **127**, 665 (1991).
28. M. Inomata, A. Miyamoto and Y. Murakami, *J. Catal.*, , **62**, 140 (1980).

Res. Chem. Intermed., Vol. 24, No. 4, pp. 439-448 (1998)

PARTICIPATION OF MOLECULAR HYDROGEN-ORIGINATED PROTONIC ACID SITES IN ACID-CATALYZED REACTIONS

HIDESHI HATTORI*, TAKAYUKI YAMADA, and TETSUYA SHISHIDO
Center for Advanced Research of Energy Technology, Hokkaido University, Sapporo 060, Japan

Received 22 September 1997; accepted 15 December 1997

Abstract--The formation of protons from molecular hydrogen was observed by IR of adsorbed pyridine for Pt/SO_4^{2-}-ZrO_2 when heated in the presence of hydrogen. The promotive effects of hydrogen on skeletal isomerization of alkanes and cumene cracking are rationalized by the formation of protons from hydrogen. The formation of protons and the promotive effects of hydrogen were also observed for other catalysts such as Co.Mo/SiO_2-Al_2O_3 and a physical mixture of Pt/SiO_2 and H-ZSM-5. The concept "molecuar hydrogen-originated protonic acid site" is proposed as an important and widely applicable concept in acid-catalyzed reactions.

INTRODUCTION

The acid sites on the catalysts have been believed to be intrinsically generated during preparation of the catalysts. Although a proton is formed by heterolytic dissociation of a hydrogen molecule, participation of molecular hydrogen in the formation of the protonic acid sites has never been taken into account. This is partly because a definite promoting effect of hydrogen had not been reported. Most of the studies concerning hydrogen effect on acid-catalyzed reaction reported suppressive effect of hydrogen on the reaction rate [1-6].

One more reason why participation of molecular hydrogen in the acid-catalyzed reaction had not been considered may arise from the fact that the hydrogen molecule is easily dissociated into atomic hydrogen on a certain metal catalysts. The homolytic dissociation of molecular hydrogen into atomic hydrogen occurs even below room temperature on a transition metal catalyst. There had been no room to consider heterolytic dissociation of molecular hydrogen which requires a high energy.

In recent years, several papers have reported promotion effects of hydrogen. Baba *et al.* reported that the promotion effects of hydrogen on ethylbenzene disproportionation over Ag-Y zeolite are caused by the formation of protons from hydrogen molecules accompanied by reduction of Ag^+ ions to Ag^0 metals in the presence of hydrogen [7-9]. The protons thus formed are eliminated by the reverse reaction of Ag^0 to Ag^+ if the gaseous hydrogen is removed from the reaction mixture. Although the formation of a proton by reduction of metal ion by hydrogen is generally observable, the unique feature with Ag-Y is the occurrence of the reverse reaction as the gas phase hydrogen is removed. A prominent effect of hydrogen was also reported by Sachtler *et al.* with Pd-Y zeolite for isomerization of methylcyclopentane in the presence of hydrogen [10-12]. They proposed "electron-deficient" $Pd\text{-}H^+$ adducts as the active sites in the presence of hydrogen.

In 1988, Hosoi *et al.* reported that zirconium oxide modified with platinum and sulfate ion ($Pt/SO_4^{2-}\text{-}ZrO_2$) persists at a high activity for a long period in skeletal isomerization of alkanes when the reactions were carried out in the presence of hydrogen [13]. They reported that a high and stable activity in the presence of hydrogen resulted from the removal of the carbonaceous deposit on the catalyst surface by hydrogenation. However, through our studies of hydrogen effects, we have reached a different conclusion. The promotion effects of hydrogen on the skeletal isomerization of alkanes are rationalized by the generation of protonic acid sites from molecular hydrogen [14,15].

In the present paper, I wish to report that the formation of protonic acid sites from hydrogen molecules is not a phenomenon restricted to observation for $Pt/SO_4^{2-}\text{-}ZrO_2$, but is observable for many other catalysts, and to assert the importance of the protonic acid sites originated from molecular hydrogen in acid-catalyzed reactions.

EXPERIMANTAL

Preparation of Catalysts

The catalysts examined are $Pt/SO_4^{2-}\text{-}ZrO_2$, $Co.Mo/SiO_2\text{-}Al_2O_3$, a physical mixture of Pt/SiO_2 and H-ZSM-5, $SiO_2\text{-}Al_2O_3$, H-ZSM-5, and $SO_4^{2-}\text{-}ZrO_2$. The preparative methods for these catalysts are as follows.

For $Pt/SO_4^{2-}\text{-}ZrO_2$, the sulfated ion treated $Zr(OH)_4$ was prepared by impregnation of $Zr(OH)_4$ with a 1N H_2SO_4 aq. solution followed by filtration and drying at 383K. The $Zr(OH)_4$ was obtained by the hydrolysis of $ZrOCl_2 \cdot 8H_2O$ with aqueous ammonia. The $Pt/SO_4^{2-}\text{-}ZrO_2$ sample (0.5 wt%Pt) was prepared by impregnation of $SO_4^{2-}\text{-}ZrO_2$ with a 1% H_2PtCl_6 aq. solution followed by drying at 383K and calcination at 873K in air. The amount of S remaining in the resulting catalyst was 1.5 wt% as determined by XRF. The catalyst was pretreated in a hydrogen stream at 623 K for 2 h before use for reaction. For IR and temperature-programmed desorprion (TPD) measurements, the catalyst was treated further in a vacuum at 623 K.

Pt/SiO_2 was prepared by impregnation of silica supplied from the Catalysis Society, Japan (JRC-SIO-1) with a H_2PtCl_6 aq. solution followed by drying, and calcining at 773 K for 5 h. The content of Pt was adjusted to be 2 wt%. H-ZSM-5 was prepared by ten times ion-exchange of Na-ZSM-5 supplied by TOSOH with a 1N NH_4Cl aq. solution followed by calcination at 803 K for 3 h. The physical mixture of the Pt/SiO_2 and H-ZSM-5 was prepared by mixing the two components in the ratio 1:10 (Pt/SiO_2 : H-ZSM-5) in an agate mortar.

For Co.Mo/SiO_2-Al_2O_3, into a solution of cobalt nitrate, aqueous ammonia was added to form an aqueous solution of cobalt ammonium complex. To the solution, aqueous ammonium molybdate was added to make a mixed solution containing cobalt and molybdenum. Silica-alumina supplied from the Catalysis Society, Japan(JRC-SAL-2) was impregnated with the mixed solution followed by drying at 353K for 36 h, and calcining at 773 K for 16 h. The contents of Co and Mo were 4 and 8 wt% as CoO and MoO_3, respectively.

IR study of Adsorbed Pyridine

A self-supported wafer of the sample was placed in an in situ IR cell with CaF_2 windows, and pretreated. The pretreatment procedures varied with the catalyst samples. For Pt/SO_4^{2-}-ZrO_2, the calcined sample was pretreated in a hydrogen flow at 623 K for 2 h followed by outgassing at 623 K for 2 h. For Co.Mo/SiO_2-Al_2O_3, the calcined sample was pretreated in a hydrogen flow at 673 K for 2 h followed by outgassing at 773 K. For presulfided Co.Mo/SiO_2-Al_2O_3, the calcined sample was treated in a 3%H_2S in H_2 at 673 K for 2 h followed by outgassing at 773 K for 17 h. For the physical mixture of Pt/SiO_2 and H-ZSM-5, the calcined sample was outgassed at 773K and cooled to 673 K at which the sample was treated in a hydrogen flow for 2 h followed by outgassing at 773 K for 17 h. For H-ZSM-5, the calcined sample was outgassed at 773 K for 17 h.

After the pretreatment, pyridine was introduced into the cell at a pressure of ca. 2 Torr for 15 min at 423 K, and then outgassed normally at 673 K for 20 min. In the case of Co.Mo/SiO_2-Al_2O_3, the final outgassing was done at 423 K. To examine the effects of hydrogen on the adsorbed pyridine, the pyridine-covered sample was exposed to 500 Torr of hydrogen at room temperature, and heated stepwise from room temperature to 573 K or 673 K by 50 K increments. Following the processes of heating in the presence of hydrogen, the sample was outgassed stepwise from 373 K to 673 K by 50 K increments. All spectra were recorded on an FT/IR-5300 infrared spectrometer (JASCO) at room temperature.

For determination of the number of protonic sites and Lewis acid sites on the surface, the integrated absorbances of the bands at 1450 cm^{-1} (due to pyridine chemisorbed on Lewis acid sites, L-Py) and 1490 cm^{-1} (due to both the L-Py and pyridine chemisorbed on protonic acid sites, B-Py) were used for all samples. The values obtained were normalized to the weight of the sample wafer. To obtain the apparent absorption coefficients of the bands, a known amount of pyridine was adsorbed on the sample, and the absorbance of each band was measured. Then, a small quantity of water which is sufficient to convert all the pyridine adsorbed on Lewis acid sites into pyridinium ion was introduced into the IR-cell. The apparent absorption coefficients were calculated from the changes of the absorbances at 1450 and 1490 cm^{-1} on exposure to water vapor. As the apparent absorption coefficient for a certain band is different for a different wafer, the calibration of the apparent absorption coefficient was performed for each wafer.

Temperature-Programmed Desorption of Hydrogen and/or Deuterium

The pretreatment conditions were the same as those for IR study of adsorbed pyridine. The sample pretreated in a vacuum was first exposed to ca. 200 Torr H_2 at a certain temperature for 15 min and evacuated for 15 min. Then the sample was exposed to ca. 200 Torr D_2 for 15 min After cooling to room temperature, the sample was outgassed for 15 min prior to TPD run. TPD was run at a heating rate of 10 K min^{-1}, and the desorbed gases were analyzed by mass spectrometry.

Reaction Procedures for Heptane Skeletal Isomerization and Cumene Cracking

A pressurized flow reactor was employed for carrying out heptane isomerization. For cumene cracking, a pressurized flow reactor and an atmospheric pulse reactor were employed. For the reaction over Co.Mo/SiO_2-Al_2O_3 and SiO_2-Al_2O_3, the flow reactor was used with a carrier flow rate of 100 ml min^{-1} and a cumene feed rate of 14 ml min^{-1} under the pressure of 30 atm.

For the reaction over Pt/SO_4^{2-}-ZrO_2, the physical mixture of Pt/SiO_2 and H-ZSM-5, and H-ZSM-5, the pulse reactor was used. A dose of cumene, 0.5 ml (0.036 mmol), was passed over the catalyst in a carrier flowing at 50 ml min^{-1}, and the products were trapped at 77 K before being flash-evaporated into gas chromatographic columns (PEG-20M and VZ-7).

RESULTS and DISCUSSION

Skeletal Isomerization

The promoting effect of hydrogen on the catalytic activity of Pt/SO_4^{2-}-ZrO_2 for skeletal isomerization of heptane was observed as shown in Fig. 1. The activity was high and persisted for a long time on stream in the presence of hydrogen, while the activity was low and decreased rapidly with the time on stream when the reaction was carried out in the stream of nitrogen.

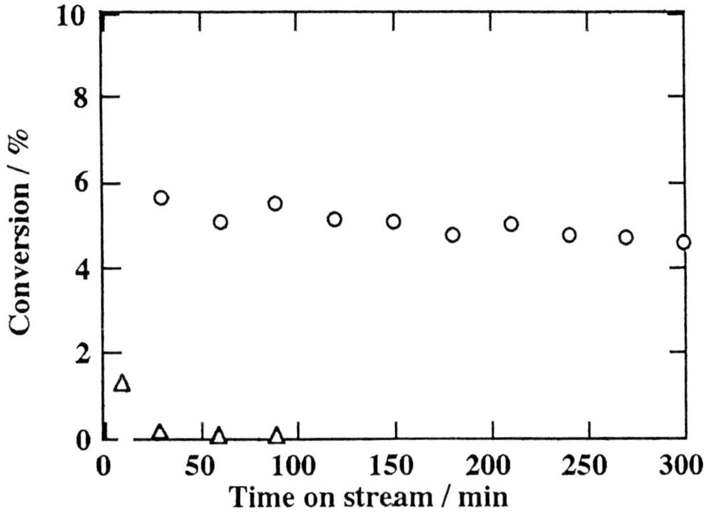

Figure 1. Effects of hydrogen on heptane isomerization over Pt/SO_4^{2-}-ZrO_2. Under a hydrogen stream (○), and a nitrogen stream (Δ). Reaction temperature; 523 K. Total pressure; 3 MPa.

IR of Adsorbed Pyridine

To examine the acidic properties of catalysts in the presence and in the absence of hydrogen, IR studies of adsorbed pyridine were performed. Variations of the amounts of protonic acid site and Lewis acid site on Pt/SO_4^{2-}-ZrO_2 with heating in the presence of hydrogen and successive outgassing are shown in Fig. 2. The IR results show that heating in the presence of hydrogen results in the formation of protonic acid sites in compensation of Lewis acid sites. Outgassing the gas phase hydrogen results in the restoration of Lewis acid sites and the decrease in protonic acid sites.

The formation of protonic acid sites with concomittant decrease of Lewis acid sites by heating in the presence of hydrogen was observed not only for the Pt/SO_4^{2-}-ZrO_2 but also for all the catalysts examined, though easiness of the interconversion between protonic acid sites and Lewis acid sites was dependent on the type of catalyst. It is noted that the formation of protonic acid sites was also observed for the metal free acidic catalysts such as SiO_2-Al_2O_3 and H-ZSM-5.

Cumene Cracking

In the presence of hydrogen, the activity for skeletal isomerization was enhanced and the formation of protonic acid sites was observed. It is suggested that the activity enhancement is caused by the formation of protonic acid sites on the Pt/SO_4^{2-}-ZrO_2. Alkane, however, undergoes skeletal isomerization by two different mechanisms.

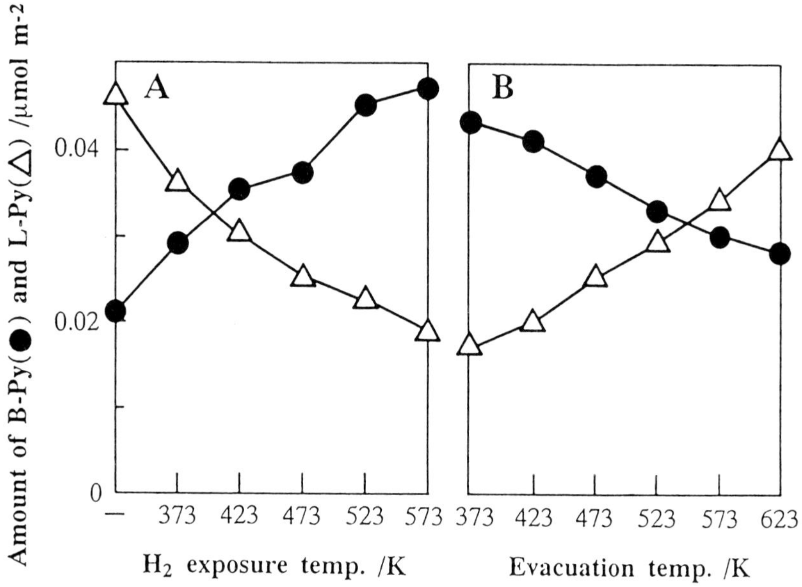

Figure 2. Change in protonic acid site (B-Py ●) and Lewis acid site (L-Py Δ) with heating in the presence of hydrogen (A) and with evacuation (B) for Pt/SO_4^{2-}-ZrO_2.

One is a metal-acid bifunctional mechanism and the other is an acid monofunctional mechanism. It is not certain which mechanism is operating in the skeletal isomerization of heptane over the Pt/SO_4^{2-}-ZrO_2 in the presence of hydrogen. Therefore, it is not certain which step the promotive effect of hydrogen is involved in. It would become much clearer if the hydrogen effect is examined on the reaction which proceeds only by an acid-catalyzed mechanism.

Cumene cracking is one of the typical acid-catalyzed reactions. The active sites for the reaction are protons on the surface. The effects of hyrogen on the cumene cracking are examined for Pt/SO_4^{2-}-ZrO_2, Co.Mo/SiO_2-Al_2O_3, Co.Mo/Al_2O_3, physical mixture of Pt/SiO_2 and H-ZSM-5, metal free H-ZSM-5, and SiO_2-Al_2O_3. The results for Pt/SO_4^{2-}-ZrO_2, are shown in Fig. 3. The conversions were about 50% in the hydrogen stream. The conversions decreased as the carrier of hydrogen was switched to helium. As the carrier helium was switched again to hydrogen, the conversions began to increase to recover almost the original value. The increase and decrease in the conversion in response to the presence and absence of hydrogen correspond to the formation and elimination of protonic acid sites as evidenced by IR spectroscopy of adsorbed pyridine. It is suggested that the high conversions observed in the presence of hydrogen are due to the formation of protonic acid sites from molecular hydrogen.

The promotive effects of hydrogen on cumene cracking were observed not only for Pt/SO_4^{2-}-ZrO_2, but also for Co.Mo/SiO_2-Al_2O_3, and a physical mixture of

Res. Chem. Intermed., Vol. 24, No. 4, pp. 449-459 (1998)

ISOBUTANE ALKYLATION OVER SOLID ACID CATALYSTS UNDER SUPERCRITICAL CONDITIONS

GENKI FUNAMOTO[1], SHIGETOSHI TAMURA[1], KOHICHI SEGAWA[1], KAM T. WAN[2], and MARK E. DAVIS[2]

[1]*Department of Chemistry, Faculty of Science and Technology, Sophia University, 7-1 Kioi-cho, Chiyoda-ku, Tokyo 102, Japan*

[2]*Division of Chemistry and Chemical Engineering, California Institute of Technology, Pasadena, California 91125, U.S.A.*

Received 19 September 1997; accepted 27 December 1997

Abstract--Solid catalyzed isobutane alkylation has been investigated for decades, but it has not yet been applied in any commercial uses because of the rapid deactivation of the catalyst. Here, the alkylation reaction has been studied under supercritical conditions using metal-promoted and unpromoted sulfated zirconia as catalysts. The catalytic activity at the supercritical condition of 5.0 MPa, 423K was significantly higher than at lower reaction pressure conditions and the deactivation rate was clearly reduced, independent on the catalyst. Iron- and manganese-promoted sulfated zirconia (SFMZ) showed higher activities under all conditions than unpromoted sulfated zirconia (SZ).

INTRODUCTION

Aromatics and olefins in gasoline have been claimed as harmful substances for humans. Thus, the removal of these components from gasoline has been required recently. However, these components usually have high octane numbers, and a blending component should be added to gasoline in order to maintain its octane number. Alkylate gasoline, which is formed by highly branched paraffins, has a high octane number and a clean combustion nature. It is therefore thought to be an ideal blending component. Sulfuric or hydrofluoric acid has been used as a commercial catalyst [1]. These liquid acids are toxic and have safety issues. Furthermore, there are some economical disadvantages to these catalyst systems such as erosion of the apparatus, separation of the spent catalyst from the products, and disposal of the spent catalyst. Because of all these problems, there has been a strong demand for replacing the liquid acid with a solid acid catalyst. If it can be replaced, the use of the alkylate gasoline will increase drastically. Many solid acids have been studied for quite some time [2-18]. However, most of the solids have a problem of rapid catalyst deactivation due to the coke formation on the catalyst surface and have not yet been applied for commercial use. Therefore, the reduction of the coke formation would be the largest key for the solid acid- catalyzed isobutane alkylation.

Supercritical fluids have been widely studied as extraction media [18-20]. A supercritical fluid has several unique properties, such as wide density change with slight pressure or temperature change and easy mass transfer due to low viscosity and high diffusivity [19]. Therefore, there are frequent reports of supercritical fluid as extraction or reaction media in the fields of medicine, food, cosmetics, energy, and new materials [20]. There are also some reports about supercritical fluid extraction of coke formed on solid catalysts [21]. Thus, it is expected that a supercritical fluid could extract the coke formed on the catalyst and keep the catalyst surface fresh during the isobutane alkylation reaction.

In this work, we have studied the alkylation reaction of isobutane with 1-butene under the supercritical conditions of the reaction mixture. As solid acid catalysts, sulfated zirconia (SZ) and iron- and manganese-promoted sulfated zirconia (SFMZ) were tested.

EXPERIMENTAL SECTION

Catalyst Preparation and Characterization

The iron- and manganese-promoted zirconia and unpromoted sulfated zirconia catalysts were prepared from $Zr(OH)_4$ by stepwise impregnation. The iron- and manganese-promoted sulfated zirconia is labeled SFMZ, and unpromoted sulfated zirconia is labeled SZ. The detailed procedure of the preparation of SFMZ and SZ has been described elsewhere [22]. The elemental composition of these catalysts is shown in Table 1.

Table 1
Elemental composition of the catalysts.

Catalyst	Composition/wt%		
	Sulfur	Iron	Manganese
SFMZ	1.4	1.2	0.9
SZ	1.2	-	-

In order to study the acidity of the catalysts, IR spectra were taken with ammonia adsorption. Spectra were recorded on a Shimadzu FTIR-8200PC equipped with a vacuum system. Prior to the measurement, the pelletized samples were calcined *in-situ* overnight at 773 K under O_2 flow and evacuated at the same temperature. Then IR spectra were taken as dry bases. The samples were cooled down to 373 K and were exposed to 670 Pa of ammonia for 30 min. After evacuation, the temperature was raised and IR spectra were taken at the corresponding temperatures.

In order to study the acid amount of the catalysts, the uptakes of chemisorbed ammonia were taken by a vacuum balance. The samples were pelletized, crushed, and placed in a sample basket. Prior to the measurement, the samples were calcined *in-situ* overnight at 773 K under O_2 flow and evacuated. After the evacuation, samples were cooled down to 473 K and fresh sample weights were measured as dry bases. Then the samples were exposed to 1300 Pa of ammonia for 2 hours and evacuated at 473 K. The weights of the catalysts were measured when the weights became stable and the amounts of the chemisorbed ammonia were calculated.

Nitrogen adsorption measurements were performed at 77 K on an automatic gas adsorption apparatus (BEL JAPAN, BELSORP 28SA) to determine the specific surface area and the pore volume. As pretreatment, 250 mg of sample were placed in a quartz tube and evacuated at 573 K for 6 hours. The specific surface was obtained by BET method, and the pore volume was obtained by t-plot method.

Alkylation of Isobutane

Alkylation of isobutane with 1-butene was carried out in a fixed bed flow reactor, which is shown in Figure 1, at a reaction pressure of 423 K, 1.0-5.5 MPa of a total pressure, an isobutane/1-butene molar ratio of 4, and a total WHSV of 50 h^{-1}. The catalysts were pelletized and crushed before being introduced into the reactor. Prior to the reaction, the catalysts were calcined *in-situ* at 773 K for 8 hours under O_2 flow, cooled to the reaction temperature, and purged with He for 1 hour. The gas and liquid products were separated before being analyzed. The gas products were analyzed by an on-line TCD gas chromatograph (Shimadzu GC-8AIT), where the different compounds were separated in a column (Unicarbon A-400, 4 m, GL Sciences). The liquid products were taken out from the sample cylinder and analyzed by a FID gas chromatograph (Shimadzu GC-14B), where the different compounds were separated in a capillary column (fused silica SPB-1, 60 m length, SUPELCO).

RESULTS AND DISCUSSION

Critical Points of the Feed Stock Mixture

As was mentioned in the introduction section, we studied the alkylation reaction under supercritical conditions, expecting reduction of the catalyst deactivation. Calculation of the critical points of the reaction mixtures is required, since the density of the fluid changes drastically by a slight change of temperature or pressure around the critical point. Although both of the reactants have their own critical

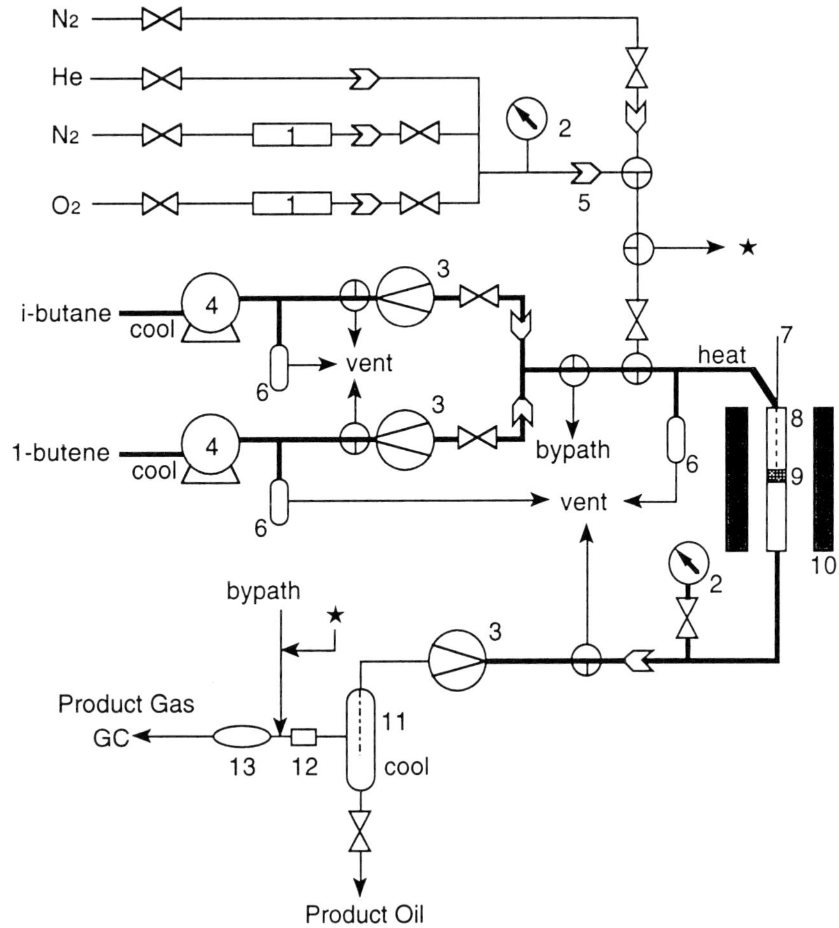

Figure 1. Reaction apparatus of alkylation of isobutane with 1-butene: (1) mass flow controller, (2) pressure gauge, (3) back pressure regulator, (4) feed pump, (5) check and relief valve, (6) safety valve, (7) thermo couple, (8) reactor, (9) catalyst bed, (10) furnace, (11) sample cylinder, (12) filter, (13) flow meter.

points, the critical point of the reaction mixture changes depending on the ratio of the reactants. Figure 2 shows the critical point as a function of the ratio of the reaction mixture based on the calculation of J. M. Prausnitz *et al.* [21,22] Although the critical temperatures of reaction mixtures tend to be between those of the two reactants, the critical pressures tend to be higher than those of the two reactants. The critical pressure is the highest at the isobutane/1-butene ratio of 4. In this work we chose the ratio of 4, where the critical temperature is 414.9 K and the critical pressure is 4.945 Mpa.

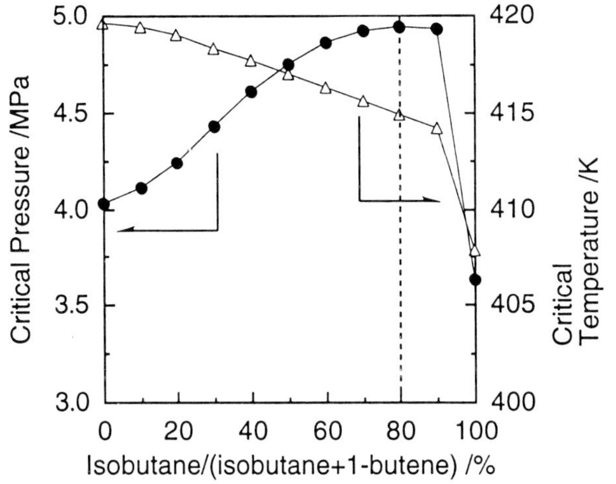

Figure 2. Critical point of the reaction mixture as a function of the isobutane/(isobutane+1-butene) rate: (●) critical pressure, (△) critical temperature.

Reaction Study

We studied the alkylation reaction at the condition of 423K and 5.0 MPa, which is slightly higher than the critical point of the reaction mixture, over SZ and SFMZ as catalysts. The catalytic activity as a function of time on stream is summarized in Figure 3. Since 1-butene is much more reactive than isobutane, oligomerization of 1-butene occurs and it causes the coke formation. Therefore, the catalytic activity for isobutane and 1-butene should be discussed separately. However, isobutane

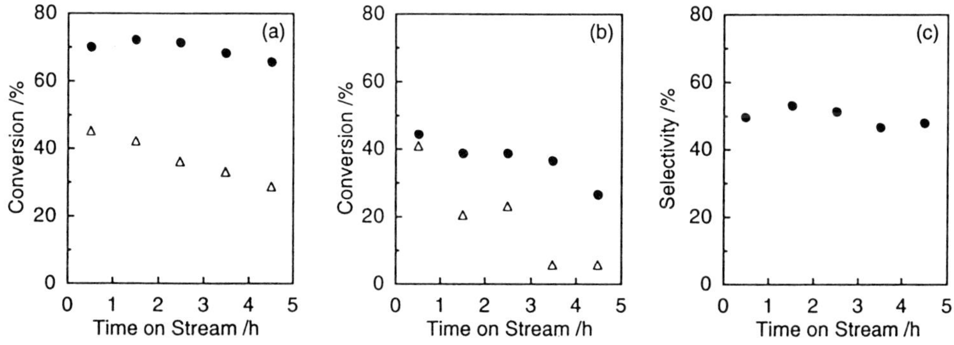

Figure 3. Catalytic activity as a function of time on stream: (a) conversion of 1-butene, (b) conversion of isobutane, (c) C8 selectivity in liquid products; (●) SFMZ, (△) SZ; reaction conditions: 423 K, 5.0 MPa, isobutane/1-butene = 4, total WHSV = 50 h^{-1}.

conversion can not be compared with 1-butene conversion simply because the isobutane/1-butene ratio is 4. Thus, we calculated the conversion for each reactant and normalized the isobutane conversion. At the condition of 423 K and 5.0 MPa, SFMZ showed a higher activity than SZ on both isobutane and 1-butene conversions. SZ was not active enough to give liquid products at this condition. Furthermore, the deactivation rate of SFMZ seems to be smaller than that of SZ. It is clear that the difference of the catalytic activity between SFMZ and SZ is due to the effect of Fe and Mn promotion. This effect has been widely studied, and various interpretations have been suggested by many researchers [23-25]. For example, there is a report assuming that Mn plays a role of dispersing Fe highly on the catalyst surface [25]. On the other hand, there is also a report that a supercritical fluid such as n-pentane could be an hydrogen donor with existence of a small amount of alkaline metals, alkaline earth metals, or transition metals [20]. However, the role of the promoted metals has not yet been clearly established. For both catalysts, 1-butene conversion is always higher than isobutane conversion indicating that 1-butene is consumed not only by alkylation but also by dimerization and oligomerization. The deactivation rate of 1-butene also seems to be smaller than that of isobutane. It is thought that the active sites responsible for alkylation deactivate much faster than the sites responsible for dimerization and oligomerization. This is also clear from the result that both the selectivity for C8 in liquid products and that for C8 paraffins in C8 products decrease with time on stream. It becomes more selective for heavy ends formed by oligomerization with longer time on stream.

We studied the influence of the reaction pressure on the catalytic activity of SFMZ. The results are shown in Figure 4. In this work, at the conditions of 5.0 and 5.5 MPa are the supercritical conditions. Comparing all the reaction pressures, the conversions are the highest and the deactivation rate seems to be the smallest when the reaction was carried out at the pressure of 5.0 MPa. No liquid products were obtained at the reaction of 3.0 MPa. The effect of the supercritical condition is clearly seen as compared with the reaction of 3.0 and 4.0 MPa. Furthermore, the initial conversions at 5.0 MPa are relatively higher than those of 3.0 and 4.0 MPa. However, despite the fact that it is a supercritical condition, it did not show a high activity at the pressure of 5.5 MPa. Although the reason for this low activity at the pressure of 5.5 MPa has not been clarified, one possible reason is that the density of the reaction mixture increased and the diffusivity becomes dominant at the reaction pressure of 5.5 MPa since the density of fluid changes drastically with a slight change of pressure. The low diffusivity probably retarded the mass transfer, and, as a consequence, the catalytic activity appeared to be low. With regard to the selectivity for C8 products in the liquid products, it showed the highest selectivity at the condition of 5.0 MPa. At the conditions of 4.0 and 5.5 MPa, it became more selective for heavier products. The catalytic activity at the condition of 3.0 MPa was low and no liquid products were obtained. No big difference was observed for the

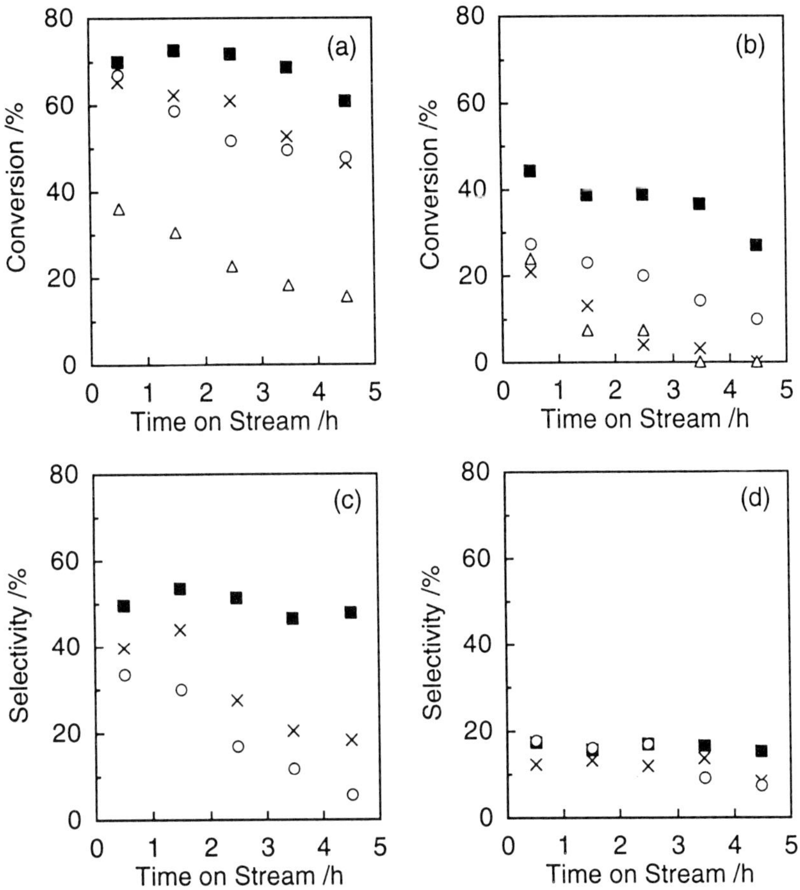

Figure 4. Influence of the reaction pressure on the catalytic activity of SFMZ: (a) conversion of 1-butene, (b) conversion of isobutane, (c) C8 selectivity in liquid products, (d) C8 paraffin selectivity in C8 fraction; (△) 3.0 MPa, (○) 4.0 MPa, (■) 5.0 MPa, (×) 5.5 MPa; reaction conditions: 423K, 3.0-5.5 MPa, isobutane/1-butene = 4, total WHSV = 50 h^{-1}.

C8 paraffin selectivity in C8 products.

Deactivation of the Catalysts

As was stated previously, the deactivation rate is the key for the solid acid catalyzed isobutane alkylation, since the rapid deactivation is a serious problem for this reaction. In this work, the catalytic deactivation can also be seen clearly from the results of the catalytic activity even under supercritical conditions (Figure 4). Thus, we calculated the deactivation rate for each reactant with the equation below, based on the results of the conversion:

$$\ln C = -at + b$$

(C, conversion; a, deactivation rate; t, time on stream; b, constant)

The deactivation rates for each reactant at various reaction pressures over SZ and SFMZ are summarized in Figure 5. It is obvious that the deactivation rates of both isobutane and 1-butene conversion for SFMZ are the smallest at the pressure of 5.0 MPa, due to the effect of the supercritical condition. It is noteworthy that the supercritical condition is effective even for SZ which did not show a high activity, as shown in Figure 5 (b). Indeed, the supercritical condition was valid for all the catalysts tested, though the degree of its effect appeared to be dependent on the catalysts.

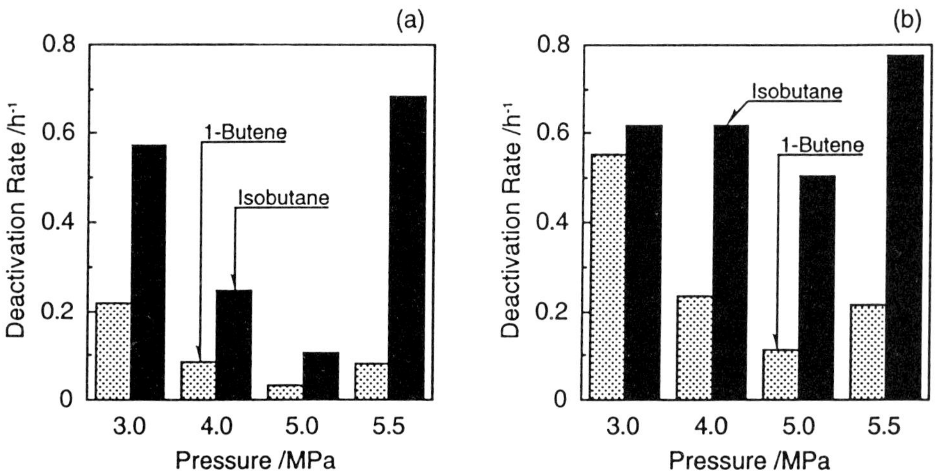

Figure 5. Influence of the reaction pressure on the deactivation rate: (a) SFMZ, (b) SZ.

For all reaction conditions, the deactivation rate of 1-butene conversion is always smaller than that of isobutane conversion, due to the difference of the reactivity between 1-butene and isobutane. Since a strong acid is required for alkylation reaction because isobutane must be activated, while 1-butene is still consumed by dimerization or oligomerization even on rather weak acid sites. Thus, it is clear from the results that the catalysts lose the alkylation activity much faster than the dimerization and oligomerization activity.

In order to study the effect of coke formed on the catalysts, the catalysts used for 5 hours of the reaction were collected from the reactor as spent catalysts, and acid amount and surface properties of spent catalysts were measured as well as the catalysts before the reaction (fresh catalysts). The results are summarized in Table 2. Before the reaction, SFMZ has larger acid amount than SZ. All the catalysts after

Table 2
Surface properties of the catalysts before and after the reaction.

Catalyst	Pressure/ MPa	NH_3 Uptake/ mmol g^{-1}	Surface Area/ m^2 g^{-1}	Pore Volume/ cm^3 g^{-1}
SFMZ	(fresh)	0.171	116	0.062
	1.0	0.068	88	0.043
	3.0	0.095	91	0.048
	5.0	0.129	105	0.050
	5.5	0.082	96	0.047
SZ	(fresh)	0.125	59	0.032
	1.0	0.030	6	0.002
	3.0	0.043	9	0.007
	5.0	0.066	19	0.009
	5.5	0.047	9	0.003

the reaction lost a part of their acid amount due to the coke formation. Nevertheless, SFMZ was less influenced by the coke formation on the acid amount than SZ. Furthermore, the decrease rate of the acid amount depends on the reaction pressure, and the amount of coke formed on the catalyst was most suppressed at the pressure of 5.0 MPa. With regard to the specific surface area, the values after the reaction were smaller than those of the fresh catalysts because of the pore blockage by the coke. When the reaction was carried out at the pressure of 5.0 MPa over SFMZ, the decrease of the specific surface area was most suppressed as the case of the acid amount. Besides, the same tendency was observed in the case of the pore volume. These results agree well with the results of the catalytic activity and the deactivation rate (Figures 4 and 5).

Acidic Nature of the Catalysts

In order to study the surface acidity of the catalysts, IR spectra of adsorbed ammonia were taken. The spectra of the N-H region over SZ and SFMZ are shown in Figure 6. The peak at 1612 cm^{-1} corresponds to δNH_3 and it indicates the existence of Lewis acid sites [27]. The other at 1446 cm^{-1} corresponds to δNH_4^+ indicating the existence of Brönsted acid sites [27]. After NH_3 was adsorbed on the catalysts at 773 K and evacuated, a large peak of δNH_3 indicating the existence of Lewis acid sites was seen for both catalysts. SFMZ had a small peak of δNH_4^+ indicating the existence of Brönsted acid sites, while SZ had no Brönsted acidity. Adsorbed NH_3 starts to desorb as the temperature is raised, and no adsorbed NH_3 could not be observed when the temperature was raised to 673 K.

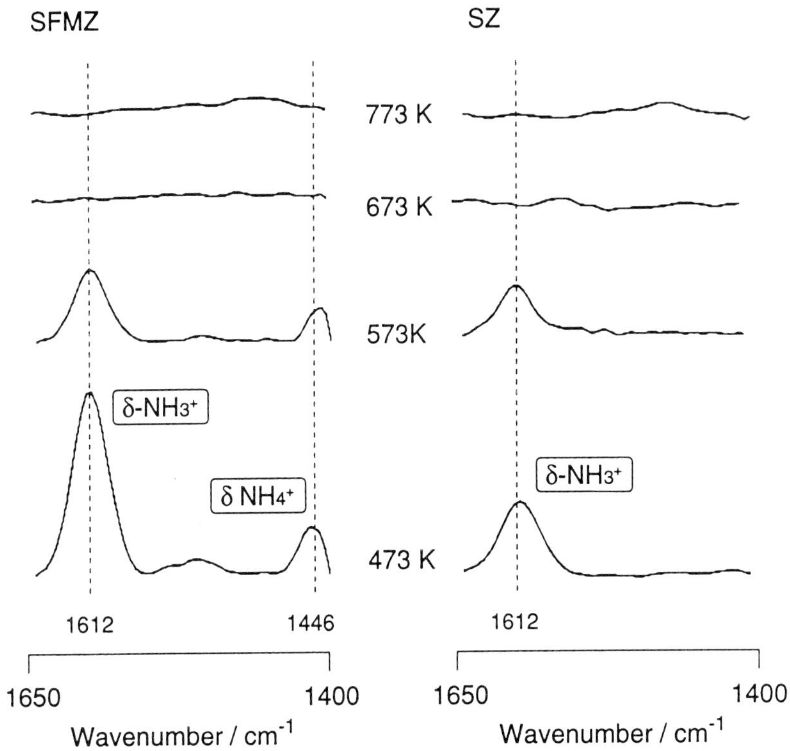

Figure 6. IR spectra of adsorbed ammonia on the catalysts.

CONCLUSION

Rapid deactivation of the catalysts is a serious problem for alkylation of isobutane over solid acid catalysts. We have studied this reaction under supercritical conditions, expecting that the supercritical fluid would extract the coke precursor formed on the catalyst surface. At a condition of a slightly higher than the critical point of the reaction mixture (423 K, 5.0 MPa), the catalyst deactivation was the most reduced, especially when iron- and manganese-promoted sulfated zirconia (SFMZ) was used as a catalyst. SFMZ mainly has Lewis acid sites. However, the correlation between the acidity and the activity of the catalyst has not been clarified yet.

REFERENCES

1. J.A. Cusmano, *CHEMTECH* **48**, 2 (1992).

2. A. Corma, V. Gomez, and A. Martinez, *Appl. Catal. A: General* **119**, 83 (1994).
3. A. Corma, A. Martinez, and C. Martinez, *Catal. Lett.* **28**, 187 (1994).
4. A. Corma, A. Martinez, and C. Martinez, *J. Catal.* **146**, 185 (1994).
5. F. Cardona, N.S. Gnep, M. Guisnet, G. Szabo, and P. Nascimento, *Appl. Catal. A: General* **128**, 243 (1995).
6. J. Weitkamp and P.A. Jacobs, *New Frontiers in Catalysis, Proceedings of the 10th International Congress on Catalysis*, 19-24 July 1992, Budapest, Hungary.
7. Y.F. Chu and A.W. Chester, *ZEOLITES* **6**, 195 (1986).
8. M. Stocker, H. Mostad, and T. Rorvik, *Catal. Lett.* **28**, 203 (1994).
9. S. Unverricht, S. Ernst, and J. Weitkamp, *Stud. Surf. Sci. Catal.* **84**, 1693 (1994).
10. T. Okuhara and K. Fujimori, *PETROTECH*, Vol. **18**, No. 10, 838 (1995).
11. A. Corma, M.I. Juan-Rajadell, J.M. Lopez-Nieto, A. Martinez, and C. Martinez, *Appl. Catal. A: General* **111**, 175 (1994).
12. C. Guo, S. Yao, J. Cao, and Z. Qian, *Appl. Catal. A: General* **107**, 229 (1994).
13. K. Hatakeyama, T. Suzuka, and M. Yamane, *Sekiyugakkaishi* **34** (3), 267 (1991).
14. Mobil, USP 4,956,518 (1990).
15. CATALYTICA, WO 92/04977 (1992).
16. T. Rorvik, I.M. Dahl, H.B. Mostad, and O.H. Ellestad, *Catal. Lett.* **33**, 127 (1995).
17. T.J. Huang and S. Yurchak, *Preprints of Symposium on Recent Adv. in Alkylation* **22**, 301 (1977).
18. T. Okuhara, M. Yamashita, K. Na, and M. Misono, *Chem. Lett.* 1451 (1994).
19. Y. Minamino and T. Aki, *KAGAKU KOGAKU* Vol. **58**, No. 9, 689 (1994).
20. K. Arai, *KAGAKU KOGAKU* Vol. **59**, No. 6, 373 (1995).
21. D.M.G. and B. Subramanian, *Stud. Surf. Sci. Catal.* **88**, 327 (1994).
22. D.R. Lide, *HANDBOOK of CHEMISTRY and PHYSICS 73rd Edition*, 1992-1993.
23. J.M. Prausnitz and P.L. Chuah, *Computer Calculations for High-Pressure Vapor-Liquid Equilibra*, Prentice-Hall Inc.
24. V. Adeeva, J.W. de Haan, J. Janchen, G.D. Lei, V. Schunemann, L.J.M. van de Ven, W.M.H. Sachtler, and R.A. van Santen, *J. Catal.* **151**, 364 (1995).
25. C. Lin and C. Hsu, *J. Chem. Soc.* **20**, 1479 (1992).
26. M.A. Coehlo, D.E. Resasco, E.C. Sikabwe, and R.L. White, *Catal. Lett.* **32**, 253 (1995).
27. N.S. Kotsarenko, L.G. Karakchiev, and V.A. Dzisko, Kinetika *I Kataliz* **9**, 158 (1968).

Res. Chem. Intermed., Vol. 24, No. 4, pp. 461-471 (1998)

RECENT ADVANCES IN IMMOBILIZATION OF HETEROPOLYACIDS

YUSUKE IZUMI*

Department of Molecular Design and Engineering, School of Engineering, Nagoya University, Furo-cho, Chikusa-ku, Nagoya 464-01, Japan

Received 3 September 1997; accepted 12 December 1997

Abstract-- Freely water-soluble heteropolyacid could be made insoluble either by supporting on active carbon or by forming acidic cesium salts. In addition, 12-tungstophosphoric acid and its acidic cesium salt could be effectively immobilized in silica matrices by means of a sol-gel technique which involves the hydrolysis of ethyl orthosilicate to give insoluble, readily recoverable, and highly active solid acid catalysts for the hydrolysis of ethyl acetate.

INTRODUCTION

The Keggin-type heteropolyacid (hereafter abbreviated as HPA) is a unique and practical catalyst material since it possesses dual catalytic functions of strong acidity and high oxidizing ability [1-5]. Although its debut in the field of catalyst chemistry and technology is rather recent, HPA is now commercially applied mainly in Japan as an efficient catalyst to several petrochemical processes. These include the direct hydration of propene (1972) [6,7], isobutene (1984) [8] and *n*-butenes (1989) [9], the oxidation of methacrolein to methacrylic acid (1982) [10], and the oligomerization of tetrahydrofuran to polymeric diols (1985) [11].

HPA is a typical strong Brönsted acid like perchloric and sulfuric acids. It is very soluble in oxygen-containing polar solvents such as water, alcohol, ether, and ketone. In these polar solvents, HPA works as an efficient acid catalyst for alkene hydration, Prins reaction, nucleophilic cleavage of alicyclic and cyclic ethers, esterification, hydrolysis, transesterification, and acetalization. HPA exhibits much higher catalytic activities for these reactions than ordinary protonic acids such as sulfuric and *p*-toluenesulfonic acids, because the activation energies of the HPA-catalyzed reactions are remarkably lowered owing to stabilization of the cationic reaction intermediates by heteropoly anion [1,4].

Since environmentally benign solid acid catalysts have recently been awaited in place of sulfuric acid and aluminum chloride, it is an interesting and significant research target to immobilize such catalytically highly efficient HPA into an insoluble, readily separable solid acid. In this paper, several studies on the immobilization of HPA are described mainly based on our recent work.

IMMOBILIZATION ON ACTIVE CARBON

The authors have first pointed out that active carbon tightly immobilizes or entraps HPA inside its pores and that the carbon-entrapped HPA catalyzes liquid-phase organic reactions in polar media as an insoluble solid acid [12].

The carbon-entrapped HPA was prepared by impregnating an active carbon (preferably coconut-derived carbon) of 0.4-0.7 mm particle size with a methanol solution of HPA, followed by removing methanol at 40 °C under evacuation. The non-entrapped HPA was removed by extracting with water or methanol. As shown in Figure 1, the fraction of the HPA dissolved out of the carbon support changed with the initial HPA loading and with the kind of solvent served for extraction. Leakage of HPA ceased after a certain period of extraction and the entrapped HPA with a maximum loading of 7.2-13.9 wt% was obtained. The maximum HPA loading depended on pore volume and pore size of carbon support. The chemical nature of the carbon surface such as the amounts of surface carboxyl or hydroxyl groups had little influence on its entrapping ability. Diatomaceous earth and silica, which are often employed as supports for HPA in vapor-phase reactions, were both

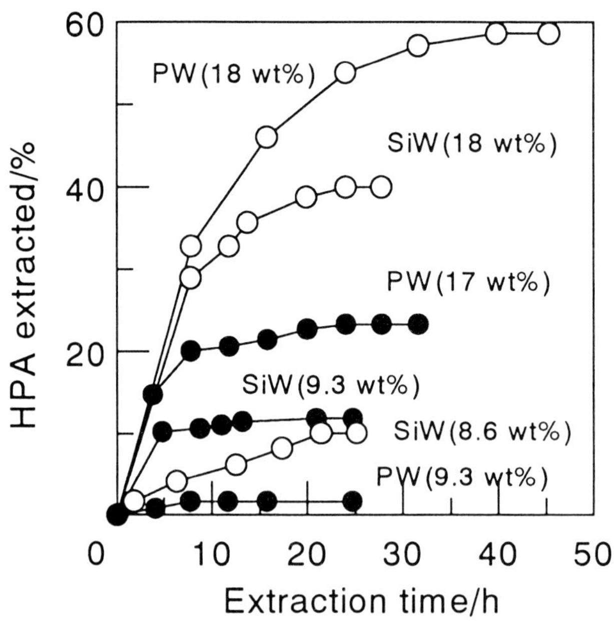

Figure 1. Extraction of heteropolyacid supported on carbon with water (○) and methanol (●). The values in parentheses denote the initial HPA loading (wt%) before extraction.

incapable of immobilizing HPA in polar liquid media because the HPA supported on them was almost completely removed by solvent extraction.

Table 1 demonstrates an application of the carbon-entrapped $H_4PW_{12}O_{40}$ catalyst to the intermolecular dehydration between butanol and *t*-butyl alcohol into butyl *t*-butyl ether. Under these reaction conditions, neither a self-condensation product (dibutyl ether) nor an intramolecular dehydration product (isobutene) was formed. When the reaction was repeated several times using the recovered catalyst by renewing the reactants, no leakage of HPA from the carbon support was observed, and the catalytic activity remained unchanged. Heteropolymolybdic acids such as $H_3PMo_{12}O_{40}$ and $H_4SiMo_{12}O_{40}$ were inadequate to be trapped in carbon because these HPA were susceptible to reduction with carbon, resulting in decrease in acid strength.

Table 1
Dehydration between butanol and *t*-butyl alcohol into butyl *t*-butyl ether catalyzed by carbon-entrapped $H_3PW_{12}O_{40}$[a].

Reaction repeated	*t*-BuOH conversion (%)			
	30min	60min	90min	120min
1	12	22	29	35
2	14	24	31	36
3	13	23	30	36
4	14	22	29	36

[a]Carbon-entrapped $H_3PW_{12}O_{40}$ catalyst (0.58 g) with an HPA loading of 13.9 wt%, butanol 5 ml, *t*-butyl alcohol 2 ml, 106 °C.

HPA supported on active carbon worked as an insoluble solid acid catalyst also in liquid-phase esterification [13]. The catalytic efficiency of carbon-entrapped $H_4PW_{12}O_{40}$ was compared with a cation-exchange resin and *p*-toluenesulfonic acid for the esterification of acetic acid with butanol (Table 2). Non-polar solvents such as hexane and benzene were used to minimize the leakage of HPA from the carbon support. Since active carbon strongly adsorbed ester, a work-up procedure of extraction with an appropriate solvent such as methanol was necessary to recover the product completely from the catalyst. The carbon-entrapped $H_4PW_{12}O_{40}$ was not so active for the esterification as Amberlyst 15 and *p*-toluenesulfonic acid because of its lower acid strength. Carbon-entrapped HPA, which is heat-resistant, may be effectively applicable to esterification processes operated at relatively higher temperatures at which the product water can be continuously removed from the reactor.

Table 2
Liquid-phase esterification of acetic acid with butanol catalyzed by carbon-entrapped $H_3PW_{12}O_{40}$[a].

Catalyst	AcOBu yield/%	
	1.0[b]	3.0[b]
$H_4PW_{12}O_{40}$/C[c]	57	89
Amberlyst 15	92	-
p-TsOH[d]	91	-

[a]Acid amount: 0.1 mmol, AcOH: 1 mmol, hexane: 5 ml, 60 °C, 6 h.
[b]BuOH/AcOH mole ratio.
[c]Active carbon: Kuraray GC (1656 m^2/g).
[d]*p*-toluenesulfonic acid.

IMMOBILIZATION BY SALT FORMATION

It has been well known that cesium salts and ammonium salts of HPA are insoluble in water. Misono and his co-workers have recently demonstrated that acidic Cs salt $Cs_{2.5}H_{0.5}PW_{12}O_{40}$, which is insoluble in water and has a large surface area of 136 m^2/g, showed good catalytic activity for the decomposition of cyclohexyl acetate at 100°C in liquid phase [14]. Such insoluble acid salts of HPA may be effective candidates for solid acid catalysts in heterogeneous liquid-phase reactions.

We have pointed out that $Cs_{2.5}H_{0.5}PW_{12}O_{40}$ is very stable against hydrothermal treatment at 120°C, and applied it to several water-participating reactions such as esterification and ester hydrolysis in which ordinary inorganic solid acids cannot be employed [15]. The results are summarized in Table 3. Acidic Cs salts of HPA were more active than H-ZSM-5 and silica-alumina for the hydrolysis of ethyl acetate and the esterification of acetic acid with ethanol, because they showed higher acid strength ($\underline{H}_0 > 0.8$) than H-ZSM-5 ($\underline{H}_0 > 1.5$) in both water and alcohol. In terms of turnover frequency based on the total acid amount, $Cs_{2.5}H_{0.5}PW_{12}O_{40}$ was more active than Amberlyst 15.

Figure 2 illustrates the changes in hydrolysis activity and surface area of the salt with the Cs content in terms of x in the formula of $Cs_xH_{3-x}PW_{12}O_{40}$. The hydrolysis activity decreased monotonously with increasing Cs content or decreasing acidity, whereas the surface area increased abruptly from less than 10 to 130 $m^2\ g^{-1}$ at above x = 2.0. $CsH_2PW_{12}O_{40}$ dissolved during the hydrolysis reaction, and $Cs_2HPW_{12}O_{40}$ partly dissolved. $Cs_{2.5}H_{0.5}PW_{12}O_{40}$, however, remained as an insoluble solid. The activity was related directly to the total acid amount of the salt, which suggests that the hydrolysis took place throughout the bulk of the salt. This catalytic behavior shown in the hydrolysis was quite different from that observed in the liquid-phase Friedel-Crafts reaction where catalytic efficiency was greatly dependent

Table 3
Catalytic activity of solid acids in hydrolysis and esterification reactions.

Solid acid	Surface area (m^2/g)	Acid amount (meq/g)	Hydrolysis[a]		Esterification[b]	
			Rate[c]	TOF[d]	Rate[c]	Tf[d]
$Cs_{2.5}H_{0.5}PW_{12}O_{40}$	102	0.156[h]	2.04	13.1	0.52	33.3
$Cs_{2.5}H_{1.5}SiW_{12}O_{40}$	49	0.468[h]	4.56	9.7	-	-
$Cs_3HSiW_{12}O_{40}$	83	0.305[h]	1.76	5.8	-	-
Amberlyst-15	-	4.9[i]	7.80	1.6	1.64	3.5
H-ZSM-5[e,f]	-	0.234[h]	0.43	1.8	0.03	6.6
SiO_2-Al_2O_3[e,g]	-	-	0	0	-	-

[a]Ethyl acetate/water/catalyst = 1080/27.8/0.1 mmol, 60 °C.
[b]Acetic acid/ethanol/catalyst = 100/100/0.1 mmol, 60 °C.
[c]Specific rate/10^{-3} $min^{-1}g^{-1}$.
[d]Turnover frequency/10^{-3} mmol $meq.^{-1}$ min^{-1} (based on the total acid amount).
[e]321 mg.
[f]Si/Al = 70.
[g]Si/Al = 6.25.
[h]Calculated value based on chemical composition.
[i]Determined by titration.

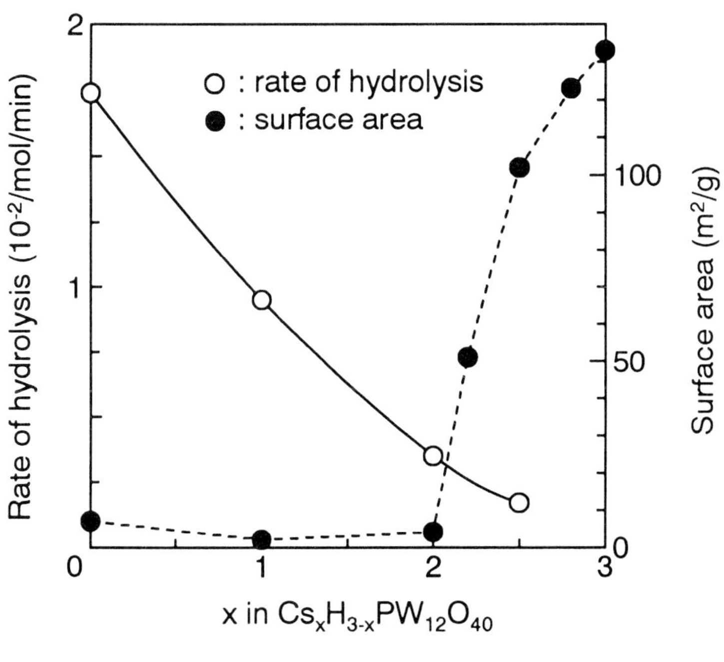

Figure 2. Hydrolysis of ethyl acetate over $Cs_xH_{3-x}PW_{12}O_{40}$. AcOEt = 9.5 g, H_2O = 0.5 g, catalyst = 0.1 mmol, 60 °C.

on surface area because the Friedel-Crafts reaction proceeded exclusively on the surface owing to low polarity of the reaction media [15]. Other acidic salts of HPA such as $(NH_4)_2HPW_{12}O_{40}$, $Cs_2H_2SiW_{12}O_{40}$, and $Cs_{2.5}H_{0.5}PMo_{12}O_{40}$, which worked as efficient and insoluble solid acid catalysts for the liquid-phase Friedel-Crafts reaction, were unfavorable for the hydrolysis of ethyl acetate because they partly dissolved in the course of the reaction.

IMMOBILIZATION WITHIN SILICA MATRIX

Immobilization of Acidic Cesium Salt

The salt of $Cs_{2.5}H_{0.5}PW_{12}O_{40}$ consists of relatively uniform very fine crystal particles with ca. 10nm diameter on average. In the hydrolysis of ethyl acetate with a low concentration of water, the salt particles could be separated by filtration as large-size aggregates from the liquid medium. However, if applied to the hydrolysis under the commonly employed reaction condition using a large excess of water, i.e. in a highly polar reaction medium, the salt particles formed a colloidal solution and became inseparable. In order to make the salt recoverable from water, we tried to include the salt with silica matrix into a composite by means of a sol-gel technique which involved the hydrolysis of ethyl orthosilicate [16a]. The ratios of $SiO_2/Cs_{2.5}H_{0.5}PW_{12}O_{40}$, $H_2O/(EtO)_4Si$, and $EtOH/(EtO)_4Si$ were the most important criteria in including the salt with silica. Higher SiO_2 content brought about smaller leakage of the salt. Higher water content accelerated much the hydrolysis of $(EtO)_4Si$ to cause more leakage of the salt, probably due to the formation of a rough silica network [17]. Ethanol was necessary to make ethyl orthosilicate and water miscible each other. In the present preparative method for silica inclusion, the salt to be included promoted the hydrolysis of ethyl orthosilicate by its own acid catalysis. The composite was a porous material with mesopores and a large surface area of 364 m^2/g.

Table 4 lists the catalytic efficiency of silica-included $Cs_{2.5}H_{0.5}PW_{12}O_{40}$ for the hydrolysis of ethyl acetate with excess water; the activity of non-modified $Cs_{2.5}H_{0.5}PW_{12}O_{40}$ is given for comparison. Silica itself was inactive for the hydrolysis. At a silica weight ratio of less than 1.2, considerable leakage of $Cs_{2.5}H_{0.5}PW_{12}O_{40}$ as colloid from the silica composite was observed during the reaction, but the leakage could be effectively suppressed when the ratio was increased to 8. The silica composites were comparable with the non-modified salt in catalytic activity. They were less active than Amberlyst-15, and more active than H-ZSM-5 on the basis of the rate per unit catalyst weight. However the values of their turnover frequency based on the total acid amount were larger than for Amberlyst-15 and H-ZSM-5. This means that the silica-included $Cs_{2.5}H_{0.5}PW_{12}O_{40}$ shows higher acid strength in water than the resin and the zeolite. A silica

Table 4
Hydrolysis of ethyl acetate over silica-included $Cs_{2.5}H_{0.5}PW_{12}O_{40}$[a].

$SiO_2/Cs_{2.5}$[b]	Surface area (m^2/g)	Catalytic activity		Leakage[e] (%)
		Rate[c]	TOF[d]	
0.0[f]	105	2.80	17.9	-
1.2	598	3.06	17.7	>1
2.0	-	1.94	12.4	0.77
4.0	-	2.78	17.8	0.62
8.0	364	2.07	13.2	0.20
SiO_2[g]	865	0.0	0.0	-
Amberlyst-15[h]	-	4.9[j]	3.9	-
H-ZSM-5[i]	-	0.23[j]	8.1	-

[a]H_2O = 9.5g, AcOEt = 0.5 g, catalyst = 0.05 mmol on the basis of $Cs_{2.5}H_{0.5}PW_{12}O_{40}$, 60 °C. Catalyst was prepared under the conditions of $H_2O/(EtO)_4Si$ = 5, $EtOH/(EtO)_4Si$ = 1.3 (mole ratio), and calcined at 30 °C for 3 h.
[b]Weight ratio of SiO_2 to $Cs_{2.5}H_{0.5}PW_{12}O_{40}$ calculated from the amounts of reagents charged at the preparation.
[c]First-order specific rate per unit weight of $Cs_{2.5}H_{0.5}PW_{12}O_{40}$ (10^{-3} min^{-1} g^{-1}).
[d]Turnover frequency/10^{-3} mmol meq.$^{-1}$ min^{-1} (based on the total acid amount).
[e]Leakage of $Cs_{2.5}H_{0.5}PW_{12}O_{40}$ during the hydrolysis reaction (3 h), estimated by UV spectrometry.
[f]Non-modified $Cs_{2.5}H_{0.5}PW_{12}O_{40}$.
[g]Prepared by hydrolysis of $(EtO)_4Si$.
[h]160 mg, dried at 100 °C.
[i]160 mg, calcined at 300 °C. [j]First-order specific rate per unit weight.

composite with a silica weight ratio of 8 was readily recoverable by simple filtration, and it could be reused for the hydrolysis reaction repeatedly (4 times) without deactivation. The composite was less stable than the salt itself, but it worked as a solid acid catalyst even after being treated at 450 °C, although there was a gradual deactivation due to the reduction in surface area. IR spectroscopy indicated that the Keggin structure of the $PW_{12}O_{40}^{3-}$ anion was retained completely in the composite until 450°C. No IR absorption bands characteristic of $SiW_{12}O_{40}^{4-}$ anion were observed, which suggests that the exchange between P and Si in the Keggin anion did not occur in the course of the procedure for silica inclusion.

Immobilization of Free Acid

In silica-included $Cs_{2.5}H_{0.5}PW_{12}O_{40}$, Cs ions had no direct effect on acid catalysis. This problem was solved by inclusion of free acid itself with silica. In this case, higher silica contents were required to make the silica network denser than in the case of salt inclusion because heteropoly anions with a very small molecular size of 1 nm must be caged. A preparative method for silica-included 12-tungstophosphoric acid ($H_3PW_{12}O_{40}$) with good reproducibility is shown in Figure 3. According to this method, $H_3PW_{12}O_{40}$ was completely immobilized in silica matrix, and no leakage of $H_3PW_{12}O_{40}$ was observed in the course of extraction with hot water. In addition, since tetraethyl orthosilicate charged for inclusion was also totally hydrolyzed, silica-included $H_3PW_{12}O_{40}$ composite was yielded quantitatively. The content of $H_3PW_{12}O_{40}$ included in silica matrix was 11 wt%. A Langmuir-type nitrogen adsorption isotherm was observed for this composite. The silica-included $H_3PW_{12}O_{40}$ was a microporous material having a surface area of 554 m^2/g and a pore volume of 0.2 ml/g.

Table 5 summarizes the catalytic activity of silica-included $H_3PW_{12}O_{40}$ in the hydrolysis of ethyl acetate with excess water, comparing with the activities of an ion-exchange resin, Amberlyst 15, and H-ZSM-5. As for the activity per unit weight of solid acid, Amberlyst-15 was the most active because the acid amount is

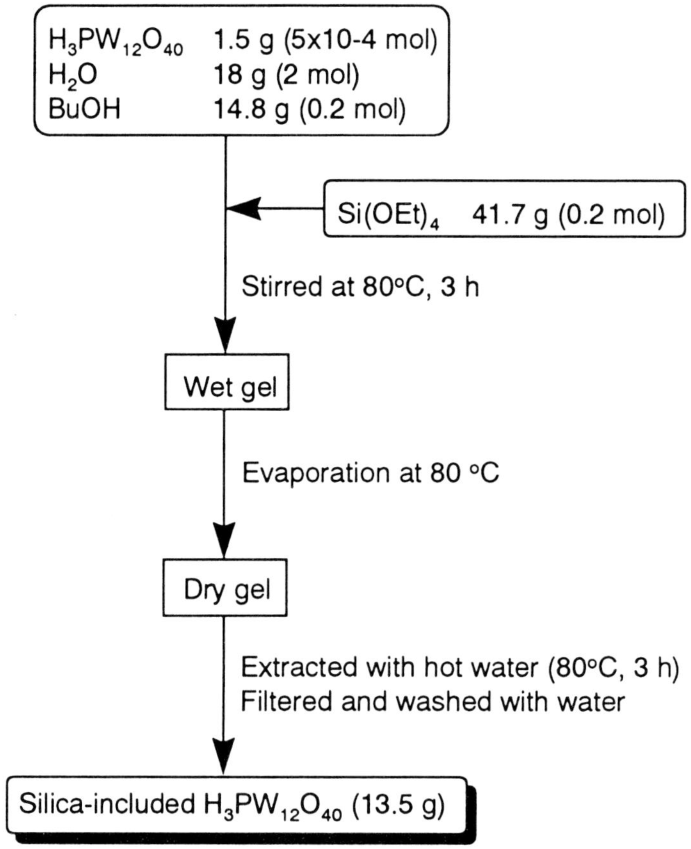

Figure 3. Standard preparative method of silica-included $H_3PW_{12}O_{40}$.

Res. Chem. Intermed., Vol. 24, No. 4, pp. 473-480 (1998)

VAPOR PHASE NITRATION OF BENZENE OVER SOLID ACID CATALYSTS (1): NITRATION WITH NITROGEN DIOXIDE (NO_2)

H. SATO* and K. HIROSE
Organic Synthesis Research Laboratory, Sumitomo Chemical Co., Ltd., 10-1, 2-chome, Tsukahara, Takatsuki-city, Osaka 569, Japan

Received 19 September 1997; accepted 22 December 1997

Abstract--Vapor phase nitration of benzene over acidic catalysts is expected to be a clean process with no sulfuric acid waste. We investigated this process over acidic catalysts utilizing nitrogen dioxide (NO_2) as a nitrating agent, and found that several mixed metal oxides, such as silica-alumina, zinc oxide-titania, and tungsten oxide-molybdenum oxide, exhibited a fairly good activity. Among them, WO_3-MoO_3 is the most active; however, only this catalyst deviated from the linear relationship between the activity and the acidity. In order to explain this discrepancy, several factors such as acid amount, acid strength, and BET surface area, are examined in detail.

INTRODUCTION

Vapor phase nitration of benzene to nitrobenzene over solid acid catalysts is expected to be a clean process without sulfuric acid waste, in contrast to the conventional liquid process which uses a mixed acid (concentrated nitric acid and sulfuric acid) as a nitrating agent, accompanied by a large amount of diluted sulfuric acid waste (Eq. 1). Many efforts devoted to develop a vapor phase process have been unsuccessful, because either activity or life of catalysts is not satisfactory [1,2].

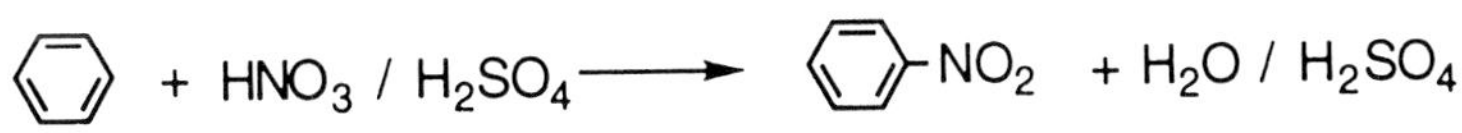

(1)

Previously we reported in a piece of patent literature [3] that a mixed metal oxide comprising WO_3 and MO_3 exhibited a fairly high and stable activity.

In this paper we will report the results of our detailed study on the catalytic performances of various solid acid catalysts for benzene nitration with nitrogen dioxide (NO_2) as a nitration agent (Eq. 2).

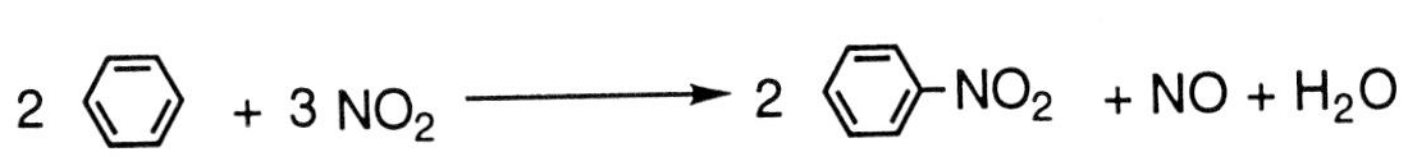

(2)

EXPERIMENTAL

Mixed metal oxides were prepared in either a co-precipitation method, a co-grinding method, or an impregnation method. In particular, a WO_3-MoO_3 system was prepared by mixing and grinding the two oxides in a wet state. The mixed oxides were then calcined in air at 500°C for 3 h. Vapor phase nitration was conducted in a flow reactor made of quartz with a fixed catalyst bed at atmospheric pressure. A nitrogen dioxide vapor was introduced into the flow reactor from a container with a nitrogen carrier gas. The nitration products were cold-trapped, analyzed, and identified by gas chromatography.

RESULTS AND DISCUSSION

Investigation of Active Catalysts for Nitration

The results of benzene nitration with nitrogen dioxide over various solid acid catalysts are summarized in Table 1. Among many catalysts examined, the mixed

Table 1

Nitration of benzene with NO_2 over solid acid catalysts.

Catalyst	Reaction Conditions			Yield[a]/% of Nitrobenzene	
	NO_2/ Bz.[b]	SV/h^{-1}	Temp./ °C	after 1 h	after 1 h
WO_3 (75%)-MoO_3 (25%)	2.3	13,500	200	24.5	25.8
TiO_2 (90%)-MoO_3 (10%)	1.0	12,000	150	0.8	-
Silica-alumina[c]	1.0	2,600	150	18.3	8.0
Nafion-H	2.3	19,500	200	23.6	1.9
Silicotungustoic acid[d]	1.0	2,800	150	15.4	1.6
25% H_3PO_4/Silica-alumina[c]	1.0	3,600	150	3.6	-
30% H_2SO_4/Silica-alumina[c]	1.0	3,800	150	17.6	15.4
20% $NiSO_4$/Silicagel	1.0	2,000	200	3.8	1.7
18.5% $Al_2(SO_4)_3$/Silicagel	1.0	3,500	150	1.3	0.7
10% (WO_3 (80%)-MoO_3 (20%))/Silica-alumina[c]	1.0	1,300	1505.6	6.8	
10% WO_3/Silicagel	1.0	2,240	160	2.3	1.3

[a]Nitrobenzene yield based on NO_2.
[b]Molar ratio of NO_2/benzene.
[c]N631L; Al_2O_3 = 8%.

oxide catalyst comprising WO_3 (75 mol%)-MoO_3 (25 mol%) afforded a fairly good yield of nitrobenzene with almost no decay of activity for 5 h. The sulfuric acid catalyst supported on silica-alumina also exhibited a good yield with a slight decay of activity. Other solid acid catalysts such as TiO_2 (90 mol%)-MoO_3 (10 mol%), silica-alumina, Nafion-H, silico-tungstoic acid, 25 wt% H_3PO_4/silica-alumina, 30 wt% H_2SO_4/silica-alumina, 20 wt% $NiSO_4$/silicagel, 18.5 wt% $Al_2(SO_4)_3$/silicagel, 10 wt% (WO_3 (80 mol%)-MoO_3 (20 mol%))/silica-alumina, and 10 wt% WO_3/silicagel did exhibit low yields or rapid decay of activity.

Relation Between Nitration Activity and Acidity of the Catalysts

In order to elucidate the properties of the active site for nitration, acidities of the mixed oxides were determined by the amine titration method utilizing Hammett indicators. The results are shown in terms of the first order rate constant (k) for nitration (Figures 1 and 2). These results show that the nitration activity was almost proportional to both the acid amount (Figure 1) and the acid strength (Figure 2) of the catalysts. This tendency is in accordance with the conventional nitration mechanism by a nitronium cation ($HNO_3 + 2H_2SO_4 \rightarrow NO_2^+ + H_3O^+ + 2HSO_4^-$) in the liquid phase. However it is noteworthy that the most active mixed oxide comprising WO_3 (95%)-MoO_3 (5%) exhibited the least amount of acid and the lowest acid strength. Therefore, the active site in the WO_3-MoO_3 system seems to differ from the ordinary acidic site.

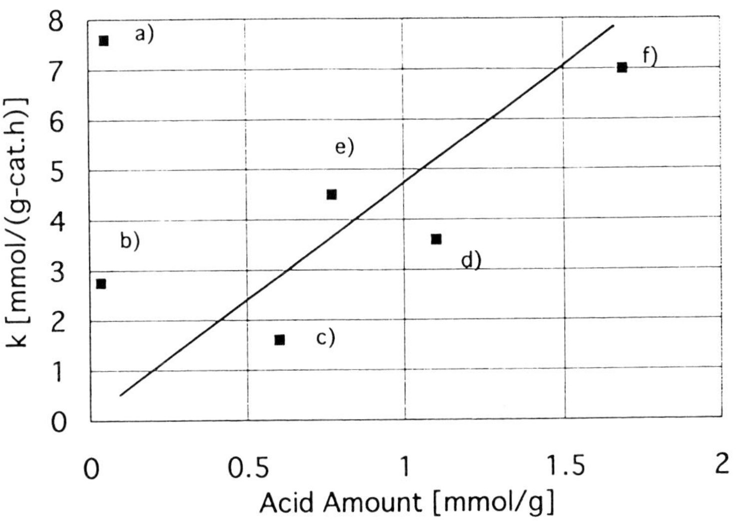

Figure 1. Relation between the rate constant (k) of nitration and the acid amount of the catalyst. Reaction conditions: NO_2/benzene/N_2 = 0.2/1.0/0.7 (molar ratios), Temp. = 150°C, SV = 220 h^{-1}. Catalysts (molar ratio): a) WO_3(95)-MoO_3(5), b) WO_3(90)-TiO_2(10), c) WO_3(10)-TiO_2(90), d) ZnO(7)-TiO_2(93), e) SiO_2(92)-Al_2O_3(8), f) SiO_2(81)-Al_2O_3(19).

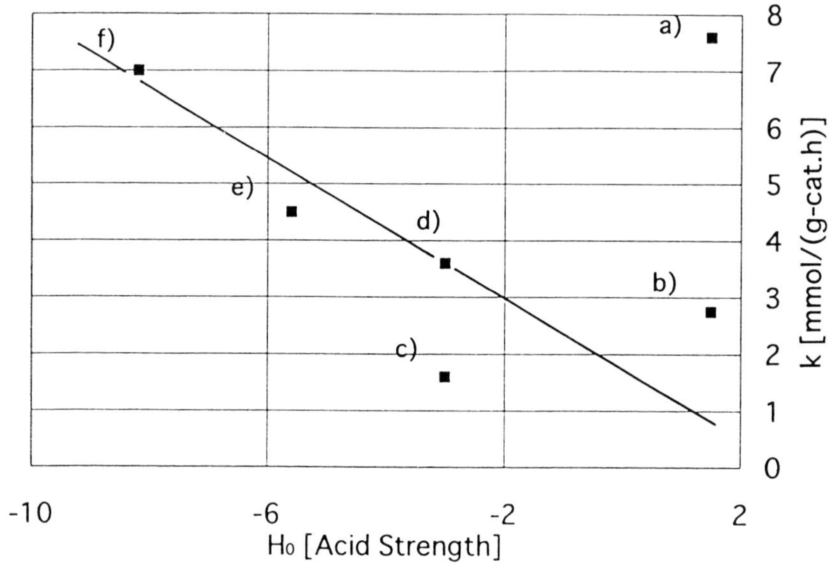

Figure 2. Relation between the rate constant (k) of nitration and the acid strength of the catalyst. Reaction conditions: see captions in Figure 1. Catalysts (molar ratios): a) WO_3(95)-MoO_3(5), b) WO_3(90)-TiO_2(10), c) WO_3(10)-TiO_2(90), d) ZnO(7)-TiO_2(93), e) SiO_2(92)-Al_2O_3(8), f) SiO_2(81)-Al_2O_3(19).

Time-on-stream of the Nitration Reaction Over the WO_3-MoO_3 Catalyst

The time-on-stream of the nitration of benzene over the WO_3 (95%)-MoO_3 (5%) catalyst is shown in Figure 3. A high yield of 93% (versus nitrogen dioxide) of nitrobenzene was obtained for more than 12 hours time-on-stream.

Factors Influencing the Catalytic Activity of the WO_3-MoO_3 System

The nitration activity increases with the content of MoO_3 in the WO_3-MoO_3 system, and it reaches a maximum at 5 mol% of MoO_3 (Figure 4). However no proportionality was observed between the content of MoO_3 and the acidity of the WO_3-MoO_3 system. The rate constant of the nitration reaction was plotted against the BET values of the WO_3 (95%)-MoO_3 (5%) catalysts prepared in various conditions (Figure 5), and an almost linear relationship was obtained between them. Therefore other factors than acidity, e.g. density of hydroxyl groups on the surface of the WO_3-MoO_3 system, might be considered important.

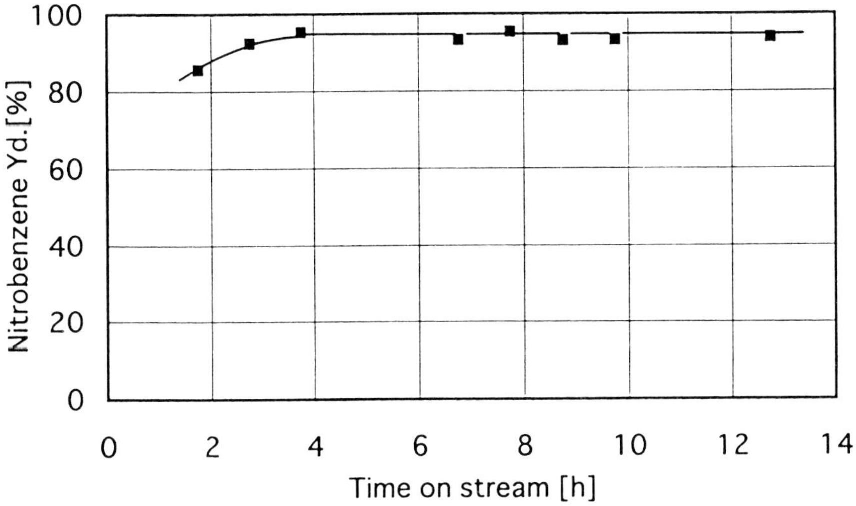

Figure 3. Time on stream of nitration reaction. Reaction conditions: NO_2/benzene/N_2 = 0.2/1.0/0.7 (molar ratios), Temp. = 150°C, SV = 100 h^{-1}. Catalyst; WO_3(95)-MoO_3(5).

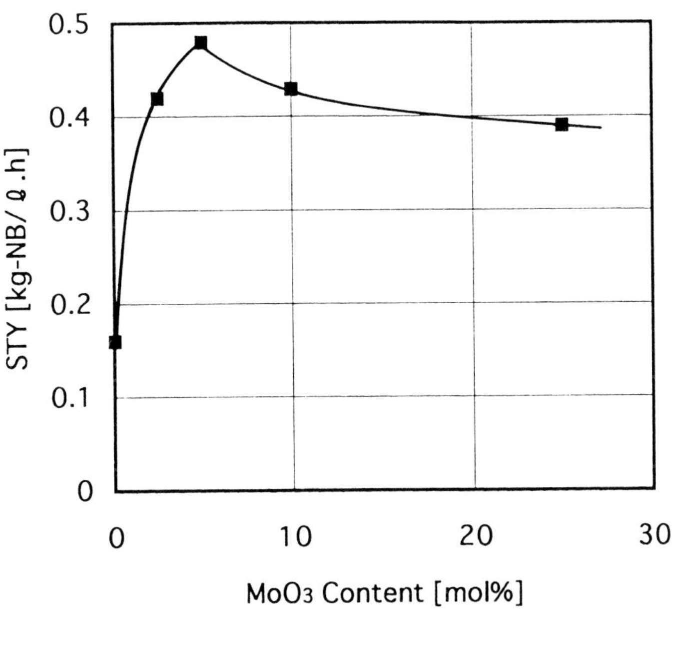

Figure 4. Influence of MoO_3/WO_3 molar ratios on the STY of nitrobenzene. Catalyst; WO_3(95)-MoO_3(5). Reaction conditions: NO_2/benzene/N_2 = 1.0/1.0/3.0 (molar ratios), Temp. = 150°C, SV = 2,000 h^{-1}.

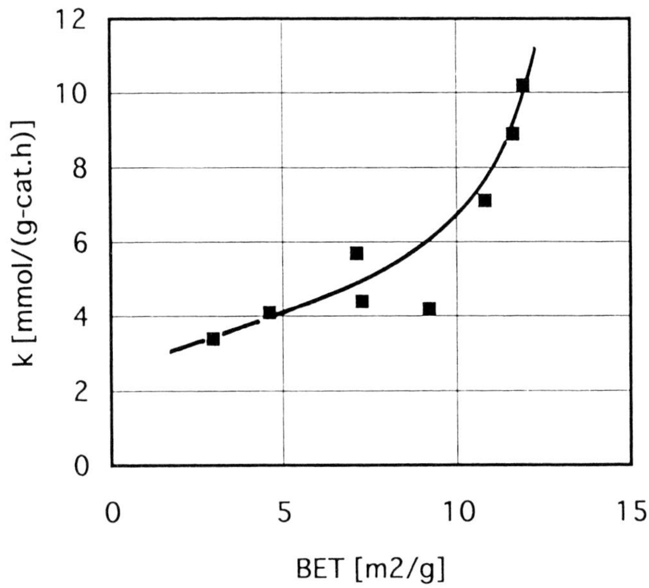

Figure 5. Dependence of the rate constant (k) on the BET surface area of the catalyst. Catalyst: WO_3(95)-MoO_3(5). Reaction conditions: NO_2/benzene/N_2 = 0.2/1.0/0.7 (molar ratios), Temp. = 150 °C, SV = 680 h^{-1}.

Nitration by Nitrogen Dioxide in the Presence of Oxygen Over the WO_3 (95%)-MoO_3 (5%) Catalyst

In the vapor phase nitration by nitric oxide, one third of nitric oxide is consumed as nitric oxide (NO) according to the stoichiometry of Eq. 2. Therefore in order to reuse nitric oxide by in situ oxidation, nitration in the presence of oxygen was examined (Eq. 3).

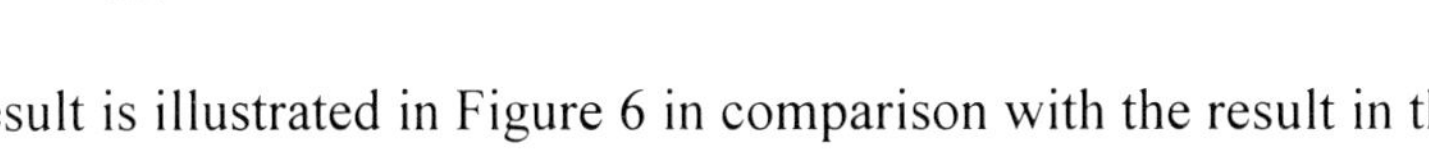

(3)

The result is illustrated in Figure 6 in comparison with the result in the absence of oxygen. In the presence of oxygen and at a long contact time such as 40~45 seconds (V/F = 0.27 ml-cat.h/mmol), the yield of nitrobenzene based on benzene increased from 11.5% to 15.8%. However, theoretical yields based on nitrogen dioxide correspond to 86.6% and 79.0%, respectively. Therefore, the theoretical yield based on nitrogen dioxide rather decreased in the presence of oxygen. This means that in situ oxidation of nitric oxide into nitrogen dioxide is not so efficient as to attain a high theoretical yield of nitrobenzene. Therefore, the complete utilization of nitrogen dioxide is not possible within a reasonable contact time over this catalyst.

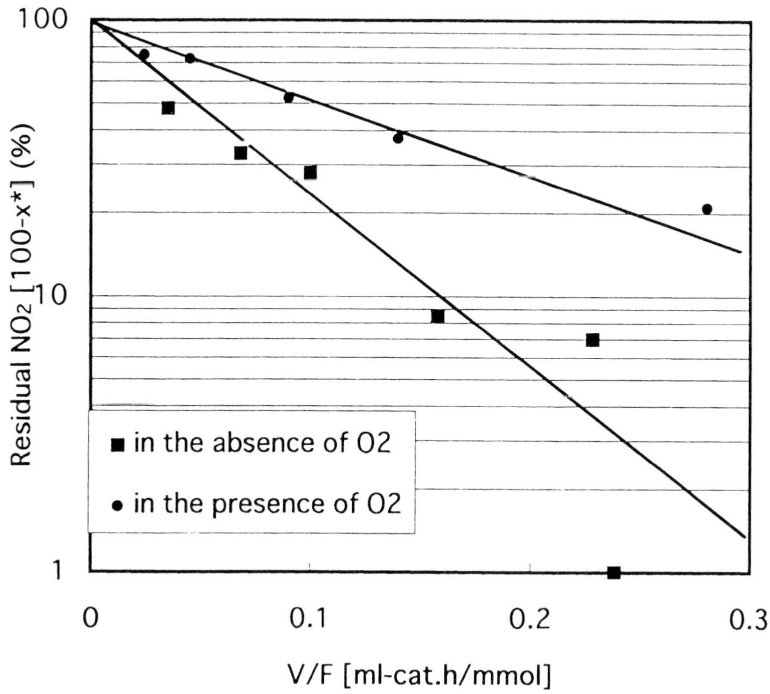

Figure 6. Comparison of the first order plots in the absence and in the presence of oxygen for in situ oxidation of NO to NO_2. Reaction conditions: NO_2/benzene/N_2 = 0.2/1.0/0.17-0.7 (molar ratios), NO_2/benzene/O_2 = 0.2/1.0/0.17-0.7 (molar ratios), Temp. = 150 °C, Catalyst: WO_3(95)-MoO_3(5).

Comparison of Benzene Nitration Activities Between Vapor Phase Nitration by NO_2 and Conventional Liquid Phase Nitration by Mixed Acid

Nitration activities of benzene are compared between the vapor phase nitration by NO_2 over WO_3 (95 mol%)-MoO_3 (5 mol%) catalyst and a conventional liquid phase nitration by mixed acid, as shown in Table 2. The activity in the vapor phase without oxygen (k_1 = 0.0948 sec^{-1}) is about one-fifth of that in the liquid phase (k_1 = 0.5 sec^{-1}). Furthermore, in the presence of oxygen, the activity in the vapor phase (k_1 = 0.0389 sec^{-1}) is almost one-tenth of that in the liquid phase. Therefore, the catalytic activity in the vapor phase nitration has to be improved more than ten times in order to replace the conventional liquid process.

CONCLUSIONS

We investigated the vapor phase nitration of benzene by nitrogen dioxide over various solid acid catalysts, and found that the mixed metal oxide comprising WO_3-MoO_3 affords the highest catalytic activity. This high activity can not be explained

Table 2
Comparison of benzene nitration activities between vapor phase nitration by NO_2 and conventional liquid phase nitration by mixed acid.

Activity Parameters	Vapor Phase Nitrations[a]		Liquid Phase Nitration[d]
	without O_2[b]	with O_2[c]	
ΔEa ($kcal.mol^{-1}$)	7.22	-	15.3
k1 [sec^{-1}]	0.948	0.0389	0.5
STY [$(kg\text{-}NB).L^{-1}.h^{-1}$][e]	138	45.8	385

[a]Cat.; WO_3 (95 mol%)-MoO_3 (5 mol%), Temp. = 150 °C.
[b]NO_2/Benzene/N_2 = 0.2/1.0/0.17-0.68 (molar ratios).
[c]NO_2/Benzene/O_2 = 0.2/1.0/0.17-0.68 (molar ratios).
[d]Mixed acid: $HNO_3/H_2SO_4/H_2O$ = 32/60/8 (wt %), Temp. = 50 °C.
[e]Space Time Yield (STY).

simply by the solid acidity (acid strength and acid amount) of the catalyst. Other factors such as the density of hydroxyl groups on the surface of the mixed oxide might be considered. The complete utilization of nitrogen dioxide by in situ oxidation of by-produced nitric oxide (see the following reaction sequences) within a reasonable contact time could not be attained.

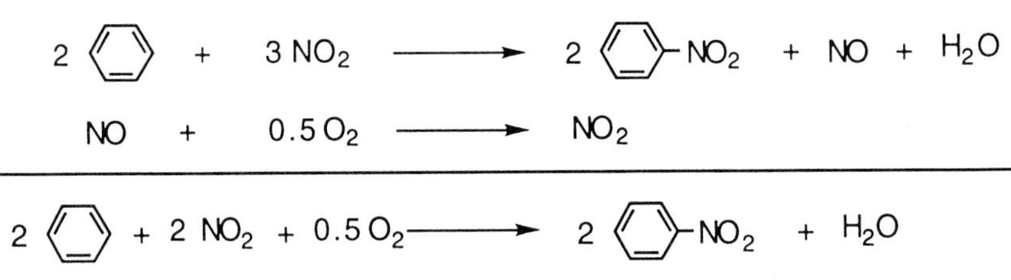

(2)

(4)

(3)

Therefore, more efficient catalysts have to be further investigated for the industrial use.

Acknowledgements
Authors wish to express their sincere thanks to Ms. A. Sakamoto for her contribution in characterizations of catalysts and to Mr. H. Tojima for his help with reaction techniques.

REFERENCES

1. S. Suzuki, K. Tohmori, and Y. Ono, *Chem. Lett.* 747 (1986).
2. U.S. Patent No. 4660702 (1986) (to Monsanto Co.).
3. H. Sato, K. Hirose, and S. Nakamura, U.S. Patent No. 4551568 (1985).

Res. Chem. Intermed., Vol. 24, No. 5, pp. 481-494 (1998)

MICROSTRUCTURE OF SILICA MONOLAYER SOLID ACID CATALYSTS DETERMINED BY ^{29}Si NMR SPECTROSCOPY

NAONOBU KATADA and MIKI NIWA

Department of Materials Science, Faculty of Engineering, Tottori University, 4-101, Koyama-cho Minami, Tottori 680, Japan

Received 8 September 1997; accepted 22 November 1997

Abstract--The nuclear magnetic resonance (NMR) of ^{29}Si showed that the silica monolayer, which was prepared by chemical vapor deposition (CVD) of $Si(OCH_3)_4$ at 593 K on alumina, mainly consisted of $Si(OAl)_1(OSi)_3$ and $Si(OAl)_1(OSi)_2(OH)_1$ species, in agreement with the mechanism of formation of monolayer. The Brønsted acid site is suggested to be the latter species. On the other hand, such an isolated species as $Si(OAl)_3(OH)_1$ was formed from $Si(OCH_3)(CH_3)_3$. Lack of acidity on this species indicates that the acidity requires the siloxane network.

INTRODUCTION

The mixing of two oxides creates a novel catalytic activity. In particular, the combination of silica and alumina is known to form the silica-alumina catalyst, widely applied as a solid acid catalyst to the petrochemical industries and organic syntheses [1]. The acidic property, *i.e.*, the acid amount (number or concentration of acid sites), acid strength and identification of Brønsted or Lewis acidity, is significantly changed by the composition and preparation technique. Several models of the acid sites have therefore been proposed, and explanations about the generation of the acidity have been made on the basis of various concepts [2,3], including the quantum-chemical [4-6] and molecular dynamics [7] calculations. The origin of acidity is however confused, because the conventional silica-alumina catalyst prepared by the liquid-phase procedure usually possesses a complicated structure and hence multiple types of acid sites, which will hardly confirm the proposed models.

In order to obtain a model compound, and to obtain a novel functionalized material, we showed that the chemical vapor deposition (CVD) of $Si(OCH_3)_4$ on γ-alumina formed the silica monolayer, and the Brønsted acid sites were generated on it [8,9]. The acid strength of the silica monolayer on alumina was weak compared to those on zeolite or conventional silica-alumina catalyst. The silica monolayer was applied to Beckmann rearrangement of cyclohexanone oxime [10]. Subsequently we clarified the mechanism of CVD [11], and the effect of the Si precursor on the network structure and acidity [12]. High thermal stability was found on the silica monolayer [13], as reviewed [14].

From the analysis of products of CVD of $Si(OCH_3)_4$, it was suggested that the network of siloxane was well developed on the alumina surface at such a high temperature as 593 K [11]. The ^{29}Si nuclear magnetic resonance (NMR) spectroscopy can directly analyze the microstructure of the silicon species [15-22]. The present paper will confirm the proposed model by means of the NMR. Moreover, the difference in the microstructure, which was suggested to be caused by changing the deposition temperature and silicon precursor [12], will be clarified.

Independently or referring to our study, some research groups are now studying the CVD of silica on metal oxide [23-28]. Sato *et al.* [29] and Sheng *et al.* [30] have already utilized the NMR spectroscopy for the analysis of silica layer prepared on γ-alumina by the CVD method, but these authors utilized a continuous flow method as the CVD procedure, while we utilize a static method. This probably gives a variation in the structure. The present study will confirm the homogeneous structure realized by the static method [14].

EXPERIMENTAL

Preparation of Sample

According to the previous papers [8 - 12], the CVD of silicon alkoxide was carried out on a reference catalyst JRC-ALO4 (γ-Al_2O_3, 161 m^2 g^{-1}) supplied by Catalysis Society of Japan. Silica layers deposited from $Si(OCH_3)_4$ at 473 and 593 K, from $Si(OCH_3)(CH_3)_3$ at 473 K on alumina were analyzed in this paper. The loading of silica was determined by gravimetry and usually shown by the surface concentration of silicon atom (Si nm^{-2}), which was determined based on the BET surface area measured by the nitrogen adsorption at 77 K.

NMR Spectroscopy

The NMR measurements were done after the sample was exposed to a humid atmosphere (relative humidity *ca.* 80 %) or was pretreated at 673 K in flowing nitrogen for 1 hour followed by quickly transporting the sample into the NMR tube. Magic-angle spinning (MAS) NMR spectra were observed for ^{29}Si by a Varian UNITY plus-400 spectrometer at 79.456 MHz of the magnetic field frequency, 5.5 μsec of the pulse width, 51 msec of the acquisition time, 4.0 sec of the residence time and about 4.2 kHz of the spinning rate. Cross-polarization (CP) spectra were measured with 0.5 - 6 msec of the contact time.

RESULTS AND DISCUSSION

Influence of ^1H Cross-Polarization

Figure 1 (a) shows the NMR spectrum of silica monolayer fully covering the alumina surface (19.6 wt% of SiO_2, *i.e.*, 11.4 Si nm^{-2} deposited from $Si(OCH_3)_4$ at 593 K) measured without the CP. The peak was observed on a wide chemical shift range from -80 to -120 ppm. The main peak appeared at *ca.* -100 ppm, but this noisy spectrum is hardly analyzed, although more than 100,000 free induction decays (FID) were accumulated for more than 5 days to improve the signal to noise ratio. Under such conditions or even in a lower magnetic field, clear spectra were obtained in the cases of silica gel and zeolites [18 - 22]. It was observed on zeolites [31] that the ^{29}Si NMR spectrum was broadened by the presence of aluminum due to the magnetic interaction. The present low signal to noise ratio, namely a large broadening of spectrum, is also supposed to be due to the aluminum atoms which

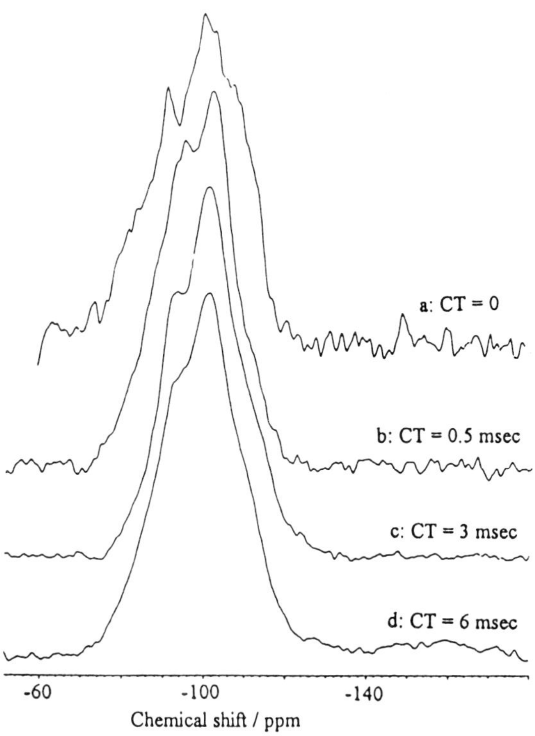

Figure 1. ^{29}Si NMR spectra of SiO_2 / Al_2O_3 (11.4 Si nm^{-2}) deposited from $Si(OCH_3)_4$ at 593 K. The spectra were measured in humid conditions without CP (a), and with CP at 0.5 (b), 3 (c) and 6 msec (d) of the contact time.

located near the silicon atoms. In other words, almost all of silicon atoms of the present sample reside close to aluminum, namely just onto the alumina surface, with distances short enough to induce the magnetic interaction. This behavior, obviously different from the cases of such bulky silica compounds as silica gel and zeolites, supports the ultra thin layer structure loaded on alumina.

However, the structure cannot be discussed based on such a noisy spectrum. Therefore, the ^1H cross-polarization (CP) technique was applied. As shown in Figure 1 (b) - (d), the CP improves the signal to noise ratio, but does not seem to change the peak shape and position. This is quite different from the cases of silica gel and zeolites; usually, the peak at large chemical shift (low negative value) is enhanced by the CP, because such a peak is interacting with hydroxyl group while the peak at small chemical shift (high negative value) is due to the silicon atoms $Si(OSi)_4$ located inside the bulk. The simplest explanation for this phenomenon is that most of the silicon atoms locate near the surface and are equally enhanced by surface hydroxide and/or adsorbed water. In other words, this supports the validity of monolayer structure. In the following description, the measurements were mainly carried out with the CP technique at 3 msec of the contact time.

Silica Monolayer Deposited from $Si(OCH_3)_4$ at 593 K on Alumina

Figure 2 shows the spectra of SiO_2 / Al_2O_3 prepared from $Si(OCH_3)_4$ at 593 K on alumina measured under the dried conditions, *i.e.*, after it was pretreated at 673 K for 1 hour. The monolayer fully covering the alumina surface (12.2 Si nm^{-2}) [32] is shown as (c). Since the spectrum seemed to consist of multiple fragments, the spectrum was deconvoluted as shown in Figure 3 c. Four Lorentzian fragments were obtained at -86 (D), -93 (C), -101 (B) and -108 (A) ppm. The half-height width of the fragment was not fixed, and adjusted to fit the deconvoluted and observed curves, although the actual width of each fragment, which must be affected by the magnetic interaction between the silicon species and alumina substrate, was unknown. The present deconvolution has this shortcoming, however, it is presumed that the present spectrum mainly consists of the (B) and (C) species, and relatively small amount of (A) and (D) species. The peak (A) at -108 ppm is ascribable to the $Si(OSi)_4$ species, because materials consisting of SiO_2 usually show the resonance at *ca.* -105 to 115 ppm [18]. At the least, a doubly accumulated layer is necessary to form such an $Si(OSi)_4$ species (Scheme 1). Since the peak intensity corresponds to *ca.* 10 %, it is pointed out that a small fraction of silicon formed such an accumulated layer with *ca.* 12 Si nm^{-2} of silica.

On the other hand, it is known that one-atom substitution of silicon by aluminum and hydrogen makes the chemical shift 6 and 10 ppm higher (smaller negative value), respectively [18,22]. From this principle, the chemical shift of various silicon species are predicted as shown in Table 1. According to this table, the main peak (B, -101 ppm) has two possibilities; it can be ascribed to $Si(OAl)_1(OSi)_3$ or $Si(OSi)_3(OH)_1$.

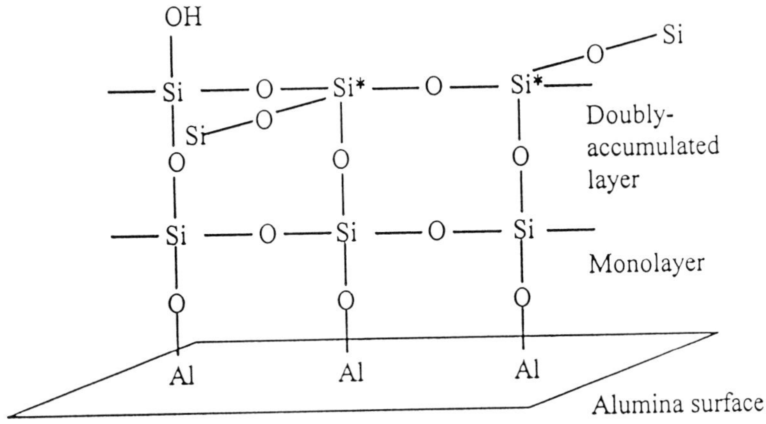

Scheme 1

Model of $Si(OSi)_4$ (A) species (marked by *) in a doubly-accumulated layer.

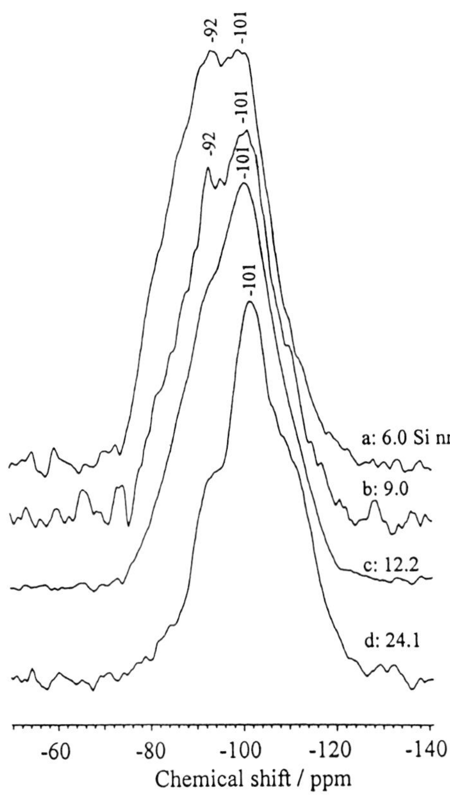

Figure 2. Spectra of SiO_2 / Al_2O_3 (a: 6.0 Si nm^{-2}, b: 9.0 Si nm^{-2}, c: 12.2 Si nm^{-2} and d: 24.1 Si nm^{-2} deposited from $Si(OCH_3)_4$ at 593 K) dried at 673 K. The contact time was 3 msec.

The former $Si(OAl)_1(OSi)_3$ species is consistent with the proposal on the structure of silica monolayer which consists of 1 : 1 bondings of Al-O-Si and a

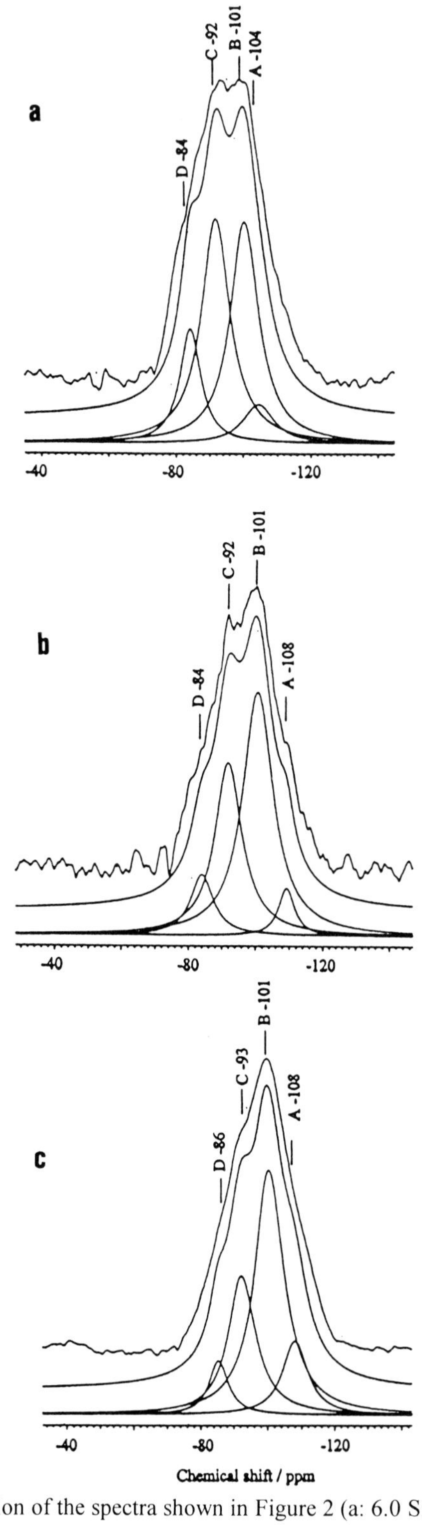

Figure 3. Deconvolution of the spectra shown in Figure 2 (a: 6.0 Si nm^{-2}, b: 9.0 Si nm^{-2} and c: 12.2 Si nm^{-2}). The spectrum (d) could not be analyzed quantitatively because this spectrum was affected by the CP as shown in Figure 4.

Table 1.
Predicted species at from -80 to -108 ppm estimated from the assumptions that the $Si(OSi)_4$ species appears at -108 ppm, and one-atom substitution of silicon by aluminum and hydrogen makes the chemical shift into 6 and 10 ppm higher (smaller negative value), respectively.

Chemical shift / ppm	Species classified as $Si(OAl)_0$	$Si(OAl)_1$	$Si(OAl)_2$	$Si(OAl)_3$
-108	$Si(OSi)_4$			
-102		$Si(OAl)_1(OSi)_3$		
-98	$Si(OSi)_3(OH)_1$			
-96			$Si(OAl)_2(OSi)_2$	
-92		$Si(OAl)_1(OSi)_2(OH)_1$		
-90				$Si(OAl)_3(OSi)_1$
-88	$Si(OSi)_2(OH)_2$			
-86			$Si(OAl)_2(OSi)_1(OH)_1$	
-82		$Si(OAl)_1(OSi)_1(OH)_2$		
-80				$Si(OAl)_3(OH)_1$

dimensional siloxane (Si-O-Si) network [11]. On the contrary, if the latter species was assumed, most of the surface of this sample should be covered by the hydroxyl species. However, such a high coverage with the hydroxyl groups is incredible, since this sample had been pretreated at 673 K, and the IR spectroscopy showed the silanols isolated from other hydroxyl groups [9]. Therefore the former $Si(OAl)_1(OSi)_3$ species is more probable. Of course it is possible that the peak (B) contains a small amount of the $Si(OSi)_3(OH)_1$ species. It is thus concluded that the peak (B) is mainly due to the $Si(OAl)_1(OSi)_3$ species.

The peak (C) is ascribable to the $Si(OAl)_1(OSi)_2(OH)_1$ species. The silica monolayer showed weak Brønsted acidity [9]. Because the Brønsted acid site is presumed to be the surface hydroxyl group, this hydroxyl species (C) is considered as the Brønsted acid site on the monolayer. The small peak (D) is probably due to the $Si(OAl)_1(OSi)_1(OH)_2$ species.

Since the peaks (B) and (C) occupy more than 80 % of the whole peak area, the structure of silica monolayer with *ca.* 12 Si nm^{-2} of silicon, which fully covers the alumina surface, is concluded to mainly consist of a network, as proposed by the previous study [11], containing the $Si(OAl)_1(OSi)_3$ and the Brønsted acidic $Si(OAl)_1(OSi)_2(OH)_1$ species. Since the ratio of peak area of (C) to (B) species is almost 1 : 2, the concentrations of the (B) and (C) species are estimated to be approximately 8 and 4 nm^{-2}, respectively, although some uncertainties are supposed to be due to the CP. This approximation leads to a model of monolayer as shown in Scheme 2 which consists of 2 : 1 of the $Si(OAl)_1(OSi)_3$ and $Si(OAl)_1(OSi)_2(OH)_1$ species, and these species form a two-dimensional network of siloxane. It was observed by means of gravimetry and gaseous product analysis that, during the deposition of $Si(OCH_3)_4$, the two-dimensional network of siloxane was formed [11]. The present NMR spectroscopy confirms this speculation.

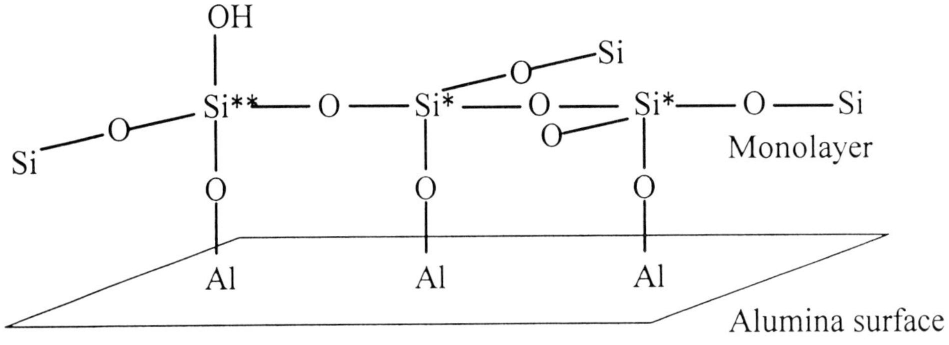

Scheme 2

Proposed structure of silica monolayer fully covering alumina surface. The symbol Si* shows the $Si(OAl)_1(OSi)_3$ (B) species, and the symbol Si** shows the $Si(OAl)_1(OSi)_2(OH)_1$ (C) species.

From the relationship between chemical shift of $Si(OSi)_4$ species and bond angle proposed by Thomas *et al.* [19], the bond angle Si-O-T (T = Al or Si) is estimated to be 143°. From the chemical shift of $Si(OAl)_1(OSi)_3$ species, the angle is estimated as 140°.

Figure 2 (a) - (d) compares the spectra of samples with various silicon concentrations. In the region from 6 to 12 Si nm^{-2}, the $Si(OAl)_1(OSi)_3$ and $Si(OAl)_1(OSi)_2(OH)_1$ species were main composites, but the peak shoulder at -108 ppm ($Si(OSi)_4$ species) obviously increased at 24 Si nm^{-2}. Moreover, varying the contact time of CP significantly changed the peak shape at 24 Si nm^{-2} as shown in Figure 4; this shows that the silica did not form a thin layer. These results indicate the structure of multiply accumulated layer or particle of 24 Si nm^{-2} of silicon, as predicted from the silicon concentration which was twice higher than the full-coverage monolayer.

On the sample with the low Si concentration (Figure 3a), the ratio of peak area of (C) to (B) was higher than that of the monolayer fully-covering the surface. It can be explained that, at the beginning of deposition, a small patch consisting of siloxane network is supposed; the smaller patch has the higher ratio of (C) to (B) species, because the hydroxyl (C) species is considered to be enriched at the edge of the proposed siloxane network. However, also the selective deposition of silica on the specific plane is possible to be considered. We will next clarify this point by the measurement of concentration of hydroxyl groups [33].

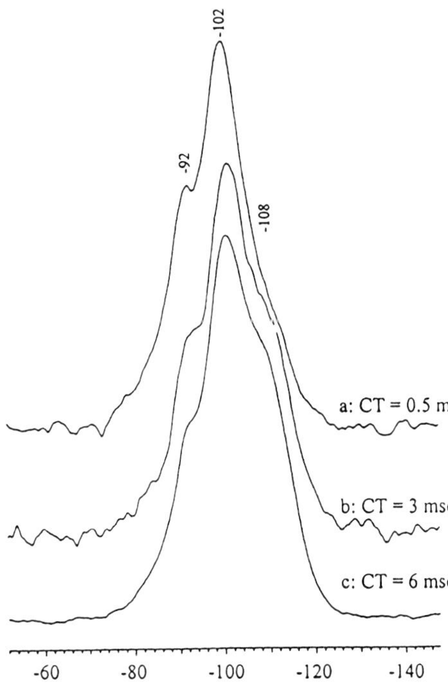

Figure 4. Spectra of SiO_2 / Al_2O_3 (24.1 Si nm^{-2}) dried at 673 K with CP at 0.5 (a), 3 (b) and 6 (c) of the contact time.

Sato *et al.* [29] and Sheng *et al.* [30] have utilized the NMR to analyze the silica layers loaded on γ-alumina by the CVD method. They reported different spectra for the silica layers prepared in flowing oxygen and nitrogen. Sato *et al.* proposed a doubly-accumulated structure for the layer deposited from $Si(OC_2H_5)_4$ in oxygen, based on the spectrum showing a large amount of $Si(OSi)_4$ species (-107 ppm) [29]. The reported spectrum is obviously different from the present spectra, indicating the deposition of Si alkoxide in oxygen causes the accumulation of silica layer probably on the external surface of alumina particle because of the fast decomposition rate in oxygen. On the other hand, Sheng *et al.* observed a relatively homogeneous and thin layer formed from $Si(OCH_3)_4$ in an inert atmosphere (nitrogen), but proposed a random build-up mechanism in flowing nitrogen [30]. The reported spectrum was similar to the present results, but not a negligible amount of $Si(OSi)_4$ was observed at *ca.* 12 Si nm^{-2}. Moreover, Sheng *et al.* ascribed the peak at *ca.* -100 ppm as $Si(OSi)_3(OH)_1$ species, because the surface of the accumulated layer, which was suggested to be formed as above, should be covered with this hydroxyl species. On the contrary, we consider that the peak at -101 ppm is mainly due to the $Si(OAl)_1(OSi)_3$ species, because the monolayer structure of silica is reliable in our case. It seems that the reason for the disagreement is the difference in methods used by Sheng *et al.* (flow method) and the present study (static method), and the static method is probably needed to form a homogeneous monolayer [14]. We have compared the NMR spectra of the silica layers prepared by the flow and static methods, and will report the influence of preparation procedure and conditions [34].

Thermal Stability

Figure 5 shows the spectra of the silica layer with 9 Si nm^{-2} calcined at 673 (a) and 1493 K (b). The calcination at the very high temperature changed the spectrum.

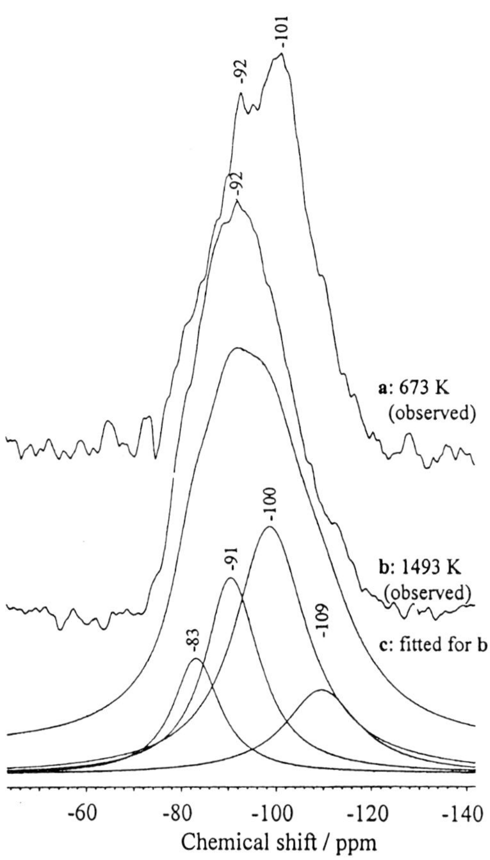

Figure 5. Spectra of SiO_2 / Al_2O_3 (14.9 % of the weight ratio of SiO_2 to Al_2O_3) calcined at 673 (a) and 1493 K (b). The surface concentration of Si was changed by the high temperature calcination because of the decrease of surface area; 9.0 Si nm^{-2} before and 20.3 Si nm^{-2} after calcination at 1493 K, respectively. The sample was exposed once to a humid atmosphere for several days after the calcination, and then dried at 673 K before the NMR measurement. The best fitted curve for (b) was overlapped, and that for (a) is shown in Figure 3 (b).

Table 2:
Change of peak intensity of (A) to (D) species on SiO_2 / Al_2O_3 (14.9 % of weight ratio of SiO_2 to Al_2O_3, deposited from $Si(OCH_3)_4$ at 593 K) by calcination at 1493 K.

Calcination temperature / K	Fraction of peak area relative to the total area / %			
	$Si(OSi)_4$ (A)	$Si(OAl)_1(OSi)_3$ (B)	$Si(OAl)_1(OSi)_2(OH)_1$ (C)	$Si(OAl)_1(OSi)_1(OH)_2$ (D)
673	5.3	51.5	33.4	9.8
1493	15.5	44.4	26.4	13.8

After the calcination, the main composites are still the $Si(OAl)_1(OSi)_3$ (B) and $Si(OAl)_1(OSi)_2(OH)_1$ (C) species, qualitatively same as before the high temperature calcination. This confirms the thermal stability [13] of the silica monolayer: the structure of the monolayer consisting of these species is qualitatively maintained.

From the deconvolution (shown in Figure 5c) of this peak, it is shown that the $Si(OSi)_4$ (A) species increased up to *ca.* 15 %, as shown in Table 2. Simultaneously the species (B) and (C) decreased, and the (D) species increased. Moreover, for this sample, the CP changed the peak shape. The degree of thermal sintering was small on the silica monolayer, but the surface area was reduced by the sintering from *ca.* 150 to 60 m^2 g^{-1} at 1493 K. It is presumed that the sintering formed a multiply accumulated layer or particle of silica, and this causes the increase of $Si(OSi)_4$ (A) species. We have reported that the catalytic activity for the double-bond isomerization of butene per surface area was maintained by the high temperature calcination, but this does not seem to correspond with the change in the concentration of species (C) which is supposed to be the acid site as described in the previous section.

Isolated Species on Alumina

From the previous study, it was proposed that an isolated species of Si, like $Si(OAl)_3(OH)_1$, was selectively formed from $Si(OCH_3)_n(CH_3)_{4-n}$ (n = 1 - 3) or from $Si(OCH_3)_4$ at low temperature (< 473 K) [12]. Figure 6 shows the change of spectra (measured under humid conditions) by varying the deposition temperature and Si precursor in the low Si concentration region (2 - 6 Si nm^{-2}). The sample prepared from $Si(OCH_3)(CH_3)_3$ (Figure 6 a) showed the resonance at -70 to -90 ppm, while the silica monolayer prepared from $Si(OCH_3)_4$ at 593 K (Figure 6 c) showed the resonance at -75 to -120 ppm. It is clear that silica deposited from $Si(OCH_3)(CH_3)_3$ mainly consists of the species with no or only one siloxane bonding, *i.e.*, $Si(OAl)_3(OH)_1$, $Si(OAl)_3(OSi)_1$, $Si(OAl)_2(OSi)_1(OH)_1$ and / or $Si(OAl)_1(OSi)_1(OH)_2$, because only these species can show peaks in the region from -80 to -90 ppm as shown in Table 1; the $Si(OSi)_2(OH)_2$ is also possible to appear, but this species can be formed only on a doubly or further accumulated layer, and the possibility is therefore ignored. The isolated structure formed from $Si(OCH_3)(CH_3)_3$ was thus clear.

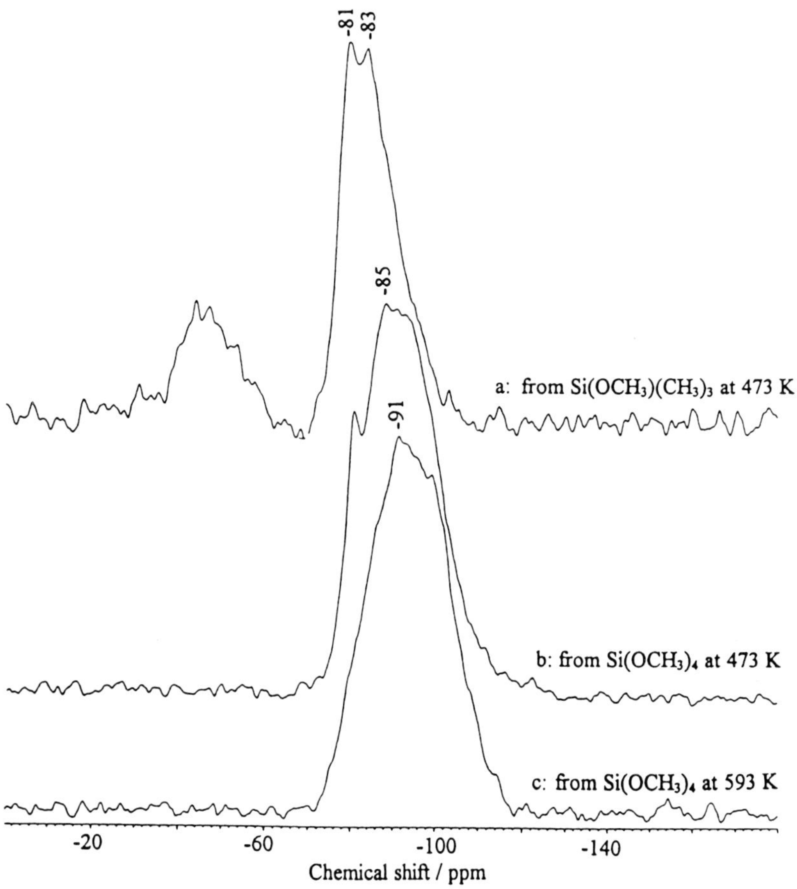

Figure 6. Spectra of SiO_2 / Al_2O_3 (a: 2.0 Si nm^{-2} from $Si(OCH_3)(CH_3)_3$ at 473 K, b: 3.0 Si nm^{-2} from $Si(OCH_3)_4$ at 473 K and c: 6.0 Si nm^{-2}, deposited from $Si(OCH_3)_4$ at 593 K) measured under humid conditions.

On the other hand, the change in structure by varying the deposition temperature was rather equivocal. The sample deposited from $Si(OCH_3)_4$ at the low temperature (473 K, Figure 6b) showed the resonance at -75 to -120 ppm, like the sample deposited at 593 K (Figure 6c). However, the peak maximum was observed at -85 ppm on the former sample, whereas that on the latter one showed the smaller chemical shift (higher negative value), -91 ppm. It is presumed that the content of the isolated species like $Si(OAl)_3(OH)_1$ (-80 ppm) was relatively high on the silica deposited at lower temperature, while the main composites of the sample deposited at 593 K were the $Si(OAl)_1(OSi)_3$ (*ca.* -100 ppm) and $Si(OAl)_1(OSi)_2(OH)_1$ (-91 ppm) species, as shown above. The difference in degree of siloxane network due to the difference of Si precursor and deposition temperature [12] was thus clarified.

Because the acidity was observed only on the sample from $Si(OCH_3)_4$ at high temperatures (> 493 K) [12], it is pointed out that the Brønsted acidity requires the two-dimensional network consisting of the $Si(OAl)_1(OSi)_3$ and $Si(OAl)_1(OSi)_2(OH)_1$ species. As proposed previously [12], the strained structure of the siloxane network developed on alumina is suggested to generate the acidity on the surface hydroxyl species. The acidity of the silica monolayer was weak in comparison with zeolite [9]. The origin of acidity is considered to be different from the case of zeolite, in which the isomorphous substitution of aluminum atom into the silicate matrix generates the strong acidity. Therefore, the silica monolayer can be considered as a model of another type of acid site. An amorphous mixed oxide such as commercially utilized silica-alumina catalyst usually has a wide distribution of acid strength, and hence the multiple types of acid sites are suggested. The acidic species the same or similar to that observed on the present silica monolayer may also exist on the amorphous material.

CONCLUSION

1. The silica monolayer on alumina, which was deposited from $Si(OCH_3)_4$ at 593 K, mainly consisted of the $Si(OAl)_1(OSi)_3$ and $Si(OAl)_1(OSi)_2(OH)_1$ species. This agrees well with the mechanism of growth of the silica monolayer. The Brønsted acid site is suggested to be the latter species. This structure was qualitatively maintained at quite high temperature (1493 K).
2. The structure formed from $Si(OCH_3)(CH_3)_3$ is confirmed to be such an isolated species as $Si(OAl)_3(OH)_1$. Because this species showed no acidity, the two-dimensional network structure consisting of the $Si(OAl)_1(OSi)_3$ and $Si(OAl)_1(OSi)_2(OH)_1$ species is considered to generate the acidity.

Acknowledgment

The authors are grateful to Mr. Toshimi Sawano, Nippon Shokubai Co. Ltd. for his cooperation for the NMR measurement. This study was partly supported by a Grant-in-Aid from the Ministry of Education, Science, Sports and Culture, Japan (No. 08232257 and 08750910).

REFERENCES

1. K. Tanabe, *Solid Acids and Bases*, Kodansha-Academic Press. Tokyo, 1970, p. 58.
2. C.L. Thomas, *Ind. Eng. Chem.*, **41,** 2564 (1949).
3. K. Tanabe, in: *Catalysis: Science and Technology Vol. 2*, J.R. Anderson and M. Boudart (Eds.), Springer-Verlag, Berlin, 1981, p. 231.
4. W. Grabowski, M. Misono, Y. Yoneda, *J. Catal.*, **61,** 103 (1980).
5. H. Kawakami, S. Yoshida, T. Yonezawa, *J. Chem. Soc., Faraday Trans. 2*, **80,** 205 (1984).
6. G.M. Zhidomirov, V.B. Kazanski, *Adv. Catal.*, **34,** 131 (1984).
7. S. Blonski and S. H. Garofalini, *J. Phys. Chem.*, **100,** 2201 (1996).
8. M. Niwa, T. Hibino, H. Murata, N. Katada and Y. Murakami, *J. Chem. Soc., Chem. Commun.*, **1989,** 289.
9. M. Niwa, N. Katada and Y. Murakami, *J. Phys. Chem.*, **94,** 6441 (1990).
10. N. Katada, T. Tsubouchi, M. Niwa and Y. Murakami, *Appl. Catal., A: General*, **124,** 1 (1995).
11. N. Katada, T. Toyama and M. Niwa, *J. Phys. Chem.*, **98,** 7647 (1994).
12. N. Katada, T. Toyama, M. Niwa, T. Tsubouchi and Y. Murakami, *Res. Chem. Intermed.*, **9,** 137 (1995).
13. N. Katada, H. Ishiguro, K. Muto and M. Niwa, *Adv. Mater. Chem. Vap. Deposition*, **1**, 54 (1995).
14. N. Katada and M. Niwa, *Adv. Mater. Chem. Vap. Deposition*, **2**, 125 (1996).
15. G.E. Maciel and D.W. Sindorf, *J. Am. Chem. Soc.*, **102,** 7606 (1980).
16. J.M. Thomas, C.A. Fyfe, S. Ramdas, J. Klinowski and G.C. Gobbi, *J. Phys. Chem.*, **86,** 3061 (1982).
17. D.W. Sindorf and G.E. Maciel, *J. Am. Chem. Soc.*, **105,** 1487 (1983).
18. J. Klinowski, *Prog. NMR Spectrosc.*, **16,** 268 (1984).
19. J.M. Thomas and J. Klinowski, *Adv. Catal.*, **33,** 199 (1985).
20. K. Segawa, Y. Nakajima, S. Nakata, S. Asaoka and H. Takahashi, *J. Catal.*, **101,** 81 (1986).
21. G. Engelhardt and D. Michel, *High resolution Solid-State NMR of Silicates and Zeolites*, J. Wiley & Sons, New York, 1987, p. 1.
22. . T. Bein, R.F. Carvaer, R.D. Farlee and G.D. Stucky, *J. Am. Chem. Soc.*, **110,** 4546 (1988).
23. T. Jin, T. Okuhara and J. M. White, *J. Chem. Soc., Chem. Commun.*, **1987,** 1248.
24. Y. Imizu, A. Tada and I. Toyoshima, *Shokubai (Catalyst)*, **30,** 388 (1988).
25. B.-Q. Xu, T. Yamaguchi and K. Tanabe, *Chem. Lett.*, **1989,** 149.
26. S. Sato, M. Toita, T. Sodesawa and F. Nozaki, *Appl. Catal.*, **62,** 73 (1990).
27. N. Okazaki, T. Kohno, R. Inoue, Y. Imizu and A. Tada, *Chem. Lett.*, **1993,** 1195.
28. T.-C. Sheng and I. D. Gay, *J. Catal*, **145,** 10 (1994).
29. S. Sato, T. Sodesawa, F. Nozaki and H. Shoji, *J. Mol. Catal.*, **6,** 343 (1991).
30. T.-C. Sheng, S. Lang, B.A. Morrow and I.D. Gay, *J. Catal.*, **148,** 341 (1994).
31. C.A. Fyfe, H. Strobl, G.T. Kokotailo, G.J. Kennedy and G.E. Barlow, *J. Am. Chem. Soc.*, **110,** 3375 (1988).
32. For the dried conditions, we have not confirmed the unchanging of the peak shape by the CP technique, while it was confirmed for the humid conditions as shown in Figure 1. However, the peak shape in the CP NMR spectrum was unchanged when the contact time was varied from 0.5 to 6 msec. Therefore, we judged that the CP did not modified the peak shape also in the dried conditions, and the measurement without CP, which must take a very long time, was skipped in this case.
33. N. Katada, T. Fujii, K. Iwata and M. Niwa, *3rd Conf. Adv. Catal. Sci. Tech.*, submitted.
34. N. Katada, H. Fukui, H. Ishiguro and M. Niwa, *J. Catal.*, submitted.

Res. Chem. Intermed., Vol. 24, No. 5, pp. 495-505 (1998)

CATALYTIC PROPERTY OF EUROPIUM SUPPORTED ON ALUMINA FROM ITS AMMONIACAL SOLUTION

TOSHIHIDE BABA, SATORU HIKITA, HARUHISA HANDA, and YOSHIO ONO*

Department of Chemical Engineering, Tokyo Institute of Technology, Ookayama, Meguro-ku Tokyo 152, Japan

Received 1 October 1997; accepted 17 December 1997

Abstract--Europium supported on alumina (Eu/Al_2O_3) was prepared by impregnation from an ammoniacal solution of the metal. The catalytic activity of Eu/Al_2O_3 for the isomerization of 2,3-dimethylbut-1-ene to 2,3-dimethylbut-2-ene was examined. Eu/Al_2O_3 showed the highest catalytic activity for the isomerization when heated under a vacuum at 523 K, the yield of 2,3-dimethylbut-2-ene in 20 min being 81.6% at 314 K. IR spectrum shows the presence of two kinds of Eu amides. The change of the IR spectra and the catalytic activities with evacuation temperature suggest that only one of the two Eu amides is catalytically active for this base-catalyzed isomerization.

INTRODUCTION

Eu and Yb metals dissolve in liquid ammonia to yield a blue solution [1], characteristic of the solvated electron (e_{am}) [2]. It is well-known that the Eu and Yb amides are formed in their ammoniacal solution [3-5]. The solution processes of lanthanide metals, Ln (Yb or Eu) in liquid ammonia with subsequent reactions are described through the following equations;

$$Ln \rightarrow Ln^{2+} + 2e_{am}$$

$$e_{am} + NH_3 \rightarrow NH_2^- + 1/2\ H_2$$

$$Ln^{2+} + 2\ NH_2^- \rightarrow Ln(NH_2)_2$$

$$Ln^{2+} + 3NH_2^- \rightarrow Ln(NH_2)_3 + e_{am}$$

We have reported the catalytic and physico-chemical properties of low-valent Yb or Eu species which were introduced into K-Y zeolite by an impregnation method from their ammoniacal solution [6-8]. The catalytic properties were examined by the isomerization of alkenes such as 1-butene to 2-butenes, the Michael reaction of cyclopent-2-enone with diethylmalonate and hydrogenation of alkenes such as ethylene. The chemical states of the surface species were investigated by infrared (IR), X-ray absorption fine structure (XAFS), and X-ray photoelectron (XPS) and temperature programmed desorption method. On the basis of these experimental results, the nature of Yb or Eu supported on K-Y zeolites (Yb/K-Y, Eu/K-Y) can be summarized as follows.

(1) Yb/K-Y and Eu/K-Y heated under a vacuum at ca. 500 K were active for the isomerization of alkenes and the Michael reaction, while the catalysts evacuated at a temperature higher than ca. 600 K were inactive for the isomerization.

(2) The catalytically active species for these base-catalyzed reactions were Yb(II) amide or Eu(II) imide formed on K-Y.

(3) Yb/K-Y and Eu/K-Y heated under a vacuum at ca. 900 K were active for the hydrogenation of ethylene and but-1,3-diene, while the catalysts evacuated at a temperature lower than ca. 500 K, were inactive for the hydrogenation.

(4) Yb and Eu nitrides, which were formed by the decomposition of Yb amide and Eu imide, respectively, were active species for the hydrogenation of ethylene and but-1,3-diene.

Recently Imamura and co-workers also reported that Yb and Eu nitride species which were prepared by the decomposition of Yb and Eu amide supported on activated carbon, respectively showed high catalytic activities for the hydrogenation of alkenes such as ethylene, and that their catalytic activities strongly depended on the temperature of decomposition under a vacuum [9].

We found that Eu supported on alumina by impregnation from an ammoniacal solution of the metal followed by heating it under a vacuum around 500 K, showed much higher catalytic activity for the isomerization of 2,3-dimethylbut-1-ene, as compared with Yb/K-Y and Eu/K-Y. In this work, the catalytic properties of the Eu supported on alumina as solid base were examined by the isomerization of 2,3-dimethylbut-1-ene and the chemical states of the Eu species on alumina were investigated by IR measurements.

EXPERIMENTAL

Materials

Ingots of Yb and Eu, whose purities were 99.9% , were obtained from Soekawa Kagaku Inc. K-Y (SiO_2/Al_2O_3 = 5.8) was prepared from Na-Y with a conventional ion exchange method. The degree of K^+ cations exchanged was determined by atomic absorption analysis to be 100%. Alumina obtained from Marubi Lid., had a surface area of 125 m^2 g^{-1} and a average pore diameter of 15 nm. Silica obtained from Fujidavison Chemical Ltd., had a surface area 296 m^2 g^{-1}.

Preparation of Catalysts

K-Y, alumina and silica supports were heated in a quartz reactor under vacuum to 10^{-3} Pa at 773 K for 3 h. The ingot of Yb or Eu was placed in the quartz reactor under nitrogen. After evacuating the system to 10^{-3} Pa at room temperature, NH_3 was liquefied into the quartz reactor, which was cooled with a mixture of dry ice and ethanol, in order to dissolve Yb or Eu metal. The metal ammoniacal solution was kept in contact with the support for 1 h. The support was warmed to room temperature removing ammonia and was heated under a vacuum at the prescribed temperature for 1 h.

Reaction Procedure

The isomerization of 2,3-dimethylbut-1-ene was carried out in a conventional gas-circulating system of volume 422 cm^3. 2,3-Dimethylbut-1-ene was purified by distillation before the reaction. The reaction temperature was 314 K and the initial pressure of 2,3-dimethylbut-1-ene was 10.5 kPa. The reaction mixture was collected by a gas sampler every few minutes, and was analyzed by a gas chromatograph with a 50 m column of OV-101.

The isomerization was also carried out in a liquid phase as follows; 2,3-dimethylbut-1-ene purified by distillation was put into a glass tube, which was then attached to the side arm of the quartz reactor and it was degassed with a freeze-thaw method. The reaction was started by transferring the 2,3-dimethylbut-1-ene into the reactor containing a catalyst prepared as described above, without solvent. The reaction temperature was 313 K and the reaction time was 20 h.

IR Measurements

IR spectra were recorded on JASCO FTIR 7000 spectrometer at room temperature by using a home-made IR cell [7]. The most important characteristic of the IR cell is that a self-supported disk placed on the holder can be impregnated in a metal-ammonia solution and then heat-treated in situ.

The sample for IR measurements was prepared as follows: Alumina was pressed into a thin self-supporting disk (10 mg cm^{-2}). The disk was placed in a small holder and then heated under a vacuum at 773 K for 3 h. The IR cell was removed from the vacuum system to put a piece of Eu metal (0.20 g or 0.008 g) into the bottom of the IR cell under nitrogen. After evacuating the IR cell, ammonia (3.8 cm^3) was liquefied into the bottom of the cell cooled with a mixture of dry-ice and ethanol to form europium-liquid ammonia. The alumina disk was then immersed in the europium-liquid ammonia solution for 1 h, and withdrawn from the solution. Most of the ammonia in the IR cell was removed by evacuation at room temperature. The disk was again moved down and heated under a vacuum at a prescribed temperature. The sample was then pulled up into the IR beam for IR measurements.

RESULTS AND DISCUSSION

Isomerization of 2,3-dimethylbut-1-ene over Eu/Al_2O_3

The isomerization of 2,3-dimethylbut-1-ene proceeded over Eu/Al_2O_3 at 314 K, when the catalyst was heated under a vacuum at 523 K for 1 h. Only the double-bond migration of 2,3-dimethylbut-1-ene to 2,3-dimethylbut-2-ene proceeded, no skeletal isomerization being observed. No gaseous hydrocarbons other than 2,3-dimethylbut-1-ene and 2,3-dimethylbut-2-ene were detected in the reaction mixture. The change in the yield of 2,3-dimethylbut-2-ene with reaction time over Eu/Al_2O_3 (the loading amount of Eu was 5 wt%, Eu (5 wt%)/Al_2O_3) is shown in Figure 1. The yield of 2,3-dimethylbut-2-ene increased with reaction time and reached 81.6% in 20 min.

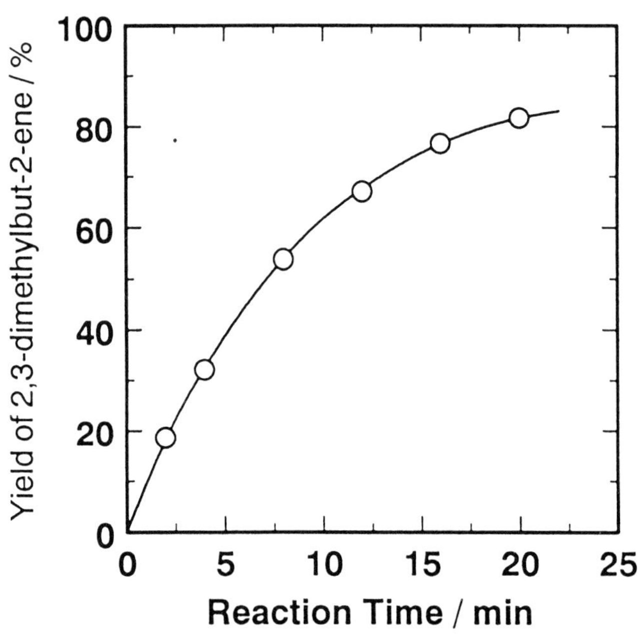

Figure 1. Yield of 2,3-dimethylbut-2-ene with reaction time. Catalyst: 0.108 g of Eu (8 wt%)/Al_2O_3 evacuated at 523 K for 1 h. Reaction temperature: 314 K, Pressure of 2,3-dimethylbut-1-ene: 10.5 kPa.

Influence of Support on the Isomerization Activity

Alumina, K-Y and silica were used as supports for Eu or Yb and the initial rates of isomerization of 2,3-dimethylbut-1-ene were measured at 314 K. The amount of Eu or Yb as a metal was 8 wt% and the catalyst was heated under a vacuum at 473 K for 1 h. The pressure of 2,3-dimethylbut-1-ene was 10.5 kPa.

Alumina, K-Y and silica showed no isomerization activity without Eu or Yb under the reaction conditions. As shown in Table 1, the catalytic activities depended strongly on the support, and alumina was most effective as the support. On every support, Eu is more active than Yb. The catalytic activity of Eu/Al_2O_3 was by far the highest. For example, Eu (8 wt%)/Al_2O_3 showed ca. 15 and 35 times higher isomerization rate than Yb (8 wt%)/Al_2O_3 and Yb (8 wt%)/K-Y, respectively. Thus, Eu/Al_2O_3 is more effective catalyst than Eu or Yb supported on alkali-exchanged zeolites, which have reported previously [6].

Table 1
Rate of isomerization of 2,3-dimethylbut-1-ene to 2,3-dimethylbut-2-ene.

Catalyst	Rate / 10^{-5} mol g-cat^{-1} min^{-1}
Eu/Al_2O_3	88.6
Yb/Al_2O_3	6.4
Eu/SiO_2	3.5
Yb/SiO_2	0.7
Eu/K-Y	2.5
Yb/K-Y	2.8

The amount of metal: 8 wt%, Reaction temperature: 314 K, Pressure of 2,3-dimethylbut-1-ene: 10.5 kPa.

Influence of Evacuation Temperature on the Catalytic Activity

The influence of evacuation temperature on the catalytic activity of Eu (8 wt%)/Al_2O_3 for the isomerization is shown in Figure 2. The initial rate of isomerization increased with increasing evacuation temperature and reached a maximum around 520 K. At higher evacuation temperature, it declined. Thus, the optimum evacuation temperature lies around 520 K.

We have already reported the dependence of the evacuation temperature on the catalytic activities of Eu/K-Y for the olefin isomerization [8], whose temperature dependence for the isomerization of 2,3-dimethylbut-1-ene was similar to that on Eu/Al_2O_3. However, the catalytic activity of Eu/Al_2O_3 is much higher than that of Eu/K-Y.

The catalytic activity of Eu (8 wt%)/Al_2O_3 was also examined by the liquid phase isomerization of 2,3-dimethylbut-1-ene at 313 K. As shown in Table 2, the catalytic activity was strongly influenced by the evacuation temperature. The optimum evacuation temperature lies also around 520 K, the highest yield of 2,3-

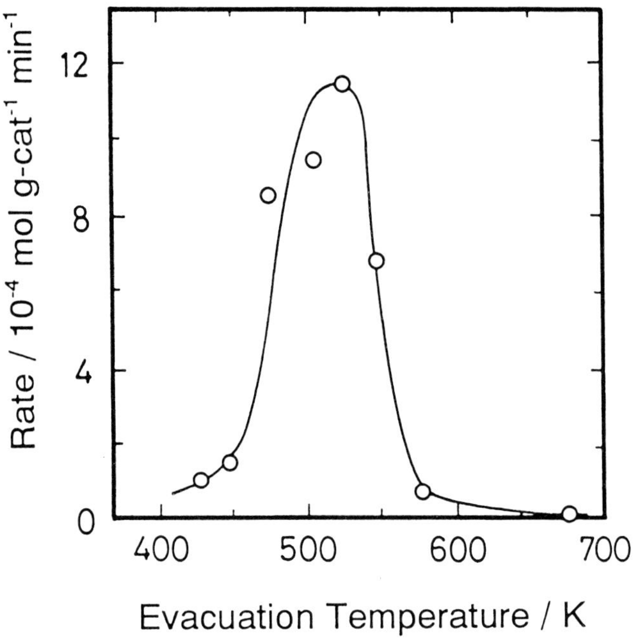

Figure 2. Inference of evacuation temperature on the catalytic activity of Eu (8 wt%)/Al_2O_3. Reaction temperature: 314 K, Pressure of 2,3-dimethylbut-1-ene: 10.5 kPa.

Table 2
Influence of the evacuation temperature on the catalytic activity of Eu/Al_2O_3 for the liquid phase isomerization of 2,3-dimethylbut-1-ene.

Catalyst	Evacuation temperature/K	Yield of 2,3-dimethylbut-2-ene/%
Eu (8 wt%)/Al_2O_3	423	51
	473	59
	523	83
	573	72
	873	21
Yb (8 wt%)/Al_2O_3	423	13
	523	19

Reaction temperature: 313 K, Reaction time: 20 h, 2,3-dimethylbut-1-ene: 24 mmol, Catalyst weight : 0.25 g.

dimethylbut-2-ene being 83% in 20 h when Eu (8 wt%)/Al_2O_3 was heated under a vacuum at 523 K.

When Yb (8 wt%)/Al_2O_3 was evacuated at 423 and 473 K, the yields of 2,3-

dimethylbut-2-ene were 13% and 29%, respectively (Table 2). Eu/K-Y and Eu/SiO_2 showed very low catalytic activity as in the case of the gas-phase isomerization, the yields being 2.4% and 0.2%, respectively.

Effect of the Supported Amount of Eu on Alumina on the Catalytic Activity

The effect of the amount of Eu supported on alumina on the rate of isomerization of 2,3-dimethylbut-1-ene was examined at 314 K (Figure 3). The catalytic activity depended strongly on the amount of Eu. As mentioned above, alumina did not have any activity. The rate of isomerization was very low when the loaded amount was about 4 wt% or below. At loading higher than ca. 5 wt%, the rate of isomerization increased with increasing the amount of Eu.

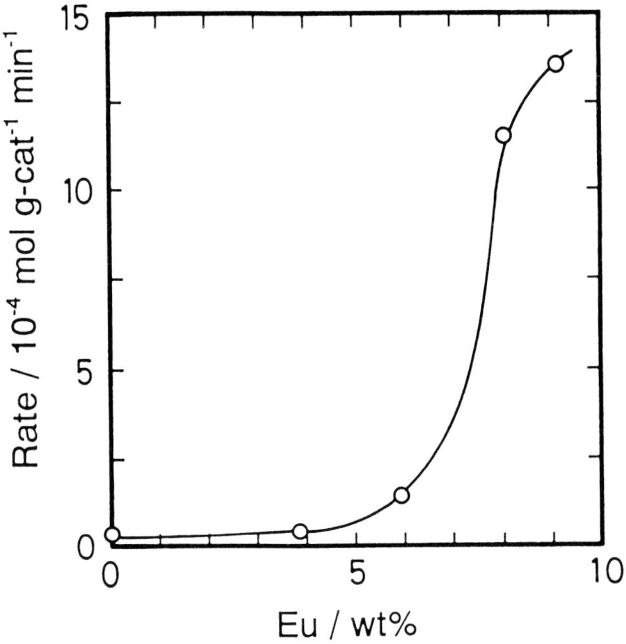

Figure 3. Effect of the amount of Eu supported on alumina on the rate of isomerization of 2,3-dimethylbut-1-ene. Eu/Al_2O_3 evacuated at 523 K for 1 h. Reaction temperature: 314 K, Pressure of 2,3-dimethylbut-1-ene: 10.5 kPa.

IR Study on Eu Species

As mentioned above, there is a possibility that the state of Eu species on alumina varies with the amount of Eu and the evacuation temperature. To get information on surface Eu species on alumina, IR spectra of the catalyst were recorded. The exact amount of Eu loaded on alumina could not be quantified.

When alumina was heated under a vacuum at 773 K for 3 h, three IR bands were observed at ca. 3700 cm^{-1} (Figure 4 a). These bands are due to OH groups on alumina [9]. The alumina pellet was immersed in Eu ammoniacal solution after dissolving 0.008 g of Eu in 3.8 cm^{-3} of ammonia. The IR spectrum was measured after the sample was evacuated at 298 K for 20 min. It showed three bands at 3302, 3242 and 1520 cm^{-1} (Figure 4 (b)). The bands at 3302 and 3242 cm^{-1} are attributed to asymmetrical and symmetrical stretching vibrations of N-H, respectively and the band at 1526 cm^{-1} is attributed to the bending vibration of H-N-H. Linde and Juza reported that IR bands due to the $Eu(NH_2)_2$ were observed at 3263, 3200 and 1503 cm^{-1} [10]. Three IR bands as shown in Figure 4 (b) agree reasonably well with those of europium amide, indicating that Eu supported on alumina exists as the amide upon evacuating the sample at 298 K. The Eu amide observed in this sample will be denoted as Eu amide-I.

The IR bands due to OH groups at ca. 3700 cm^{-1} disappeared upon supporting Eu. Moreover, XANES spectra of Eu (5 wt%)/Al_2O_3 showed the presence of both Eu(II) and Eu(III) [11]. The fraction of Eu(II) and Eu(III)

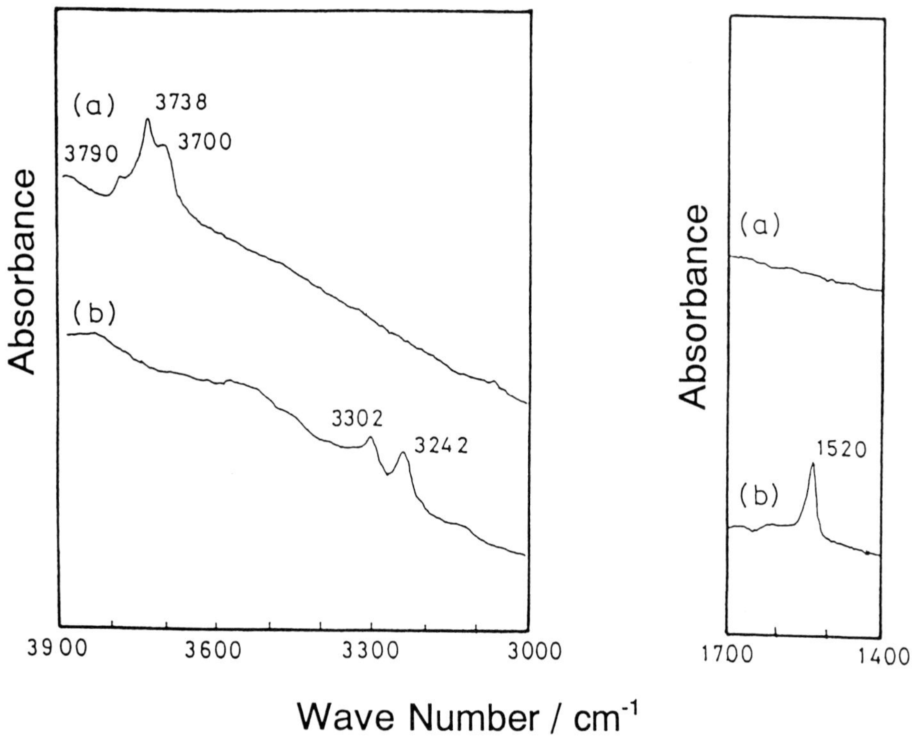

Figure 4. IR spectra of Eu/Al_2O_3. (a) Al_2O_3 evacuated at 773 K for 3 h. (b) Al_2O_3 wafer was immersed in Eu ammoniacal solution dissolving 0.008 g of Eu in 3.8 cm^{-3} of liquid ammonia, and then was evacuated at 298 K for 20 min.

estimated by the deconvolution of XANES spectra was 92:8, when Eu (5 wt%)/Al_2O_3 was heated under a vacuum at 523 K for 1 h [11]. Therefore, there is possibility that the OH groups on the surface of alumina react with Eu amides as follows;

$$Al\text{-}OH + Eu(NH_2)_2 \rightarrow Al\text{-}O\text{-}EuNH_2 + NH_3$$

Imamura *et al.*, reported the impregnation of Yb and Eu metal on silica from their ammoniacal solution of the metal results in the formation of Si-O-Ln-NH_2 when the amounts of their lanthanide metals were lower than ca. 15 wt% [12,13].

The second sample was prepared by immersing the alumina disk in a more concentrated solution of 0.20 g of Eu in 3.8 cm^3 of liquid ammonia. Its IR spectrum was measured after evacuating it at 298 K for 20 min (Figure 5a). The new three bands were observed at 3268, 3218 and 1560 cm^{-1} together with the bands observed in the spectrum of the first sample at 3320, 3242, and 1520 cm^{-1}. The new bands are also attributed to Eu amide, which is denoted as Eu amide-II.

The intensity of the peaks due to Eu amide-I decreased by increasing the evacuation temperature and the band disappeared at 673 K (Figure 5). On the other hand, the intensity of the peak due to Eu amide-II started to decrease around 623 K, and the bands disappeared at 773 K. Thus, Eu amide-II is thermally more stable than Eu amide-I.

The band was observed at 1970 cm^{-1}, when the second sample was evacuated at 423 K. The intensity of the band increased with increasing the evacuation temperature. Moreover, the peak at 2020 cm^{-1} was also observed by evacuating the second sample at temperatures above 773 K. These two bands around 2000 cm^{-1} may be assigned to a N=N stretching vibration or (N=N=N)-anions stretching vibration in an azide like species [14]. Thus, the Eu amide decomposed to azide like species.

Catalytically Active Species for the Isomerization of 2,3-dimethylbut-1-ene

The characteristics of Eu/Al_2O_3 catalyst for the isomerization of 2,3-dimethylbut-1-ene are as follows;

(1) The catalytic activity sharply increased by increasing the evacuation temperature to reach a maximum around 523 K, and decreased.

(2) The catalytic activity appeared when the amount of Eu is more than ca. 5 wt%. The amount of 5 wt% Eu is almost equal to a half the amount of OH groups on the surface alumina, which was estimated by thermogravimetric analysis up to 1273 K.

IR spectra of Eu/Al_2O_3 showed the formation of two kinds of Eu amides as

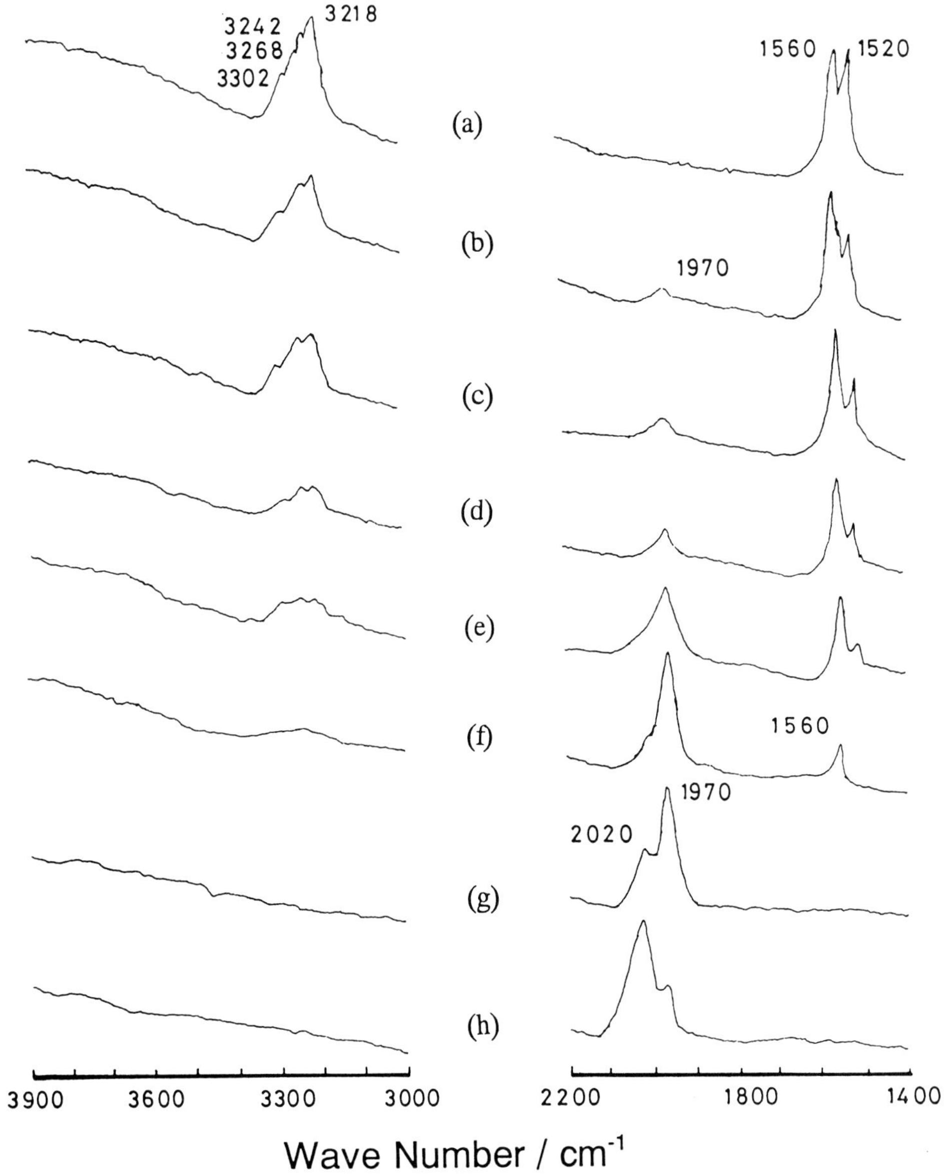

Figure 5. IR spectra of Eu/Al_2O_3. The sample was prepared by immersing the Al_2O_3 wafer in a solution of 0.20 g of Eu in 3.8 cm^{-3} of liquid ammonia. The sample was evacuated at (a) 298 K for 20 min, (b) 423 K for 1 h, (c) 473 K for 1 h, (d) 523 K for 1 h, (e) 573 K for 1 h, (f) 673 K for 1 h, (g) 773 K for 1 h, (h) 973 K for 1h.

described above. Eu amide-II was observed only in the second sample which was prepared in higher Eu concentration. Moreover, the intensity of the peaks due to Eu amide-I decreased with the evacuation temperature, and the peak was very weak at the evacuation temperature of 523 K. On the other hand, the significant decrease of the amount of Eu amide-II occurs only above 574 K. Therefore, Eu amide-II is a major species after heating at 573 K where the maximum catalytic activity was observed. From these findings, we conclude that Eu amide-II is a catalytically active species in Eu/Al_2O_3 for the base-catalyzed isomerization.

We reported that Eu imide supported on K-Y is catalytically active species for the isomerization of 2,3-dimethylbut-1-ene and the Michael reaction when Eu was introduced into K-Y zeolite by impregnation method [6]. In the case of Eu/ Al_2O_3, however, the Eu imide was not observed by the IR spectroscopy. Moreover, the catalytic activity of Eu/ Al_2O_3 much higher than that of Eu/K-Y. Theses results suggest that the catalytically active species of Eu/ Al_2O_3 is different from that of Eu/K-Y.

REFERENCES

1. J.C. Warf and W.L. Krost, *J. Phys. Chem.* **60**, 1590 (1956).
2. M.C. Symons, *Quart Rev. (London)* **13**, 99 (1959).
3. A. Iandelli and A. Dalenzona, *J. Less-common Metals.* **29**, 293 (1972).
4. E.E. Havinga, K-H.J. Buschow, and H.J. van Dual, *Solid State Commun* **13**, 621 (1973).
5. J.C. Warf and V. Gutmann, *J. Inorg. Nucl. Chem.* **33**, 1538 (1971).
6. T. Baba, G.J. Kim, and Y. Ono, *J. Chem. Soc., Faraday Trans.* **88**, 891 (1992).
7. T. Baba, S. Hikita, R. Koide, T. Hanada, T. Tanaka, S. Yoshida, and Y. Ono, *J. Chem. Soc., Faraday Trans.* **89**, 3177 (1993).
8. T. Baba, S. Hikita, R. Koide, T. Hanada, T. Tanaka, S. Yoshida, and Y. Ono, *J. Mol. Catal., A* **98**, 49 (1995).
9. H. Imamura, T. Konishi, E. Suda, Y. Sakata, and S. Tsuchiya, *Bull. Chem. Soc. Jpn.* **69**, 77 (1996).
10. V.G. Linde and R. Juza, *Z. Anorg. Allg. Chem.* **409**, 199 (1974).
11. T. Yoshida, T. Tanaka, S. Yoshida, T. Baba, and Y. Ono, *Catal. Lett.* submitted.
12. H. Imamura, T. Konishi, Y. Tokunaga, Y. Satake, and S. Tsuchiya, *Bull. Chem. Soc. Jpn.* **65**, 244 (1992).
13. H. Imamura, T. Konishi, Y. Satake, and S. Tsuchiya, *J. Chem. Soc., Faraday Trans.* **88**, 275 (1992).
14. J. Kritzenberger, E. Jobsen, A. Wokaun, and A. Baiker, *Catal. Lett.* **5**, 73 (1990).

Res. Chem. Intermed., Vol. 24, No. 5, pp. 507-514 (1998)

INEVITABLE REVIVAL OF C_1 CHEMISTRY IN THE FIRST HALF OF THE TWENTY-FIRST CENTURY

ZHAN-SHENG HAN, WEI PAN, JIN-LU LI* and QI-MING ZHU*

The State Key Laboratory of C_1 Chemistry and Technology, Department of Chemistry, Tsinghua University, Beijing 100084, China

Received 29 September 1997; accepted 24 December 1997

Abstract --The future of fuel production via C_1 chemistry is discussed in this paper. Four different processes from coal or natural gas to methanol or gasoline are compared with the present commercialized process from crude oil to gasoline on the aspects of economics, environmental friendliness and energy. Considering the future trend of crude oil price, it is believed that C_1 chemical technologies, especially indirect conversion technologies, will become economically feasible by the year 2030.

The oil crises between early 1970's and early 1980's led to a sharp rise in oil price from US$ 1.67 to 32/bbl (current US$). This triggered major world-wide inflation and a severe recession lasting at least ten years. The effects are still with us, indicated by the company downsizing and unemployment levels as we see today. The lesson is that mankind relied on oil excessively. During that period, people took every measure to contract oil consumption, including seeking for ways to substitute oil products with alternatives which could be derived from resources other than oil, e.g. coal, natural gas (NG), and biomass etc. As a result, C_1 chemistry received great attention and developed rapidly. These measures and the worldwide recession resulted in a declined demand for oil and a steep fall in oil price from early 1980's to mid-1980's down to US$ 16/bbl. From second half of 1980's until present, the oil price has shown a modest rise though in most cases highly unstable and fluctuating. Along with the fall in oil price, the enthusiasm for C_1 chemistry has been decreased.

Is it realistic to be so optimistic about oil sources? Should we prepare for ways to cope with exhaustion of oil? Let's have a look at the prospect of the oil supply and demand.

According to the estimation of experts [1], the limitation of the world oil reserve is approximately 1750 Gb and the annual increment of the oil production will be 1.5~3% in the next decades. Based on these estimations, the evolution of the annual oil production is plotted in Figure 1 (for the rising rate of the oil production, we took a middle value 2%). Based on this figure and assuming a 2% annual

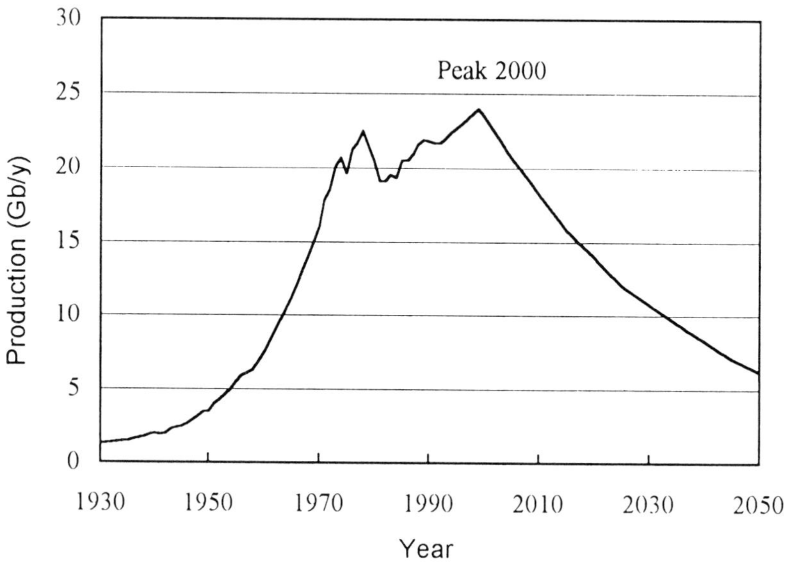

Figure 1. Evolution of annual oil production in teh world (with projection) [1].

increment of the oil production, its output will reach a maximum at the end of this century (see Figure 1), after which there will be a steep drop. At the oil consumption rate of 1996, the remaining oil can only be sustained for about 45 years. This figure for coal and natural gas is about 240 and 70 respectively. The production contract will certainly cause an abrupt increase in oil price.

In view of the above estimation, it is evident that mankind will have to go through an era in which coal and natural gas play a major part in supplying mankind with energy in the next half century.

In order to make use of coal and natural gas as raw materials to meet the demand of mankind for clean fuels (especially liquids) and various chemicals, people naturally turn to C_1 chemistry. A variety of technological routes and products are commercially available at present or in the future by means of C_1 chemistry starting from coal and natural gas and other carbon containing materials even garbage.

A limited amount of oil is used for overall chemicals production (*ca.* 8.5 %) so that even a substantial substitution would not exceed a few percent of the total oil consumption. Therefore, this paper will focus on the fuel uses of the products derived via C_1 chemistry. The principal processes for fuel production via C_1 chemistry are given in Figure 2. The processes on the right are commercially available at present, while those on the left, that is, the processes involved in the direct conversion of methane are under development.

We have recently made a study on 3E (Economic, Environmental and

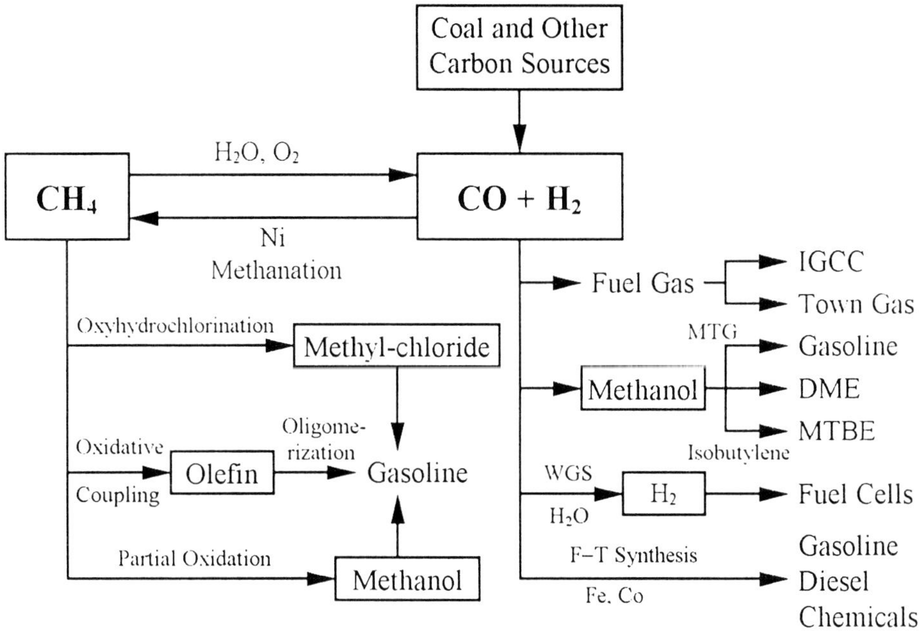

Figure 2. The principal processes for fuel production via C_1 Chemistry.

Energy) life cycle assessment of some alternative automotive fuels. Although this study is done based on the background of China, the general conclusions may be useful for the whole world.

Four technologies were selected for evaluation in the study: 1. Methanol from coal - Texaco coal slurry gasification technology combined with the Lurgi methanol synthesis technology represented coal-to-methanol synthesis; 2. Methanol from NG - the ICI natural gas-to-methanol synthesis technology; 3. Gasoline from coal - A modified Fischer-Tropsch technology developed in China was modeled for the coal-to-gasoline scenario. 4. Gasoline from NG - Methane reforming combined with modified Fischer-Tropsch synthesis; 5. Gasoline from crude oil is taken as a base case for comparison. The results are summarized in Table 1.

The production cost of gasoline from coal and NG is much higher than gasoline from oil. Taking into account of the consumption ratio for same driving distance (about 1.7 for methanol/gasoline in volume), the cost of methanol from coal is considerably higher than the base case. It seems that methanol from natural gas is competitive with gasoline from oil. However, if other cost is added for running automobiles, methanol from natural gas is still less economical now.

Figure 3 shows the cost increment of one vehicle during its lifetime using various fuels versus the oil price. The cost includes the manufacturing cost of the

Table 1
Production costs of several indirect conversion technologies of coal and NG to liquids fuels.

Technology No.	1	2	3	4	5
Feedstock	Coal	NG	Coal	NG	Oil
Major Product	Methanol	Methanol	Gasoline	Gasoline	Gasoline
Production Rate (tons/stream day)	1000	1000	1834	1834	3030
Total Capital Investment (million US$)	324	100	1525	8235	288
Raw Material Consumption (tons/stream day)	1436	676	7523	6018	18206
Capital Cost per Gallon (US$)	0.33	0.09	0.38	0.21	
Resource Cost per Gallon (US$)	0.08	0.22	0.28	0.55	
Operating Cost per Gallon (US$)	0.14	0.06	0.84	0.54	
Total Cost per Gallon (US$)	0.55	0.37	1.50	1.30	0.63
Total Cost per Barrel(US$)	23.1	15.5	63.6	54.6	26.5
Relative Mileage Cost	1.48	1.00	2.40	2.06	1.00

vehicle, fuel cost at the filling station and maintenance cost. It's obvious that no alternative fuels are economically feasible at present.

In his review [2], J.M. Fox III provides an economic comparison among six indirect and direct conversion technologies for producing liquid fuels from methane. Based on his data and assuming a gas price of $ 1.0 / GJ, the production costs of the fuels from various technologies are calculated and listed in Table 2. Each of the three direct technologies also involves an oligomerization step in which the primary products (methanol, olefins, and methyl chloride respectively) are converted to liquid hydrocarbons. It is seen from Table 2 that all the six technologies are not

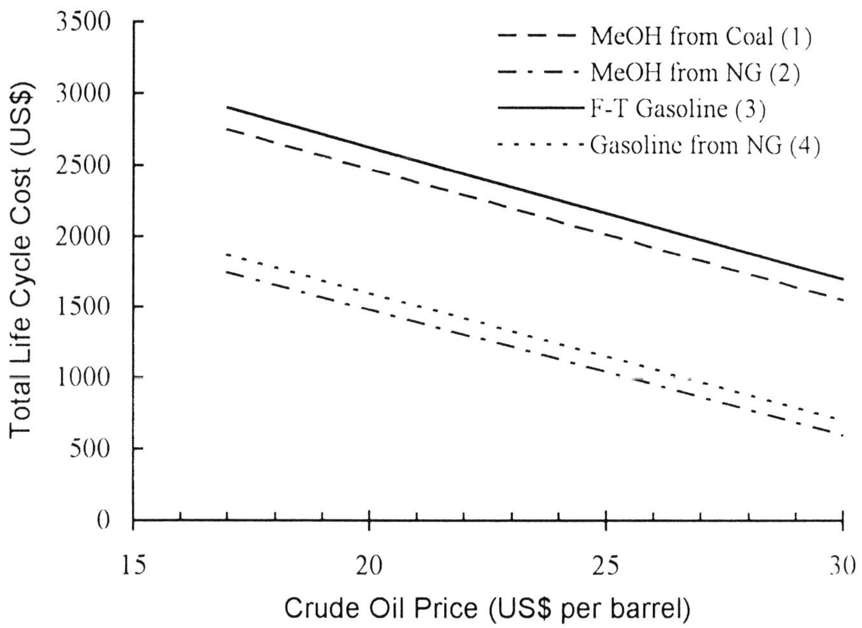

Figure 3. Cost increment from basecase for one vehicle during its lifetime.

Table 2
Production cost comparison for methane conversion technologies. US$/bbl liquid HCs, 1996.

Partial oxidation	Oxidative coupling	Oxyhydro-chlorination	Fischer-Tropsch synthesis	Fixed-bed MTG	Fluid-bed MTG
77	66	61	55	54	52

competitive either and that direct conversion technologies are even worse as long as economy is concerned.

Although all the alternative fuels discussed above are not competitive with oil products at present, they have bright prospects. On one hand, the rate of increase in oil price is much higher than for natural gas and coal. On the other hand, the room for improvement in the alternative fuel technologies is much larger than for gasoline from oil. If we prolong the lines in Figure 3, we will get their intercepts with the X coordinate. They are $36, 37, 46, 48 for technologies No. 2, 4, 1, 3, respectively. Assuming an extra increase rate of 3% per year in oil price starting from $18 in 1996, the corresponding years are 2019, 2020, 2028, and 2029, respectively. That is to say, by the year 2030, all the four alternative fuel technologies will be competitive with oil products. If we take into account of the improvement in these technologies, the years when C_1 chemical technologies become competitive will be earlier.

As for the other two E (Environment and Energy), the results of assessment are given in Table 3. Generally, technologies from coal result in more pollution than the base case, while those from NG cause less pollution problems. The energy efficiency of gasoline from oil is higher than all the fuels from alternative technologies. This is due to fact that the process of oil refining is simpler than the processes via C_1 chemistry.

In order to enhance the economic feasibility of the C_1 chemical technologies, development or improvement work in selectivity of desired product, single-pass conversion, equipment, and reaction conditions should be done. In particular for indirect methanol synthesis, the key is to conduct the reaction with more favorable thermodynamics. One way is the development of more active catalysts to lower the reaction temperature. For example, the Brookhaven liquid low temperature process takes place at 80-120°C, yielding >90% syngas conversion per pass with over 95% selectivity to methanol [3]. Another way is to remove methanol from the reaction system by absorbing with a high boiling inert solvent [4], adsorbing with an adsorbent [5], condensing under proper conditions [6] and further reacting to other compounds such as dimethyl ether in the presence of an added acid catalysts such as γ-Al_2O_3 [7]. By using these methods, 90-100% syngas conversion per pass has been achieved.

The key to the improvement of the Fischer-Tropsch synthesis is to circumvent the limitation of Florry-Schulz chain growth kinetics to raise the selectivity of liquid hydrocarbons. This can be done by means of zeolite encaged FTS catalysts or ultrafine catalysts.

The improvement of the first step of direct conversion of methane is a challenge. High selectivity and high conversion are contradictory. There is still a long way to go before they become economically competitive. Better catalysts providing high conversion and selectivities are still the vital factor for the improvement. Good combinations of components for methane activation and by product suppression may be a choice [8]. Reactors which permit a sudden cooling of the product stream or separation of desired products from oxygen may suppress deep oxidation of useful products.

In conclusion, because of the oil exhaustion in future decades, C_1 chemical technologies will play more and more important roles in the production of liquid fuels and chemicals. Although C_1 chemical technologies are not competitive with oil refining at the present time, by the year of 2030, all the indirect conversion technologies will become economically feasible. With significant improvements, the direct technologies for methane conversion will also become competitive.

Table 3
Lifecycle emission (kilograms per vehicle) and energy efficiency and their ranking (numbers in brackets).

Feedstock	Fuel	Carbon Dioxide	Sulfur Dioxide	Oxides of Nitrogen	Carbon Monoxide	Hydrocarbon	Particulate Mater	Energy Efficiency
NG	Methanol	45000(1)	60(1)	50(1)	60(2)	40(1)	5(1)	10%(2)
Petroleum	Gasoline	59000(3)	210(4)	90(3)	140(4)	90(3)	5(1)	12%(1)
Coal	Methanol	76000(4)	160(3)	60(2)	330(5)	60(2)	10(2)	8%(3)
Coal	Gasoline	87000(5)	240(5)	110(5)	110(3)	90(3)	20(3)	8%(3)
NG	Gasoline	50400(2)	90(2)	95(4)	40(1)	60(2)	10(2)	10%(2)

REFERENCES

1. C.J. Campbell, *Energy World*, No. **240**, June, 7 (1996).
2. J.M. Fox III, *Catal. Rev. - Sci. Eng.* **35(2)**, 169 (1993).
3. A. Sapienza, W.A. Slegeir, T.W. O'Hare, and D. Mahajan, *US Patent* 4,614,749 (1986).
4. M. Berty, C. Krishman, and I.R. Elliot, Jr., *Chemtech* **20**(October), 624 (1990).
5. R. Westerterp and M. Kuczynski, *Hydrocarbon Processing*, 80 (1986).
6. B. Hansen and F. Joensen. In: *Proceedings of Natural Gas Conversion Symposium*, A. Holman, J.J. Jens, and S. Kolboe (Eds.), Elsevier, Amsterdam, 457 (1991).
7. D.M. Brown, B.L. Bhatt, T.H. Hsiung, J.J. Lewnard, and F.J. Waller, *Catal. Today* **8**, 279(1991).
8. Z.S. Han, W. Pan, J.L. Li, and Q.M. Zhu, *Tsinghua Science and Technology*, **1(4)**, 420 (1996).

Res. Chem. Intermed., Vol. 24, No. 5, pp. 515-527 (1998)

ACTIVITIES OF POLYHEDRAL VANADIUM-CONTAINING SILSEQUIOXANE-BASED CATALYSTS FOR PHOTO-ASSISTED OXIDATION OF HYDROCARBONS

K. WADA, M. NAKASHITA, A. YAMAMOTO, H. WADA and T. MITSUDO*
Department of Energy and Hydrocarbon Chemistry, Graduate School of Engineering, Kyoto University, Sakyo-ku, Kyoto 606-01 Japan

Received 11 September 1997; accepted 10 November 1997

Abstract--The catalytic activity of a polyhedral vanadium-containing silsesquioxane ((c-$C_5H_9)_7Si_7O_{12}VO$, **1b**) with a cubic core structure towards the selective photooxidation of benzene, cyclohexane, and adamantane was examined under O_2 (1 atm) at around 300 K. The reaction of benzene in the presence of **1b** afforded phenol selectively, but the TON was low (~1.4). Complex **1b** showed high activity towards the reaction of cyclohexane (TON ~16) and adamantane (TON ~7.0). The order of the reactivity of the C-H bonds in adamantane was tertiary C-H > secondary C-H. In addition to **1b**, the mixtures of vanadyl bisacetylacetonate and incompletely-condensed silsesquioxanes with silanol functionality showed relatively high activities, although vanadyl bisacetylacetonate itself was not effective. The degradation of the catalysts in the course of the reaction was partly suppressed in the presence of molecular sieves-4A.

INTRODUCTION

Supported transition metal oxide catalysts are now widely used in chemical industries. Eminently, supported vanadium oxide is one of the most important catalysts effective for various selective oxidation reactions and de-NO_x reactions [1]. On the other hand, we have examined the selective photo-assisted catalytic oxidation of methane and other light alkanes into corresponding aldehydes at elevated reaction temperatures as high as 493 K [2]. Especially, silica-supported vanadium oxide catalysts were found to be quite effective. This reaction is, however, very sensitive to the surface structure of the catalyst. Although similar high sensitivities are often observed with silica supported vanadium oxide catalysts, the strict control of the surface active species is generally extremely difficulty [3].

On the other hand, silsesquioxane is a compound which is composed of a silicon-oxygen framework with organic substituent groups on the silicon atoms, of which the composition formula is $RSiO_{3/2}$ [4]. Among them, incompletely condensed oligomer with silanol groups is very interesting as a model compound of a silica support [5,6]. Oligometallasilsesquioxanes including transition-metal species also have been synthesized (see Scheme 1) [6-14]. Feher and coworkers established the chemistry of corner-capped type complexes [6-9]. Roesky's group developed heterosiloxanes which can be regarded as models for zeolites [10]. Although their potential abilities as catalysts have been often advocated, studies on their catalytic activities are limited [6,11-13]. Feher *et al.* reported that vanadium and aluminum-containing silsesquioxanes showed activities towards ethylene polymerization [11]. Molybdenum-containing complexes were also reported to be good catalysts for olefin metathesis [12]. Recently, van Santen *et al.* reported the effectiveness of a titanium-containing silsesquioxane for the epoxidation of olefins [13].

From these points of view, we have prepared a new cyclopentyl derivative **1b** of Feher's vanadium-containing silsesquioxane (**1a**) [8], and examined its activity towards the photooxidation of benzene and cyclohexane. The preliminary results were reported in the communication [15]. In the present study, the ability of a vanadium-containing silsesquioxane as an organic solvent-soluble silica supported oxide catalyst was investigated in full detail. The photo-assisted catalytic oxidations of benzene and alkanes were performed as probe reactions to examine the nature of the catalyst [16]. In addition to effects of the reaction conditions, effects of various incompletely-condensed silsesquioxanes as ligands for homogeneous vanadyl complexes are also discussed.

EXPERIMENTAL SECTION

Materials

Vanadyl triisopropoxide (Azumax) and vanadyl bisacetylacetonate (Aldrich) were used as received. Silica (Alfa, large pore 70 micron) was evacuated ($<10^{-2}$ Torr) at 423 K for 4 h before use. Dioxygen gas of high purity was purchased from Kyoto Teisan Co., Ltd. Molecular sieves-4A (MS-4A, granular) was purchased from Nakarai Tesque Co., Ltd. Other organic reagents were delivered from Wako Co., Ltd., and distilled just before use.

A vanadium-containing silsesquioxane **1b** was prepared in accordance with Feher's method [8]. Water of 146 cm^3 was added dropwise into a mixture of acetone (520 cm^3) and cyclopentyltrichlorosilane (Aldrich, 28.8 g, 141 mmol) in a 1000 cm^3 flat bottom flask, followed by reflux for 65 h. After the workup white powder of trisilanol **2b** was obtained in 37% yield (6.52 g) [17]. After 0.50 g (0.57 mmol) of **2b** in 30 cm^3 of benzene was agitated for 30 min at room temperature, 0.57 mmol of vanadyl triisopropoxide was added. After agitation for 2 h and vacuum evaporation

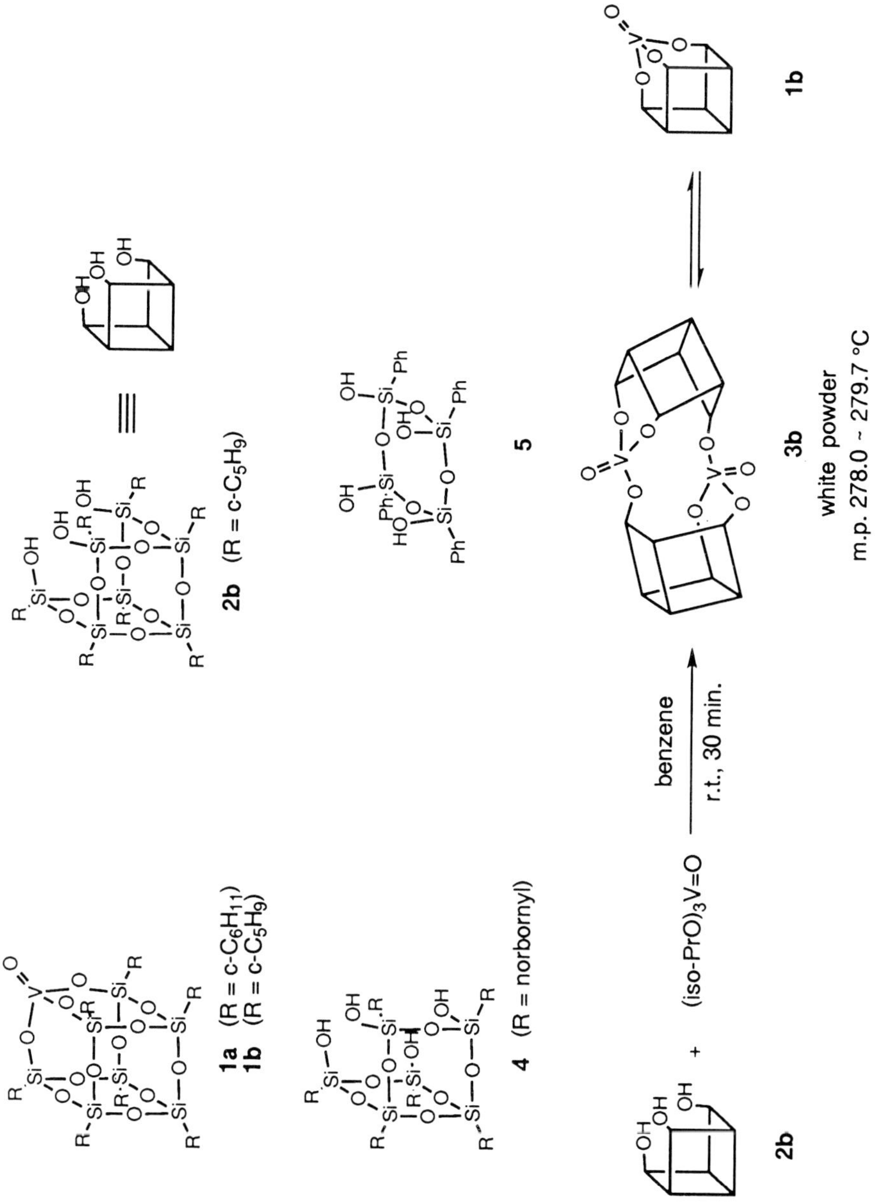
1a (R = c-C6H11)
1b (R = c-C5H9)
2b (R = c-C5H9)
4 (R = norbornyl)
5
2b
(iso-PrO)3V=O
benzene
r.t., 30 min.
3b
white powder
m.p. 278.0 ~ 279.7 °C
1b

Scheme 1

of the solvent, the white powder of vanadium-containing silsesquioxane (0.23 g, yield 44%) was obtained. Although we have not obtained sufficient X-ray crystallographic data until now, characteristic behaviour similar to **3a**, a well analyzed cyclohexyl derivative [8], could be observed in the temperature-variable ^{13}C-NMR study. Namely, in toluene at room temperature there were three predominant peaks for the *ipso*-carbons of cyclopentyl groups, indicating the presence of monomer **1b** of C_{3v} symmetry. At lower temperature (253 K), five peaks of dimer **3b** were predominant. Details of the preparative procedure and spectroscopic data of the complex has been described elsewhere [15]. Since the reactions were performed at around 300 K, and **1b** acted as a homogeneous catalyst in organic solvents, we have hereafter designated it as **1b**. Other incompletely-condensed silsesquioxanes with silanol functionality, **4** [18] and **5** [19], were prepared by the reported methods. The silica-supported catalysts were prepared by usual wet incipient impregnation using an oxalic acid solution of ammonium metavanadate [2].

Figure 1 shows the ultraviolet absorption of **1b** and related complexes in cyclohexane solution. For **1b**, λmax was 252 nm ($\varepsilon = 4840\ M^{-1}\ cm^{-1}$). On the other hand, the parent trisilanol **2b** showed no absorption in the range from 190 nm to 900 nm.

Procedure

For the photooxidation of liquid hydrocarbons, a 50 cm^3 Pyrex reactor or a quartz glass reactor equipped with a gas buret was used. UV irradiation was performed using a 200 W high pressure mercury vapor lamp with a water filter from the outer side of the reactor. The reaction temperature was monitored by a thermometer directly inserted into the reaction mixture. Typical reaction conditions are as follows: For the photooxidation of benzene, benzene 10 cm^3, solvent 10 cm^3, catalyst 26 μmol, dioxygen 1 atm, reaction time 6 h, temperature 305 K. For the reaction with cyclohexane, substrate 20 cm^3 without solvent, catalyst 13 μmol. For the oxidation of adamantane, substrate 1.0 mmol, 1,2-dichloroethane (solvent) 20 cm^3, catalyst 13 μmol.

Analysis

The products were analyzed by GC-MS (Shimadzu QP 2000 S), gas chromatography, ^{1}H-NMR and ^{13}C-NMR. For G.C. analyses n-propylbenzene or ethylbenzene was used as an internal standard. Details of the analytical conditions are as follows: For CO and CO_2, activated carbon column (stainless steel column, i.d. 3.0 mm x 3.0 m, at 323 K); for gaseous alkanes, Porapak Q column (stainless

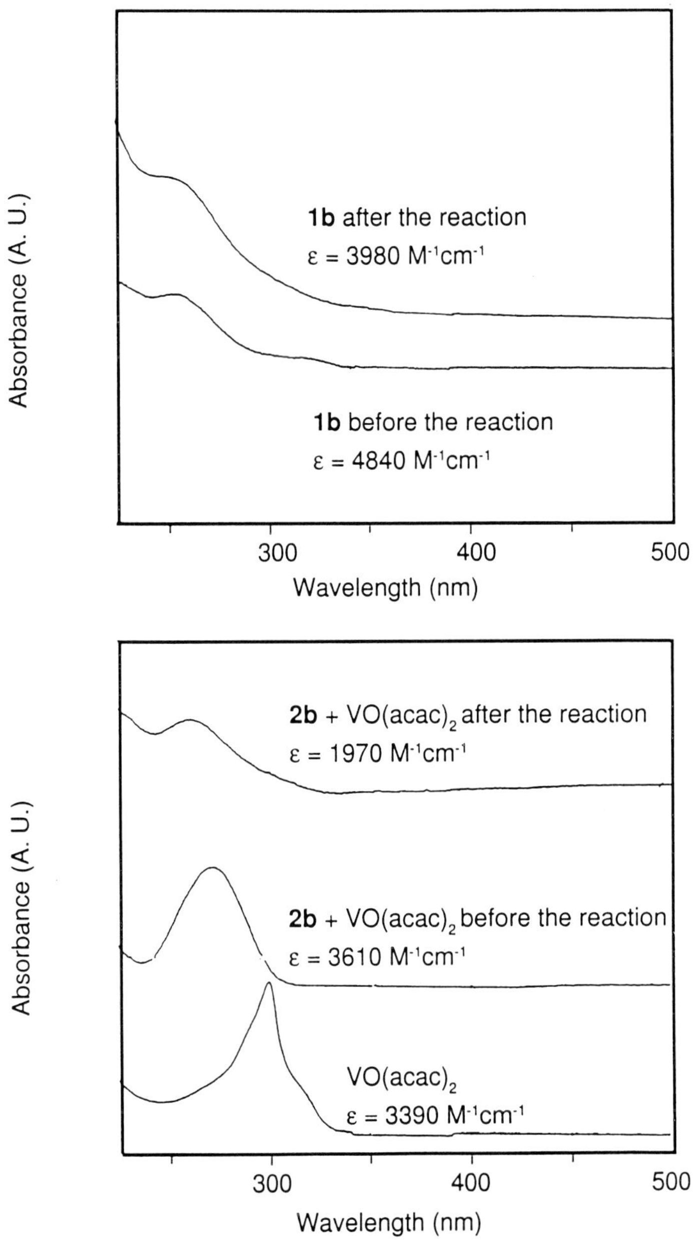

Figure 1. UV spectra of **1b** and related complexes in cyclohexane solutions.

steel column, i.d. 3.0 mm x 3.0 m, at 353 K); for phenol, alcohols and ketones, CBP10-S25-050 (Capillary column, i.d. 0.33 mm, length 25 m, o.d. 0.43 mm, at 323~473 K). UV-visible absorption spectra in solutions were measured by using a

Hitachi U-3210 spectrophotometer. ^{1}H-NMR (400 MHz) and ^{13}C-NMR (100 MHz) spectra were measured using a JEOL EX-400.

RESULTS AND DISCUSSION

Photooxidation of Benzene

Phenol was selectively formed by the photo-assisted oxidation of benzene catalyzed by a vanadium-containing silsesquioxane **1b**. To prevent photochemical reactions of benzene in which the catalysts are not get involved, a Pyrex reactor was used. It should be noted that light through a Pyrex reactor (passes $\lambda > 290$ nm) is hardly able to activate the catalysts efficiently (see Figure 1). Whereas both heterogeneous silica supported vanadium oxide catalysts and homogeneous vanadyl bisacetylacetonate did not show significant activities as described in the communication [15], the reaction in the presence of **1b** together with acetonitrile afforded 36 μmol of phenol. Formation of CO and CO_2 was not observed. Only a trace amount of biphenyl was detected. As shown in Table 1, the solvent drastically affected the reaction. Among the solvents examined, acetonitrile was found to be most effective. The yield of phenol decreased remarkably when the reaction was performed in the absence of the solvent. Whereas addition of 1,2-dichloroethane slightly increased the yields, the addition of dichloromethane decreased the yield of phenol.

On the other hand, vanadyl bisacetylacetonate (26 μmol) showed a comparatively high activity to produce 13 μmol of phenol under coexistence with **2b**, although neither vanadyl bisacetylacetonate nor **2b** showed any activity. Effects of the addition of these silanols will be discussed in detail in the later section. When the reaction was carried out in an argon atmosphere in place of O_2, phenol was hardly formed, indicating the involvement of vapor phase dioxygen. It could be seen that the reactions with benzene generally resulted in low turnover numbers (TON). The TON was at best 1.4 even in the presence of **1b**.

Photooxidation of Cyclohexane

The reaction of cyclohexane in the presence of **1b** afforded cyclohexanol and cyclohexanone in high yields. Again, to prevent photochemical reactions in which the catalysts are not get involved, a quartz glass reactor with a Pyrex filter or a UV-27 filter and a lamphouse was used. It should be noted that the use of a lamphouse greatly decreases the number of photons irradiated into the reactor. As shown in Table 2, much higher TON was recorded than for those in the reactions of benzene. The use of a UV-27 filter in place of a Pyrex filter greatly increased the yields.

Table 1
Effects of solvents on photo-assisted catalytic oxidation of benzene.[a]

Run	Catalyst	Solvent	Phenol Yield (μmol)	TON[b]
1	none	CH_3CN	0	-
2	**1b**	CH_3CN	36	1.4
3	none	CH_2Cl_2	trace	-
4	**1b**	CH_2Cl_2	trace	0
5	none	$C_2H_4Cl_2$	trace	-
6	**1b**	$C_2H_4Cl_2$	7.0	0.2
7[c]	**1b**	none	6.4	0.2
8[d]	**1b**	CH_3CN	0	0
9	**2b**	CH_3CN	trace	0
10	V_2O_5/SiO_2[e]	CH_3CN	4.5	0.6
11	$VO(acac)_2$	CH_3CN	trace	0
12	$VO(acac)_2$ + **2b**	CH_3CN	13	0.5

[a]Catalyst 26 μmol, benzene 10 cm^3, solvent 10 cm^3, O_2 1 atm, temp. 305 K, 6 h, using a 200 W high pressure Hg-lamp and a Pyrex reactor. Formation of carbon oxides was not observed.
[b]Turnover number (mol-products/mol-catalyst used).
[c]Benzene 20 cm^3.
[d]Ar 1 atm without O_2.
[e]V 2.0 mol%, 25 mg.

Table 2
Photo-assisted catalytic oxidation of cyclohexane.[a]

Run	Catalyst	Filter[b]	Yield (μmol)		Total TON
			6	**7**	
13	**1b**	Pyrex	44	32	5.8
14	none	UV-27	trace	trace	-
15	**1b**	UV-27	120	86	16
16	V_2O_5/SiO_2[c]	UV-27	trace	trace	0
17	$VO(^iPrO)_3$	UV-27	trace	trace	0
18	$VO(acac)_2$	UV-27	11	11	1.7
19	$VO(acac)_2$ + **2b**	UV-27	85	55	11
20	$VO(acac)_2$ + **4**	UV-27	23	21	3.4
21	$VO(acac)_2$ + **5**	UV-27	10	9	1.5

[a]Catalyst 13 μmol, cyclohexane 20 cm^3, O_2 1 atm, temp. 305 K, 3 h, using a 200 W high pressure Hg-lamp. Formation of carbon oxides was not observed.
[b]A quartz glass reactor was used. Irradiation using a lamphouse.
[c]V 6.7 mol%, 25 mg.

Again, there was no formation of carbon oxides. A silica supported vanadium oxide catalyst and a homogeneous V(V)=O complex did not show significant activity.

In runs 18 to 21, effects of several incompletely-condensed silsesquioxanes with silanol functionality onto the reaction using vanadyl bisacetylacetonate are shown. Although vanadyl bisacetylactonate itself showed relatively low activity, addition of **2b** greatly increased the yields of the desired products. The addition of **4** showed a moderate effect, whereas **5** was not effective. As shown in Figure 1, by mixing vanadyl bisacetylacetonate with **2b**, the UV spectra immediately changed to show new absorption at around 275 nm. This also differs from that of **1b**. Attempts to isolate and identify the products were still unsuccessful. It should be noted that there were no changes in UV spectra by the addition of **5** to the solution of vanadyl bisacetylactonate.

One of the problems we come up against is the degradation of the catalysts. Figure 2(a) shows the effects of the reaction time. A Pyrex reactor without a lamphouse was used. The rate of oxidation decreased in the course of the reaction. Just after the completion of the reaction using **1b**, the solution was clear and colorless. But afterwards, greenish colloidal matter was formed. This is probably due to the hydrolysis of metallasilsesquioxane by water produced by the oxidation of the substrate. Regeneration of silanol functionality may decrease its solubility in organic solvents. Then we tried to prevent the degradation of the catalysts by the addition of MS-4A. As shown in Figure 2(b), this attempt was partly successful. Although the initial activity decreased, the deactivation was fairly suppressed in the presence of MS-4A (130 mg). But too much MS-4A (550 mg) conversely decreased the yields (cyclohexanol 149 μmol, cyclohexanone 117 μmol, reaction time 6 h, other reaction conditions as in Figure 2). It should be noted that the ratio of cyclohexanone to cyclohexanol increased by the addition of MS-4A.

Figure 3 shows the effects of the amount of the catalyst **1b**. The yields increased monotonously along with an increase in the amount of the catalyst up to 13 μmol. But further addition of the catalyst did not increase the yields. This is probably due to the limited number of photons irradiated into the reactor.

Photooxidation of Adamantane

The main product of the photooxidations of adamantane was 1-adamantanol, the results of which can be seen in Table 3. Because of the limited solubility of the substrate, 1,2-dichloroethane was used as a solvent. A Pyrex reactor without a lamphouse was used in the present reaction. The reaction using **1b** at 299 K for 6 h afforded 1-adamantanol, 2-adamantanol and 2-adamantanone in yields of 27 μmol, 3.3 μmol, and 7.1 μmol, respectively. On the other hand, the products were obtained in higher yields in the presence of both vanadyl bisacetylacetonate and **2b** (run 26).

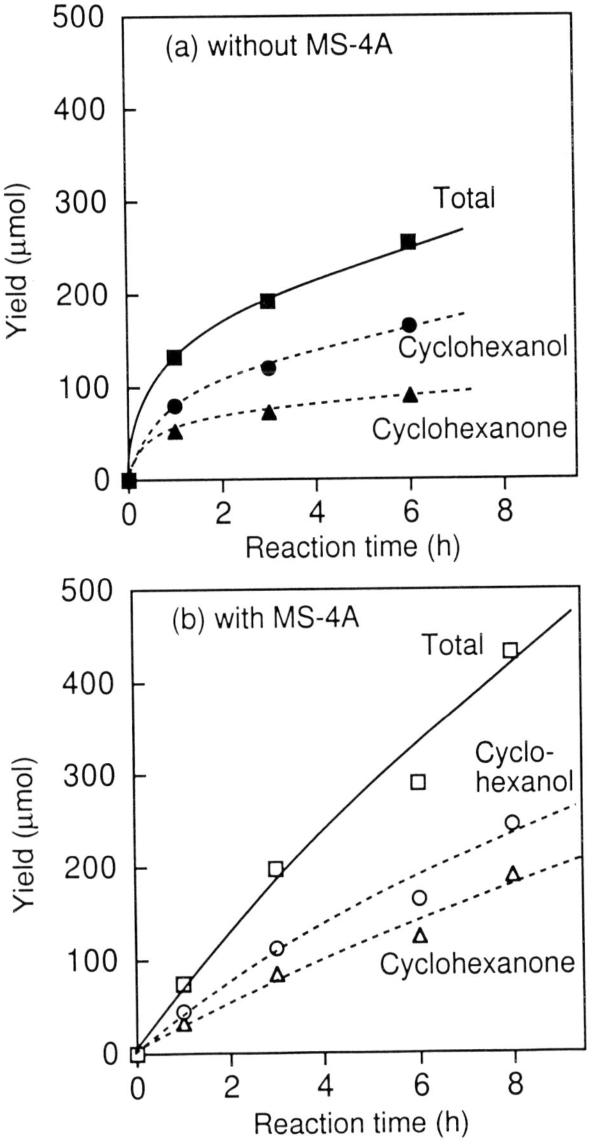

Figure 2. Time dependence of photo-assisted catalytic oxidation of cyclohexane. Catalyst **1b** 13 μmol, cyclohexane 20 cm^3, O_2 1 atm, 305 K, MS-4A 130 mg, using a 200 W high-pressure Hg-lamp and a Pyrex reactor.

Since this catalytic system has an absorption band of longer wavelength than that of **1b** (see Figure 1), the high activity of this mixed catalyst seems partly due to its ability in utilizing light of longer wavelengths more efficiently.

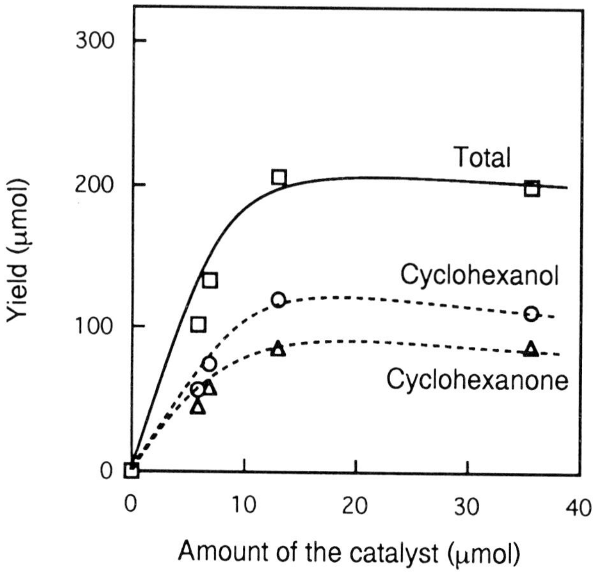

Figure 3. Effects of the amount of the catalyst **1b** on photo-assisted catalytic oxidation of cyclohexane. Cyclohexane 20 cm^3, O_2 1 atm, 305 K, 3 h, using a 200 W high-pressure Hg-lamp and a quartz glass reactor with a UV-27 filter and a lamp house.

Table 3
Photo-assisted catalytic oxidation of adamantane.[a]

Run	Catalyst	Yield (μmol)			Total TON
		8	**9**	**10**	
22	none	8.4	3.0	2.3	-
23	**1b**	27	3.3	7.1	2.1
24	V_2O_5/SiO_2[b]	3.1	trace	trace	0.2
25	V_2O_5/SiO_2[c]	8.0	trace	2.7	0.6
26	$VO(acac)_2$	13	trace	2.1	1.0
27	$VO(acac)_2$ + **2b**	70	21	trace	7.0

[a]Catalyst 13 μmol, adamantane 1.0 mmol, 1,2-dichloroethane 20 cm^3, O_2 1 atm, temp. 299 K, 6 h, using a 200 W high pressure Hg-lamp and a Pyrex reactor. Formation of carbon oxides was not observed.
[b]V 6.7 mol%, 12.5 mg.
[c]V 2.0 mol%, 25 mg.

Reaction Pathways

Although we have no experimental evidence to propose a clear-cut reaction pathway, on the basis of earlier studies on highly dispersed silica-supported vanadium oxide catalysts, we propose one of the possible pathways as follows: First, the vanadyl group in tetrahedral co-ordination is photo-activated via charge transfer to produce strongly electrophilic radicallic active oxygen species. The cleavage of the C-H bond of the substrate by the oxygen species and simultaneous oxygen transfer may form an intermediate species, probably an alkoxide species. The recombination of the proton would lead to the formation of phenol or alcohols. Besides the further oxidation of the alcohols, abstraction of another hydrogen, if possible, from the intermediate species would also form ketones. This explanation is based on that proposed by Kazansky's group for heterogeneous silica supported oxide catalysts [20], but details are still unclear. Detailed studies on the activated complex and/or active oxygen species is now in progress.

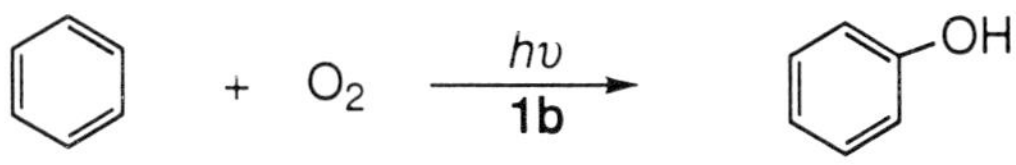

(1)

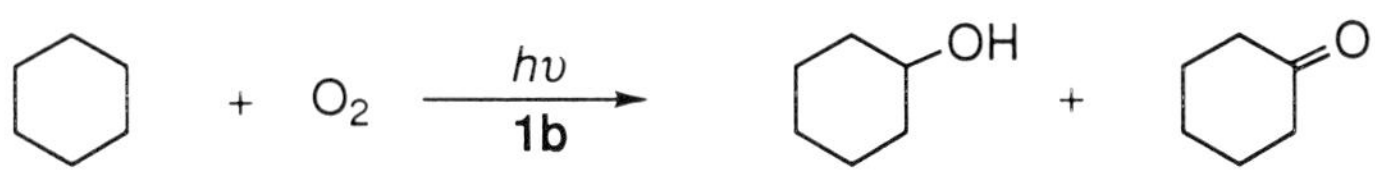

6 7

(2)

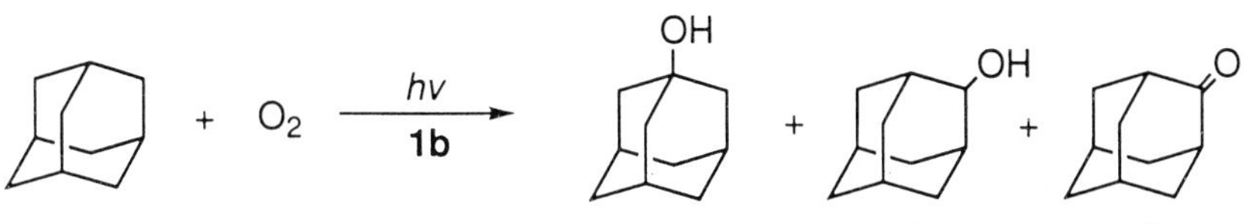

8 9 10

(3)

If the reaction proceeds in accordance with the above mechanism, electrophilicity of activated oxygen species would be the key which determines the activity. According to precedent studies on ^{13}C chemical shift and Hammett substitution parameters, the electron-withdrawing properties of the cubic-type siloxane frameworks of polyhedral silsesquioxanes were recognized. For example, the electron-withdrawing tendency of the Si_8O_{12} core was estimated to be almost equal to that of a trifluoromethyl group [21]. Such properties would not only increase the Lewis acidity of a vanadium-containing complex, but also enlarge the electrophilicity of the photo-activated oxygen species. Besides their high solubility, this would be one of the reasons why **1b** showed high activity.

CONCLUSION

A polyhedral vanadium-containing silsesquioxane **1b** showed high activities for the selective photooxidation of benzene, cyclohexane, and adamantane under O_2 (1 atm) at around 300 K. Phenol was formed from benzene without significant formation of carbon oxides and other byproducts, but the TON of this reaction was low. The oxidation of cyclohexane into cyclohexanol and cyclohexanone with **1b** proceeded more efficiently. The main product of the oxidation of adamantane was 1-adamantanol. The reactivity order was tertiary C-H > secondary C-H. In addition to **1b**, the mixtures of vanadyl bisacetylacetonate and some incompletely-condensed silsesquioxanes with silanol groups such as **2b** and **4** also showed relatively high activities, although vanadyl bisacetylacetonate itself did not show the significant catalytic activity. The degradation of the catalysts was partly suppressed by the addition of molecular sieves-4A.

The results in the present study indicate the potential abilities of oligometallasilsesquioxanes as "organic solvent soluble silica supported catalysts".

REFERENCES

1. G.I. Golodets, *Heterogenous Catalytic Reactions Involving Molecular Oxygen*, Elsevier, New York, 1983.
2. K. Wada and Y. Watanabe. In: *Methane and Alkane Conversion Chemistry*, M.M. Bhasin and D.W. Slocum (Eds.), Plenum, New York, 1995, p. 179.
3. G. Poncelet, P. Grange, and P.A. Jacobs (Eds.), *Preparation of Catalysts III*, Elsevier, Amsterdam, 1983.
4. R.H. Baney, M. Ito, A. Sakakibara, and T. Suzuki, *Chem. Rev.* **95**, 1409 (1995).
5. J.F. Brown, Jr., *J. Am. Chem. Soc.* **87**, 4317 (1965).
6. For a review, F.J. Feher and T.A. Budzichowski, *Polyhedron* **14**, 3239 (1995).
7. F.J. Feher, *J. Am. Chem. Soc.* **108**, 3850 (1986).
8. F.J. Feher and J.F. Walzer, *Inorg. Chem.* **30**, 1689 (1991).
9. F.J. Feher and R.L. Blanski, *J. Am. Chem. Soc.* **114**, 5886 (1992).
10. For example, V. Chandrasekhar, R. Murugavel, A. Voigt, H.W. Roesky, H. Schmidt, and M. Noltemeyer, *Organometallics* **15**, 918 (1996).
11. F.J. Feher and R.L. Blanski, *J. Chem. Soc., Chem. Commun.* 1614 (1990).
12. F.J. Feher and T.L. Tajima, *J. Am. Chem. Soc.* **116**, 2145 (1994).
13. H. Abbenhius, S. Krijnen, and R. van Santen, *Chem. Commun* 331 (1997).
14. W.A. Herrmann, R. Anwander, V. Dufaud, W. Scherer, *Angew. Chem. Int. Ed. Engl.* **33**, 1285 (1994).
15. K. Wada, M. Nakashita, A. Yamamoto, H. Wada, and T. Mitsudo, *Chem. Lett.* (1997), in press.
16. F. Fujihira, Y. Satoh, and T. Osa, *Nature* **293**, 206 (1981).
17. F.J. Feher, T.A. Budzichowski, R.L. Blanski, K.J. Weller, and J.W. Ziller, *Organometallics* **10**, 2526 (1991).
18. T.W. Hambley, T. Maschmeyer, and A.F. Masters, *Appl. Organomet. Chem.* **6**, 253 (1992).

19. J.F. Brown, Jr., and L.H. Vogt, Jr., *J. Am. Chem. Soc.* **87**, 4313 (1965).
20. S.L. Kaliaguine, B.N. Shelimiov, and V.B. Kazansky, *J. Catal.* **55**, 384 (1978).
21. F.J. Feher and T.A. Budzichowski, *J. Organomet. Chem.* **379**, 33 (1989).

Res. Chem. Intermed., Vol. 24, No. 5, pp. 529-540 (1998)

PROBABLE MECHANISM FOR ALKANE OXIDATION WITH H_2O_2 OVER VS-2

TAKASHI TATSUMI, YOSIAKI WATANABE, YOSHIRO HIRASAWA and JUNICHI TSUCHIYA

Engineering Research Institute, Faculty of Engineering, The University of Tokyo, Yayoi, Tokyo 113, Japan

Received 20 October 1997; accepted 5 December 1997

Abstract--Metallosilicate zeolites containing Ti and V, TS-2 and VS-2, have been synthesized using tetrabutylammonium cation as the template. Although both Ti and V exist in zeolite framework in a highly dispersed state, V easily leaches by K_2CO_3 or H_2O_2 treatments in contrast to Ti. Spin trapping experiments and relative reactivity of linear/cyclic alkanes indicate that the oxidation of alkanes over TS-2 and VS-2 proceed by different mechanisms; it is conceivable that the oxidation occurs by way of the attack of •OH species produced from the V species in the zeolite framework with H_2O_2.

INTRODUCTION

In recent years increasing attention has been directed toward the study of new zeolitic materials, especially metallosilicate analogs of zeolites containing transition metal ions with variable valence in the framework because of their interesting catalytic properties. The presence of transition metals in the framework of a zeolite lattice can impart oxidation activity which can be combined with shape selectivity of the molecular sieve. Titanium silicate molecular sieves, TS-1 and TS-2, which are titanium analogs of ZSM-5 and ZSM-11, respectively, have been more extensively studied [1]. Vanadium silicate, VS-2, a vanadium analog of ZSM-11, catalyzed the oxidation of alkanes at mild temperature in a similar manner to TS-1 and TS-2 [2]. In the oxidation of hexane, the formation of primary alcohols and aldehydes was observed with VS-2, in contrast with TS-1 and TS-2, which gave only secondary alcohols and ketones.

Compared to TS-1 and TS-2, few studies have been conducted on the microenvironments of the active species and reaction mechanism for the liquid phase oxidations over VS-2 [3]. The purpose of this study is to investigate the structure of the V species and the reaction pathway in the oxidation of hydrocarbons with H_2O_2 on VS-2.

EXPERIMENTAL SECTION

Catalysts

VS-2 was prepared by the reported method [4]. K-VS-2 was prepared by treating calcined VS-2 in aqueous K_2CO_3 (0.5 M) solution at room temperature for 2 h followed by calcination at 763 K for 3 h. H-VS-2 was obtained by treating K-VS-2 in aqueous HCl (1.0 M) solution at room temperature for 18 h and the subsequent calcination at 823 K for 3 h. VS-2-H_2O_2 denotes VS-2 treated with 30 wt% aqueous solution of H_2O_2 at 333 K for 2 h. V-imp.S.-2 was prepared by impregnation of silicalite-2 with aqueous solution of NH_4VO_3 containing H_2O_2. TS-1 was synthesized according to a literature method using tetrapropylammonium hydroxide as template [5]. TS-2 was prepared by the method reported by Reddy and Kumar [6].

Characterization

X-ray diffraction (XRD) powder patterns were collected on a Rigaku RINT 2400 X-ray diffractometer. FT-IR were recorded on a Perkin Elmer 1600 spectrometer. Electron spin resonance spectra were recorded at room temperature on a JEOL JM-EFIX spectrometer at 9.3 GHz (X-band).

Reactions

Oxidation of hexane and cyclohexane was conducted in a flask with 30 % H_2O_2. The products were analyzed on a Shimadzu GC 14 gas chromatograph equipped with a Nukol capillary column.

Spin Trapping

Spin trapping agent phenyl N-*tert*-butyl nitrone (PBN) was used as purchased. Highly dispersed Mn^{2+} on MgO was used as a marker. To detect radicals in the presence of hexane, to the mixture of 1.0 ml of 30% aqueous H_2O_2 and 3.9 ml of hexane was added a catalyst (VS-2 and V-imp.S.-2 50 mg, TS-2 73 mg), and the resulting mixture was stirred for 15 minutes at 333 K. Then 1.0 ml of the supernatant solution was taken out and 4 mg of PBN was added to this solution.

RESULTS AND DISCUSSION

The XRD patterns of VS-2 closely matched with those reported for MEL structures [7]. There was no change in the XRD patterns with the treatment with aqueous solution of K_2CO_3, HCl or H_2O_2. The ratios of Si to V are shown in Table 1. The Si/V ratios of K-VS-2 and VS-2-H_2O_2 decreased to 1510 and 1250, respectively, indicating that the treatments with K_2CO_3 or H_2O_2 leached almost all the vanadium from the zeolite framework. This is in contrast to the stability of TS-1 and TS-2 to aqueous solution of K_2CO_3. No leaching of Ti was observed while the IR band at 960 cm^{-1} owing to the incorporation of Ti into the zeolite framework was weakened as shown in Table 1.

Table 1
Characterization of VS-2 and its related materials.

Sample	Si/Metal ratio	I_{960}/I_{550} ratio
VS-2	58	0.13
K-VS-2	1510	0.12
H-VS-2	1792	0.35
VS-2-H_2O_2	1250	-
TS-1	92	0.20
K-TS-1	80	0.14

Although it is well known that the ratio of the intensities of the 960 to 550 cm^{-1} band is proportional to the Ti content of the zeolite framework in TS-1 and TS-2, the assignment of the band at 960 cm^{-1} is still in dispute. It is proposed that this band is assigned to Ti-O-Si stretching mode [8], Si-OH stretching mode in the neighborhood of titanium [9] or Ti=O stretching mode [10]. The XANES spectra showed no evidence for the Ti=O structure [11]. Khouw and Davis observed that treatment of TS-1 with aqueous NaOH solution resulted in the decrease in the intensity of IR band at 960 cm^{-1} and the appearance of a weak shoulder band at 985 cm^{-1} [12]. They suggested the interconversion as illustrated in Scheme 1. It is considered that the I_{960}/I_{550} ratio can be taken as a measure for the number of silanol groups (Si-O-H in Scheme 1) in the neighborhood of titanium.

The IR spectrum of calcined VS-2 also exhibited a band in the similar region as shown in Figure 1. Photoluminescence studies indicated the presence of tetrahedrally coordinated V-oxide species; a vibrational fine structure owing to the vanadyl (V=O) band was clearly observed [13]. The energy gap between the ($0 \rightarrow 0$) and ($0 \rightarrow 1$) transition bands of the V=O vibration was found to be 956 cm^{-1}, in good agreement with the IR frequency. Thus the band at ca. 960 cm^{-1} on VS-2 seems to be due to the V=O stretching mode. The presence of vanadyl (V=O) species in VS-2 was also revealed by a XAFS study; the best fitting of the XAFS spectrum of the

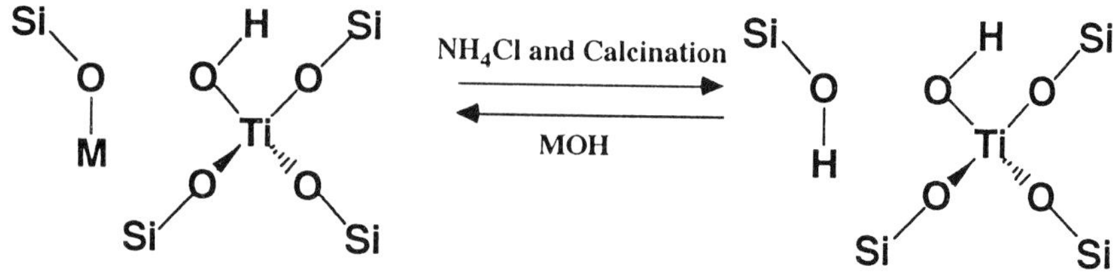

Scheme 1
Ion exchange of silanol group.

VS-2 sample was obtained with one oxygen atom in the short V-O distance of 1.61 Å and three oxygen atoms in the long V-O distance of 1.77 Å [13]. However, a weak band at ca. 960 cm^{-1} was also observed for silicalite-2 as well as for V-imp.S-2, where no evidence for the V=O structure was obtained from photoluminescence studies. These facts suggest that the IR band at ca. 960 cm^{-1} observed with V-imp.S-2 and silicalite-2 was assigned to Si-OH at zeolite defects. Hence it is not clear whether the band near 960 cm^{-1} in the IR spectra of VS-2 is assigned to V=O or Si-OH.

As shown in Table 1, the relative intensity of the ca. 960 cm^{-1} band (I_{960}/I_{550}) was slightly decreased upon K_2CO_3 treatments, while the V content drastically decreased. The sample obtained by the subsequent acid treatment, H-VS-2, showed relative intensity higher than the starting VS-2, in spite of the very low concentration of vanadium. Thus the 960 cm^{-1} band is apparently independent of vanadium in VS-2. Figure 2 is the magnification of the IR spectra around 960 cm^{-1}. The spectra revealed that the band at 960 cm^{-1} actually consists of several bands. The band at 970 cm^{-1} in VS-2 disappeared after K_2CO_3 treatment and never recovered by the subsequent HCl treatment, suggesting the 970 cm^{-1} peak is directly related to the vanadium in framework. Thus the 970 cm^{-1} peak is assigned to the V=O stretching mode. The peak at 974 cm^{-1} also disappeared after K_2CO_3 treatment and a peak at 978 cm^{-1} appeared instead. It was reported that the Si-O-H group in TS-1 was converted to Si-O-K by K_2CO_3 treatment and that Si-O-K was reconverted to Si-O-H by HCl treatment [14]. Treatment of K-VS-2 with HCl resulted in recovery of the peak at 974 cm^{-1} associated with emergence of a peak at 976 cm^{-1}. As shown in Figure 1(b), the peak at 3740 cm^{-1} ascribable to silanol groups was stronger in H-VS-2 than in VS-2 and absent in K-VS-2. Hence it is suggested that alkaline metals (M = Na and/or K) originating from template impurities is contained in the form of Si-O-M in the as-synthesized and calcined VS-2. Further Si-O-K groups could be formed by the K_2CO_3 treatment and this Si-O-K linkage should have been changed into Si-O-H by the subsequent HCl treatment. These findings suggest that the peaks

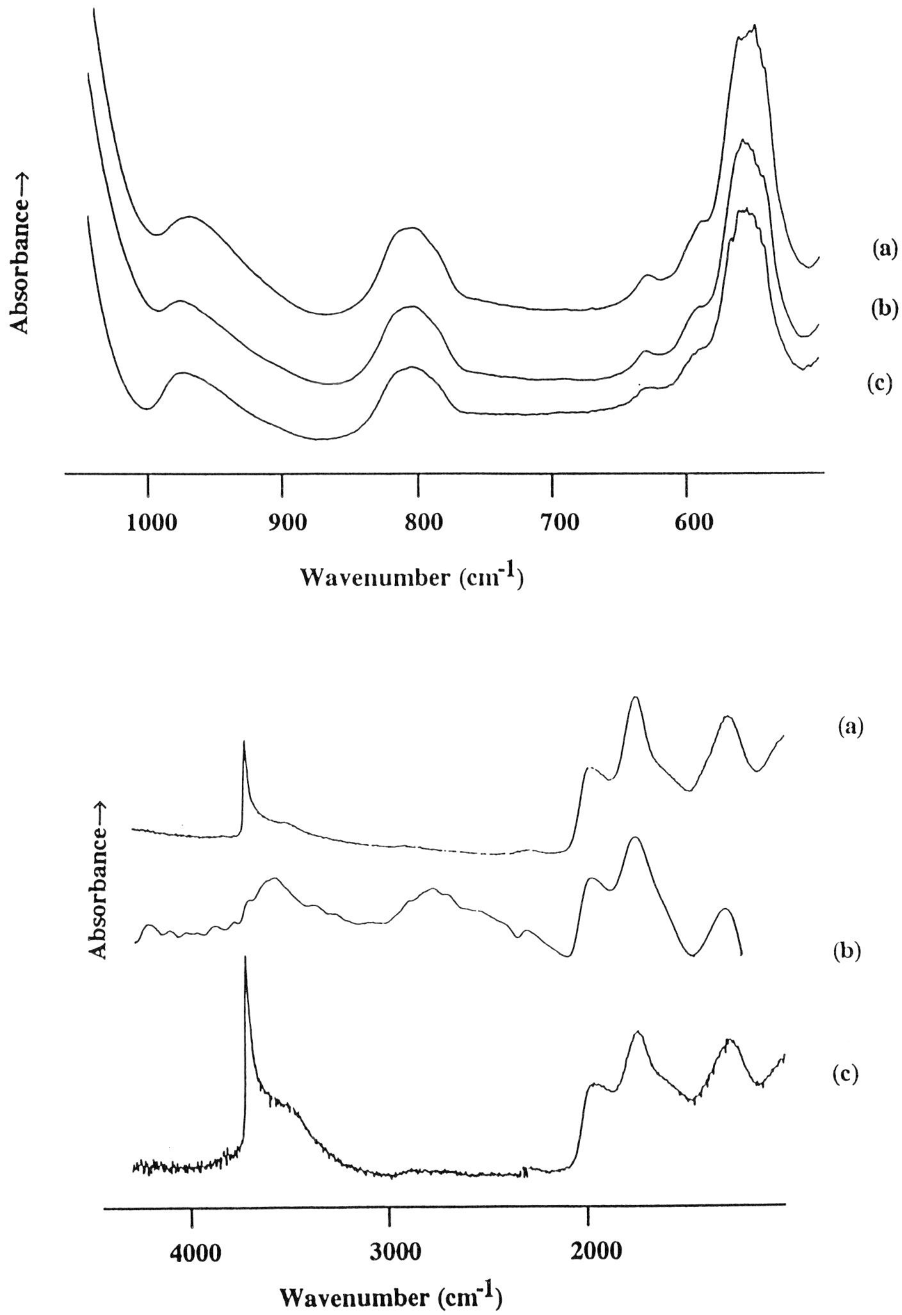

Figure 1. IR spectra of VS-2(a), K-VS-2(b) and H-VS-2(c).

at 974 and 976 cm^{-1} are assignable to Si-O-H and that the peak at 978 cm^{-1} is assignable to Si-O-K as depicted in Figure 3.

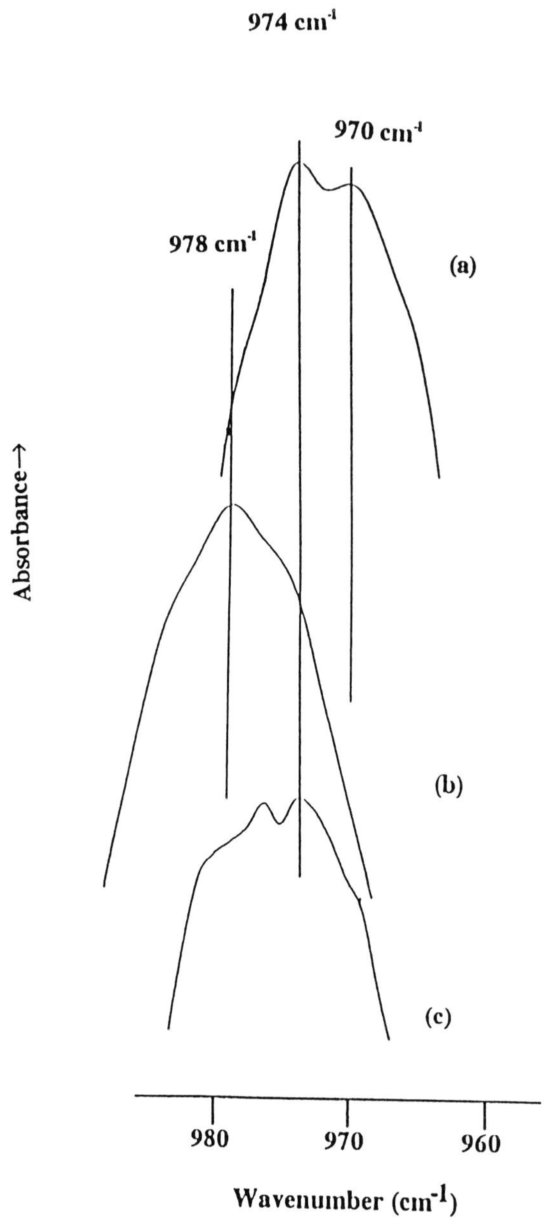

Figure 2. IR spectrum around 960 cm^{-1} of VS-2(a), K-VS-2(b) and H-VS-2(c).

Table 2 compares the results of oxidation of hexane over TS-2 and VS-2. The oxidation of the terminal carbon did not proceed on TS-2 in contrast to on VS-2, where oxidation of the terminal carbon took place. The turnover number on VS-2 without acetonitrile was much lower than that on TS-2 and the difference in turnover

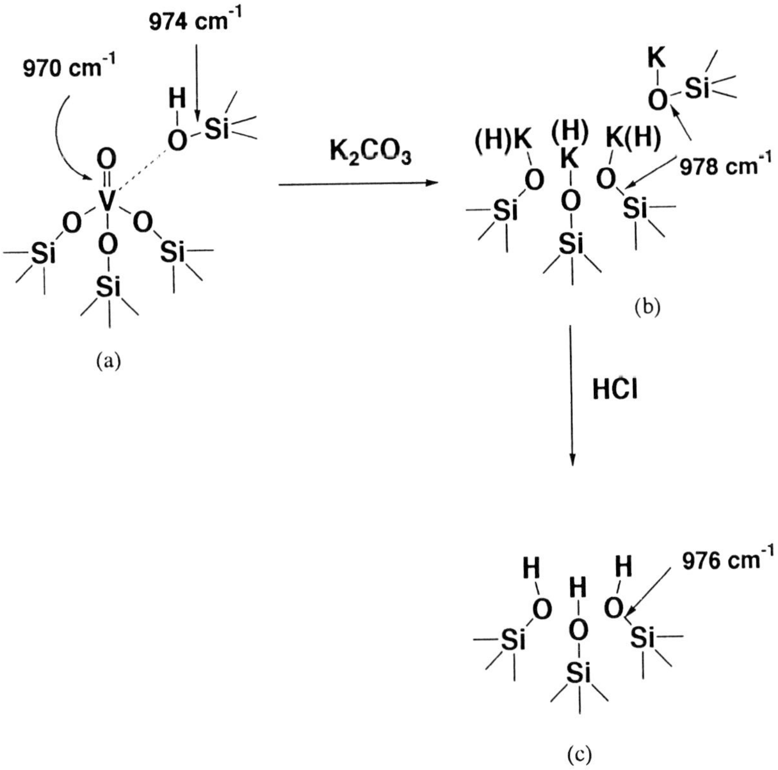

Figure 3. Structures of VS-2(a), K-VS-2(b) and H-VS-2(c).

number between after 2 h and after 3 h was only 0.3. The turnover frequency in this case decreased with increasing reaction time. When acetonitrile was used as a solvent, however, the turnover number after 2 h was about four times higher than that in the absence of acetonitrile and furthermore the turnover number increased in proportion to reaction time. These results suggest that acetonitrile might mitigate the leaching of vanadium out of zeolite framework during the oxidation reaction. As described above, the treatment of VS-2 with aqueous H_2O_2 in the absence of hexane or acetonitrile gave rise to VS-2-H_2O_2 with extremely low V content, which negligible activity.

Figure 4 shows the ESR spectra of the PBN adduct of radical species evolved in the reaction of TS-2, VS-2 and V-impregnated silicalite-2 with H_2O_2 in the presence of hexane. In all three spectra, the coupling constants are $A_N = 12.9$ G and $A_H = 1.89$ G. Although the formation of active •OH species is also expected, the A_N smaller than 15 G indicates that the trapped radical was •OOH species [15]. The concentration of •OOH radical generated in the VS-2-H_2O_2-hexane system was much higher than in the TS-2- and V-imp.S.-2- H_2O_2-hexane system.

Table 2
Oxidation of hexane over TS-2, VS-2 and V-imp.S-2.

Catalyst	Reaction time (h)	Turnover number (mol/mol-Ti or V)	Turnover frequency (mol/(mol*h))	Selectivity (%)	
				Terminal oxidation	Internal oxidation
TS-2	2	15	-	0.0	100
VS-2[a)]	2	1.1	0.55	31	69
	3	1.4	0.47	32	68
VS-2[b)]	2	4.4	2.2	32	68
	3	6.3	2.1	29	71
	4	8.9	2.2	23	77
V-imp.S-2	2	0.003	-	-	-

TS-2(Si/Ti = 85) 50 mg, H_2O_2 (30% aq.) 2.5 ml, substrate 2.5 ml, 323 K.
VS-2[a)](Si/V = 58) 50 mg, H_2O_2 (30% aq.) 1.0 ml, substrate 3.9 ml, 333 K.
VS-2[b)](Si/V = 58) 50 mg, H_2O_2 (30% aq.) 1.0 ml, substrate 3.9 ml, acetonitrile 12.5 ml, 333 K.
V-imp.S-2 (Si/V = 58) 50 mg, H_2O_2 (30% aq.) 1.0 ml, substrate 3.9 ml, 333 K, 2h.

The presumable scheme of hexane oxidation over VS-2 is presented in Figure 5. It is conceivable that •OH and •OOH radials are derived from the reaction of V-O-O• with H_2O_2. Highly active •OH radical would be able to attack both terminal and internal carbons of alkane. The fact that the spin adduct of •OH species was not observed can be accounted for by the very high activity and too short life of •OH radical. This mechanism results in the consumption of two moles of H_2O_2 per one mole of oxidized products. This is consistent with the low H_2O_2 selectivity with vanadium silicate catalysts compared to metallosilicates containing Ti or Sn [16].

Table 3 shows the results of oxidation of hexane and cyclohexane over VS-2. It is expected that the conversion of cyclohexane over VS-2 is lower than that of hexane because it is difficult for cyclohexane to enter the pore channels of the 10-membered oxygen ring. This is the case for TS-2, which showed remarkable shape selectivities in the oxidation of alkanes and cycloalkanes. However, with VS-2 the turnover number of cyclohexane was about three times higher than that of hexane. This suggests that the oxidation of cyclohexane mainly proceeded outside of VS-2 by •OH radical derived from H_2O_2 and V-O-O• generated inside of VS-2. We speculate that there is another mechanism of oxidation: the direct reaction of V-O-O• inside the zeolite crystals with alkane leading to the formation of V-O-O-H and alkyl radical (R•) followed by production of ROH and regeneration of vanadyl

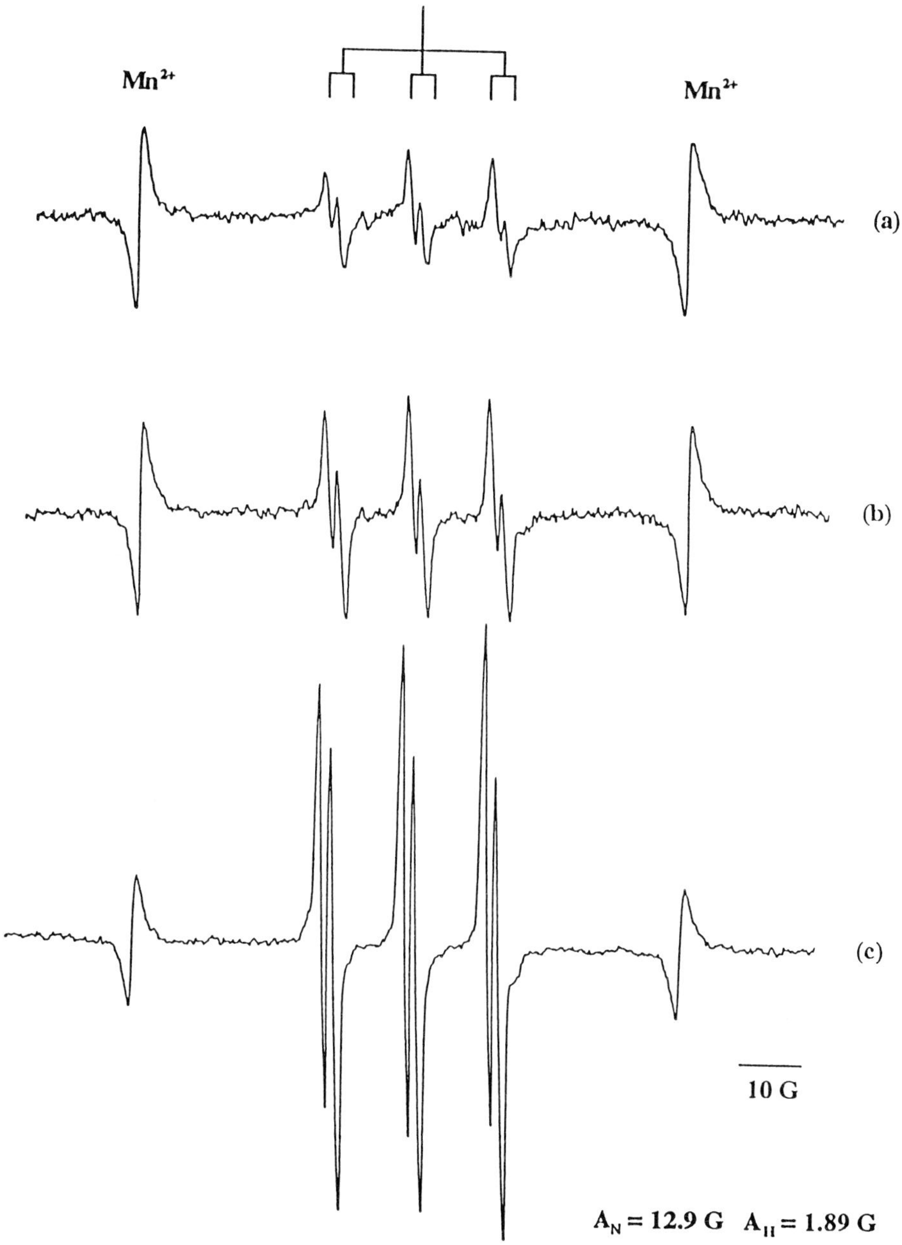

Figure 4. ESR spectra of spin adduct of PBN in the reaction of TS-2(a), V-imp.S.-2(b) and VS-2(c) with hexane and aqueous H_2O_2.

species. However, the high reactivity of cyclohexane compared to hexane suggests that this route plays a minor role.

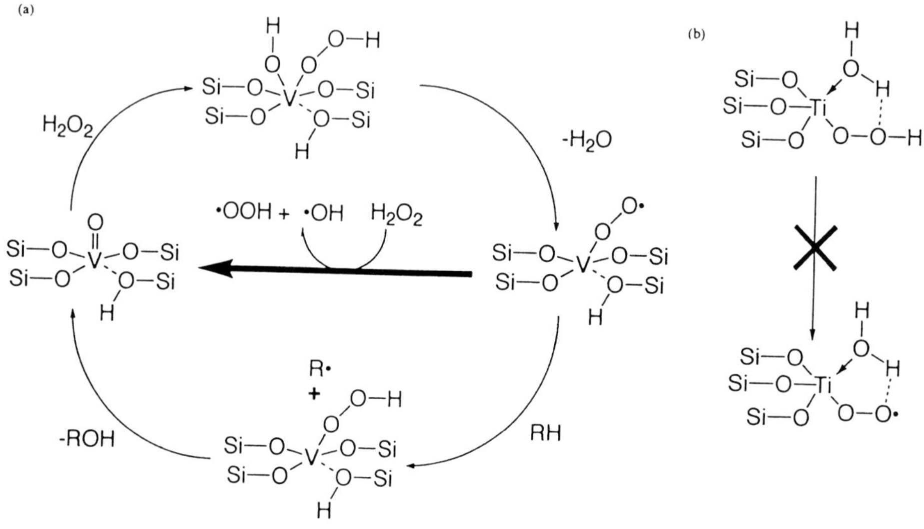

Figure 5. Scheme of hexane oxidation over VS-2 (a) and intermediate of TS-1 with H_2O_2 (b).

Table 3
Oxidation of hexane and cyclohexane over TS-2 and VS-2.

Catalyst	Turnover number (mol/mol-Ti or V)		L/C ratio[a]
	Hexane	Cyclohexane	
TS-2	15	0.34	44
VS-2	8.9	28	0.32

TS-2(Si/Ti = 85) 50 mg, H_2O_2 (30% aq.) 2.5 ml, substrate 2.5 ml, 323 K, 2h.
VS-2(Si/V = 58) 50 mg, H_2O_2 (30% aq.) 1.0 ml, substrate 3.9 ml for Hexane and 3.0 ml for cyclohexane (hexane/H_2O_2 molar ratio = 3), Acetonitrile 12.5 ml, 333 K, 4h.
[a](linear hydrocarbon conversion)/(cyclo hydrocarbon conversion) ratio.

The lower concentration of •OOH radical species with TS-2 than with VS-2 might be also due to the relatively high redox potential of the Ti^{4+}/Ti^{3+} couple compared to that of the V^{5+}/V^{4+} couple.

$$(\equiv SiO)_3 Ti^{4+}OOH \rightleftarrows (\equiv SiO)_3 Ti^{3+} + \bullet OOH$$

$$(HO)(\equiv SiO)_3V^{5+}OOH \rightleftarrows (HO)(\equiv SiO)_3V^{4+} + \bullet OOH$$

On TS-2 the oxidation of hexane may proceed by a mechanism involving a metal coordinated OOH group [17]. Ti-O-O• could not be generated from the Ti-O-O-H group (Figure 5(b)) and hence the formation of primary alcohols and aldehydes could not proceed. As shown in Table 2, V-imp.S-2 hardly promoted the oxidation of hexane. This may be due to the absence of V=O species which leads to the formation of V-O-O• by way of a metal coordinated OOH group. On VS-2, the oxidation of internal carbons might proceed not only by •OH radical but also by the metal coordinated OOH group inside the zeolite. However, this route might not be major because the oxidation of internal carbon of cyclohexane proceeded very quickly, as described above.

Vanadium easily leached from VS-2 during the oxidation reaction using H_2O_2 as described above. Figure 6 shows the ESR spectrum of radical adduct of

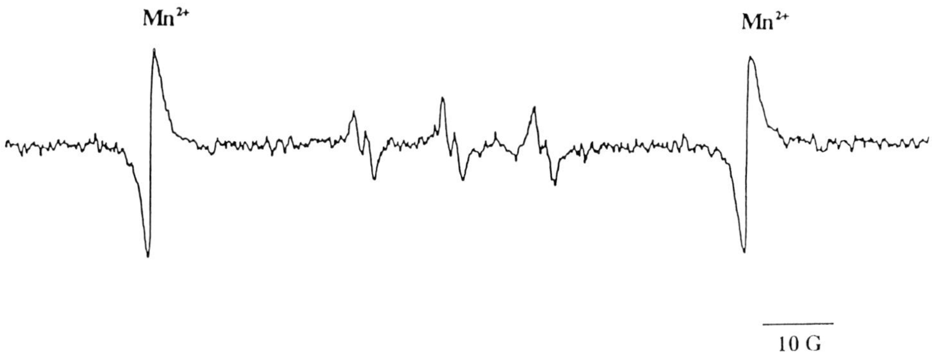

Figure 6. ESR spectrum of spin adduct of PBN in the reaction of dissoluble vanadium leached from VS-2 with hexane and aqueous H_2O_2.

PBN in the reaction of hexane and aqueous H_2O_2 in the presence of dissolved vanadium leached from VS-2. It can be seen that the dissolved vanadium could not generate •OOH radicals effectively. In agreement with this, the dissolved vanadium showed no activity for oxidation of hexane. Thus vanadium incorporated into zeolite framework was indispensable for the alkane oxidation, whereas V-impregnated silicalite-2 was virtually inactive in this oxidation.

CONCLUSIONS

The oxidation of alkanes catalyzed by VS-2 proceeded mainly by way of the attack of •OH radical derived from the reaction of V species incorporated into zeolite framework with H_2O_2.

Acknowledgements
We thank Prof. M. Anpo and H. Yamashita for measuring photoluminescence spectra of TS-2 and VS-2.

REFERENCES

1. B. Notari, *Stud. Surf. Sci. Catal.* **37**, 133 (1987).
2. A.V. Ramaswamy, S. Sivasanker, and P. Ratnasamy, *Microporous Mater.* **2**, 451 (1994).
3. P.R. Hari Prasada Rao, A.V. Ramaswamy, and P. Ratnasamy, *J. Catal.* **141**, 604 (1993).
4. T. Tatsumi, Y Hirasawa, and J. Tsuchiya, *Heterogeneous Hydrocarbon Oxidation*, American Chemical Society, Washington DC, 1996, p. 374.
5. A. Thangaraj, R. Kumar, S.P. Mirajkar, and P. Ratnasamy, *J. Catal.* **130**, 1 (1991).
6. J.S. Reddy, R. Kumar, and P. Ratnasamy, *Appl. Catal.* **57**, L1 (1990).
7. P.R. Hari Prasada Rao, A.V. Ramaswamy, and P. Ratnasamy, *J. Catal.* **137**, 225 (1992).
8. M.R. Boccuti, K.M. Rao, A. Zecchina, G. Leofanti, and G. Petrini, *Stud. Surf. Sci. Catal.* **48**, 133 (1989).
9. M.A. Camblor, A. Corma, and J. P rez-Pariente, *J. Chem. Soc. Chem. Commun.* 557 (1993).
10. G. Deo, A.M. Turek, I.E. Wachs, D.R.C. Huybrechts and P.A. Jacobs, *Zeolites* **13**, 365 (1993).,
11. S. Bordiga, S. Coluccia, C. Lamberti, L. Marcese and A. Zecchina, *J. Phys. Chem.* **98** 4125 (1994).,
12. C.B. Khouw and M.E. Davis, *J. Catal.* **151**, 77 (1995).
13. M. Anpo, S.G. Zanf, H. Yamashita, *Stud. Surf. Sci. Catal.* **101**, 941(1996). ; M. Anpo, S. Zanf, H. Yamashita, Y. Hirasawa, and T. Tatsumi, *Abstract of the 7th Japan-China-USA Synposium on Catalysis*, 1992, p. 55.
14. T. Tatsumi, Y. Watanabe, and K.A. Koyano, *Chem. Commun.* 2281 (1996).
15. G.R. Buettner, *Free Radical Biology & Medicine* **3**, 259 (1987)
16. N.K. Mal and A.V. Ramaswamy, *Appl. Catal. A* **143**, 75 (1996)
17. A. Corma, P. Esteve, and A. Martínez, *J. Catal.* **161**, 11 (1996)

Res. Chem. Intermed., Vol. 24, No. 5, pp. 541-550 (1998)

OXIDE NANOPARTICLES WITHIN A HOST MICROPOROUS MATRIX: POLYNUCLEAR COPPER SPECIES IN Cu-ZM5 AND THEIR ROLE IN THE REDUCTION OF NO

G. CENTI,[1,2] F. FAZZINI[2] and A. GALLI[2]
[1]*Dept. Industrial Chemistry, Univ. Messina, Salita Sperone 31, 98166 Messina, Italy*
[2]*Dept. Ind. Chem. and Materials, V.le Risorgimento 4, 40136 Bologna, Italy*

Received 16 September 1997; accepted 25 November 1997

Abstract--The role of copper-oxide nanoparticles inside a host ZSM5 zeolite, identified by UV-Visible-NIR diffuse reflectance spectroscopy on the reactivity in NO reduction to N_2 in the presence of propane/O_2 or ammonia/O_2 is discussed. These species lead to an enhancement in the rate of alkane depletion in the reduction of NO in the presence of propane/O_2, whereas they cooperate with isolated ions at exchangeable sites in determining the overall catalytic behaviour in NO reduction in the presence of ammonia/O_2. Also the stability of these species depends on the type of reductant.

INTRODUCTION

Increasing attention has been given in recent years to the study of the characteristics and reactivity of microporous materials containing transition metals with redox character in light of the possibility to have well defined isolated ions anchored at specific sites (mainly Brønsted acid sites) of the host matrix [1]. Much less attention instead has been given to the analysis of the nature, characteristics and especially reactivity of small oxide particles within the pores of the structured host microporous material, even though some evidence indicates that in several cases polynuclear oxide species form together with isolated transition metal ions. For example, in copper containing zeolites active for NO conversion to N_2, there is evidence indicating the formation of bi- or poly-nuclear copper oxide species together with isolated copper species [2,3], but their nature and catalytic role has not been analyzed in detail.

It is known that oxide nanocrystals often posses quite different and peculiar reactivity properties in comparison with the corresponding bulk oxide, because changes are found in the crystalline habits, morpholopy, bulk oxygen transport properties, electronic properties, etc. Although various methodologies have been developed to synthetize these oxide nanocrystals [4], it is usually very difficult to control the characteristics of the nanocrystals obtained and especially to avoid recrystallization during catalytic runs. Both these problems can be overcome in principle when these oxide crystals are stabilized inside a solid ordered matrix such as zeolites which control both the growth and crystalline habit of the guest nanoxide. Furthermore, the presence of a strong electrostatic field inside the zeolite channels, the possibility of synergetic cooperation between isolated transition metal ions and oxide nanoparticles and the possible presence of shape selectivity effects are further reasons indicating that the catalytic behaviour of oxide nanoparticles within the cavities of host microporous materials can be significantly different from those of the same oxide, either in the form of bulk materials or supported over conventional oxide materials.

As a part of a more general interest for the investigation of the catalytic properties of oxide nanoparticles within a host microporous matrix, we report in this work a study on the role of copper oxide nanoparticles within a host ZSM-5 type zeolite on the catalytic behaviour and stability in the reduction of NO in the presence of O_2 using two types of reductants: (1) propane and (ii) ammonia. The formation of copper oxide nanoparticles depends on several factors, such as (i) the Si/Al ratio, (ii) the exchange level and (iii) the method of addition of copper to host zeolite. It is well known that the catalytic behaviour of Cu/ZSM-5 catalysts in NO reduction depends on the ion exchange level and Si/Al ratio [3] and thus the catalytic role of copper oxide nanoparticles cannot be clearly evidenced when these factors are changed. For this reason, in this work different Cu/ZSM-5 samples were prepared changing the method of addition of copper, but using the same ZSM-5 host zeolite and adjusting the conditions of preparation as to have a nearly equivalent amount of copper in the various samples and thus an analogous "formal" level of ion exchange. However, as a consequence of the modality of preparation, a different distribution of copper species was obtained for the various samples.

The UV-Visible-NIR diffuse reflectance technique was used to characterize the various copper species, although this technique does not allows one to obtain their detailed structural determination, because other techniques of characterization such as XRD, IR, XPS, EPR, do not allow clear evidence on the presence of the copper oxide nanoparticles especially to be obtained.

EXPERIMENTAL SECTION

Sample Preparation and Characterization

Cu/ZSM-5 samples were prepared using the same batch of microcrystalline high silica ZSM-5 ($SiO_2/Al_2O_3 = 90$, sodium form) synthesized according to US Patent 4112056 (1978). Copper was introduced by an ion exchange method using a dilute Cu-acetate solution and variable temperature and pH of the solution as indicated in Table 1. The solution concentration was adjusted to have, after 24 h of exchange

Table 1
Code name and main characteristics of the Cu/ZSM-5 samples prepared.

Code	% wt CuO	Exchange Level[a], %	Temp. of the ion exchange, °C	pH of the ion exchange solution
CuZ5-*a*	1.93	131	25	5
CuZ5-*b*	1.82	124	25	7
CuZ5-*c*	1.70	115	50	5

[a]See text in the experimental part for definition of the exchange level.

under continuous mixing and reflux, an amount of copper deposited of around 1.7-1.9% wt. (as CuO) corresponding to an exchange level of around 115-130% (100% corresponds to a 0.5 ratio between Cu^{2+} and Al^{3+} ions in the zeolite). After ion exchange, the samples were washed, dried at 120°C and calcined up to 550°C using a slow temperature increase (about 13.5 h from room to final temperature). Complete elimination of Na in the samples after ion exchange and the level of Cu^{2+} introduced were determined by atomic absorption spectroscopy. The crystallinity of the samples and absence of formation of CuO particles were checked by infrared spectroscopy and X-ray diffraction analysis. The same methods were used to check the stability of the samples after the catalytic tests. Samples were checked also by scanning electron microscopy before and after the catalytic tests to verify the absence of CuO deposition on the outer surface of the zeolite crystals.

UV-Vis-NIR diffuse reflectance spectra were recorded in the 170-2400 nm range using a Perkin-Elmer Lamda 19 instrument and a $BaSO_4$ coated integrating sphere. Reflectance (R) values were converted using the Kubelka-Munk function: $F(R) = (1-R)^2/2R_\infty$. Spectra were recorded in air. Spectra in vacuum were also recorded, but are not reported here for brevity, because they do not give relevant information regarding the questions discussed in this work.

Catalytic Tests

Catalytic tests for NO reduction on Cu/ZSM-5 samples were made in a flow quartz reactor with on-line continuous analysis of products and feed composition using a mass quadrupole detector as described in detail elsewhere [5]. Tests were made using 0.3 g of sample (30-60 mesh), a space-velocity of 15,500 h^{-1} and the following feed compositions: (a) NO reduction with propane/O_2; NO = 0.1%, C_3H_8 = 0.2%, O_2 = 3%, remainder He; (b) NO reduction with NH_3/O_2; NO = 0.1%, NH_3 = 0.1%, O_2 = 3%, remaining He. The following procedure was used for the catalytic tests: (1) the calcined sample was loaded into the reactor and the catalytic behaviour in the 150-500°C temperature range analyzed for increasing reaction temperatures; for each temperature a stationary activity was reached usually in 0.5-1.0 h; (2) the catalyst was cooled down to room temperature in an inert flow; (3) the catalytic behaviour was tested again at increasing temperatures; (4) after cooling in an inert flow, the catalyst was discharged from the reactor and characterized. If the catalytic data of the 2nd run differed significantly from those of the 1st run, the procedure was repeated until reproducible behaviour was obtained.

RESULTS AND DISCUSSION

The role of copper oxide nanoparticles within a host ZSM-5 zeolite on the catalytic reactivity of Cu/ZSM-5 samples in the selective conversion of NO to N_2 was studied preparing by ionic exchange a series of samples (i) using the same host ZSM-5 and (ii) containing nearly the same amount of copper (around 2% wt.), but (iii) modifying the nature and relative amount of copper species by a change in the pH and temperature during the ion exchange procedure (Table 1). Due to the known dependance of the catalytic behaviour of Cu/ZSM-5 samples from the Si/Al ratio and the exchange level, clear indication of the role of copper oxide nanoparticles within a host ZSM-5 cannot be derived when these parameters are also varied. Therefore, in this work it was chosen to maintain these parameters constant, but vary the nature and distribution of the copper species changing the method of copper addition to host zeolite, although a limited number of samples can be prepared in this way.

The catalytic behaviour of Cu/ZSM-5 samples was studied using two different types of selective reductants for NO: (i) propane and (ii) ammonia. Furthermore, due to the possible changes of copper species during the catalytic tests, the reactivity of the samples was analyzed for the catalysts "as prepared" and after "in situ" transformation up to obtain a constant catalytic activity. The change in the distribution of copper species during "in situ" modifications and the parallel change in the catalytic behaviour was used as an additional way to modify the characteristics of copper species in Cu/ZSM-5 samples without changing the global exchange level, in order to more clearly derive the catalytic role of copper oxide nanoparticles within a host zeolite matrix.

Characterization of Cu-ZSM5 Samples by UV-Vis-NIR Diffuse Reflectance

The analysis by UV-Vis-NIR diffuse reflectance (UVN-DR) spectra of these samples (Figure 1) shows the considerable influence of the procedure of ion exchange on the type of copper species present in the Cu/ZSM-5 samples. The UVN-DR spectra are characterized by two intense bands at about 210 and 254 nm which can be attributed to low energy charge transfer (LCT) $Cu^{2+} \leftarrow O$ bands of copper ions in a highly distorted octahedral environment [6]. The zeolite oxygen

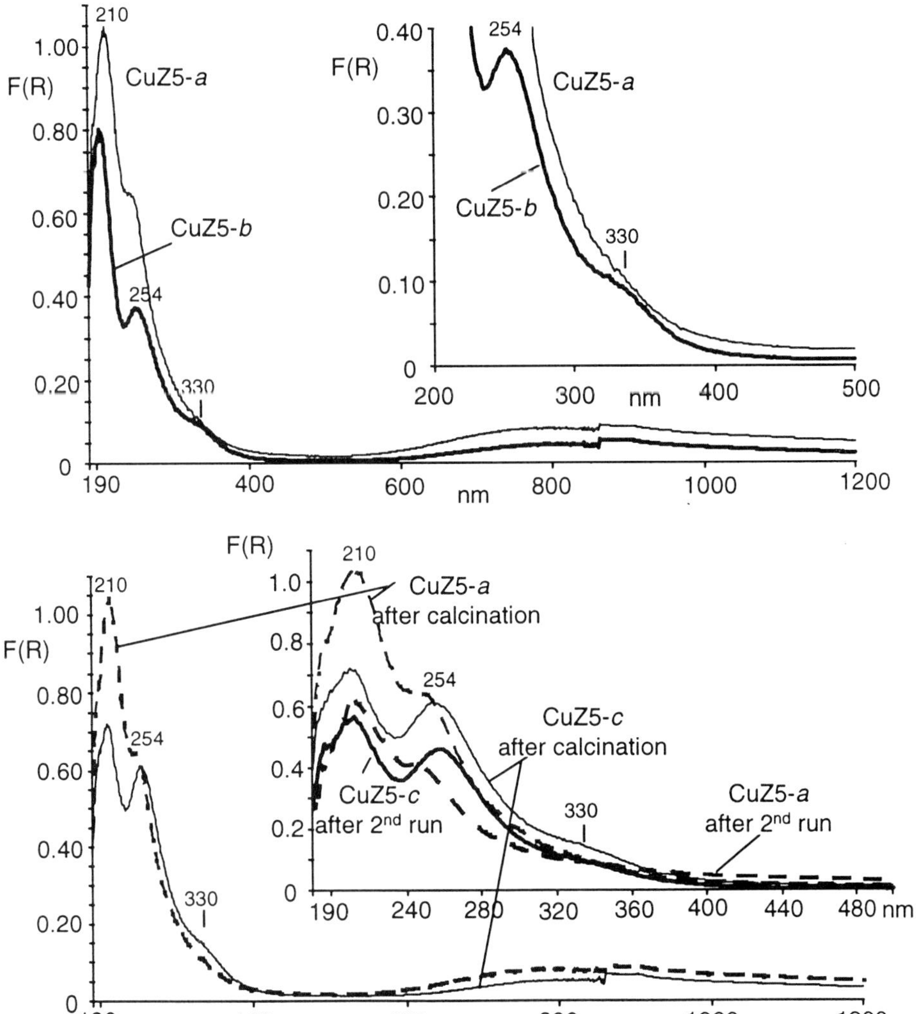

Figure 1. UV-Visible-NIR diffuse reflectance spectra of CuZ5-*a* and CuZ5-*b* after calcination (upper spectra; in the inset, expansion of the region between 200 and 500 nm), and CuZ5-*a* and CuZ5-*c* after calcination (lower spectra; in the inset, expansion of the region between 190 and 500 nm of the same samples after calcination and after 2nd runs of catalytic tests).

environment around copper ions at the possible exchange sites in ZSM-5 is not symmetrical [7] (two short and two longer O •••Cu bonds are present), and thus for a single site two LCT $Cu^{2+} \leftarrow O$ bands are expected. On the other hand, preferential coordination sites for Cu^{2+} ions do not exist in ZSM-5 structure differently from other zeolites such as mordenite [7], because the coordination environment is relatively similar. However, a slight modification is expected in position and especially in relative intensity of the two bands at about 210 nm with respect to the band at 254 nm as a function of the localization of copper ions. These two bands therefore indicate a family of isolated distorted octahedral Cu^{2+} species anchored at exchangeable sites of ZSM-5, but an increase in the ratio of the intensities of the bands at 210 and 254 nm is indicative of an increased fraction of sites in more open positions at the zeolite cage walls [7]. The presence of isolated distorted octahedral Cu^{2+} sites is confirmed by a broad, but low intensity band centred near 780 nm, due to *d-d* transitions.

Together with these bands, a shoulder centred at 330 nm is also clearly visible. This band falls at energies too low for oxygen to copper charge-transfer transitions, neither could it be attributed to the presence of CuO crystalline particles, for which copper ions are in nearly tetrahedral coordination (corresponding *d-d* transitions above about 1000 nm were not observed). Furthermore, UVN-DR spectra of mechanical mixtures of CuO and ZSM-5 confirm the possibility of noting the species of CuO particles, when formed, and their different spectra. It is interesting to note that a band at about 350 nm is observed in dimeric Cu^{2+}-acetate and not in the analogous monomeric form. This band was attributed to copper to copper electron hopping [8]. It is thus reasonable to attribute the band at 330 nm to electron hopping in polynuclear copper-oxide rows deposited inside the zeolitic channels. In this species the coordination environment of copper ions is different from that in CuO and the presence of a nearly bi-dimensional chain favours the delocalized hopping of electrons responsible for the band at about 330 nm, not possible in crystalline or amorphous CuO. This suggested structure for the polynuclear copper-oxide type species is confirmed by the fact that when copper hydroxide is precipitated over zeolite particles, a broad tail from about 250 to 1000 nm is instead observed. This species rapidly sinters during the catalytic reaction of NO reduction with propane/O_2, differently from the species characterized by the band at 330 nm.

Based on the above attributions, it is possible to conclude that UVN-DR spectra indicate that although the amount of copper in the three Cu/ZSM-5 samples analyzed is similar (Table 1), a different distribution of species is present. Together with isolated Cu^{2+} species in distorted octahedral coordination (probably more than one species and in different amounts in the three zeolites as indicated by the change in the relative intensities of 210 to 254 nm bands), a polynuclear copper-oxide type species (tentatively a nearly bidimensional copper-oxide chain) is also present. The amount of this species decreases in the CuZ5-*b* ≈ CuZ5-*c* > CuZ5-*a* series.

Catalytic Behaviour in the Reduction of NO to N_2

All these Cu-ZSM5 samples were active in the reduction of NO in the presence of O_2 and propane or ammonia as reductants, but the nature of the active species influences their reactivity. In order to compare their catalytic reactivity, a reference temperature of 350°C for tests using propane and 250°C using ammonia was chosen. This reaction temperature corresponds to a conversion of reactants in the 40-85% range for all samples and thus allows a good comparison without the typical errors when tests are made at too low or too high conversion. A plug flow type reactor model and a first order reaction rate were used to derive the values of the rate constants. In the range of concentrations used in these tests, it has been shown that this simplified model correctly describes the kinetic behaviour [3,5]. The results of this comparison of the catalytic behaviour are reported in Figure 2 (Figure 2a for NO reduction with propane and O_2; Figure 2b for NO reduction with NH_3 and O_2) as a plot of first order rate constant of NO conversion to N_2 versus first order rate constant of reductant depletion (propane or ammonia). The oxygen concentration is instead in excess and does not convert significantly. The results refer to the behaviour observed in the 1st and 2nd runs of catalytic tests (see experimental). While in the case of propane the catalytic behaviour does not change significantly from the first to the second run (apart from an improvement in catalytic behaviour for CuZ5-*c*), a considerable change is observed in the case of ammonia as reductant. In Figure 2b the arrows indicate the direction of the change which is different for the three samples: CuZ5-*a* shows an activation (increase in the rate of NO depletion) form the first to the second run, whereas the other two samples deactivate. Also UVN-DR spectra after tests using ammonia indicate a change in the distribution of copper species in the three Cu-ZSM5 samples, whereas modifications in the spectra are only minor after tests using propane.

Role of Copper Oxide Nanoparticles on the Catalytic Behaviour

The data reported in Figure 2 together with UVN-DR spectra (Figures 1 and 3, the latter figure refers to the change in the spectra after tests with ammonia) provide indications of the role of copper-oxide type species inside zeolitic channels on the catalytic behaviour of Cu-ZSM5 samples in NO reduction. Using propane as the reductant (Figure 2a), comparison of the results of CuZ5-*a* and CuZ5-*b* samples is interesting. These two samples show an analogous rate of NO reduction to N_2, but a considerable difference in the rate of propane depletion. The comparison of UVN-DR spectra (Figure 1, upper spectra) clearly shows the presence in the sample CuZ5-*b* of a large amount of polynuclear copper-oxide species (band at 330 nm) and as a consequence less intense bands of isolated octahedral Cu^{2+} species (region 200-300 nm for LCT $Cu^{2+} \leftarrow O$ bands and 600-1200 nm for *d-d* transitions). This suggests that the species with the band at 330 nm is responsible for the enhanced rate of propane conversion.

This species is also present in CuZ5-*c* samples, but it was found that these samples activate during the catalytic reaction both in terms of rate of NO depletion

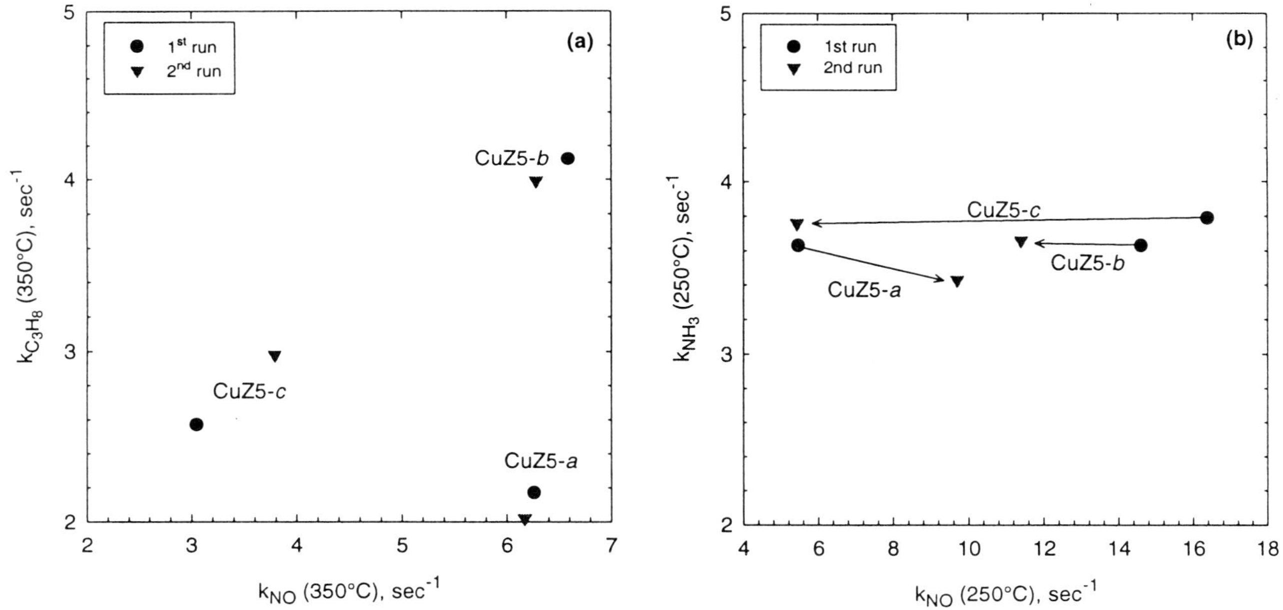

Figure 2. Comparison of the first order rate constant of NO reduction to N_2 versus first order rate constant of reductant depletion (propane, Figure 2a; ammonia, Figure 2b) on Cu-ZSM5 samples during first and second runs of tests of catalytic reactivity. Data refer to a reaction temperature of 350 °C (a) or 250 °C (b). See experimental part for conditions of the tests.

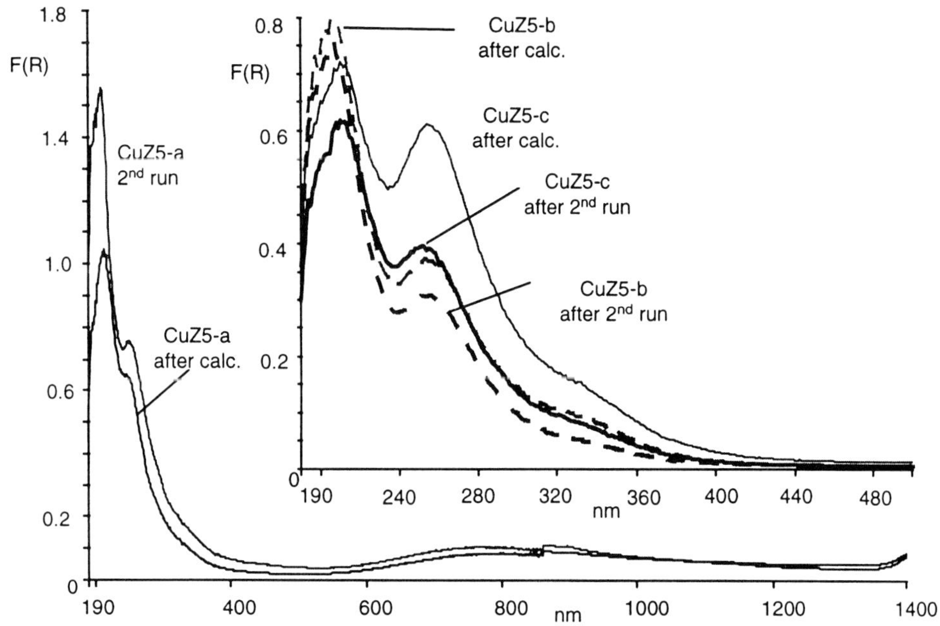

Figure 3. UV-Visible-NIR diffuse reflectance spectra of CuZ5-*a* before and after catalytic tests of NO reduction with ammonia and oxygen. In the inset, corresponding spectra for CuZ5-*b* and CuZ5-*c*.

and rate of propane depletion. After the 2nd run of the catalytic tests, the intensity of the band at 330 nm in CuZ5-*c* decreases, but the band does not completely disappear. At the same time, a broad low intensity band is detectable near 1400 nm indicating the partial formation of microcystalline CuO particles not detectable by XRD or SEM analyses. Probably due to the higher temperature of ion exchange, in this case deposition of copper occurs mainly near the outer surface of crystals, and thus during the catalytic reaction part of the copper may crystallize forming small CuO particles which grow on the outer surface of the zeolite crystallites, and part of this deposited copper redisperses throughout the whole zeolite crystal. This explains the increase in catalytic behaviour of CuZ5-*c*, as compared with CuZ5-a and CuZ5-b, while the different relative intensities of the bands at 210 and 254 nm (indicative of a different localization of isolated copper ions) explain the lower activity in NO conversion.

Changes in both reactivity and UVN-DR spectra are more evident in the case of ammonia as the reductant (Figure 2b and Figure 3). Samples CuZ5-*b* and CuZ5-*c* both deactivate considerably during the catalytic reaction. The rate of NO depletion decreases to about one third in the case of CuZ5-*c*, whereas the change is less marked for CuZ5-*b*. The analysis of the spectra after catalytic reaction using ammonia as the reductant (inset in Figure 3) clearly indicates considerable reduction of the intensity of the band at 330 nm after the catalytic tests (especially in CuZ5-*c*), whereas there is no marked change in the ratio of the bands at 210 and 254 nm. The absolute intensity of these bands before and after the catalytic tests is instead not indicative, because part of the isolated Cu^{2+} ions reduced during the catalytic reaction.

In fact, the intensity of these bands increases with oxidative treatment, but that of the band at 330 nm does not, indicating an effective transformation of the latter species and not only a reversible reduction.

Sample CuZ5-*a*, differently from the other two, activates during the catalytic reaction (Figure 2b). In this sample the intensity of the band at 330 nm increases slightly, but the ratio of the intensities of the bands at 210 and 254 nm increases considerably. As discussed above, we generally observe an increase in the rate of NO depletion when this ratio increases, not only in these, but also in other Cu-ZSM5 samples with different characteristics [9].

It may be concluded that whereas the presence of small polynuclear copper-oxide type particles within the pores of host zeolite leads mainly to an enhancement in the rate of alkane depletion in the reduction of NO in the presence of propane/O_2, the cooperation between isolated copper ions and these polynuclear species is instead responsible for the catalytic behaviour in NO reduction in the presence of ammonia/O_2 on the same catalysts. It is also shown that not only the presence of these species, but their stability during the catalytic reaction should be considered and that the stability of these species depends considerably on the modality of preparation.

Finally, it can be noted that data on propane combustion indicate that zeolites containing small and stable polynuclear copper-oxide type species can be used for the synthesis of highly active materials for hydrocarbon combustion, in substitution for noble-metal based catalysts.

REFERENCES

1. J.N. Armor. In: *Science and Technology in Catalysis 1994*, Y. Izumi, H. Arai, and M. Iwamoto (Eds.), Elsevier Science Publishers, Amsterdam, 1995, p. 51.
2. J. Sarkany, J.L. d'Itri, and W.M.H. Sachtler, *Catal. Letters* **16**, 241 (1992).
3. G. Centi and S. Perathoner, *Appl. Catal. A: General* **132**, 179 (1995).
4. W.R. Moser (Ed.), *Advanced Catalysts and Nanostructured Materials - Modern Synthetic Methods*, Academic Press San Diego, CA, 1996.
5. G. Centi, S. Perathoner, and L. dall'Olio, *Appl. Catal. B: Environmental* **7**, 359 (1996).
6. N.S. Hush and R.J.M. Hobbs. In: *Progress in Inorganic Chemistry*, F.A. Cotton (Ed.), Interscience Publishers, New York, Vol. 10, 1968, p. 259.
7. R.A. Schoneydt, *Catal. Rev.-Sci. Eng.* **35**, 129 (1993).
8. L.S. Foster and C.J. Ballhausen, *Acta Chem. Scand.* **16**, 1385 (1962).
9. A. Galli, Doctoral Thesis, Univ. Bologna, 1997.

Res. Chem. Intermed., Vol. 24, No. 5, pp. 551-560 (1998)

HYDROCRAKING OF DIBENZOTHIOPHENE OVER METAL-LOADED ZEOLITES CATALYST

KAZUYOSHI KANEDA, TAKEMA WADA, SHIN-ICHI NAKAGAWA, SATORU MURATA, TERUO SUZUKA, and MASAKATSU NOMURA*

Department of Applied Chemistry, Faculty of Engineering, Osaka University, 2-1 Yamada-oka, Suita, Osaka 565, Japan

Received 17 September; accepted 10 November 1997

INTRODUCTION

Sulfur oxides produced in the combustion of gas oil are known to cause air pollution. The reduction of sulfur content in gas oil is an important subject to solve these issue. It has already been found that several percent of sulfur is contained in gas oil and more than sixty kinds of sulfur-containing organic compounds exist. Dibenzothiophene (DBT) especially is believed to be one of the tough materials to be eliminated and many researchers have been studying the hydrodesulfurization (HDS) reaction of DBT [1]. In the petroleum industry, Ni-Mo/Al_2O_3 or Co-Mo/Al_2O_3 are usually employed as HDS catalysts for gas oil [2,3]. These catalysts are not sufficient for deep HDS reaction. Therefore, the development of more active catalysts is expected.

We have found that, for the hydrocracking reaction of polycyclic aromatic compounds, palladium- and nickel- co-loaded Y-type zeolite catalyst (Pd-Ni-Y) showed higher activity than the palladium- or nickel-loaded one (Pd-Y, Ni-Y). In our recent study, we applied the Pd-Ni-Y catalyst for hydrocracking reaction of DBT and found that DBT could be effectively hydrocracked and simultaneously desulfurized. We also obtained an interesting result, namely that the catalytic activity directed to HDS of DBT was improved by means of loading of palladium and nickel on Y-type zeolite in a reverse order. We also investigated the HDS reaction using ternary metal-loaded Y-type zeolites, that is, adding Mo or La as a third component to these binary metal-loaded ones. In this report, the details of HDS of DBT using these catalysts and characterization of the catalysts using FT/IR are described.

EXPERIMENTAL SECTION

Samples

NH_4-substituted Y-type zeolite (No. 23650) was purchased from Catalysts and Chemicals Industries Co. Ltd., Japan. The other reagents employed in this study were commercially available and purified by conventional distillation or recrystallization before use.

Preparation of the Metal-loaded Zeolites

(Ni-Y, Pd-Y, Pd-Ni-Y, Ni-Pd-Y). The NH_4-substituted Y-type zeolite (50 g) was stirred in 1000 mL of an aqueous solution of $Ni(NO_3)_2$ (0.25 M) at 90 °C for 24 h, then filtered and dried at 110 °C to obtain the nickel cation substituted zeolite (Ni-Y). For the Pd supported one (Pd-Y), NH_4-substituted Y-type zeolite (15 g) was treated in aqueous solution (200 mL) of $[Pd(NH_3)_4](NO_3)_2$ (0.025 M) at 40 °C for 24 h. For the palladium and nickel co-loaded one (Pd-Ni-Y), NH_4-substituted Y-type zeolite was treated with an aqueous $[Pd(NH_3)_4](NO_3)_2$ solution (0.025 M), which was filtered and dried, and then, treated with an aqueous $Ni(NO_3)_2$ solution (0.25 M). For the the Pd supported Ni-Y (Ni-Pd-Y), NH_4-substituted Y-type zeolite was treated with an aqueous $Ni(NO_3)_2$ solution (0.25 M), which was filtered and dried, and then, treated with an aqueous $[Pd(NH_3)_4](NO_3)_2$ solution (0.025 M). The resulting cation exchanged zeolites were calcined in a stream of air (52 mL/min) at 450 °C for 4 h, then submitted to reduction with an H_2 flow (52 mL/min) at 450 °C for 1 h. The contents of nickel and palladium in each zeolite were determined by using a Rigaku Denki System 3270 type fluorescence X-ray analyzer. Amounts of metal loading were as follows: Ni 3.7 wt% (Ni-Y); Pd 5.5 wt% (Pd-Y); Pd 3.6 wt% and Ni 3.3 wt% (Pd-Ni-Y); Ni 3.0 wt% and Pd 3.8 wt% (Ni-Pd-Y).

(Pd-Ni-Mo-Y, Ni-Pd-Mo-Y, Pd-Ni-La-Y, Ni-Pd-La-Y). Ternary-metal-loaded Y-type zeolites were prepared by means of impregnating Pd-Ni-Y or Ni-Pd-Y in aqueous $(NH_4)_6Mo_7O_{24}$ or aqueous $La(NO_3)_3$ solutions at 50 °C for 1 h. Impregnated zeolites were calcined in a stream of air (52 mL/min) at 450 °C for 4 h, and then submitted to reduction with an H_2 flow (52 mL/min) at 450 °C for 1 h. The amount of molybdenum or lanthanum in the ternary catalyst was about 3 wt%.

Hydrocracking Reaction of Dibenzothiophenes

The substrate (1 g) and the catalyst (0.5 g) were placed into a 70 mL SUS 316 autoclave, which was pressurized to 70 kg/cm^2 with hydrogen, and then heated up to 320 °C at a rate of 8 °C/min. Reaction time here means the duration at a desired temperature. The gaseous products were passed through an aqueous iodine solution (1 mol/L, 20 mL) to recover the hydrogen sulfide produced. The collected gaseous products from which hydrogen sulfide was eliminated were analyzed quantitatively by a Shimadzu GC-3BT (active carbon column, 2 m) and a Shimadzu GC-8AIT (silica gel column, 60/80 mesh, 3 m) gas chromatographs. The resulting iodine solution was diluted to 200 mL by deionized water, 20 mL of which was submitted to titration with sodium thiosulfate solution (0.1 mol/L) by using aqueous starch solution as an indicator. The liquid and solid products were recovered by washing the inside of the autoclave with CH_2Cl_2. The components of liquid products (CH_2Cl_2 soluble fraction) were identified using a JEOL JMS-DX-303HF type GC-MS, and quantified using a Shimadzu GC-14APFSC (CBP-1 capillary column, i.d. 0.5 mm x 25 m) gas chromatograph after adding an appropriate internal standard (usually dicyclohexyl phthalate). Coke was defined as the carbonaceous materials remaining on the surface or inside of the pores of the catalyst after washing with dichloromethane. The yield was calculated by taking the carbon weight from the microanalysis of the catalyst recovered into consideration.

Measurement of FT/IR Spectra of Pyridine-sorbed Zeolites

After calcination and reduction of each zeolite catalyst, it was placed in a glass vessel. Then, a pyridine-saturated nitrogen stream was passed through the glass tube for 3 min, being kept for 1 h. Then, the apparatus was heated at 150 °C (bath temperature) under nitrogen stream to remove the excess amount of pyridine. The resulting zeolite (5 mg) was mixed with potassium bromide (500 mg), ground in an agate mortar, extruded to a disk (10 mm diameter), and submitted to FT/IR measurement. The spectra were recorded on a Shimadzu FTIR-8100M spectrometer and processed with the computer equipped with it.

RESULTS AND DISCUSSION

Hydrocraking of Dibenzothiophenes

Three kinds of metal-loaded zeolites, such as nickel-loaded (Ni-Y), palladium-loaded (Pd-Y), and palladium- and nickel-co-loaded Y-type zeolite (Pd-Ni-Y), were examined as a hydrocracking catalyst of DBT, the results being shown in Figure 1. The reaction was conducted at 300 °C under 70 kg/cm^2 of hydrogen pressure for 1 h.

With Pd-Y as a catalyst, about 87% of DBT was converted to alkylbenzenes, alkylcyclohexanes, and hydrocarbon gases (ethane, propane, *i*-butane, and *n*-butane) along with H_2S. However, unreacted DBT and hydrogenated DBT derivatives, still

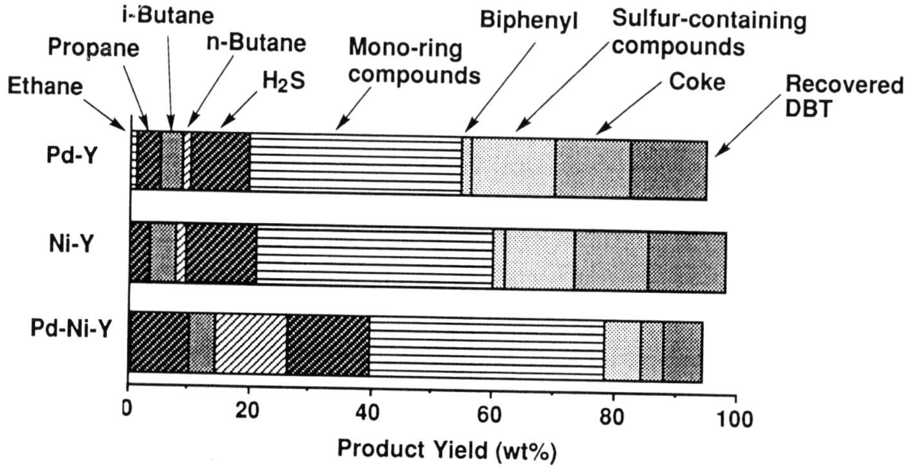

Figure 1. Hydrocracking reaction of DBT in the presence of metal-loaded zeolite catalysts. DBT 1.0 g, catalyst 0.5 g, H_2 70 kg/cm^2, 300 °C, 1 h.

remain (13 wt% and 12 wt%, respectively). Carbonaceous materials causing deterioration of activity corresponded to 12 wt% on the catalyst. The catalytic performance of the Ni-Y catalyst was similar to that of the Pd-Y catalyst. On the other hand, when using Pd-Ni-Y, the conversion of DBT increased up to 94 wt% and only 4 wt% of sulfur containing compounds were detected in the products extracted with dichloromethane. Formation of coke also considerably decreased to as low as 6 wt%. Yields of hydrogen sulfide were 13.3 wt%, these values corresponding to 76 wt% of the amount of sulfur contained in the starting DBT. In the case of these zeolite catalysts, biphenyl and cyclohexylbenzene were produced only in very small amounts, although they were predominantly obtained in the HDS reaction of DBT with Ni-Mo/Al_2O_3 and Co-Mo/Al_2O_3. These results suggest that the catalysts employed in this study have higher hydrocracking activity than the alumina supported catalyst. In particular, Pd-Ni-Y shows the highest activity among the three catalysts.

To optimize reaction conditions the effects of both reaction time and temperature on the extent of conversion were examined with the Pd-Ni-Y catalyst indicating the highest activity for HDS. The results are shown in Figures 2a and 2b. Either increase of the reaction time to 2 h or increase of the reaction temperature up to 320 °C, brought about more than 99 wt% conversion of DBT and trace amounts of sulfur-containing compounds were detected. In these reaction conditions, yields of hydrogen sulfide reached 13.8 wt% (at 320 °C for 1 h) and 15.1 wt% (at 300 °C for 2 h), these values corresponding to 80% and 89% of the amount of sulfur contained in the starting DBT, respectively. Moreover, coke formation was

acid sites corresponding to the C=C stretching in aromatic rings. Generally, Brønsted acid sites are thought to be important for hydrocracking of DBT. In Figure 5, relative intensity (Brønsted type) in FT/IR vs. conversion of DBT using seven kinds of metal-loaded zeolites was plotted in order to make clear the correlation between Brønsted acid sites with conversion of DBT. Conversion of DBT corresponded to the results of the reaction conducted at 300 °C. This Figure implies that Brønsted acid sites on zeolites play a very important role for the hydrocracking activity of DBT because a clear correlation between hydrocracking activities of catalysts and Brønsted acid sites is observed. On the other hand, compared with the number in parenthesis shown in Table 2, Lewis acid sites that those were believed to cause coking of catalysts did not show any obvious relationship to the yield of coke with each catalyst.

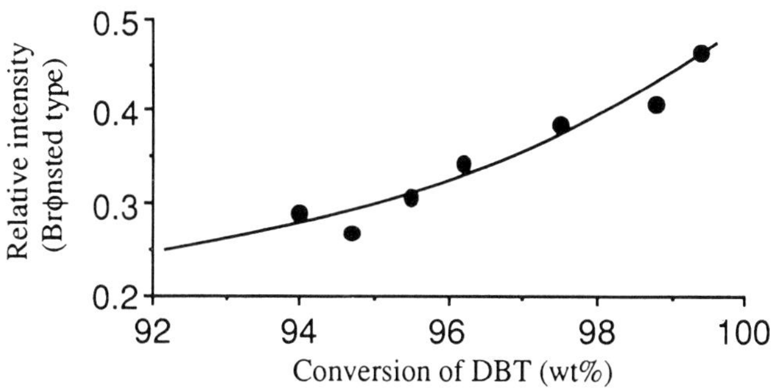

Figure 5. Plot of relative intensity (Brønsted type) in FT/IR *versus* conversion of DBT for seven kinds of metal-loaded zeolites.

CONCLUSIONS

- DBT could be almost completely converted over Pd-Ni-Y catalyst at 320 °C in 1 h or at 300 °C in 2 h.
- Catalytic activity for hydrocracking of DBT was improved by means of loading of palladium and nickel on Y-type zeolites in a reverse order. Moreover, the addition of La or Mo as a third component on the catalysts increased catalytic activity.
- In the case using mixture of *n*-dodecane and DBT as a model of gas oil, Pd-Ni-Y catalyst could selectively react with DBT.
- The results of FT/IR measurement for pyridine-absorbed zeolites indicated that the larger amount of Brønsted acid sites was closely related to the higher activity for HDS.

REFERENCES

1. X. Ma, *Ind. Eng. Chem. Res.* **34**, 748 (1995).
2. C. W. Curtis, J. H. Chen, and Y. Tang, *Energy Fuels* **9**, 195 (1995).
3. T. Kabe, W. Qian, and A. Ishihara, *J. Catal.* **149**, 171 (1994).
4. E. P. Parry, *J. Cat.* **2**, 371 (1963).
5. T. R. Hughes and H. M. White, *J. Phy. Chem.* **71**, 2192 (1967).

Res. Chem. Intermed., Vol. 24, No. 5, pp. 561-570 (1998)

EFFECT OF AUXILIARY CHEMICALS ON PREPARATION OF SILICA MCM-41

S. NAMBA* and A. MOCHIZUKI

Department of Materials, Teikyo University of Science & Technology, Uenohara-machi, Kitatsuru-gun, Yamanashi 409-01, Japan

Received 3 October 1997; accepted 21 November 1997

Abstract--Silica MCM-41 materials were prepared hydrothermally by using 1,3,5-trimethylbenzene (MS) or 1,3,5-triisopropylbenzene (TI) as an auxiliary chemical. The BJH pore size of MCM-41 increased up to 12 nm with increasing amounts of MS. However, MCM-41 materials prepared with MS displayed irregular pore arrangements and a half of these materials exhibited low thermal and hydrothermal stabilities. On the other hand, MCM-41 materials prepared with TI as an auxiliary chemical displayed regular pore arrangements and high thermal and hydrothermal stabilities, but their BJH pore sizes did not go over 4.0 nm (intrinsic value; *ca.* 5 nm). MCM-41 prepared with TI together with a small amount of MS (TI/MS/SiO_2 = 1.5/0.5/1) displayed regular pore arrangements and high thermal and hydrothermal stabilities. Its BJH pore size was 4.7 nm (intrinsic value; *ca.* 6 nm).

INTRODUCTION

Recently mesoporous silica molecular sieves have been developed [1-3]. One of them is called "MCM-41", which has a hexagonal structure, uniform one-dimensional mesopores and a very high surface area [2,3]. It has been also reported that pore size of MCM-41 materials can be tailored in the range of 1.6~10 nm through the choice of surfactant agents as templates and the addition of auxiliary chemicals [2,3]. However, larger pore materials prepared with 1,3,5-trimethylbenzene (MS) as an auxiliary chemical display XRD patterns exhibiting a single broad line in the extremely low angle region and also display somewhat irregular, yet essentially hexagonal, pore arrangements and a broad pore size distribution [2]. Such materials may exhibit lower thermal and hydrothermal stabilities.

Hue *et al.* have also reported the preparation of silica MCM-41 materials with MS as a auxiliary chemical [4]. They observed that the expansion of the pore size by addition of MS could be achieved only by introducing MS at low temperature and/or short reaction times before significant silicate condensation took place. The d_{100} value increased with increasing MS/SiO_2 ratios and reached almost the ceiling value at MS/SiO_2 = 1. The largest d_{100} value of their MCM-41 was *ca.*

7 nm. However, they did not mention the structural regularity of their MCM-41 materials.

Recently Ulagappan and Rao have reported the preparation of silica MCM-41 materials with various kinds of *n*-alkanes (C_5-C_{18}) [5]. The observed d_{100} value of MCM-41 materials increased with increasing carbon numbers of alkanes and was maximized at a carbon number of 14 (d_{100} = 7.4 nm). However, XRD spectra and TEM images suggest that their MCM-41 materials with wide pores may display irregular pore arrangements.

We have reported that highly ordered silica MCM-41 materials prepared without auxiliary chemicals exhibit high thermal and hydrothermal stabilities and are excellent adsorbents for recovery of organic solvent vapors by a pressure swing adsorption process [6].

In this study we aim to prepare larger pore MCM-41 materials with regular pore arrangements and to clarify the relationship between regularity of pore arrangements and thermal and hydrothermal stabilities.

EXPERIMENTAL

MCM-41 materials with larger pore sizes were synthesized hydrothermally at an initial pH value of 10.0 and at 373 K for 10 days by using waterglass, sulfuric acid, water, hexadecyltrimethylammonium bromide (C_{16}TMABr) as template, and MS or TI as auxiliary chemicals. The molar gel composition for hydrothermal synthesis was SiO_2•0.31 Na_2O•0.14-0.21 H_2SO_4•0.51 C_{16}TMABr•0-6.00 MS (or 0-6.00 TI)•75-95 H_2O.

The synthesis procedure was as follows. C_{16}TMABr (50.3 g) was dissolved in pure water (150.7 g) at 323 K. A certain amount of auxiliary chemical was added to the resulting solution. The mixture was vigorously stirred (Solution 1). Waterglass (SiO_2;28.9 wt%, Na_2O;9.3 wt%) was added dropwise to pure water (120 g) containing sulfuric acid (3.6 g) which had been added with stirring (Solution 2). Solution 2 was added to Solution 1. The mixture was stirred vigorously for 30 min and then pure water (60 g) was added to the mixture with stirring. The pH value of the mixture was adjusted to 10.0 with diluted sulfuric acid. The resulting suspension was poured into a polytetrafluoroethylene vessel and allowed to stand for 10 days at 373 K. The solid product was washed with large amounts of warm water (500 ml x 6 times), dried at 373 K, heated to 813 K in air at a rate of 1K/min, and then calcined in air at 813 K for 6 h. The calcined samples were characterized by TEM, XRD and N_2 adsorption for determination of specific surface area and pore size distribution.

The MCM-41 materials were also calcined in air at 973-1273 K for 2 h to examine their thermal stability and treated in a flowing steam (0.5 atm, balance; He) at 923 and 1073 K for 2 h to examine their hydrothermal stability. Samples treated were characterized by XRD and N_2 adsorption to determine their surface area and pore size distribution.

RESULTS AND DISCUSSION

MCM-41 Materials Prepared with 1,3,5-trimethylbenzene (MS)

MCM-41 materials were prepared with various amounts of MS. These materials are denoted in MCM-41-MS(x), where x is a MS/SiO_2 ratio. Surface areas and BJH pore sizes of MCM-41-MS and a "blank" MCM-41 prepared without auxiliary chemicals (MCM-41-16) are given in Table 1. The surface area of every MCM-41 sample was very high. The BJH pore size of MCM-41 prepared with MS increased up to 12 nm with increasing MS/SiO_2 ratios. The BJH pore size distributions of MCM-41-16 and MCM-41-MS loaded with various amounts of MS are shown in Figure 1. The pore size distribution of every MCM-41-MS was broad, compared with that of MCM-41-16.

XRD spectra of MCM-41-16 and three MCM-41-MS are shown in Figure 2. All the MCM-41-MS exhibited a single broad line in the extremely low angle region, as observed by Beck *et al.* [3]. The θ_{100} value seems to decrease with increasing MS/SiO_2 ratios. These XRD spectra and the pore size distribution shown in Figure 1 indicate irregular pore arrangements of MCM-41-MS materials. Moreover little structural regularity was observed on TEM images of MCM-41-MS.

From these results it is apparent that in spite of an obvious pore size increase, the structural regularity of MCM-41 was reduced by the addition of MS.

MCM-41 Materials Prepared with 1,3,5-triisopropylbenzene (TI)

MCM-41 materials were loaded with various amounts of TI. XRD spectra of these MCM-41-TI together with the parent MCM-41-16 are shown in Figure 3. The θ_{100} values decreased with the addition of TI, indicating pore enlargement. However the amount of TI added little influenced the θ_{100} values. Every spectrum exhibited the typical four peaks attributable to hexagonal structure. From these results it is suggested that the MCM-41-TI materials remain highly ordered and may have almost the same pore size.

The pore size distribution of MCM-41-TI was measured. Every MCM-41-TI studied here exhibited a similar BJH pore size of *ca.* 4 nm (Figure 4) and a similar pore size distribution. The BJH pore size distributions of MCM-41-TI(2)and MCM-41-16 are shown in Figure 5. The MCM-41-TI(2) exhibited a sharp pore size distribution. Thus the pore size of MCM-41 prepared with TI did not extend over 4.0 nm, but pore size distributions and XRD spectra indicate that MCM-41-TI

Table 1
Effect of calcination and steaming on surface area and pore size distribution of various MCM-41 sample.

Sample	BJH Pore Size/nm	BJH Pore Size Distribution	Specifice Surface Area/m^2g^{-1}					
			Parent	Calcination/K			Steaming/K	
				1073	1173	1273	923	1073
MCM-41-16	3.0	Sharp	1020	973	926	478	984	858
MCM-41-MS (0.7)	3.7	Broad	1078	923	721	196	-	181
MCM-41-MS (1)	4.5	Broad	980	657	157	25	-	34
MCM-41-MS (1)*	4.5	Broad	1093	1021	957	755	-	744
MCM-41-MS (2)	7.7	Broad	1125	1024	1011	736	-	753
MCM-41-MS (2)*	7.0	Broad	1146	1059	966	691	-	857
MCM-41-MS (4)	8.0	Broad	1120	953	827	431	-	658
MCM-41-MS (6)	10.0	Broad	1000	427	25	3	-	10
MCM-41-MS (6)*	12.0	Broad	1000	435	79	7	357	-
MCM-41-TI (2)	4.0	Sharp	1073	1000	856	621	997	755
MCM-41-TI,MS (1.5,0.5)	4.7	Sharp	1101	1031	902	598	-	704

Each MCM-41-MS (X) sample was prepared in a similar manner to that for the corresponding MCM-41-MS(X) sample.

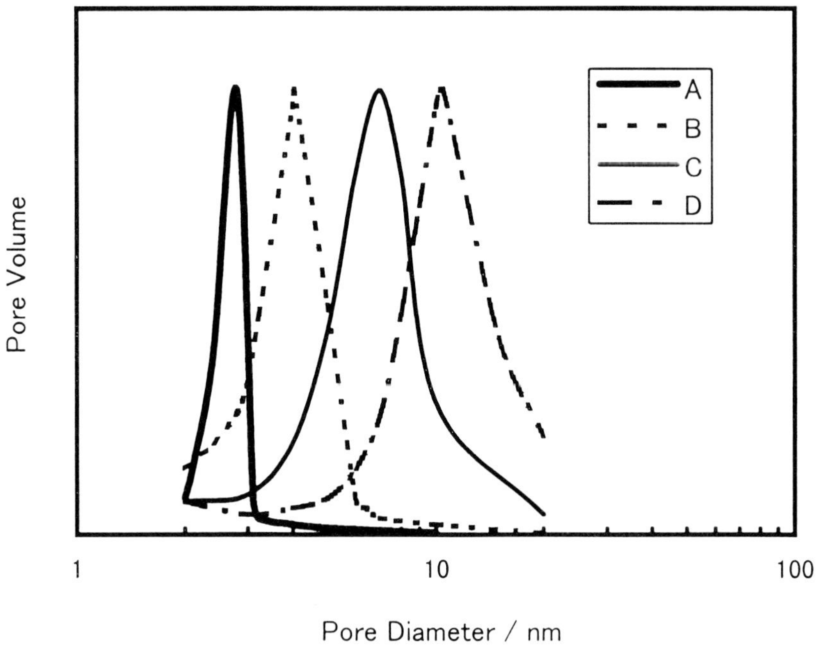

Figure 1. BJH pore size distribution of (A) MCM-41-16, (B) MCM-41-MS(1), (C) MCM-41-MS(2), and (D) MCM-41-MS(6).

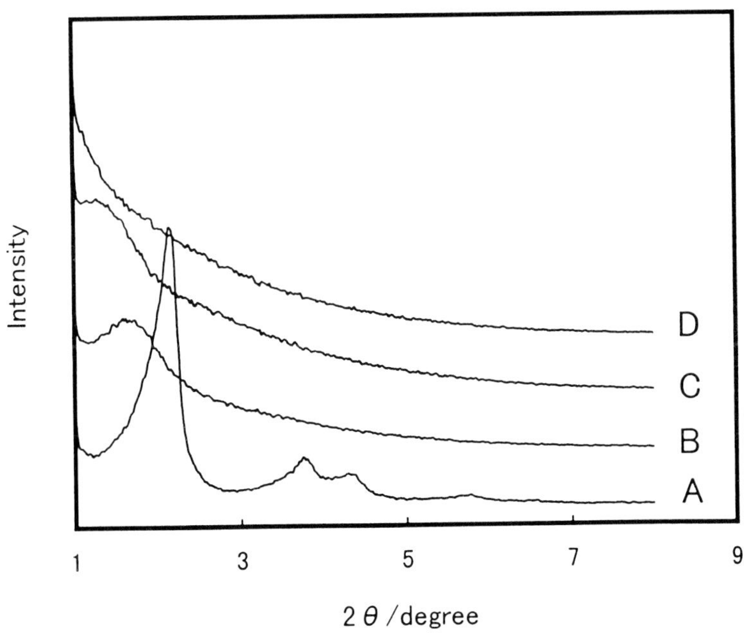

Figure 2. XRD spectra of (A) MCM-41-16, (B) MCM-41-MS(1), (C) MCM-41-MS(2), and (D) MCM-41-MS(6).

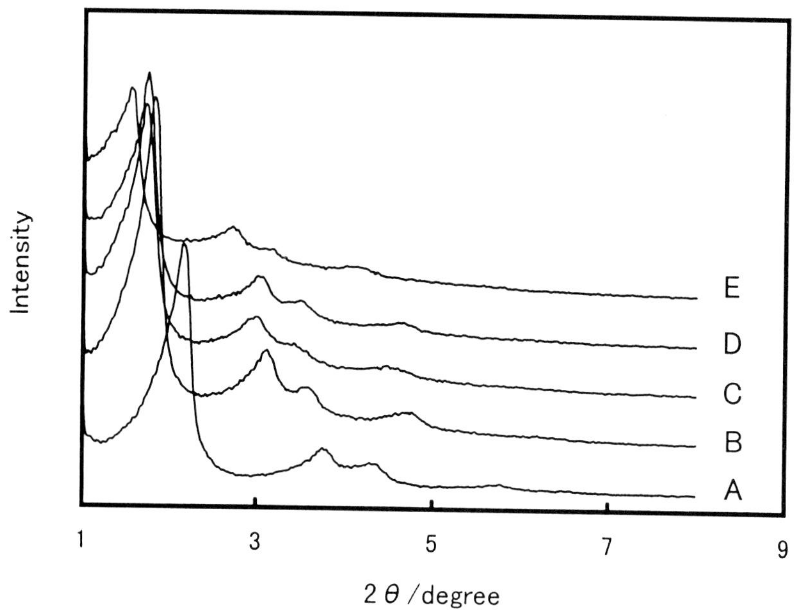

Figure 3. XRD spectra of (A) MCM-41-16, (B) MCM-41-TI(2), (C) MCM-41-TI(4), (D) MCM-41-TI(6), and MCM-41-TI,MS(1.5,0.5).

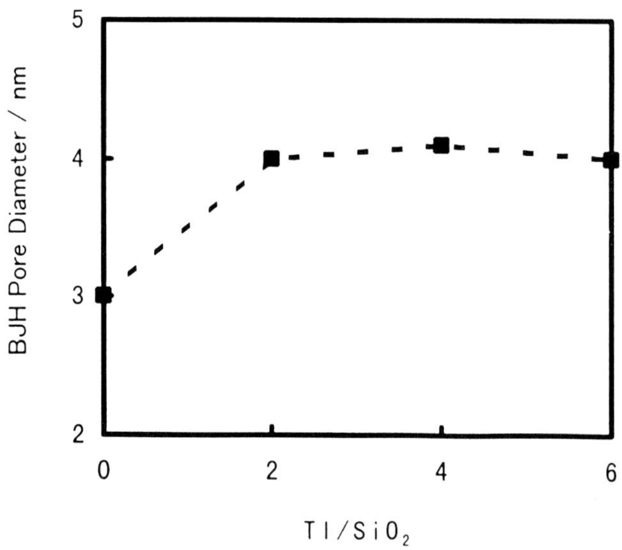

Figure 4. Relationship between BJH pore size and TI/SiO_2 for various MCM-41-TI samples.

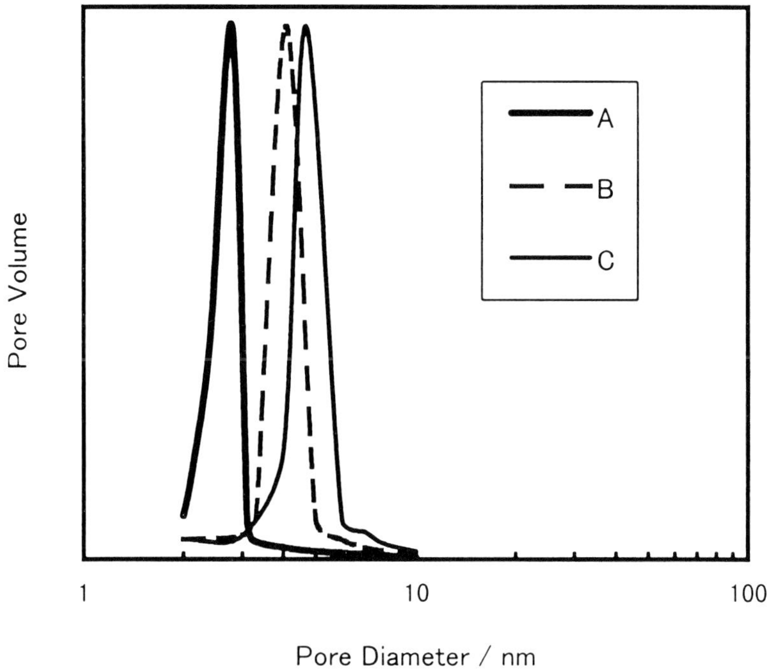

Figure 5. BJH pore size distribution of (A) MCM-41-16, MCM-41-TI(2), and MCM-41-TI,MS(1.5,0.5).

materials display regular pore arrangements.

MCM-41 was prepared by using both TI and MS as auxiliary chemicals. The maximum BJH pore size of MCM-41-TI,MS with regular pore arrangements was attained at TI/SiO_2 = 1.5 and MS/SiO_2 = 0.5 and was 4.7 nm. The highest ratio of MS/SiO_2 provided the widest pore, but the less regular pore arrangements.

The BJH pore size distribution of MCM-41-TI,MS(1.5,0.5) is shown in Figure 5. A sharp distribution was observed for MCM-41-TI,MS(1.5,0.5). The XRD spectrum of MCM-41-TI,MS(1.5,0.5) is shown in Figure 3. All the four peaks attributable to the hexagonal structure were observed, although the peak intensity was a little weaker, compared to MCM-41-TI materials.

TEM images of MCM-41-16, MCM-41-TI(2), and MCM-41-TI,MS(1.5,0.5) are shown in Figure 6. These images indicate regular pore arrangements not only for MCM-41-16 but also for MCM-41-TI(2) and MCM-41-TI,MS(1.5,0.5).

Schacht *et al.* have reported that the formal application of BJH method leads to an under estimation of the pore diameter of the MCM-41 materials [7]. The difference between the intrinsic pore diameter and the BJH pore diameter may be *ca.* 1 nm. Therefore the intrinsic pore size of MCM-41-16 should be of 4 nm. This

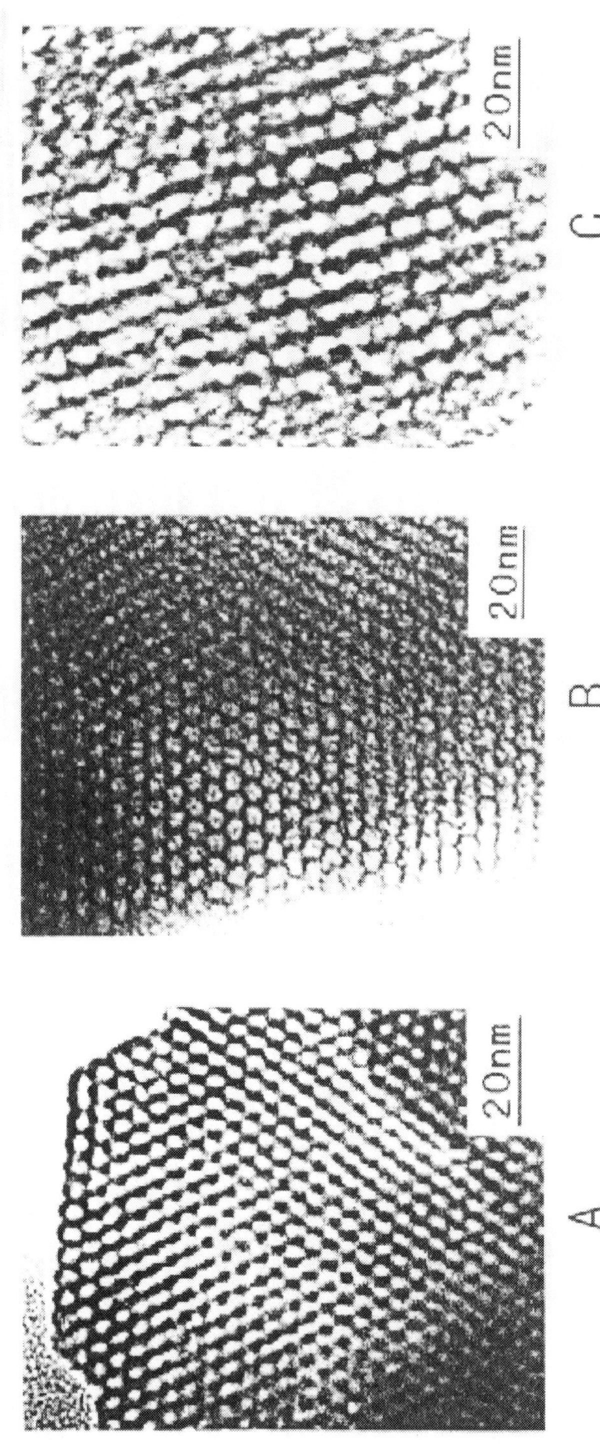

Figure 6. TEM images of (A) MCM-41-16, (B) MCM-41-TI(2), and (C) MCM-41-TI,MS(1.5,0.5).

value coincides with that reported by Beck *et al.* [3]. The addition of TI or TI with a small amount of MS enlarges the pore size without reducing the regularity of the pore arrangements. The intrinsic pore sizes of MCM-41-TI and MCM-41-TI,MS(1.5,0.5) are 5 nm and 6 nm, respectively.

Thermal and Hydrothermal Stabilities of MCM-41 Materials

The thermal and hydrothermal stabilities of MCM-41 materials were examined. The change in surface area by calcination or steaming is shown in Table 1. In the case of MCM-41-16 the surface area decreased to 90% of that of the parent material, upon calcination at 1173 K and to about 50% upon calcination at 1273 K. The 10% loss in surface area may be due to the shrinkage by the condensation reaction of hydroxyl groups in MCM-41-16. However by calcination at 1273 K the hexagonal structure destroyed partially. XRD data confirmed these results. The loss in surface area of MCM-41-16 by steaming at 923 K was very small and that by steaming at 1073 K was of only 15%. The other MCM-41 materials with regular pore arrangements exhibited similar thermal and hydrothermal stabilities as MCM-41-16, even though the pore sizes were enlarged. On the other hand, a half of eight MCM-41-MS materials with irregular pore arrangements exhibited lower thermal and hydrothermal stabilities. That is, MCM-41 materials with irregular pore arrangements do not always exhibit high thermal and hydrothermal stabilities. At the present time we cannot clarify what factor governs thermal and hydrothermal stabilities.

CONCLUSION

The BJH size of MCM-41 increased up to 12 nm with increasing amounts of MS. However, MCM-41 materials prepared with MS displayed irregular pore arrangements and some of them exhibited low thermal and hydrothermal stabilities. On the other hand, MCM-41 materials prepared with TI displayed regular pore arrangements and high thermal and hydrothermal stabilities, but their BJH pore sizes did not extend over 4.0 nm of BJH pore size (intrinsic value; *ca.* 5 nm). MCM-41 prepared with TI together with a small amount of MS ($TI/MS/SiO_2$ = 1.5/0.5/1) exhibited a highly ordered hexagonal structure and 4.7 nm of BJH pore size (intrinsic value; *ca.* 6 nm) and showed high thermal and hydrothermal stabilities.

Acknowledgements

This work has been carried out as a research project of The Japan Petroleum Institute commissioned by the Petroleum Energy Center with the support of the

Only a few studies have dealt with the use of organic media in place of water for hydrothermal method. Thus, synthesis of silica-sodalite in nonaqueous systems [24], synthesis of fine-particle-size transition alumina by the treatment of aluminum *sec*-butoxide in *sec*-butanol [25], and synthesis of group 8~10 metal powders by the treatment of metal hydroxide in glycol [26] have been reported.

We have explored the synthesis of inorganic materials in organic media at elevated temperatures (200-300°C) under autogenous pressure of the organics [27-37], and showed that α-alumina can be crystallized in glycol at 300°C, much lower temperature than the α-alumina formation temperature in hydrothermal reaction [28], and that microcrystalline double oxides such as yttrium aluminum garnet were directly formed when a stoichiometric mixture of two starting materials such as aluminum isopropoxide and yttrium acetate was heated in 1,4-butanediol at 300°C (glycothermal method) [29,30]. We also examined the reaction of aluminum and titanium alkoxides in inert organic solvents and found that these reactions gave thermally stable χ-alumina [31] and microcrystalline titania (anatase) [32], respectively.

In this paper, we have examined the reaction of zirconium alkoxides (or acetylacetonate) in various organic media, and found the formation of microcrystalline tetragonal zirconia which maintains a relatively large surface area even after calcination at 500°C.

EXPERIMENTAL SECTION

Zirconium *n*-propoxide (ZNP), isopropoxide (ZIP) or acetylacetonate (ZAA) (12.5 g) was added in 130 ml of an organic solvent (glycols or toluene) and the mixture (or solution) was placed in a test tube serving an autoclave liner. The test tube was then set in a 300 ml autoclave, and in the gap between the test tube and the autoclave wall, an additional 30 ml of the solvent was added. The autoclave was thoroughly purged with nitrogen, heated to a desired temperature at a rate of 2.5°C/min, and held at that temperature for 2 h. After the mixture was cooled, the resulting powders were washed repeatedly with methanol or acetone and air dried. The product was calcined for 1 h in a box furnace.

Powder X-ray diffraction (XRD) was measured on a Rigaku Geigerflex-2013 diffractometer using Ni-filtered CuKα radiation. Crystallite size of the product was calculated from the half-height width of the 111 diffraction peak of tetragonal zirconia using the Scherrer equation; the value of the shape factor, *K*, was arbitrarily taken to be 0.9. The specific surface area was calculated by the BET single-point method on the basis of the nitrogen uptake measured at 78 K on a Shimadzu ADS-1B sorptograph equipped with a gas-chromatograph (GC-3BT, Shimadzu). Morphologies of the particles were observed with a transmission electron microscope, Hitachi H-800, operated at 200 kV. Infrared (IR) spectra were obtained on a Shimadzu IR-430 spectrometer with the usual KBr-pellet technique. Simultaneous thermogravimetric and differential thermal analyses were performed on a Shimadzu TG-30 thermal analyzer at a heating rate of 10°C/min in a 40 ml/min-flow of dried air.

RESULTS AND DISCUSSION

The reaction of ZNP in 1,4-butanediol (1,4-BG) at 300°C yielded tetragonal zirconia (Fig. 1c), while the reaction of ZNP in ethylene glycol (EG) yielded poorly crystallized tetragonal zirconia (Fig. 1b). The crystallite size of the products increased in the following order (carbon number of glycol): 2 <6 <4 (Table 1).

In the previous papers, we examined the reaction of aluminum isopropoxide (AIP) in various glycols and found that the crystallite size of the product, the glycol derivative of boehmite, was affected by the carbon number of the glycols and increased in the following order [33, 34]:

$$HO(CH_2)_2OH < HO(CH_2)_3OH < HO(CH_2)_6OH < HO(CH_2)_4OH$$

These results suggested that the development of the product structure was controlled by the heterolytic cleavage of O-C bond of the glycoxide intermediate, >Zr-O-$(CH_2)_n$OH, formed by transesterification of zirconium alkoxide with the glycol. Because of the inductive effect of the intramolecular hydroxyl group, heterolytic cleavage of the O-C bonds of the glycoxide becomes more difficult with decreasing the carbon number of the glycol and therefore, the product obtained by the reaction in EG had the smallest crystallite size. The largest crystallite size, obtained by the use of 1,4-BG can be explained by facile cleavage of the O-C bonds due to the participation of the intramolecular hydroxyl group [38], yielding a zirconate ion and protonated tetrahydrofuran (Scheme 1). Tetrahydrofuran was actually detected by gas-chromatographic analysis of the supernatant after the reaction in 1,4-BG.

The reaction of ZIP in an inert organic solvent (toluene) at 300°C also yielded tetragonal zirconia (Fig. 1a). Mazdiyasni *et al.* [39, 40] reported that thermal decomposition of zirconium alkoxides in vapor phase at 325-400°C yielded zirconia which was essentially amorphous to X-ray. Therefore, the presence of the organic solvent molecules somehow facilitated the development of he crystal structure of the product.

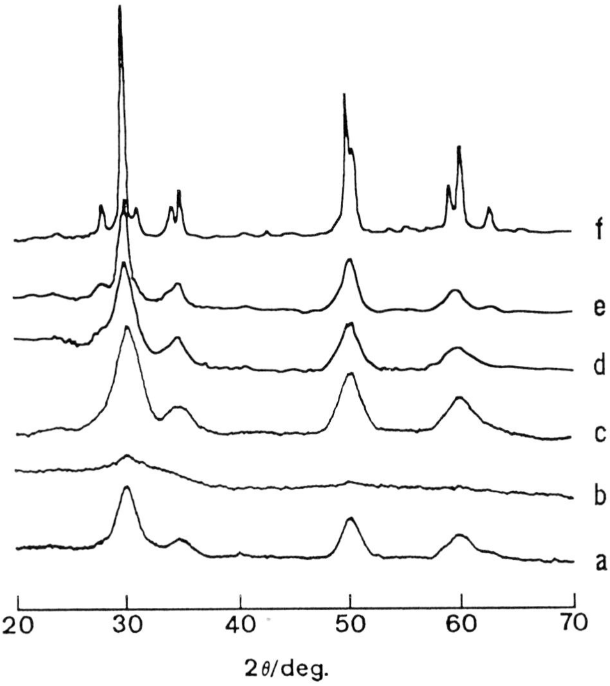

Figure 1. X-ray diffraction patterns of the products obtained by the reaction of: (a), zirconium isopropoxide in toluene; (b), zirconium *n*-propoxide in ethylene glycol; (c), zirconium *n*-propoxide in 1,4-butanediol; each at 300 °C for 2 h, and XRD patterns of the samples obtained by calcination of product (c) at: (d), 500 °C; (e), 700 °C; (f), 900 °C; for 1 h.

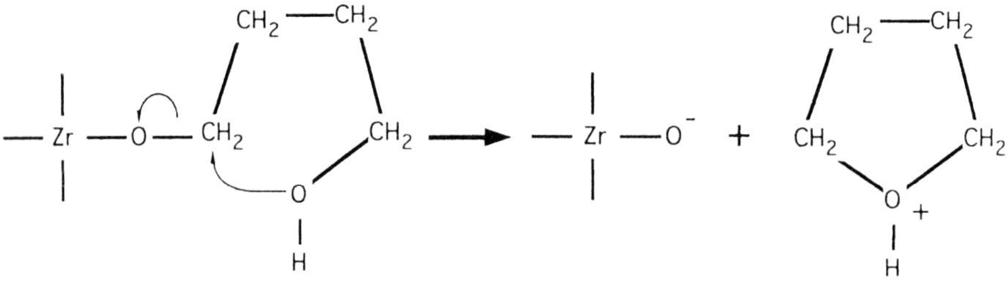

Scheme 1
Decomposition of the ω-hydroxybutyl group.

ZIP also decomposed at 265 °C, although the product was amorphous. On the other hand, ZNP did not decompose even at 300 °C and no solid product was obtained, while zirconium *tert*-butoxide decomposed at as low as 200 °C yielding amorphous zirconia. These results suggest that direct cleavage of the O-C bonds of zirconium alkoxides yielding alkyl cation and zirconate anion was the prime factor for the decomposition of zirconium alkoxides. Similar effect of the alkyl group of alkoxides was observed for the thermal decomposition of aluminum or titanium alkoxides in inert organic solvents [31,32].

Table 1
Typical properties of the products obtained by the thermal decomposition of zirconium alkoxides in organic media and of the calcined samples

Zirconium source[a]	Solvent	Temperature (°C)	Product		Sample calcined at 500°C		
			Phase[b]	d_{111}[c] (nm)	Phase[b]	d_{111}[c] (nm)	BET surface area (m^2/g)
ZNP	Ethylene glycol	300	T	*ca.* 1.5	T, (M)	8.2	91
ZNP	1,4-Butanediol	300	T	3.2	T, (M)	4.1	157
ZNP	1,6-Hexanediol	300	T	2.9	T, (M)	4.8	16
ZNP	Toluene	300	No reaction				
ZIP	1,4-Butanediol	300	T, (M)	5.4	T, (M)	5.8	141
ZIP	Toluene	300	T	4.3	T, (M)	6.9	92
ZIP	Toluene	265	Amorphous	-	T, M	6.6	107
ZIP	Toluene	250	No reaction				
ZTB	Toluene	200	Amorphous	-	T, (M)	10	68
ZAA	1,4-Butanediol	300	T	3.6	T, (M)	6.1	99
ZAA	Toluene	300	T	*ca.* 2.0	T	5.0	4

[a] ZNP; *n*-propoxide, ZIP; isopropoxide, ZTB; tert-butoxide, ZAA; acetylacetonate.
[b] T; tetragonal, M; monoclinic. Minor phases detected in the product are shown in parentheses.
[c] Crystallite size calculated from 111 diffraction peak of tetragonal zirconia.

It must be noted that the decomposition temperatures of aluminum, titanium, and zirconium alkoxides are quite similar to each other, when the alkyl moiety is the same. These results suggest that the decomposition rate of metal alkoxide in inert organic solvent is essentially determined by the structure of alkyl group, irrespective of the metal ions. In other words, metal alkoxides show the following general order of reactivity:

tertiary > secondary > primary

This argument leads to an idea that a homogeneous product would be formed when two alkoxides are decomposed in an inert organic solvent. In fact, thermal decomposition of a mixture of aluminum and silicon alkoxides yielded an amorphous product from which mullite was crystallized at 1000°C [35]. It is well known that the crystallization temperature of mullite is determined by the mixing state of aluminum and silicon atoms: From the homogeneous mixture, mullite crystallizes at 1000°C while heterogeneous mixing of aluminum and silicon atoms resulted in crystallization of mullite at 1300°C.

The reaction in 1,4-BG proceeds more easily than the reaction in the inert organic solvents. This is not due to the polarity of the solvent, but due to transesterification of alkoxide with glycol prior to the formation of zirconia. In fact, the IR spectra of the products obtained in glycols exhibited bands corresponding to liquid glycol and the tetragonal zirconia (480 cm^{-1}, broad), which suggests that the alkyl moieties of zirconium alkoxide are completely expelled from the product.

In the glycothermal reaction of aluminum alkoxide [34], however, the structure of the alkyl group of the alkoxide did not have an essential effect on the physical properties of the products, *i.e.*, glycol derivatives of boehmite. In the present reaction, however, this was not the case; crystallite size of the product increased in the following order (starting material): ZNP < ZAA < ZIP. Because ZAA and ZIP also decomposed in the inert organic solvent, direct decomposition of alkoxide moieties yielding zirconate ion seems to occur concurrently with the glycoxide route and facilitate the growth of the crystals. The fact that the products obtained in toluene always had smaller crystallite sizes than that obtained in 1,4-BG also supports this argument.

The TEM micrograph of the product obtained in 1,4-BG (Fig. 2a) indicated that the product was comprised of agglomerates of primary particles having an average diameters of 4 nm. This value is in good agreement with the crystallite size (3.2 nm) calculated by the line-broadening technique suggesting that each particle observed by TEM is a single crystal of zirconia.

Thermal analysis of the product obtained in 1,4-BG showed 20% weight loss from 200 to 500°C which is associated with a large, broad exothermic peak in DTA. Because physisorbed 1,4-BG can easily be detected by the weight decrease

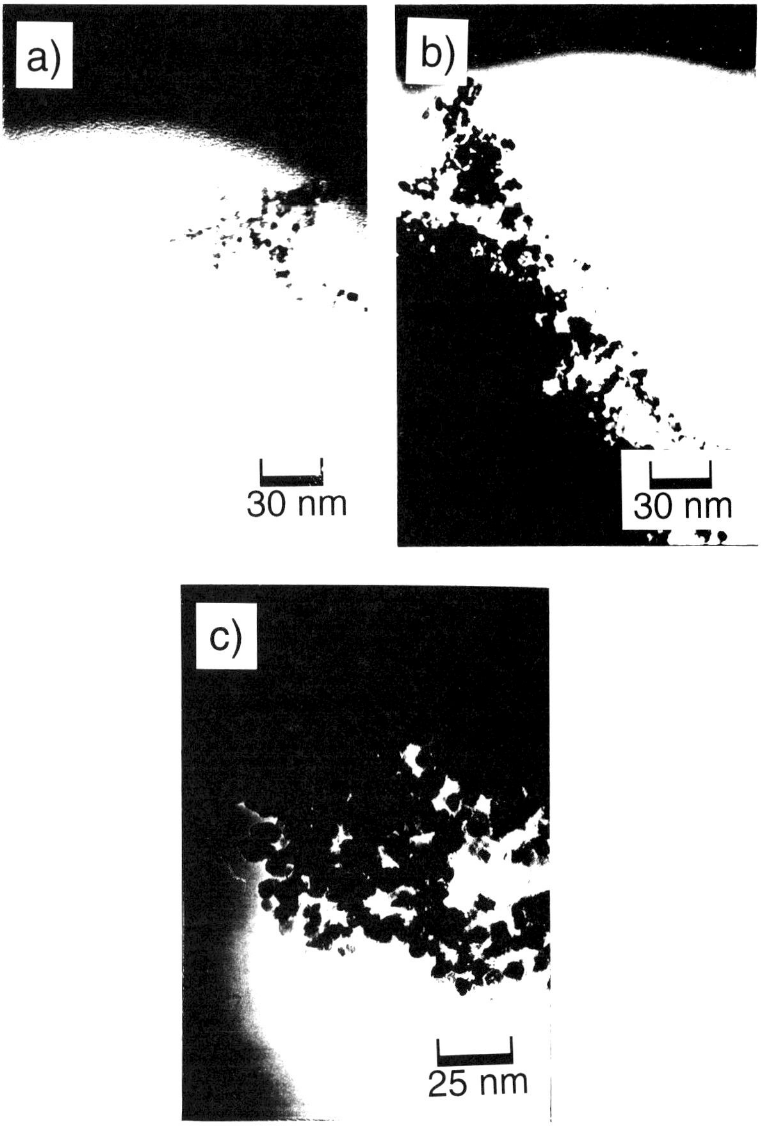

Figure 2. Transmission electron micrographs of: (a), the products obtained by the reaction of zirconium *n*-propoxide in 1,4-butanediol at 300°C for 2 h; (b), the sample obtained by calcination of product (a) at 500°C; (c), the sample obtained by 500°C-calcination of the product obtained by the reaction of zirconium isopropoxide in toluene at 300°C for 2 h.

below 150°C associated with the endothermic response in DTA [27], the organic residue is not due to simple adsorption on the product surface. Small amounts of alkyl (*i.e.*, ω-hydroxybutyl) groups covalently bonded to the surface oxygen atoms of the particles were always detected in the products obtained by the glycothermal reactions [27,33,34]. Assuming that each zirconium atom in the surface of spherical particles of zirconia (specific gravity, 6.1 g/cm^3) with a 3.2 nm diameter is bound to a glycol moiety forming Zr-O$(CH_2)_4$OH bond, one can calculate the theoretical weight loss of the product to be 18%, which is in fairly good agreement with the observed weight loss. If one assumes further that the zirconium atom in corners or edges of the particle possesses two glycol moieties, the agreement between observed and calculated weight losses becomes much better. Therefore, we concluded that the glycol moieties were bound to the surface zirconia atoms of the product.

The product obtained in EG contained a larger amount of organic matters than the product obtained in 1,4-BG; thermal analysis of the former product showed 27% weight loss from 200 to 500°C, which is in accord with the argument that the difficulty of the cleavage of O-C bond of zirconium glycoxide is the prime factor giving the product with quite small crystallite size.

The tetragonal phase was stable up to 700°C (See, Fig. 1d-f). Above this temperature it gradually transformed to the thermodynamically stable monoclinic form and transformation was completed at around 1000°C. All the samples calcined at 500°C maintained relatively large surface areas and the product obtained by the reaction of ZNP in 1,4-BG maintained the largest surface area (157 m^2/g). The TEM observation of this sample showed that the average particle size was 4.8 nm (Fig. 2b), which is in good agreement with the crystallite size (4.1 nm) but is smaller than the particle size calculated from the surface area (6.2 nm). The average particle size of the sample from the reaction in toluene was 6.4 nm (Fig. 2c). Again this value is in good agreement with the crystallite size (6.9 nm) but is smaller than the particle size calculated from the surface area (10.5 nm). These results indicate that coagulation of primary particles took place during the calcination and that a part of the surface of the primary particles is not accessible for nitrogen molecules so that the BET surface area is smaller than that expected from the particle size.

In conclusion, microcrystalline tetragonal zirconia was formed by the reaction of zirconium alkoxide in organic media. The products showed high thermal stability and maintained relatively large surface areas (90-160 m^2/g) even after calcination at 500°C and therefore, they have a potential for the catalyst use. The method reported here is unique in that the product free from the influence of counter anions can be obtained without any precaution or careful control of the processes such as hydrolysis, aging and drying just as required in the sol-gel method.

Acknowledgments
The present work was partly supported by a Grant-in-Aid for Scientific Research on Priority Areas "Dynamics of Fast Ions in Solids and Its Evolution for Solid State Ionics" No. 09215221 from The Ministry of Education, Science, Sports and Culture, Japan.

REFERENCES

1. Y. Nakano, T. Iizuka, H. Hattori, and K. Tanabe, *J. Catal.*, **57**, 1 (1979).
2. K. Arata, S. Akutagawa, and K. Tanabe, *Bull. Chem. Soc. Jpn.*, **49**, 390 (1976).
3. K. Arata, K. Kato, and K. Tanabe, *Bull. Chem. Soc. Jpn.*, **49**, 563 (1976).
4. T. Yamaguchi, H. Sasaki, and K. Tanabe, *Chem. Lett.*, 1017 (1973).
5. B.H. Davis and P. Ganesan, *Ind. Eng. Chem., Prod. Res. Dev.*, **18**, 191 (1979).
6. Y. Nakano, T. Yamaguchi, and K. Tanabe, *J. Catal.*, **80**, 307 (1983).
7. R. Bird, C. Kemball, and H.F. Leach, *J. Chem. Soc., Faraday Trans. I*, **83**, 3069 (1987).
8. T. Yamaguchi and J.W. Hightower, *J. Am. Chem. Soc.*, **99**, 4201 (1977).
9. Y. Tanaka, H. Hattori, and K. Tanabe, *Bull. Chem. Soc. Jpn.*, **51**, 3641 (1978).
10. B.-Q. Xu, T. Yamaguchi, and K. Tanabe, *Appl. Catal.*, **64**, 41 (1990).
11. H. Fujii, N. Mizuno and M. Misono, *Chem Lett.*, 2147 (1987).
12. T. Iizuka, Y. Tanaka and K. Tanabe, *J. Catal.*, **76**, 1 (1982).
13. T. Iizuka, Y. Tanaka and K. Tanabe, *J. Mol. Catal.*, **17**, 381 (1982).
14. T. Iizuka, M. Kojima and K. Tanabe, *J. Chem. Soc., Chem. Commun.*, 638 (1983).
15. P. Turlier, J.A. Dalmon, G.A. Martin, and P. Vergnon, *Appl. Catal.*, **29**, 305 (1987).
16. M.S. Scurrell, *Appl. Catal.*, **34**, 109 (1987).
17. P.D.L. Mercera, J.G. van Ommen, E.B.M. Doesburg, A J. Burggraff, and J.R.H. Ross, *Appl. Catal.*, **57**, 127 (1990).
18. P.D.L. Mercera, J.G. van Ommen, E.B.M. Doesburg, A.J. Burggraff, and J.R.H. Ross, *Appl. Catal.*, **71**, 363 (1991).
19. P.D.L. Mercera, J.G. van Ommen, E.B.M. Doesburg, A.J. Burggraff, and J.R H. Ross, *Appl. Catal.*, **78**, 79 (1991).
20. M.S. Scurrell, *Appl. Catal.*, **34**, 109 (1987).
21. E. Tani, M. Yoshimura, and S. Somiya, *J. Am. Ceram. Soc.*, **64**, C-181 (1981); **66**, 11 (1983).
22. H. Nishizawa, N. Yamasaki, K. Matsuoka, and H. Mitsushio, *J. Am. Ceram. Soc.*, **65**, 343 (1982).
23. B. Mottet, M. Pichavant, J.-M. Bény, and J.-A. Alary, *J. Am. Ceram. Soc.*, **75**, 2515 (1992).
24. D.M. Bibby and M.P. Dale, *Nature (London)*, **317**, 157 (1985).
25. A.J. Fanelli and J.V. Burlew, *J. Am. Ceram. Soc.*, **69**, C-174 (1986).
26. M. Figlarz, F. Fievet, and J.P. Lagier, *International Meeting on Advanced Materials,* Tokyo, May 30 to June 3, C5.1, 1988; Nikkan Kogyo Shinbun, Ltd., Tokyo, Japan, 1988.
27. M. Inoue, Y. Kondo, and T. Inui, *Inorg. Chem.*, **27**, 215 (1988).
28. M. Inoue, H. Tanino, Y. Kondo, and T. Inui, *J. Am. Ceram. Soc.*, **72**, 352 (1989).
29. M. Inoue, H. Otsu, H. Kominami, and T. Inui, *J. Am. Ceram. Soc.*, **74**, 1452 (1991).
30. M. Inoue, H. Otsu, H. Kominami, and T. Inui, *J. Alloys Comp.*, **226**, 146 (1995).
31. M. Inoue, H. Kominami, and T. Inui, *J. Am. Ceram. Soc.*, **75**, 2597 (1992).
32. M. Inoue, H. Kominami, H. Otsu, and T. Inui., *Nippon Kagakukai-shi*, 1364 (1991).
33. M. Inoue, H. Kominami, and T. Inui, *J. Am. Ceram. Soc.*, **73**, 1100 (1990).
34. M. Inoue, H. Kominami, and T. Inui, *J. Chem. Soc., Dalton Trans,* 3331 (1991).
35. M. Inoue, H. Kominami, and T. Inui, *J. Am. Ceram. Soc.*, **79**, 793 (1996).
36. M. Inoue, H. Otsu, H. Kominami, and T. Inui, *Ind. Eng. Chem., Res.*, **35**, 295 (1996).
37. M. Inoue, T. Nishikawa, T. Nakamura, and T. Inui, *J. Am. Ceram. Soc.,* **80**, 2157 (1997).
38. S. Winstein, E. Allred, R. Heck, and R. Glick, *Tetrahedron,* **3**, 1 (1958).
39. K.S. Mazdiyasni, C.T. Lynch and J.S. Smith, *J. Am. Ceram. Soc.*, **48**, 372 (1965).
40. K.S. Mazdiyasni, C.T. Lynch and J.S. Smith, *J. Am. Ceram. Soc.*, **49**, 286 (1966).

Res. Chem. Intermed., Vol. 24, No. 5, pp. 581-592 (1998)

PEROVSKITE OXIDES IN CATALYSIS: PAST, PRESENT AND FUTURE

E. A. LOMBARDO AND M. A. ULLA
Instituto de Investigaciones en Catálisis y Petroquímica - INCAPE (FIQ,UNL-CONICET) - Santiago del Estero 2829 - 3000 - Santa Fe - ARGENTINA. Tel.: 54-42-536861. Fax: 54-42-571162. e-mail: nfisico@fiqus.unl.edu.ar

Received 13 October 1997; accepted 20 November 1997

Abstract-- The application of perovskite oxides in catalysis is reviewed. Selective and total oxidation, nitrogen oxides decomposition and reduction as well as other miscellaneous uses of mixed oxide catalysts are discussed. A final section describes the most promising areas of research to increase the number of commercial uses of perovskite oxides.

I. INTRODUCTION

Perovskite oxides with the general formula ABO_3 have been tried as catalysts for a large variety of reactions. The ideal perovskite structure is cubic although it may be somewhat distorted according to the type of the A and B cations involved. Typical A elements are rare earths, alkaline and alkaline earths while the B sites are usually occupied by transition metals.

These oxides exert a double research attraction. On the one hand, they offer the possibility of preparing a large series of isomorphic solids for the study of the effect of partial substitution of A and B cations ($A_{1-X}A'_xB_{1-y}B'_yO_3$) upon catalytic behavior. On the other hand, these oxides are stable at elevated temperatures and may be used in many areas where other formulations fail, e.g., high temperature catalytic combustion. However, one of the shortcomings of these mixed oxides is their low surface area. Many preparation methods have been reported in the literature aimed at increasing the surface area of the massive mixed oxide [1,2] and to speed up the synthesis [3-5]; even so, surface areas between 5 and 50 m^2/g at most are achieved in massive oxides. This is why in the last decade many efforts have been devoted to the preparation of perovskites supported on high surface area solids [6-9].

We will review here the use of perovskite oxides to advance the knowledge of the catalytic phenomena and more commercially oriented applications. With this goal in mind, we have chosen to group the perovskite studies by the type of catalyzed reaction.

concentration of the active sites for methyl radical formation and limiting the concentration of weakly bonded oxygen associated with the total oxidation of methane.

The dispersion and the location of the promoter components in a host mixed oxide are not readily controlled, moreover high temperature could cause their partial loss. Taking into account that the structure of layered perovskites, $A_2R_2Ti_3O_{10}$ (A = K, Na; R = La, Ce, Pr), allows layers of alkali ions dispersed between lanthanide perovskite layers, these solids seem to be a good system to fix the alkali in a perovskite environment. Some of the results obtained by Campbell *et al.* [20] are shown in Table 2. The C_2 yield obtained using $K_2La_2Ti_3O_{10}$ is comparable to many other good OCM catalysts. The important feature being that this layered perovskite did not show any significant deactivation after 149 h on stream.

IV. PARTIAL OXIDATION OF METHANE TO SYNTHESIS GAS (SYNGAS)

Syngas, CO and H_2, can be obtained as equilibrium product of the reaction between CH_4 and O_2. A good catalyst should activate methane and oxygen under conditions that favor syngas products and avoid total oxidation and coke formation.

The syngas catalyst works under reducing conditions and high temperature. Thus, it is expected that during the reaction, the catalytic surface will be reduced and finely dispersed metal particles will act as active sites. Perovskites are potential catalysts for this reaction since the B element can be reduced to the metallic state, yielding a highly dispersed metal on oxide solid [23,24].

Different La-based perovskites were tested for the partial oxidation of methane to syngas [25,26]. $LaNiO_3$ was active and selective; however deactivation was observed during the catalytic process, probably due to coke formation. This solid showed drastic structural changes after the reaction, only $La(OH)_3$ and Ni^o were identified using XRD by Slagtern and Olsbye [25]. Ni^o sintered during the reaction, favoring the formation of coke.

Choudhary *et al.* [26] found out that the following reactions occurred simultaneously over $LaNiO_3$: partial and total oxidation of methane (1, 2), shift gas reaction (3), and CO_2 and steam reforming of methane (4, 5).

Exothermic reactions

$$CH_4 + 1/2\,O_2 \rightarrow CO + 2\,H_2 \qquad \textbf{(1)}$$

$$CH_4 + 2\,O_2 \rightarrow CO_2 + 2\,H_2O \qquad \textbf{(2)}$$

$$CO + H_2O \rightarrow CO_2 + H_2 \qquad \textbf{(3)}$$

Endothermic reactions

$$CH_4 + CO_2 \rightarrow 2\,CO + 2\,H_2 \qquad \textbf{(4)}$$

$$CH_4 + H_2O \rightarrow CO + 3\,H_2 \qquad \textbf{(5)}$$

Coupling these reactions over the same catalyst, the catalytic process could be carried out at slightly endothermic or exothermic regime changing the reaction conditions. Choudhary *et al.* [26] reported that $LaNiO_3$ at 850°C yielded high conversions for methane and oxygen, almost 100% selectivity to CO and H_2 and positive conversions for CO_2 and H_2O. Since exothermic reactions produced heat, which could be used by endothermic reactions, the overall process could be conducted in a more energy-efficient manner.

V. CATALYTIC COMBUSTION

Perovskite-type oxides have been attractive catalysts for hydrocarbon combustion since Voorhoeve *et al.* [27] in 1972 reported their high oxidation activity.

In order to evaluate the effect of A- and B-site elements, Nitadori *et al.* [28] studied the catalytic activities of $LnBO_3$ (Ln = lanthanides; B = Cr, Mn, Fe, Co, Ni) for propene oxidation at 230°C. They observed that the specific activities decreased according to $LnCoO_3 > LnMnO_3 > LnNiO_3 >> LnFeO_3 \sim LnCrO_3$. Similar results were reported by Kremenić *et al.* [10] for C_3 and C_4 oxidation over $LaBO_3$ (B = Cr, Mn, Fe, Co, Ni), showing that the transition metal in the B-site plays an important catalytic role as active site. Co and Mn are the most active 3d-transition metals. The role of the lanthanides is secondary as long as they are trivalent.

Arai *et al.* [29] studied the catalytic behavior of $LaBO_3$ (B = Co, Mn, Fe, Cu, Ni, Cr) for methane combustion in comparison to Pt(1%)/Al_2O_3. The temperature at which the conversion of methane was 50% ($T_{50\%}$) was used to rate the oxidation activity. Co, Fe and Mn based perovskites were the most active in this series. Their activities, $T_{50\%}$, were 525, 571, and 579°C, respectively, quite close to that of Pt(1%)/Al_2O_3 ($T_{50\%}$: 518°C).

The catalytic properties of these perovskite-type oxides are modified by the partial subtitution of A- and/or B-sites. The catalytic activity for C_4H_8 oxidation over $Ln_{1-x}Sr_xCoO_3$ (Ln = lanthanides) increased several times upon the Sr subtitution for Ln [28]. However, a maximum was observed as the x value increased. Yamazoe and Teraoka [30] summarized the specific rate for oxidation of hydrocarbons over Co, Fe and Mn La-based perovskites partially substituted by Sr showing the same tendency.

$La_{0.8}Sr_{0.2}MnO_3$ was tested as a catalyst for combustion of acetone and compared to $CuMn_2O_4$ [31]. The latter solid is a commercial catalyst for VOCs removal. The perovskite was more active than the copper manganate.

Strong synergic effect in the catalytic behavior was observed as Co was partially substituted by Fe in $La_{0.2}Sr_{0.8}Co_{1-y}Fe_yO_3$ for n-butane combustion [32]. The activity increased as y = 1 < 0.6 < 0.8 < 0.4. However, the activity for methane combustion over these samples was hardly affected by this subtitution.

A number of patents dealing with perovskite oxidation catalysts were summarized by Quinlan *et al.* [33], Zwinkels *et al.* [34] and M. Misono *et al.* [35]. $La_{0.9}Ce_{0.1}CoO_3$ was the first perovskite-type catalyst for complete oxidation in domestic emissions (U.S. Patent 4748143 (1988)) in commercial use [35].

An interesting application of perovskite-type oxides is as catalysts for high temperature combustion (gas turbine, catalytic burners) because of their thermochemical stability in atmosphere containing oxygen and steam at high temperature. However, the catalytic system must be adequately shaped to achieve low pressure drop, mechanical strength and chemical stability. Isupova *et al.* [36] tried to form rings of different perovskites without any addition of binders or peptizers, but these rings showed low mechanical strength. An option is to incorporate a catalytic component on a support, which satisfies those criteria mentioned above. Note that an important bonus of catalytic combustion is to decrease the formation of NO at the source, the reason being the sharp decrease in the appearance of high temperature spots typical of non-catalytic combustion.

VI. NITROGEN OXIDES DECOMPOSITION

Every year, new, more rigorous legislation appears all over the world imposing lower limits to the nitrogen oxides content in exhaust gases. This is then a large driving force to search for new ways to eliminate these oxides. In theory, the best way to eliminate the thermodynamically unstable NO is by catalyzing its decomposition. Although there have been some breakthroughs no catalysts so far can operate in the presence of excess oxygen as is usually the case in combustion gases from diesel and lean burning Otto engines and power plants. The most recent study that throws some light into the way perovskite and related solids act has been published by Zhao *et al.* [37]. They found out that La_2NiO_4 is much more active than $LaNiO_3$. They also reported that the oxides with oxygen vacancies produced via the partial substitution of La by Sr are also more active. In view of this, they concluded that both Ni^{2+} and oxygen vacancies are key players in the redox reaction mechanism they propose. However, the performance of these materials falls short from industrial application.

A similar picture appears in the case of N_2O decomposition. The most recent work done by Misono and co-workers [38] indicates that N_2O decomposition on a variety of mixed oxides is first order in nitrous oxide and negative order in oxygen.

In brief, to decompose both NO and N_2O there is a delicate balance between the ability of vacancies to adsorb N_2O with enough strength to break the N=O bond, and that of retaining the chemisorbed oxygen weakly enough to allow the rapid regeneration of the active site. This is very difficult to achieve in practice over all kinds of solids.

VII. NO_X REDUCTION

In principle, the presence of a reducing agent should speed up the "cleaning" of the catalyst surface. Of course, this is a valid conclusion if the same mechanism operates in the presence of the reducing agent, be it CO or hydrocarbons.

The first impulse on perovskite catalysis took place in the '70s when industry was hastily searching for materials to replace noble metals in the manufacturing of what are now called three-way catalysts. These efforts collapsed when it was discovered that the perovskites tried by the Voorhoeve team were easily poisoned by SO_2.

During this decade the study of NO reduction on perovskites took-off again. Most articles are concerned now with multicomponent mixed oxides but only a few of them report the effect of excess oxygen. Misono and co-workers [39] studied the reduction of NO with CO. They also investigated the reaction of CO with either NO_2 or O_2. Their prime goal was to elucidate the controlling factor in NO reduction. They concluded that the reaction occurs at least in two steps with the formation of N_2O as an intermediate. Öcal *et al.* [40] and Forni *et al.* [41] using different mixed oxides agree with the role played by N_2O as an intermediate.

Swedish and Japanese researchers [42,43] tried a variety of complex supported and unsupported perovskite-type oxides. They are also among the few groups to use hydrocarbons as reducing agents and to explore the effect of excess oxygen. Skoglund *et al.* [42] prepared nine alumina washcoated mixed oxide catalysts containing La, Sr, Cu and Ru. These samples contained small particles (60-200 μm) that sometimes were accompanied by larger (1-10 μm) flake-like formations. The surface area of the impregnated washcoat calcined at 900°C varied between 40 and 60 m^2/g. The nanoparticles had the chemical composition of perovskites, while the flakes were closer to La_2CuO_4. Table 3 shows the data obtained by Skoglund *et al.* [42] over five of their supported catalysts. Note the key role of both Ru and Sr to achieve high conversions of NO to N_2. These data were obtained using a stoichiometric reactant mixture. When they modified the feedstream composition from overall reducing to oxidizing, the NO conversion

Table 3.

NO reduction with CO and C_3H_6 under stoichiometric conditions at 450°C [a]

Catalyst [b]	NO	CO	C_3H_6
		Conversion % [c]	
LaCu	4	76	71
$LaCu_{0.8}Ru_{0.2}$	59	86	76
$La_{0.9}Sr_{0.1}Cu_{0.8}Ru_{0.2}$	75	89	91
$La_{0.8}Sr_{0.2}Cu$	4	78	81
$La_{0.8}Sr_{0.2}Cu_{0.8}Ru_{0.2}$	73	89	91

[a]Stoichiometric gas mixture $(2(O_2) + [NO])/([CO] + 9\ [C_3H_6]) = 1$; SV = 100000 h^{-1}. Data from Skoglundh *et al.* [42].
[b]The alumina washcoat contained 10.6 - 12.0 % (wt.%) of metal oxide.
[c]NO to N_2 and CO and C_3H_6 to CO_2 and H_2O.

drastically decreased. This behavior is in fact very close to that observed with commercial three-way catalysts. Of course, a pending matter is to test how these solids behave in the presence of ca. 50 ppm SO_2 and 10-15% water in the feed stream.

The X-ray diffraction patterns of the nanoparticles clearly show the reflections of a perovskite structure. However, it is still risky to adscribe a unique catalytic function to this phase. More studies are required to investigate the possible role of other species that might be present in all these washcoated formulations.

Teraoka *et al.* [43] reported data obtained for massive $La_{0.8}Sr_{0.2}Co_{1-2y}Cu_yRu_yO_3$ with $0 \leq y \leq 0.2$. They used them to catalyze both the CO oxidation and the CO + NO reaction. The partial substitution of Co by Cu/Ru decreases the rate of both reactions but sharply increases the selectivity to N_2. Since these reactions have been proposed to occur via an intrafacial mechanism, the negative effect of the Co substitution by Cu/Ru may be explained by the decrease in weakly bonded oxygen reported by the authors during the TPD experiments. Teraoka *et al.* [43] further emphasized the importance of Cu/Ru incorporation in the perovskite matrix. They reached the conclusion that $La_{0.8}Sr_{0.2}Co_{1-2y}Cu_yRu_yO_3$ has an activity similar to both $La_{0.8}Sr_{0.2}CoO_3$ and 0.5% Pt/Al_2O_3 for the NO + CO reaction. In all cases N_2O appears at ca. 200°C and disappears above 400°C. This behavior is again consistent with the role of intermediate assigned to nitrous oxide.

$LaNiO_3/Al_2O_3$ and La_2NiO_4/ZrO_2 were tested for the NO + CO reaction by Ladavos and Pomonis [44]. They compared the activity of the supported mixed oxides with massive ones. The alumina support was not suitable for this perovskite. This is not surprising due to the known preference of Ni to occupy the octahedral positions within the alumina framework to form the spinel. A similar phenomenon occurred when trying to support $LaCoO_3/Al_2O_3$. To overcome the formation of

$CoAl_2O_4$, Nudel *et al.* [6] preimpregnated the alumina with lanthanum to block the octahedral sites. La_2NiO_4/ZrO_2 was about four times more active than massive La_2NiO_4 on a per-gram basis. Further comparison is not possible since the authors were not able to measure the dispersion of the mixed oxide over the zirconia support.

In this new round of studies using a large variety of mixed oxides formulations, a large effort is being made to prepare, characterize and advantageously use supported catalysts. It is still too early to foresee the possible commercial applications of these catalysts.

VIII. MISCELLANEOUS APPLICATIONS

$LaM_{1-x}Cu_xO_3$ (M = Ti, Mn) have been used to catalyze the conversion of methanol to methyl formate by Rodriguez-Ramos *et al.* [45]. Particularly using $LaMn_{0.5}Cu_{0.5}O_3$ the yield averaged 25% between 200° and 260°C. A key point here is to decrease the reaction temperature for the product readily decomposes at 230°C and above. No significant effect was observed when Mn was replaced by Ti. The pre-reduction of the solids did not affect their catalytic behavior. In fact, by means of XPS analysis they found no difference between pre-reduced fresh and used catalysts. They both contained either Cu^+ or Cu°, or both types of reduced species. XPS is not a good tool to differentiate Cu^+ or Cu° but a closer look to the Auger parameters corresponding to $L_3M_{4.5}M_{4.5}$ transition undoubtedly confirmed that only Cu° was present in the used catalysts. XPS measurements indicate that some migration of copper from the subsurface to the topmost layer seems to occur on pre-reduced and/or used catalysts.

Other miscellaneous applications are summarized in Table 4. Besides the methyl formate reaction the disproportionation observed with toluene and the liquid phase hydroxylation of phenol are striking systems that deserve further research.

Table 4

Miscellaneous catalytic applications of perovskite-type oxides.

Reactants	Main product	Catalyst	Reference
CH_3OH/N_2	Methyl formate	$LaM_{1-x}Cu_xO_3$ (M = Mn, Ti)	Rodriguez-Ramos *et al.* [45]
Toluene + O_2	disproportionation products [a]	$La_{1-x}Sr_xNiO_3$ [b]	Liang and Weng [46]
Phenol + H_2O_2	[c]	$La_{2-x}Sr_xCuO_{4\pm\lambda}$	Liu *et al.* [47]

[a]Ethylbenzene and xylenes.

[b]The authors concluded that La_2NiO_4 and NiO are the active phases under reaction conditions.

[c]Catechol, hydroquinone, benzoquinone.

IX. FUTURE OF PEROVSKITE CATALYSIS

After this brief review of perovskite applications we conclude that the use of these and related mixed oxides in combustion is already established and has reached in some cases, commercial status. Many potential applications have also been reported in the patent literature.

Turning our attention to the academic media one sees many attempts to find new uses of these ceramics in NO_x abatement, catalytic combustion, syngas production and other miscellaneous albeit interesting applications.

There are three key points that require much effort if new commercial applications of these materials are to be achieved. They are:

i) The development of new methods or improvement of existing ones to prepare supported mixed oxides.

ii) Related to the previous point but standing by itself is the preparation of membranes containing active perovskite materials, or dense ceramic membranes made of these mixed oxides.

iii) The study of chemical transformations under conditions very close to those found in practice.

The first item requires the use of preparation procedures such as chemical vapor deposition, atomic layer epitaxy, etc. These methods are potentially able to yield well dispersed perovskites on high surface area single oxides such as ZrO_2 and Al_2O_3. The use of sophisticated surface instrumentation is needed to characterize the solids.

Non-catalytic materials may be used to make membranes that support active ingredients in their channels. An interesting option is the use of dense ceramic membranes with catalytic properties shaped into a hollow-tube reactor. Such membranes manufactured from certain perovskites are permeable to O_2 and not to N_2 at elevated temperatures. This may be an excellent solution for partial oxidation and in fact it has already been tried. Balachandran *et al.* [48] used this approach for partial oxidation of methane to syngas. A very promising field opens up for the use of membrane technology in perovskite catalysis.

The last proposition is illustrated by a classical situation occurring in environmental catalysis. The exhaust gases from different types of engines usually contain large concentrations of water vapor and SO_2. Therefore, any realistic approach to the development of catalysts to eliminate pollutants from these streams need to include SO_2 and water in the feed of the experimental reactors.

In brief, we believe that there are many potential practical applications for these materials that will be fully realized in the near future.

Acknowledgments
We are indebted to all the people in our group who worked and still investigate perovskite catalysis. To Elsa Grimaldi for her help in editing the manuscript in English. To CONICET and UNL for their sustained support to our research.

REFERENCES

1. C.D. Chandler, C. Roger and M.J. Hampden-Smith, *Chem. Rev.* **93**, 1205 (1993).
2. S. Moreau, J. Choisnet and F. Beguin, *J. Phys. Chem. Solids,* **57**, 1049 (1996)
3. A.G. Merzhanov and I.P. Boroviskaya, *Dokl. Acad. Navk.* **204**, 366. (1972).
4. V.G. Milt, R. Spretz, M.A. Ulla and E.A. Lombardo, *J. Mat. Sci.* **14**, 428 (1995)
5. P. Gordes, N. Christiansen, E.J. Jensen and J. Villadsen, *J. Mat. Sci.* **30**, 1053 (1995).
6. J.N. Nudel, B.S. Umansky and E.A. Lombardo, *Appl. Catal.* **31**, 275 (1987).
7. N. Mizuno, *Catal. Today* **8**, 221 (1990).
8. N. Mizuno, H. Fujii, H. Igarashi and M. Misono, *J. Am. Chem. Soc.* **114**, 7151 (1992a).
9. G. Saracco, G. Scibilia, A. Iannibello and G. Baldi, *Appl. Catal. B: Environmental* **8**, 229 (1996).
10. G. Kremenić, J.M.L. Nieto, J.M.D. Tascón and L.G. Tejuca, *J. Chem. Soc., Faraday Trans. I,* **81**, 939 (1985).
11. K.L. Madhock, *React. Kinet. Catal. Lett.* **30**, 185 (1985).
12. D.D. Agarwal and H.S. Goswami, *Kinet. Catal. Lett.* **53**, 441 (1994).
13. G. Yi, T. Hayakawa, A.G. Andersen, K. Suzuki, S. Hamakawa, A.P.E. York, M. Shimizu and K. Takehira, *Catal. Lett.* **38**, 189 (1996).
14. T. Shimizu, *Catal. Rev. Sci. Eng.* **34**, 337 (1992).
15. S. Hansen, J. Otamini, J.O. Bonvin and A. Andersson, *Nature* **334**, 143 (1988).
16. C. de Leitenburg, A. Tovarelli, J. Llorca, F. Cavani and G. Bini, *Appl. Catal. A: General* **139**, 161 (1996).
17. T. Shimizu, *Appl. Catal.* **28**, 81 (1986).
18. R. Radha and C.S. Swamy, *Surface Technology* **24**, 157 (1985).
19. N.D. Spencer, *J. Catal.* **109**, 512 (1988).
20. K.D. Campbell, *Catal. Today* **13**, 245 (1992).
21. R.V. Siriwadarne and A. Shamsi, *Appl. Catal.* **60**, 119 (1990).
22. M. Yamamura, N. Tsuzuki, H. Okado, T. Wakatsuki and K. Otsuka, *Appl. Catal. A: General* **115**, 269 (1994).
23. J. Marcos, R.H. Buitrago and E.A. Lombardo, *J. Catal.* **105**, 95 (1987).
24. a) M.A. Ulla, R.A. Migone, J.O. Petunchi and E.A. Lombardo, *J. Catal.* **105**, 107 (1997); b) J.O. Petunchi and E.A. Lombardo, *Catal. Today* **8**, 201 (1990).
25. Å. Slagtern and U. Olsbye, *Appl. Catal. A: General* **110,** 99 (1994).
26. V.R. Choudhary, B.S. Uphade and A.A. Belhekar, *J. Catal.* **163**, 312 (1996).
27. R.J.H. Voorhoeve, J.P. Remeika, P.E. Freeland and B.T. Mathias, *Science* **177**, 353 (1972).
28. T. Nitadori, T. Ichiki and M. Misono, *Bull. Chem. Soc. Japan* **61**, 621 (1988).
29. H. Arai, T. Yamada, K. Eguchi and T. Seiyama, *Appl. Catal.* **26**, 265 (1986).
30. N. Yamazoe and Y. Teraoka, *Catal. Today* **8**, 175 (1990).
31. H. G. Lintz and K. Wittstock, *Catal. Today* **29**, 457 (1996).
32. H.M. Zhang, Y. Shimizu, Y. Teraoka, N. Miura and N. Yamazoe, *J. Catal.* **121**, 432 (1990).
33. M.A. Quinlan, H. Wise and J.G. McCarty, *Basic Research on Natural Gas Combustion*

Phenomena - Catalytic Combustion, GRI-89/0141, SRI International, Menlo Park, CA, 1989.

34. M.F.M. Zwinkels, S.G. Järås and P.G. Menon, *Catal. Rev. - Sci. Eng.* **35**, 319 (1993).
35. M. Misono and N. Nojiri, *Appl. Catal.* **64**, 1 (1990).
36. L.A. Isupova, V.A. Sadykov, L.P. Solovyova, M.P. Andrianova, V.P. Ivanov, G.N. Kryukova, V.N. Kolomiichuk, E.G. Avvakumov, I.A. Pauli, O.V. Andryushkova, V.A. Poluboyarov, A.Ya. Rozovskii and V.F. Tretyakov, *Preparation of Catalysts VI*, 637 (1995).
37. Z. Zhao, X. Yang and Y. Wu, *Appl. Catal. B: Environmental* **8**, 281 (1996).
38. J. Wang, H. Yasuda, K. Inumaru and M. Misono. *Bull. Chem. Soc. Jpn.* **68**, 1226 (1995).
39. N. Mizuno, M. Tanaka and M. Misono, *J. Chem. Soc. Faraday Trans.* **88**, 91 (1992).
40. M. Öcal, R. Oukaci and G. Marcelin, *Ind. Eng. Chem. Res.* **33**, 2930 (1994).
41. L. Forni, C. Oliva, F.P. Vatti, M.A. Kandala, A.M. Ezerets and A.V. Vishniakov. *Appl. Catal. B: Environmental* **7**, *269* (1996).
42. M. Skoglundh, L. Löwendahl, K. Jansson, L. Dahland, M. Nygren, *Appl. Catal. B: Environmental* **3**, 259 (1994).
43. Y. Teraoka, H. Nii, S. Kagawa, K. Jansson and M. Nygren, *J. Mat. Chem* **6**, 97 (1996).
44. A.K. Ladavos and P.J. Pomonis, *Catal. Today* **17**, 181 (1993).
45. I. Rodriguez-Ramos, A. Guerrero-Ruiz, M.L. Rojas and J.L.G. Fierro, *Appl. Catal.* **68**, 217 (1991).
46. J.J. Liang and H.S. Weng, *J. Catal.* **140**, 302 (1993).
47. Ch. Liu, Z. Zhao, X. Yang,, X. Ye and Y. Wu, Chem. Commun. 1019 (1996).
48. U. Balachandran, J.T. Dusek, R.L. Mieville, R.B. Poeppel, M.S. Kleefisch, S. Pei, T.P. Kobylinski, C.A. Udovich and A. Bose, *Appl. Catal. A: General* **133**, 19 (1995).

Res. Chem. Intermed., Vol. 24, No. 5, pp. 593-603 (1998)

THE PROMOTER EFFECT OF LANTHANA ON MgO SUPPORTED RUTHENIUM CATALYSTS FOR AMMONIA SYNTHESIS

YUSUKE NIWA and KEN-ICHI AIKA*

Department of Environmental Chemistry and Engineering, Interdisciplinary Graduate School of Science and Engineering Tokyo Institute of Technology, 4259 Nagatsuta, Midori-ku, Yokohama 226, Japan

Received 24 September 1997; accepted 7 January 1998

Abstract--When activated at high temperature (873 K) doped lanthana such as La_2O_3, Ce_2O_3, Sm_2O_3 derived from nitrate showed remarkable effect on the MgO-supported ruthenium activity for ammonia synthesis. The activity treated with hydrogen at 873 K became 5 times as active as that treated at 623 K. The activity levels were almost the same as those of the Ru-lanthana (support) system under the same activation condition. It was thought that the lanthana was partly reduced and a strong metal support interaction occurred. The partially reduced lanthana was thought to donate electrons to the contacting Ru atoms forming the active centers. The other characters such as the stability were also discussed by comparing with alkali promoters.

INTRODUCTION

The promoter effect on ammonia synthesis ($N_2+3H_2 = 2NH_3$) is remarkable in the ruthenium catalysts compared with the iron catalysts [1,2]. Ruthenium catalysts had been expected as the second-generation ammonia catalyst after the iron catalysts. A new type of promoted ruthenium catalyst has been used in a commercial ammonia plant [3]. Considering energy and environmental problems in the future, effective ruthenium catalysts will become necessary for the future ammonia plants.

Our systematic studies of Ru catalysts for ammonia synthesis have revealed that the combination of three components (Cl-free ruthenium source, basic oxide support, and an alkali nitrate promoter) is the most effective for the synthesis under atmospheric pressure [4]. It is thought that the basic supports and alkali promoter donate electrons to Ru; as a result, the dissociation of dinitrogen on the Ru surface is promoted [5]. On the contrary, alkali-promoted Ru catalysts suffer from the strong adsorption of hydrogen that retards the activation of dinitrogen [6,7]. The strong hydrogen adsorption must retard the ammonia synthesis under industrial conditions (high pressure).

Recently, lanthanide oxides (lanthana) were found to be effective promoters for the Ru catalyst [8]. A small amount of these are as effective as larger amounts of cesium hydroxide on Ru/Al_2O_3 [8]. The characteristics of lanthana are not only electronic donation but also weakening of the hydrogen poisoning [9]. In the last paper the authors tried to clarify the interaction between Ru and lanthana "support" (two component system) because of simplicity, and found the remarkable effect of hydrogen-treatment-temperature (probably SMSI effect) [10].

In this paper the authors tried to investigate the interaction between Ru and lanthana "promoter". Although, the Ru-lanthana/"MgO" had been once studied and turned out to be a less active catalyst [8], those were not treated at high temperature. When the lanthana was used as a support, the nitrate was once decomposed to oxide (support) which was again treated with hydrogen at higher temperature after Ru is added. In this work, the nitrate which had been added to Ru/MgO was reduced (accompanying decomposition) by hydrogen at high temperature all at once.

EXPERIMENTAL SECTION

Several reference catalysts were prepared. MgO (Ube Co.) baked in air at 773 K ($130 m^2 g^{-1}$, pressed and crushed into 8 to 10 mesh size) was impregnated with $Ru_3(CO)_{12}$ (Tanaka Noble Metals Co.) in tetrahydrofuran (THF). $Ru_3(CO)_{12}$ is a better precursor than $RuCl_3$ because it gives high dispersion and is free from poisonous chlorine for N_2 activation [8,11]. After evaporation and drying, a sample was heated in vacuo at 623 K for 2 h to decompose the carbonyl compound to the metal cluster (Ru/MgO). Ru/MgO was transferred into an aqueous solution of the alkali or lanthanide nitrate and was dried (Ru-promoter/MgO). These samples were treated with $N_2 + 3H_2$ at 623 K or at a given temperature for 4 h before ammonia synthesis.

Lanthana-supported Ru catalysts were prepared as was described in the last paper [10]. For these catalysts $Ru_3(CO)_{12}$ was decomposed under N_2 gas flow or in vacuo at 773 K for 3 h and reduced with $N_2 + 3H_2$ for 4 h at a given temperature between 623 and 923 K, up to which the temperature was increased at a heating rate of 10 K min^{-1}. To investigate the SMSI effect, Ru/CeO_2 reduced at 773 K was cooled to room temperature, oxidized for 1 h, and subsequently, reduced again at the given temperature. The ceria support is baked once in air and transformed to CeO_2 [10].

The rate of ammonia synthesis was usually measured at 588 K with a flow rate of 60 mL min^{-1} using a conventional flow system containing 1g of the catalyst [10]. Gas mixture: $N_2 + 3H_2$ was used without further purification (Takachiho Shoji Co.). The reaction rate was determined through the decreasing rate of electric conductivity of diluted sulfuric acid solution, which fixed the produced ammonia. Nitrogen and hydrogen gas adsorption were measured at 75 K and 298 K in a closed circulation system, respectively. Details are shown in the last paper [10].

Dispersions of typical Ru-Promoter/MgO and Ru/Lanthana were 0.40 (Ru-CsOH/MgO) and 0.35 (Ru/CeO_2), respectively.

RESULTS

Effect of Lanthanide Oxides as "Promoter" on Ru/MgO Catalysts

Rates of ammonia synthesis on promoted Ru/MgO catalysts were measured and compared with the rate on promoted Ru/Al_2O_3 [8]. The results are shown in Table 1. In the case of "alumina"-supported Ru catalysts, lanthana promoters were as effective as CsOH [8]. Smaller amounts of lanthana were enough to get the maximum activity compared with the CsOH promoter [8].

Table 1

NH_3 synthesis rates under 101kPa of $N_2 + 3H_2$ at 588 K on 3 wt% Ru/MgO Ube product) catalysts and 2 wt% Ru/Al_2O_3 catalysts with different promoters.

Promoter	Mol ratio M/Ru	Rate mmol NH_3 h^{-1} g-cat^{-1}	E^a kJ mol^{-1}
for 3 wt% Ru/MgO			
Cs^+	1.0	2772 (1445)[b]	96.6
La^{3+}	1.0	340 (115)[b]	78.3
Ce^{3+}	1.0	424 (100)[b]	63.0
Sm^{3+}	1.0	335 (86)[b]	64.4
none	0.0	218 (60)[b]	91.5
for 2 wt% Ru/Al2O3			
Cs^+	10.1	214	98.6
La^{3+}	2.8	226	44.7
Ce^{3+}	3.1	216	48.1
Sm^{3+}	3.0	204	60.6
none	0.0	13	62.7

[a]From reference 8; $N_2 + 3H_2 = 80$ kPa.
[b]From reference 5; $N_2 + 3H_2 = 80$ kPa ; 2 wt% Ru/MgO (Soekawa product).

Since for "magnesia"-supported Ru catalysts lanthana were not as effective as with Ru/Al_2O_3, no detailed work has been carried out [8]. Each result obtained here with high purity and high surface area MgO was slightly different from the previous results with MgO (Soekawa Chemical) [4]. Most of the g-base activities of the MgO samples used here were about 3 times higher than the previous one, for which the Ru content and pressure were lower (2/3 and 4/5, respectively) than the present samples. As has been reported [4] Cs promotes the Ru/MgO most effectively: promotion was more than 10 fold. On the other hand, lanthana such as La_2O_3 and Sm_2O_3 promote the Ru/MgO catalyst less effectively; the promotion was approximately 1.5 times. These tendencies were similar to the reported results using a different MgO source [4]. However, Ce_2O_3-promoted catalyst has been found to be moderately active: approximately 2 times promotion and having a low activation energy.

Thus, we investigated the dependence of Ce addition on the activity of the Ru/MgO catalyst. The results are shown in Figure 1. The activity was increased up

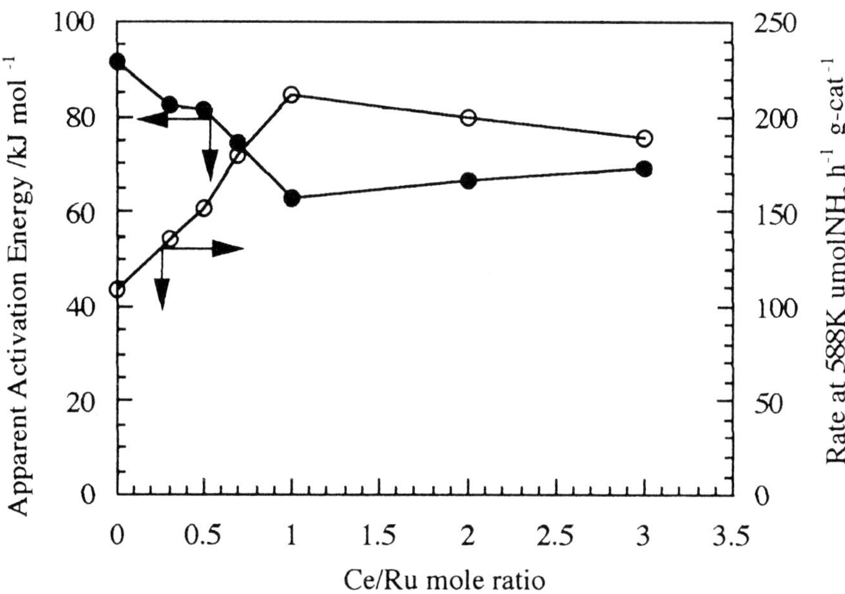

Figure 1. Ammonia synthesis rate at 588 K and apparent activation energy (573-623 K) as a function of Ce/Ru mole ratio in 3 wt% Ru-Ce_2O_3/MgO.

to 2 times and the apparent activation energy was decreased with an increase of Ce up to 1 or 2 mol ratio (Ce/Ru), with fall of activity after that (more than 2 mol ratio). The small decrease of activity is considered to be due to the excess covering of ceria. Judging from the change of apparent activation energy, the active site is thought to be changed from Ru-MgO to Ru-Ce_2O_3 above the Ce/Ru mole ratio of 1. The apparent activation energy on 3wt% Ru/CeO_2 treated at 623 K was 45 kJ mol^{-1}[10], lower than that for Ru-Ce_2O_3/MgO (Ce/Ru = 1~2). Although both energies (65 and 45kJ mol^{-1}) are lower than that on Ru/MgO, the reason for the discrepancy (Ru-Ce_2O_3/MgO, and Ru-CeO_2) has not been well studied.

Dependency of Reduction Temperature with Hydrogen

The authors once studied Ru-$Sm(NO_3)_3$/"Al_2O_3", and found that the high temperature reduction over 623 K "decreased" the activity [8]. A similar tendency has also been obtained for Ru-CsOH/MgO [8]. In this work, a different trend was found for Ru-lanthana/"MgO". Figure 2 shows the reduction temperature dependency on the activity of Ru-lanthana/MgO. Here, the activity even increases by 5 times when they are treated at 873 K before use, and the activity levels become comparable with those of Ru/lanthana [10] and Ru-CsOH/MgO (Table 1). These temperature dependencies were similar to that of Ru/lanthana [10]. Lanthana "promoters" were found to behave like lanthana "supports" on MgO but not on Al_2O_3. The SMSI phenomena have been proposed to occur for Ru/"lanthana" [10]. The reduction temperature dependency on the activity of Ru-CsOH/MgO was also studied here and was shown in Figure 3. As has been reported [8], the activity increased with the treatment temperature up to 673 K, after which it decreased drastically.

Activity Change With Time

Time courses of ammonia synthesis activity at (588 K) of three Ru-lanthana/MgO catalysts prereduced at 873 K are shown in Figure 4. The activities were found to drop with time-on-stream, but recovered when they were treated again by hydrogen

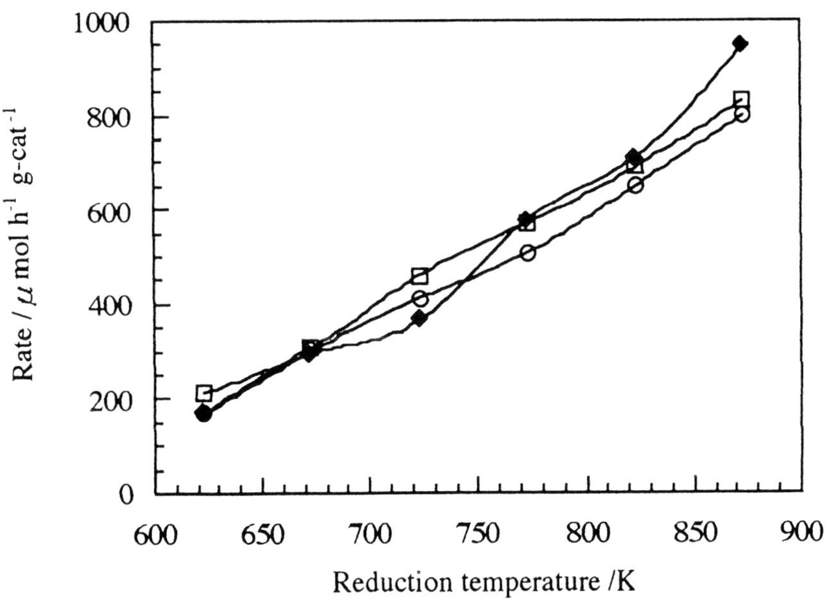

Figure 2. Ammonia synthesis rates at 588 K over lanthana promoted 3wt% Ru/MgO catalysts as a function of hydrogen treatment temperature. (□) Ru-La_2O_3/MgO, (O) Ru-Ce_2O_3/MgO, (◆) Ru-Sm_2O_3/MgO.

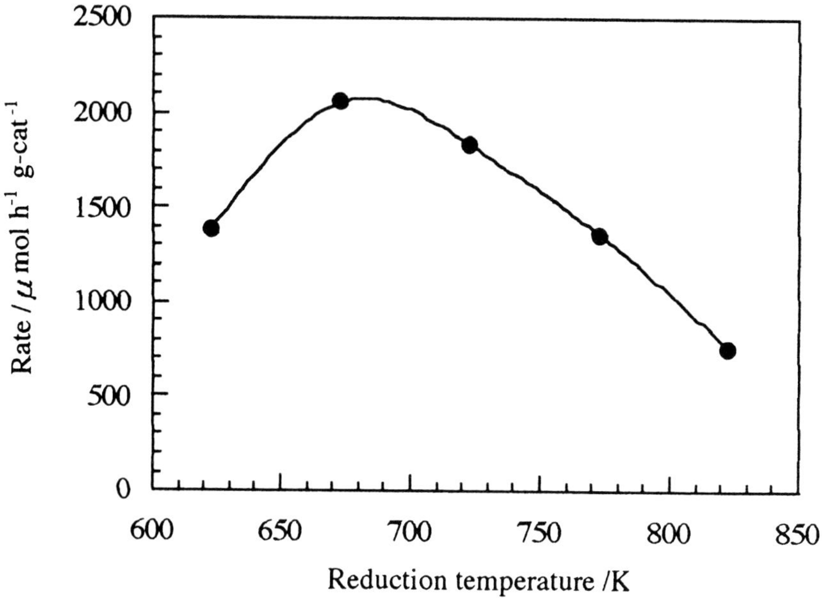

Figure 3. Ammonia synthesis rates at 588 K over 3wt% Ru-CsOH/MgO catalysts as a function of hydrogen treatment temperature.

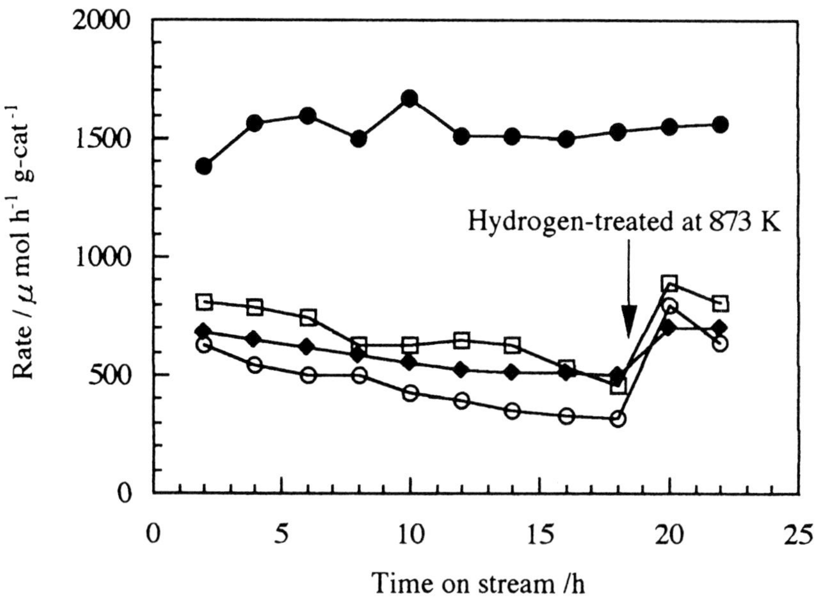

Figure 4. Time course of ammonia synthesis rates at 588 K over rare earth-promoted Ru catalysts (□) Ru-La_2O_3/MgO, (○) Ru-Ce_2O_3/MgO, (◆) Ru-Sm_2O_3/MgO that has been reduced at 873 K and over (●)Ru-CsOH/MgO catalysts that have been reduced at 623 K.

at 873 K for 2-4 h. It was also found that the activity never dropped below the lowest level (activated at 623 K in Figure 2). These characters are all similar to those of Ru/lanthana [10]. On the other hand the activity of Ru-CsOH/MgO reduced at 623 K was stable for 24 h as is shown in Figure 4.

DISCUSSION

Interaction Between Ruthenium and Lanthana "Promoter"

Lanthana oxide can be used both as an additive (third component) and as a support for ruthenium. The former is a three component system (Ru-lanthana/MgO) and the latter is a two component system (Ru/lanthana). Ru can be influenced both by the support and the promoter, for example MgO and Ce_2O_3 in the former case. The apparent activation energy gradually decreases from 100 to 65 kJ mol^{-1} by increasing Ce content on Ru/MgO. The activity also increases up to Ce^{3+}/Ru ratio of 1. This suggests that the active sites change from MgO-promoted Ru to Ce_2O_3-promoted Ru, when the Ce/Ru mole ratio increased to unity. Such a small amount of promoter is enough to interact with ruthenium, because the promoter decomposes at the Ru surface by activated hydrogen.

A similar trend has been observed for Ce^{3+}/Ru on Al_2O_3, Ba^{2+}/Ru on MgO, and Cs^+/Ru on MgO [8]. However, it was quite different for Cs^+/Ru on Al_2O_3. On the last catalyst, the activity was maximum when Cs^+/Ru ratio was more than 10 [8]. If the support is acidic like Al_2O_3, the alkali promoter may spread over whole the support surface independent of the Ru particles. As has been discussed in the previous paper, the promoter nitrate is decomposed on the Ru surface (hydrogen source) and may migrate to the support depending on the acid-base nature or melting point of promoters [4,8]. Considering this result, it is thought that lanthanide nitrate is also decomposed into oxide over the Ru surface, stays around the Ru particle, and the excess part migrates to the support surface.

Strongly Reduced Lanthana Oxide as Electron-Donor

The basic support or promoter has been found to be an effective component of Ru catalyst for ammonia synthesis because basic compounds have been thought to give electrons to the Ru surface, which activates the N-N bond adsorbed [4]. Sanderson's electronegativity of compounds can be used as relative electron accepting standards as follows; 2.63 (TiO_2), 2.37 (CeO_2), 2.23 (Sm_2O_3), 2.18 (La_2O_3), 2.12 (MgO), and 1.73(CsOH). It can be reasonably understood that Ru-CsOH/MgO is very active, Ru/MgO is active but, Ru/TiO_2 is extraordinarily less active [4]. In this sense, Ru/lanthana or Ru-lanthana/MgO is expected to give similar activity to Ru/MgO.

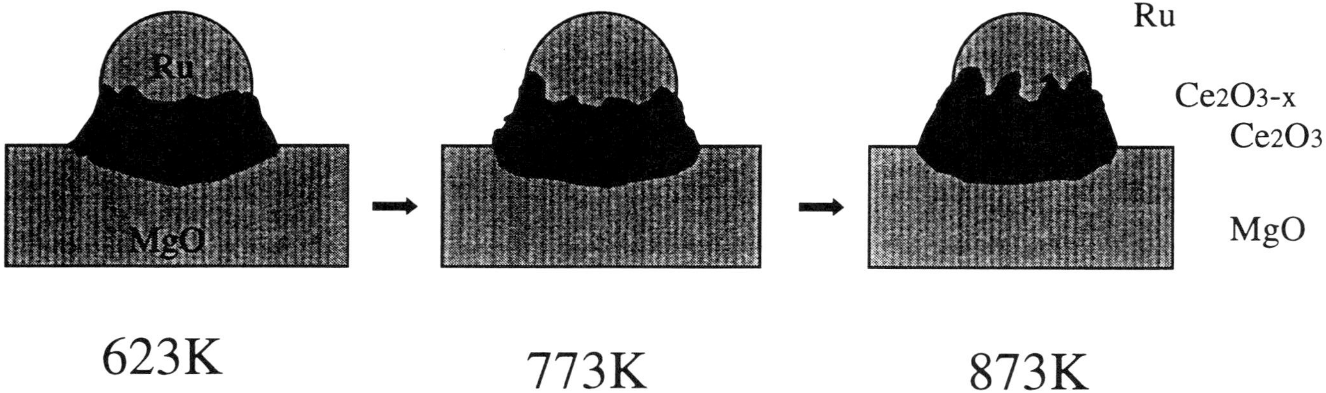

Figure 5. Active site model: Ru atoms attaching to the reduced ceria increases with the reduction temperature.

In this and the last study, however, Ru-lanthana/MgO (and Ru/lanthana) was found to give higher activity than Ru/MgO (Table 1 and Figure 1) and even to increase the activity when they are reduced by hydrogen at high temperature (Figure 2). As has been suggested partially reduced lanthana can donate an electron to Ru through the oxygen anion owing to the tendency to return to the stable and high oxidation state. Because the electronegativity of Ce_2O_3 (2.20) is still higher than that of MgO (2.12), the driving force of electrodonation must not be the bulk property but come from the special local (molecular) structure. The electron-donating effect of similar systems has been reported for the following examples. In the Ce-Ru (0001) system, it is presumed from the Auger peak ratio that metallic Ce donates electrons to Ru in the region less than 0.5 monolayer of Ce coverage [12]. In the highly reduced Ni/CeO_2 catalyst, the interaction between Ni and CeO_2 increases Ni electron density accompanied by the crystal-field effect [13]. If the cerium promoter exists as CeO_2 like the support, the reduced state might be Ce_2O_3 and the explanation might be easy.

Any stable lower valance states are not known for La_2O_3, Ce_2O_3, and Sm_2O_3, however, the activated hydrogen on the Ru surface may possibly migrate to the above oxides to form partially reduced states (SMSI): La_2O_{3-x}, Ce_2O_{3-x} or Sm_2O_{3-x} [14]. Such partially reduced lanthanide oxides tend to donate electrons to Ru. The high activity for CO, CO_2, and acetone hydrogenation has been explained as a result of interaction between the metal and cerium cation (Ce^{3+}) or oxygen anion vacancy [15,16,17]. N_2 dissociation might be also activated by a similar interaction. An SMSI model is proposed in Figure 5, where the Ru atoms near the reduced lanthana are considered to be the active site [18,19,20].

Deactivation and Regeneration of Active Site of Ru-Lanthana/MgO

The catalytic activity of Ru-lanthana/MgO gradually decays for a long time as Ru/lanthana does [10]. However, it can be recovered after hydrogen treatment at 873 K (Figure 4). Thus, the deactivation is not due to the ruthenium metal sintering. Two reasons are discussed here. One of the causes might be impurities included in the reactant, O_2, H_2O, CO and CO_2, which might oxidize the SMSI state (see the next section). N_2 and H_2 gas were used in ammonia synthesis without purification. To clarify the relation between Ru and CeO_2, the catalyst was treated by oxygen (at 298 K) and hydrogen (623-823 K), and checked the activity of Ru/CeO_2 (Figure 6). The oxidation treatment at room temperature turns surface ruthenium to oxides [21] and a part of the ceria must be oxidized again. The second hydrogen treatment restores Ru surface to metallic above 373 K. The activity was restored to about 80% of original value as is shown in Figure 6. These results may support the deactivation by impurities.

Another possibility is the surface reconstruction during ammonia synthesis,

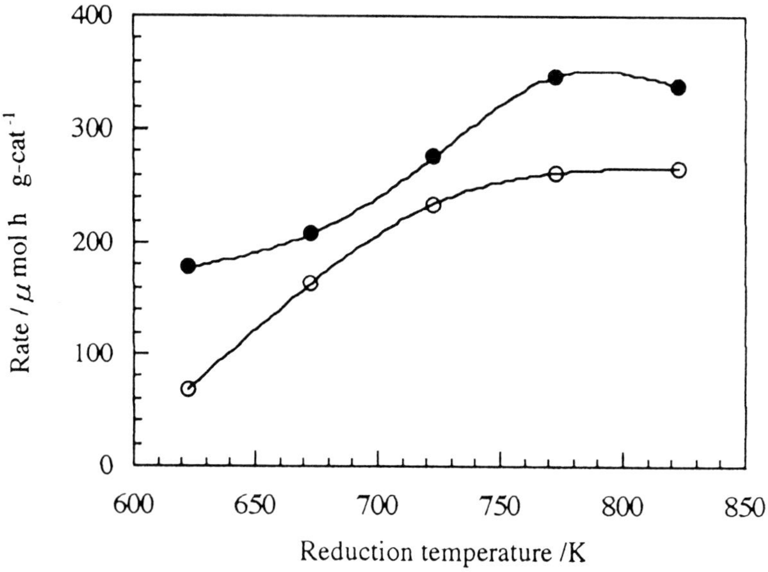

Figure 6. Ammonia synthesis rates over 1wt% Ru/CeO_2 at 588 K as a function of reduction temperature; comparison of the rate after first reduction (●) and the rate after second reduction (○). The sample was oxidized in O_2 at 298 K between the two reduction process. First and second hydrogen treatment were done at the same temperature.

which is known on iron surfaces [1]. Unstable active sites might be stabilized spontaneously at low reaction temperature (588 K) or through the adsorption-desorption cycle during the reaction. The active site "structure" can be recovered by hydrogen reduction at high temperature (873 K). At this moment, the deactivation mechanism (or active site structure) is not fully understood.

Comparison of Lanthana and Alkali as a Promoter

As is shown in Figure 3 the maximum activity of Ru-CsOH/MgO activated at 673K decreased to one third when it is treated at 823K. This is considered to be due to CsOH evaporation. It is also interesting to point out the CsOH promoted system is quite stable during the reaction at 588K as in shown in Figure 4. The Ru-CsOH system may not be affected by impurities such as H_2O or the active structure may be thermally stable. Promotion of Ru by CsOH is not considered to be due to SMSI, but due to electron donation of stable CsOH to Ru. The promotion mechanisms (or interacting mechanism with Ru) are different between lanthana and alkali.

CONCLUSIONS

Lanthana "promoter" increases the activity of Ru/MgO catalyst approximately 2 times for ammonia synthesis, when the catalyst is treated at low temperature (623 K). However, the activity increases 5 times when reduced at 873K compared to that reduced at 623 K. These phenomena were the same as that for Ru/lanthana but different from that for Ru-lanthana/Al_2O_3. A morphological structure change was suggested during hydrogen treatment for Ru-lanthana/MgO and Ru/lanthana (SMSI). It may increase the electron donating effect to the Ru. The highly active structure is not stable. The two reasons are considered: reoxidation with the impurity such as H_2O or the thermal reconstruction. A belt area model of active site was proposed: Ru atoms of the spherical cluster connecting to the reduced lanthana. CsOH promoter was also compared with lanthana, the former is stable for long runs but evaporated at high temperature. Both promoters have different interaction mechanisms. Although the highly active structure of Ru-lanthana (reduced) is not stable, the stabilized (low) activity is still attractive for the second generation ammonia catalyst because of the small hydrogen poisoning.

REFERENCES

1. K. Aika and K. Tamaru. In: *Ammonia*, A. Nielsen (Ed.), Springer-Verlag, Berlin, 1995.
2. K. Aika, A. Ohya, A. Ozaki, Y. Inoue, and I. Yasumori, *J. Catal.* **92**, 305 (1985).
3. Report article, *Appl. Catal.* **67**, N18 (1991).
4. K. Aika, T. Takano, and S. Murata, *J. Catal.* **136**, 126 (1992).
5. J. Kubota and K. Aika, *Chem. Commun.* 1554 (1991); *ibid, J. Phys. Chem.* **98**, 11293 (1994).
6. T. Hikita, Y. Kadowaki, and K. Aika, *J. Phys. Chem.* **95**, 9396 (1991).
7. K. Aika, M, Kumasaka, T. Oma, O. Kato, H. Matsuda, N. Watanabe, K. Yamazaki, A. Ozaki, and T. Onishi, *Appl. Catal.* **28**, 57 (1986).
8. S. Murata and K. Aika, *J. Catal.* **136**, 118 (1992).
9. Y. Kadowaki and K. Aika, *J. Catal.* **161** 178(1996).
10. Y. Niwa and K. Aika, *J. Catal.* **162** 138(1996).
11. P. Moggi, G. Predieri, G. Albanesi, S. Papadopulos, and E. Sappa, *Appl. Catal.* **53**, (1989).
12. A.P. Walker and R.M. Lambert, *J. Phys. Chem.* **96**, 2265 (1992).
13. E. Ramaroson, J.F. Tempere, M.F. Guilleux, F. Vergand, H. Roulet, and G. Dufour, *J. Chem. Soc. Faraday Trans.* **88**, 1211 (1992).
14. S.J. Tauster, S.C. Fung, and R.L. Garten, *J. Am. Chem. Soc.* **100**, 170 (1978).
15. A. Trovarelli, G. Dolcetti, C. Leitenburg, J. Kaspar, P. Finetti, and A. Santoni, *J. Chem. Soc. Faraday Trans.* **88**, 1311 (1992).
16. L. Mendelovici and M. Steinberg, *J. Catal.* **93**, 353 (1985).
17. L. Mendelovici and M. Steinberg, *J. Catal.* **96**, 285 (1985).
18. G. Wrobel, M.P. Sohier, A. D'Huysser, J.P. Bonnelle, and J.P. Marcq, *Appl. Catal.* A **101**, 73 (1993).
19. R. Burch and A.R. Flambard, *J. Catal.* **78**, 389 (1982).
20. J.B. Bracey and R. Burch, *J. Catal.* **86**, 384 (1984).
21. S.K. Masthan, K.V.R. Chary, and P.K. Rao, *J. Catal.* **124**, 289 (1990).

Res. Chem. Intermed., Vol. 24, No. 5, pp. 605-612 (1998)

COMPARATIVE STUDY OF COKE DEPOSITION ON CATALYSTS IN REACTIONS WITH AND WITHOUT OXYGEN

P. PRASERTHDAM, C. CHAISUK and P. KANCHANAWANICHKUN
Petrochemical Research Laboratory, Department of Chemical Engineering, Chulalongkorn University, Bangkok 10330, Thailand

Received 1 December 1997; accepted 27 November 1997

Abstract--Two types of catalysts, i.e. Pt/γ Al_2O_3 and Cu/Na-ZSM-5, were used to investigate the catalyst activity and amount of coke formation on the spent catalysts. The reactions of particular interest were the hydrocarbon oxidation and the SCR of NO with and without O_2. Propane and propene were used as the hydrocarbon sources. The reaction conditions were as follows: reaction temperature = 170-500°C, GHSV = 4,000 hr^{-1}, TOS = 2 hr, feed composition depending on each reaction, but the composition of gases were fixed as HC = 3,000 ppm, NO = 1,000 ppm and O_2 = 2.5%, using He balance. It was found that both the case of Pt/γ Al_2O_3 and the case of Cu/Na-ZSM-5, propene provided higher conversion and coke deposition than propane in the presence or the absence of O_2 and/or NO. For Pt/γ Al_2O_3 catalyst, in case of the absence of oxygen reactions, the propene conversion dropped more rapidly than the propane conversion. Finally the reaction of propene gave a lower percent of hydrocarbon conversion than the reaction of propane. Additionally, propene had a higher percent selectivity of coke formation for the reaction with the absence of oxygen, but propane had a higher percent selectivity of coke formation for the reaction with the presence of oxygen. For Cu/Na-ZSM-5, in the system with absence and presence of oxygen, the addition of oxygen caused a significant change in % coke selectivity. With the presence of NO_x, the percent conversion of both propane and propene decreased and that the % coke selectivity of propane decreased, whereas that of in propene increased.

INTRODUCTION

Coke deposition is an important deactivation mode in the hydrocarbon conversion process In general, most researchers [1-5] have emphasized only coke formation in reducing atmospheres e.g. dehydrogenation, cracking, reforming etc. Nevertheless, in oxidizing atmospheres, coke deposition can also take place on the catalyst surface. Therefore, in the present work, it is set up to compare the amount of coke formed on Pt/Al_2O_3 and Cu/Na-ZSM-5 catalysts for reactions with the absence and presence of oxygen. The catalytic activity and the amount of coke on the spent Pt/Al_2O_3 and Cu/Na-ZSM-5 catalysts were investigated for reactions with oxygen, *i.e.* hydrocarbon combustion and reduction of NO_X, and without oxygen, *i.e.* dehydrogenation or aromatization and reduction of NO_X. Propane and propene were used as the hydrocarbon sources for all four reactions.

EXPERIMENTAL

Pt/Al_2O_3 (0.3 wt.% Pt) catalyst in this study was prepared by a dry impregnation method using H_2PtCl_6 as the salt precursor. The parent Na-ZSM-5 zeolite with Si/Al ratio of 50 was hydrothermally synthesized from gel and decant solution in an autoclave. The structure of ZSM-5 was confirmed by X-ray diffraction (XRD). Cu/Na-ZSM-5 zeolite was prepared by exchanging Cu^{2+} into Na-ZSM-5 sample in the aqueous solution. The catalytic reactions were carried out at atmospheric pressure in a fixed bed reactor. A 0.5 g of catalyst was packed in a quartz tube reactor. In the case of Cu/Na-ZSM-5, before the reaction, the catalyst was heated under He flow from room temperature to 500°C in 1 hr., and held for 1 hr. before being cooled down. In the case of Pt/Al_2O_3, the catalyst was reduced to 500°C for 1 hr using hydrogen as the reductant gas. The 50 cc./min. of mixed gas feed consisting of 100 ppm NO, 3000 ppm hydrocarbon, 2.5% vol. oxygen and He was introduced to the reactor at a space velocity of 4000 hr^{-1} . The temperatures of the reaction with and without oxygen were 170°C and 350°C (in the case of Pt/Al_2O_3) or 500 °C (in the case of Cu/Na-ZSM-5) respectively. The outlet gases were analyzed by SHIMADZU GC-8APT gas chromatograph with MS-5A column for nitrogen, oxygen and carbon monoxide and with SHIMADZU GC-8AIT porapak QS column for carbon dioxide, propane and propene. Coke deposited on the catalysts was characterized by temperature programmed oxidation (TPO). Before starting the TPO, 0.5 g of the spent catalysts was heated to 130°C at 10°C/min. under He atmosphere and held for 3 hr. The heat treatment removed any air and water that was adsorbed on the catalyst. Then the pretreated sample was heated from 50°C to 700°C with a heating rate of 5°C/min. then in a 30 cc./min. stream of 1% O_2 in helium gas. The carbon dioxide formed was determined by SHIMADZU GC-8AIT gas chromatograph using a thermal conductivity detector with parapak QS column. The percentage of carbon in coke can be calculated from TPO curves. CO_2 area is divided by a internal time in which CO_2 flows through the sampling loop (1 cc.). The rate of CO_2 formation is, hence, obtained. The area under the curve of CO_2 formation rate versus time gives the value of total CO_2 formation. Finally, this value is converted to milligram carbon or percentage of carbon by using a calibration curve.

RESULTS AND DISCUSSION

For the Case of Pt/Al_2O_3

The experimental results were shown in Table 1, Figure 1 and Figure 2. In all four reactions, it was observed that the reactions with propene as a reactant gave a larger percent of hydrocarbon conversion in the initial interval of time on stream and also

Table 1
Hydrocarbon conversion, amount of coke and coke selectivity of 0.3% Pt/Al_2O_3.

Reaction	Reactant	Temperature (°C)	% HC conversion (at 5 min.)	% Carbon in coke	% Selectivity of coke formation
HC	Propane	350	59.36	0.17	0.22
	Propene	350	98.88	0.32	0.27
HC+NO	Propane	350	46.17	0.13	0.19
	Propene	350	66.14	0.16	0.34
$HC+O_2$	Propane	170	11.71	0.13	1.13
	Propene	170	100.00	0.48	0.29
$HC+NO+O_2$	Propane	170	5.59	0.12	1.07
	Propene	170	40.12	0.33	0.54

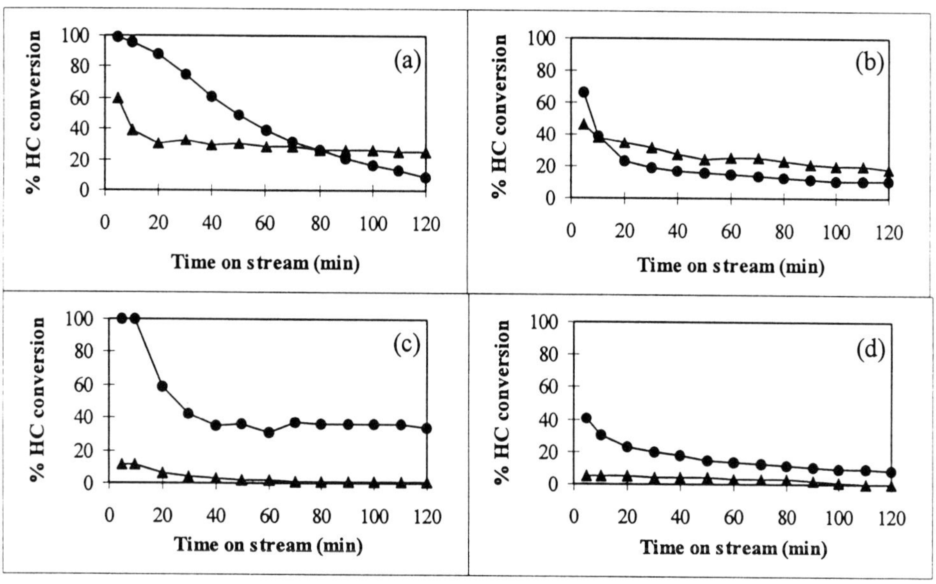

Figure 1. Relationship between % HC conversion versus time on stream. (a) Feed: 3000 ppm HC + He balance, Temperature = 350°C. (b) Feed: 3000 ppm HC + 1000 ppm NO + He balance, Temperature = 350°C. (c) Feed: 3000 ppm HC + 2.5 vol.% O_2 + He balance, Temperature = 170°C. (d) Feed: 3000 ppm HC + 1000 ppm NO + 2.5 vol.% O_2 + He balance, Temperature = 170°C.: (♦) Propane, (●) Propene.

had a larger amount of coke than the reactions with propane as a reactant. However, in the dehydrogenation reaction and NO_X reduction under the absence of oxygen condition, the propene conversion dropped more rapidly than the propane

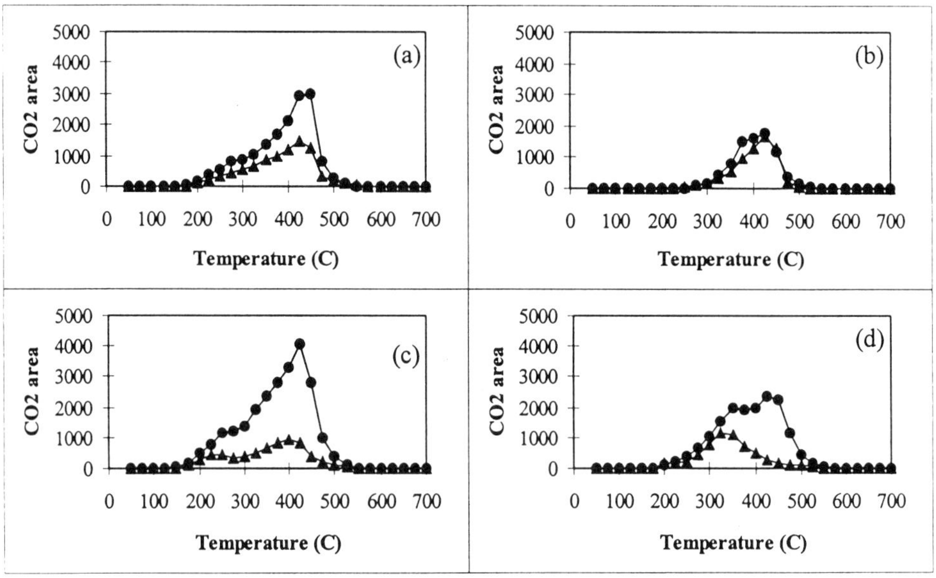

Figure 2. TPO profiles of 0.3 g coked catalyst from reaction. (a) Dehydrogenation, Temperature = 350°C. (b) Reduction of NO_x under absence of oxygen condition, Temperature = 350°C. (c) Combustion, Temperature = 170°C. (d) Reduction of NO_x under absence of oxygen condition, Temperature = 170°C.: (♦) Propane, (●) Propene.

conversion. Finally the reaction of propene gave a lower percent of hydrocarbon conversion than the reaction of propane. It meant that propene is more active than propane [6]. Since propene reacted more, there was more opportunity to convert to coke precursor. Additionally, it was clarified that the product of propene from the reaction with the absence of oxygen, propadiene which is converted irreversibly to ethylidyne leading to coke deposition [7,8], was more reactive than propene [6]. Thus the catalysts in the reaction using propene as a reactant were rapidly covered by carbonaceous deposits. The TPO profiles of 0.3% Pt/Al_2O_3 catalysts are indicated in Figure 2. Each sample showed a TPO peak around 425°C except the TPO profiles of spent catalyst from reduction of NO_X in the presence of oxygen condition. This TPO profiles showed a TPO peak at a temperature of around 325°C for the reaction with propane as a reactant and two peaks at the temperature around 325°C and 425°C for that with propene as a reactant. Barbier *et al.* [3] suggested that the first peak is the coke on metal and the second peak is the coke on the support. In this paper, we define the selectivity of coke formation as the ratio of carbon atom in coke to carbon atom of feed hydrocarbon converted. It was found that propene had a higher percent of selectivity of coke formation for the reaction with the absence of oxygen, but propane had a higher percent of selectivity of coke formation for the reaction with the presence of oxygen. Under the presence of oxygen, it suggested that coke is formed in parallel with carbon dioxide formation. Propene or coke precursor is more effectively reacted with oxygen to carbon dioxide [9]. Additionally, for the reaction with the presence of oxygen using propane as a reactant, it was observed that the reaction with the absence of NO gave a higher percent of selectivity of coke formation than the reaction with the presence of NO. On the other hand, in the case of propene, the reactant with the presence of NO gave higher percent selectivity of coke formation. It suggests that propane or propene is first reacted with adsorbed oxygen to be converted to intermediates [10,11]. In the case of propane [10,11], these intermediates are preferably reacted with NO which result in, when NO was added in feed, propane producing less selectively of coke formation. However, in the case of propene [12], NO hardly reacted with the intermediates but it preferred to dissociate into dinitrogen. Thus, since the dissociation of NO hinders the reaction of carbon dioxide formation, the intermediates prefer produce coke rather than carbon dioxide for the case of the presence of NO reaction.

For the Case of Cu/Na-ZSM-5

The experimental results were summarized in Figure 3. It was found that the propene conversion was higher than that of propane. This is particularly obvious for the reaction with the presence of oxygen and absence of NO_x as shown in Figure 3(b). Propane conversion was only 10 % ,whereas propene conversion is about 100 %. Figure 3(c),(d) (reduction of NO_x with and without O_2) exhibited the effect of NO_x on the reaction. It was found that the % of propene conversion decreased when NO_x was added, however the % of propene conversion was still higher than that of propane. It means that propene is more active than propane [6]. The temperature program oxidation (TPO) results were shown in Figure 4. It was found that the amount of coke from the four reactions of propene was greater than propane over Cu/Na-ZSM-5 zeolite. The results in Table 2 can suggest that in aromatization, the percentage of propene conversion was greater than that of propane; however, on the other hand, the % coke selectivity of propene was less than that of propane, because propene was converted to product more than to coke while propane formed product less than coke. In the system with absence and presence of oxygen it was found that the addition of oxygen caused a significant change in % of coke selectivity [13-15]. With the presence of NO_x, it was found that the percent conversion of both propane and propene decrease and that the % of coke selectivity of propane decreased whereas that of in propene increased. From this result in the case of propene, we can propose that NO_x was adsorbed on the surface of the catalyst to form an intermediate which is strongly adsorbed and, hence, the desorption rate is slow [16-21]. As a result, the intermediate can not selectively form the product but can be further

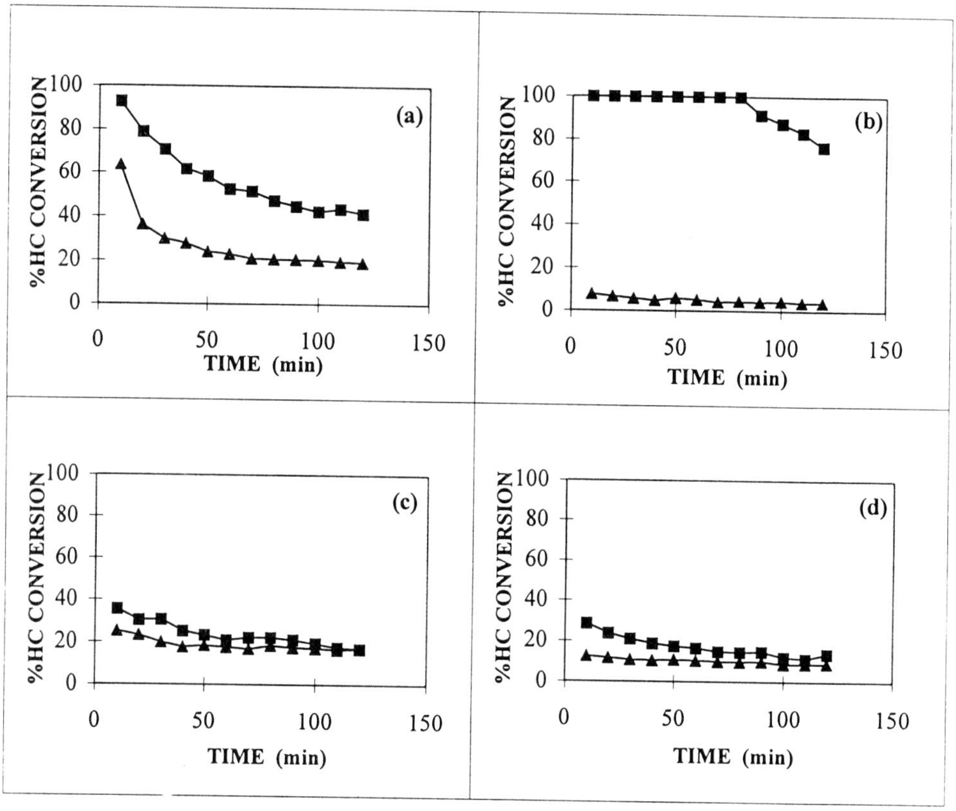

Figure 3. Relationship between % HC conversion versus time over 1.58% Cu/Na-ZSM-5; (a) Feed: 3000 ppm HC + He balance, Temperature = 500°C. (b) Feed: 3000 ppm HC + 2.5 vol.% O_2 + He balance, Temperature = 170°C. (c) Feed: 3000 ppm HC + 1000 ppm NO + He balance, Temperature = 500°C. (d) Feed: 3000 ppm HC + 1000 ppm NO + 2.5 vol.% O_2 + He balance, Temperature = 170°C: (▲) C_3H_8, (■) C_3H_6

converted to coke.

CONCLUSION

Both in the case of Pt/γ Al_2O_3 and in the case of Cu/Na-ZSM-5, propene provided both higher conversion and coke deposition than propane in the presence or the absence of O_2 and/or NO. For Pt/γ Al_2O_3 catalyst, in the case of the absence of oxygen reactions, the propene conversion dropped more rapidly than the propane conversion. Finally, the reaction of propene gave a lower percentage of hydrocarbon conversion than the reaction of propane. Additionally, propene had a higher

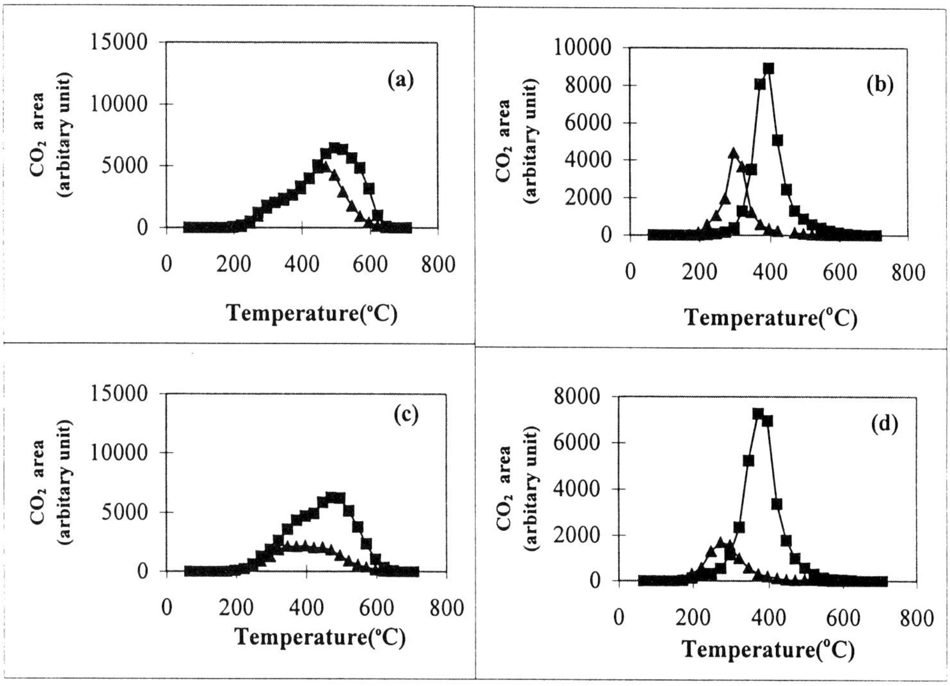

Figure 4. TPO curve of 1.58% Cu/Na-ZSM-5 after the reaction; (a) HC at 500 °C, (b) HC+O_2 at 170 °C, (c) HC+NO at 500 °C, (d) HC+NO+O_2 at 170 °C; (▲) C_3H_8, (■) C_3H_6; In the case of propene (b,d), amount of coked catalyst for TPO is five times than that of propane.

Table 2
Percentage HC conversion and selectivity of coke over 1.58%Cu/Na-ZSM-5

Reaction	Reactant	Temperature (°C)	% HC conversion (at 5 min.)	% Carbon in coke	% Selectivity of coke formation
HC	propane	500	75	0.45	0.74
	propene	500	98	0.68	0.51
HC+O_2	propane	170	9	0.17	1.43
	propene	170	100	2.03	0.86
HC+NO_x	propane	500	26	0.25	0.45
	propene	500	38	0.67	1.51
HC+NO_x+O_2	propane	170	15	0.10	0.39
	propene	170	38	1.82	2.80

percentage of selectivity of coke formation for the reaction with the absence of oxygen but propane had higher percent selectivity of coke formation for the reaction with the presence of oxygen. For Cu/Na-ZSM-5, in the system with absence and presence of oxygen, the addition of oxygen caused a significant change in the % of coke selectivity. With the presence of NO_x, the percent conversion of both propane and propene decreased and that of the % coke selectivity of propane decreased, whereas that in propene increased.

REFERENCES

1. W.G. Appleby, J.W. Gibson, and G.M. Good, *I & EC Process Design and Development* **1**, 102 (1962).
2. J.N. Beltramini, E.E. Martinelli, E.J. Churin, N.S. Figoli, and J.M. Parera, *Applied Catalysis* **7**, 43 (1983).
3. J. Barbies, *Catalyst Deactivation* 1 (1987).
4. J. Biswas, P.G. Gray, and D.D. Do, *Applied Catalysis* **32**, 249 (1987).
5. Mikael Larsson, Magnus Hulten, Edd A Blekkan, and Bengt Andersson, *Journal of Catalysis* **164**, 44 (1996).
6. Morrison and Boyd, *Organic Chemistry (sixth edition)*: 122, 287, 410.
7. Gabor A. Somorjai, *Introduction to Surface Chemistry and Catalysis*: 420.
8. B.J. McIntyre, M. Salmeron, and G.A. Somorjai, *Journal of Catalysis* **164**, 184 (1996).
9. R. Burch and T.C. Watling, *Catalysis Letters* **43**, 19 (1997).
10. Motoi Sasaki, Hideaki Hamada, Yoshiaki Kintaichi, and Takehiko Ito, *Catalysis Letters* ***15***, 297 (1992).
11. Megumu Inaba, Yoshiaki Kintaichi, and Hideaki Hamada, *Catalysis Letters* **36**, 223 (1996).
12. R. Burch, P.J. Millington, and A.P. Walker, *Applied Catalysis* **B4**, 65 (1994).
13. J.L. d' Itri and W.M.H. Sachtler, *Catal. Lett.*. **15**, 289 (1992).
14. H. Hamada, Y. Kintaichi, M. Sasaki, T. Ito, and M. Tabata, *Appl. Catal*. **64**, L1 (1990).
15. A.Yu Stakheev, C.W. Lee, S.P. Park, and P.J.Chong, *Progess in Zeolite and Microporous Materials.*, Vol **105** , 1579 (1997).
16. H. Hamada, Y. Kintaichi, M. Sasaki, M. Tabata, and T. Ito, *Appl. Catal.,* **70**, L15 (1991).
17. C.J. Bennett, P.S. Bennett, S.E. Golunski, J.W. Hayes, and A.P. Wallker, *Appl. Catal. A,* **86** L1 (1992).
18. R.Burch and P.J. Millington , *Appl. Catal B:Env, 2.* 101 (1993).
19. M. Iwamoto, H. Yahiro, H. Khin, M.Watanabe, J Gue, M. Konno, T. Chikabisa and T. Murayama, *Appl. Catal* **B5**., L1 (1994).
20. Chikafumi Yokotama and Makoto Misono, *Journal of Catalysis.* **160**, 95 (1996).
21. Janos Szanyi and Mark T. Paffett, *Journal of Catalysis.* **164** 232 (1996).

Res. Chem. Intermed., Vol. 24, No. 5, pp. 613-623 (1998)

A COMPARATIVE STUDY ON THE ACTIVITY OF METAL EXCHANGED MCM22 ZEOLITE IN THE SELECTIVE CATALYTIC REDUCTION OF NO_X

A. CORMA*, A.E. PALOMARES+ and V. FORNÉS*

* *Instituto de Tecnología Química, UPV-CSIC, Universidad Politécnica de Valencia, Avda. de los naranjos s/n, 46022 Valencia SPAIN*

+ *Departamento de Ingeniería Química y Nuclear, Universidad Politécnica de Valencia, Avda. de los naranjos s/n, 46022-Valencia, SPAIN*

Received 8 December 1997; accepted 26 December 1997

Abstract --A series of MCM22 zeolites exchanged with copper and cobalt are studied for the selective catalytic reduction (SCR) of NO with C_3H_8 and the behavior compared with that of Cu and Co-beta and ZSM5 zeolites. The results show that Co and Cu-MCM22 samples are stable SCR catalysts. These zeolites give the maximum activity at 450°C. Their behavior towards oxygen content in the reactant phase and exchange level of the metal in the catalyst, is qualitatively similar to that of metal exchanged beta and ZSM5 zeolite, but the yields obtained with this zeolite are lower in any case. The infrared studies of adsorbed NO show, contrary to what is occurring in ZSM5 in which only Cu^+ sites are observed at low NO partial pressure, that in this condition, Cu^+ and Cu^{2+} species are formed on MCM22. The results indicate that in MCM22, the copper located in the 10 member ring (MR) circular channels behaves similarly to that present in ZSM5, while the Cu present in the 12 MR cavities has a strong tendency to agglomerate forming non active CuO clusters.

INTRODUCTION

The demand for environmentally friendly technologies has shown the enormous potential of catalysis to help in reducing waste and undesired emissions. Indeed, more efficient catalysts increase the selectivity to the desired products maximizing the profitability of the natural raw materials, while decreasing the formation of unwanted and sometimes difficult to store subproducts. Moreover, when the formation of undesired or contaminant products cannot be avoided, catalysis may allow the reuse of the unwanted emissions or at least it can transform them in less harmful products. This is the case for NO_x emissions which can be now treated by using transition metal containing zeolite catalysts [1]. Unfortunately the NO_x decomposition activity is strongly diminished by the presence of O_2 and SO_2 [2,3]. Nevertheless the possibilities of this type of catalysts are still high taking into account that they also catalyze the selective reduction of NO_x with hydrocarbons in the presence of O_2 [4,5]. This type of reaction could replace the traditional removal of NO using NH_3 as reductant [6,7], overcoming the safety problems associated with the use of NH_3 [8].

Large and medium pore zeolite structures exchanged with different cations, i.e., Cu, Ga, Ce, Fe, Rh, etc. have shown activity for reducing NO_x with hydrocarbons such as propane, propylene, ethylene and other C_3^+ hydrocarbons [3, 9-18]. The best results were achieved with Cu exchanged ZSM5 zeolite [9,10] and Cu exchanged beta zeolite [14], but these results are not still good enough to allow its industrial use, therefore improved catalysts should be designed. It has been recently shown [19,20] that Co exchanged ZSM5 and beta zeolites are also able to remove NO and N_2O simultaneously at temperatures in the 400-450°C range and in the presence or not of oxygen [21-25]. It is also interesting to point out that with this type of catalysts, methane (not considered traditionally as a selective reductant) can be used as reductant, even though the conversion achieved is not very high and the system is very sensitive to the O_2 concentration [21].

The reason why these zeolites are active for the removal of NO, is still a matter of discussion and active research. The aim of this work is to compare the activity of Cu and Co containing ZSM5 and beta zeolites with a MCM22 zeolite exchanged with Cu and Co. This zeolite which can be synthesized in a large range of compositions has a very peculiar structure. Thus, the study of the catalytic activity of Cu and Co containing MCM22 for the selective catalytic reduction (SCR) of NO should allow us, not only to find the intrinsic possibilities of this new material, but also to bring some light on the influence of the zeolite structure on the SCR ability of zeolites.

EXPERIMENTAL

Materials.

MCM22 synthesized in our laboratory following [26] and beta and ZSM5 zeolites obtained from PQ Zeolites have been used as starting materials. The metal exchange was done in an aqueous solution containing $Co(CH_3COO)_2 \cdot 4H_2O$ or $Cu(CH_3COO)_2 \cdot 4H_2O$ with adequate concentrations to achieve the desired amount of metal on the zeolite, and with a solid/liquid ratio = 10 $g \cdot l^{-1}$. The exchange was carried out at 80°C under stirring for 24 hours. After that, the zeolite was filtered and washed, and then calcined at 450°C for 4 hours. The composition of the samples was analyzed by atomic absorption, and the results are given in Table 1.

In Table 1 zeolites are named starting with the name of zeolite structure, followed by numbers indicating the Si/Al ratio and the level of metal exchange (100% atomic exchange is considered when the metal/Al ratio is 0.50).

TABLE 1
Chemical analysis of the zeolites.

Catalyst	SiO_2	Al_2O_3	Na_2O	CoO	CuO	Si/Al	Met/Al	%
			w%					exch.
ZSM5-Cu-27-74	95.3	3.0			1.7	27	0.37	74
beta-Cu-12-106	87.9	6.6	0.18		5.4	11.4	0.53	106
MCM22-Cu-23-88	93.9	3.5	0.16		2.4	22.7	0.44	88
Y-Cu-13-207	85.7	5.4	0.27		8.7	13.5	1.04	207
beta-Co-15-119	90.0	5.3		4.6		14.6	0.6	119
ZSM5-Co-17-178	90.0	4.3	0.02	5.7		17.6	0.89	179
MCM22-Co-14-145	88.8	5.4	0.07	5.8		13.9	0.72	145

Catalytic measurements

The NO_x reduction experiments were carried out in a fixed bed, quartz tubular reactor of 2.2 cm of diameter and 53 cm of length. In a typical experiment, 3.0 g of a zeolite catalyst, as particles of 0.25-0.42 mm size, were introduced into the reactor and were heated up to 450°C under nitrogen flow. The flow was maintained at this temperature for 6 hours. After that, the desired reaction temperature was set and the reaction feed admitted. This consists of 1500-2000 ml.min^{-1} of a mixture composed of 650 ppm of NO, 450 ppm of C_3H_8 and variable quantities of oxygen. The NO_x present in the outlet gases from the reactor was continuously analyzed by means of a chemiluminescence detector.

Infrared study on NO adsorption

For i.r. measurements the samples were compressed into self supporting wafers of 10 mg.cm^{-2} and placed in a vacuum i.r. cell. After treating the samples to 400°C at 10^{-4} Torr during 10 hours, background spectra were recorded. Afterwards, increasing amounts of NO were introduced in the cell and the spectra were registered at room temperature.

RESULTS AND DISCUSSION

In a preliminary experiment the stability of the catalysts was studied by carrying out the NO_x reduction at 450°C, 3% of oxygen, 650 ppm of NO and 425 ppm of C_3H_8. No significant changes in activity were observed at least during the two hours duration of the experiment (Figure 1). This observation allowed us to perform several experiments on the same catalyst sample. Nevertheless, and in all the cases, the stability of the catalyst was checked by repeating at the end of the last run, the first experiment. No changes due to the catalyst deactivation were observed in any of the samples studied.

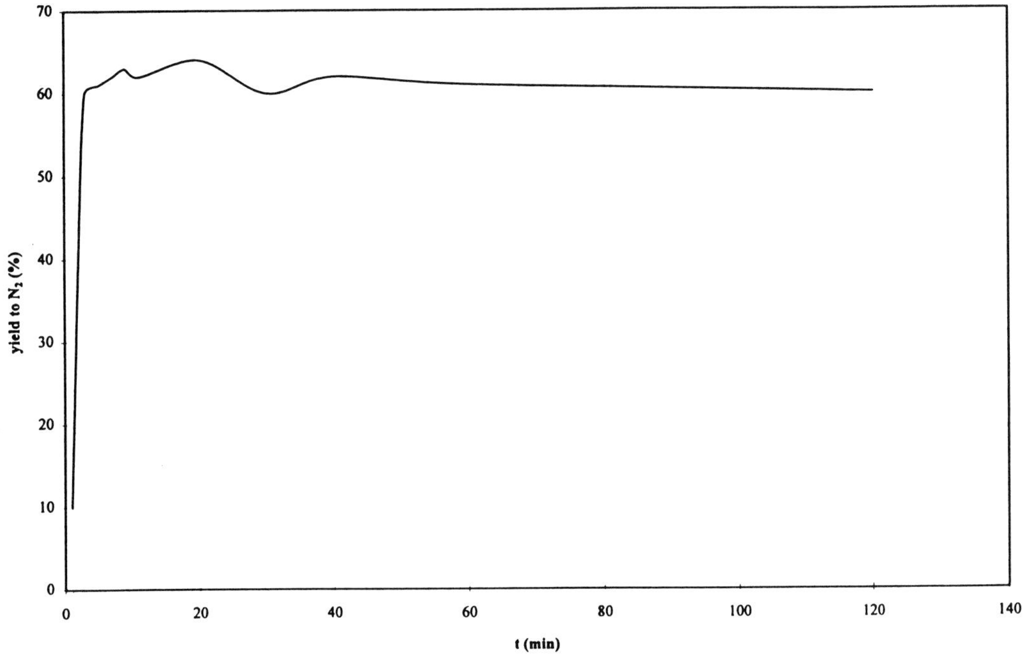

Figure 1. Stability of MCM22-Cu-23-88 in the course of the reaction, T=450°C, 3% O_2, 650 ppm NO, 425 ppm C_3H_8.

The impact of the reaction conditions on the catalytic activity was studied for the MCM22 exchanged with copper and with cobalt. In Figure 2, the influence of the reaction temperature on the activity of Co-MCM22 for the reduction of NO with propane is presented and compared with that of the Co-beta and Co-ZSM5 zeolites. The maximum activity for the catalysts is achieved at 450°C, observing a strong increase in the conversion when the temperature goes from 350 to 450°C. At higher temperatures than the above, the conversion decreases as a consequence of the higher activation energy of the competing reaction, i.e., the oxidation of propane, which produces a depletion of reductant with the corresponding decrease in the efficiency of NO_x reduction:

$$2NO + C_3H_8 + 4O_2 \rightarrow N_2 + 3CO_2 + 4H_2O$$

$$C_3H_8 + 5O_2 \rightarrow 3CO_2 + 4H_2O$$

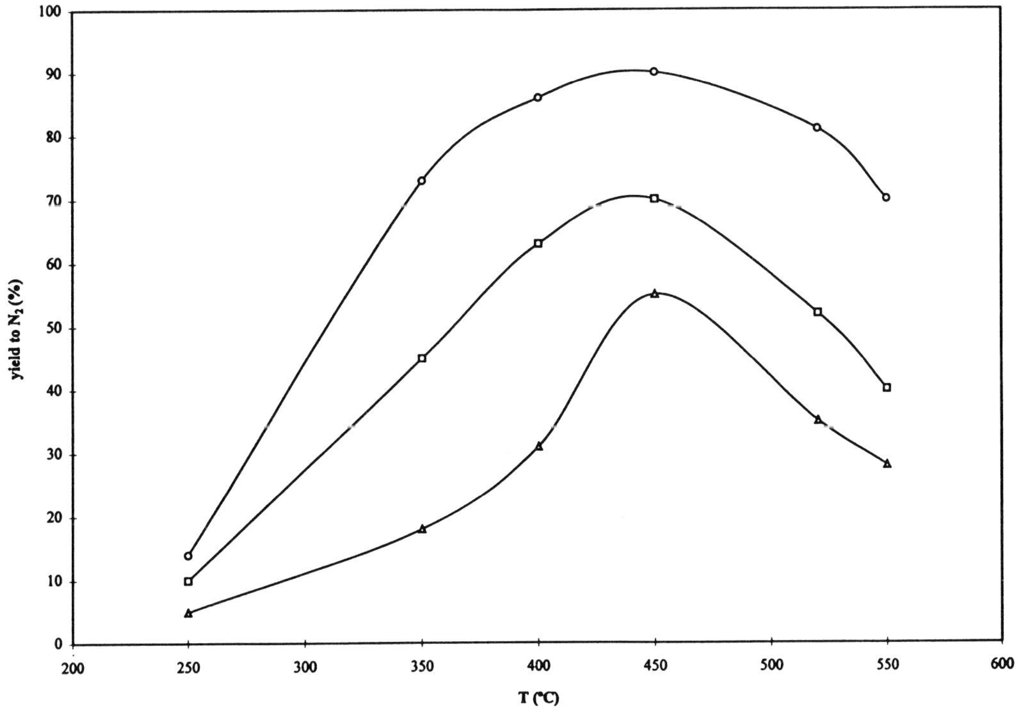

Figure 2. Influence of the temperature in the SCR of NO with propane on cobalt exchanged zeolites. (❑) ZSM5-Co-17-178, (○) beta-Co-15-119, (Δ) MCM22-Co-14-145, 3% O_2, 650 ppm NO, 425 ppm C_3H_8.

Similar results were obtained when copper instead of cobalt was exchanged with MCM22, as can be seen in Figure 3. The maximum activity is achieved at 450°C in MCM22, while in the case of Cu-ZSM5 and Cu-beta, the maximum occurs at temperatures close to 350°C. In MCM22, as well as in beta and ZSM5, a further increase of the temperature leads to a decrease in the activity of the catalyst. It should be pointed out that an increase in reaction temperature does not cause a permanent loss in catalyst activity, since when the reaction temperature was restored to 450°C, the maximum conversion was again observed. The activity of the copper exchanged MCM22 was also compared with that of a zeolite with large cavities such as a copper containing dealuminated faujasite with a unit cell of 24.33 Å. It can be seen in Figure 3, that MCM22 is more active than the faujasite at temperatures higher than 375 °C, but it is less active than ZSM5 or beta. In other words, the MCM22 zeolite is showing an intermediate behavior between faujasite and beta or ZSM5 zeolites.

In figures 4 and 5 the influence of the oxygen concentration in the selective catalytic reduction of NO with propane over Co and Cu-MCM22 is shown and the results are compared with those obtained on the other zeolites. In all cases, when oxygen is not present in the reaction mixture, the conversion is very low. However it increases rapidly after adding small quantities of O_2. The maximum activity is

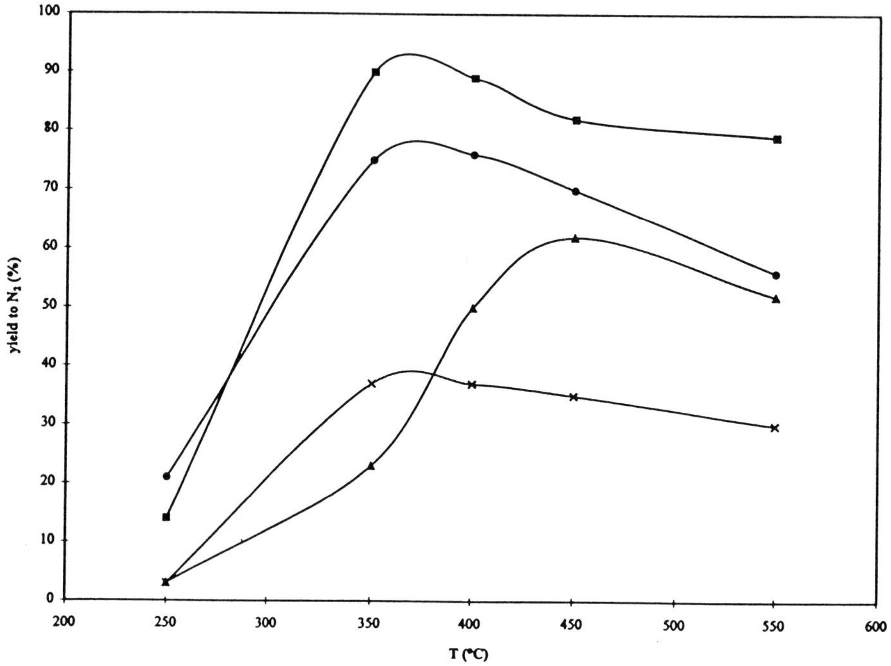

Figure 3. Influence of the temperature in the SCR of NO with propane on copper exchanged zeolites. (■) ZSM5-Cu-27-74, (●) beta-Cu-12-106, (▲) MCM22-Cu-23-88 and (x) Cu-Y-13-207, 3% O_2, 650 ppm NO, 425 ppm C_3H_8.

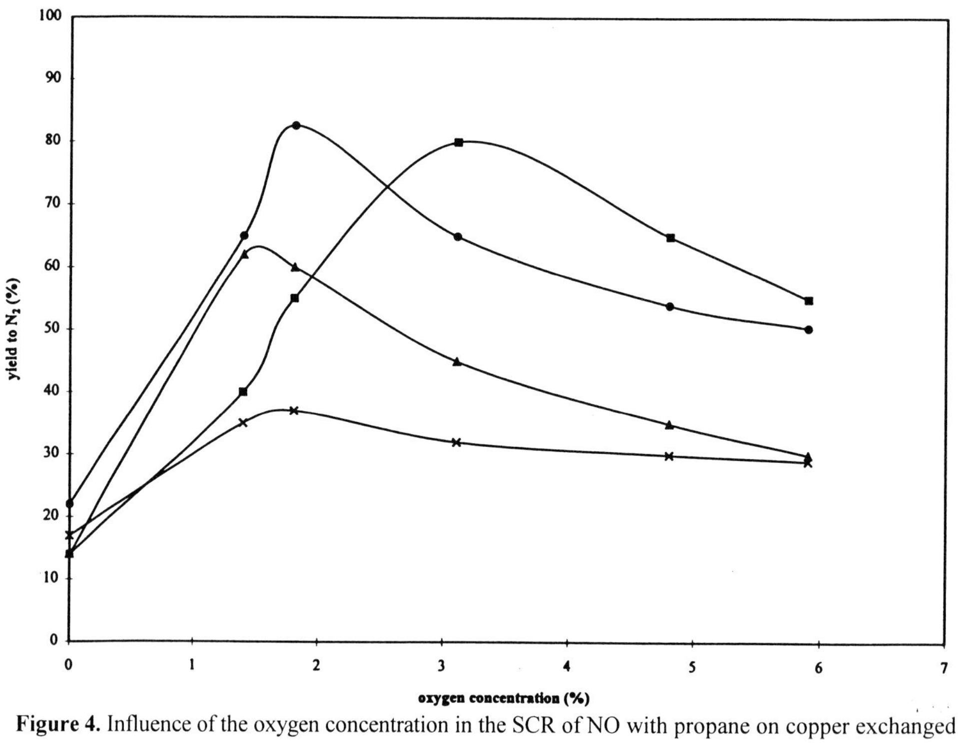

Figure 4. Influence of the oxygen concentration in the SCR of NO with propane on copper exchanged zeolites. (■) ZSM5-Cu-27-74, (●) beta-Cu-12-106, (▲) MCM22-Cu-23-88 and (x) Cu-Y-13-207, T=450°C, 650 ppm NO, 450 ppm C_3H_8.

obtained at different oxygen concentrations, depending on the metal exchanged. Cobalt exchanged MCM22 achieves the maximum activity at higher oxygen concentration (2-3%) than copper exchanged MCM22 (1-2%). A further increase in the oxygen concentration results in a decrease of the activity up to 5-6% oxygen concentration where a stable conversion is obtained. These results are qualitatively similar to those obtained with the beta and ZSM5 zeolite, but the activity of MCM22 zeolite is lower. On the other hand when the activity of MCM22 is compared with that of faujasite (Figure 4), the former zeolite catalyst is more active, and again shows an intermediate behavior between faujasite and beta and ZSM5 zeolites. In all cases the decrease in conversion when increasing the oxygen content, does not lead to a permanent loss of the catalyst activity and the activity is restored to its

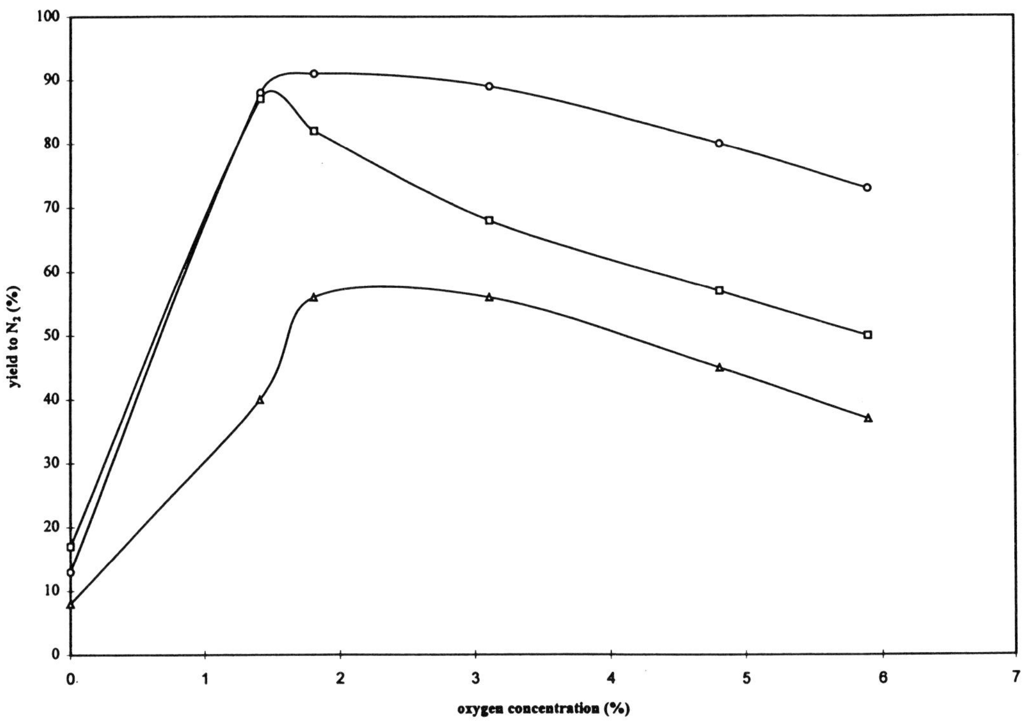

Figure 5. Influence of the oxygen concentration in the SCR of NO with propane on cobalt exchanged zeolites. (❑) ZSM5-Co-17-178, (○) beta-Co-15-119, (Δ) MCM22-Co-14-145, T=450°C, 650 ppm NO, 450 ppm C_3H_8.

original value when the initial concentration of oxygen is restored. Then, it appears that as occurred for Cu-beta zeolite [14], a redox process is established on the surface of MCM22 and the ratio of O_2/alkane, controls the proportion of the metal surface species with different oxidation states.

It is known that in SCR of NO on metal containing zeolites, the metal content plays a very important role on the activity shown by the catalyst. Indeed, for Cu-ZSM5, as well as for Cu-beta zeolite catalysts, the optimum activity is obtained for Cu contents slightly above the exchange capacity of the zeolite [4,14]. This is also the case of the MCM22, (figure 6) which shows the best results on samples containing enough metal to exchange all the hydroxyl sites, probably with an excess over the stoichiometric exchange ratio (0.5 atom.atom^{-1}).

From the results presented up to now it appears that Co and Cu MCM22 zeolites are active catalysts for the NO selective reduction with propane, being the copper exchanged MCM22 more active than cobalt exchanged MCM22 (figures 2 and 3). The behavior of these zeolites is qualitatively similar to that observed for beta and ZSM5 zeolites, but the activity achieved on MCM22 is lower. In contrast, if MCM22 is compared with faujasite, the former is more active and presents an intermediate behavior between fauajsite and ZSM5 and beta zeolites.

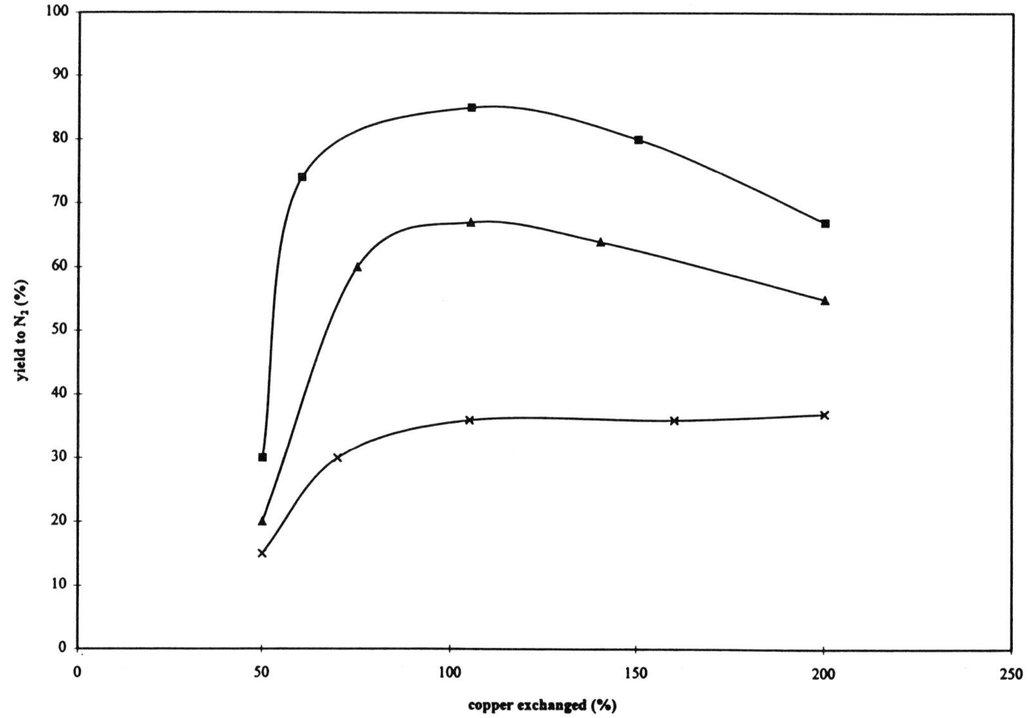

Figure 6. Influence of catalyst composition in the SCR of NO with propane on copper exchanged zeolites. (■) ZSM5, (▲) MCM22 and (x) Y zeolites, T=450°C, 1.5% O_2, 750 ppm NO, 480 ppm C_3H_8.

The catalytic observation is in agreement with the NO adsorption results obtained on these samples. Indeed, the results presented in Figure 7A shows that Cu-ZSM5 adsorbs NO at very low pressure (1 Torr) leading to the appearance of a band at 1810 cm^{-1} assigned to NO adsorbed on Cu^+. When increasing the amount of NO adsorbed, the intensity of this band increases, while two new bands with a maximum at 1896 and 1630 cm^{-1} appear. These bands have been assigned to NO adsorbed on Cu^{2+} and to NO_2 [27-28]. All these bands increase with increasing the amount of NO adsorbed up to NO pressures of ca. 10 torr, where the 1810 cm^{-1} band starts to decrease. Then two new bands appear at 2135 and 1734 cm^{-1} and a shoulder emerges at 1824 cm^{-1}. These bands are associated to NO_2^+, (NO_2)-Cu^+ (asymmetric vibration) and $(NO)_2$-Cu^+ (symmetric vibration), respectively [27-29]. These results show that when increasing the NO pressure, some Cu^+ centers are oxidized to Cu^{2+} while monomeric species are dimerized.

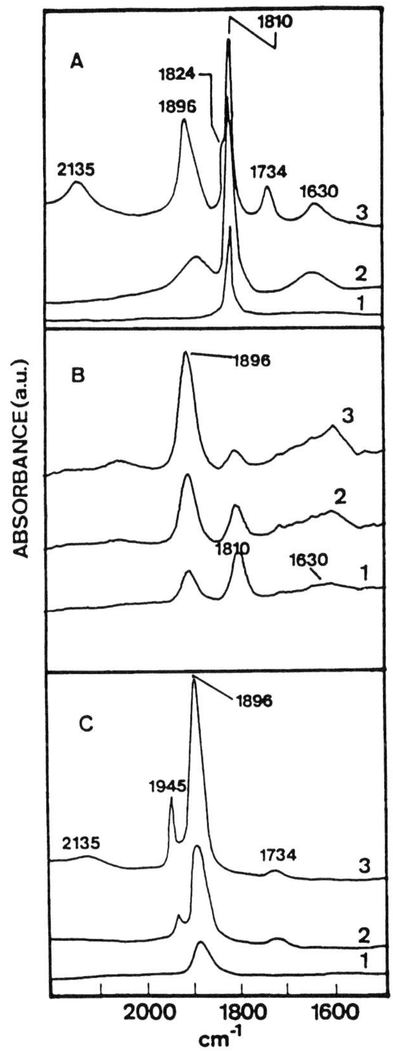

Figure 7. I.R. spectra of NO adsorbed on zeolite (a) ZSM5-Cu, (b) MCM22-Cu, (c) faujasite-Cu at different partial pressures of NO (1) 1 Torr, (2) 5 Torr, (3) 10 Torr.

From figure 7C we can see that some different results are achieved when NO is adsorbed on copper exchanged Y zeolite. When increasing the amount of NO adsorbed some important bands are observed at 1896 and 1945 cm^{-1}, assigned to Cu^{2+}-NO [30] and only a small shoulder at 1734 cm^{-1} suggests the presence of some $(NO)_2$-Cu^+. In this case practically no Cu^+ appears in the sample even at low NO pressure.

When the NO adsorption was done on Cu-MCM22, (Figure 7B) an intermediate behavior between the faujasite and the ZSM5 is observed. In this zeolite Cu^+ and Cu^{2+} species are formed at low NO concentration (1 Torr), as is indicated by the bands at 1810 and 1896 cm^{-1}. This is in contrast with ZSM5 and faujasite zeolites, where mainly Cu^+ or Cu^{2+} centers are respectively observed. These results suggest that some of the Cu atoms present in the big cavities of MCM22 (as occurs with faujasite zeolites) can sinterize, forming CuO clusters, something that will not occur in the circular 10 member ring channels with dimensions close to those of the ZSM5 zeolite [31]. These CuO clusters are not active for the NO selective catalytic reduction and this is consistent with the catalytic results obtained.

From this work it can be concluded, that copper or cobalt exchanged MCM22 is a stable catalyst for NO reduction with propane, but less active than ZSM5 and beta zeolites. The best results are obtained at 450°C and 1.5% oxygen concentration, if copper is exchanged slightly over the 100% ratio. The behavior of the MCM22 zeolite is intermediate between ZSM5 and beta zeolites, and faujasite. The activity results are consistent with the results obtained in the i.r. study of NO adsorption, which indicates that in faujasite and MCM22 zeolites, the presence of large cavities favors the formation of CuO clusters, which are not active for the SCR of NO.

REFERENCES

1. M. Iwamoto, H. Yahiro, S. Shundo, Y. Yu-u, N. Mizuno, *Shokubai (Catalyst)*, **32,** 430 (1990).
2. M. Iwamoto *et al.*, *Stud. Surf. Sci. Catal.*, **28**, 943 (1986); M. Iwamoto in: *Future Opportunities in Catalytic and Separation Technology, Stud. Surf. Sci. Catal.*, M. Misono *et al.* (Eds.), Vol. 54, Elsevier, Amsterdam, 1990, p. 121.
3. M. Iwamoto, H. Hamada, *Catal. Today*, **10**, 57 (1991).
4. M. Iwamoto, H. Hidenori, *Catal. Today*, **22**, 5 (1994).
5. W. Held, A. Koening,. *Ger Offen.* DE 3642018 (1987).
6. F.P. Boer, L.L. Hegedus, T.R. Gouber, K.P. Zak, *Chem. Tech.* 312, May 1990.
7. G.L. Baverle, S.C. Wu, K. Nobe, *Ind. Eng. Chem. Prod. Res. Dev.*, **14**, 268 (1975).
8. S. Matsuda, T. Kamo, A. Kato, F. Nakajima, T. Kumura, H. Kuroda, *Ind. Eng. Chem. Prod. Res. Dev.*, **21**, 48 (1982).
9. M. Iwamoto, H. Yahiro, *Catal. Today*, **22**, 5 (1994).

10. T.Tabata, M. Kokitzu, O. Okada, *Catal. Today*, **22**, 147 (1994).
11. R. Burch, J. Millington, *Appl. Catal. B*, **2** , 101 (1993).
12. M. Iwamoto, H. Yahiro, S. Shundo, Y. Yu-u, N. Mizuno, *Appl. Catal.*, **69**, L15 (1991).
13. M. Iwamoto, H. Yahiro, *Catal. Today*, **22**, 5 (1994).
14. A. Corma, A.E. Palomares, V. Fornes, *Appl. Catal. B*, **11**, 233 (1997).
15. K. Yogo, S. Tanaka, M. Ihara, T. Hishiki, E. Kikuchi, *Chem. Lett.*, 1025 (1992).
16. M. Misono, K. Kondo, *Chem. Lett.*, 2001 (1991).
17. Y. Li, J.N. Armor, *Appl. Catal. B*, **1**, L21 (1992).
18. Y. Kintaichi, H. Hamada, M. Tabata, M. Sasaki, T. Ito, *Catal. Lett.*, **6**, 239 (1990).
19. Y. Li, J.N. Armor, *US Patent* 5,171,553 (1992).
20. Y. Li, J.N. Armor, *J. Catal.*, **142**, 561 (1993).
21. R. Burch, S. Scire, *Appl. Catal. B*, **3**, 295 (1994).
22. Y. Li, J.N. Armor, *Appl. Catal. B*, **1**, L31 (1992).
23 K. Yogo, M. Ihara, I. Terasaki, E. Kikuchi, *Appl. Catal. B*, **2**, L1 (1993).
24. T. Tabata, M. Kokitsu, H. Ohtsuka, O. Okada, L.M.F. Sabatino, G. Bellussi, *Catal. Today* **27**, 91 (1996).
25 H. Ohtsuka, T. Tabata, O. Okada, L.M.F. Sabatino, G. Bellussi, *Catal. Letters* **44**, 265 (1997).
26 A. Corma, C. Corell, J. Perez-Pariente, *Zeolites* **15**, 2 (1995).
27. M. Iwamoto, N. Mizuno, H. Yahiro, W.X. Zhang, Y. Mine, H. Furukawa, S. Kagawa, *J. Phys Chem.*, **96**, 9360 (1992).
28. E. Giamello, C. Murphy, G. Magnacca, C. Morterra, Y. Shioya, T. Nomura and M. Anpo, *J. Catal.*, **136**, 510 (1992).
29. J. Valyon and W. K. Hall, *J. Phys. Chem.* **97**, 1204 (1993).
30. J. Valyon and W. K. Hall, *J. Phys. Chem.* **97**, 7054 (1993).
31. A. Corma, A.E. Palomares, F. Marquez, *J. Catal.* **170**, 132 (1997).

S. Suga · N. Watabe (Eds.)

Hard Tissue Mineralization and Demineralization

With 106 Figures

Springer-Verlag
Tokyo Berlin Heidelberg New York London
Paris Hong Kong Barcelona Budapest

SHOICHI SUGA, D.D.S., D.M.S.
Professor, Department of Pathology,
The Nippon Dental University School of Dentistry at Tokyo,
1-9-20 Fujimi, Chiyoda-ku, Tokyo, 102 Japan

NORIMITSU WATABE, D.SC.
Director of Electron Microscopy Center,
Professor of Biological Sciences and Marine Science Program,
University of South Carolina,
Columbia, SC 29208, USA

ISBN 4-431-70092-7 Springer-Verlag Tokyo Berlin Heidelberg New York
ISBN 3-540-70092-7 Springer-Verlag Berlin Heidelberg New York Tokyo
ISBN 0-387-70092-7 Springer-Verlag New York Berlin Heidelberg Tokyo

Library of Congress Cataloging-in-Publication Data. Hard tissue mineralization and demineralization/S. Suga, N. Watabe (eds.). p. cm. "Papers mainly presented at the open seminar held in Tokyo, October 13, 1990, under the auspices of the Society of the Nippon Dental University." Includes bibliographical references and index. ISBN 4-431-70092-7 (Springer-Verlag Tokyo Berlin Heidelberg New York). — ISBN 3-540-70092-7 (Springer-Verlag Berlin Heidelberg New York Tokyo) — ISBN 0-387-70092-7 (Springer-Verlag New York Berlin Heidelberg Tokyo). 1. Biomineralization—Congresses. I. Suga, S. (Shōichi), 1927– . II. Watabe, Norimitsu, 1922–
QH512.H37 1992. 591.19'214—dc20. 92-17466

Printed in Hong Kong

Typesetting, printing, and binding: Best-set Typesetter Ltd., Hong Kong.

Preface

Mineral formation occurs commonly in many organisms ranging from bacteria to humans within and/or outside the living tissues. More than forty types of such biominerals have been reported to date, and two of the most common are calcium carbonates and calcium phosphates. Much work has been carried out recently to elucidate the mechanisms of biomineralization. The studies are concentrated in the areas of calcium phosphate mineral formation in vertebrate bone and teeth and calcium carbonate mineral formation in coral skeletons and molluscan shells. Demineralization of bone and teeth has also been extensively investigated as it relates to human health, and this phenomenon has become recognized also in invertebrate skeletal systems. The discussions presented here are also concerned with those areas. We hope that this volume will stimulate interactions and cooperation between the investigators of invertebrate and vertebrate biomineralization to increase our understanding of the mechanisms of biological mineral formation and demineralization.

This volume is a collection of papers mainly presented at the open seminar held in Tokyo, October 13, 1990, under the auspices of The Society of The Nippon Dental University. The objective of the seminar was to introduce the status of biomineralization research to investigators in many different fields who had not been able to participate in the Odawara Symposium (the Sixth International Symposium on Biomineralization, October 9–12, 1990).

We are grateful to The Society of The Nippon Dental University and Lion Corporation, Japan whose generous financial support made the open seminar and publication of this volume possible. We also thank the staff of Springer-Verlag Tokyo for their constant assistance in the editing and production of this volume.

SHOICHI SUGA
NORIMITSU WATABE

Contents

List of Contributors

Towards an Understanding of Tissue-Specific Biomineralization

HAROLD C. SLAVKIN[1], CHING C. HU[1], YASUYUKI SASANO[2], SASSON DAVID[1], PABLO BRINGAS, JR.[1], YASUNORI SAKAKURA[1], VALENTINO SANTOS[1], and MASANORI NAKAMURA[2]

Summary. We propose to test the hypothesis that time and position-restricted growth factor expression regulates dentine and enamel biomineralization. We predict that a developmental program which defines timing and position for the putative dentine-enamel junction (pDEJ) serves to initiate the nucleation process which propagates either as (1) dentine mineral deposition, or (2) enamel mineral deposition patterns. A simple organ culture system, devoid of exogenous serum or plasma supplementation, provides a controlled model to test our hypothesis and these predictions. Dentine- and enamel-specific biomineralization were observed when E15 mouse cap stage tooth organs were cultured for 3 weeks, and root and alveolar bone formations were observed within 4 weeks using serumless, chemically defined medium. Under these permissive conditions, a developmental program expressed the DEJ and biomineralization. The initial DEJ constituents appeared to provide instructions for nucleation and specificity for dentine and enamel crystal size, orientation, number, and growth patterns. In tandem, EGF and EGF receptors were expressed by the same cells in this model system, suggesting autocrine regulation. These results have provided a rationale towards understanding tissue-specific biomineralization.

Key words: Enamel—Dentine—Grwoth factors—In vitro—Mouse

Introduction

Cells and their extracellular matrices (ECM) control tissue-specific biomineralization [1–2]. Anionic macromolecules (e.g., phospholipids and phosphoproteins) are associated with biomineralization throughout vertebrate evolution and are implicated in the control processes for biomineralization [1–4] (Fig. 1). However, the determinants for the timing, position, and unique specificities

[1] Center for Craniofacial Molecular Biology, School of Dentistry, University of Southern California, Los Angeles, CA 90033, USA
[2] Department of Anatomy, School of Dentistry, Tohoku University, Sendai, 980 Japan

Cell- and extracellular matrix-mediated biomineralization

Secretion of extracellular matrix constituents

↓

Intra- and intermolecular interactions

↓

Induction and propagation of tissue-specific biomineralization (e.g., position, size, and patterns of crystal morphogenesis)

Fig. 1. Cell- and extracellular matrix-mediated biomineralization

for structural gene products associated with biomineralization are not as yet known.

Calcium phosphate formation resulting in calcium hydroxyapatite crystal (HAP) morphogenesis in bone, dentine, cementum, and enamel are associated with heterogeneous nucleation processes, controlled by specific regulatory and structural gene products within temporally and positionally restricted extracellular microenvironments. It is assumed that ECM organic components (e.g., type I collagen, phosphoproteins, phospholipids, etc.) function as nucleators or substrates [1–4]; chemical groups (e.g., amino acid side chains, glycosylations, and/or phosphorylations in the case of proteins) can produce intermolecular three-dimensional stereochemical geometry with attendant electrical charge distribution which replicates specific crystal planes of the HAP crystal to be nucleated [1–2]. The nucleation substrate for HAP crystal formation is assumed to be a solid state aggregate. Therefore, the specific physicochemical properties of the solid state aggregate, such as structural details characteristic for each of the dental tissue ECM proteins, provides tissue-specific biomineralization (e.g., number, size, and patterns of crystal growth). How these solid aggregates are formed and how they induce and control tissue-specific patterns of HAP crystal growth are not known.

The mouse tooth organ provides an excellent model for a number of questions related to the timing, position, and sequential inductive processes which result in cellular differentiation to become specific phenotypes related to biomineralization [5–11]. Enamel and dentine provide two curiously different and unique tissue examples of ECM protein-mediated biomineralization. Ectodermally derived epithelia synthesize and secrete enamel proteins to form an ECM in juxtaposition to and coordinated with ectomesenchymally derived odontoblasts which synthesize and secrete dentine proteins to form and dentine ECM [3, 8]. At the interface between dental inner enamel epithelia and adjacent odontoblasts, hole zone regions formed within type I collagen fibrils, as well as matrix vesicles, may provide *de novo* nucleation sites for initial HAP crystal formation [4, 12]. These initial processes result in the position of the putative dentine-enamel junction (pDEJ) [12]. From the initial DEJ, relatively small and randomly arranged HAP crystals increase in number (multiplication) but not in size, and propagate dentine-specific biomineralization. In juxtaposition, HAP crystals from the DEJ grow in length along their *c*-axis and become highly ordered, enamel-specific biomineralization. Two different tissue-specific patterns of biomineralization appear to be initiated at the same interface (DEJ), yet grow to different sizes and patterns in opposite directions, presumably regulated by two very different cell/tissue phenotypes. Whereas many ultrastructural and physicochemical details of this process have been described [reviews 1–4], tissue specific mechanisms remain to be elucidated.

Recently, dentine and enamel tissue-specific biomineralization were demonstrated in a simple in vitro model without endocrinic or long-range humoral supplements. Embryonic mouse molar tooth organs (E15/E16, early cap stage) expressed dentine- and enamel-specific patterns of HAP crystal biomineralization in a serumless, chemically defined medium [5–6]. When these cultures were extended for periods up to 31 days in vitro, alveolar bone was also expressed in association with the forming tooth organ roots [7]. The culture medium was devoid of transferrin, serum, or other exogenous growth/regulatory factors (e.g., epidermal growth factor, transforming growth factors alpha or beta, nerve growth factor, insulin-like growth factors, vitamin A, vitamin D, etc.). Therefore, endocrine-type, long-range regulatory molecules (e.g., steroid or polypeptide hormones, vitamin D, serum or plasma factors, etc.) were not pre-requisite for dentine- or enamel-specific biomineralization.

The present contribution summarizes and reviews the results of these experimental strategies and attempts to identify potential rules for biomineralization. We anticipate that multiple regulatory factors determine when, where, and how tissue-specific biomineralization takes place. It is anticipated that the suggested criteria might serve to further catalyze efforts towards defining the mechanisms which regulate the determinants for biomineralization.

Material and Methods

Embryonic Mouse Molar Dissections and a Simple Organ Culture System

Embryonic mandibular first molars (M_1) were dissected from 15-days gestation (E15), timed-pregnant Swiss Webster strain mice and then cultured for periods up to 31 days in vitro using a serumless, chemically defined medium, as previously described [5–7].

Von Kossa Histochemistry

Non-demineralized cultured explants and control in vivo specimens (birth, days 2 and 7 postnatal) were processed for histochemistry to determine the deposition of calcium phosphate salts, as previously described [5–7].

Anhydrous Fixation, Transmission Electron Microscopy, and Electron Diffraction Studies

Non-demineralized specimens were processed by the Landis method as previously reported [13] and physical studies as previously described [6].

Immunocytochemistry

Anti-amelogenin antibody and antibodies designed to identify motifs (-LPPHPGHPGYI-) within the amelogenin primary structure were used for immunohistochemistry by methods previously described [14, 15]. In addition, polyclonal anti-EGF antibodies were used to identify the translation product within odontogenic cells.

In Situ Hybridization

AMEL (amelogenin) transcripts were identified and localized using ^{35}S-labeled sense and antisense RNA probes by methods previously reported [16, 17].

EGF mRNA Phenotyping

The RT-PCR strategy (reverse transcriptase, polymerase chain reaction) for mRNA phenotyping, as recently described [18–20], was used to detect relatively low levels of EGF transcripts during in vitro tooth development.

Results

Summary of Tooth Development In Vivo

The salient descriptions of mouse tooth organ development are readily found in a number of very useful reviews [e.g., 8–11]. Figure 2 outlines the sequence and timing of major stages in mouse mandibular first molar tooth organ (M_1) development. Dentine phosphoprotein was identified within odontoblasts prior to biomineralization, but was not secreted until after the DEJ was formed. Initial biomineralization was defined using ultrastructural criteria for HAP crystal formation associated with matrix vesicles and collagen. In M_1, this was observed along the major buccal cusp 24 h prior to birth. Anionic enamel protein(s) represented 46 kd (pI, 5.5) and 72 kd (pI, 5.8) proteins which were first expressed prior to AMEL and shared one or more epitopes with enamel proteins.

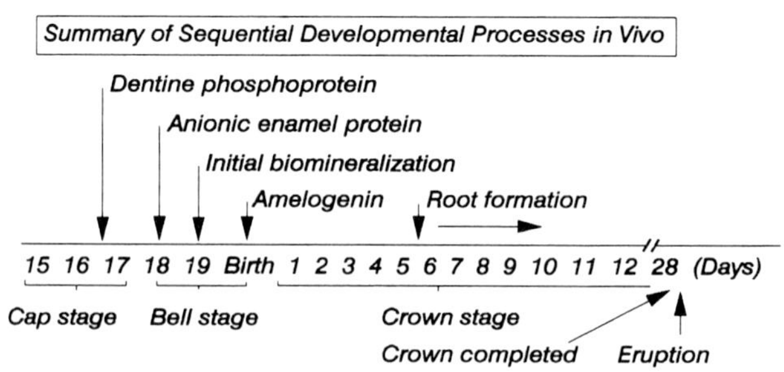

Fig. 2. Summary of sequential developmental processes in vivo

Fig. 3. Summary of sequential developmental processes in vitro

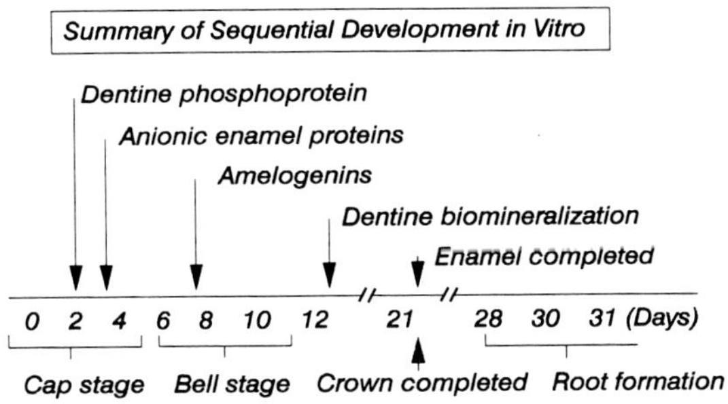

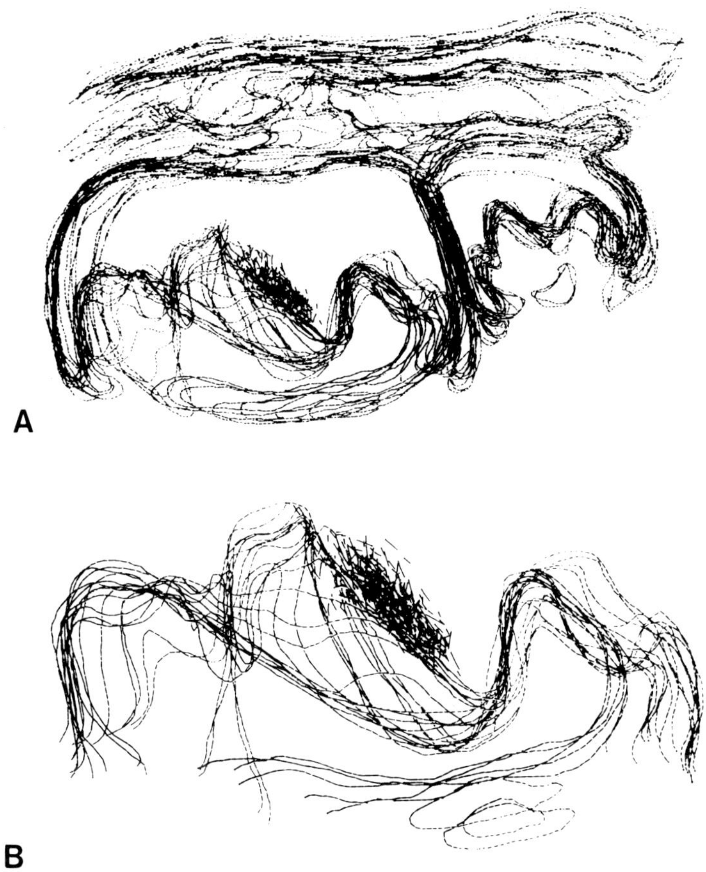

Fig. 4. Computer-assisted, three-dimensional reconstructions of AMEL localization using in situ hybridization. **A** AMEL transcripts were first observed along the distal surface of the major buccal cusp in M_1 at 6 days in vitro. The E15 cap stage explant with oral epidermis and enamel organ epithelia produced both M_1 and M_2 molars. **B** AMEL transcript localization patterns were comparable to those described for in vivo development

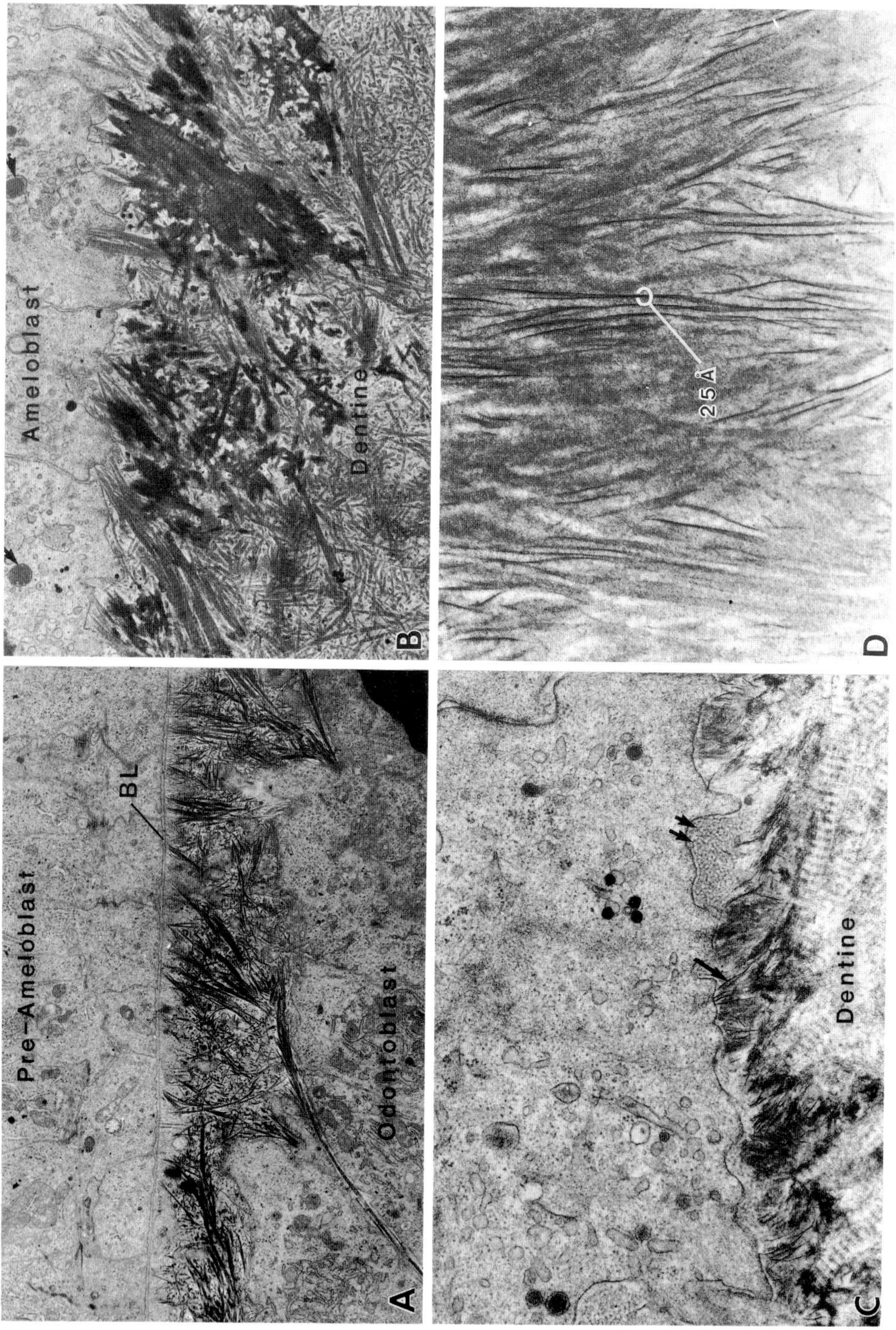
Pre-Ameloblast
BL
Odontoblast
A
Ameloblast
Dentine
B
Dentine
C
25 Å
D

Fig. 5. Ultrastructural survey of dentine-enamel junction (DEJ) formation and tissue-specific biomineralization. **A** Preameloblasts with a continuous basal lamina (*BL*) expressed anionic enamel proteins prior to biomineralization and AMEL expression. Odontoblasts produced matrix vesicles, type I collagen and dentine phosphoprotein prior to biomineralization. **B** In tandem with basal lamina degradation and removal from ameloblasts, biomineralization along the putative DEJ appeared as disparate aggregates of HAP crystals associated with either matrix vesicles or type I collagen. AMEL translation products were packaged within secretory vesicles (*arrows*). **C** Enamel crystals (*arrow*) appeared to form at the DEJ. Electron-dense, granular materials (AMEL) were secreted from ameloblasts beginning at birth (*double arrows*). **D** Enamel crystals were approximately 25 Å in diameter

Summary of Tooth Development In Vitro

The embryonic mouse molar tooth organ has been used in a number of in vitro studies [reviews, 8–11]. Figure 3 outlines the sequence and timing of M_1 development using serumless, chemically defined medium [5–7]. Albeit slower and smaller in scale, the sequence of the observed tooth development was comparable to the sequence observed for in vivo development.

The Sequence and Timing of Tissue-Specific Biomineralization

The major buccal cusp of the M_1 region demonstrated the initial ultrastructural evidence for tissue-specific biomineralization. A detailed description of this process has recently been reported by our laboratory [14, 16–17]. HAP crystals were first observed within matrix vesicles in the position of the future DEJ region, in juxtaposition to inner enamel epithelia with continuous basal lamina and adjacent odontoblast cell processes. Small HAP crystals were observed in association with type I collagen fibrils. HAP crystals formed aggregates, the aggregates coalesced and subsequently produced the mineralizing dentine ECM.

In the M_1 model, initial biomineralization was indicated approximately 24 h before birth. The timing and position of AMEL gene expression in vivo and in vitro were determined using in situ hybridization. Inner enamel epithelia initially expressed amelogenin (AMEL) transcription and translation at birth [16], *after* initial biomineralization [14]. When M_1 E15/16 explants were cultured, AMEL was detected at 6 days along the distal buccal surface of the major buccal cusp (Fig. 4A,B). Subsequent patterns for AMEL expression were comparable to those observed for in vivo amelogenesis [16–17].

Figure 5 provides an ultrastructural survey of the formation of the DEJ. A continuous basal lamina was characteristic of the post-mitotic preameloblasts prior to initial biomineralization (Fig. 5A). Preameloblasts expressed non-collagenous, presumably anionic enamel proteins prior to AMEL expression at birth. The basal lamina was removed from the preameloblasts coincident with the formation of the DEJ and biomineralization (Fig. 5B). AMEL was first expressed at birth and characteristic enamel-specific HAP crystal patterns were detected (Fig. 5C).

Anhydrous fixation methods without subsequent staining provided a methodology to observe electron-dense images representing the inorganic constituents of forming enamel and dentine (Fig. 6). Our observations of the DEJ were interpreted to suggest that enamel HAP crystals appeared to grow from the dentine collagen HAP crystals. These observations confirmed those previously reported by Arsenault and Robinson [12].

Endogenous EGF Expression

Recently, Snead and colleagues reported that EGF transcripts were identified in E14 embryonic mouse odontogenic tissues [21]. Earlier, Adamson et al. reported that EGF receptors (EGFR) were first identified at 9-days gestation in

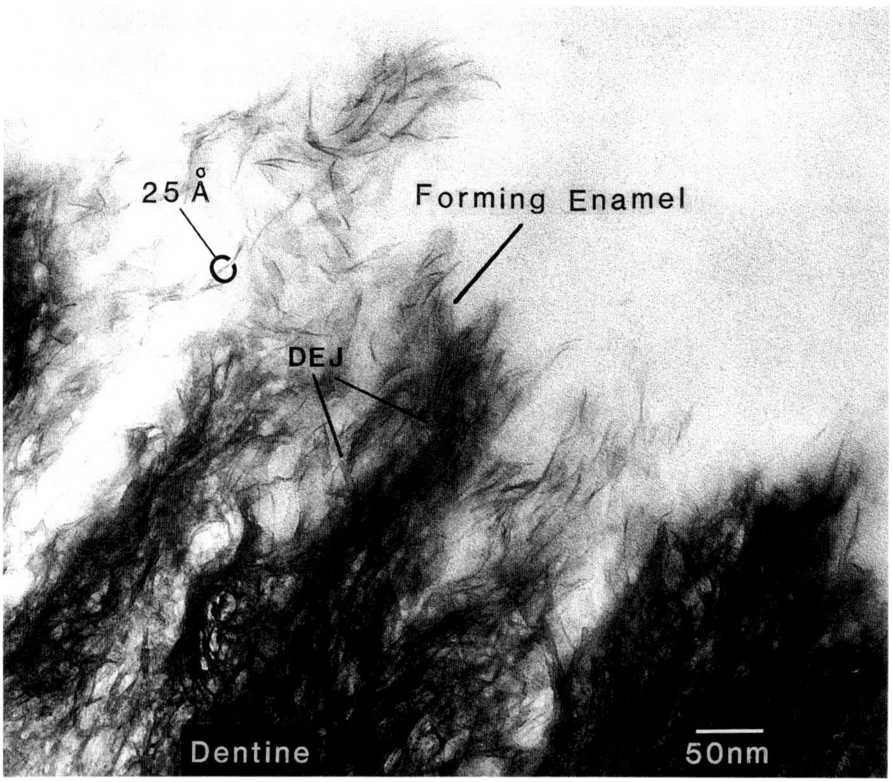

Fig. 6. Enamel HAP crystals appeared to form from dentine HAP crystals along the DEJ when observed in M_1 specimens processed by anhydrous fixation and without staining

Fig. 7. EGF mRNA phenotyping during M_1 development in vivo and in vitro. *PN*, Postnatal development selected was at birth

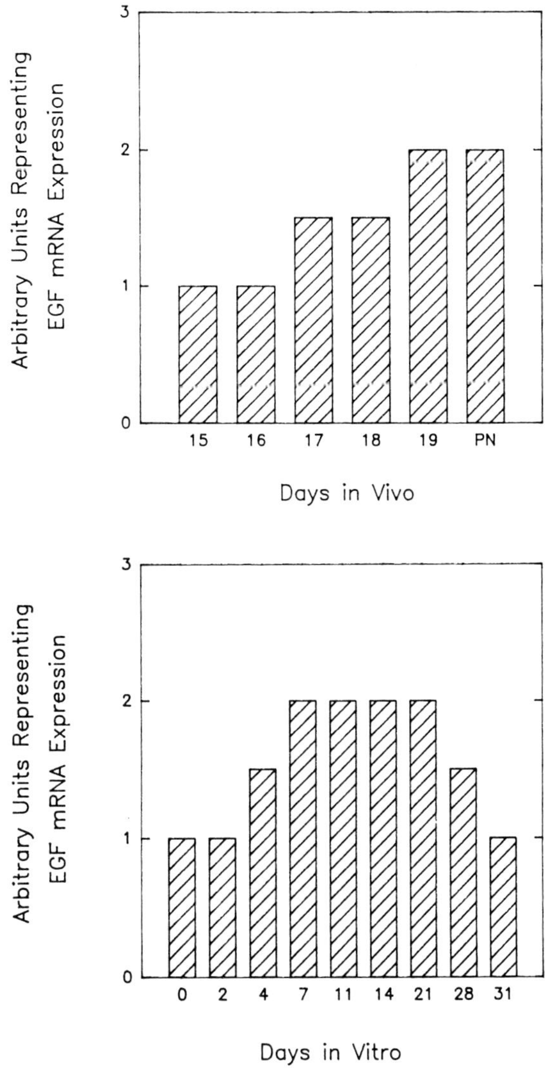

mouse embryos [22]; EGFR were also identified on early embryonic mouse dental tissues [23].

These results were used to formulate our hypothesis that endogenous EGF, and possibly other growth factors, function as either autocrine and/or paracrine regulatory molecules for biomineralization during tooth development. More recent studies in our laboratory demonstrated that EGF transcripts and translation products were expressed during tooth development: the relative levels and patterns of expression changed during development [20, 24]. Figure 5 provides a survey of relative EGF mRNA levels of expression during tooth organ development in vivo and in vitro. A two-fold increase in EGF transcription was observed between the cap and bell stages (e.g., dentine and enamel biomineralization). Figure 6 summarizes the distribution of EGF translation product and EGFR during tooth development. Immunocytochemical evidence showed that cells which produced EGF also contained EGFR (Fig. 7), suggesting that these cells may function by autocrinic regulation [24].

Complementary studies using tyrphostin (10–60 μM) inhibited signal transduction mediated by EGF binding to EGFR with subsequent autophosphorylation of tyrosine residues, produced significant reductions in DNA synthesis, and retarded or inhibited odontogenic cell/tissue differentiations [24]. E15/16 M_1 responded to exogenous EGF (2–20 ng/ml) by a dose-response increase in DNA synthesis [24], whereas E17–E19 M_1 explants responded by increased total protein synthesis and selective inhibition of type I collagen and enamel proteins [25]. In tandem, preliminary studies found that M_1 explants cultured in serumless medium also expressed TGF-alpha, TGF-beta$_1$, IGF-I, IGF-II, bFGF and a number of other putative regulatory factors possibly associated with tissue-specific biomineralization.

Discussion

This analysis provides evidence in support of three potentially important predictions related to tissue-specific biomineralization:

1. A genetically-defined developmental program controls morphogenesis, differentiation, and tissue-specific biomineralization.
2. A series of sequential epithelial-mesenchymal interactions, mediated through autocrine and/or paracrine controls, regulates the timing and position for the future DEJ.
3. The putative DEJ is producted by both epithelial- and mesenchymal-derived ECM products. These products serve as nucleation mediators for HAP crystal formation, and tissue-specific microenvironments are either permissive for multiplication of small and randomly arranged HAP crystals as in dentine, or for increased growth along the *c*-axis resulting in large, ordered HAP crystals associated with enamel.

The following is selected evidence which further substantiates the rationale for these assertions.

Developmental Genetic Program For Tissue-Specific Biomineralization

Regulatory and/or structural genes and their products required for biomineralization have not yet been identified. Whereas ultramicromethods have provided detection for relatively few transcripts and their translation products in embryonic and fetal mouse dental tissues, direct evidence implicating one or more regulatory factors with specific processes related to biomineralization is not yet available.

However, three lines of evidence should be considered towards understanding the genetic program for tissue-specific biomineralization. First, a recent report has demonstrated that a homeotic gene transcript, (i.e., Hox 7.1, a mouse homologue for the *Drosophila* bicoid homebox), was uniquely expressed and coordinated with each mouse tooth organ during the cap stage of development in vivo as well as in vitro [27]. Second, Hox 7.1 expression appeared to be coordinated with endogenous growth factor and receptor expressions prior to and during dental tissue-specific biomineralization (e.g., EGF, TGF-alpha, TGF-beta$_1$, IGF-I, IGF-II, bFGF) [20–21, 23–24, 26]. Third, differential expression of substrate adhesion molecules (e.g., tenascin/cytotactin and syndecan) as well as ECM constituents (e.g., fibronectin and types I and III collagens) were also coordinated with early to late cap stages of tooth development [reviews 8–11, 23, 28]. These three complementary lines of evidence suggest that coordinated and hierarcal gene regulation provides the controls for the timing and position of tissue-specific biomineralization. We suggest that a sequence of multiple and coordinated genetically controlled processes are required to enable cell- and/or ECM-mediated biomineralization.

Highly specific inhibition or perturbation studies, using either antisense or immunoperturbation strategies, might be useful to correlate a specific growth factor, or pattern of growth factor expressions, with a specific developmental process related to biomineralization. The experimental strategy of constructing the so-called "knock-out" or deletion experiments to prevent gene expression within a transgenic mouse model system could significantly enhance an understanding of structure/function during tissue-specific biomineralization.

Criteria Towards Understanding the Molecular Biology of Tissue-Specific Biomineralization

International agreement for criteria is essential towards defining when, where, what, and how cell- and/or ECM-derived constituents regulate tissue-specific biomineralization. The following criteria are suggested for studies designed to determine genes controlling biomineralization: (1) Gene(s) must be expressed during development at precise times and positions concurrent with initiation and propagation of tissue-specific biomineralization (e.g. enamel, dentine, bone, and cementum); and (2) Gene mutations must produce abnormalities in biomineralization (e.g., amelogenesis imperfecta, dentinogenesis imperfecta, and osteogenesis imperfecta).

DEJ Is Formed by both Epithelia and Mesenchyme-Derived Products Which Control Tissue-Specific Patterns of Biomineralization

Prior to and during the formation of the ECM interface, interposed between inner enamel epithelia and adjacent preodontoblast/odontoblasts (e.g., late

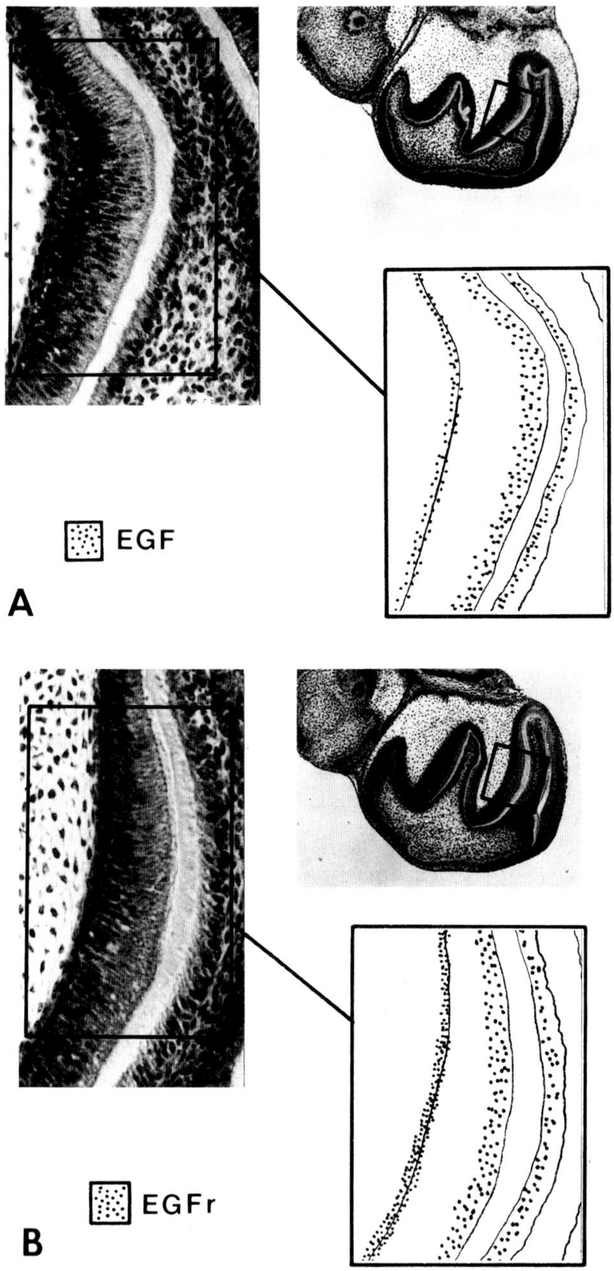

Fig. 8. *EGF* presumably functions in an autocrine mechanism during M_1 development. **A** Immunolocalization of EGF during cap stage development. The region within the rectangle indicating the mesial buccal cusp of the M_1 is shown at low magnification, enlargement, and schematically with immunostained cells indicated by *stippled material*. **B** Immunolocalization of *EGFR* (receptor) in comparable specimens as shown in **A**

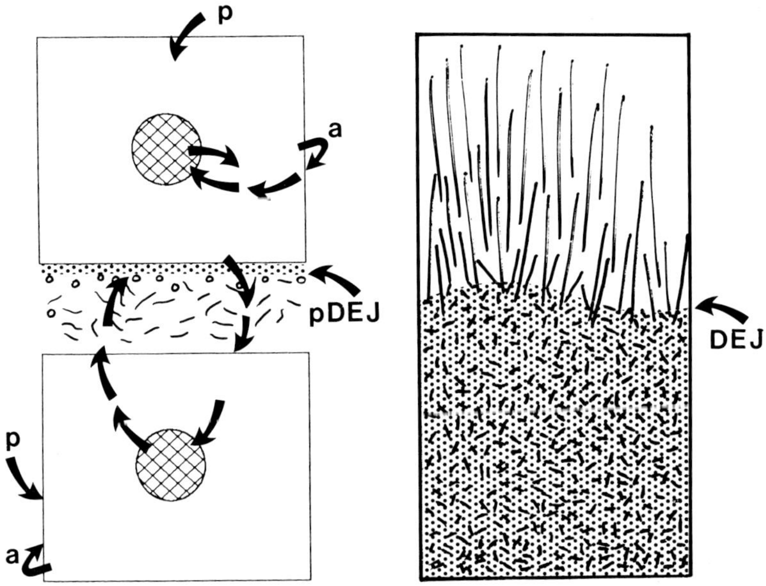

Fig. 9. Evidence supports the hypothesis that both autocrine (*a*) and/or paracrine (*p*) mechanisms possibly control formation of the putative dentine-enamel junction (*pDEJ*). The pDEJ appeared to provide nucleation for HAP crystal formation which proceeds as either (1) small, randomly arranged crystals in dentine, or (2) large, highly ordered HAP crystals in enamel

cap stage of tooth development), both tissue types synthesize and secrete regulatory factors as well as macromolecular products into the forming ECM interface (Fig. 8). Growth factors (e.g., EGF, TGF-beta$_1$, IGF-I, and IGF-II) were sequentially produced and regulated developmental processes through either autocrine or paracrine mechanisms. ECM products include preodontoblast-derived matrix vesicles, types I and III collagens, proteoglycans, substrate adhesion molecules, sequestered growth factors, and epithelial-derived anionic enamel proteins [14]. We suggest that the initial nucleation of HAP crystals within matrix vesicles and that associated with collagen fibrils along the putative DEJ were the resultants of a microenvironment produced and propagated by both epithelial and mesenchymal tissues, and instructive for biomineralization. We further suggest that the putative DEJ, a macromolecular aggregate precipitated at the interface between preameloblasts and odontoblasts, functions as the nucleation site for both enamel and dentine, as previously suggested by Arsenault and Robinson [12]. We assume that subsequent AMEL expression *following the initiation of biomineralization* may function to regulate the size, pattern, and direction of enamel HAP crystal growth.

Acknowledgements. Often in science a hypothesis and supporting evidence which can facilitate the realization and significance of the original hypothesis is published well in advance of other discoveries. An example for this is found in the original hypothesis, which predicted that procaryotic as well as eucaryotic cells interpret and function within their microenvironments through a metabolic code. This idea has been essential for our recent research studies. Therefore, the authors wish to especially acknowledge the intellectual legacy provided through the original hypothesis for a metabolic code put forth by the late Professor Gordon Tomkins (The Metabolic Code. *Science* 189: 760–768, 1975) preceding the discoveries of cytokines, growth factors, and autocrine or paracrine regulation during development. These studies were supported in part by research funds granted by the NIH, USPHS (DE-02848, DE-06425).

References

1. Lowenstam HA, Weiner S (1989) On biomineralization. Oxford University Press, New York, pp 1–324
2. Mann S, Webb J, Williams RJP (1989) Biomineralization. VCH, New York, pp 1–541
3. Veis A (1989) Biochemical studies of vertebrate tooth mineralization. In: Mann S, Webb J, Williams RJP (eds) Biomineralization. VCH, New York, pp 189–222
4. Glimcher MJ (1989) Mechanism of calcification: Role of collagen fibrils and collagen-phosphoprotein complexes in vitro and in vivo. Anat Rec 224(2):139–153
5. Bringas P, Nakamura M, Nakamura E, Evans JC, Slavkin HC (1987) Ultrastructural analysis of enamel formation during in vitro development using chemically-defined medium. Scan Electron Microsc 1:1103–1108
6. Evans J, Bringas P, Nakamura M, Nakamura E, Santos V, Slavkin HC (1988) Metabolic expression of intrinsic developmental programs for dentine and enamel biomineralization in serumless, chemically-defined, organotypic culture. Calcif Tissue Int 42:1–12
7. Slavkin HC, Bringas P, Bessem C, Santos V, Nakamura M, Hsu M-Y, Snead ML, Zeichner-David M, Fincham AG (1988) Hertwig's epithelial root sheath differentiation and initial cementum and bone formation during long-term organ culture of mouse mandibular first molars using serumless, chemically-defined medium. J Periodont Res 23:28–40
8. Slavkin HC (1990) Molecular determinants of tooth development: A review. Crit Rev Oral Biol 1:1–16
9. Thesleff I, Hurmerinta K (1981) Tissue interactions in tooth development. Differentiation 18:75–88
10. Kollar EJ (1983) Epithelial-mesenchymal interaction in the mammalian integument. In: Sawyer RH, Fallon FJ (eds) Epithelial-mesenchymal interaction in development. Praeger, New York, pp 27–50
11. Lumsden AGS (1988) Spatial organization of the epithelium and the role of neural crest cells in the initiation of the mammalian tooth germ. Development 103:155–169
12. Arsenault LR, Robinson BW (1989) The dentino-enamel junction: A structural and microanalytical study of early mineralization. Calcif Tissue Int 45:111–121

13. Landis WJ (1983) Special techniques for problem tissues. In: Jones BR (ed) Electron microscopy. Academic, New York, pp 459–476
14. Slavkin HC, Bessem C, Bringas P, Zeichner-David M, Nanci A, Snead ML (1988) Sequential expression and differential function of multiple enamel proteins during fetal, neonatal and early postnatal stages of mouse molar organogenesis. Differentiation 37:26–39
15. Nanci A, Bendayan M, Slavkin HC (1985) Enamel protein biosynthesis and secretion in mouse incisor secretory ameloblasts as revealed by high resolution immunocytochemistry. J Histochem Cytochem 33:1153–1160
16. Snead ML, Luo W, Lau EC, Slavkin HC (1988) Spatial- and temporal-restricted pattern for amelogenin gene expression during mouse molar tooth organogenesis. Development 104:77–85
17. David S, Bringas P, Santos V, Luo W, Slavkin HC (1991) Individual molar cusp units express independent amelogenin (AMEL) transcription. J Dent Res 70:271a
18. Rappolee DA, Mark D, Banda MJ, Werb Z (1988) Wound macrophages express TGF-alpha and other growth factors in vivo: Analysis by mRNA phenotyping. Science 241:708–712
19. Rappolee DA, Brenner CA, Schultz R, Mark D, Werb Z (1988) Developmental expression of PDGF, TGF-alpha and TGF-beta genes in preimplantation mouse embryos. Science 241:1823–1826
20. Slavkin HC, Snead ML, Luo W, Bringas P, Kikunaga S, Sasano Y, Bessem C, Mayo M, MacDougall M, Rall LB, Rappolee D, Werb Z (1990) Analysis of embryonic cartilage and bone induction in a defined culture system. In: Pecile A, de Bernard B (eds) Bone regulatory factors. Plenum, New York, pp 55–68
21. Snead ML, Luo W, Oliver P, Nakamura M, Don-Wheeler G, Bessem C, Slavkin HC (1989) Localization of epidermal growth factor precursor in tooth and lung during embryonic mouse development. Dev Biol 134:420–429
22. Adamson ED, Deller MJ, Warshaw JB (1981) Functional EGF receptors are present on mouse embryo tissues. Nature 291:656–659
23. Partanen AM, Thesleff I (1989) Growth factors and tooth development. Int J Dev Biol 33:165–172
24. Hu CC, Slavkin (1991) Mitogenic effect of exogenous EGF on developing mouse tooth buds. J Dent Res 70:270a
25. Hata R, Bessem C, Bringas P, Hsu M-Y, Slavkin HC (1990) Epidermal growth factor regulates gene expression of both epithelial and mesenchymal cells in mouse molar tooth organs in culture. Cell Biol Int Rep 14:509–519
26. D'Souza RN, Happonen RP, Ritter NM, Butler WT (1990) Temporal and spatial patterns of transforming growth factor-beta$_1$ in developing rat molars. Arch Oral Biol 35:957–965
27. Mackenzie A, Leeming GL, Jowett AK, Ferguson MWJ, Sharpe PT (1991) The homeobox gene Hox 7.1 has specific regional and temporal expression patterns during early murine craniofacial embryogenesis, especially tooth development in vivo and in vitro. Development 111:269–285
28. Slavkin HC, MacDougall M, Zeichner-David M, Oliver P, Nakamura M, Snead ML (1988) Molecular determinants of cranial neural crest-derived odontogenic ectomesenchyme during dentinogenesis. Am J Med Genet 4:7–22

Fluoride and Iron Concentrations in the Teeth of Primitive Bony Fishes (Osteichthyes)

SHOICHI SUGA[1] and YASUHIKO TAKI[2]

Summary. Previous studies by the authors have revealed that fluoride and iron concentrations in the enameloid of the teeth of some teleostean fishes are very high compared with those in human enamel, and that their concentrations are related to the phylogeny of the fishes rather than to the environmental water, tooth morphology, and feeding habits. In order to trace the origin of the concentration-mechanism of these elements into the enameloid in the evolutionary history of the bony fishes, electron microprobe investigations of tooth fluoride and iron were made in this study in non-teleostean bony fishes which are ancestral to or distantly diverged from the teleostean lineage, i.e., the dipnoan lungfishes, the crossopterygian coelacanth, the primitive antinopterygian polypterids (bichers and reedfishes), gars, and bowfin.

A high fluoride concentration in the entire layer of the enameloid, as seen in many teleostean fishes, was found only in the bowfin, in which the concentration was about 3.5% in the surface layer and about 2.2% in the inner layer. Fluoride concentration was from 0.17% to 0.70% in the enamel of lungfishes and coelachanth and in the enameloid of polypterids and gars. On the other hand, a high iron deposition at the outer layer of enameloid (0.5%–1.3%) was detected in polypterids, gars, and bowfin, while the enamel in lungfishes and coelacanth did not contain any detectable amounts of iron under the analytical conditions of this study.

Based on these results, the mechanisms of high fluoride and iron concentrations seem to have been acquired only in actinopterygians, and the origin of the high fluoride concentration could be traced back to the pre-teleostean stage of actinopterygian evolution while the high iron concentration mechanism seemed to originate as far back as the cladistian (polypterid) stage.

Key words: Teeth—Enameloid—Primitive bony fishes—Fluoride—Iron

[1] Department of Pathology, The Nippon Dental University School of Dentistry at Tokyo, Tokyo, 102 Japan
[2] Tokyo University of Fisheries, Tokyo, 108 Japan

Introduction

Previous investigations on the fluoride (F) concentration in fish teeth using the electron microprobe analysis first indicated that the teleostean fishes may be classified into 2 groups in terms of the fluoride concentration in the enameloid: 1 group whose enameloid contains more than about 2.50% fluoride and the other which contains less than about 0.17%. Such a remarkable difference in fluoride concentration in the enameloid has been found to be related to the phylogeny of the fishes rather than to their feeding habits and the fluoride concentration in the environmental water [1–6], although marine water everywhere in the world contains about 1.3 ppm fluoride [7], while fresh water has usually less than about 0.3 ppm.

Iron (Fe) is also deposited at a high level in the enameloid of some teleostean fishes [8–14]. It was found by the quantitative electron microprobe analysis that the iron concentration in the enameloid of some teleostean fishes ranges from 0.4% to 13.5% [13, 15], whereas that in human enamel from 0.02% to 0.028% in surface enamel and from 0.012% to 0.022% in subsurface enamel [16]. It was suggested that iron concentration is also related to the phylogeny of teleostean fishes [15], although it has been proposed that it is related to the feeding habits of fishes [14].

A comparative investigation using the electron microprobe made on the developing tooth germs of a perciform fish, whose enameloid contains high fluoride and iron concentrations, indicated that the timing and mode of deposition of fluoride and iron into the developing enameloid are different [13].

In order to verify the relationship between the fluoride and iron concentrations in the enameloid and fish evolution more precisely, a quantitative analysis of the fluoride and iron of the teeth of primitive bony fishes was conducted in this study. The classification and interrelationships of major groups of bony fishes (class Osteichthyes) have long been a subject of debate [17–22]. In this study, we tentatively classified the osteichthyans into three subclasses, i.e., the Dipnoi (lungfishes), the Crossopterygii (lobefins, including coelacanths), and the Actinopterygii (ray-finned fish), with the Actinopterygii being further divided into the Cladistia (the Polypteridae or bichers and reedfish, regarded by some authors as representing a distinct subclass, the Brachiopterygii), Chondrostei (sturgeons) and Neopterygii [19–22]. The Neopterygii contains the Ginglymodi (gars) and the Halecostomi, consisting of the Halecomorphi (bowfin) and Teleostei (a major group of advanced actinopterygians, typical familiar fishes ranging from herrings to perches, and the subject of our previous investigations).

Materials and Methods

The teeth examined in the present study were obtained from 3 species of dipnoan lungfishes (Ceratodontidae, Lepidosirenidae, and Protopteridae), 1 species of the crossopterygian coelacanths (Latimeriidae), 5 species of the

Table 1. Fluoride and iron concentrations in the teeth of major groups of bony fishes (class Osteichthyes) other than the Teleostei. Taxa lower than subclass are not given taxonomic categories

Classification/species	*Fluoride (%)*		*Iron (%)*	
Subclass Dipnoi	*(Enamel)*	*(Dentin)*	*(Enamel)*	*(Dentin)*
Neoceratodus forsteri (Australian lungfish)	0.18 ± 0.01	0.20 ± 0.02 p 0.20 ± 0.02	ND	Tr p ND
Lepidosiren paradoxa (South American lungfish)	0.21 ± 0.02	0.19 ± 0.03 p 0.32 ± 0.02	ND	ND p ND
Protopterus dolloi (African lungfish)	0.17 ± 0.01	0.24 ± 0.01 p 0.19 ± 0.02	ND	ND p ND
Subclass Crossopterygii	*(Enamel)*	*(Dentin)*	*(Enamel)*	*(Dentin)*
Latimeria Chalumnae (Coelacanth)	0.34 ± 0.04	0.37 ± 0.01	ND	ND
Subclass Actinopterygii *Cladistia*	*(Enameloid)*	*(Dentin)*	*(Enameloid)*	*(Dentin)*
Polypterus ornatipinnis (Bicher)	a 0.22 ± 0.02 b 0.18 ± 0.01	0.19 ± 0.02	a 0.5 ± 0.3 b ND	ND
Polypterus delnezi (Bicher)	a 0.18 ± 0.02 b 0.19 ± 0.03	0.25 ± 0.02		ND
Polypterus senegalus (Bicher)	a 0.20 ± 0.01 b 0.18 ± 0.02	0.20 ± 0.01	a 2.1 ± 0.9 b ND	ND
Polypterus palmas (Bicher)	a 0.24 ± 0.04 b 0.20 ± 0.02	0.22 ± 0.01	a 1.2 ± 0.2 b ND	ND
Erpetoichthys calabaricus (=*Calamoichthys*, reedfish)	a 0.27 ± 0.08 b 0.21 ± 0.02	0.27 ± 0.02	a 1.0 ± 0.2 b ND	ND
Chondrostei Sturgeons	*Teeth rudimentary or absent*			
Neopterygii Ginglymodi *Lepisosteus spatula* (Gar)	a 0.80 ± 0.01 b 0.20 ± 0.02	0.24 ± 0.01	a 0.7 ± 0.2 b ND	ND
Lepisosteus osseus (Gar)	a 0.74 ± 0.18 b 0.35 ± 0.12	0.25 ± 0.02	a 0.7 ± 0.3 b ND	ND ND
Halecostomi Halecomorphi				
Amia calva (Bowfin)	a 3.38 ± 0.06 b 2.21 ± 0.03	0.34 ± 0.02	a 0.6 ± 0.2 b ND	ND

a, The outer layer; *b*, the inner layer; *ND*, not detectable; *p*, Pleromin

cladistian bichirs and reedfish (Polypteridae), 2 species of the neopterygian gars (Lepisosteidae), and 1 species of neopterygian bowfin (Amiidae) (Table 1). All these species are freshwater in distribution except for the coelacanth, *Latimeria chalumnae*. They were stored in isopropyl alcohol and/or 10% formalin. Adult sturgeons (Chondrostei) do not have teeth, although young ones have very small teeth which could not be used for the present analysis.

All the specimens were embedded in polyester resin after dehydration in an ascending concentration series of alcohol. Plano-parallel ground sections about 50–60 μm in thickness were made using a grindstone after sectioning with a diamond saw. In order to visualize the mineralization pattern, the ground sections were first microradiographed using soft X-ray (Softex type-CMR, Softex Co., Tokyo). Then, they were mounted on a polished surface of an aluminum block with epoxy resin. The electron microprobe (quantitative and line-scan analyses were performed on the polished surface of the ground sections after applying a thin coating of carbon to form a conducting layer by evaporation in a vacuum.

The electron microprobe apparatus used for the present investigation was a Shimadzu-ARL, type EMX-SM, equipped with a wave length dispersive monochrometer. Simultaneous line-scan analysis of fluoride, calcium, and magnesium with an accelerating voltage of 15 kV and a specimen current of 0.15 μA was carried out on the line which ran perpendicularly towards the enameloid or enamel surface from the dentinal side. Following that, another line-scan analysis for iron, calcium, and magnesium with 20 kV and 0.05 μA was performed along the same line, in order to compare the patterns of fluoride and iron distributions. A 10.16 cm RAP curved single crystal was used for the detection of the FKα and CaKβ emissions, a 10.16 cm LiF for the FeKα emission, and a 10.16 cm ADP for CaKα and MgKα emissions.

Quantitative analysis was performed by measuring the intensity of FKα and FeKα emissions obtained by the mode of point analysis using fluorapatite (Durango, F: 3.5%) and the glass beads of anhydrous lithium tetraborate ($Li_2B_4O_7$) containing various amount of iron (III) oxide (Fe_2O_3) (Fe: 5.0%, 3.0%, 1.0%, 0.5%, 0.1%, and 0.05%), which were fused in a furnace (1100°C)

a b

Fig. 1. Microradiograms of a longitudinal ground section of the tooth of a coelacanth, *Latimeria chalumnae*. **a** Low magnification showing the entire view of a tooth. **b** High magnification of the upper one-third of **a**. The tooth surface is covered by very thin highly mineralized enamel

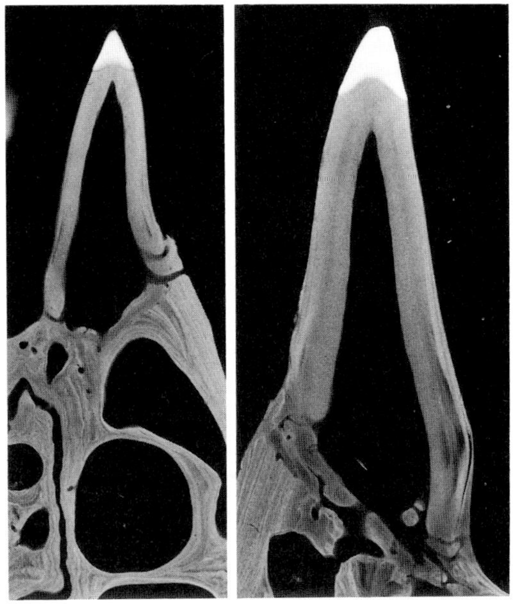

Fig. 2. Microradiograms of longitudinal ground sections of the teeth of cladistian bicher and reedfish. **a** *Polypterus senegalus* (bicher). **b** *Calamoichthys calabaricus* (rccdfish)

[23], as standard samples. A point analysis was performed at certain points in the enameloid or enamel and the dentin. The patterns of line-scan analysis made from the dentin layer towards the enameloid or enamel surface were compared with the mineralization pattern revealed by microradiography.

Results

Form, Structure and Mineralization Pattern of Teeth

Coelacanth, polypterids, gars, and bowfin have separate chisel-like teeth. In the microradiogram of longitudinal ground sections of the teeth of these fishes, the hard tissue, which can be very easily distinguished from the dentin by its very high degree of mineralization, was observed to cover the coronal surface of teeth (Figs. 1–3). In the teeth of polypterids, gars and bowfin, the arrowhead-shaped highly mineralized tissue, which has been considered to be the enameloid, covered only the cuspal tip of the teeth (Figs. 2, 3). Alternatively, in the teeth of the coelacanth, a very thin highly mineralized tissue, which has been identified as the enamel [24–26], covered the entire coronal surface of the teeth (Fig. 1).

The 3 species of lungfishes have plate-like teeth. In the microradiograms of longitudinal ground sections, a wide central part of dentin adjacent to the pulpal space was more highly mineralized than the rest of the dentin and bone. It has been suggested that this highly mineralized part be named "pleromin" or "petrodentin" because of the peculiar structure of the organic matrix and the mineralization pattern [27]. On the other hand, the lateral thin layer of

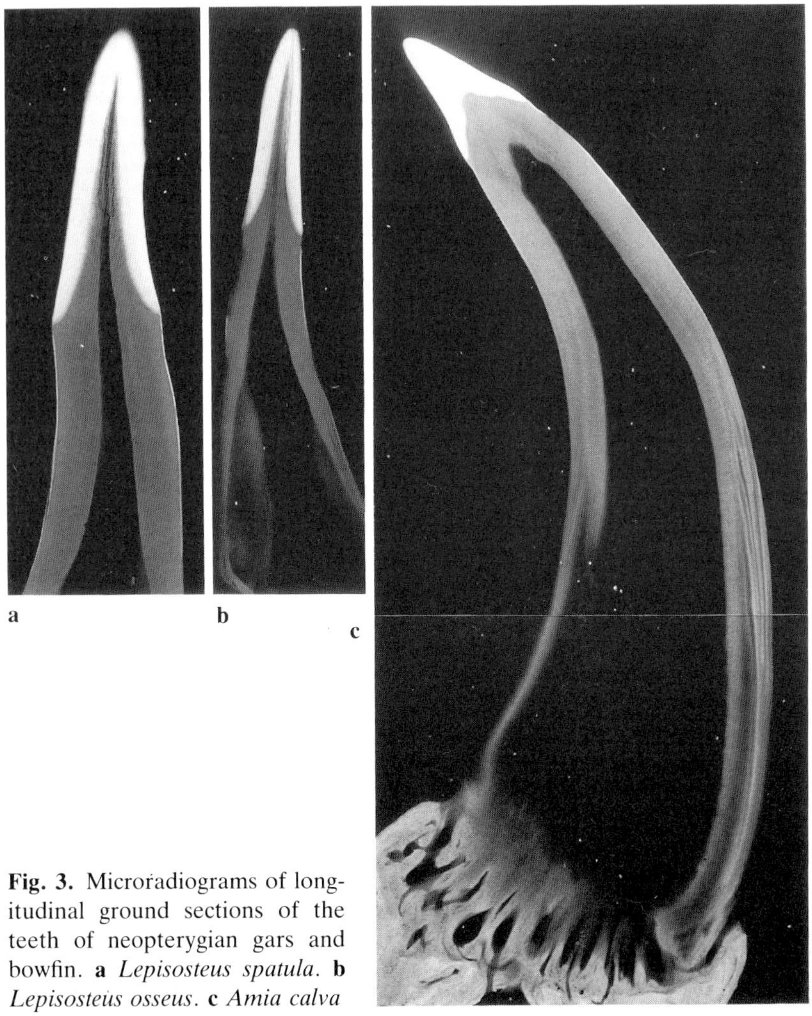

Fig. 3. Microradiograms of longitudinal ground sections of the teeth of neopterygian gars and bowfin. **a** *Lepisosteus spatula*. **b** *Lepisosteus osseus*. **c** *Amia calva*

the tooth plate is also very highly mineralized (Fig. 4). This layer has been suggested to be enamel by some investigators [28–30], although others considered it to be enameloid [31, 32]. In the present investigations, this highly mineralized layer is tentatively called the "enamel."

Elemental Analysis by the Electron Microprobe

In all the fishes examined, the line-scan analysis performed along a line crossing the enameloid or enamel and dentin layers of the fully mineralized teeth showed that the calcium concentration is higher in the enameloid or enamel than in the dentin (Figs. 5–9).

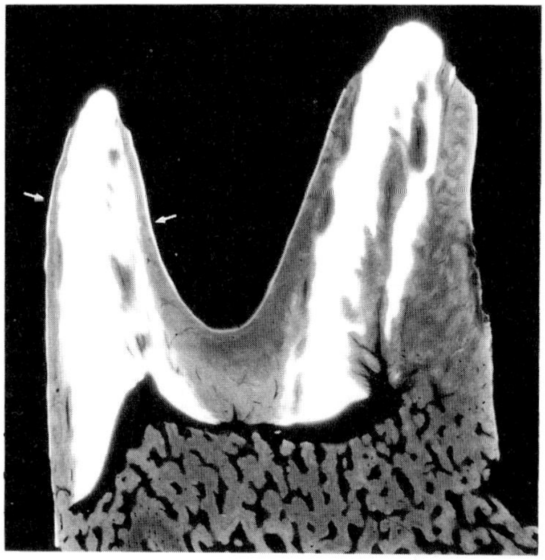

Fig. 4. Microradiogram of a longitudinal ground section of the tooth plate of a dipnoan lungfish, *Lepidosiren paradoxa* (South American lungfish). *Arrow* indicates the highly mineralized thin enamel

Magnesium concentration is much lower in the enameloid and enamel than in the dentin, and a sharp decline of magnesium concentration was observed at the enameloid (or enamel)-dentin junction, followed by a gradual decrease towards the enameloid or enamel surface (Figs. 5–9). In the pleromin of the lungfishes, the calcium concentration was higher and the magnesium concentration much lower than in the rest of the dentin (Fig. 5).

Fluoride concentration was very low and showed an almost even distribution throughout the entire layer of enamel or enameloid and dentin of the teeth of the lungfishes, coelacanth, and polypterids (Figs. 5a, 7a); it was also very low in the pleromin of lungfishes (Fig. 5a). In the teeth of the gars, the fluoride concentration was very low in the dentin and enameloid, except for the narrow outer layer in which fluoride showed a slight increase of its concentration towards the surface (Fig. 8a). In the teeth of the bowfin, fluoride concentration was also very low in the dentin; however, as it entered into the enameloid, the fluoride at first increased its concentration very sharply and highly and then increased gradually towards the surface (Fig. 9a).

A quantitative point analysis performed at the outer and inner layers of the enameloid or enamel and at the middle layer of dentin indicated that the fluoride concentration ranged from about 0.17% to 0.34% in the entire layer of the enameloid or enamel, from about 0.19% to 0.37% in the dentin of the lungfishes, coelacanth, and polypterids, and from about 0.19% to 0.32% in the pleromin of the lungfishes. However, in the enameloid of the gars, the fluoride concentration was about 0.80% and 0.74% in the outer layer and about 0.20% and 0.35% in the inner layer, and, in the enameloid of the bowfin, it was about 3.38% in the outer layer and about 2.21% in the inner layer. In the dentin of the gars and bowfin, the fluoride concentration ranged from about 0.24% to 0.34% (Table 1).

Line-scan analysis indicated that the iron concentration was very low throughout the entire layer of enamel and dentin of the lungfishes (Fig. 5b) and

a b

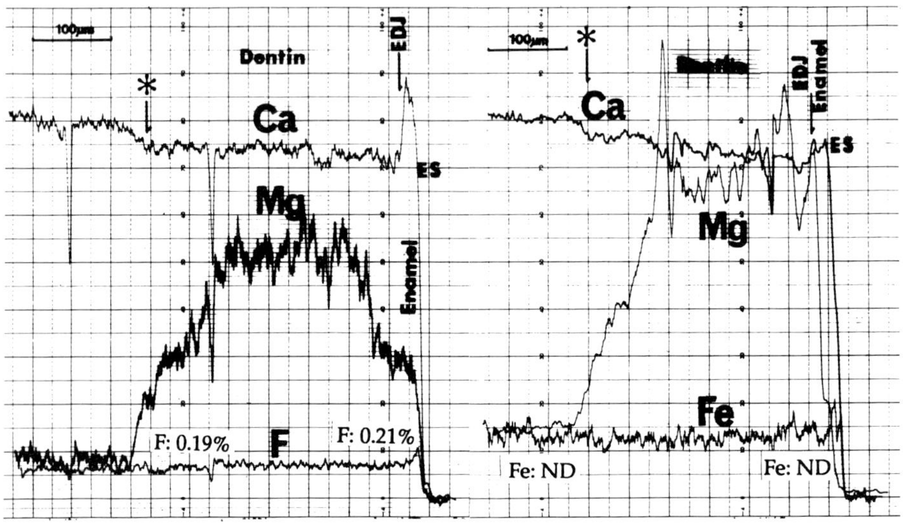

a b

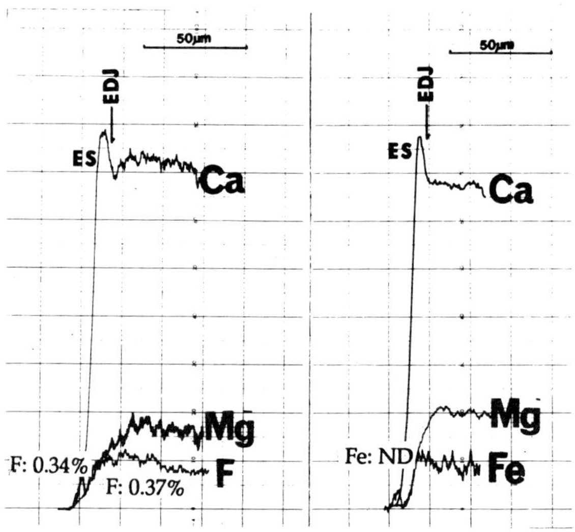

Fig. 5. (*Top*) The distributions of fluoride, calcium, magnesium, and iron in the tooth of a lung fish, *Lepidosiren paradoxa*, revealed by line scan analyses by means of the electron microprobe, which were performed along the same line crossing the enamel, dentin and pleromin. The results of a quantitative point analysis of fluoride and iron at the certain points are indicated. **a** The distribution of fluoride, calcium and magnesium. **b** The distribution of iron, calcium and magnesium. *EDJ*, Enamel-dentin junction; *ES*, enamel surface, *, dentin-prelomin junction. Pleromin is to the *left* of the *

Fig. 6. (*Bottom*) The distribution of fluoride, calcium, magnesium, and iron in the tooth of a coelacanth, *Latimeria chalumnae*, revealed by line-scan analyses by means of the electron microprobe, which were performed along the same line crossing the enamel and dentin. Results of a quantitative point analysis of fluoride and iron at the certain points are indicated. **a** Distribution of fluoride, calcium, and magnesium. **b** Distribution of iron, calcium, and magnesium. *EDJ*, Enamel-dentin junction; *ES*, enamel surface. The enamel is between ES and EDJ and the dentin is *to the right* of EDJ the coelacanth (Fig. 6b), and in the pleromin of the lungfishes (Fig. 5b). According to a quantitative point analysis, the emission intensity of iron could not be discriminated from the background value (Table 1). However, in the enameloid, especially in its outer layer, of polypterids, gars, and bowfin, iron concentration showed a steep increase from the side of the middle layer towards the surface, whereas, in the rest of enameloid and dentin, it was very low (Figs. 7b–9b). A quantitative point analysis indicated that iron concentration ranged from 0.5% to 1.3% at the outer layer and from about 0.1% to 0.2% at the middle layer of the enameloid of these fishes, whereas, in the inner layer of enameloid and in the dentin, it was not detectable under the present analytical condition (Table 1).

Discussion

A quantitative electron microprobe fluoride analysis of the teeth of the Perciformes, which is one of the largest orders of teleostean fishes and includes a considerable number of freshwater species in addition to the predominant marine forms, showed that, while the dentin and bone contained fluoride lower than about 0.50%, the enameloid of all of the 78 species examined contained in its surface layer a rather high fluoride level, ranging mainly from 2.10% to 5.45% [6]. There was no significant difference in the fluoride concentration associated with the saline environments of the water in which these fishes live.

The teeth of marine species of the Tetraodontiformes, which are considered to have evolved from a perciform-like stem and are classified into 2 suborders, i.e., the more generalized Balistoidei and the more specialized Tetraodontoidei, were also examined in the same way [4]. The results indicated that the enameloid of the balistoids studied contained a very high fluoride concentration, ranging from 2.31% to 5.16%, whereas the concentration in the tetraodontids was only 0.14%–0.28%. In the perciform family Acanthuridae (marine fishes) from which the Teraodontiformes may have stemmed, the fluoride concentrations were higher than 3.0% [4]. These facts suggest that the fluoride concentration in the enameloid of teleostean fishes is related to the phylogeny of the fishes rather than to the fluoride concentration in the environmental water, and that the fluoride concentration mechanisms may have been lost secondarily in some groups. However, the biological significance of the high concentration of fluoride into the enameloid is not clearly understood.

Crystallographic investigations indicated that fluoride concentration into the enameloid results in a greater crystal size, reduction in carbonate content, and systematic decrease in the *a*-axis lattice parameter [33].

Balistoid, tetraodontoid, and acanthurid fishes have also been subjected to iron analysis using the electron microprobe [15]. The results showed that the balistoids and acanthurids contain a rather high iron concentration in the enameloid, in the former at the surface layer (0.4%–13.5%) and in the latter throughout the entire layer (2.7%–3.9%), but the tetraodontoids contain such

a b

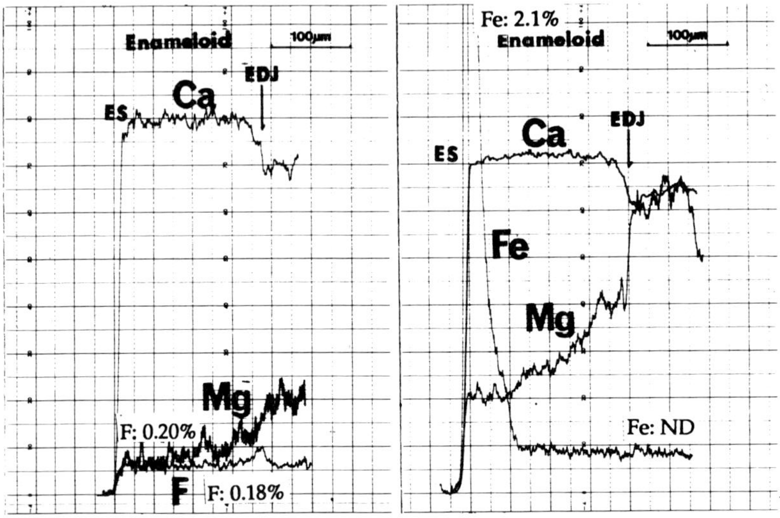
Enameloid
100μm
ES
Ca
EDJ
Mg
F: 0.20%
F
F: 0.18%
Fe: 2.1%
Enameloid
100μm
ES
Ca
EDJ
Fe
Mg
Fe: ND

a b

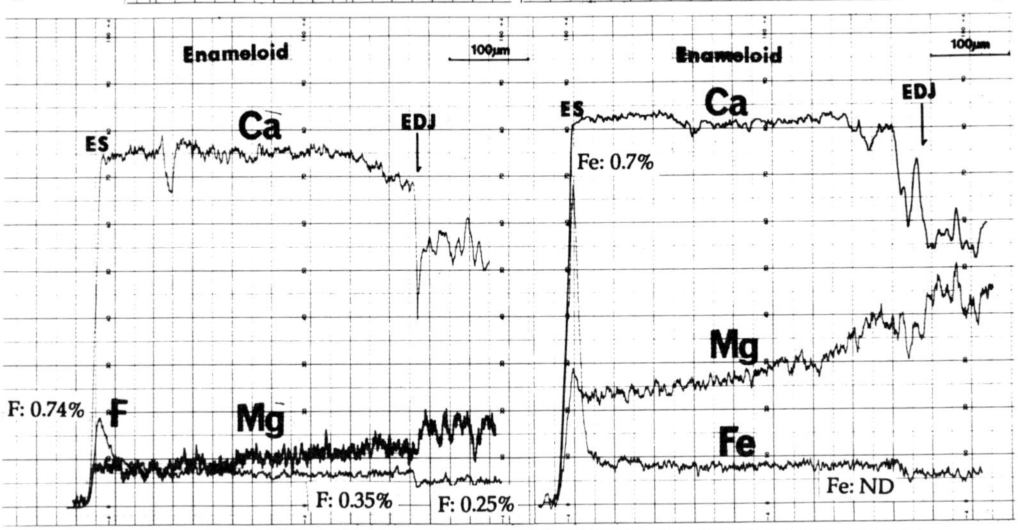
Enameloid
100μm
ES
Ca
EDJ
F: 0.74%
F
Mg
F: 0.35%
F: 0.25%
Enameloid
100μm
ES
Ca
EDJ
Fe: 0.7%
Mg
Fe
Fe: ND

a b

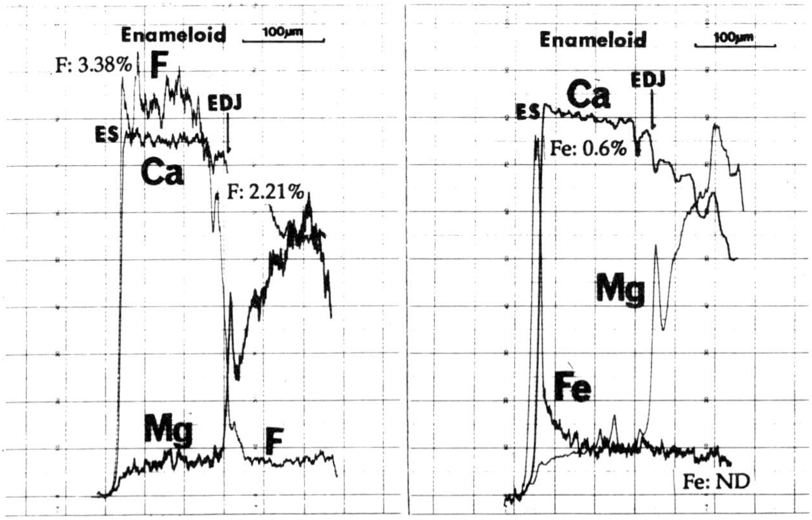
Enameloid
100μm
F: 3.38%
F
EDJ
ES
Ca
F: 2.21%
Mg
F
Enameloid
100μm
ES
Ca
EDJ
Fe: 0.6%
Mg
Fe
Fe: ND

a very low iron concentration that it could not be discriminated from the background value. It is inferred from these results that the iron concentration into the enameloid is also related to the phylogeny of these fishes and that the mechanism has been lost in the course of differentiation of the tetraodontoids. However, some investigators consider that the iron deposition hardens the enameloid and reduces abrasion and cracking, and is related to the feeding habits of fishes [12, 14].

The chemical state and significance of the iron deposition in the enameloid have not been elucidated. However, the presence of $Fe3+$ ions is known to induce a high crystallinity of apatite [34]. Electron microscopic investigations of heavily deposited iron at the narrow surface layer of enamel of the rat incisor showed that it is not present as a separate particulate phase but is mainly bound to the surface of apatite crystals [35]. A Mössbauer effect study also suggested that iron is mainly attached on the surface of crystals [36].

It is interesting to note that both fluoride and iron, which originally have no particular chemical correlation, are deposited together in some heavily mineralized tissues, such as the enameloid of some teleostean fishes [4, 13, 15], and the teeth of the radula of the chiton (Polyplacophora: Mollusca) [13, 37], although their depositions into the enameloid are not fully synchronized with each other in the process of mineralization of the developing enameloid and their final distributional patterns do not coincide [13]. At least in the enameloid of teraodontiform fishes such a phenomenon seems to be related to the phylogeny of the fishes [15], in view of their concurrent disappearance in the course of differentiation of these fishes.

Fig. 7. (*Top*) The distribution of fluoride, calcium, magnesium, and iron in the tooth of a bicher, *Polypterus senegalus*, revealed by line-scan analyses by means of the electron microprobe, which were performed along the same line crossing the enameloid and dentin. Results of a quantitative point analysis of fluoride and iron at the certain points are indicated. **a** Distribution of fluoride, calcium and magnesium. **b** Distribution of iron, calcium and magnesium. *EDJ*, Enameloid-dentin junction; *ES*, enameloid surface. The dentin is *to the right* of EDJ

Fig. 8. (*Middle*) The distribution of fluoride, calcium, magnesium, and iron in the tooth of a gar, *Lepisosteus spatula*, revealed by line-scan analyses by means of the electron microprobe performed along the same line crossing the enameloid and dentin. Results of a quantitative point analysis of fluoride and iron are indicated. **a** The distribution of fluoride, calcium, and magnesium. **b** The distribution of iron, calcium, and magnesium. *EDJ*, Enameloid-dentin junction; *ES*, enameloid surface. The dentin is *to the right* of EDJ

Fig. 9. (*Bottom*) The distribution of fluoride, calcium, magnesium, and iron in the tooth of bowfin, *Amia calva*, revealed by line-scan analyses by means of the electron microprobe, which were performed along the same line crossing the enameloid and dentin. Results of a quantitative point analysis of fluoride and iron are indicated. **a** The distribution of fluoride, calcium and magnesium. **b** The distribution of iron, calcium and magnesium. *EDJ*, Enameloid-dentin junction; *ES*, enameloid surface. The dentin is *to the right* of EDJ

All the above observations point to the existence of certain relationships between fluoride and iron concentrations and the phylogeny of teleostean fishes. In order to elucidate these relationships more precisely and trace the origin of the concentration mechanisms, the teeth of non-teleostean bony fishes ancestral to or distantly divergent from the present-day predominating teleostean fishes were examined in this study using the same methods as in our previous studies [1–6, 13, 15].

The coronal portion of the teeth of all bony and cartilaginous fishes had been considered to be covered with enameloid, which is very highly mineralized, as observed in the enamel of tetrapods, and in which very fine filaments of collagen formed by odontoblasts persist; there is also a complex woven structure with varying orientations of the crystals [12, 29, 38]. However, recent histological investigations of fish teeth using transmitted, polarizing, and scanning electron microscopies, and microradiography have revealed that the teeth of the coelacanth are covered with true enamel, in which the crystals are arranged perpendicularly to the surface without any trace of collagen and with incremental lines growing away from the side of the dentin and running throughout the entire layer [24, 25]. This fact was considered to be of great importance since some crossopterygians have been considered by many authors as being ancestral to tetrapods [24]. Properties similar to the enamel, rather than the enameloid, containing crystals that run perpendicularly to the surface, have also been observed in lungfishes [28–30], although other investigators considered them to be enameloid or highly mineralized dentin [31, 32, 39]. On the basis of cladistic analyses of anatomical characters, it was indicated that lungfishes and tetrapods are sister groups [21, 40, 41].

The present study revealed that no high concentrations of fluoride and iron, as observed in the enameloid of some teleosteans, were found in the enamel of the coelacanth and lungfishes, whereas, calcium and magnesium in these 2 groups showed a distributional pattern similar to that in the enamel of mammals and in the enameloid of fishes [4–6, 15, 42].

The relationships of the cladistians, or the polypterids, have been the subject of a century-long dispute. They have been considered to represent either a highly specialized lineage of the subclass Actinopterygii (ray-finned fish) or a group sufficiently distinct from other osteichthyans to warrant their separation into a subclass, the Brachiopterygii [18, 22, 43]. The fact that the polypterids have enameloid is in favor of, or at least not inconsistent with, their allocation to the Actinopterygii.

In the present study, high concentrations of iron in the enameloid, especially in its surface layer, were observable in the polypterids, gars, and bowfin, whereas the adjacent dentin contained a very low iron concentration which was not detectable under the analytical conditions of this study. The concentration of the iron in the enameloid of these fishes increased steeply from the middle layer towards the surface, as seen in many teleosteans [13, 15]. On the other hand, very high concentrations of fluoride throughout the entire layer of enameloid, as seen in many teleosteans, were found only in the bowfin. In the other examined species, only the neopterygian gars showed a slight increase in the narrow outer layer.

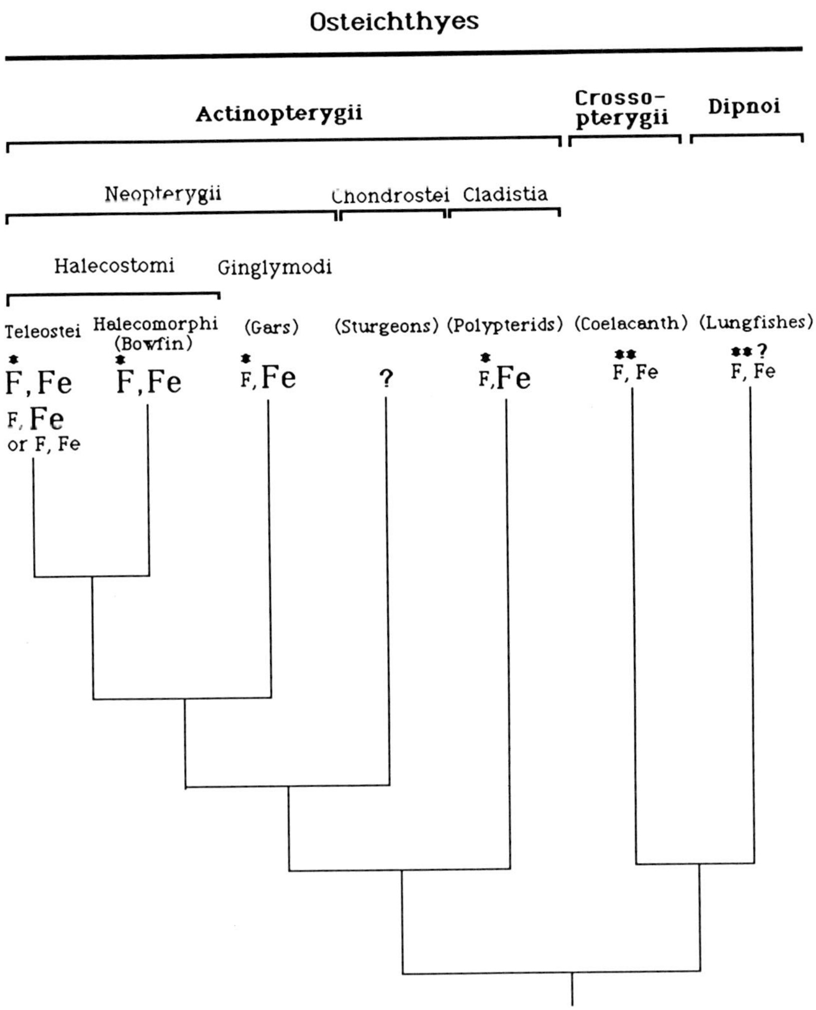

Fig. 10. A possible phylogenetic diagram of major groups of extant bony fishes (Osteichthyes) with summarized results of the quantitative analysis of fluoride and iron in the enameloid and enamel. * Enameloid, ** Enamel, ***F, Fe***, High concentration in the enamel and enameloid (fluoride = >3.0%; iron = >0.5%); *F, Fe*, low concentration in the enamel and enameloid (fluoride = <0.4%; iron = not detectable under the present analytical conditions)

Incorporating all the observations made in this study, the mechanisms of high fluoride and iron concentrations into the developing enameloid are considered to have occurred only in the Actinopterygian lineage. The high iron concentration in the polypterids supports their inclusion in the Actinopterygii. The mechanism of high iron concentration can be traced back to the cladistian stage of actinoperygian evolution, while that of high fluoride concentration to the neopterygian stage (Fig. 10). The two mechanisms are probably independent of each other and have different evolutionary backgrounds. This idea is supported by the fact that the iron concentration is very low in the enameloid of sharks and rays of cartilaginous fishes, whereas the fluoride concentration is very high [44]. Since, in addition to the present results, it was found that there are some teleostean fishes of the Characiformes whose enameloid contains very high iron and low fluoride concentrations [44], the high iron concentration mechanism seems to have appeared earlier and disappeared later than the high fluoride concentration mechanism in the actinopterygian evolution.

Deposition of histologically or histochemically detectable amounts of iron has been found in the enameloid of some teleostean fishes and in the enamel of some amphibians, reptiles, and mammals, although no investigation has been made on its biological and phylogenetic significance [8–10, 12]. In view of the absence of detectable amounts of iron in the enamel of the Dipnoi and Crossopterygii, the iron deposition mechanisms of tetrapods may have appeared in parallel with that of actinopterygian fishes, in the former group after they established themselves as terrestrial animals and in the latter from the cladistian stage.

High fluoride concentrations are not found in amphibians, reptiles, or mammals [45] nor in dipnoans or crossopterygians. The high concentration of fluoride seen in the enameloid of neopterygian fishes is probably due to a certain mechanism specially designed for advanced actinopterygians for some unknown biological reasons.

Acknowledgements. We thank Dr. T. Ueno, National Science Museum, Tokyo, and Dr. D.E. McAllister, National Museum of Natural Sciences, Ottawa, for supplying many specimens.

References

1. Suga S, Wada K, Ogawa M (1977) Fluoride concentration in the enameloid of fishes (abstract). J Dent Res 56(A):54
2. Suga S, Wada K, Ogawa M (1978) Possible relationship between the fluoride concentration in the enameloid and phylogeny of fishes (abstract). J Dent Res 57(A):280
3. Suga S, Wada K, Ogawa M (1980) Fluoride concentration in the enameloid of fishes. In: Omori M, Watabe N (eds) The mechanisms of biomineralization in animals and plants. Tokai University Press, Tokyo, pp 229–240
4. Suga S, Wada K, Ogawa M (1981) Fluoride concentration in teeth of tetraodontiform fishes and its phylogenetic significance. Jpn J Ichthyol 28:304–312
5. Suga S, Wada K, Ogawa M (1981) Fluoride concentration in the enameloid of fish teeth and its relationship to the phylogeny of fishes. In: Binder K, Hohenegger M (eds) Fluoride metabolism. Whilhelm Maudrich, Wien, pp 79–88

6. Suga S, Taki Y, Wada K (1983) Fluoride concentration in the teeth of perciform fishes and its phylogenetic significance. Jpn J Ichthyol 30:81–93
7. Greenhalgh R, Riley JP (1963) Occurrence of abnormally high fluoride concentration at depth in the ocean. Nature 197:371–372
8. Schmidt WJ (1958) Natürliche Färbung von Reptilien- und Fischzähnen durch Eisenoxyd. Zool Anz 161:168–178
9. Schmidt WJ (1969) Die Schmelznatur der eisenoxidhaltigen Kappe auf Teleostier-Zähnen. Z Zellforsch Mikrosk Anat 93:447–450
10. Miles AEW (1963) Pigmented enamel. Proc R Soc Med 56:918–920
11. Garant PR (1970) Observations on the ultrastructure of the ectodermal component during odontogenesis in *Helostoma temmincki*. Anat Rec 166:167–188
12. Shellis RP, Berkovitz BKB (1976) Observations on the dental anatomy of piranhas (Characidae) with special reference to tooth structure. J Zool 180:69–84
13. Suga S (1984) The role of fluoride and iron in mineralization of fish enameloid. In: Fearnhead RW, Suga S (eds) Tooth enamel IV. Elsevier, Amsterdam, pp 472–477
14. Motta PJ (1987) A quantitative analysis of ferric iron in butterflyfish teeth (Chaetodontidae, Perciformes) and the relationship to feeding ecology. Can J Zool 65:106–112
15. Suga S, Wada K, Taki Y, Ogawa M (1989) Iron concentration in teeth of tetraodontiform fishes and its phylogenetic significance. J Dent Res 68: 1115–1123
16. Torell P (1957) Determination of iron in dental enamel. Odont Tidskr 65:20–23
17. Nelson GJ (1969) Gill arches and the phylogeny of fishes, with notes on the classification of vertebrates. Bull Am Mus Nat Hist 141:475–552
18. Jessen HL (1973) Interrelationships of actinopterygians and brachiopterygians: Evidence from pectral anatomy, In: Greenwood PH, Miles RS, Patterson C (eds) Interrelationships of fishes. Academic, London, pp 227–232
19. Patterson C (1973) Interrelationships of holosteans. In: Greenwood PH, Miles RS, Patterson C (eds) Interrelationships of fishes. Academic, London, pp 233–305
20. Shaeffer B (1973) Interrelationships of chondrosteans. In: Greenwood PH, Miles RS, Patterson C (eds) Interrelationships of fishes. Academic, London, pp 207–226
21. Rosen DE, Forey RL, Gardiner BG, Patterson C (1981) Lungfishes, tetrapods, paleontology, and plesiomorphy. Bull Am Mus Nat Hist. 167: 159–276
22. Lauder GV, Liem KF (1983) The evolution and interrelationships of the actinopterygian fishes. Bull Mus Comp Zool 150:95–197
23. Stephenson DA (1969) An improved flux-fusion technique for X-ray emission analysis. Anal Chem 41:966–967
24. Smith MM (1978) Enamel in the oral teeth of *Latimeria chalumnae* (Pisces: Actinistia): A scanning electron microscope study. J Zool Lond 185:355–369
25. Shellis RP, Poole DFG (1978) The structure of the dental hard tissues of the coelacanthid fish *Latimeria Chalumnae* Smith. Arch Oral Biol 23:1105–1113
26. Sasagawa I, Ishiyama M, Kodera H (1984) Fine structure of the pharyngeal teeth in the coelacanthid fish (*Latimeria chalumnae*). In: Fearnhead RW and Suga S (eds) Tooth enamel IV. Elsevier, Amsterdam, pp 462–466
27. Orvig T (1976) Palaeohistological notes. 4. The interpretation of osteodentine, with remarks on the dentition in the devonian dipnoan griphognathus. Zool Scripta 5:79–96
28. Peyer B (1968) Comparative odontology, University of Chicago Press, Chicago, pp 110–118
29. Schmidt WJ, Keil A (1971) Polarizing microscopy of the dental tissues. Pergamon, Oxford, pp 240–244

30. Ishiyama M, Ogawa T (1983) Existence of true enamel on tooth plate in the lungfish *Lepidosiren paradoxa* (in Japanese). Acta Anat Nipponica 58:157–161
31. Smith MM (1977) The microstructure of the dentition and dermal ornament of three dipnoans from the Devonian of Western Australia. Phil Trans R Soc (B) 281:29–72
32. Kemp A (1979) The histology of tooth formation in the Australian lungfish, *Neoceratodus forsteri Krefft*. Zool J Linn Soc 99:251–287
33. LeGeros RZ, Suga S (1980) Crystallographic nature of fluoride in enameloids of fish. Calcif Tissue Int 32:169–174
34. LeGeros RZ, Tsheri MH, Quirolgico CB, LeGeros JP (1980) Formation and stability of apatite: Effects of some cationic substituents. Proc 2nd International Congress on Phosphorus Compounds, Boston, pp 89–103
35. Selvig AK, Halse A (1975) The ultrastructural localization of iron in rat incisor enamel. Scand J Dent Res 83:88–95
36. Bauminger E, Ofer S, Gedalia I, Horowitz G, Mayer I (1985) Iron uptake by teeth and bones: A Mössbauer effect study. Calcif Tissue Int 37:386–389
37. Lowenstam HA (1967) Lepidocrocite, an apatite mineral, and magnetite in teeth of chitons (Polyplacophora). Science 156:1373–1374
38. Shellis RP (1981) Dental tissues evolution In: Osborn JW (ed) Dental anatomy and embryology. Blackwell Scientific, Oxford, pp 155–165
39. Denison RH (1974) The structure and evolution of teeth in lungfishes. Fieldiana Geology 33:31–58
40. Gardiner BG (1980) Tetrapod ancestry: A reappraisal. In: Panchen AL (ed) The terrestrial environment and the origin of land vertebrates. Academic, London, pp 175–185
41. Forey PL (1986) Relationships of lungfishes. In: Bemis WE, Burggren WW, Kemp NE (eds) The biology and evolution of lungfishes. Alan R Liss, New York, pp 75–91
42. Suga S, Kondo M, Onodera A, Kubota Y, Ohtuka M (1971) Electron microprobe analysis on the distributions of Cl, Mg and Na, in the enamels of various animals (in Japanese). Jpn J Oral Biol 13:85–94
43. Patterson C (1982) Morphology and interrelationships of primitive actinopterygian fishes. Am Zool 22:241–259
44. Suga S, Taki Y, Wada K, Ogawa M (1991) Evolution of fluoride and iron concentrations in the enameloid of fish teeth, In: Suga S, Nakahara H (eds) Mechanisms and phylogeny of mineralization in biological systems. Springer, Tokyo, pp 439–446
45. Suga S, Ogawa M (1989) Comparative observations on trace element distribution in enamel and enameloid, In: Fearnhead RW (ed) Tooth enamel V. Florence, Yokohama, pp 353–359

Fish Tooth Formation: An Assessment of Biological Factors Affecting the Fluoride Content of Enameloid

Kenneth S. Prostak, Philip Seifert, and Ziedonis Skobe[1]

Summary. The fluoride content in fish tooth enameloid is dependent on species phylogeny, not on its food or habitat. The source of osteichthian enameloid collagen, formed from both the ectoderm and mesoderm, has no relationship to the fluoride content of the enameloid. Similarities in the electrophoretic profiles, amino acid composition and the extent of collagen cross-linking between bony fish enameloids having a high or low fluoride content make it unlikely that specific proteins of the enameloid matrix avidly bind fluoride. Chondrichthian enameloid is distinct from that of bony fishes: it contains no ectoderm collagen and little mesodermal collagen, but is enriched in carbohydrate and non-collagenous mesodermal proteins. On the other hand, the unique chondrichthian enameloid suggests a fluoride-binding property.

The serum fluoride levels in all species tested were low, apparently not related to the fluoride content of their enameloid. Therefore, it seems unlikely that there is a systemic mechanism concentrating fluoride into enameloid.

We propose that a triad of cellular morphologies, found exclusively in bony fish tooth buds forming a high fluoride enameloid, provides the structural framework for fluoride concentration into the enameloid. First, a rich vascular network surrounding the enameloid organ in the mineralization stage allows site-specific enrichment of ions, including fluoride, from fenestrated capillaries into enameloid organ cells. Second, the specialized morphology of outer dental epithelial cells in species with high fluoride levels consistent with the hypothesis that fluoride ions are preferentially transported across cell membranes as are chloride ions in "chloride cells" of fish gills. Third, the intercellular junctions of the enameloid organ serve as barriers to the outward diffusion of transported fluoride.

Key words: Enameloid—Development—Fluoride—Ion Transport—Matrix

Introduction

The homology of fish tooth enameloid with mammalian enamel has been debated considerably over the past century. The presence of collagen in the enameloid matrix has led to the perception that this tissue was a modified

[1] Forsyth Dental Center, Boston, MA 02115, USA

dentin [1]. Subsequently, it has been suggested that an epithelial protein secreted into the mesodermal collagen modified the collagen cross-linking, allowing large apatite crystals to form [2]. However, we showed that collagen of the enameloid matrix was not mesodermal in origin, but that the inner dental epithelial cells (IDE) produced a large quantity of type I collagen in the enameloid matrix [3, 4]. IDE cells of several bony fish also secrete an amorphous material on the surface of mineralizing enameloid. Here, we provide additional evidence that the enameloid matrix of the bony fish enameloid is composed primarily of ectodermally derived collagen.

It is known that bony fishes may be classified into 2 groups: the fishes whose enameloid contains high fluoride (higher than about 2.0%) and those whose enameloid is low in fluoride (lower than about 0.4%). The fluoride content in the enameloid is dependent on species phylogeny, not fluoride concentration in the environmental water or the feeding habits of the fishes [5–7; Chap. 2] We demonstrated that the fluoride content of the surface layer reflects the high or low fluoride content of its underlying enameloid. Thus, it is unlikely that these amorphous proteins have specific fluoride-concentrating properties. The observation of mineralizing amorphous material adjacent to the pulpal aspect of enameloid suggests that odontoblasts associated with the enameloid matrix synthesis also synthesize and secrete non-collagenous proteins into the enameloid matrix. The homology of these and other surface layers to mammalian dental enamel is explored.

Little is known concerning the biochemistry of collagen and other proteins of the enameloid matrix, especially in relation to the fluoride content of enameloid. In this study, we determined whether potential differences in electrophoretic gel patterns, amino acid composition, and collagen cross-linking between high and low fluoride-containing enameloid matrices account for the different fluoride content in enameloid.

Finally, data are presented supporting a novel hypothesis of fluoride regulation through the enameloid organ of bony fish containing high fluoride in enameloid. Unique cytostructural features of the outer dental epithelial cells (ODE) are compared to the anion regulating "chloride cells" of fish gills. We suggest that fluoride is transported by specialized ODE cells found in tooth buds forming high fluoride enameloid.

Materials and Methods

Numerous specimens representing the following families were routinely processed for either TEM [4, 8] or freeze-dried for elemental analysis by electron microprobe: Skate-*Rajidae*; Blue Shark-*Galeidae*; Tautog-*Labridae*; Sea Robin-*Triglidae*; Parrotfish-*Scaridae*; Cichlid-*Cichlidae*; Triggerfish-*Balistidae*; Pufferfish-*Tetraodontidae*; Alligator Gar-*Lepisosteidae*; Bluefish-*Pomatomidae*; and Goldfish-*Cyprididae*.

Some fish were injected with either colchicine [4, 8], insulin, or glucagon (2 mg/kg), killed at various intervals post-injection, and then processed for TEM.

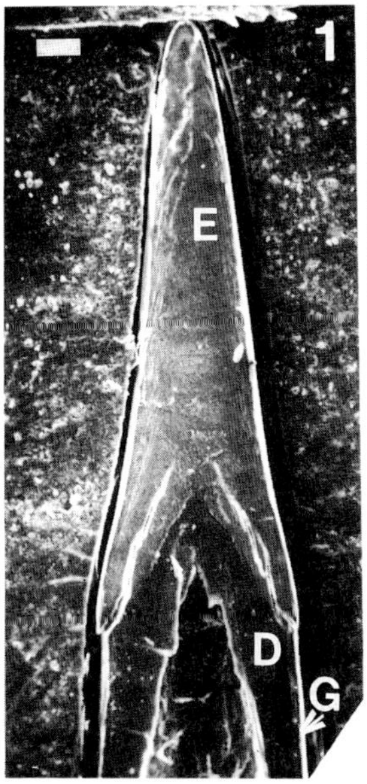

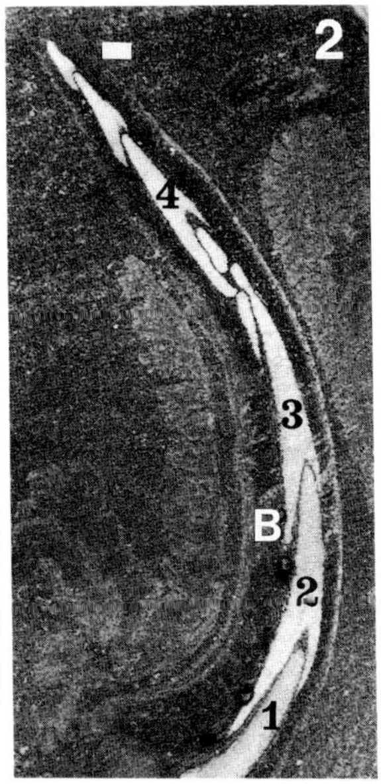

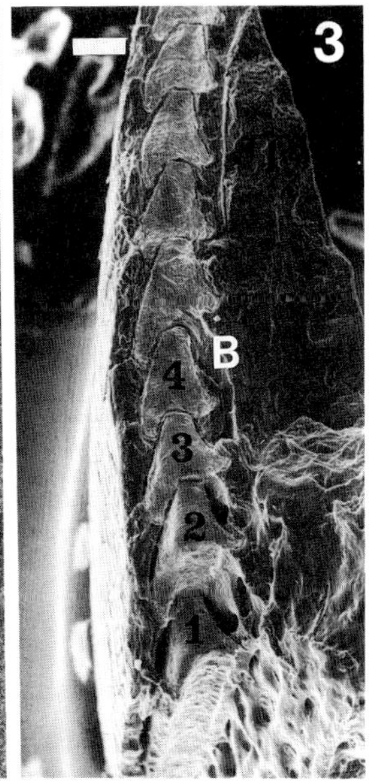

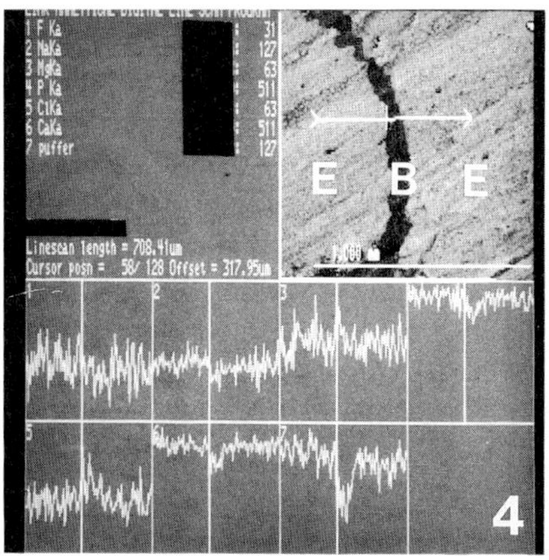

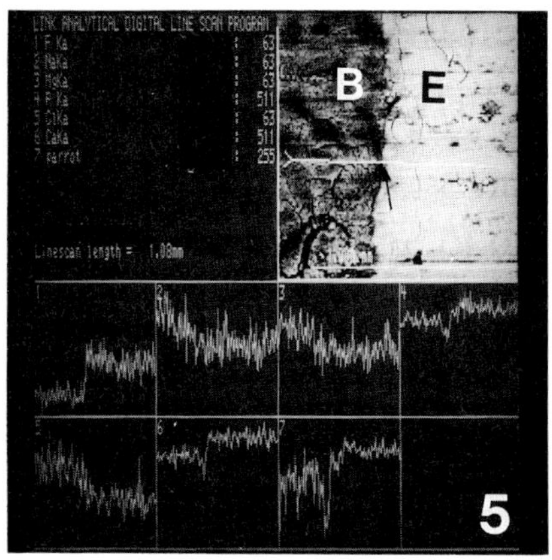

Fig. 1. Secondary electron image of longitudinally sectioned, mature gar tooth depicting enameloid (*E*), dentin (*D*), and ganoid/enamel (*G*). *Scale* = 100 μm

Fig. 2. Backscattered electron image of a parrotfish jaw section showing arrangement of spade-shaped teeth. Bone (*B*), but not dentin, is interspersed between the enameloid pieces (*1–4*) of successive tooth rows. *Scale* = 100 μm

Fig. 3. Secondary electron image of a fractured sharpnose pufferfish jaw showing the eruptive pattern of the plate-like enameloid pieces (*1–4*). Dentin is not present in this species. *B* = Bone. *Scale* = 100 μm

Fig. 4. EDS linescan of mature pufferfish enameloid. The *line* on the digitized image indicates the path of the linescan. The numerical values for each element listed correspond to the total vertical scale CPM. There were distinct decreases in calcium (*6*) and phosphorus (*4*), but not fluoride (*1*), at the intersection of enameloid (*E*) and bone (*B*). *Block 7* is a grey-scale histogram of image. The *arrows* indicate the position of bone

Fig. 5. EDS linescan of mature parrotfish enameloid. Distinct increases in calcium (*6*), phosphorus (*4*), and fluoride (*1*) were observed in enameloid (*E*) relative to bone (*B*). Chloride (*5*) content of bone is higher relative to that in enameloid. *Block 7* is a grey-scale histogram of the image. The *arrows* indicate the position of bone

Environmental water, the developing enameloid matrix, and blood samples of each fish were taken and assayed for fluoride by microdiffusion [9]. The amino acid composition and collagen cross-link content of freeze-dried parrotfish, pufferfish, and blue shark enameloids were determined by Y. Kuboki [10]. Ten-percent polyacrylamide gel electrophoresis was performed on 0.5 M acetic acid soluble enameloid proteins of several developing tooth stages from tautog, parrotfish, pufferfish, skate, and shark. Antisera developed against the rat enamel matrix [3] was used in Western blotting and immunocytochemical [11] assays to determine which enameloid proteins were potentially homologous to those of the mammalian enamel matrix.

Morphometric analyses were performed on calibrated micrographs using either SigmaScan (Jandel Scientific; Calif.) or Link (Link Analytical; Wis.) analytical software.

Results and Discussion

Mature Dentition

Two major groups of tooth morphologies were represented in our samples. One group (sharks, skates, tautog, gar, goldfish, cichlid, bluefish, and sea robin) had enameloid caps supported by tooth shaft dentin (Fig. 1), whereas the other group had enameloid pieces with little (ballonfish, and porcupine fish) or no dentin (parrotfish and sharpnose pufferfish) (Figs. 2, 3). Tautog, cichlid, and bluefish had thin layers of collar enameloid on their tooth shaft surface, whereas the gar displayed a hypermineralized layer of ganoid/enamel containing incremental lines [11, 12] not found on collar enameloid.

Fluoride Content of Enameloid

Electron microprobe analysis (by wavelength dispersive spectroscopy; WDS) showed a species-specific, high (2.0–3.7 wt%) or low (<0.4 wt%) fluoride content in the mature enameloid. Species containing high fluoride in enameloid were the shark, tautog, cichlid, sea robin, and parrotfish whereas pufferfish, bluefish, goldfish, and gar had low fluoride. The fluoride content of the gar ganoid/enamel layer was also low. All species had consistently low fluoride (0.8–0.1 wt%) in their dentin and bone. Energy dispersive spectroscopy (EDS) linescan analysis across the tooth-bone boundary indicated no fluoride gradient between pufferfish enameloid and bone (Fig. 4), whereas a sharp fluoride gradient was observed between parrotfish enameloid and bone (Fig. 5). Previously, we demonstrated that fluoride accumulates in developing enameloid matrix, and that the final fluoride content of the enameloid is not related to serum fluoride levels [9]. Thus, a specific fluoride-binding protein or cellular process must be responsible for concentrating fluoride into enameloid.

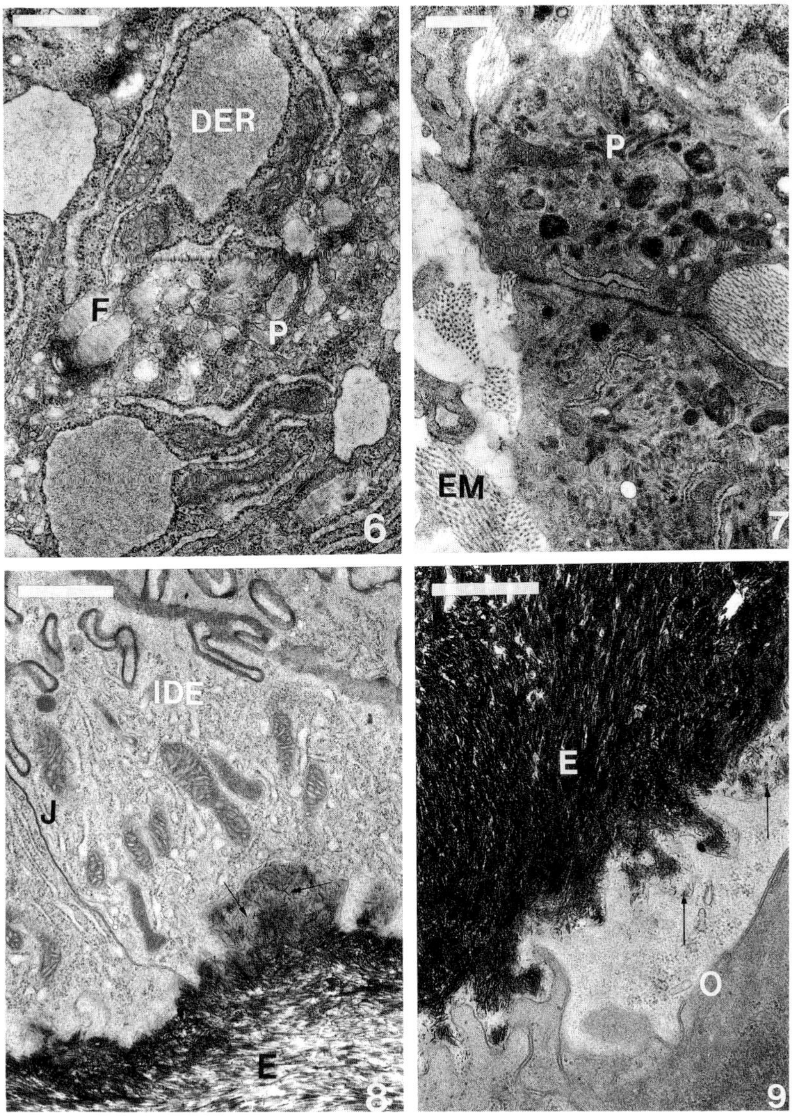

Fig. 6. TEM of colchicine-treated sharpnose pufferfish IDE cells at the enameloid matrix secretion stage. Dilated granular endoplasmic reticulum (*DER*), numerous procollagen granules (*P*), and striated fibers (*F*) were characteristic of colchicine-treated IDE cells. *Scale* = 1 μm

Fig. 7. TEM of colchicine-treated sharpnose pufferfish odontoblasts associated with enameloid matrix (*EM*). Accumulations of procollagen granules (*P*) were frequently but not consistently observed in all odontoblasts. *Scale* = 1 μm

Fig. 8. TEM of amorphous surface layer on mineralizing sharpnose pufferfish enameloid (*E*). The newly secreted material has small crystallites with random orientation (*arrows*). Intercellular junctions (*J*) are observed between neighboring *IDE* cells. *Scale* = 1 μm

Fig. 9. TEM of amorphous material (*arrows*) associated with sharpnose pufferfish odontoblasts (*O*) at the enameloid mineralization stage. *Scale* = 1 μm

Development of Enameloid Matrix

The presence of procollagen granules within both normal and colchicine-treated IDE cells from pufferfish (Fig. 6), parrotfish, cichlid, tautog, goldfish, and gar demonstrates that the bony fish enameloid matrix collagen is synthesized and secreted by the ectoderm. Procollagen granules have been seen in IDE cells of salmon [13] and eel (P. Shellis, personal communication, 1988).

It has been suggested that mesodermal collagen provides a scaffolding on which the ectodermal collagen is oriented [3]. However, the proportions of ectodermal to mesodermal collagens in enameloid remains unresolved [3, 13, 14]. Since parrotfish and sharpnose pufferfish contain no dentin (Figs. 2, 3), they are ideal models to investigate the relative contributions of the ectoderm and mesoderm to the enameloid matrix collagen.

Both pufferfish (Fig. 6) and parrotfish IDE cells responded to colchicine injection by accumulating procollagen granules in their cytoplasm at the enameloid matrix secretion stage. Puffer and parrotfish odontoblasts were confined to a small layer of cells at the base of each tooth bud. In colchicine-treated specimens, accumulations of procollagen granules were sometimes, but not consistently, observed within puffer and parrotfish odontoblasts (Fig. 7). The data suggest that the odontoblasts provide some "scaffolding" collagen which orients the ectodermally derived collagen into fibers [3]. Since this "scaffolding" collagen is present in both high and low fluoride-containing enameloid matrices, it is unlikely that the origin of collagen influences the fluoride content of enameloid.

Non-Enameloid Surface Layers

Pufferfish and parrotfish IDE cells contained non-collagenous, electron-dense secretory granules which were released onto the surface of the mineralizing enameloid (Fig. 8). Nascent crystallites subsequently formed in this material, but had no particular orientation and were quite different in size from the subjacent enameloid crystallites. At later stages of mineralization, pufferfish surface crystallites could not be distinguished from those of its enameloid, whereas parrotfish surface crystallites were somewhat shorter than those in the underlying enameloid.

Amorphous material was also observed between pufferfish odontoblasts and the mineralizing enameloid matrix (Fig. 9). This collagen-free odontoblast product also contained crystallites which, at later stages, were indistinguishable from enameloid crystallites. Possibly, odontoblasts can also express non-collagenous matrix proteins which modulate apatite formation during the mineralization stage.

Previous light and scanning electron microscopic studies suggested that in crossopterygian coelacanth, the IDE cells secrete a true enamel layer [15, 16]. An ectodermally derived enamel-like layer has also been described on lungfish (*Protopterus sp*) teeth [17]. However, the ultrastructure of the formation of these enamel-like layers has yet to be investigated. We have found that the

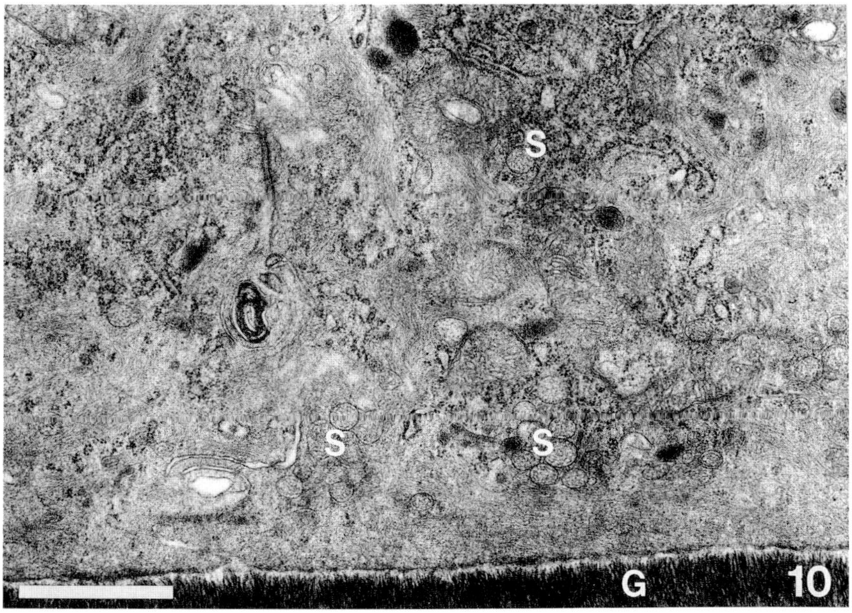

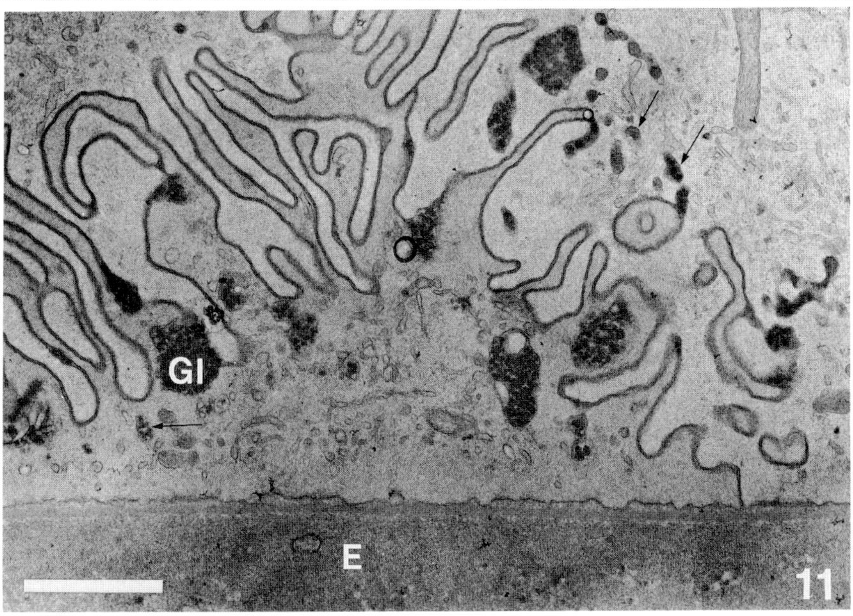

Fig. 10. TEM of ganoid/enamel formation in the alligator gar. Newly secreted enamel-like matrix (*G*) contains crystallites oriented normal to the IDE cell surface. Characteristic granules (*S*) contain "stippled" material. *Scale* = 1 μm

Fig. 11. TEM of glucagon-treated skate IDE cells, 1 h post-injection. As in colchicine-treated material, increased glycogen (*Gl*) is observed within intracellular vesicles (*arrow*) and between neighboring IDE cells. *Scale* = 1 μm tooth shaft of the alligator gar also contains a similar, non-collagenous, ectodermally derived enamel-like layer [11, 12].

The formation of ganoid/enamel is distinct from the parrotfish and pufferfish surface layers, which mineralize to become indistinct from the underlying enameloid. First, ganoid/enamel is found on the tooth shaft dentin and not on the enameloid crown itself [12] (Figs. 1, 10). Second, crystallites form immediately within the matrix and are oriented perpendicular to the tooth surface (Fig. 10). Third, characteristic secretory granules are observed within IDE cells associated with ganoid/enamel secretion but not within IDE cells associated with enameloid formation. The presence of incremental lines parallel to the tooth surface in ganoid/enamel bear a remarkable resemblance to mammalian aprismatic enamel, and supports its true ectodermal origin [11]. However, the biochemical properties of the various surface layers should be characterized to determine whether they do, in fact, represent a simple enamel.

Chondrichthian Enameloid

Chondrichthian enameloid has long been known to differ from osteichthian enameloid [18–21]. The ultrastructure of the matrix is unique in that the IDE cells do not synthesize large quantities of proteins during its formation, and the characteristic association of "giant" fibers with odontoblast processes strongly suggest their mesodermal and non-collagenous nature [8, 19]. The IDE cells store large quantities of glycogen [8, 18, 19, 21] which is secreted and then degraded in intercellular spaces [8, 19]. Our preliminary evidence using insulin and glucagon to interfere with glycogen metabolism in skates indicated that glucagon, but not insulin, produced effects similar to colchicine [19], causing increased exocytosis of glycogen-containing vesicles into intercellular spaces (Fig. 11). Glycogen (or its breakdown products) may possibly play a role in osmoregulation of the enameloid matrix, or provide a source of phosphate for growing apatite crystallites.

Tubular, membrane-like material in the elasmobranch enameloid matrix has frequently been observed and thought to arise from odontoblast processes [20]. The different enameloid matrix morphologies observed between skates [8, 19] and sharks [20] may be due to species differences, chemical processing, or temporal differences. We have observed similar tubules in well-fixed skate specimens, but mostly in samples which have been demineralized. Changes in the ionic composition of the enameloid fluids may affect the conformation of the matrix materials, leading to altered matrix morphologies. Additionally, the removal of material from the matrix by prolonged fixation may unmask the tubular structures. Future studies are being geared towards resolving these questions.

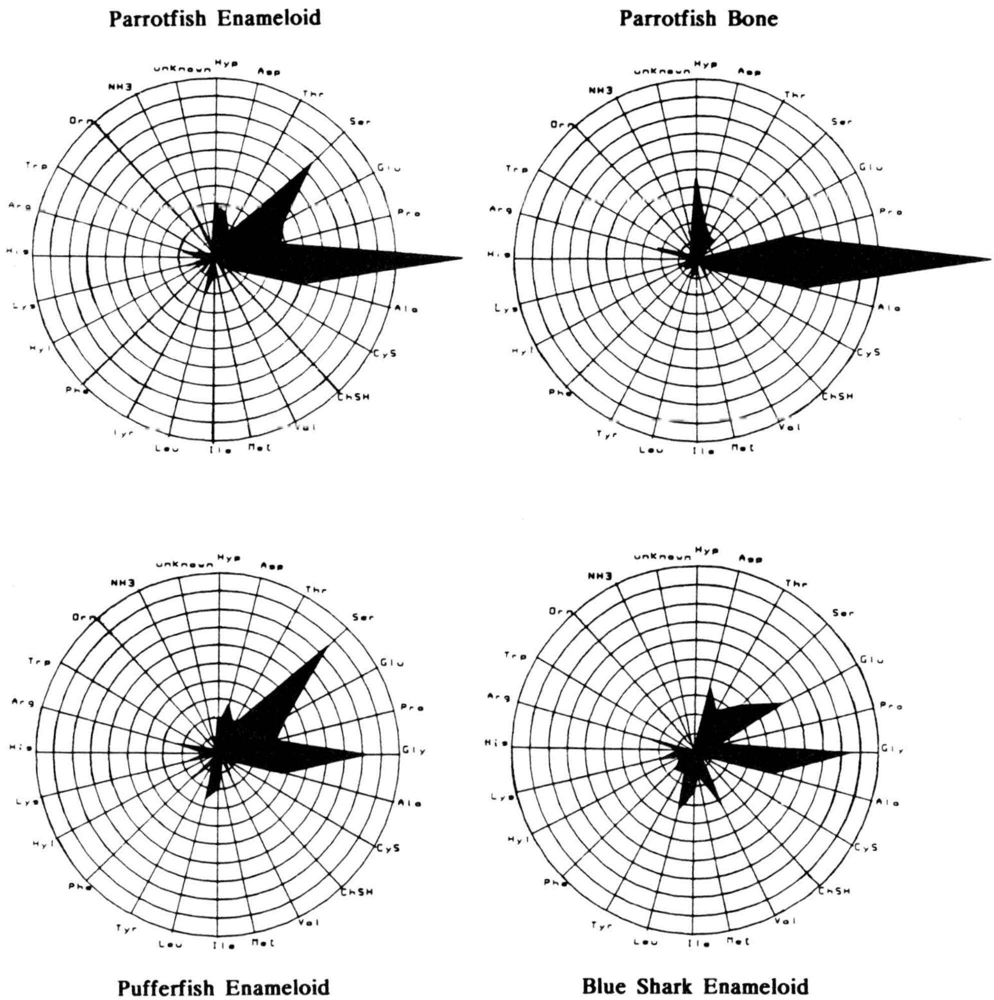

Fig. 12. Rose diagrams depicting the amino acid composition of parrotfish, pufferfish, and blue shark enameloid and parrotfish bone. Large collagen and acidic protein components are present in bony fish having both high (parrotfish) and low (pufferfish) fluoride in their enameloid. Blue shark enameloid, from which the collagen has been subtracted, contains high proportions of basic amino acids relative to bony fish enameloid proteins

Enameloid Matrix Biochemistry

Amino acid analyses of parrotfish and pufferfish enameloid matrices demonstrate the presence of collagen and other acidic proteins (Fig. 12). The bony fish enameloid matrix was enriched in acidic proteins compared to mature fish bone (Fig. 12) and had phosphorylated serine residues (100 res/1000) which

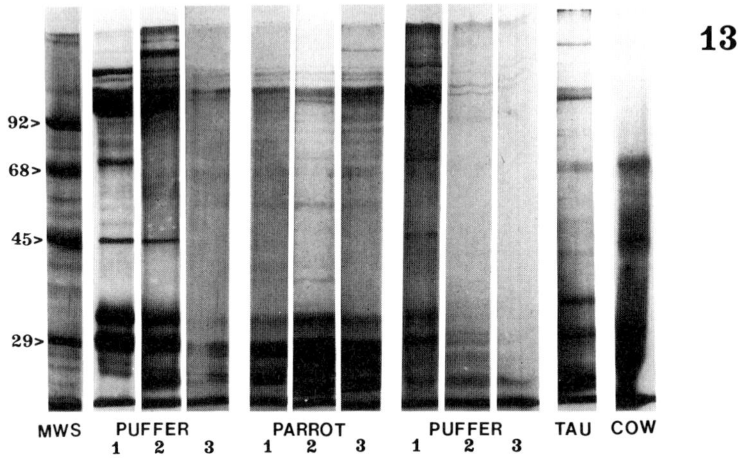
13
92>
68>
45>
29>
MWS
PUFFER
1 2 3
PARROT
1 2 3
PUFFER
1 2 3
TAU
COW

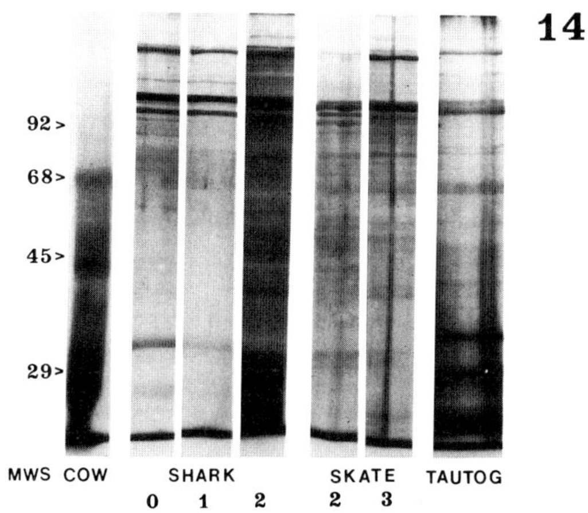
14
92>
68>
45>
29>
MWS
COW
SHARK
0 1 2
SKATE
2 3
TAUTOG

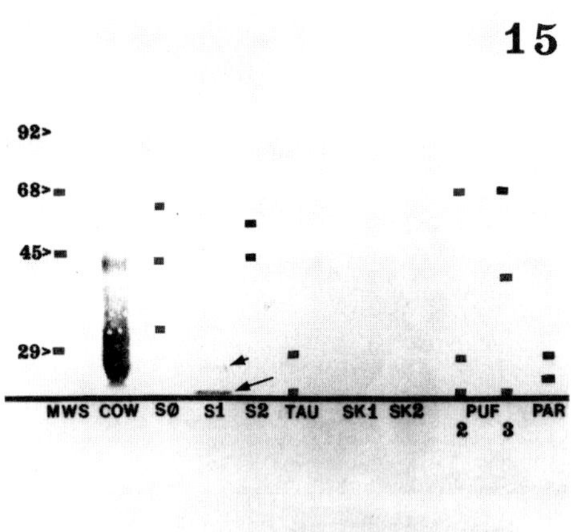
15
92>
68>
45>
29>
MWS
COW
SØ
S1
S2
TAU
SK1
SK2
PUF
2 3
PAR

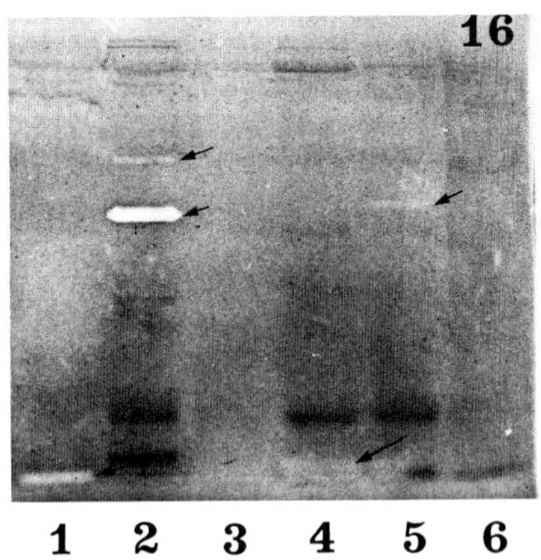
16
1 2 3 4 5 6

declined with increased mineralization (30 res/1000) in both fish enameloids. The overall amino acid compositions between the high (parrotfish) and low (pufferfish) fluoride enameloid matrices were strikingly similar. Thus, it is unlikely that a fluoride-binding protein accounts for the differences in fluoride content between the enameloids of these fish species.

The amino acid composition of the blue shark enameloid, corrected for collagen content, revealed a somewhat different profile from those of bony fish enameloid (Fig. 12). Enrichment of the basic amino acids in the shark enameloid matrix indicated a non-amelogenin type molecule [22], an admixture of actin from cytoplasmic processes, keratin-like proteins [8, 19], or contaminating serum albumin [22].

The electrophoretic patterns of acetic acid soluble proteins were quite similar between the high (parrotfish and tautog) and low (pufferfish) fluoride-containing enameloid matrices of bony fishes (Fig. 13), but different from the elasmobranchs (Fig. 14). All fish species showed characteristic Type I collagen doublets at 100 kd, a weak 68 kd band, but only bony fish enameloid had prominent 33 and 30 kd bands. With progressive mineralization (tooth rows 1, 2, and 3), the 33 and 30 kd bands decreased in intensity and whereas bands below 30 kd transiently appeared more intense. The electrophoretic fingerprints of bony fish enameloid strongly resembled peptide fragments from pollack bone [23] and are suggested to represent degradation peptides of the enameloid collagen itself. Proteases have been localized in enameloid [24], and have similar molecular weight (MW) between the high and low fluoride bony fish groups (Fig. 15). The extent of proteolysis, however, is not known.

Electrophoretic patterns of elasmobranch enameloid proteins were different from both bony fish enameloid and bovine enamel matrix proteins. Although elasmobranch enameloid matrix samples contained some dentin collagen, few

←

Fig. 13. SDS-PAGE of bony fish enameloid and cow enamel matrix proteins. Early stage matrices (*row 1*) have major bands at 100, 33, and 30 kd, whereas more mineralized matrices (*rows 2, 3*) have fewer high MW bands but more transient bands below 30 kd. All species have a 68 kd band, indicating albumin. *MWS*, Molecular weight standards; *TAU* = tautog

Fig. 14. SDS-PAGE of elasmobranch and tautog enameloid proteins. The 100 kd doublet in elasmobranch enameloid most likely represents dentin collagen. Note that the other bands, except for those at 100 and 68 kd, do not correspond with those of bony fish enameloid, in this case, tautog (high fluoride)

Fig. 15. Western blotting of cow enamel and fish enameloid proteins using rat enamel matrix antisera. Cow enamel is distinctly cross-reactive whereas tautog (*TAU*), pufferfish (*PUF*), parrotfish (*PAR*), and skate (*SK1*, *SK2*) enameloid proteins are unreactive. Only row 2 of blue shark (*S1*) had two cross-reactive proteins with this sera (*arrows*). *Rectangles* represent proteins which were positively transferred, identified by Coomassie blue staining

Fig. 16. Gelatin gel depicting gelatinase activity (*arrows*) in the low fluoride porcupine fish (*lane 1*, row 2; *lane 2*, row 1) and the high fluoride parrotfish (*lanes 3–5*, rows 4, 2, and 1) enameloid of the lower MW collagen degradation bands were observed (Fig. 14). Paucity of low MW bands reinforces the idea that elasmobranch enameloid matrix is primarily a keratin-like protein, different from bony fish enameloid collagen matrix [8, 19].

Homology

Previously, we had shown that antisera to mammalian enamel matrix proteins did not cross-react with bony fish enameloid matrix proteins [11]. Thus, the lower MW bands of bony fish enameloid are not related to the amelogenin class of enamel proteins. High resolution TEM immunocytochemical analyses of bony fish enameloid matrices, surface amorphous layers, and ganoid/enamel also produced negative results with this molecular probe [11]. However, we subsequently found that two blue shark enameloid matrix proteins cross-reacted with the enamel antisera (Fig. 16). However, our cross-reactive shark proteins had a lower MW (<20 kd) than the 66 kd protein reported in cat shark enameloid [25]. Possibly, mimeotopes may exist between structurally distinct proteins which are involved in calcification processes [26].

Other studies suggest that an enamel-like protein is present in fish enameloid [27]. If definitive evidence substantiates this hypothesis, then this protein can not be strongly related to the fluoride content of enameloid. Specifically, enamel-like antigens have been reported in both sharks of a high fluoride species, as well as in the low fluoride-containing enameloids of cod and bluefish [27]. Additionally, the mature surface layers in pufferfish (Fig. 4; low fluoride) and parrotfish (Fig. 5; high fluoride) have the same fluoride content as the underlying enameloid. Thus, even in areas enriched in non-collagenous proteins, the fluoride content of the enameloid was not affected.

Enameloid Collagen Cross-Links

The putative non-collagenous ectodermal proteins of enameloid have been suggested to interfere somehow with collagen cross-linking [2]. However, this hypothesis has not been substantiated. Two cross-link amino acids, dihydroxylysinonorleucine (DHLNL) and hydroxylysinonoleucine (HLNL), are commonly associated with bone and dentin collagens. Preliminary data (courtesy of Y. Kuboki, 1991) concerning these cross-links indicate several interesting features in fish enameloid and bone. In bone, the ratios of DHLNL/HLNL appear species specific [23], with HLNL cross-links reduced in fish compared to bovine bone. Generally, few differences were observed in the relative degree of cross-linking between high fluoride parrotfish developing enameloid and bone collagen (Fig. 17). Likewise, at similar stages of development, the collagen in low fluoride pufferfish enameloid had a nearly identical DHLNL/HLNL ratio to that in the high fluoride parrotfish enameloid (Fig. 17). Thus, the biochemical evidence strongly suggests that there is no interference in bony fish enameloid collagen cross-linking when compared to its homologous bone collagen. Likewise, the similarity in the other biochemical

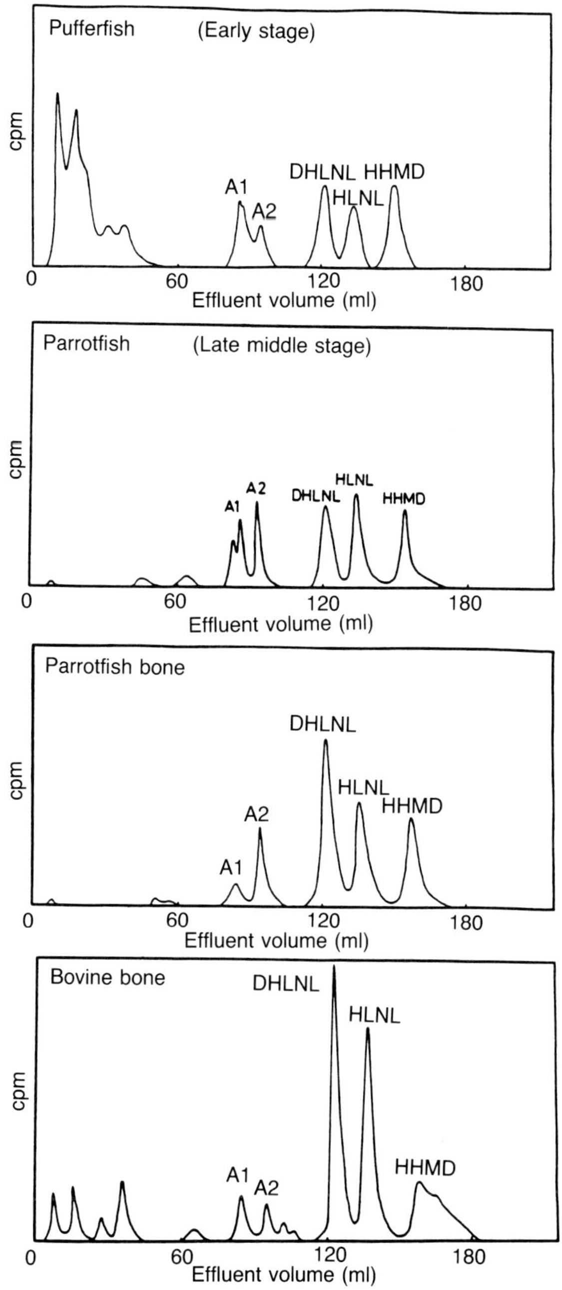

Fig. 17. Chromatograph demonstrating cross-link amino acids, dihydroxylysinonorleucine (*DHLNL*), hydroxylysinonoleucine (*HLNL*) and histidinohydroxymerodesmosine (*HHMD*), in developing parrotfish and pufferfish enameloid and in parrotfish and bovine bone. *A1* and *A2* peaks indicate the alpha 1 and 2 chains of collagen, respectively. The quantity of cross-links between both high and low fluoride enameloid and fish bone collagens are very similar

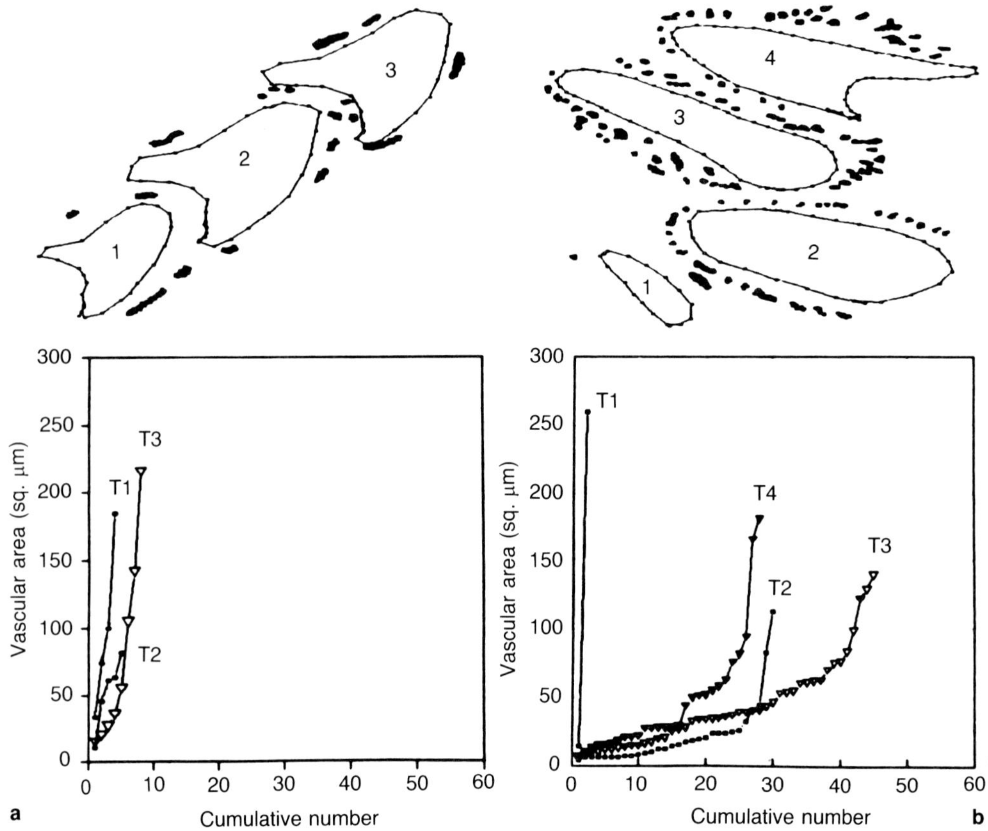

Fig. 18. Digitized images of light microscopic sections of pufferfish (**a** *rows 1–3*) and parrotfish (**b** *rows 1–4*) tooth buds. *Solid objects* represent capillaries surrounding the ODE cells. For each tooth bud and species, the vascular cross-sectional area is plotted against the cumulative number of vessels present. The high fluoride parrotfish displays numerous small capillaries throughout the mineralization stages (**b** *rows T2–4*), whereas the low fluoride, pufferfish capillaries are relatively few and large (**a**) with no change occuring during enameloid mineralization (*rows T2–3*)

features between the enameloid matrices of high and low fluoride bony fish makes it improbable that a fluoride-binding matrix protein can account for the difference in the final fluoride content of bony fish enameloid.

Interestingly, the quantity of DHLNL (0.93 mol/mol collagen) and HLNL (0.57 mol/mol collagen) cross-links in developing blue shark teeth (enameloid with adhering dentin) was approximately 3 times higher than bovine dentin (0.33 mol/mol collagen; 0.1 mol/mol collagen, respectively) (not shown). These data suggest that the collagen present in shark teeth is of odontoblastic origin. Secondly, the lack of interference with the cross-linking of the mixed enameloid-dentin sample supports the idea that chondrichthian enameloid is a

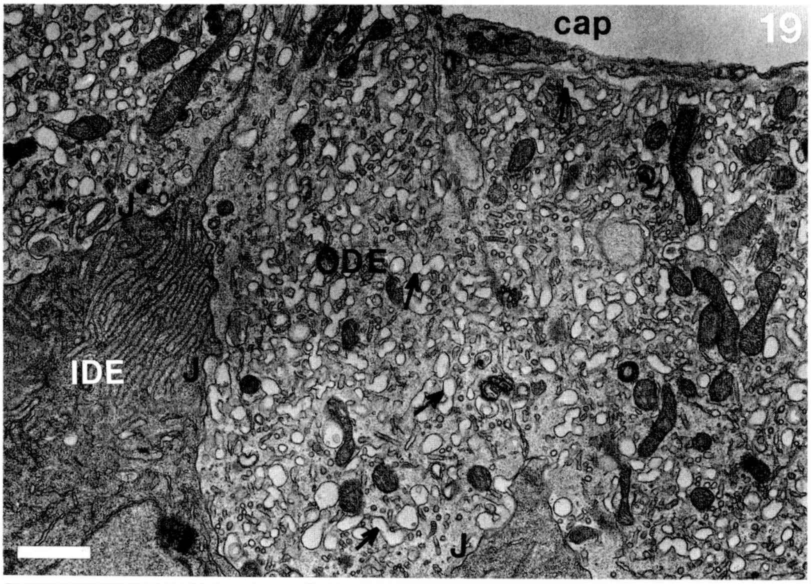

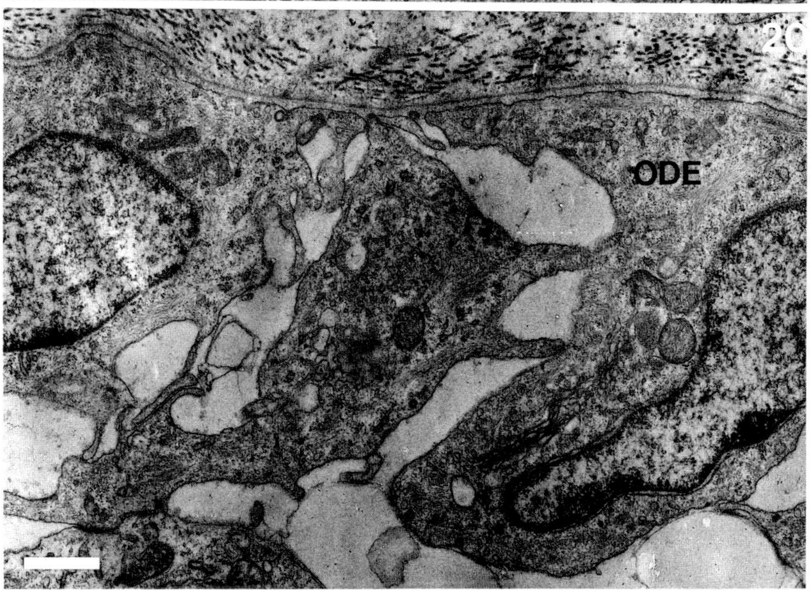

Fig. 19. High fluoride, cichlid ODE cells at the enameloid mineralization stage. Fenestrated capillaries (*cap*) are juxtaposed to the ODE basal lamina. A network of tubular structures (*arrows*), some contiguous with the plasma membrane, fill the cytoplasm of the ODE cells. Numerous mitochondria are present and extensive gap junctions (*J*) connect ODE with IDE cells. The specialized structure of ODE cells is remarkably similar to that seen in "chloride cells" of gills. *Scale* = 1 μm

Fig. 20. Low fluoride, alligator gar ODE cells at the enameloid mineralization stage. Capillaries are rarely found near the ODE cells in the low fluoride enameloid organs. ODE cells are not specialized, have few intercellular junctions, and display few tight junctions with neighboring IDE cells. *Scale* = 1 μm unique non-collagenous protein [8, 19, 20]. Possibly, this unique enameloid matrix does have fluoride-binding properties.

A Hypothesis of Fluoride Regulation During Enameloid Matrix Mineralization

The ionic composition and pH are important factors which influence the incorporation of fluoride into apatite in vitro [28]. However, our preliminary EDS data indicate that the concentrations of detected ions (Na, Cl, Ca, and PO_4) in developing enameloid are similar between high- and low-fluoride species. Additionally, high pH values (around 10.5) can decrease the fluoride content in forming apatite by fourfold [28], but are unlikely to occur in vivo [28] and would not account for the 37-fold difference in fluoride content between high and low fluoride species.

Instead, we suggest that a triad of cytostructural features, characteristic of enameloid organs in high fluoride-containing bony fish tooth buds, contributes to fluoride transport and concentration in enameloid. These features, present specifically during the enameloid mineralization stage, are: (1) an outer dental follicle which is highly vascularized with fenestrated capillaries (Fig. 18), (2) a highly invaginated, tubular network of membranes within ODE cells [3] (Fig. 19) which are remarkably similar to those observed in "chloride cells" of

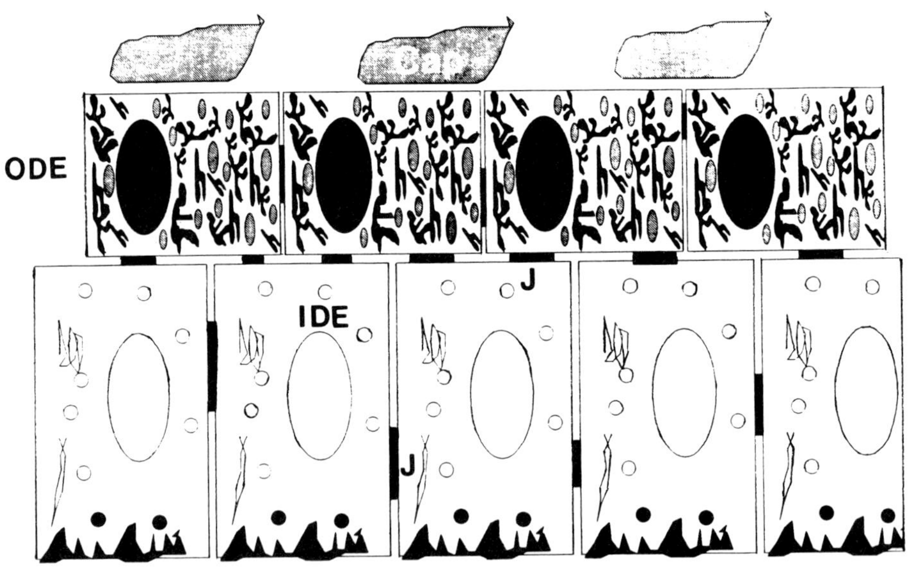

Fig. 21. Diagrammatic representation at a mineralization stage of a bony fish enameloid organ forming a high fluoride-containing enameloid. Numerous fenestrated capillaries (*Cap*) are closely juxtaposed to specialized ODE cells possessing a network of tubular membrane invaginations (*solid irregular structures*). Extensive gap junctions connect ODE to IDE cells (*heavy lines*)

fish gills [29], and (3) tight junctions or gap junctions between ODE and IDE cells which facilitate intracellular diffusion of ions towards the enameloid matrix while limiting intercellular diffusion [3]. High fluoride bony fish species which to date display all these features are cichlid, tautog, sea robin, and the Labrus and Sargus species. Low fluoride species which show neither a rich vasculature, specialized ODE cells, nor numerous intercellular junctions are the goldfish and alligator gar (Fig. 20). Although the low fluoride pufferfish have somewhat specialized ODE cells, their tubular networks are not extensive, nor are they highly vascularized. Thus, the combination of a low but continuous source of fluoride through fenestrated capillaries, specialized ODE cells, and numerous gap junctions are strongly related to the fluoride content of enameloid (Fig. 21).

We propose that the similarity in structure between the specialized ODE cells and "chloride cells" of fish gill epithelia is also reflected in their function. In "chloride cells", serum chloride enters the cell through sodium chloride symport channels [30]. Subsequently, large quantities of sodium are extruded from the "chloride cell" by the membrane Na-K ATPase, thus providing an

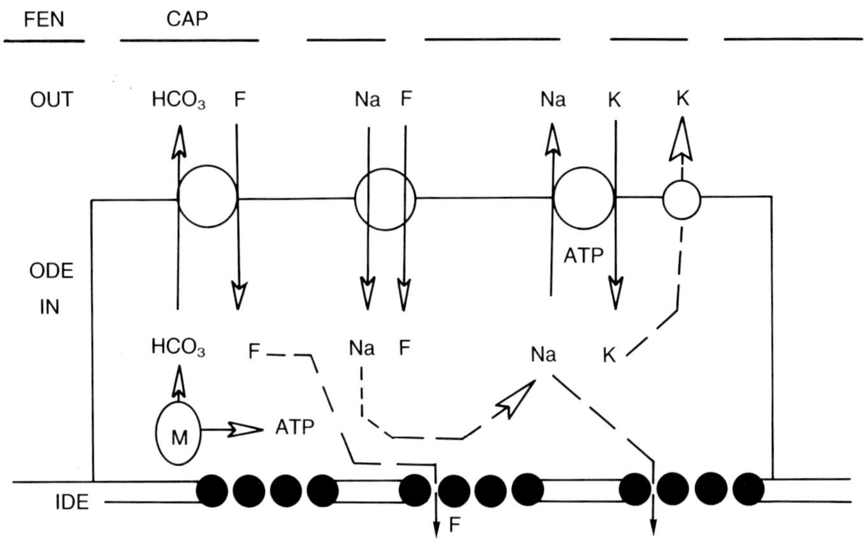

Fig. 22. Diagrammatic representation of fluoride transport through the specialized ODE cell. Fenestrated capillaries (*Fen Cap*) allow for the free diffusion of ions towards the ODE cell tubular invaginations. NaF symport channels allow these ions to preferentially diffuse into the ODE cytoplasm. The Na-K ATPase maintains the intracellular levels of Na low and K high. Another plausible mechanism for fluoride transport into ODE cells is via a bicarbonate-fluoride antiport system. Here, mitochondria release CO_2 resulting from ATP production, thus driving the bicarbonate-fluoride antiport system. Ionic fluoride, unable to diffuse through the plasma membrane, is driven down its concentration gradient through gap junctions connecting ODE to IDE cells towards the enameloid

electrochemical gradient to drive chloride into the environment [31]. In fact, "chloride cells" have the highest quantity of membrane Na-K ATPase of any tissue [31]. Since fluoride is the most electronegative ion and is a halide similar to chloride, it is reasonable that fluoride can be preferentially selected over chloride from the extracellular fluid (Fig. 22). Ionic selectivity is observed in cationic membrane pumps (Na/K ATPase) and through anionic membrane channels where chloride is preferentially secreted over fluoride [32, 33]. Presently, we are investigating the possibility that specialized ODE cells are also rich in membrane Na-K ATPase, and whether anion-selective channels are specific for fluoride. This hypothesis raises the possibility that the anion-selective membrane channels have a fluoride-"binding" property capable of distinguishing fluoride from chloride. Future studies are aimed at clarifying this proposed mechanism of fluoride transport towards the enameloid matrix.

Acknowledgements. This work was supported by NIDR Grant 07677. The authors express their gratitude to Dr. Yoshinori Kuboki and staff, Department of Biochemistry, Hokkaido University, Japan, for performing the amino acid composition and collagen cross-linking analyses of enameloid matrices.

References

1. Peyer B (1968) Comparative Odontology. Zangerl R (ed) University of Chicago Press, Chicago
2. Shellis P, Miles AEW (1974) Autoradiographic study of the formation of enameloid and dentine matrices in teleost fishes using tritiated amino acids. Proc R Soc Lond 185:51–72
3. Prostak K (1984) An ultrastructural study of enameloid matrix formation and mineralization in a teleost, *Cichlasoma cyanoguttatum*, using selected experimental systems. PhD Dissertation, Boston University, Boston, Mass.
4. Prostak K, Skobe Z (1985) The effects of colchicine on the ultrastructure of the dental epithelium and odontoblasts of teleost tooth buds. J Craniofac Genet Dev Biol 5:75–88
5. Suga S, Wada K, Ogawa M (1980) Fluoride concentration in the enameloid of fishes. In: Omori M Watabe N (eds) The Mechanisms of biomineralization in animals and plants. Tokai University Press, Tokyo, pp 229–240
6. Suga S, Wada K, Ogawa M (1981) Fluoride concentration in teeth of tetraodontiform fishes and its phylogenetic significance. Jpn J Ichthyol 28:304–312
7. Suga S, Taki Y, Wada K, Ogawa M (1991) Evolution of fluoride and iron concentrations in the enameloid of fish teeth, In: Suga S, Nakahara H (eds) Mechanisms and phylogeny of mineralization in biological systems. Springer, Tokyo, pp 439–446
8. Prostak K, Seifert P, Skobe Z (1990) The effects of colchicine on the ultrastructure of odontogenic cells in the common skate, *Raja erinacae*. Am J Anat 189:77–91
9. Prostak K, Seifert P, Skobe Z (1991) Serum fluoride level and fluoride content of enameloid. J Dent Res 70(9):1266–1271
10. Kuboki Y, Tsuzaki M, Sasaki S, Liu CF, Mechanic G (1981) Location of the intermolecular cross-links in bovine dentin collagen, solubilization with trypsin

and isolation of cross-link peptides containing dihydroxylysinonorleucine and pyridinoline. Biochem Biophys Res Comm 102:119–126

11. Prostak K, Seifert P, Skobe Z (1991) Tooth matrix formation and mineralization in extent fishes. In: Suga S, Nakahara H (eds), Mechanisms and phylogeny of mineralization in biological systems. Springer, Tokyo, pp 465–469
12. Prostak K, Seifert P, Skobe Z (1989) Ultrastructure of developing teeth in the gar pike, (*Lepisosteus*). In: Fearnhead R (ed) Tooth enamel V, Florence, Yokohama, Japan
13. Sasagawa I (1984) Formation of cap enameloid in the jaw teeth of dog salmon, *Oncorhynchus keta*. Jpn J Oral Biol 26:477–495
14. Sasagawa I, Ishiyama M (1988) The structure and development of the collar enameloid in two teleost fishes, *Halichoreses poecilopterus* and *Pagrus major*. Anat Embryol 178:499–511
15. Miller W (1969) Tooth enamel of *Latimeria chalumnae* (Smith). Nature 221:1244
16. Smith M (1978) Enamel in the oral teeth of *Latimeria chalumnae* (Pisces: Actinistia): A scanning electron microscope study. J Zool Lond 185:355–369
17. Ishiyama M, Ogawa T (1983) Existence of true enamel on tooth plate in the lungfish, *Lepidosiren paradoxa*. Acta Anat Nippon 58:157–161
18. Tomes CS (1898) Upon the structure and development of the enamel of elasmobranch fishes. Philos Trans R Soc Lond [Biol] 190:443–466
19. Prostak K, Skobe Z (1988) Ultrastructure of odontogenic cells during enameloid matrix synthesis in tooth buds from an elasmobranch, *Raja erinacae*. Am J Anat 182:59–72
20. Sasagawa I (1989) The fine structure of initial mineralization during tooth development in the gummy shark, *Mustelus manazo*, Elasmobranchia. J Anat 164:175–187
21. Kerr T (1955) Development and structure of the teeth in the dogfishes, *Squalus acanthias*, L. and *Scyliorhynus caniculus* (L.). Proc Zool Soc Lond 125:95–114
22. Robinson C, Kirkham, Fincham A (1989) The enamelin/non-amelogenin problem. A brief review. In: Glimcher M, Lian J (eds) The chemistry and biology of mineralized tissues. Gordon and Breach, New York, pp 719–726
23. Kuboki Y, Watanabe T, Tazaki M, Takita H (1991) Comparative biochemistry of bone matrix proteins in bovine and fish. In: Suga S, Nakahara H (eds) Mechanisms and phylogeny of mineralization in biological systems. Springer, Tokyo, pp 495–499
24. Kawasaki K, Shimoda S, Fukae M (1987) Histological and biochemical observations of developing enameloid of sea bream. Adv Dent Res 1:191–195
25. Deutsch D, Palmon A, Defni L, Shenkman A, Sherman J, Fisher L, Termine J, Young M (1991) Enamelin and enameloid. In: Suga S, Nakahara H (eds) Mechanisms and phylogeny of mineralization in biological tissues. Springer, Tokyo, pp 73–77
26. Kakei M, Nakahara H, Takeyama H (1991) Immunoblotting study of developing enamel matrix proteins. In: Suga S, Nakahara H (eds) Mechanisms and phylogeny of mineralization in biological tissues. Springer, Tokyo, pp 95–99
27. Herold R, Rosenbloom J, Granovsky M (1989) Phylogenetic distribution of enamel proteins: Immunohistochemical localization with monoclonal antibodies indicates the evolutionary appearance of enamelins prior to amelogenins. Calcif Tissue Int 45:88–94
28. Shimoda S, Aoba T, Moreno E, Miake Y (1990) Effects of solution composition on morphological and structural features of carbonated calcium apatites. J Dent Res 69:1731–1740

29. Pisam M, Prunet P, Boeuf G, Rambourg A (1988) Ultrastructural features of chloride cells in the gill epithelium of the atlantic salmon, *Salmo slar*, and their modifications during smoltification. Am J Anat 183:235–244
30. Potts W (1977) Fish Gills. In: Gupta B, Moreton R, Oschman J, Wall B (eds) Transport of ions and water in animals. Academic, New York, pp 453–480
31. Rankin J, Davenport J (1981) Movement between fresh water and sea water. In: Rankin J, Davenport J (eds) Animal osmoregulation. Wiley, Chichester, pp 83–100
32. Hanrahan J, Alles W, Lewis S (1985) Single anion-selective channels in basolateral membrane of mammalian tight epithelium. Proc Natl Acad Sci 82:7791–7795
33. Valdivia H, Dubinsky W, Coronado R (1988) Reconstruction and phosphorylation of chloride channels from airway epithelium membranes. Science 242:1441–1444

Amelogenin Hydrolysis in Isolated Enamel

IOANNA S. IATRIDI and MILES A. CRENSHAW*[1]

Summary. The purpose of this research was to determine if amelogenin hydrolysis in enamel can occur in the absence of an enamel organ and to determine whether protease inhibitors or fluoride have an effect on the hydrolysis. Fetal bovine secretory stage enamel was incubated in a buffer alone or in the same buffer containing either a serine protease inhibitor, a sulfhydryl protease inhibitor, or sodium fluoride. Hydrolysis was assessed by quantitative protein determinations and by gel electrophoresis. The results indicated that about 25% of the amelogenins were hydrolyzed within 5 days at 37°C, and that this hydrolysis was reduced by serine and sulfhydryl protease inhibitors. Fluoride at 1 mM had no apparent effect on the degradation of amelogenins but appeared to reduce their release from the enamel.

Key words: Enamel—Amelogenins—Proteases—Fluoride

Introduction

During the secretory stage of enamel development, ameloblasts deposit a protein matrix that is mineralized to a limited extent. The proteins consist mainly of hydrophobic, proline-rich amelogenins with a smaller fraction of acidic enamelins [1, 2]. A gradual loss of the amelogenins appears to begin within hours of their secretion [3]. In the maturation stage, the remaining amelogenins are rapidly lost while the enamelins are partially conserved [1, 2, 4]. During this stage, additional mineral is deposited until the mature enamel is about 95% mineral and 2% matrix by weight.

The loss of amelogenins appears to result from proteolytic activity present in developing enamel which was first demonstrated by Suga [5]. The mechanism involves active hydrolysis of a parent amelogenin into smaller polypeptides by proteases [4]. These enzymes probably are derived from the ameloblasts [1, 5–10].

[1] Department of Pediatric Dentistry and Dental Research Center, University of North Carolina, Chapel Hill, NC 27599–7455, USA
* Corresponding author

It has been estimated by electrophoresis that the parent amelogenin is 25–30 kd and that, during enamel development, it is cleaved in steps toward a <10 kd end product [4, 7]. These small polypeptides are more soluble and mobile than their parent amelogenin and are able to migrate out of the forming enamel. If this hypothesis is correct, protein hydrolysis should take place in secretory enamel in the absence of ameloblasts and without further mineral deposition. Preliminary observations by Robinson and Kirkham [2] indicate that the breakdown of amelogenins does continue after the enamel organ is removed from rat incisors.

Protease activities, which have been identified in developing enamel by various investigators, are mainly attributed to serine proteases [6–10]. Carter et al. [10] isolated and purified a serine protease from developing enamel and showed its ability to hydrolyze enamel proteins. Recently, the presence of metalloproteinase activity has been reported. Gelatin was used to demonstrate this activity by Overall and Limeback [11]. DenBesten and Heffernan [12] used the 27 kd amelogenin, purified from secretory bovine enamel, as the substrate for these kinds of proteases. They concluded that metalloproteinases have relatively weak activity against amelogenin.

Fluorosed enamel has a higher protein content than normal enamel [12, 14–16]. DenBesten and Crenshaw [16] found that the increased protein content was first evident in the maturation stage of incisor enamel from rats that were given fluorotic levels of fluoride. They suggested that a reduction or delay in amelogenin loss from enamel may be the initial effect of fluorosis in developing enamel. Because an enamel protease was found to be indirectly inhibited by fluoride [8], it was suggested that this inhibition may account for the higher protein content in maturing enamel. Subsequently, DenBesten [17] showed that in developing fluorosed enamel there was, first, a delay in the hydrolysis and, second, a delay in the loss of amelogenins from the enamel in vivo. However, it is not known whether this effect can occur in the absence of ameloblasts. The demonstration of an inhibition of amelogenin hydrolysis by fluoride in isolated enamel would support the mechanism proposed by DenBesten and Crenshaw [16]. This would provide additional insight as to whether fluoride can produce fluorosis by affecting the enamel directly.

Materials and Methods

Enamel

Near-term fetal calves were obtained from freshly slaughtered cows at a local abattoir. The mandibles were quickly removed, packed in ice, and returned to the laboratory. Each mandibular tooth was dissected free, and the soft tissues covering the enamel carefully wiped away with cotton sponges. While it was still moist, the soft secretory-stage enamel apical to the white opaque boundary was removed, using a blunt spatula. A thin layer of enamel was left over the dentin to avoid contamination with dentin.

Incubation of Enamel

Hydrolysis of proteins in isolated enamel and inhibition of this hydrolysis was observed by the following method. Samples of 100 mg from pooled collections of secretory stage enamel were placed into separate plastic microcentrifuge tubes. A 1-ml volume of 50 mM tris acetate buffer, pH 7.40, was added to each enamel sample with or without the addition of a protease inhibitor or fluoride. The protease inhibitors and concentrations that were used were: 80 μg/ml aprotinin (used primarily as a serine protease inhibitor) [18], 2 mM phenylmercuric acetate (PMA) and 100 mM N-ethylmaleimide (NEM) (sulfhydryl protease inhibitors) [19, 20]. Preliminary experiments using the serine protease inhibitor, phenylmethylsulfonyl fluoride (PMSF), had no significant effect on the release of amelogenins into the incubation buffer. A separate study, conducted to determine whether or not the presence of this mineral affects the inhibitory activity of PMSF, indicated that this inhibitor was too unstable for these experiments [21]. The fluoride samples contained 1 mM sodium fluoride (NaF).

After the addition of the incubation solutions, the enamel was dispersed with a glass rod and placed on a rotating wheel at 37°C for 5 days. Other samples, incubated for the same period of time (without the addition of inhibitors or fluoride) at −20°C, served as controls (frozen controls). It was anticipated that the primary, high-molecular weight amelogenin would be hydrolyzed. The low molecular weight amelogenins produced by this hydrolysis would subsequently dissolve from the enamel into the tris acetate buffer.

After the incubation, each sample was centrifuged, and the supernatant was collected and stored at −20°C (tris extracts). The enamel pellet was dispersed and extracted overnight by stirring at 4°C with 1 ml 4 M guanidine hydrochloride (GuHCl) in tris acetate buffer, pH 7.40. This procedure extracted the amelogenin fraction that was not released into the tris buffer during the incubation [1]. The samples were centrifuged and the supernatants were collected and then stored at −20°C (GuHCl extracts).

Quantitative Protein Determination

Two protein assays were used to determine (1) the amount of protein and (2) the number of free amino groups in each extract. The bicinchromic acid protein assay (BCA) [22] was used to determine the amount of protein. Bovine serum albumin was used to construct the standard curve. The *o*-phthalaldehyde (OPA) assay was used to determine the number of free amino groups [23]. Leucine was used to construct the standard curve. An increase in the ratio of free amino groups to the amount of protein in the tris extracts, compared to the frozen controls, would indicate hydrolysis. A low ratio in the tris extracts incubated with inhibitors or fluoride, compared to the tris extracts incubated alone, would indicate inhibition of the hydrolysis.

Gel Electrophoresis

After completion of the quantitative protein determinations, all samples were dialyzed (3500 mw cut-off) and lyophilized. SDS polyacrylamide gel electrophoresis (SDS-PAGE) was subsequently used to separate the amelogenins and assess their hydrolysis [2–4]. The method described by Laemmli [24] was followed, using 13% polyacrylamide gels. Approximately 20 µg protein was applied in each well. The molecular weight standards used were: carbonic anhydrase (30 kd), soybean trypsin inhibitor (20.1 kd) and α-lactalbumin (14.4 kd). Peptides smaller than approximately 10 kd were not detectable in this electrophoresis system. The gels were fixed with 1% glutaraldehyde for 1 hour and then stained with Coomassie brilliant blue. Decreases in the relative abundance of high molecular weight amelogenins in the tris extracts incubated without inhibitors or fluoride, when compared to the frozen controls, would indicate hydrolysis. In the samples incubated with inhibitors or fluoride, the presence of amelogenins with molecular weights higher than those of the samples incubated with tris alone would indicate reduction of hydrolysis.

Statistical Analysis

Statistical significance of the quantitative differences between the means of samples was evaluated by using the Student's t-test. Values of t indicating a probability of less than 0.05 were regarded as statistically significant [25].

Results

Amelogenin Hydrolysis in the Absence of Enamel Organ

After the incubation of the samples containing no inhibitors or fluoride, approximately 25% of the amelogenins were extracted by the tris acetate buffer. Compared to a 2% release of amelogenins in the frozen controls, this difference was significant ($P < 0.05$) (Table 1).

The ratio of free amino groups to the amount of protein in the tris extracts was significantly greater, $P < 0.005$, in the incubated samples compared to the

Table 1. Percentage of total protein in tris extract

	n	*Mean*	*S.D.*	*Compared to tris control*
Frozen control	8	2.00	1.62	
Tris control	9	12.7	3.2	$P < 0.001$*
Aprotinin	5	12.6	5.1	$P < 0.5$
PMA	4	11.2	4.5	$P > 0.5$
NEM	4	14.6	4.5	$P > 0.5$
NaF	5	6.66	2.81	$P < 0.005$

* Compared to frozen control

Table 2. Moles free amino groups/gm protein in tris extract

	n	*Mean*	*S.D.*	*Compared to tris control*
Frozen control	8	0.478	0.220	
Tris control	9	2.156	1.196	$P < 0.001^*$
Aprotinin	5	0.153	0.066	$P < 0.001$
PMA	3	0.987	0.317	$P < 0.025$
NEM	4	0.179	0.039	$P < 0.0001$
NaF	5	1.825	0.621	$P < 0.2$

* Compared to frozen control

Table 3. Moles free amino groups/gm protein in GuHCl extract

	n	*Mean*	*S.D.*	*Compared to tris control*
Frozen control	8	0.028	0.020	
Tris control	9	0.036	0.024	$P < 0.4^*$
Aprotinin	5	0.031	0.020	$P < 0.5$
PMA	3	0.028	0.013	$P < 0.4$
NEM	4	0.018	0.013	$P < 0.3$
NaF	5	0.015	0.010	$P > 0.4$

* Compared to frozen control

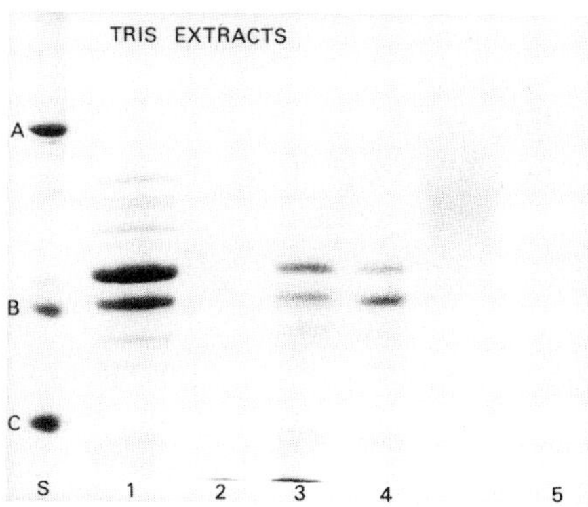

Fig. 1. SDS-PAGE of Tris Extracts. *Lane S*, Molecular size standards (A = 30 kd, B = 20.1 kd, C = 14.4 kd); *lane 1*, frozen control; *lane 2*, tris control; *lane 3*, aprotinin; *lane 4*, PMA; *lane 5*, NaF

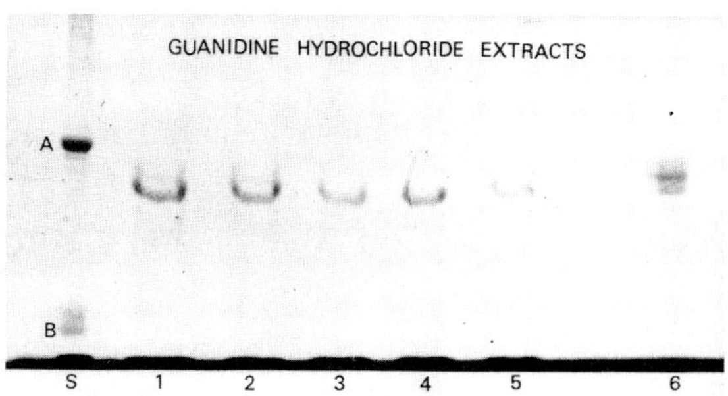

Fig. 2. SDS-PAGE of GuHCl Extracts. *Lane S*, Molecular size standards (A = 30 kd, B = 20.1 kd); *lane 1*, frozen control; *lane 2*, tris control; *lane 3*, aprotinin; *lane 4*, PMA; *lane 5*, NaF

frozen controls (Table 2). There were no differences in the amount of protein or in the ratio of free amino groups to the amount of protein between the guanidine-extracted controls and the guanidine-extracted experimentals (Table 3).

The electrophoretograms of the tris extracts showed that the 18 kd and 22 kd amelogenins were the most abundant proteins in the frozen controls. In the incubated tris extracts, no amelogenins greater than 10 kd were detected (Fig. 1). The guanidine extracted controls and experimentals were essentially identical, with a predominance of the 27 kd amelogenin and lesser amounts of the 28 kd amelogenin. No smaller amelogenins were detected (Fig. 2).

Effect of Protease Inhibitors on Amelogenin Hydrolysis

Sulfhydryl Inhibitors

Neither phenylmercuric acetate (PMA) nor N-ethylmaleimide (NEM) significantly inhibited the release of amelogenins into the medium (Table 1).

The ratio of free amino groups to the amount of protein in the tris extracts was significantly less for both PMA and NEM samples, $P < 0.05$ and $P < 0.005$, respectively (Table 2). In the guanidine extracts, this ratio was not significantly different between the samples containing either inhibitor and the incubated control (Table 3).

The electrophoretograms of the PMA and NEM tris extracts showed that the predominant amelogenins were 18 kd and 22 kd in these samples (Fig. 1). The GuHCl extracts of these samples were identical to the GuHCl extracts of the frozen controls and the GuHCl extracted samples containing no inhibitors or fluoride (Fig. 2).

Serine Protease Inhibitor

The amount of protein that was released into the tris acetate medium for the samples incubated with 80 μg/ml aprotinin was not significantly different from the samples incubated with tris alone (Table 1).

The ratio of free amino groups to the amount of protein in the tris extracts of these samples was significantly less at the $P < 0.001$ level when compared to the samples incubated in tris alone (Table 2). The difference in the ratios was not significant for the GuHCl extracts of these samples (Table 3).

The SDS-PAGE revealed that in the tris extracts of the aprotinin samples there was a predominance of the 18 kd and 22 kd amelogenins (Fig. 1). The guanidine extracts of the inhibited and control samples were identical (Fig. 2).

Effect of Fluoride on Amelogenin Hydrolysis

The inclusion of 1 mM NaF in the incubation buffer significantly diminished the amount of the amelogenin released into this medium, $P < 0.05$. Approximately 6% of the amelogenins were detected in this medium (Table 1).

The ratio of free amino groups to amount of protein was not significant for either the tris or the GuHCl extracts of the experimental samples when compared to their respective controls (Tables 2, 3).

No amelogenins above the detection limit of approximately 10 kd were observed in the tris extracts by SDS-PAGE (Fig. 1). The GuHCl extracts of the experimental and control samples were identical (Fig. 2).

Discussion

The results show that release of amelogenins from the enamel into the incubation buffer occurs during incubation of isolated secretory enamel at 37°C for 5 days. Approximately 25% of amelogenins were released from the developing enamel of the incubated samples. This release was significant when compared to the frozen controls. The hydrolysis of the released amelogenins was of primary interest since it is generally thought that the amelogenins are lost from the enamel after degradation of the primary 28.4 kd amelogenin into smaller polypeptides [4]. Our results (quantitative protein determination and SDS-PAGE) support this hypothesis. No amelogenins above 10 kd were detected in the tris extracts of the incubated enamel.

It is widely accepted that the primary amelogenins are hydrolyzed by proteases that are secreted from the ameloblasts together with the enamel matrix [5–9, 11, 12]. Our results clearly show that, indeed, amelogenins can be hydrolyzed in the absence of enamel organ and without further mineral deposition.

Several proteolytic activities have been identified in developing enamel by various investigators. These include serine proteases [6–10] and metalloproteinases [11, 12]. The roles of these enzymes in amelogenin loss has not been proven in vitro even though some have been shown to hydrolyze amelogenins [6, 9, 12]. Our results indicate that serine protease inhibitors are capable of reducing the hydrolysis of the amelogenins once they are released into the tris medium. The effect of the sulfhydryl protease inhibitors may indicate that this enzyme activity is responsible, in part, for amelogenin hydrolysis. However, sulfhydryl protease inhibitors may also inhibit some serine proteases by disrupting disulfide bonds necessary for the activity of these enzymes [9, 10]. Neither serine nor sulfhydryl protease inhibitors reduced the release of amelogenins from the enamel.

The model system for the study of amelogenin hydrolysis used in this research had a large fluid to enamel volume ratio, opposite from the ratio in vivo. For this reason, the model allowed the study of intermediate stages of amelogenin hydrolysis. The 28.4 kd primary amelogenin was found only in the guanidine extracts of the enamel. This observation confirmed that this protein has a significant affinity for the mineral. The partial hydrolysis of this protein to the 18–21 kd amelogenins eliminated this affinity, and this group of proteins was released from the mineral into the tris buffer. This step was evident in the tris extract containing inhibitors. In the absence of inhibitors, the 18–21 kd amelogenins were hydrolyzed into peptides too small to be detected by the electrophoretic system used.

The first step in amelogenin hydrolysis, the production of the 18–21 kd amelogenins, is the rate-limiting step. It appears that this step is specific for amelogenin hydrolysis. None of the protease inhibitors used in this study affected this first step. The proteolytic activity could be reduced only after the release of the amelogenins into the incubation medium. Therefore, we cannot conclude that the serine and sulfhydryl protease activities observed in this study are necessarily unique in the developing secretory enamel; any number of enzymes may digest the amelogenins after their release into the medium.

Suga [5] demonstrated inhibition of proteolytic activity in enamel by 5 mM NaF. Crenshaw and Bawden [8] found that the serine protease activity present in decalcified embryonic bovine secretory enamel could be inhibited indirectly by 5 μM fluoride at low calcium concentrations. In the current study of developing enamel, we elected to use a high NaF concentration, 1 mM. We reasoned that fluoride at this concentration would dramatically affect amelogenin loss from the enamel. The results indicated that fluoride did not inhibit the hydrolysis of amelogenins released into the tris medium.

Fluoride at this concentration did reduce the release of amelogenins from the enamel. It has been shown that the delay in the loss of amelogenins from fluorosed enamel [16] may be due to a direct effect of the fluoride on the enamel. Fluoride may have altered the enamel crystal. This would make the adsorbed 28.4 kd amelogenin unavailable for the primary hydrolysis. Alternatively, fluoride may inhibit this primary hydrolysis.

Acknowledgements. This work was supported, in part, by USPHS grant #DE05504.

References

1. Termine JD, Belcourt AB, Christner PJ, Conn KM, Nylen MU (1980) Properties of dissosiatively extracted fetal tooth matrix proteins. I. Principal molecular species in developing bovine enamel. J Biol Chem 255:9760–9768
2. Robinson C, Kirkham J (1985) Dynamics of amelogenesis as revealed by protein compositional studies. In: Butler WT (ed) The chemistry and biology of mineralized tissues. Ebsco Media, Birmingham, pp 248–263
3. Smith CE, Pompura JR, Borenstein S, Fazel A, Nanci A (1989) Degradation and loss of matrix proteins from developing enamel. Anat Rec 224:292–316
4. Fincham AG, Belcourt AB, Termine JD (1982) Changing patterns of enamel matrix proteins in the developing bovine tooth. Caries Res 16:64–71
5. Suga S (1970) Histochemical observations of proteolytic enzyme activity in the developing dental hard tissues from the rat. Arch Oral Biol 15:555–558
6. Shimizu M, Tanabe T, Fukae M (1979) Proteolytic enzyme in porcine immature enamel. J Dent Res 58:782–789

7. Moe D, Birkedal-Hansen H (1979) Proteolytic activity in developing bovine enamel. J Dent Res 58:1012–1013
8. Crenshaw MA, Bawden JW (1984) Proteolytic activity in embryonic bovine secretory enamel. In: Fearnhead RW, Suga S (eds) Tooth enamel IV. Elsevier, Amsterdam, pp 109–113
9. Carter J, Smillie AC, Sheperd MG (1984) Proteolytic enzyme in developing porcine enamel. In: Fearnhead RW, Suga S (eds) Tooth Enamel IV. Elsevier, Amsterdam, pp 229–233
10. Carter J, Smillie AC, Sheperd MG (1989) Purification and properties of a protease from developing porcine enamel. Arch Oral Biol 34:195–202
11. Overall CM, Limeback H (1988) Identification and characterization of enamel proteinases isolated from developing enamel. Biochem J 256:965–972
12. DenBesten PK, Heffernan LM (1989) Separation by polyacrylamide gel electrophoresis of multiple proteases in rat and bovine enamel. Arch Oral Biol 34:399–404
13. Aoba T, Fukae M, Tanabe T, Shimizu M, Moreno EC (1987) Selective adsorption of porcine amelogenins onto hydroxyapatite and their inhibitory activity on hydroxyapatite growth in supersaturated solutions. Calcif Tissue Int 41:281–289
14. Silness J, Gustavsen F (1970) Some observations on the fine structure of fluorosed enamel. Acad Odontol Scand 28:701–720
15. Shinoda H (1983) Effects of long-term administration of fluoride on enamel formation in rats. In: Suga S (ed) Mechanisms of tooth enamel formation. Quintessence, Tokyo, pp 273–284
16. DenBesten PK, Crenshaw MA (1984) The effects of chronic high fluoride levels on forming enamel in the rat. Arch Oral Biol 29:675–679
17. DenBesten PK (1986) Effects of fluoride on protein secretion and removal during enamel development in the rat. J Dent Res 65:1272–1277
18. Trautschold I, Werle E, Zickgraf-Rüdel G (1967) Trasylol. Biochem Pharmacol 16:59–72
19. Scocca J, Lee YC (1969) The composition and structure of the carbohydrate of pineapple stem bromelain. J Biol Chem 244:4852–4863
20. Bobb D (1972) Isolation of stem bromelain and its partial characterization by gel electrophoresis. Prep Biochem 2:347–354
21. James GT (1978) Inactivation of the protease inhibitor phenylmethylsulphonyl fluoride in buffers. Anal Biochem 86:574–579
22. Sorensen K, Brodbeck U (1986) A sensitive protein assay method using microtiter plates. Experientia 42:161–162
23. Church FC, Porter DH, Gatignani GL, Swaiswood, HE (1985) An *o*-phthalaldehyde spectrophotometric assay for proteinases. Anal Biochem 146:343–348
24. Laemmli UK (1970) Cleavage of structural proteins during the assembly of the head of bacteriophage T4. Nature 227:680–685
25. Snedecor W (1959) Statistical methods. Iowa State College Press, Ames

Enamel Formation During Porcine Amelogenesis

Takaaki Aoba[1]

Summary. The present work was aimed at elucidating the mechanism of enamel mineralization during porcine amelogenesis by specifically investigating (1) the driving force for precipitation in the fluid microenvironment from which enamel crystals form, (2) changes in the property of matrix proteins, particularly the amelogenins, originating from their post-secretory degradation, and (3) regulatory mechanisms of crystal growth related to the interaction between these proteins and enamel mineral. Analyses of ionic concentrations (and activities) of the fluid separated from secretory porcine enamel indicated that enamel mineralization during the secretory stage of porcine amelogenesis occurs in a specific microenvironment segregated from the circulating blood. It was also suggested that the driving force for precipitation at this stage is controlled by the cellular transport of lattice ions (particularly Ca ions) and the presence of Ca-binding ligands (regulating calcium ion activity) in the fluid. Our in vitro work also showed that the originally secreted porcine amelogenin (25 kd by SDS-PAGE), as well as 60–90 and 32 kd non-amelogenins, can adsorb selectively onto apatitic surfaces. Importantly, this adsorption of amelogenin is intimately related to the inhibitory activity of crystal growth of hydroxyapatite in supersaturated solutions and regulation of isotopic exchange of ions (e.g., calcium) on the crystal surfaces in saturated conditions. This potential function of amelogenins in situ seems to be modulated by enzymatic cleavages of specific segments at the N- and C-termini.

Key words: Biomineralization—Amelogenesis—Enamel Mineral—Crystal Growth—Amelogenins

Introduction

Enamel mineralization during amelogenesis is generally divided into the secretory, transitional, and maturational stages on the basis of changes in mineral and protein contents and cellular structures [1–3]. Considerable work using various techniques and animal models has been conducted on the

[1] Forsyth Dental Center, 140 Fenway, Boston, MA 02115, USA

individual stages of amelogenesis. The present consensus is that mineralization during amelogenesis is controlled by cellular activities (transport of mineral lattice ions and secretion and removal of organic matter). Autoradiographic studies have shown that ionic components in a circulating blood, e.g., Ca ions, can be transported rapidly into the forming enamel tissue [4]. Bawden et al. [5] demonstrated that the calcium transport in the secretory stage was regulated by the ameloblast, in contrast to the transport of phosphate, while no regulation of transport of either ion was observed at the maturational stage. However, there is a paucity of information about the microenvironment surrounding the precipitating enamel crystals and the driving force for precipitation during amelogenesis. Also, little is known about the nature, stoichiometry, and thermodynamic property of enamel crystals formed at various developmental stages.

The enamel matrix proteins, secreted by the ameloblasts during early stages of amelogenesis, are generally categorized into 2 groups, amelogenins and enamelins (or non-amelogenins), based on their amino acid compositions, solubility properties, and apparent molecular weights [6–8]. Studies using molecular biological approaches also suggested that the amelogenins and the enamelins are distinct classes of proteins [9]. The group of enamelins is minor matrix constituents and, recently, questions have been raised about the nature and origin of this group of proteins [10]. In contrast, the amelogenins are major matrix constituents and the primary structure of the secreted amelogenins has already been determined in various mammalian species [9, 11–13].

The most distinctive feature of amelogenesis of mammals, in contrast to their dentinogenesis and osteogenesis, is that the secreted matrix proteins are removed, almost completely, during the developmental stages, leaving large apatite crystallites in a well oriented fashion. It is most likely that the remarkable length and orientation of enamel crystallites are the results of the ameloblast-mediated matrix-mineral interaction. Recently, Weiner [14] postulated, on the basis of features in tooth, bone, and mollusc shell formation, that 2 kinds of proteins, hydrophilic (acidic in nature, aspartic acid-rich and, in some cases, phosphoserine-rich) and hydrophobic (proline or hydroxyproline-rich), are always involved in the process of "organic matrix-mediated" biomineralization. As to the function of these molecules, it has been proposed that the acidic proteins, bound to the hydrophobic proteins (or rigid substrates in vitro), play a role as nucleators but can also act as crystal growth inhibitors when free in solution [15]. Analogous roles have also been advanced for enamel proteins. In the literature, it is widely accepted that the enamelins are intimately associated with enamel apatite, while the amelogenins occupy the space between the crystals [3]. It is also proposed that enamelin-like proteins are highly conserved during vertebrate evolution [16] and that these anionic macromolecules may serve a primary function in the initiation of calcium apatite formation [17]. However, the initial precipitation during mammalian amelogenesis is induced following dentin mineralization. Furthermore, there is no direct evidence for the binding of enamelins to some of the hydrophobic proteins (amelogenins) or substrates, thus playing a role as nucleators during the secretory stage of amelogenesis. On the basis of the foregoing consideration, it is most likely that the secreted matrix proteins, including amelogenins and enamelins, function in a different way in enamel mineralization than do matrix proteins in dentinogenesis or osteogenesis.

Our working hypothesis about the mechanism of enamel mineralization at early stages of amelogenesis is that: (1) the nature and precipitation kinetics of enamel mineral are primarily determined by the degree of saturation of the fluid phase, which is controlled by the supply of lattice ions (especially calcium transport) through the ameloblast layer, (2) with the intermittent supply of calcium ions from the cell layer, the driving force in the fluid phase becomes high enough to induce rapid precipitation of calcium phosphate salts in the vicinity of the ameloblast, (3) the secretion of amelogenins and enamelins prevent further precipitation through their adsorption onto the formed crystal surfaces, (4) their functions as inhibitors of crystal growth are modulated through the degradation (i.e., cleavages of specific segments) of these macromolecules prior to their massive removal, and (5) the Ca-ligands complex in the fluid acts as a reservoir of the ion (through dissociation), maintaining a modest degree of supersaturation for a slow growth of the formed apatite crystals in the underlying secretory enamel, concomitant with the degradation of the inhibitory proteins. The present work was planned to gain experimental evidence for this overall hypothesis.

Materials and Methods

Preparation of Enamel Samples

The enamel used in this work was obtained from freshly slaughtered piglets. The main reason for the selection of this animal model is that the permanent teeth yield sufficient amounts of enamel tissue at various stages of amelogenesis. The incisors and molars (with the attached enamel organ tissues to prevent contact with blood) were dissected out of the mandible of 5- to 6-month-old animals. The incisors were used to dissect the secretory enamel layer-by-layer from the outermost (young) toward the innermost (older) zones, as reported previously [18]. Prior to the enamel dissection, the covering soft tissues were removed. The exposed enamel surface was wiped gently with a paper tissue to remove the cellular debris and traces of blood. The outermost enamel, about 100 μm in thickness, was scraped carefully from the labile surface of the secretory tissue with the aid of a razor blade. Following this separation, the underlying secretory enamel tissues were separated into four layers (each 200–300 μm in thickness) from the outer toward the dentin-enamel junction (DEJ). The enamel samples thus dissected into 5 layers were individually weighed and then freeze-dried. The molars were used for dissection of the secretory enamel and then in the separation of the fluid.

Separation of Enamel Fluid

The fluid separation was commenced immediately after the dissection of the enamel organs. The covering soft tissues were removed as soon as possible to minimize the flux of ions into (or out of) the enamel tissue through the cellular membrane. The exposed surface of the secretory enamel was covered with mineral oil. The mineral oil used was presaturated with water vapor by bubbling 1.8/98.2 CO_2/N_2 gas that had passed through a 160 mM NaCl solution. The secretory enamel was dissected with the aid of an excavator and pooled under the mineral oil in a 5 mL-volume ultracentrifuge tube (Polyallomer, Calif.). Separation of the fluid from the solid tissue was carried out by ultracentrifugation at 2.4×10^5 g. The yield of the fluid was about 45 to 55 μL from 1 g of secretory enamel. Pooled fluid samples were used for measurements of the concentrations of total Ca, Mg, Na, K, total P, HCO_3 (ion chromatography, Dionex 4010i with a conductivity detector), Cl, and F (in conjunction with ion-selective electrodes). The pH of the fluid was measured in an oil bath at room temperature (about 25°C), using needle pH and reference electrodes (Microelectrodes Inc. N.H.), as reported previously [19]. The calcium activity of the fluid was measured with a specific Ca ion microelectrode (Ionetics Inc., Calif.). In order to assess the ionic strength of the fluid, the osmolality was determined in a Nanoliter Osmometer (Clifton, N.Y.).

Preparation of Enamel Matrix Proteins

The preparation of enamel proteins was conducted in 3 different ways for specific purposes. (1) To collect whole matrix proteins from secretory porcine enamel, the dissected tissues were dissolved in 0.5 M acetic acid containing protease and phosphatase inhibitors [20]. The proteins extracted in the acid solution were desalted and concentrated by ultrafiltration using a YM-5 membrane (Amicon; M.W. cutoff 5 kd). (2) To separate the matrix proteins based on their hydrophobic/hydrophilic properties, the extraction was conducted using chloroform/methanol 3:1. The dissected secretory enamel was added to a centrifuge tube containing the organic solvent (solid/solution = 1/40 wt/v) and then well dispersed with the aid of a sonic dismembrator. The slurry was separated by centrifugation. The supernatant was removed carefully; the proteins extracted in the solvent were concentrated and then freeze-dried. The sedimented solid was dissolved in 0.5 M acetic acid and the proteins dissolved in the acid solution were recovered as described above. (3) To purify the amelogenins having various molecular weights, the matrix proteins extracted according to (1) were further fractionated using gel filtration chromatography through a Sephadex G-100 column equilibrated with 50 mM carbonate buffer (pH 10.8) in the presence of protease- and phosphatase-inhibitors. Each protein fraction was concentrated by ultrafiltration. Further purification of the amelogenins was carried out by recycling the chromatography 5 times under the same conditions or using ion-exchange columns [21]. Following the final chromatography, eluent fractions corresponding to the desired proteins were pooled, desalted using the ultrafiltration membranes, and then lyophilized. The purity of the separated protein samples was verified by SDS-PAGE and amino acid analysis, and by determining their N- and C-terminal sequences.

Adsorption Experiments

The adsorbent used was hydroxyapatite (6.4 m^2/g) which was prepared and characterized in our laboratories. The matrix proteins separated from the whole secretory enamel were used as adsorbates. Adsorbate-adsorbent equilibration was conducted in acetate buffer (pH 6.0) at 0°C, as reported previously [20]. The adsorbates were solubilized in the buffer to yield a 0.1% wt/v protein solution. Accurately weighed amounts of the adsorbent, 5–100 mg, were added to 2 mL of the adsorbate solution. The slurry was mixed continuously in a rotator for 1 h at the specified temperature and then separated by centrifugation at 10,000 g for 10 min. The supernatant was recovered carefully and then another 2 mL of the same solvent used for equilibration was added to the centrifuge tube containing the pellet. The crystals were dispersed with a vortex mixer, centrifuged again, and then the supernatant was removed. This washing procedure was repeated 3 times to minimize the contamination of unadsorbed macromolecules or to remove species non-specifically adsorbed onto the crystals. Following the final washing, the sedimented crystals were completely dissolved in 0.5 M acetic acid. All procedures for protein recovery were conducted at 0°C unless otherwise specified. The acid solution containing the adsorbed proteins was desalted by passing it through a PD-10 column (Pharmacia-LKB, N.J.), using 0.5 M acetic acid as an eluent, and then freeze-dried. The unadsorbed proteins remaining in the equilibrating solvent and the proteins in washing solutions were also desalted in the same way and then freeze-dried. The amounts of proteins adsorbed were determined by dissolving the freeze-dried samples in 20 mM acetic acid and analyzing the total protein concentration by Lowry's method or amino acid analysis. The protein moieties recovered from the crystal surface and the proteins remaining in the equilibrium solution (i.e., the first supernatant) were characterized by SDS-PAGE using 15% to 18% polyacrylamide gels.

Effect of Protein-Crystal Interaction on Ca Incorporation on the Crystals

In this work, our interest was focused on the coatings of apatite crystals with enamel proteins in dispersed forms. Since the amelogenins were sparingly soluble under physiological conditions but became soluble in low ionic strength solutions at cool temperatures [8], the whole matrix proteins extracted from the secretory enamel (mainly the amelogenins) were solubilized in deionized water at 0°C to yield 0.2% wt/v adsorbate solution. The protein solution was mixed with the same volume of 0.32 M NaCl solution; the resulting solution (0.1% wt/v proteins) had an ionic strength of 160 mM. In this solution and at room temperature, the proteins commenced to aggregate, resulting in a clouding of the solution. Hydroxyapatite crystals (the same batch as used in the adsorption experiments), in various amounts to provide total surface areas covering the range of 0.01 to 0.2 m^2, were introduced into the protein solution and well mixed with a vortex. The protein-crystal equilibration was continued for 1 h at room temperature. Following the equilibration, the slurry was separated by centrifugation, as described above. The sedimented (protein-coated) crystals were used to investigate the incorporation of calcium ions onto the protein-coated crystals in a well-defined solution. The experimental solution (100 mL in volume) had an ionic composition similar to that of the enamel fluid (pH 7.3, total phosphate = 3.9 mM, total CO_3 = 10 mM, and background electrolyte = 150 mM as NaCl) but with a lower total calcium concentration (0.10 mM) to yield about the same calcium activity of the fluid and saturation with respect to hydroxyapatite. The solution was maintained at 37° ± 0.2°C in a water bath and well stirred with a magnetic stirrer. Trace amounts of $^{45}CaCl_2$ (10 μCi) were added to the solution. The prepared protein-coated crystals were dispersed in 1 mL of 160 mM NaCl and then introduced into the experimental solution (time zero). Time courses of the changes in radioactivity in solution were monitored by taking aliquots (1 mL) before the addition of the crystals and at timed intervals over the experimental period (5 h). The withdrawn aliquot was transferred to a microcentrifuge tube and then centrifuged at 10,000 g for 1 min. The radioactivity in the supernatant was counted using a liquid scintillation counter. The results (kinetics of incorporation and the incorporated amount at equilibrium) were compared between the experiments using the protein-free crystals and the protein-coated crystals. The experimental solution used was close to saturation with respect to the crystals. Nevertheless, in all experiments, it was verified that no appreciable changes in the total Ca concentration of the experimental solution occurred before and after equilibration. In parallel experiments, but without the addition of the

Table 1. Ionic composition and degree of saturation of enamel fluid

Analytical values		*Calculated values*		*Degree of saturation*[a]
Total Ca	0.50 mM	Free Ca^{2+}	1.53 mM	$(IP/K_{HA})^{1/9}$
Ca^{2+} Activity	0.053 mM	$CaHPO_4^0$	1.18×10^{-2} mM	= 3.56
Total P	3.90 mM	$CaH_2PO_4^+$	2.38×10^{-4} mM	
Total Mg	0.83 mM	$CaHCO_3^+$	8.33×10^{-3} mM	$(IP/K_{OCP})^{1/8}$
Total HCO_3	10 mM	$CaCO_3^0$	5.97×10^{-4} mM	= 0.59
Na + K	160 mM	Ca-lactate^0	1.71×10^{-2} mM	
Cl	130 mM	Bound Ca^0	0.31 mM	$(IP/K_{EN})^{1/7.8^b}$
Lactate	11 mM	Free HCO_3^-	8.64 mM	=0.93 to 1.21
pH	7.3	Pco_2	0.0184 atm	
Osmolality[c]	312	Ionic strength	165 mM	

[a] IP stands for the ionic activity product of the salt concerned and the degree of saturation is expressed in terms of the mean ionic activity products

[b] The adopted stoichiometries of the outer and inner secretory porcine enamel were $[Ca]_{4.16}[HPO_4]_{0.41}[CO_3]_{0.42}[PO_4]_{2.16}[OH]_{0.16}$ and $[Ca]_{4.29}[HPO_4]_{0.31}[CO_3]_{0.40}[PO_4]_{2.29}[OH]_{0.29}$, respectively

[c] mosmol/kg H_2O

radioactivity, the protein-coated crystals were incubated for 5 h in the experimental solution and, at the end, the slurry was separated by centrifugation. The proteins remaining on the crystal surfaces and the proteins dissolved in the experimental solution were individually recovered, according to the same procedures as used in the adsorption studies, and then characterized using SDS-PAGE.

Electron Microscopy

Part of the secretory enamel tissue was lyophilized immediately after dissection and then embedded with POLY/BED 812. Ultra-thin sections were prepared with a diamond knife and the morphology of the crystals, without staining, was observed in either a JEM 1200EX electron microscope at 100 kV or a JEM-2010 electron microscope at 200 kV.

Results and Discussion

Table 1 shows the results of the chemical composition of the separated fluid. A significant finding was that the ionic calcium activity, $(Ca^{2+}) = 0.053$ mM, was one order of magnitude lower than the total calcium concentration, suggesting the presence of ligands complexing the calcium ions. Additionally, the fluid contained about 10 mM total HCO_3, which was much higher than the total concentrations of lattice ions, [Ca] = 0.5 mM and total $[PO_4] = 3.9$ mM. At the determined pH value of the fluid, it was found that the partial pressure of CO_2 of the fluid corresponded to 0.018 to 0.02 atm which is less than that (0.05 atm) of porcine serum. The foregoing results indicate that enamel mineralization during the secretory stage of amelogenesis occurs under a distinct, compartmentalized microenvironment, isolated from the circulating blood (at least in relation to some of the common constituents), although the fluid was kept isosmotic with serum, thus preventing undesirable water flows across tissues. Previously, we reported the estimated degree of saturation (DS) of the fluid with respect to calcium phosphate salts of interest, which were calculated using analytical data available at that moment [19]. Since the concentration of bicarbonate and organic acids (e.g., the total concentration of lactate of the fluid was 11 mM [22]) were analytically determined after that publication, the DS of the fluid was recalculated by taking into consideration the pertinent ion pairs with these species. A new mass balance for calcium considered was:

$$\text{Total Ca} = \text{free Ca}^{2+} + \text{CaHPO}_4^0 + \text{CaH}_2\text{PO}_4^+ + \text{CaHCO}_3^+ + \text{CaCO}_3^0 + \text{Ca-Lactate}^+ + \text{bound Ca}^0$$

Since the measurement of pH of the fluid was conducted at 25°C, the numerical values of the solubility product for hydroxyapatite (K_{HA}) and octacalcium phosphate (K_{OCP}) are 2.63×10^{-60} and 1.05×10^{-47}, respectively (detailed in [19]). Recently, we reported the solubility product constants for secretory porcine enamel (K_{EN}), which were determined under controlled $Pco_2 = 0.018$ atm (N_2-CO_2 mixtures). The values of K_{EN} obtained for the outer and inner secretory enamel on the basis of the stoichiometry model $[Ca]_{5-x}[HPO_4]_v[CO_3]_w[PO_4]_{3-x}[OH]_{1-x}$, were 1.45×10^{-40} and 1.08×10^{-43}, respectively [23]. In Table 1 are also shown the calculated ionic composition of the fluid and the DS values with respect to HA, OCP, and the enamel minerals. The results indicate that the calcium-ligand complex accounts for about 62% of the total calcium, significantly affecting the DS of the fluid. The DS of the fluid separated from the secretory enamel is essentially in equilibrium with the forming enamel apatites, and slightly supersaturated and undersaturated with respect to HA and OCP, respectively.

Concerning the Ca-ligands, our previous studies [24] showed that the apparent dissociation constant of the assumed complex was about 7.5×10^{-5} mol/L. This value indicates that the strength of the Ca-ligand binding is rather modest; the determined value was significantly greater than the dissociation constant (1.95×10^{-6} mol/L) of Ca-binding by the whole organic matrix of fetal bovine enamel, reported by Drinkard et al. [25]. This feature seems advantageous if complexed calcium acts as a reservoir (a fine regulatory system through association-dissociation) of calcium levels in the fluid during the secretory stage of amelogenesis. Our computer simulation of changes of fluid composition taking into consideration the presence of Ca-ligands, clearly showed that free Ca^{2+} concentrations in fluid can increase from 0.15 mM to 0.2 mM with an increase in total calcium from 0.5 mM to 1 mM (Fig. 1.A); this increase in free Ca^{2+} concentration is enough to induce supersaturated conditions with respect to OCP (Fig. 1.B).

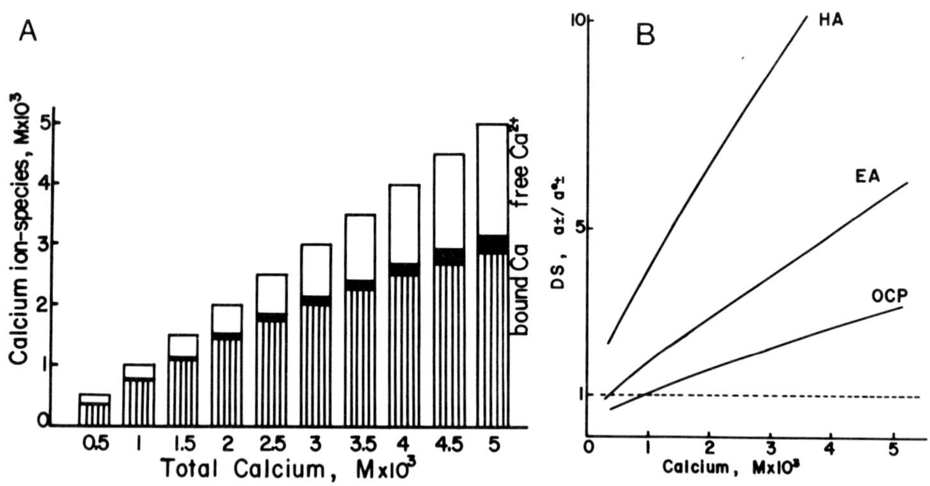

Fig. 1. A Results of calculation of calcium ionic species in fluid and **B** the degree of saturation (*DS*) of fluid with respect to hydroxyapatite (*HA*), octacalcium phosphate (*OCP*), and enamel mineral at the secretory stage (*EN*) as a function of the increment of the total calcium concentration of the fluid

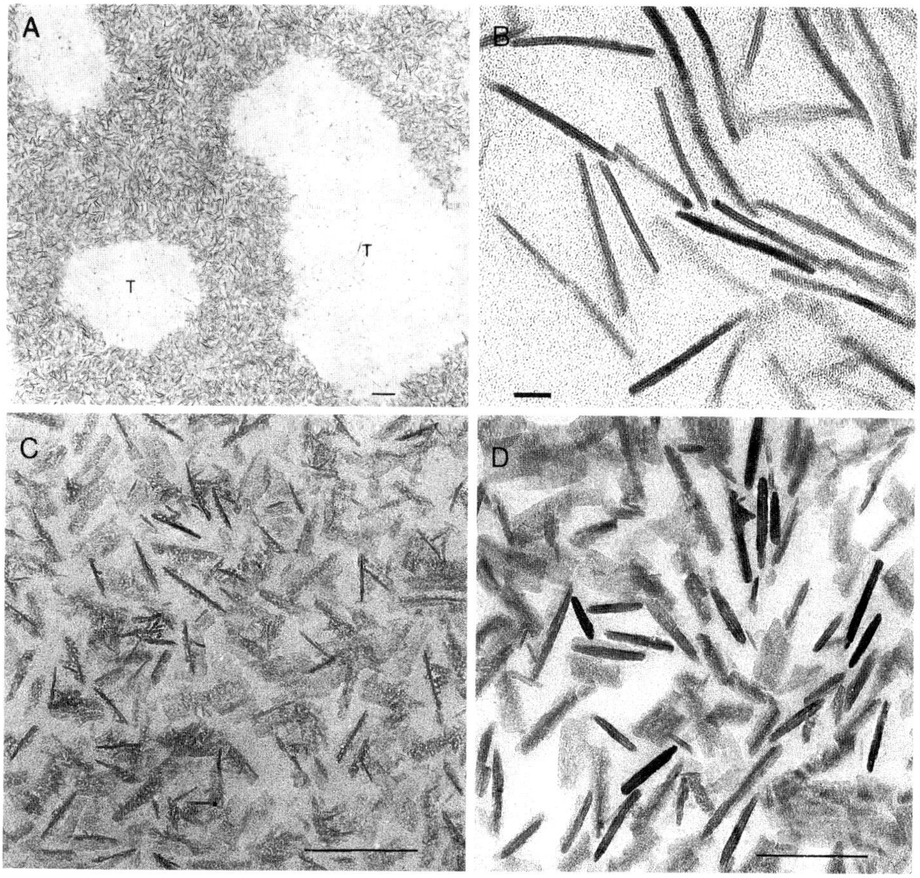

Fig. 2. Electron micrograms of porcine enamel crystals. **A** Thin ribbons precipitated around Tomes processes (*T*) of the ameloblast. **B** High resolution images of these crystals having thicknesses corresponding to several unit cells of apatite crystals. **C** The crystals found in the outer region, displaying non-uniform deposition of aptitic unit cells on the lateral planes of thin ribbons. **D** The grown crystals found in the middle region of secretory enamel; note the presence of a *central dark line. Thin and thick horizontal bars* represent 100 nm and 10 nm, respectively

Microradiographic studies by Suga [1] clearly show that mineralization within the secretory stage of mammalian amelogenesis progresses from the outer toward the inner regions. This advancement of mineralization is due to an increase in the size of the crystallites (i.e., crystal growth) (Fig. 2). The crystals precipitated in the vicinity of the ameloblast were thin ribbons in appearance (Fig. 2.A,B) and their initial growth process was characterized by a non-uniform accretion of apatitic unit cells on the lateral planes of a thin ribbon (Fig. 2.C). The grown crystallites, found in underlying zones, retained defects in their core region, the so-called central dark-line images (Fig. 2.D). Several investigators [26] have advocated that these features of enamel crys-

```
        1         45 46  62 63                  148   161              173
25 kd M----------W-L---H-A-------------------S-----W-PATDKTKREEVD

23 kd M-------------------------------------------------W

20 kd M-------------------------------------------S

      M---- 7 kd ------H A------ 11 kd -----S

      M-5 kd -W L---------- 13 kd ---------S
```

Fig. 3. The 25 kd porcine amelogenin secreted by the ameloblast and its degraded products

tallites (i.e., thin ribbons and central defects) may be ascribed to possible involvement of OCP-like phases as precursors. Interestingly, the observation of initial precipitates in in vitro carbonatoapatite preparations [27] disclosed that all the features (i.e., thin-ribboned in appearance, the central dark lines, and the non-uniform deposition) were reproduced in the neutral solutions, whereas rod-like apatite crystals, having hexagonal cross sections and without any vestiges of the central dark lines, were formed in the presence of high concentrations of fluoride or at pH 10, where the precipitation of OCP becomes kinetically unfavorable. However, no crystallographic evidence for OCP precipitation in mammalian amelogenesis is reported in the literature.

A central issue in enamel mineralization during amelogenesis is to understand how the observed precipitation and subsequent growth of enamel crystals are regulated through the protein-mineral interaction. Our interest was focused on the role played by the amelogenins. The whole sequence of the intact porcine amelogenin (25 kd by SDS-PAGE) was reported by Yamakoshi et al. [13]. This amelogenin protein, comprising 173 amino acid residues, has a unique sequence, i.e., all the charged residues are located within the N-terminal 40 and C-terminal 15 residues; particularly, the highly charged segment composed of 2 Glu, 2 Asp, 2 Lys, and 1 Arg is found at the C-terminus. It is also established that this intact protein is degraded specifically by proteolytic enzymes shortly after secretion [7], giving rise to a complex picture of protein constituents in the forming tissue, such as the moieties of 23, 20, 13, 11, 7, and 5 kd. In Fig. 3 are illustrated the sites where the enzymatic cleavages of the 25 kd porcine amelogenin takes place, according to the work by Shimizu and Fukae [8] at the Tsurumi University, Yokohama, Japan.

Figure 4 shows SDS-electrophoretograms of the proteins separated from the outermost (close to the ameloblast) through innermost (close to DEJ) zones of secretory porcine enamel. A comparison of the SDS-PAGE patterns indicated that appreciable changes in the protein composition take place within the secretory stage. The 25 kd amelogenin, as well as high molecular weight proteins (60–90 kd), were concentrated in the outermost secretory enamel. Most of the 60–90 kd proteins were assigned as being non-amelogenins [28]. In the underlying inner zones, the 20 kd amelogenin increased in relative terms as

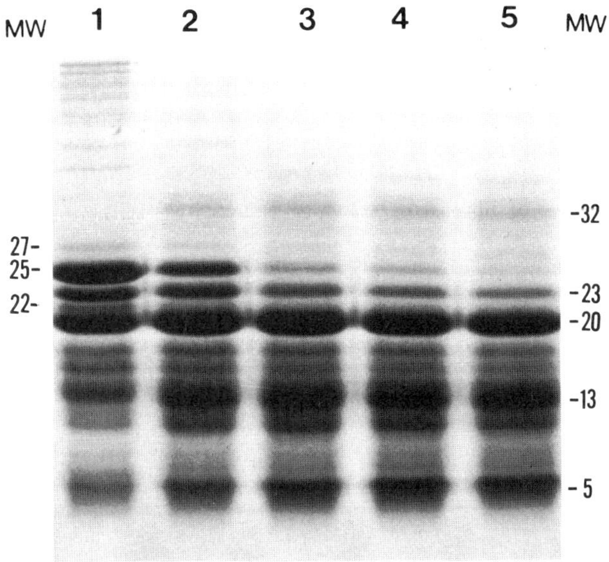

Fig. 4. SDS-electrophoretograms of enamel matrix proteins extracted from the various layers (I through V) of secretory porcine enamel. Note the presence of 27, 25, and 22 kd amelogenins, as well as the 60–90 kd non-amelogenins, in the outermost zone (close to the ameloblast). With substantial decreases of these moieties, the 20 kd and lower molecular weight amelogenins, as well as the 32 kd non-amelogenins, increased in relative terms in the underlying zones.

the 25 kd amelogenin decreased. The 11–13 and 5–7 kd amelogenin fragments also increased in the inner zones. In addition, as the 60–90 kd non-amelogenins decreased substantially, the 32 kd band became prominent. This 32 kd group (at least 7 constituents having pl = 3–4.5 were distinguished) has a unique N-terminal sequence, LxHVPGRIPPGYGRPPT-, which did not correspond to the sequence of porcine amelogenin [29]. It is notable that two faint bands at 27 and 22 kd were discerned from the proteins separated from the outermost secretory enamel and that the corresponding proteins were not discernible in the underlying tissue. In studies using antibodies raised against a synthetic peptide corresponding to the C-terminal 12 residue segment of the 25 kd porcine amelogenin, it was found that these 27 and 22 kd moieties reacted with the antibodies as the 25 kd protein did, indicating the presence of the common epitope at the C-terminus [30]. This finding supports the view that the ameloblast secrete amelogenin families through splicings of DNA information [9, 11].

The foregoing results on protein localization in secretory porcine enamel indicate that the 25 kd intact amelogenins (and the 27 and 22 kd proteins) are unstable in situ. Our recent studies using ^{1}H-NMR and photo-CIDNP (chemically induced dynamic nuclear polarization) [31] showed that the C-terminus of the 25 kd porcine amelogenin is exposed on the molecular

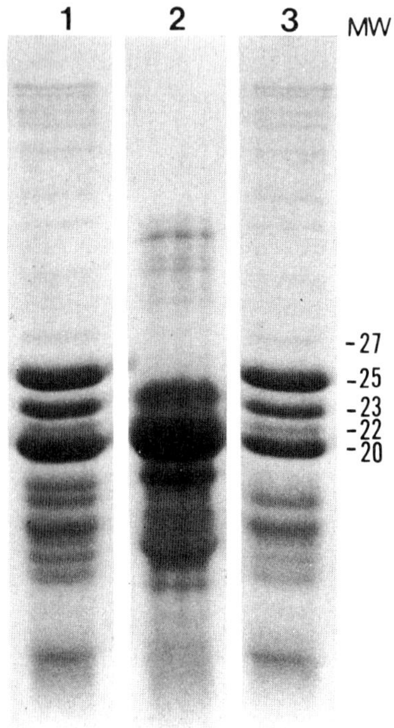

Fig. 5. SDS-electrophoretograms showing the extraction of enamel matrix proteins with chloroform/methanol 3:1 (v/v). *Lane 1*, the whole matrix proteins isolated from the outer secretory enamel; *lane 2*, proteins extracted in the organic solvent; *lane 3*, proteins recovered on the crystals after extraction. Note the remarkable difference in the hydrophobic/hydrophilic nature between the 25 kd intact amelogenin (and 27 and 22 kd proteins) and the degraded products (23 and 20 kd) lacking the hydrophilic segment at the C-terminus

surface, supporting the contention that the segment at the C-terminus is easily accessible to enzymatic cleavage. An intriguing question is whether the presence of the hydrophilic segment at the C-terminus and its cleavage are essential parts of the regulatory mechanism of enamel mineralization at early developmental stages. In the present studies, we first tested a possible influence of the C-terminal segment on the hydrophilic/hydrophobic nature of the amelogenin molecules by investigating the solubility of the proteins in the organic solvent (chloroform-methanol 3:1). The pertinent SDS-PAGE patterns are shown in Fig. 5. The most prominent findings were that (1) the 25 kd amelogenin, as well as the high molecular weight non-amelogenins, was sparingly soluble in this solvent, (2) the 27 and 22 kd moieties having a common epitope at the C-terminus were also sparingly soluble, and (3), in contrast, the 23 and 20 kd amelogenins, lacking the hydrophilic segment at the C-terminus, were extracted preferentially in the solvent. These results provide evidence that the property of the amelogenins is actually modified by the cleavage of the C-terminal segment.

As cited in the Introduction, it is widely accepted that the enamelins bind tightly onto the enamel mineral, while the amelogenins are segregated from the crystal surface during mineralization [3]. However, based on the published primary structures of the parent amelogenins showing the presence of a hydrophilic sequence at the C-terminus (and the observed difference in

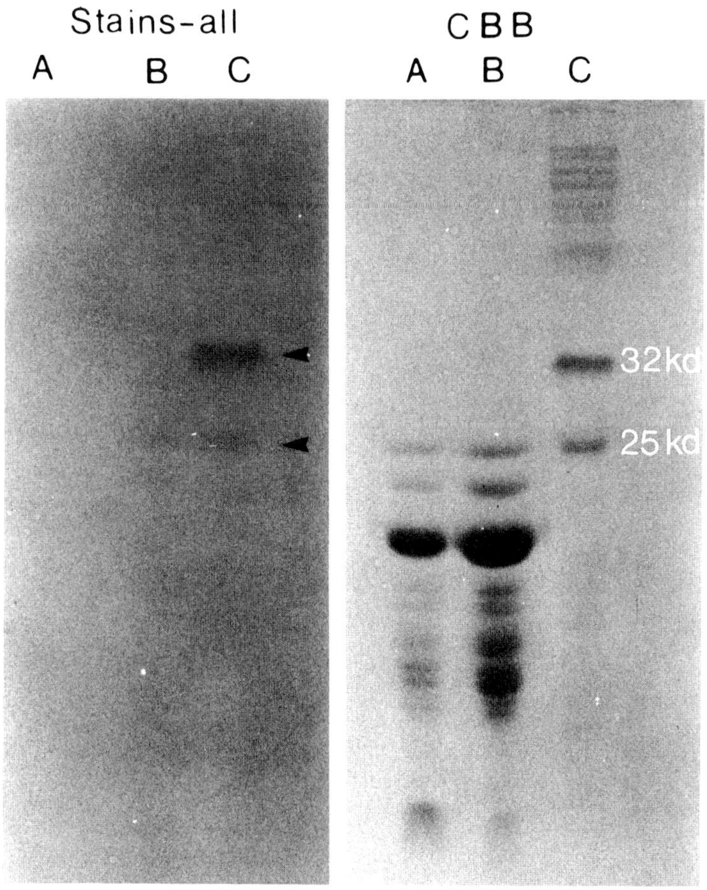

Fig. 6. SDS-electrophoretograms showing the selective adsorption of porcine enamel proteins onto hydroxyapatite. *Lanes A*, the original enamel proteins used as adsorbates, which were isolated from the secretory enamel; *lanes B*, proteins remaining in solution after equilibration with the adsorbent crystals; *lanes C*, proteins adsorbed onto the crystals. The left- and right-side lanes A through C were stained with Stains-all and Coomassie brilliant blue (*CBB*), respectively. Note the staining (blue in color) of the 32 and 25 kd proteins with Stains-all

hydrophilic/hydrophobic property between the intact amelogenin[s] and its degraded products), we hypothesized that the intact amelogenins may play a different role in the process of enamel mineralization. This hypothesis was tested in in vitro adsorption studies. Figure 6 shows typical results of the selective adsorption of enamel proteins onto hydroxyapatite. The original adsorbate shown in lane A was the whole matrix proteins separated from secretory porcine enamel. As shown in lanes B and C, whereas the degraded products of the 25 kd amelogenin (e.g., 23 and 20 kd proteins) were not adsorbed appreciably onto the crystals and most of them remained in the solution

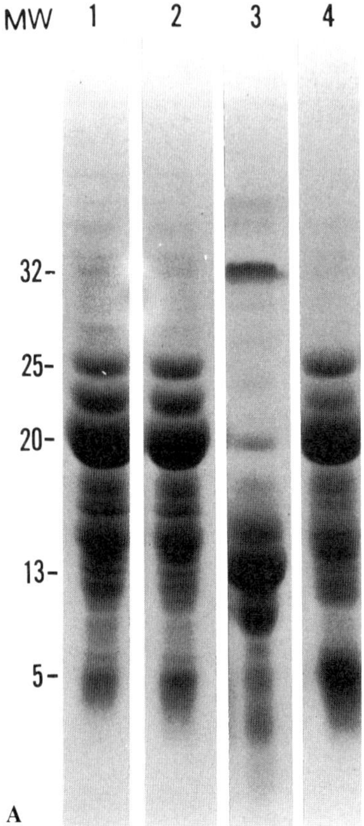

Fig. 7. A SDS-electrophoretograms showing coating hydroxyapatite with enamel matrix proteins in dispersed forms. *Lane 1*, proteins used as adsorbates; *lane 2*, proteins sedimented with the crystals in 160 mM NaCl solution at room temperature; *lane 3*, proteins dissolved in the fluid-simulated solution during the isotopic exchange reaction with calcium ion; *lane 4*, proteins remaining on the crystal surfaces at the end of the reaction. **B** Changes of the exchangeable Ca pool on the crystal surface as a function of the protein-coating of the crystals

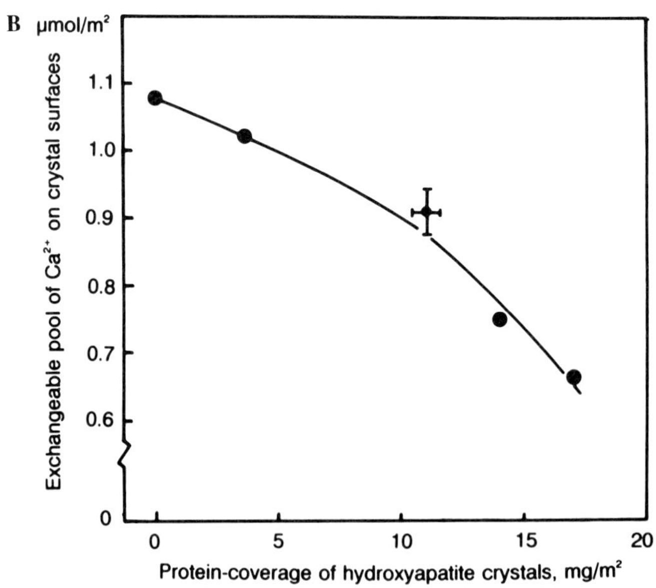

(lane B), the 25 kd intact amelogenin, as well as the 32 kd and 60–90 kd non-amelogenins, were concentrated on the crystals after equilibration (lane C). The left-side lanes of Fig. 6 were stained with Stains-all. Interestingly, the band corresponding to 32 kd proteins was stained most extensively (blue in color), while the 25 kd amelogenin was also stained faintly. The obtained staining of the 32 kd proteins is consistent with the finding that this protein group was highly phosphorylated (about 7 mols P per mol protein) [29]. Similarly, the present results suggest that some of the 25 kd proteins may be phospholyrated. In this connection, previous studies using two-dimensional electrophoresis [32] demonstrated that the 25 kd porcine amelogenin is separated into charge families, one major moiety at about pI = 6.2 and at least three or four variants having different pI values between 6 and 6.5. However, no quantitative data about the phospholyration of porcine amelogenins (and evidence for the phospholyration site) are available at present.

An important finding relevant to possible functions of the amelogenins in situ [18, 20] is that the observed adsorption of the 25 kd amelogenin onto hydroxyapatite is related intimately to the inhibition of precipitation kinetics of the crystals in supersaturated solutions. Furthermore, the present studies provided an insight into possible functions of enamel matrix proteins (in aggregated forms) in solution having a composition similar to that found in the enamel fluid. Figure 7 shows the effect of the protein-coating on the Ca incorporation onto the crystals. Following the proteins-crystal equilibration in solution having an ionic strength of 160 mM at room temperature, most of the proteins were sedimented with the crystals (lane 2). When the protein-coated crystals were added to the fluid-simulated solution for investigating Ca exchange at the crystal-solution interface, it was found that appreciable amounts of the proteins (approximately 20% of the total proteins originally coated on the crystals) were dissolved in the solution during the experimental period. The moieties which dissolved corresponded to the 32 kd non-amelogenin and the 20 and 13–11 kd amelogenins (lane 3), but the 25 and 23 kd amelogenins were barely dissolved. As illustrated in Fig. 7.B, the exchangeable pool of calcium on the crystal surface decreased with an increase of the protein coating per unit surface area of the crystals. These results, as well as previous findings with respect to fluoride incorporation into apatite lattice [33], indicate that coating hydroxyapatite crystals with enamel matrix proteins brings about a retardation of ion incorporation into the crystals. It is pertinent to point out that the expected ratio of protein/mineral in the secretory enamel is substantially higher than those ratios tested in the present study. Therefore, very significant reductions in the effective surface areas of enamel crystals are anticipated in situ, thereby retarding ion incorporation into the crystalline lattice and, by exchange, on the crystal surface.

In relation to the mechanism of the amelogenin-crystal interaction shown above, it is a logical proposition that the obtained properties (i.e., adsorption affinity and inhibitory activity) of the intact amelogenins are related to their molecular structures (i.e., presence of the functional sequences and radicals, such as phosphorylation, and the molecular conformations). In this connection, it is a significant feature of mammalian amelogenins that their primary structures display a high degree of homology, particularly the first 30 and last 10 residues at the N- and C-termini [7, 11]. The observation that the 23 kd amelogenin, lacking the C-terminal 12 residues, already substantially decreases the adsorption affinity onto the crystals (Fig. 6), suggests that the cleaved segment at the C-terminus may be the functional sequence in the adsorption reaction. However, a recent study using a synthetic decapeptide corresponding to the last 10 residue hydrophilic sequence at the C-terminus [21], demonstrated that the partial sequence in itself has only a very low adsorption affinity (of the same order of magnitude as those of single amino acids, such as aspartic acid and phosphoserine) and, as expected by its low adsorption affinity, shows a weak inhibitory activity on the growth of apatite crystals in dilute supersaturated solutions. Similar results were obtained using other amelogenin fragments [34]. Furthermore, previous studies [32] showed that all charge families of the 25 kd amelogenin adsorbed onto the crystals, indicating that the small difference in net electric charges among the 25 kd isomorphs do not significantly affect their adsorption affinity onto apatitic surfaces. Based on the foregoing results and consideration, our current conclusion is that no segmental sequences are enough to explain the properties of the intact amelogenin, and that the properties of the amelogenins (and their functional roles in enamel mineralization) may be determined by the whole molecular structure, including both the highly conserved segments at the N- and C-termini. It is also deemed likely that the molecular structure of the amelogenin in solution may change substantially by the cleavage of, first, the C-terminal hydrophilic segment and then the subsequent cleavage of the N-terminus; our studies on porcine amelogenins using NMR gave supportive evidence for this concept [31].

The overall results suggest that the nature and properties of the biominerals is primarily determined by the solution composition or, more accurately, the driving force (DS) in the media. It is also probable that the kinetics of mineralization (and thereby the nature and properties of formed biominerals) at early stages of amelogenesis is modulated substantially by the presence of physiological inhibitors, e.g., organic matrix. The secreted intact amelogenins have different properties from those of the degraded amelogenins. It is an intriguing hypothesis that the behavior of amelogenins in situ and their postulated functions are most likely related to their unique molecular structure. There is still need for further investigations to elucidate what is the molecular conformation of amelogenins in solution and in the adsorbed state, how amelogenetic mineralization is regulated through protein-mineral interaction in situ, and what is the nature and property of enamel minerals formed at various stages of amelogenesis.

Acknowledgements. The author is grateful to Drs. M. Fukae, T. Tanabe, M. Shimizu, and E.C. Moreno for their stimulating discussion and collaboration in conducting the present work. He is also grateful to Drs. S. Shimoda and Y. Miake for their technical help. This work was supported by grants DE07623 and DE08670 from the National Institute of Dental Research.

References

1. Suga S (1983) Comparative histology of progressive mineralization pattern of developing enamel. In: Suga S (ed) Mechanisms of tooth enamel formation. Quintessence, Tokyo, pp 167–203
2. Robinson C, Fuchs P, Deutsch D, Weatherell JA (1978) Four chemically distinct stages in developing enamel from bovine incisor teeth. Caries Res 12:1–11
3. Warshawsky H (1984) Ultrastructural studies on amelogenesis. In: Butler WT (ed) The chemistry and biology of mineralized tissues. Birmingham, Ala, pp 33–45
4. Takano Y, Hanawa M, Yamamoto T, Domon T, Fujinami H, Hanaizumi Y, Wakita M (1990) Time-related changes in the distribution of ^{45}Ca in the developing enamel of rat incisors as revealed by radioautography. J Biol Buccale 18:135–147
5. Bawden JW, Merriff DH, Deaton TG (1981) In vitro study of calcium 45 and phosphorus-32 uptake in developing rat molar enamel using quantitative methods. Arch Oral Biol 26:477–482
6. Termine JD, Belcourt AB, Christner PJ, Conn KM, Nylen MV (1980) Properties of dissociatively extracted fetal tooth matrix proteins. I. Principal molecular species in developing bovine enamel. J Biol Chem 255:9760–9768
7. Fincham AG, Belcourt AB, Lyaruu DM, Termine JD (1982) Comparative protein biochemistry of developing dental enamel matrix from five mammalian species. Calcif Tissue Int 34:182–189
8. Shimizu M and Fukae M (1983) Enamel Proteins. In: Suga S (ed) Mechanisms of tooth enamel formation. Quintessence, Tokyo, pp 125–141
9. Shimokawa H, Sobel ME, Sasaki S, Termine JD, Young MF (1987) Heterogeneity of amelogenin mRNA in the bovine tooth germ. J Biol Chem 262:4042–4047
10. Limeback H, Sakarya H, Chu W, Mackinnon M (1989) Serum albumin and its acid hydrolysis peptites dominate preparations mineral-bound enamel proteins. J Bone Mineral Res 4:235–241
11. Shimokawa H, Tamura M, Ibaraki K, Sasaki S (1989) Human amelogenin gene. In: Fearnhead RW (ed) Tooth enamel V. Florence, Yokohama, pp 301–305
12. Snead ML, Lau EC, Zeichner-David M, Fincham AG, Woo SLC, Slavkin HC (1985) DNA sequence for cloned cDNA murine amelogenin reveals the amino acid sequence for enamel-specific protein, Biochem Biophys Res Commun 129:812–818
13. Yamakoshi Y, Tanabe T, Fukae M, Shimizu M (1989) Amino acid sequence of porcine 25 kDa amelogenin. In: Fearnhead RW (ed) Tooth enamel V. Florence, Yokohama, pp 314–318
14. Weiner S (1986) Organization of extracellularly mineralized tissues: A comparative study of biological crystal growth. CRC Crit Rev Biochem 20:365–408
15. Linde A, Lussi A, Crenshaw MA (1989) Mineral induction by immobilized polyanionic proteins. Calcif Tissue Int 44:286–295
16. Herold R, Rosenbloom J, Granovsky M (1989) Phylogenetic distribution of enamel proteins: Immunohistochemical localization with monoclonal antibodies indicates the evolutionary appearance of enamelins prior to amelogenins. Calcif Tissue Int 45:88–94
17. Slavkin HC, Bessem C, Bringas P Jr, Zeichner-David M, Nanci A, Snead ML (1988) Sequential expression and differential function of multiple enamel proteins during fetal, neonatal, and early postnatal stages of mouse molar organogenesis. Differentiation 37:26–39
18. Aoba T, Tanabe T, Moreno EC (1987) Function of amelogenins in porcine enamel mineralization during the secretory stage of amelogenesis. Adv Dent Res 1:252–260

19. Aoba T, Moreno EC (1987) The enamel fluid in the early secretory stage of porcine amelogenesis: Chemical composition and saturation with respect to enamel mineral. Calcif Tissue Int 41:86–94
20. Aoba T, Fukae M, Tanabe T, Shimizu M, Moreno EC (1987) Selective adsorption of porcine amelogenins onto hydroxyapatite and their inhibitory activity on seeded crystal growth of hydroxyapatite. Calcif Tissue Int 41:281–289
21. Aoba T, Moreno EC, Kresak M, Tanabe T (1989) Possible roles of partial sequences at N- and C-termini of amelogenin in protein-enamel mineral interaction. J Dent Res 68:1331–1336
22. Aoba T, Moreno EC (1989) Mechanism of amelogenetic mineralization in minipig secretory enamel. In: Fearnhead RW (ed) Tooth enamel V. Florence, Yokohama, pp 163–167
23. Aoba T, Moreno EC (1992) Changes in the solubility of enamel mineral at various stages of porcine amelogenesis. Calcif Tissue Int 50:266–272
24. Moreno EC, Aoba T (1987) Calcium binding in enamel fluid and driving force for enamel mineralization in the secretory stage of amelogenesis. Adv Dent Res 1:245–251
25. Drinkard C, Gibson L, Crenshaw MA, Bawden JW (1981) Calcium binding by organic matrix of developing bovine enamel. Arch Oral Biol 26:483–485
26. Nelson DGA, Barry JC, Shields CP, Glena R, Featherstone JDB (1989) Crystal morphology, composition, and dissolution behavior of carbonated apatites prepared at controlled pH and temperature. J Colloid Interface Sci 130:467–479
27. Shimoda S, Aoba T, Moreno EC, Miake Y (1990) Effect of solution composition on morphological and structural features of carbonated calcium apatites. J Dent Res 69:1731–1740
28. Fukae M, Tanabe T (1987) Nonamelogenin components of porcine enamel in the protein fraction free from the enamel crystals. Calcif Tissue Int 40:286–293
29. Tanabe T, Aoba T, Moreno EC, Fukae M, Shimizu M (1990) Properties of phosphorylated 32 kD nonamelogenin proteins isolated from porcine secretory enamel. Calcif Tissue Int 46:205–215
30. Aoba T, Shimoda S, Akita H, Holmberg C, Taubman MA (1992) Anti-peptide antibodies reactive with epitopic domains of porcine amelogenins at the C-terminus. Arch Oral Biol 37:249–255
31. Aoba T, Kawano K, Moreno EC (1990) Molecular conformation of porcine amelogenins and its significance in protein-mineral interaction: 1H-NMR photo CIDNP study. J Biol Buccale 18:189–194
32. Tanabe T, Aoba T, Moreno EC (1988) Effect of fluoride in the apatitic lattice on adsorption of enamel proteins onto calcium apatites. J Dent Res 67:536–542
33. Aoba T, Collins J, Moreno EC (1989) Possible function of matrix proteins in fluoride incorporation into enamel mineral during porcine amelogenesis. J Dent Res 68:1162–1168
34. Aoba T, Moreno EC (1991) Structural relationship of amelogenin proteins to their regulatory function of enamel mineralization. In: Surface reactive peptides and polymers. American Chemical Society, Washington DC, pp 85–106

Recent Observations on the Formation of Carbonatoapatite In Vivo

Takaaki Aoba and Edgard C. Moreno[1]

Summary. Biominerals in tooth and bone are most adequately described as carbonatoapatites. In order to elucidate the mechanism of carbonated apatite formation in vivo, our experimental strategies have been to (1) determine changes in the nature and properties of the biominerals with advancement of mineralization or tissue development, (2) assess the driving force for precipitation of the crystals in extracellular fluids, and (3) investigate temporal and spacial modifications of mineralization kinetics caused by various native regulators. Enamel mineralization taking place during amelogenesis is a unique and most useful model to investigate carbonatoapatite formation in vivo and the mineral changes at various stages of development. It was established that (1) substantial amounts of carbonate substitute both the PO_4^{3-} and OH^- ions in enamel crystals, (2) early enamel mineralization occurs in an environment isolated from circulating blood plasma, under a partial pressure of CO_2 of around 0.018 atm, and (3) the precipitation and growth of the crystals proceed steadily, without interference due to remodeling of the formed tissue, yielding the largest crystallites in the body. In the past few years, our effort was focused on the assessment of the labile and stable pools of the crystal constituents (i.e., carbonate, acid phosphate and fluoride) in developing enamel tissues, the development of a new solubility model taking into account most of the lattice constituents, and the determination of the solubility of enamel mineral under controlled partial pressures of CO_2. Although our knowledge is still limited and further refinement of the adopted stoichiometry and solubility models is required, the information obtained is valuable to predict the driving force for carbonatoapatite precipitation, as well as the stability of enamel mineral, in normal and pathologic environments.

Key words: Biomineralization — Amelogenesis — Carbonatoapatite — Stoichiometry — Solubility—Crystal growth

Introduction

Numerous reports in the literature have provided evidence that carbonation of calcium apatites is a common phenomenon in hard tissues. It is generally considered that most of the carbonate occupies sites in the crystalline lattice,

[1] Forsyth Dental Center, 140 Fenway, Boston, MA 02115, USA

although some of the ions may exist as CO_3^{2-} or HCO_3^- adsorbed on the crystal surfaces. The carbonate incorporation into the crystalline lattice occurs at 2 sites, originating 2 types of substitution [1–4]; in type B, the carbonate substitutes for phosphate, the planar CO_3^{2-} molecule occupying one of the planes in the PO_4^{3-} tetrahedron; in type A, CO_3^{2-} replaces the hydroxyl ion. Previous studies using polarized infrared (IR) spectroscopy [2] demonstrated that the type B substitution gives rise to distinct absorption bands at 1465, 1412, and 873 cm^{-1}, and bands at 1545, 1450, and 880 cm^{-1} correspond to the type A substitution. Recent studies by Fourier transform infrared spectroscopy (FTIR), in conjunction with resolution-enhancing techniques [5, 6], distinguished an additional CO_3^{2-} band around 865 cm^{-1} in the spectra of bone and enamel minerals, which was ascribed to the labile carbonate on the crystal surface.

Concerning the effect of carbonate ions on precipitation of apatite crystals, it was shown that carbonate ions in highly supersaturated solutions inhibit apatite formation from precursors [7, 8] and also inhibit the growth of apatite crystals along the *c*-axis direction [9], thus affecting the morphology of the grown crystals [10]. It was also reported that the carbonate substitution in apatite crystals causes an increase in the number of lattice defects [11, 12], increasing the solubility of the crystals [13–15]. However, little is known about the thermodynamic properties of carbonated biominerals and the mechanism of their precipitation and growth under normal and pathological conditions. Thus, our interest in the past few years has been to investigate the nature and properties of carbonated bioapatites, in particular enamel mineral formed at various stages of amelogenesis.

Enamel Mineralization as a Unique Model for Biological Carbonatoapatite Formation

Carbonate is one of the prominent constituents (3 wt% or more) of enamel mineral [16, 17]. Recent studies by Elliott et al. [18] demonstrated that approximately 10% to 15% of the carbonate ions in mature human enamel is in the OH^- position, while the rest substitute for PO_4^{3-} ions. Most likely, these carbonate substitutions in enamel crystals take place when the mineral precipitates during amelogenesis. In fact, the relative enrichment of carbonate in early developing enamel tissue, as well as the association of carbonate with the forming enamel crystals at subsequent stages, has been reported for various mammalian species [19, 20]. There was a paucity of information, however, about the microenvironment surrounding the precipitating biominerals. In Table 1 are shown available data concerning electrolyte concentrations in various extracellular fluids relevant to mineralization in mammalian species, as well as the corresponding values for the blood plasma. Using a micropuncture procedure, Howell et al. [21] aspirated fluid samples from rat hypertrophic (normal and rachitic) cartilage and analytically determined the total concentrations of major electrolytes. Wuthier [22] also determined the electrolyte

Table 1. Electrolyte concentrations of extracellular fluids

Sample	*[Ca]* *mM*	*[total P]* *mM*	*[Mg]* *mM*	*[Na]* *mM*	*[K]* *mM*	*[Cl]* *mM*	*$[HCO_3]$* *mM*	*pH*
Hypertrophic cartilage fluid								
Rat normal[a]	1.5	2.6					36	7.58
Rat rachitic[a]	1.8	1.6					31	7.52
Chicken normal[b]	3.1	1.5	0.91	115	15	111		7.49
Bone (bovine) fluid[c]	0.48	1.8	0.4	125	25	130		
Enamel (pig) fluid[d]	0.50	3.9	0.83	139	21	146	20	7.26
Blood plasma								
Rat[a]	2.3	2.8					22	7.39
Chicken[b]	3.2	1.8	1.1	156	11	118		7.35
Bovine[c]	1.5	1.8	0.7	140	4	100		
Pig[d]	2.9	2.4	1.1	156	12	168		7.26

[a] Howell et al. [21]; [b] Wuthier [22]; [c] Neuman [23]; [d] Aoba and Moreno [24]

concentrations in chick cartilage fluid; his data were corrected by taking into account possible leakages of electrolytes from the surrounding cells. Neuman [23] assessed concentrations of electrolyte constituents in bovine "bone fluid" and compared them with the corresponding values in blood plasma. Recently, we succeeded in separating the liquid phase from the secretory enamel of porcine teeth by ultracentrifugation and determined the concentrations (and activities) of most of the lattice ionic constituents [24]. Examination of the data shown in Table 1 leads to the conclusion that (1) mineralization in cartilage, bone, and enamel occurs in fluid environments having a composition different from that of circulating blood, (2) all mineralizing systems contain sufficient carbonate ions to be incorporated into the crystal lattice, and (3) the CO_3 concentrations in these fluids are always higher than the concentrations of coexisting Ca and phosphate ions. More important, the composition of the enamel fluid is significantly different from those of the cartilage and bone fluids, specifically with respect to concentrations of calcium, phosphate and carbonate, and pH value. This difference indicates that enamel mineralization in the early developmental stages occurs in an isolated fluid environment.

In addition to the uniqueness of enamel fluid composition, mammalian amelogenesis provides an excellent model for the investigation of the carbonated hydroxyapatite formation, since the mineralization process is not intervened by any remodeling of the formed tissue, as in the case of bone (and cartilage) mineralization. Furthermore, differing from the rapid deposition of dentin crystals at the mineralization front, the growth of enamel crystals proceeds gradually, but to a significant extent, through all the developmental stages, resulting in the largest carbonatoapatite crystals in the body. An important conclusion drawn from ultrastructural studies [25–27] is that the advancement of enamel mineralization is not due to an increase in the number but in the size of the crystallites (i.e., crystal growth). Our recent studies using high resolution electron microscopy [28] showed that the enamel mineralization process is characterized by the initial precipitation of thin ribbons, the non-uniform accretion of apatite unit cells on prism planes, and the formation of flattened hexagonal crystals as revealed by cross-sectional images. Although not yet defined completely, it appears that the obtained morphology of enamel crystals and their growth kinetics at various developmental stages are controlled temporally by the modulation of cellular (ameloblast and surrounding enamel organ cells) activities and the presence of various regulators of apatite crystal growth (e.g., matrix proteins, F^-, and Mg^{2+}).

In general, the forming enamel during amelogenesis is divided into secretory, early maturing (including transitional), and late maturing (hard mature) stages on the basis of mineralization progress and associated alterations in proteinaceous matrix and tissue structures [29, 30]. Rodent incisors, growing and erupting continuously, have been widely used by many investigators, since a full spectrum of amelogenesis can be observed from the apical through the cuspal ends in a single tooth. However, it is of importance to realize that enamel mineralization during amelogenesis advances in 2 directions: from the outer to the inner zones as well as from the apical to the cuspal ends [31]. In this respect, porcine enamel is a suitable model for investigation of carbonatoapatite formation during amelogenesis, since the large tooth size facilitates dissection and recovery of sufficient amounts of enamel samples at various developmental stages, i.e., the outer (younger) secretory, inner (older) secretory, early maturing, and mature tissues. Indeed, substantial changes in the crystal morphology and protein constituents were observed between the outer (close to the ameloblasts) and inner zones of secretory porcine enamel [32]. It is also known that the chronology of enamel mineralization (and the tooth shape) of this animal is similar to that of human enamel [33]. Thus, most of our work cited in this report, i.e., changes in chemical composition [20, 32, 34, 35], sites of carbonate substitution [20, 28], morphology of enamel crystals [28, 32] and their solubility [36], was conducted with enamel samples dissected from unerupted permanent incisors of piglets.

Changes in Chemical Composition of Enamel Mineral During Amelogenesis

As mentioned above, much work has been conducted on the composition of the inorganic constituents of enamel over the last three decades. However, there remains a paucity of information about the state or environment of ionic constituents in enamel tissues at various developmental stages. Specific questions to be addressed are: (1) whether part of the ionic constituents of enamel tissues may be in labile forms (i.e., organically bound or adsorbed on crystal surfaces) rather than in a stable form (i.e., incorporated into the crystal lattice), (2) whether protonated ionic species, e.g., HCO_3^- and HPO_4^{2-}, exist in significant quantities in enamel crystals, and, as a consequence, (3) what are the true lattice stoichiometries of enamel crystals formed at various stages of enamel mineralization. The fluoride uptake into developing enamel has been studied by many investigators [37, 38] since it was disclosed that the incorporated fluoride ions are effective stabilizers of enamel mineral, whereas, at high concentrations, fluoride induces defective enamel structures. However, there still remain controversies about the state of the fluoride in developing enamel tissues or enamel proteins-fluoride binding [39, 40]. Thus, the contents of calcium, total phosphate, carbonate, acid phosphate, and fluoride were determined in the enamel samples dissected from regions of porcine incisors corresponding to the various developmental stages. Significant attention was given to assessment of the labile pools of carbonate (removal of surface CO_3 species by plasma ashing in an activated oxygen atmosphere at about 60°C [20]), acid phosphate (successive extraction with 3 mM phosphate solutions at pH 11 [35]), and fluoride (successive extractions with neutral NaCl solutions and carbonate buffers [34]).

Table 2 shows the results of chemical analyses of the total and the labile pools of carbonate, acid phosphate, and fluoride. Both the calcium and phosphorus contents (% dry weight) in porcine enamel tissue increased consistently with tissue development, from the outer secretory (Ca, 15%; P, 8%) to the

Table 2. Chemical composition of porcine enamel samples obtained at various developmental stages

Samples	*Ca* *wt%*	*P* *wt%*	*CO_3* *wt%*	*% P* *as HPO_4^{2-}*	*F* *wt% ×10^2*
Outer secretory enamel	22.8	11.6	3.8 (18)	15.6 (60)	1.0 (30)
Inner secretory enamel	25.5	12.8	4.0 (12)	10.8 (45)	0.5 (25)
Maturing enamel	30.8	15.2	3.8 (7)	8.3 (35)	1.5 (**)
Mature hard enamel	35.9	17.5	2.8 (**)	5.3 (5)	

** trace (<5%)
Figures in parenthesis represent the labile pool as percentages of the total CO_3^{2-}, HPO_4^{3-}, and F^- in each enamel tissue

mature enamel (Ca, 36%; P, 17.5%). In contrast, the total carbonate content decreased only modestly from the maximum content of 4.0 wt% in the inner secretory enamel to 2.8 wt% in the mature enamel. Importantly, significant portions (about 18% and 12%) of the carbonate in the outer and inner secretory enamel, respectively, were lost by low temperature ashing, whereas this lability of carbonate disappeared as the enamel tissue developed. The latter reduction in carbonate suggests that part of the carbonate is adsorbed on the crystal surfaces in the early stages of mineralization, while, with the advancement of mineralization, the carbonate, almost in its totality, becomes a lattice constituent in the enamel crystals. Moreover, if carbonation of enamel crystals is expressed by the stable pool of CO_3^{2-} per mol of Ca (or P), the carbonate incorporation in enamel mineral was revealed to be fairly constant (stable $CO_3/Ca = 0.09$) in the secretory enamel while it decreased from 0.07 in the maturing to 0.05 in the mature enamel. Most likely, the foregoing changes in carbonate substitution reflect compositional changes in the enamel fluid between the secretory and the further advanced stages (presumably related to the activity of the ameloblasts).

Concerning the acid phosphate content of enamel samples, it was found that (1) the outer secretory enamel contained the highest HPO_4^{2-}, corresponding to 15% or more of the total phosphate, (2) the HPO_4^{2-} content decreased with tissue development, and (3) the HPO_4^{2-} content in the mature enamel was about 5.3% of the total phosphate. The maximum value of the HPO_4^{2-} obtained thus far from the dissected porcine enamel corresponds to one-half of the content anticipated from the stoichiometry of octacalcium phosphate, $Ca_4H(PO_4)_3 \cdot 5H_2O$. Interestingly, a substantial part (60%) of the total acid phosphate in the outer (young) secretory enamel was in labile forms (i.e., lost after the extraction with the alkaline solution, due to deprotonation or exchange with PO_4^{3-} in solution [35]), whereas the corresponding pools decreased substantially to 45% in the inner (older) secretory enamel, to 35% in the maturing enamel, and to only a marginal fraction (<5%) in the mature enamel. On the basis of electron microscopic observations of enamel crystal growth [28, 32], it seems reasonable to think that the decrease of the total and the labile HPO_4^{2-} with the advancement of mineralization may result from the

Table 3. Chemical composition of rat secretory enamel from animals ingesting fluoride

Animal group	*Ca* *wt%*	*P* *wt%*	*F* *wt% ×10²*
Control	17.1	8.8	1.2 (23)
25 ppm	17.8	9.0	17.5 (**)
50 ppm	17.5	9.0	31.1 (**)
75 ppm	18.0	9.1	40.9 (**)
100 ppm	17.7	9.1	54.5 (**)

** trace (<5%)
Figures in parenthesis represent the labile pool as percentages of the total F^- in each enamel tissue

burying of an initially precipitated domain rich in HPO_4^{2-} with a subsequent deposition of an apatitic phase having different stoichiometries from that of the initially precipitated thin ribbons.

We also determined the fluoride content in porcine secretory and mature enamel samples with special attention to possible association of fluoride with the matrix proteins. Most of the amelogenins in the enamel tissue were extracted with a carbonate buffer at pH 10.8. The fluoride in the original and treated enamel samples was measured by the microdiffusion procedure, in conjunction with an ion-selective electrode [34]. The results (Table 2) indicate that (1) the total F per unit mass of Ca (or P) in the tissue was maximal at the beginning of the secretory stage and decreased (and leveled off) with the advancement of mineralization, and (2) interestingly enough, part of the fluoride in the secretory enamel, about 25%–30% of the total, was removed with the extraction of organic matter, mostly amelogenins. These results suggest that a fraction of the fluoride in the secretory enamel may be associated with the organic matter. However, the nature of that association remains obscure.

A question of interest in understanding the role of fluoride in enamel mineralization is whether the labile fluoride is simply a pool available for incorporation into enamel crystals, or if it may have other specific functions, e.g., modification of protein properties or their degradation processes in the process of enamel mineralization. A partial answer to this question was obtained in studies on the uptake of systemic fluoride by rat enamel [43]. The animals were given either deionized water or water containing 25, 50, 75, and 100 ppm fluoride as NaF. Table 3 shows the Ca, total P, and fluoride contents of the secretory enamel obtained for each of the animal groups. The results clearly showed that, whereas there were no significant differences in the Ca and P contents among the animal groups, the total fluoride content of the enamel tissue increased with increasing the concentration of fluoride in the drinking water. A conspicuous finding was that no appreciable increase in the labile fluoride was obtained in any of the animal groups. In parallel studies, it was verified that the systemic fluoride ingestion at the doses tested did not cause discernible changes in the properties of amelogenins and their degradation cascade during the secretory stage. Furthermore, in vitro crystal growth studies yielded evidence that coating hydroxyapatite seeds with enamel matrix proteins resulted in a retardation of fluoride incorporation, as a result of an inhibition of apatite precipitation. Based on the foregoing experimental data, it seems that the fluoride incorporation into enamel crystals is intimately regulated by the kinetics of mineralization but, once an excess of fluoride ions beyond the magnitude of the labile pool (depending on the amounts and properties of matrix proteins) becomes available under relatively high fluoride regimes, the excess F^- promotes the precipitation reaction, as demonstrated in vitro [44], leading to the accretion of fluoride in the forming crystals.

Structural Changes of Enamel Mineral at Early Stages of Amelogenesis

The same batches of porcine enamel samples used for the chemical analysis were examined by X-ray diffraction and FTIR spectroscopy. Figure 1 shows typical X-ray diffraction patterns recorded from the porcine enamel samples. The observed reflections, although most of them in the outer secretory enamel samples were broadened (reflecting a low degree of crystallinity of the initially precipitated mineral), match those for the apatite lattice. Table 4 shows the lattice parameters and relative crystallinity of enamel mineral. Developmental advancement from the outer secretory to the inner secretory as well as early maturation stages brought about significant changes in the unit cell parameters, namely, shrinkage and expansion of the *c* and *a* unit cell dimensions, respectively, and an improvement of the crystallinity, particularly in the *a*-axis direction. These changes in the unit cell parameters are consistent with the relative decrease of the carbonate and acid phosphate contents and the changes in CO_3^{2-} substitution sites mentioned above. Also, the increasing crystallinity is consistent with the observation that the advancement of enamel mineralization is characterized by an increase in the size of the crystallites.

For comparison, in Fig. 2 are shown typical X-ray diffraction patterns of synthetic carbonatoapatites obtained at 37°, 60°, and 80°C. It is pertinent to

Table 4. Lattice parameters and line broadening of porcine enamel apatites obtained at various developmental stages

Sample	*Lattice parameter (Å)*		*Line broadening (°(2θ))*	
	c	*a*	*(002)*	*(310)*
Outer secretory enamel	6.890	9.433	0.38	1.14
Inner secretory enamel	6.886	9.437	0.30	0.90
Maturing chalk-like enamel	6.884	9.440	0.28	0.64
Mature hard enamel	6.882	9.443	0.24	0.39

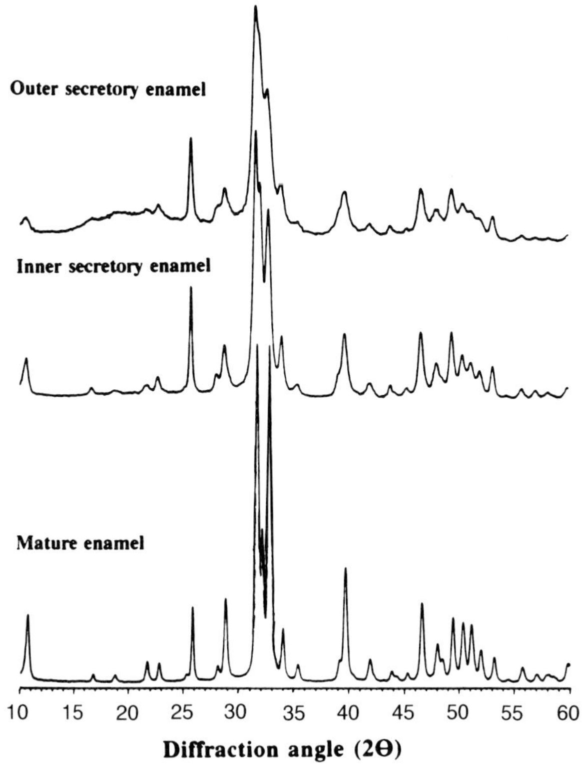

Fig. 1. X-ray diffraction patterns of porcine enamel at various developmental stages

note that the broadened pattern characteristic of the initial enamel mineral was similar to those for bone and dentin minerals and to the pattern of carbonatoapatite synthesized in aqueous solutions at 37°C [11]. In contrast, the pattern obtained for mature enamel was well resolved and comparable to those of carbonatoapatites prepared at higher temperatures [9].

Large fluoridated apatite crystals can be precipitated at 37°C in the presence of high concentrations of fluoride (1 to 10 ppm) [45]. However, it remains an enigma how mammalian amelogenesis yields large enamel crystals at the physiological temperature, with a relatively low concentration of fluoride in the enamel fluid. Presumably, one key mechanism in enamel mineralization might be the prevention of proliferation of nuclei, thus facilitating the growth of a limited number of the crystals over prolonged periods (several weeks to months). Putative regulators that may play this inhibitory role are the secreted matrix proteins. At present, it remains an intriguing and important challenge to synthesize carbonated hydroxyapatites having crystal sizes comparable to that of enamel crystals, using well-defined systems at the physiological temperature.

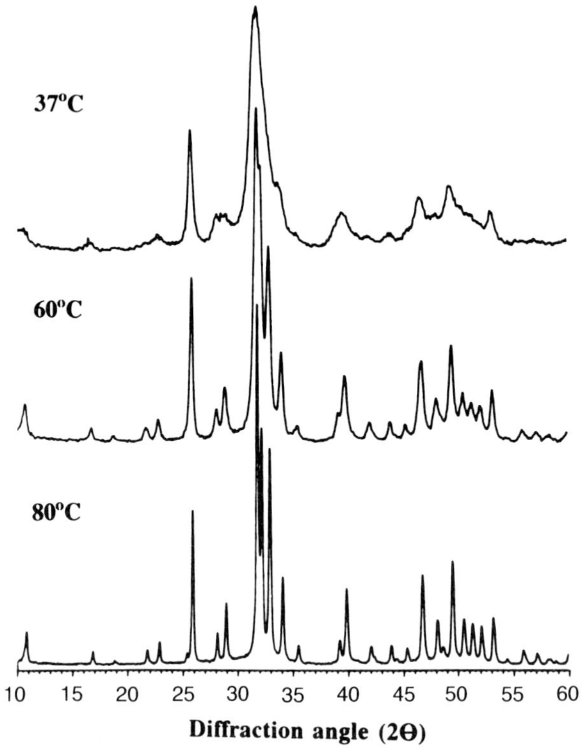

Fig. 2. X-ray diffraction patterns of carbonatoapatites precipitated at various temperatures

Figure 3 shows typical FTIR spectra obtained for the secretory and mature porcine enamel. Broad peaks at 1660, 1530, and 1450 cm^{-1} observed for the secretory enamel sample originate from the amide groups of the organic matter. These bands became obscure in the spectra obtained for the mature enamel, since little organic matter remained in the tissue. In both enamel samples, CO_3^{2-} bands were detected at 1417 and 1467 cm^{-1} corresponding to the B-type substitution. As reported previously [20], analysis of the mu_2 CO_3^{2-} band at 880–870 cm^{-1} also confirmed that the carbonate in enamel apatites formed in the early stages of porcine amelogenesis substitutes mainly for phosphate ions, while, with developmental advancement, there is an increase in the amount of CO_3^{2-} occupying OH^- sites in the crystalline lattice. Importantly, a comparison of both spectra in Fig. 3 revealed that the peak-maxima of mu_3 PO_4^{3-} bands shifted from the 1031 cm^{-1} region in the secretory enamel to the 1040 cm^{-1} in the maturing sample. This observation might be consistent with an OCP-like mineral formation at the onset of mineralization because the corresponding bands in synthetic OCP appear at 1025 and 1035 cm^{-1} [46]. In this respect, it is interesting that similar shifts in the mu_3

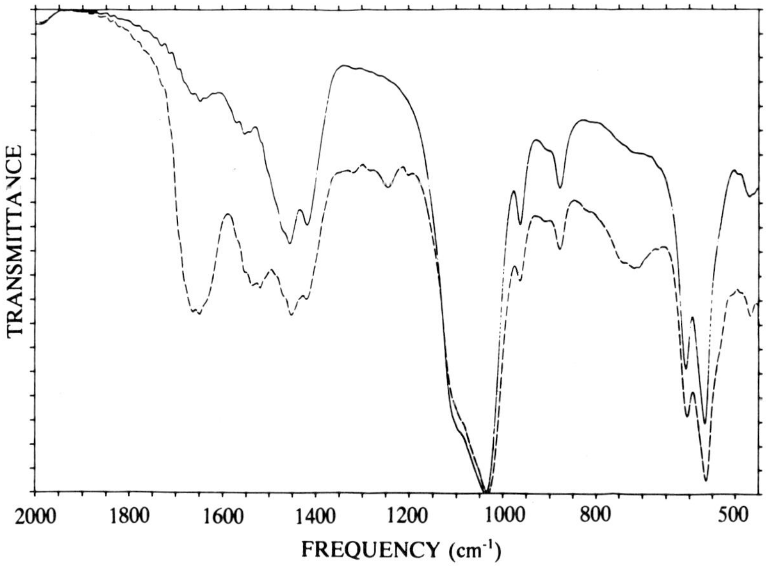

Fig. 3. FTIR spectra of secretory and mature porcine enamel. Note the shift of mu$_3$ PO_4^{3-} bands from 1031 to 1040 cm^{-1} with tissue development

PO_4^{3-} bands were observed with ageing of the mineral deposited in matrix vesicles [47] and deposits of calcifying cartilage [48].

Driving Force for Precipitation and Growth of Apatite Crystals

Previous work about crystal growth of hydroxyapatite [44, 49] demonstrated that the nature of the precipitating phase is largely dependent on the experimental conditions (e.g., degree of solution supersaturation, concentration of crystalline seeds, temperature, existing regulators, etc.). In in vivo mineralization, the degree of supersaturation (DS) is of particular importance. In the case of calcium phosphates, the DS can be expressed as the ratio of the ionic activity product (IP) of the lattice constituents in solution to the solubility product (Ksp) of the solid concerned, DS = IP/Ksp or in a form of mean activities DS = $(IP/Ksp)^{1/t}$, where t is the total sum of ions that the compound generates upon solution. In testing the dependence of the growth rate of hydroxyapatite on the DS of experimental solutions, a series of kinetic studies were conducted at 37°C with a batch of synthetic hydroxyapatite crystals (6.3 m^2/g) as seeds [50]. In those studies, the initial composition of the solution used was 1.06 mM Ca, 0.63 mM P, 50 mM NaCl with a pH value of 7.4. The DS value with respect to hydroxyapatite is 16.1. The composition ([Ca], [P] and pH) of the solution started changing immediately upon addition of the

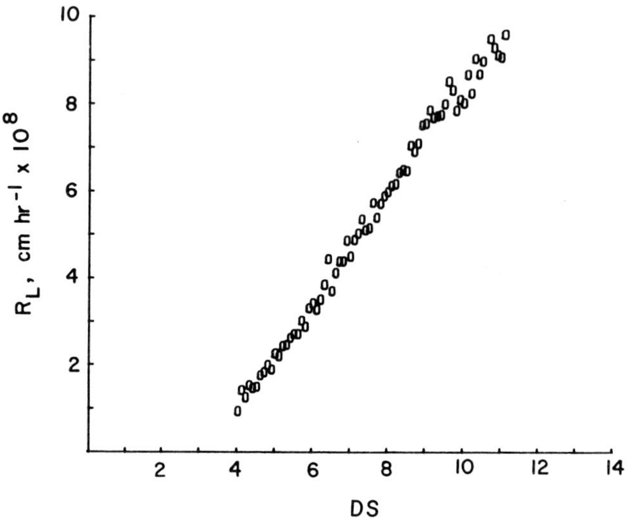

Fig. 4. Linear rate (R_L) of crystal growth of hydroxyapatite at 37°C as a function of the degree of supersaturation (*DS*) in solution with respect to hydroxyapatite

seeds (0.1 to 0.3 m^2/100 mL) at time zero and the magnitude of these changes were related to the degree of supersaturation DS. The linear growth rate, R_L, was determined from the volume of HA precipitated at any given time calculated on the basis of the phosphorus (or Ca) precipitated (using a density of 3.1 g/cm^3 for the precipitated HA); the mean linear growth up to that time is V/S where S is the surface area of the added seeds. Values of V/S were obtained for each of the experimental times. Then, a polynomial of the third degree in time was fitted to the values V/S. The linear growth rate for each experimental time was then obtained by differentiating the polynomial and using the numerical value of time in the resulting expression. In Fig. 4 are plotted the values of R_L versus their corresponding values of DS for a number of experimental systems. In this plot, the DS value was expressed by the mean activities DS = $(IP/Ksp)^{1/9}$. The results clearly showed that the growth rate of hydroxyapatite increases with increasing of the DS. The obtained relationship between the growth rates and the driving force implies a plausible mechanism for the crystal growth process [44], although further discussion in this respect is beyond the scope of this report.

In independent experiments, we also investigated the dependence of the precipitation kinetics on surface areas of the apatite seed crystals [50] and the effects of various native regulators on the growth kinetics and the nature of the precipitating crystals [51]. In Fig. 5 are shown typical results on the effects that fluoride and magnesium ions have had on crystal growth in dilute supersaturated solutions having compositions similar to that of the enamel fluid. The initial pH of the supersaturated solution was 7.3 and the initial concentrations of Ca and total P were 0.5 mM and 3.0 mM, respectively. The ionic strength

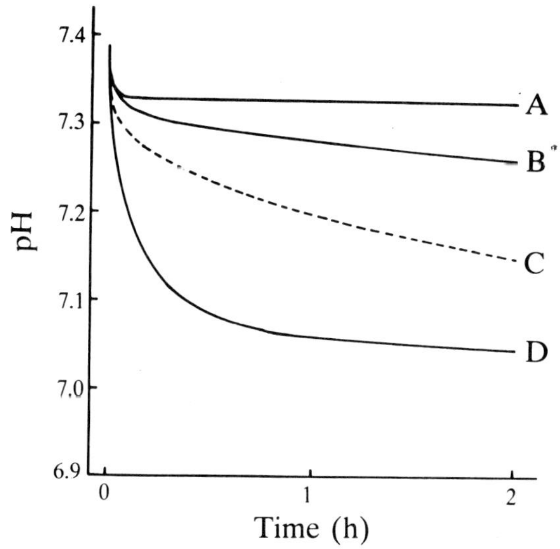

Fig. 5. Effects of fluoride and magnesium ions on the precipitation kinetics of hydroxyapatite at 37°C. Concentrations of the additives in the initial experimental solution were: *A*, [Mg] = 1 mM; *B*, [Mg] = 1 mM and [F] = 1 ppm; *C*, control ([Mg] = 0 mM and [F] = 0 ppm); *D*, [F] = 1 ppm

Table 5. Inhibition of hydroxyapatite crystal growth due to porcine amelogenins

Amelogenin MW	*Total amounts of proteins used mg*	*Amounts adsorbed onto crystals mg (mg/m^2)*	*% Inhibition of initial growth rate*
25 kd	0.33	0.26 (0.82)	22
	0.66	0.47 (1.57)	33
20 kd	0.42	0.03 (0.09)	<5
	0.64	0.04 (0.13)	5

of the solution was maintained by using 160 mM NaCl as a background electrolyte. The experimental solution was stable (no spontaneous precipitation) at 37°C over 5 h. Comparison of the plot of solution pH versus experimental time for the control run (seeding in the absence of additives) with the other curves, shows that the addition of either fluoride (1 ppm) or magnesium (1 mM) or both to the experimental solution significantly affects (acceleration by F^- and partial inhibition by Mg^{2+}) the kinetics of precipitation onto the seed crystals.

Inhibitory effects on the apatitic precipitation were also observed when the seed crystals were pretreated with the amelogenin proteins. The proteins tested were porcine 25 kd amelogenin (intact secreted protein) and its degraded product (20 kd amelogenin). The seed crystals were pretreated with specified amounts of individual proteins, as reported previously in detail [52]. Table 5 shows the effect that pretreatment of seed crystals has on the precipitation kinetics of apatite. The inhibition on the crystal growth of hydroxyapatite due to the 25 kd and 20 kd amelogenins was quantified in terms of the resulting reduction in initial precipitation rate, (Rc − Ri)/Rc × 100, where Rc and Ri were the initial precipitation rates determined using untreated and pretreated seeds with these proteins, respectively. It is apparent that (1) the 25 kd amelogenin yielded a higher inhibitory activity than the 20 kd protein, and (2) the observed inhibitory activity is related to the adsorption of the moieties onto the seed crystals. In relation to the possible inhibitory function of the intact amelogenin, it is of interest to note that the secreted 25 kd amelogenin was concentrated in the outermost layer of secretory enamel, but its amount decreased substantially in the second and underlying inner layers [32], i.e., in the older tissue. This localization of the intact protein in the forming enamel, as well as its inhibitory activity obtained in in vitro systems, may be related to the observation [32] that the crystallites existing in the outermost layer were thin plate-like in appearance but the width or thickness of the crystallites was already increased in the second layer. The foregoing results lead to the concept that enamel mineralization, in the early stages of amelogenesis, is intimately related to the secretion and degradation, i.e., enzymatic cleavages of specific segments, of amelogenins [53, 54].

Solubility of Enamel Mineral Under Controlled Pco_2

As explained in the preceding paragraphs, knowledge about the solubility product (Ksp) of the forming biomineral is essential, not only for defining the thermodynamic stability of the enamel mineral at each stage of the developmental processes, but also for assessing the driving forces involved in the precipitation of the apatite crystals. In the literature, thermodynamic solubilities of hydroxyapatite have received extensive attention. However, no definite thermodynamic approach has been applied to carbonatoapatites (better prototypes for biominerals than hydroxyapatite) or carbonated biominerals. One of the reasons for the paucity of information in the study of the thermodynamic properties of carbonated biominerals is that experimentation and interpretation of the analytical results are more complicated in the $Ca(OH)_2$-CO_2-H_3PO_4-H_2O system than those in the ternary system $Ca(OH)_2$-H_3PO_4-H_2O, or in other four-component systems without a gaseous component. In addition, the stoichiometry of the solid concerned is complicated and not well established. In some studies [55, 56], the stoichiometry of hydroxyapatite was assigned to dental enamel, and Ksp values were calculated on such a basis. However, this approach (neglecting an important constituent, carbonate) is a gross approximation of questionable validity; the obtained values of the solubility product constants are only of limited value for predicting the stability of the enamel mineral or the driving force for mineralization in physiological media.

For the last few years, our interest was focused on assessment of the solubility of porcine enamel at various developmental stages under controlled partial pressures of CO_2 (Pco_2). Powdered enamel samples (100 mg/100 mL) were equilibrated in dilute H_3PO_4 solutions (0.01 to 1.1 mM) in atmospheres of N_2-CO_2 mixtures (1.86 and 1.75% in CO_2), which bracketed the Pco_2 value (1.8%) determined in the enamel fluid [53]. Equilibrium was assumed when the solution composition (pH, concentrations of Ca, P, Mg, Na, and K) did not change for a period of 3–4 days, a procedure which took from 15 to 25 days. Possible effects of any organic components remaining in the samples particularly in the secretory enamel, on the solution composition, e.g., due to formation of metal-complexes, were ruled out because, in most cases, the value of the calculated Ca^{2+} activity was in good agreement (less than 5% discrepancy) with the value determined using specific calcium ion and reference electrodes. Since analytical work on the enamel sample disclosed that acid phosphate was also an important constituent of the crystal lattice [35], the solubility data was interpreted on the basis of the given stoichiometry $Ca_{5-x}(HPO_4)_v(CO_3)_w(PO_4)_{3-x}(OH)_{1-x}$. The stoichiometric coefficients, x, v, and w, were calculated from the analytical values of Ca/P, HPO_4/P_{total} and CO_3/Ca molar ratios [36]. On the basis of the adopted stoichiometry, the equilibrium in solutions saturated with enamel mineral can be expressed by the following equation $(Ca^{2+})^{5-x}(HPO_4^{2-})^v(CO_3^{2-})^w(PO_4^{3-})^{3-x}(OH^-)^{1-x} = K_{EN}$, in which the parentheses indicate activities of the species enclosed and K_{EN} is the solubility product constant for enamel mineral. Table 6 shows the stoichiometries of enamel samples obtained at various developmental stages and their K_{EN} values, which were calculated from the individual solution composition at equilibrium. The results indicate that (1) the outer (younger) secretory mineral was the most soluble, and (2) the solubility of enamel mineral decreased with advancing developmental stages. As reported previously [24], it is reasonable to assume that the enamel fluid is in equilibrium with the enamel mineral formed in the secretory tissue. Using the values for the stoichiometric coefficients shown in Table 1, the values of IP in the secretory enamel fluid were calculated from its composition; they are shown in the last column of Table 6. Interestingly enough, the K_{EN} and IP values exhibit an excellent agreement for the most soluble mineral (outer secretory enamel), whereas the discrepancy between the values of K_{EN} and the IP became appreciable for the samples separated at the advanced developmental stages. This fact is indirect support of the concept that the ionic composition in the enamel fluid changes at various stages of amelogenesis, probably due to changes of the cellular metabolic activities.

The Mechanism of Carbonatoapatite Formation During Amelogenesis

All the experimental data and considerations mentioned above strongly suggest that carbonate incorporation in bioapatites is directly related to the thermodynamic and kinetic aspects of mineralization processes, which, in turn, are primarily determined by the DS of the fluid environment. An important observation, made in previous solubility studies with hydroxyapatite (and enamel mineral) under controlled Pco_2 [57], was that precipitation of a car-

Table 6. Stoichiometry and solubility of enamel mineral formed at various developmental stages

Sample	*Stoichiometry*	K_{EN} [a]	*IP in fluid* [b]
Outer secretory enamel	$[Ca]_{4.16}[HPO_4]_{0.41}[CO_3]_{0.42}[PO_4]_{2.16}[OH]_{0.18}$	6.33×10^{-41}	6.47×10^{-41}
Inner secretory enamel	$[Ca]_{4.29}[HPO_4]_{0.31}[CO_3]_{0.40}[PO_4]_{2.29}[OH]_{0.29}$	1.30×10^{-43}	4.63×10^{-43}
Maturing chalk-like enamel	$[Ca]_{4.45}[HPO_4]_{0.23}[CO_3]_{0.32}[PO_4]_{2.45}[OH]_{0.45}$	1.34×10^{-46}	3.31×10^{-46}
Mature hard enamel	$[Ca]_{4.63}[HPO_4]_{0.11}[CO_3]_{0.25}[PO_4]_{2.63}[OH]_{0.65}$	1.33×10^{-49}	8.08×10^{-49}

[a] K_{EN}, Solubility product constant determined at 25°C, based on the solution composition at equilibrium
[b] *IP in fluid*, Ionic activity product calculated with the analytical values of lattice constituents in the enamel fluid shown in Table 1

bonatoapatite occurred concomitantly with the dissolution of the original solid in acidic media, meaning that carbonatoapatite is the stable mineral phase in media having certain Pco_2. This observation, as well as the ubiquitous presence of carbonate ions in extracellular fluids shown in Table 1, provides an explanation for the fact that carbonation of apatite crystals is a common, not an exceptional, phenomenon in hard tissues. However, it is not yet clear how the magnitude and location of carbonate substitution in the crystal lattice are determined in the process of biomineralization.

In order to gain an insight into the mechanism determining carbonation of biominerals, we conducted a series of carbonatoapatite preparations at various temperatures and with solutions having wide ranges of composition [9]. The characterization of the products by FTIR, X-ray diffraction, and electron microscopy, lead to the conclusion that the magnitude and sites of carbonate substitution and the growth and morphology of the formed crystals are highly dependent on both the degree of supersaturation (with respect to calcium phosphates) and the presence of "regulators" (i.e., CO_3^{2-} and F^-) in solution. Specifically, low concentrations of carbonate in solution favored the A-type substitution in the precipitated crystals [9, 58]. With increasing carbonate content, the fraction of carbonate in the OH^- site decreased substantially, whereas the carbonate in the PO_4^{3-} sites became prominent. In the presence of high concentrations of fluoride, acicular apatite crystals were precipitated initially, growing into rod-like crystals having symmetric-hexagonal cross sections and the carbonate substitution was almost exclusively of the B-type. Based on the foregoing results, we hypothesized that the selective occupation of hydroxyl sites by carbonate at the late stage of enamel formation is due to a competition for two sites, one of which (A-type) is favored energetically but is present in a limited number in the system, whereas the second site (B-type) is present in a larger number. With lower availability of carbonate, the energetic considerations prevail but, at high concentrations, the probability of occupancy in each substitution site is essentially determined by their respective number of sites. The above consideration supports the connotation that the preferential B-type substitution found in enamel crystals at early developmental stages could indicate the formation of initial biomineral in the presence of relatively high carbonate and that the subsequent apatitic growth, incorporating more carbonate in the OH^- sites, occurs in media having lower Pco_2. Similar energetic considerations may apply to the occupancy of the OH^- sites by carbonate, hydroxyl, and fluoride ions. If a true competition exists among the ions, the fluoride that introduces the least lattice disturbance [59] should be most favored. Thus, it is not surprising that, even when the carbonate concentration in solution is several times higher than the fluoride concentration, the precipitated carbonatoapatite would have all its A-sites occupied by fluoride. This is most likely the case seen for the francollite formation in some fish enameloids [28, 60].

An alternative mechanism for the incorporation of carbonate in enamel crystals, involving the initial precipitation of OCP and its hydrolysis, has also been proposed [61, 62]. Chickerur et al. [63] demonstrated that the B-type substitution of carbonate, which is the only observable substitution in the youngest (outer secretory) enamel [20], takes place in the apatite crystal lattice when OCP crystals are hydrolyzed in the presence of Na^+ and CO_3^{2-} at 37°C. As reported previously [64], our computer simulation on the basis of enamel fluid composition suggested that the fluid may become supersaturated with respect to OCP with an increase in the concentration of total Ca of more than 1 mM. This situation is likely to occur in the vicinity of the ameloblasts, when the cells transport Ca ions into the adjacent extracellular microspace. Most of the observations made in the outer secretory enamel, i.e., selective B-type carbonate substitution, thin ribbons in crystal morphology (characteristic for OCP [65]), high content of acid phosphate, and shifts of PO_4^{3-} bands in FTIR spectra, are consistent with the initial precipitation of an OCP-like phase. On the other hand, X-ray or electron diffraction work on the youngest enamel failed to discern the presence of OCP reflections [20, 66]. High resolution electron microscopy demonstrated evidence for the vestige of OCP domains in the lattice of synthetic carbonatoapatites [67, 68], while no corresponding evidence has yet been attained for enamel crystallites. Altogether, there is still a lack of conclusive evidence for precipitation of OCP during enamel mineralization.

Future Prospects

The overall results support the contention that the nature of the biomineral(s) and its precipitation kinetics are determined primarily by the degree of supersaturation in the microenvironment surrounding the precipitating crystals, although the presence of inhibitors and promoters may substantially modify the situation. In order to gain a more complete picture about the regulatory mechanism of carbonatoapatite formation in biological systems, the experimental approaches and models (e.g., solubility model) mimic more closely the in vivo situation and the actual nature of the mineral. Consideration of carbonate and acid phosphate as constituents of the biomineral is certainly better than assigning as its stoichiometry that of hydroxyapatite, $(Ca)_5(PO_4)_3(OH)$. However, it should be noted that, as shown in Table 1, the enamel fluid (and other mineralizing fluids) contains significant concentrations of Mg^{2+} and Na^+. It was reported [69, 70] that parts of the magnesium and sodium in the enamel tissue are located in the crystal lattice. Currently, little information exists on what is the extent of incorporation of these ions into the crystal lattice or if the main portion is surface bound or, as suggested by others [71, 72], present in non-apatitic phases. Similarly, Young and Spooner [73] reported that the OH^- content of human enamel apatite is at least 20%–30% below the theoretical content for stoichiometric hydroxyapatite. However, no quantitative information about the OH^- contents in enamel crystals at various developing stages is available. A better understanding of the stability of enamel mineral and the driving force for the precipitation and growth of enamel crystals will be attained through further studies taking into account most of these putative lattice constituents. A model for enamel mineral of the type $Ca_{5-x}(Mg)_q(Na)_u(HPO_4)_v(CO_3)_w(PO_4)_{3-y}(OH,F)_{1-z}$ may better reflect the chemical nature of enamel mineral and may lead to a better understanding of its properties.

Acknowledgements. The authors are grateful to Drs. S. Shimoda and Y. Miake and Ms. C. Creighton for their valuable technical help. This work was supported by Grants DE08670, DE07623, DE07009, DE03187 from the National Institute of Dental Research and a grant from the Colgate-Palmolive Company.

References

1. Bonel G, Montel G (1964) Sur une nouvelle apatite carbonatee synthetique. C R Acad Sc Paris 258:923–926
2. Elliott JC (1964) The crystallographic structure of dental enamel and related apatites. Ph.D. Thesis, University of London
3. LeGeros RZ, Trautz OR, Klein E, LeGeros JP (1969) Two types of carbonate substitution in the apatite structure. Experientia 24:5–7
4. McClellan GH, Lehr JR (1969) Crystal-chemical investigations of natural apatites. Amer Min 54:1374–1391
5. Rey C, Collins B, Goehl T, Dickson RI, Glimcher MJ (1989) The carbonate environment in bone mineral: A resolution-enhanced Fourier transformed infrared spectroscopy study. Calcif Tissue Int 45:157–164
6. Rey C, Shimizu M, Collins B, Glimcher MJ (1990) Resolution-enhanced Fourier transformed infrared spectroscopy study of the environment of phosphate ions in the early deposits of a solid phase of calcium-phosphate in bone and enamel, and their evolution with age. Investigations in the mu_4 PO_4 domain. Calcif Tissue Int 46:384–394
7. Bachra BN, Fischer HRA (1969) The effect of some inhibitors on the nucleation and crystal growth of apatite. Calcif Tissue Res 3:348–357
8. Greenfield DJ, Termine JD, Eanes ED (1974) A chemical study of apatites prepared by hydrolysis of amorphous calcium phosphate in carbonate containing aqueous solutions. Calcif Tissue Res 14:131–138
9. Shimoda S, Aoba T, Moreno EC, Miake Y (1990) Effect of solution composition on morphological and structural features of carbonated calcium apatites. J Dent Res 69:1731–1740
10. LeGeros RZ, Trautz OR, LeGeros JP, Klein E, Shirra WP (1967) Apatite crystallites: Effects of carbonate on morphology. Science 155:1409–1411
11. Aoba T (1977) X-ray diffraction study on the amorphous and crystalline components in bone mineral (in Japanese). Jpn J Oral Biol 20:616–627
12. Nelson DGA, Featherstone JDB, Duncan JF, Cutress TW (1982) Paracrystalline disorder of biological and synthetic carbonate-substituted apatites. J Dent Res 61:1274–1281
13. Hallsworth AS, Weatherell JA, Robinson C (1973) Loss of carbonate during the first stages of enamel caries. Caries Res 7:345–348
14. LeGeros RZ, Tung MS (1983) Chemical stability of carbonate- and fluoride-containing apatites. Caries Res 17:419–429

15. Mayer I, Voegel JC, Bres EF, Frank RM (1988) The release of carbonate during the dissolution of synthetic apatite and dental enamel. J Crystal Growth 87:129–136
16. Brudevold F, Soremark R (1967) Chemistry of the mineral phase of enamel. In: Structural and chemical organization of teeth, vol. 1. Miles AEW (ed) Academic New York, pp 247–277
17. Weatherell JA, Robinson C, Hiller CR (1968) Distribution of carbonate in thin sections of dental enamel. Caries Res 2:1–9
18. Elliott JC, Holcomb DW, Young RA (1985) Infrared determination of the degree of substitution of hydroxyl by carbonate ions in human dental enamel. Calcif Tissue Int 37:372–375
19. Hiller CR, Robinson C, Weatherell JA (1975) Variations in the composition of developing rat incisor enamel. Calcif Tissue Res 18:1–12
20. Aoba T, Moreno EC (1990) Changes in the nature and composition of enamel mineral during porcine amelogenesis. Calcif Tissue Int 47:356–364
21. Howell DS, Pita JC, Marquez JF, Madruga JE (1968) Partition of calcium, phosphate, and protein in the fluid phase aspirated at calcifying sites in epiphyseal cartilage. J Clin Invest 47:1121–1132
22. Wuthier RE (1977) Electrolytes of isolated epiphyseal chondrocytes, matrix vesicles, and extracellular fluid. Calcif Tissue Res 23:125–133
23. Neuman WF (1969) The *milieu interieur* of bone: Claude Bernard revisited. Fed Proc 28:1846–1850
24. Aoba T, Moreno EC (1987) The enamel fluid in the early secretory stage of porcine amelogenesis: Chemical composition and saturation with respect to enamel mineral. Calcif Tissue Int 41:86–94
25. Nylen MU, Eanes ED, Omnell KA (1963) Crystal growth in rat enamel. J Cell Biol 18:109–123
26. Kerebel B, Daculsi G, Kerebel LM (1979) Ultrastructural studies of enamel crystallites. J Dent Res 58(B):844–850
27. Daculsi G, Menanteau J, Kerebel LM, Mitre D (1984) Length and shape of enamel crystals. Calcif Tissue Int 36:550–555
28. Miake Y, Aoba T, Moreno EC, Shimoda S, Prostak K, Suga S (1990) Ultrastructural studies on crystal growth of enameloid minerals in elasmobranch and teleost fishes. Calcif Tissue Int 48:204–217
29. Robinson C, Fuchs P, Deutsch D, Weatherell JA (1978) Four chemically distinct stages in developing enamel from bovine incisor teeth. Caries Res 12:1–11
30. Warshawsky H (1984) Ultrastructural studies on amelogenesis. In: The chemistry and biology of mineralized tissues. Butler WT (ed) Birmingham, Ala, pp 33–45
31. Suga S (1983) Comparative histology of progressive mineralization pattern of developing enamel. In: Suga S (ed) Mechanisms of tooth enamel formation. Quintessence, Tokyo, pp 167–203
32. Aoba T, Tanabe T, Moreno EC (1987) Function of amelogenins in porcine enamel mineralization during the secretory stage of amelogenesis. Adv Dent Res 1:252–260
33. Fejerskov O (1979) Human dentition and experimental animals. J Dent Res 58(B):725–731
34. Aoba T, Collins J, Moreno EC (1989) Possible function of matrix proteins in fluoride incorporation into enamel mineral during porcine amelogenesis. J Dent Res 68:1162–1168
35. Shimoda S, Aoba T, Mereno EC (1991) Changes in acid phosphate content in enamel mineral during porcine amelogenesis. J Dent Res 70:1516–1523

36. Aoba T, Moreno EC (1992) Changes in the solubility of enamel mineral at various stages of porcine amelogenesis. Calcif Tissue Int 50:266–272
37. Weatherell JA, Deutsch D, Robinson C, Hallsworth AS (1975) Fluoride concentrations in developing enamel. Nature 256:230–232
38. Speirs RL (1975) Fluoride incorporation into developing enamel of permanent teeth in the domestic pig. Arch oral Biol 20:877–883.
39. Crenshaw MA and Bawden JW (1981) Fluoride-binding by organic matrix from early and late developing bovine fetal enamel determined by lowrate dialysis. Arch Oral Biol 26:473–476
40. Lussi A, Fridell RA, Crenshaw MA, Bawden JW (1988) Absence of in vitro fluoride-binding by the organic matrix of developing bovine enamel. Arch Oral Biol 33:531–533
41. Shimizu M, Fukae M (1983) Enamel proteins. In: Suga S (ed) Mechanisms of tooth enamel formation. Quintessence, Tokyo, pp 125–141
42. Aoba T, Tanabe T, Moreno EC (1987) Proteins in the enamel fluid of immature porcine teeth. J Dent Res 66:1721–1726
43. Aoba T, Moreno EC, Tanabe T, Fukae M (1990) Effects of fluoride on matrix proteins and their properties in rat secretory enamel. J Dent Res 69:1248–1255
44. Varughese K, Moreno EC (1981) Crystal growth of calcium apatites in dilute solutions containing fluoride. Calcif Tissue Int 33:431–439
45. Phakey PP, Palamara JP, Conn A, Rachinger WA, Orams HJ (1989) Artifical remineralization of tooth enamel surfaces. In: Fearnhead RW (ed) Tooth enamel V. Florence, Yokohama, pp 504–508
46. Fowler BO, Moreno EC, Brown WE (1966) Infra-red spectra of hydroxyapatite, octacalcium phosphate and pyrolysed octacalcium phosphate. Arch Oral Biol 11:477–492
47. Sauer GR, Wuthier RE (1988) Fourier transform infrared characterization of mineral phases formed during induction of mineralization by collagenase-released matrix vesicles in vitro. J Biol Chem 263:13718–13718
48. Rey C, Beshah K, Griffin R, Glimcher MJ (1991) Structural studies of the mineral phase of calcifying cartilage. J Bone Min Res 5:515–525
49. Nancollas GH, Tomazic B (1974) Growth of calcium phosphate on hydroxyapatite crystals. Effect of supersaturation and ionic medium. J Phys Chem 78:2218–2225
50. Aoba T, Moreno EC (1984) Preparation of hydroxyapatite crystals and their behavior as seeds for crystal growth. J Dent Res 63:874–880
51. Aoba T, Yoshioka C, Yagi T (1984) Seeded crystal growth of calcium apatite and inhibitory activity of small organic molecules. In: Fearnhead RW, Suga S (eds) Tooth enamel IV. Elsevier, Amsterdam, pp 3–7
52. Aoba T, Fukae M, Tanabe T, Shimizu M, Moreno EC (1987) Selective adsorption of porcine amelogenins onto hydroxyapatite and their inhibitory activity on seeded crystal growth of hydroxyapatite. Calcif Tissue Int 41:281–289
53. Aoba T, Moreno EC (1989) Mechanism of amelogenic mineralisation in minipig secretory enamel. In: Fearnheard RW (ed) Tooth enamel V. Florence, Yokohama, pp 163–168
54. Aoba T, Moreno EC (1990) Structural relationship of amelogenin proteins to their regulatory function of enamel mineralization. In: Sikes CS, Wheeler AP (eds) Surface reactive peptides and polymers. American Chemical Society, pp 85–106
55. Patel PR, Brown WE (1975) Thermodynamic solubility product of human tooth enamel: Powdered sample. J Dent Res 54:728–736

56. Larsen MJ, Jensen SJ (1989) The hydroxyapatite solubility product of human dental enamel as a function of pH in the range 4.6–7.6 at 20°C. Arch Oral Biol 34:957–961
57. Moreno EC, Aoba T (1991) Comparative solubility of human dental enamel, dentin, and hydroxyapatite. Calcif Tissue Int 49:6–13
58. Vignoles M, Bonel G, Holcomb DW, Young RA (1988) Influence of preparation conditions on the composition of type B carbonated hydroxyapatite and on the localization of the carbonate ions. Calcif Tissue Int 43:33–40
59. Kay MI, Young RA, Posner AS (1964) Crystal structure of hydroxyapatite. Nature 204:1050–1052
60. LeGeros RZ, Suga S (1980) Crystallographic nature of fluoride in enameloids of fish. Calcif Tissue Int 32:169–174
61. Brown WE, Mathew M, Tung MS (1981) Crystalchemistry of octacalcium phosphate. Prog Cryst Growth Charact 4:59–87
62. Brown WE, Chow LC, Siew C, Gruninger S (1984) Acidic calcium phosphate precursors in formation of enamel mineral. In: Fearnhead RW, Suga S (eds) Tooth enamel IV. Elsevier, Amsterdam, pp 8–13
63. Chickerur NS, Tung MS, Brown WE (1980) A mechanism for incorporation of carbonate into apatite. Calcif Tissue Int 32:55–62
64. Moreno EC, Aoba T (1987) Calcium binding in enamel fluid and driving force for enamel mineralization in the secretory stage of amelogenesis. Adv Dent Res 1:245–251
65. Terpstra RA, Bennema P (1987) Crystal morphology of octacalcium phosphate: Theory and observation. J Crystal Growth 82:416–426
66. Landis WJ, Burke G, Neuringer JR, Paine MC, Nanci A, Bai P, Warshawsky H (1988) Earliest enamel deposits of the rat incisor examined by electron microscopy, electron diffraction, and electron probe microanalysis. Anat Rec 220:233–238
67. Nelson DGA, Wood GJ, Barry JC, Featherstone JDB (1986) The structure of (100) defects in carbonated apatite crystallites: A high resolution microscope study. Ultramicroscopy 19:253–266
68. Nelson DGA, Barry JC, Shields CP, Glena R, Featherstone JDB (1989) Crystal morphology, composition, and dissolution behavior of carbonated apatites prepared at controlled pH and temperature. J Colloid Interface Sci 130:467–479
69. LeGeros RZ (1984) Incorporation of magnesium in synthetic and in biological apatites. In: Fearnhead RW, Suga S (eds) Tooth enamel IV. Elsevier, Amsterdam, pp 32–36
70. Driessens FCM, Heijligers HJM, Borggreven JMPM, Woltgens JHM (1984) Variations in the mineral composition of human enamel on the level of cross-striations and striae of Retzius. Caries Res 18:237–241
71. Casciani FS, Etz ES, Newbury DE, Doty SB (1979) Raman microprobe studies of two mineralizing tissues: Enamel of the rat incisor and the embryonic chick tibia. Scan Electron Microsc II:383–390
72. Driessens FCM, Verbeeck RMH (1985) Dolomite as a possible magnesium-containing phase in human tooth enamel. Calcif Tissue Int 37:376–380
73. Young RA, Spooner S (1969) Neutron diffraction studies of human tooth enamel. Arch Oral Biol 15:47–63

Specific Physicochemical Properties of Fluoridated Hydroxyapatites with Incorporated Trace Elements

MASAYUKI OKAZAKI[1]

Summary. The crystallinity of fluoridated CO_3-apatites showed similar patterns of variation in the degree of fluoridation irrespective of carbonate content, although, on the whole, it was lower than that of fluoridated hydroxyapatites. In Mg-containing fluoridated hydroxyapatites, the patterns of crystallinity with the increase of fluoride content were unique in each series with different magnesium contents. In contrast to these unexpected crystallinity phenomena, the apparent solubility of fluoridated CO_3-apatites and Mg-containing fluoridated hydroxyapatites decreased monotonically with the increasing degree of fluoridation, and approached that of fluoridated hydroxyapatites at a high fluoride content. Fe-containing fluoridated hydroxyapatites were more poorly crystallized than Fe-free fluoridated hydroxyapatites. Despite this inhibited crystallinity behavior, the apparent solubility decreased more than that of Fe-free fluoridated hydroxyapatites at a low fluoride content.

Key words: Fluoridated apatites—CO_3^{2-}—Mg^{2+}—Fe^{2+}

Introduction

Previous investigators [1–3] reported that, from the physicochemical standpoint, incorporation of fluoride into the crystalline lattice of apatitic tooth mineral brings about greater stability. In general, crystallinity reflects the crystal's physicochemical strength, and, therefore, an increasing fluoride content may increase crystallinity, i.e., crystal strength. Additionally, the relationship of the crystallinity of fluoridated apatites to their solubility might be closely related to the susceptibility of caries to fluoridated enamel apatites. However, as caries-preventive agents, fluoride ions often act unpredictably.

F^- ions can be almost stoichiometrically taken up into the apatite crystals. Unless it is controlled, the environment of apatite formation in vivo and in vitro is unlikely to have a stoichiometric (Ca/P) value 1.67 of the hydroxyapatite.

[1] Osaka University Faculty of Dentistry, Suita, Osaka, 565 Japan

Hydroxyapatites prepared over a wide range (0.1–6.67) of calcium-phosphate feed molar ratios, $(Ca/P)_f$, are not strongly affected by these ratios [4]. However, F^- uptake of fluoridated apatites synthesized at low $(Ca/P)_f$, i.e., below 1.67, is strongly inhibited [5]. Furthermore, the composition of the mineral phase is not exactly the same as that of the standard hydroxyapatite. There are various other constituents present. In the presence of trace elements such as CO_3^{2-}, Mg^{2+}, Fe^{2+}, etc., the formation of fluoridated hydroxyapatites may become more complex.

Computer Graphics of Fluoridated Hydroxyapatites

Information on the conformation of hydroxyapatite in relation to its crystallinity, solubility, and/or mechanical strength is useful for understanding the mechanism and prevention of dental caries. After the expenditure of a great deal of time and effort, a ball-and-stick model of hydroxyapatite using plastic spheres has been successfully devised. Recently, computer graphics have made remarkable progress: while a large computer was needed a few years ago to draw a molecular model, this can now be easily achieved with only a personal computer and a CRT [6]. The benefit of computer graphics is in making it possible to view a model from any direction, and to instantly change its elements and their sizes and positions. Therefore, hydroxyapatite substituted by a trace element can be viewed immediately. The structure of hydroxyapatite has been analyzed by using X-ray diffraction and neutron diffraction analysis [7].

Figure 1.A shows a computer graphic of fluoridated hydroxyapatite along the *c*-axis, in which F^- ions are partially substituted into OH^- positions. Two columns of Ca^{2+} ions are visible in the unit cell. If computer graphics were drawn simply using Pauling's ionic radius [8], P^{5+}(0.33Å) and H^+(≑0Å) ions would be lost. Therefore, each ionic radius was modified as follows: O^{2-}(1.40Å → 1.20Å), P^{5+}(0.33Å → 0.70Å), H^+(≑0Å → 0.50Å), F^-(1.36Å → 1.10Å). The ionic radius of Ca^{2+}(0.99Å) was not changed. By rotating this view through −90° along the X-axis of the CRT, Fig. 1.B was obtained.

The crystallographic structure of hydroxyapatite had been difficult to analyze, however, because of the shifted position of OH^- ions, which is located about 0.3Å above or below the stable position of the crystal [1]. Now the instability of the structure can be easily understood using the computer graphics of fluoridated hydroxyapatite, in which F^- ion is located just on the mirror plane at $Z = 1/4$. This computer graphics display demonstrates that fluoride ions serve to stabilize the hydroxyapatite crystal and prevent dental caries. The computer graphics are helpful in understanding the physicochemical properties of fluoridated apatites with trace elements.

CO_3-Containing Fluoridated Hydroxyapatites

A few investigators have touched briefly upon the subject of fluoridated CO_3-apatites. In most studies, however, both the synthetic and tooth apatites examined were carbonate-free fluoridated hydroxyapatite. It is well known that

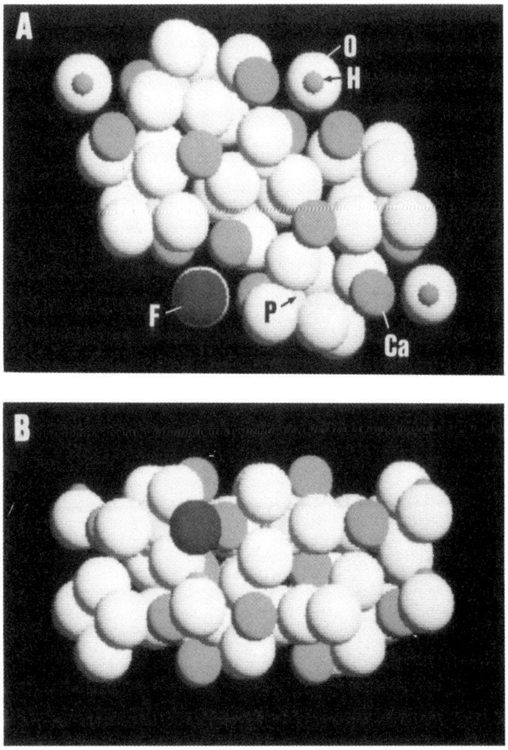

Fig. 1. Computer graphics of a fluoridated hydroxyapatite structure **A** projected on the X-Y plane and **B** rotated through −90° around the X-axis

the inorganic phases of tooth enamel are mainly carbonate-containing apatites and that carbonate in the apatites may weaken the chemical bonds in the crystals [9]. Thus, it is important to examine the interaction between the caries-preventive action of fluoride and the caries-promoting action of carbonate in apatite crystals.

We have reported that crystallinity of synthetic fluoridated hydroxyapatites showed no simple increase with the degree of fluoridation, while solubility showed a simple decrease [10]. We suspect that the fluoride content may also have a strong effect of the physicochemical properties of carbonate-containing apatites.

Fluoride contents of the fluoridated CO_3-apatites, FCO_3Ap_I and FCO_3Ap_{II}, when synthesized with carbonate-phosphate feed molar ratios, $(CO_3/P)_f = 0.5$ and 2.0, increased with the fluoride concentration in the solution [11]. Fluoride could be taken up almost stoichiometrically. Neither calcium nor phosphate contents changed significantly with changes in fluoride content. Notably, carbonate content was not affected by fluoride content. The X-ray diffraction patterns of all samples were characteristic of calcium apatites. Samples showed a lower degree of crystallization with increase of the carbonate content. No extraneous peaks due to CaF_2 were found, indicating that the fluoride content of samples did not exceed 2 mmol/g, which is equal to that of francolite (carbonate fluorapatite).

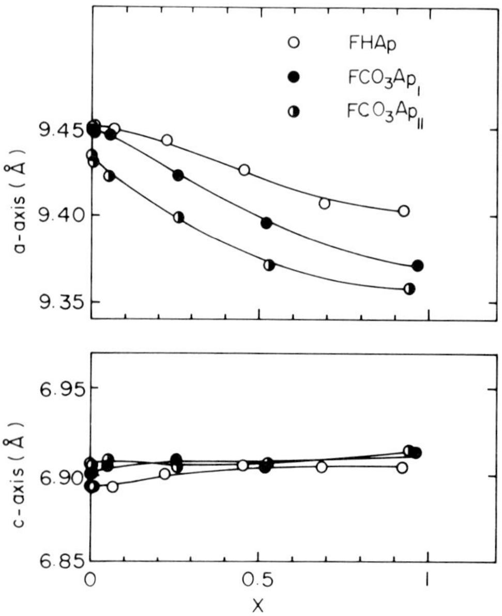

Fig. 2. *A*- and *c*-axis dimensions of fluoridated CO_3apatites as a function of the degree of fluoridation, X

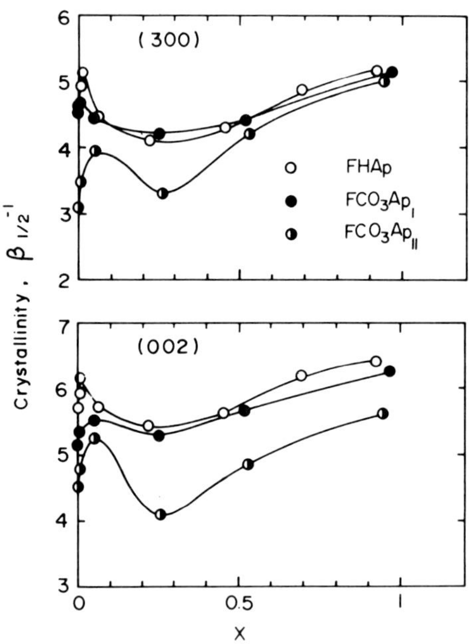

Fig. 3. Crystallinity of fluoridated CO_3-apatites shown as the inverse of the half-value breadths of the (300) and (002) reflections

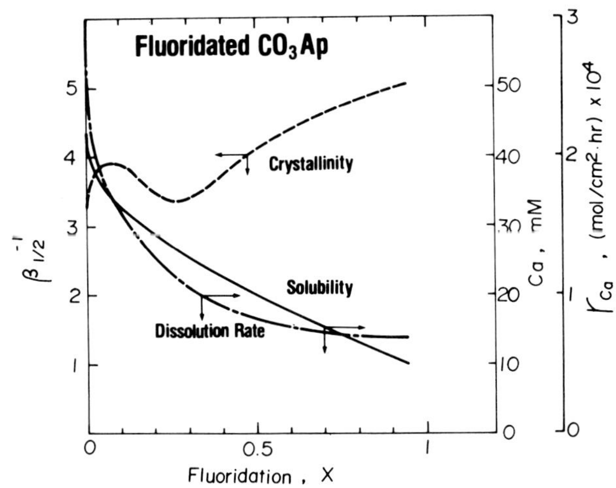

Fig. 4. Crystallinity, solubility, and dissolution rate behaviors of fluoridated CO_3-apatites

Figure 2 shows the *a*- and *c*-axis dimensions of each sample as a function of fluoride content. With an increase in the degree of fluoridation, X, the *a*-axis dimensions decreased, whereas the *c*-axis dimensions changed very little. This result agrees with the observations on fluoridated HAp by Moreno et al. [3]. These and the results of chemical analysis indicate that fluoridated CO_3-apatites were produced. In addition, the *a*-axis dimensions of samples with high carbonate content, FCO_3Ap_{II}, decreased more than those with low carbonate content, FCO_3Ap_I.

Figure 3 shows the crystallinity of fluoridated CO_3-apatites as the inverse of the half-value breadths of the (300) and (002) reflections. The crystallinity of each fluoridated CO_3-apatites initially increased slightly, then decreased with the degree of fluoridation, and finally increased greatly when the fluoride content approached that of francolite, analogous to that of the fluoridated hydroxyapatites, FHAp. It should be noted that the crystallinity of each fluoridated CO_3-apatite decreased at lower fluoride contents.

We were able to observe an interesting phenomenon in the magnified infrared absorption spectra in the region of $1000–800\,cm^{-1}$ [12]. The peak due to CO_3^{2-} ions at $875\,cm^{-1}$ shifted to a lower frequency with the degree of fluoridation. This means that the conformation of CO_3^{2-} ions in the apatite crystal may be greatly affected by the fluoride content, even though the carbonate content was not affected by the degree of fluoridation.

On the other hand, the solubility and dissolution rate of fluoridated CO_3-apatites decreased dramatically at low fluoride content levels and continued to gradually decrease monotonically. Thus, no analog to solubility behavior was found in the complex phenomena of fluctuating crystallinity (Fig. 4). The disorder and/or random distribution of F^- ions in the crystals may possibly weaken the crystallinity and/or inhibit the crystal growth at relatively low fluoride content levels.

Mg-Containing Fluoridated Hydroxyapatites

Recently, clinical attention is being focused on magnesium because of its close relation to the metabolism and physiology of the human body [13]. Bone and teeth also contain certain amounts of magnesium, which may play an important role in the initial formation of apatites and have a significant effect on their physicochemical properties. For example, magnesium may control the rate of apatite formation in the early stages of mineralization in a matrix vesicle (MV) [14]. Furthermore, it can be speculated that magnesium may associated with fluoride ions which promote the apatite formation.

The chemical analysis of Mg-containing hydroxyapatites indicated that with an increasing magnesium-calcium feed molar ratio, $(Mg/Ca)_f$, the calcium content decreased and the magnesium content increased [15]. However, the phosphate content showed no significant change below $(Mg/Ca)_f = 0.5$. The $(Ca + Mg)/P$ molar ratio of the sample of $(Mg/Ca)_f = 1$ was close to that of tricalcium phosphate, 1.5, while the other samples showed apatitic values close to the apatitic figure of 1.67 [15].

Mg-containing hydroxyapatites became poorly crystallized at higher ratios and showed an amorphous structure at $(Mg/Ca)_f = 1$. However, none of the samples gave a tricalcium phosphate pattern, even though the chemical composition of the sample of $(Mg/Ca)_f = 1$ had a similar value 1.45 of $(Ca + Mg)/P$ to tricalcium phosphate 1.5.

Although the change in *a*-axis dimensions was less distinct because of the rapid appearance of amorphous patterns, *c*-axis dimensions decreased with the increase of magnesium content. These and the results of chemical analysis suggest that Mg^{2+} ions may substitute for Ca^{2+} ions. We can then suggest that Mg^{2+} ions can substitute for calcium positions to a limited extent, and that a small amount of magnesium may decrease the crystallinity of hydroxyapatite and make it soluble.

Both the calcium and phosphate contents of Mg-containing fluoridated hydroxyapatite synthesized with the magnesium-calcium feed molar ratio, $(Mg/Ca)_f = 0.01$ (Mg-FHAp$_I$) and 0.05 (Mg-FHAp$_{II}$) [16], did not vary significantly with an increasing fluoride content (X: the degree of fluoridation). Magnesium content, on the other hand, increased with an increase in the degree of fluoridation, X (Fig. 5), indicating that magnesium uptake may be promoted by fluoride. This result agrees with the observation of McCann and Bullock [17].

The *a*-axis dimensions of both Mg-FHAp$_I$ and Mg-FHAp$_{II}$ decreased with an increased degree of fluoridation, in the same way as that of Mg-free fluoridated hydroxyapatites, FHAp, suggesting a substitution of F^- ions into the apatite crystals. The *a*- and *c*-axis dimensions of both Mg-FHAp$_I$ and Mg-FHAp$_{II}$ were lower than those of FHAp. The *c*-axis dimension, in particular, decreased with the increase of magnesium content over the whole range of fluoride content, suggesting a substitution of magnesium ions into the apatite crystals.

Figure 6 shows the crystallinity of the 2 series of Mg-containing fluoridated apatites as the inverse of the half-value breadths calculated from the (300) and

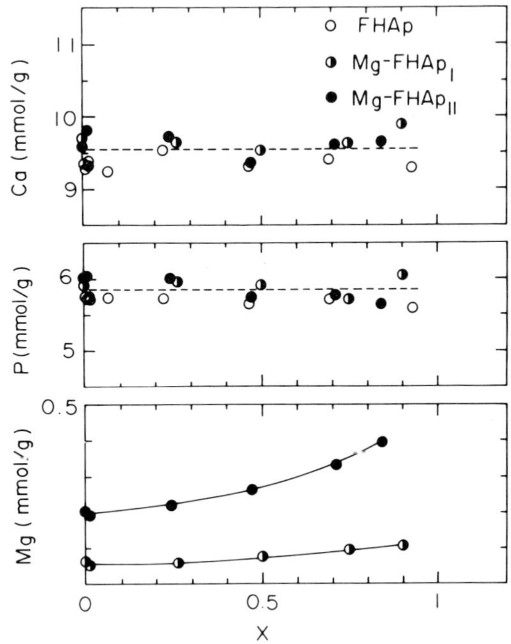

Fig. 5. Chemical compositions of Mg-containing fluoridated hydroxyapatites as a function of the degree of fluoridation, X

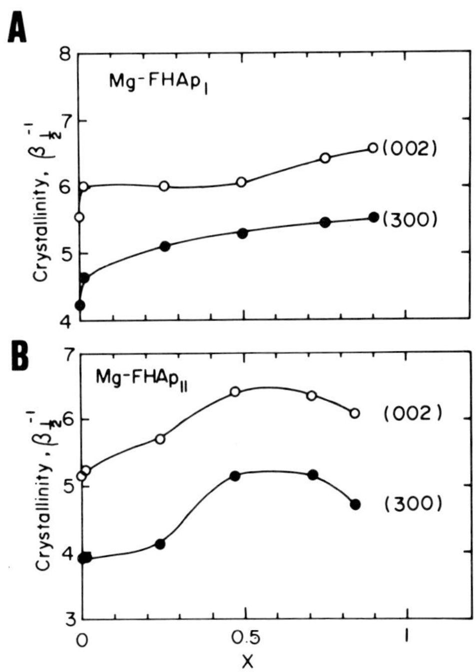

Fig. 6. Crystallinity of Mg-containing fluoridated hydroxyapatites, **A** Mg-FHAp$_I$ and **B** Mg-FHAp$_{II}$, shown as the inverse of the half-value breadths of the (300) and (002) reflections. (See Fig. 3 for Mg-free fluoridated hydroxyapatites, FHAp)

(002) reflections, which are taken as representations of the crystallinity along the *a*- and *c*-axis. In fluoridated hydroxyapatites, FHAp, the crystallinity increased initially, then decreased, and finally increased (Fig. 3). In the series with lower Mg contents, Mg-$FHAp_I$, the minimum observed at low fluoride content of Mg-free fluoridated hydroxyapatites was obliterated (Fig. 6.A); and, in the series with higher Mg-content, Mg-$FHAp_{II}$, crystallinity rose steadily to a maximum at about one-half of the maximum fluoride content (Fig. 6.B), although crystallinity of the F-free Mg-containing hydroxyapatites decreased monotonically with the increasing Mg content [15].

The apparent solubility of Mg-containing fluoridated hydroxyapatites $FHAp_I$ and $FHAp_{II}$ decreased monotonically with the increase of fluoride content in the same way as that of Mg-free fluoridated hydroxyapatites, FHAp. These 3 apatite series did not show any significant differences when approaching a high fluoride content. In the solubility behaviors, the action of magnesium is behind the strong stability action of fluoride.

Fe-Containing Fluoridated Hydroxyapatites

Iron, an indispensable element for human life, forming the core of hemoglobin molecules, Hb(Fe^{2+}), in red blood cells [13] for example, is also a trace element in tooth minerals. In East Asia, the tooth black "ohaguro," that was traditionally applied to teeth, contained iron, which could have contributed to caries prevention [18]. The effects on caries-preventive action of cations substituting Ca^{2+} sites, and of fluoride substituting OH^- sites have been studied [19]. Some fish species have a large quantity of iron and fluoride in their teeth [20]. Thus, it is important to clarify the efficiency of fluoride in combination with iron in caries prevention.

Hydroxyapatites synthesized by using $FeCl_2 \cdot nH_2O$ rapidly became poorly crystallized with the increase of ferrous ions in the solution [21] (Fig. 7). In addition to apatite, at a higher concentration of ferrous ions, a different phosphate was formed, (Fig. 7.C, D). This was identified as $Ca_3Fe_4(PO_4)_4(OH)_6 \cdot 3H_2O$, in which iron was oxidized to the ferric state. Its X-ray patterns clearly differed from those of the iron sources, $FeCl_2 \cdot nH_2O$ and $FeCl_3 \cdot 6H_2O$.

Ferrous (Fe^{2+}) ions seemed to be taken up by apatite crystals. The contraction of *a*- and *c*-axis dimensions of Fe-containing apatites with the increase of iron content is suggested, although the differences found are too small to be conclusive. However, the peak shifts of the expanded (300) and (002) reflections shown in Fig. 7.E support this suggestion.

The X-ray diffraction patterns of Fe-containing fluoridated apatites synthesized with the iron-calcium feed molar ratio $(Fe^{2+}/Ca)_f = 0.07$ [22] were similar to those of calcium apatites, but the degree of crystallization was greatly reduced in comparison with that of Fe-free fluoridated hydroxyapatites as previously reported [10, 11]. Calcium contents of samples were lower than those of Fe-free fluoridated hydroxyapatites, but the total molar contents of calcium and iron were similar to those of Fe-free fluoridated hydroxyapatites.

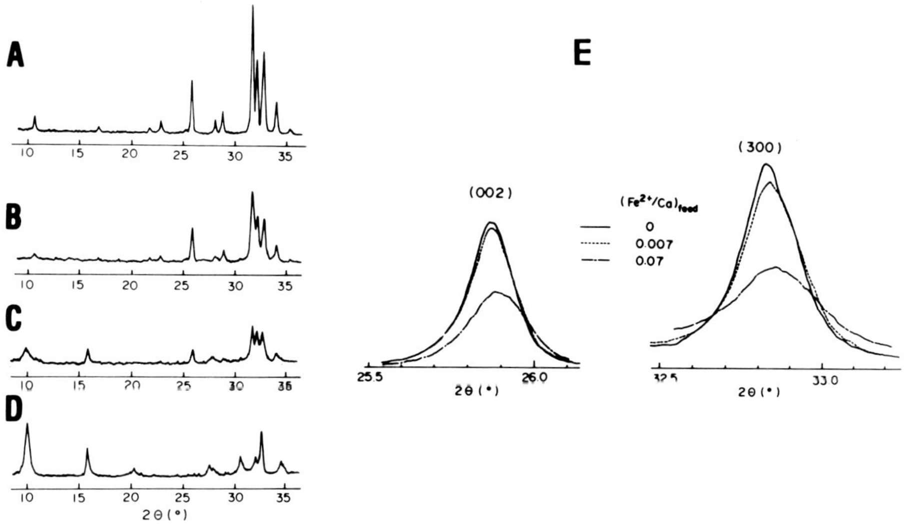

Fig. 7. X-ray diffraction patterns of Fe-containing hydroxyapatites (*A*–*D*) and their expanded (300) and (002) reflections (*E*). $(Fe^{2+}/Ca)_f$ = 0 (**A**); 0.07 (**B**); 0.37 (**C**); 0.73 (**D**)

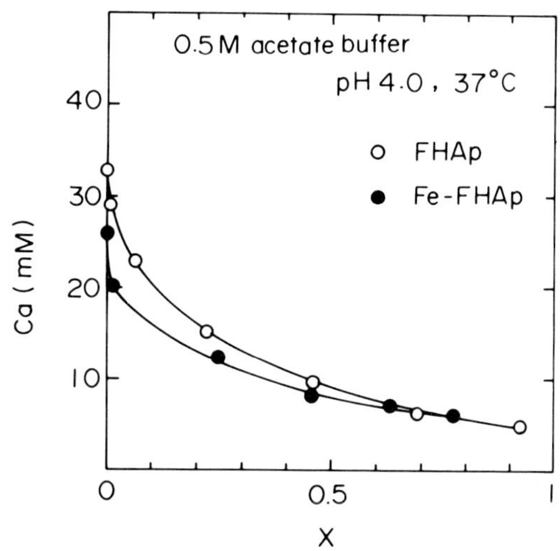

Fig. 8. Apparent solubility of Fe-containing fluoridated apatites shown as the dissolved calcium concentration in the solution

Calcium, iron, and phosphate contents were almost independent of the degree of fluoridation, X.

A-axis dimensions of Fe-containing fluoridated apatites decreased with the degree of fluoridation, in the same way as those of fluoridated hydroxyapatites [10, 11, 16]. The contraction was enhanced in both *a*- and *c*-axis dimensions, suggesting that Fe^{2+} ions may substitute into the fluoridated apatite crystals.

Figure 8 shows the apparent solubility of Fe-containing fluoridated apatites and those of Fe-free fluoridated apatites previously synthesized [10, 11]. At first, the solubility decreased greatly and then gradually with the degree of fluoridation, whereas the crystallinity showed a complex behavior. The solubility of F-free Fe-containing hydroxyapatite was significantly lower than that of F- and Fe-free hydroxyapatite in spite of the decrease in crystallinity. Such decrease in calcium concentration by the dissolution-preventive action of iron is clear at very low fluoride concentrations, even if the decrease in the relative calcium content of apatites that accompanies the substitution on Fe^{2+} ions is considered. However, at a high fluoride content, both curves seem to converge, and the additional effect of Fe^{2+} ions seems to disappear.

A number of investigators have studied the combined effects of various compounds with fluoride. Often, however, the efficiency of these compounds is masked by the dramatic caries-preventive action of fluoride. In this study, the effectiveness of iron in caries prevention has been clearly demonstrated. It is interesting that this dissolution-preventing action of iron was enhanced at very low concentrations of fluoride, irrespective of crystallinity.

References

1. Kay MI, Young RA, Posner AS (1964) Crystal structure of hydroxyapatite. Nature 204:1050–1052
2. Miles AEW (1967) Structural and chemical organization of teeth, vol. II. Academic, New York
3. Moreno EC, Kresak M, Zahradnik RT (1977) Physicochemical aspects of fluoride-apatite systems relevant to the study of dental caries. Caries Res 11 (Suppl 1): 142–171
4. Okazaki M, Aoba T, Doi Y, Takahashi J, Moriwaki Y (1979) The effect of the calcium-phosphate feed molar ratios on apatite formation. J Osaka Univ Dent Sch 19:87–93
5. Okazaki M, Takahashi J, Kimura H (1985) F^- uptake inhibition by excess during fluoridated apatite formation. Caries Res 19:342–347
6. Okazaki M, Sato M (1990) Computer graphics of hydroxyapatite and β-tricalcium phosphate. Biomaterials 11:573–578
7. Sudarsanan K, Young RA (1965) Significant precision in crystal structure details: Holly springs hydroxyapatite. Acta Cryst B25:1534–1543
8. Pauling L (1960) The nature of the chemical bond, 3rd edn. Cornell University Press, Ithaca
9. LeGeros RZ (1967) Apatite crystallites: Effects of carbonate on morphology. Science 155:1409–1411
10. Okazaki M, Aoba T, Doi Y, Takahashi J, Moriwaki Y (1981) Solubility and crystallinity in relation to fluoride content of fluoridated hydroxyapatites. J Dent Res 60:845–849
11. Okazaki M, Takahashi J, Kimura H, Aoba T (1982) Crystallinity, solubility, and dissolution rate behavior of fluoridated CO_3apatites. J Biomed Mater Res 16:851–860

12. Okazaki M (1983) F^{-}-CO_3^{2-} interaction in IR spectra of fluoridated CO_3-apatites. Calcif Tissue Int 35:78–81
13. Stryer L (1981) Biochemistry. W.H. Freeman, San Francisco
14. Anderson HC (1967) Electron microscopic studies of induced cartilage development and calcification. J Cell Biol 23:81–101
15. Okazaki M, Takahashi J, Kimura H (1986) Unstable behavior of magnesium-containing hydroxyapatites. Caries Res 20:324–331
16. Okazaki M (1987) Mg^{2+}-F^{-} interaction during hydroxyapatite formation. Magnesium 6:296–301
17. McCann HG, Bullock FA (1957) The effect of fluoride ingestion on the composition and solubility of mineralized tissues of the rat. J Dent Res 36:391–398
18. Yamaga R, Nakayama M (1975) Ohaguro (tooth black) and its prophylactic effects. Kokusai Shika J 1:401–414
19. Francis MD, Gray JA, Griebstein WJ (1968) The formation and influence of surface phases on calcium phosphate solid. Arch Oral Biol 3:83–120
20. Suga S (1984) The role of fluoride and iron in mineralization of fish enameloid. In: Fearnhead RW, Suga S (eds) Tooth enamel IV. Elsevier, Amsterdam, pp 472–477
21. Okazaki M, Takahashi J, Kimura H (1985) Iron uptake of hydroxyapatite. J Osaka Univ Dent Sch 25:17–24
22. Okazaki M, Takahashi J, Kimura H (1986) Crystallinity and solubility behavior of iron-containing fluoridated hydroxyapatites. J Biomed Mater Res 20:879–886

The Effect of Fluoride on Enamel De- and Remineralization

Jacob M. (Bob) ten Cate[1]

Summary. Fermentation of carbohydrates in the dental plaque leads to the formation of acids, causing the pH to drop and enamel to demineralize. In between such acid attacks, the condition of supersaturation of the oral fluids to hydroxyapatite results in a precipitation of calcium phosphate mineral at the sites depleted of mineral. Recent observations have shown that the processes of demineralization and remineralization are both affected significantly by fluoride, even at very low concentrations, in the oral fluids. Such fluoride levels may be achieved by topical fluoride applications, drinking water fluoridation, or the use of fluoridated dentifrices. The rationale for the inhibition of demineralization and the enhancement of remineralization is thermodynamic: when fluoride is present, mineral deposition, in the form of a fluoridated hydroxyapatite, is enhanced due to its lower solubility product. Inhibition of demineralization is the result of the reprecipitation of a fluoride-rich mineral during enamel dissolution. Moreover, the phase thus formed is less susceptible to future acid attacks.

This paper was presented during a symposium held at the Nippon Dental University, Tokyo, Japan, October 13, 1990.

Key words: Enamel—Demineralization—Remineralization—Fluoride

Introduction

For many years, caries was considered to be an inevitable disease. Once caries had begun to develop at a particular site in the dentition, it progressed rapidly, forming a cavity which needed dental treatment in the form of a restoration. However, in the 1960s evidence was presented that caries need not necessarily progress to deep cavities but, instead, that a caries defect may also be remineralized or be arrested by the influx of salivary constituents. Koulourides et al. [1] immersed enamel lesions in saliva in a laboratory experiment and showed that the uptake of calcium and phosphate from saliva results in a rehardening of the enamel lesions. This process was described as remineralization.

[1] Department of Cariology and Endodontology, Academic Centre for Dentistry, Amsterdam (ACTA), The Netherlands

At the same time, in 1966, Backer Dirks [2] presented data from a clinical study which indicated that remineralization has an impact on the rate at which caries develops in the mouth as well. In brief, these findings revealed that about 50% of the white spot lesions detected on the buccal surfaces of the first molars of 8-year-old children disappeared during the ensuing (7) years due to the remineralizing capacity of saliva, particularly in geographical areas where the water had been fluoridated.

Mechanistic Studies of Enamel Remineralization

With the clinical findings of enamel remineralization, the possible implications of this phenomenon in caries prevention were rapidly understood: if saliva was able to counteract caries attacks in a natural way, this process could possibly lead to an easy and effective method of caries inhibition. Therefore, many studies were initiated with the purpose of understanding the mechanisms of enamel remineralization, with the ultimate aim of using this knowledge to develop caries preventive treatments and regimens. Some of these studies will be discussed here, and the conclusions derived from these studies will be given.

In the 1950s and 1960s, it had become possible to mimic the process of forming white spot caries lesions in the laboratory. Several model systems had been developed to create enamel lesions with the histological characteristics of caries as it develops in the oral cavity. These included systems comprised of gelatin [3] and other gels [4] to simulate the dental plaque, and of organic acids which were similar to the acid formed in dental plaque during the fermentation of carbohydrates. Thus, different systems were available for forming white spot caries lesions in the laboratory. This obviously had the advantage that the lesions could be produced in large numbers and were easily reproducible, a necessary condition when these lesions had to be remineralized.

In studies on the mechanism of enamel lesion remineralization, artificially produced white spot lesions were generally used as the substrate material. A typical set-up for such studies involved the use of lesions and solutions mimicking saliva in its inorganic composition. Of course the organic components were considered equally important but, in the initial stages, emphasis was given to the mineral components of the saliva system. Thus, the solutions used resembled saliva in its degree of supersaturation. Nevertheless, they differed from saliva in the calcium-to-phosphate ratio. Where saliva typically is rich in phosphate and relatively low in calcium, the solutions used for remineralization studies had a calcium-to-phosphate ratio stoichiometric to hydroxyapatite (i.e., Ca/P = 1.66) [5, 6]. When solutions with a calcium content of 2.0 mMol/L, a phosphate content of 1.2 mMol/L, and a pH of 7 were used to remineralize lesions, a deposition of mineral occurred, as concluded from the depletion of ions from the remineralizing solutions [5–7] (Fig. 1).

Judging from the calcium-to-phosphorous ratio of the depositing material, this mineral was either apatite or a similar calcium phosphate mineral. Additional studies were conducted to look into the effect of adding fluoride to saliva-

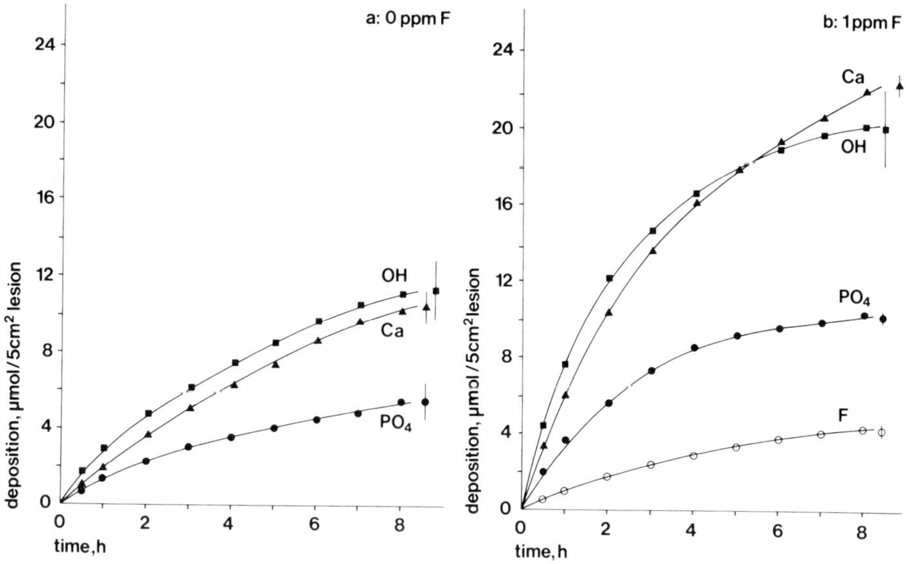

Fig. 1. Amounts of *Ca*, *P*, *OH*, and *F* deposited during 8-h immersion of 5 cm^2 of enamel lesion in 12.5 ml solution, initially containing 2.0 mMol/L $CaCl_2$, 1.2 mMol/L KH_2PO_4, and 150 mMol/L KCl at pH = 7.0 and a constant fluoride concentration of either **a** 0 (no added fluoride) or **b** 1 ppm. (With permission from [7])

like solutions, and it was found that even very low fluoride concentrations in solution would significantly speed up the apatite deposition [7–9]. The physico-chemical rationale for this was that, in the presence of fluoride, not only hydroxyapatite but also a fluoridated hydroxyapatite can form, the latter having a lower solubility and thus a higher thermodynamic driving force for precipitation.

Apart from such physico-chemical descriptive and analytical chemical studies, the lesions remineralised in vitro were analyzed using histological, histomorphometric, and tracer techniques. For instance, using the scanning electron microscope, it was found that artificial lesions were depleted of mineral, as could be seen from a thinning of crystallites in the most demineralized part of the lesions (the "lesion body"). When such lesions were subjected to remineralizing solution, it was found that, in particular, the thickness of the crystallites increased. During remineralization, new crystals were also formed in the specimens [10].

These findings indicate that not only did remineralisation occur primarily by the growth of existing crystals but also that new crystals could be formed by secondary nucleation onto existing crystals or other remnants which had remained in the lesion after demineralization.

The finding that lesion remineralization resulted in an increase of the crystallinity and of the crystallites' sizes was later confirmed in a study by Silverstone [11]. He found that, in particular, the crystallites in the surface layer and in the

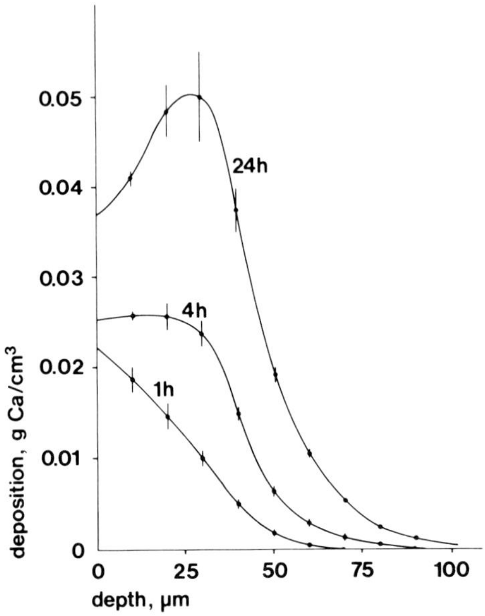

Fig. 2. Location of deposition within a lesion after 1, 4, and 24 h immersion in solutions initially containing 1.5 mMol/L $CaCl_2$, 0.9 mMol/L KH_2PO_4, and 150 mMol/L KCl at pH = 7.0. (With permission from [12])

dark zone (the region just beyond the lesion body) were increased in thickness, demonstrating that these were the regions where mineral deposition occurred during the remineralization process. Independent of these electron microscopic studies, experiments were performed with radioactive (labelled) calcium in the remineralizing solution in order to identify the sites in the lesion where deposition occurred. These studies, in which thin layers were removed by grinding off specimens which were then analyzed for radioactivity, revealed that, after an initial lag period, the radioactive deposition concentrated in the lesion body [12] (Fig. 2).

The general conclusions drawn from the numerous studies on mechanistic aspects are that remineralization is a process that can be simulated in the laboratory, that remineralization from solutions with a supersaturation similar to saliva occurs in the form of the deposition of an apatite-like mineral, and that this precipitation results in a growth of the partly demineralized crystals inside the lesion as well as the formation of some new crystals. This latter finding serves to explain why remineralized lesions never become as densely mineralized as the original, sound enamel.

This phenomenon, which could be called "scar tissue formation", is particularly evident when fluoride is provided. The rationale is that the accelerated deposition of a fluoride containing mineral occurs less specifically inside the lesion and may even occur in the narrow pores of the surface layer. This could result in lesions which are completely remineralized superficially, leaving the underlying tissue still demineralized [10, 13] (Fig. 3). This phenomenon has also been observed clinically. Backer Dirks' early observation [2] revealed that, in areas with drinking water fluoridation, many of the white spot lesions had

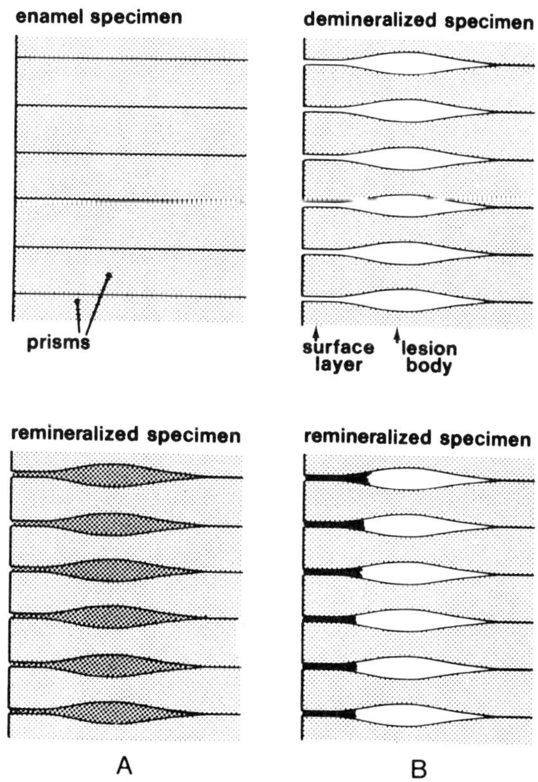

Fig. 3. **A** Schematic representation of the formation and (partial) remineralization of a caries lesion. When viewed in cross section, lesions form by the dissolution of mineral from the interprismatic regions, preferentially from sites underneath the anatomical surface. Remineralization may occur by a redeposition of mineral at the sites attacked. **B** Alternatively, the pores in the surface layer of the lesions may be filled up by mineral, causing a surface layer blocking and resulting in arrested (consolidated) lesions. (Adapted from [13])

a glossy appearance which was attributed to surface remineralization [14]. Thus, although the lesions still appeared white, it was concluded that such arrested lesions would not progress into cavities.

Another general finding of these studies was that very low fluoride concentrations in solution were able to significantly alter the rate of remineralization, and this again is of significant practical importance. When fluoride-containing dentifrices are used, this results in an elevation of the fluoride levels in the oral fluids. Obviously, the fluoride levels are increased substantially immediately after brushing, but even long after the actual brushing, a two-fold increase in fluoride baseline levels is still observed.

A similar situation exists when fluoridated drinking water is used. The physiological fluoride levels are also increased by about two-fold in those cases. The observation that remineralization is enhanced in areas with drinking water fluoridation [2, 14] serves to demonstrate that fluoride dentifrices, which yield a similar increase in fluoride levels in the oral fluids, improve the rate at which remineralization may occur. Therefore, the conclusion is justified that, for the prevention of dental caries, it is important to give remineralization an optimal chance. This is achieved by using fluoride dentifrices to increase fluoride levels in the oral fluid.

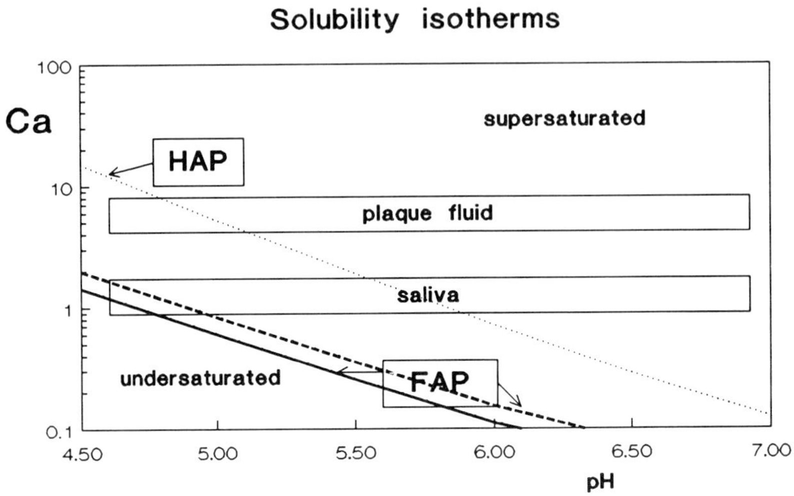

Fig. 4. Solubility isotherms for hydroxyapatite (*HAP*) and fluorapatite (*FAP*) at fluoride concentrations in solution of 1 and 10 ppm, respectively. For FAP, the *full line* represents the saturation calcium concentrations at 10 ppm F/L and the *dashed line* is the corresponding calcium levels at 1 ppm F/L. Calcium levels in saliva and plaque fluid, indicative of the degree of supersaturation with respect to the two minerals, are represented schematically

The Effect of Fluoride on Enamel Demineralization

Having established the effect of low fluoride levels in the ambient fluids on mineral deposition, a study was initiated on the effect of sub ppm levels of fluoride on enamel demineralization. Thermodynamically, the rationale for there being an effect of fluoride on enamel demineralization is very similar to its effects on remineralization. If enamel is attacked by weak acids in the presence of fluoride, a situation could arise where hydroxyapatite will dissolve but fluoridated apatite could still precipitate (Fig. 4). This would mean that while enamel dissolves, the calcium and phosphate ions could reprecipitate with the fluoride supplied from solution [15, 16]. Studies aimed to determine the effects of low fluoride levels on enamel demineralization have indicated that fluoride levels in the range of 0.05–5 ppm were very effective in inhibiting demineralization [17–19]. It was concluded that the fluoride concentration is as important as the pH in determining the rate at which a lesion develops, to such an extent that lesion formation may even be completely inhibited. The fluoride concentration at which inhibition occurs is dependent not only on the pH of the plaque, but also on the calcium and phosphate levels of the plaque. Together, they determine the driving force for the precipitation of fluoridated apatite at the sites under caries attack. Combined laboratory and clinical studies indicate that an effect of ambient fluoride on enamel demineralization is feasible at fluoride concentrations prevailing in the oral fluids.

In addition to the noted effects of fluoride on the rate of lesion formation, it was also observed that fluoride had an effect on the histology of the forming lesions. When fluoride was added to solution, lesions would form with a very thick surface layer, whereas in fluoride-depleted or fluoride-free solutions, a caries defect would form with an eroded aspect. Taking this observation one step further, it could be argued that many of the white spot lesions observed on extracted teeth are lesions which had already been remineralized to a significant extent, because such lesions would typically have quite pronounced surface layers which were rich in fluoride. Thus, the question of whether a surface layer is a phenomenon associated with caries lesion formation or whether it is the result of a remineralization process is yet unsolved, and both hypotheses still stand at present.

pH-Cycling

While the mechanistic studies on enamel demineralization and enamel remineralization were conducted, techniques were also developed which enabled the monitoring of the pH fluctuations in dental plaque in the course of time, in particular after eating fermentable carbohydrates. From the results it was concluded that caries never occurs as a continuous process, but rather during short periods of acid attack, lasting about 30 to 45 min. During major parts of the day, the dentition is bathing in saliva of near neutral pH from which mineral may be redeposited. This continuous fluctuation of the pH during the day initiated a new type of research which is now commonly referred to as "pH-cycling" [20]. In such experiments, the enamel is subjected to conditions of de- and remineralization mimicking the conditions of super- and subsaturation occurring in the mouth during the day. During the past decade, this method has been used in many laboratories [21–23] and has been advanced in different ways to such an extent that actual "Stephan curves" (giving the change in pH after acid production in the plaque) can be simulated in the laboratory [24]. These studies are of value for answering fundamental questions regarding the mineral uptake and loss by either enamel lesions or by sound enamel under different physiological conditions, as well as for studying the effects of caries preventive agents necessary to formulate optimal treatments.

Using the former approach, it was shown that low fluoride levels, present during the demineralization phase or during remineralization, were effective in shifting the net balance of demineralization and remineralization to the positive site (i.e., mineral accretion). Typically, using low fluoride levels during pH-cycling, the initial effect would be on the inhibition of demineralization during the caries attack cycles. Gradually, the fluoride also affects the mineral uptake as the inhibition of caries breakdown limits the number of sites available for regrowth of material. In other words, the amount of turnover of mineral is decreased after the addition of fluoride (Fig. 5).

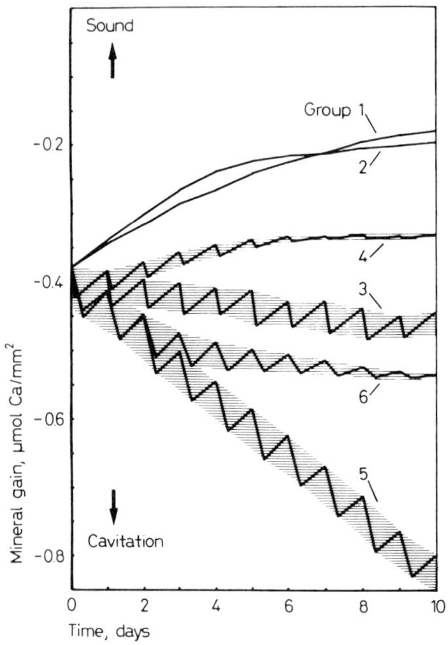

Fig. 5. Cumulative plot of calcium loss and gain for different cycling schemes with or within fluoride added at 2 ppm concentration to the remineralization cycles. In groups 1 and 2 the specimens were subjected to remineralization only. In groups 3 and 4 there was a daily 3-hour demineralization cycle and a 21-hour remineralization cycle. In groups 5 and 6, these time periods comprised 6 and 18 h, respectively. Before the start of the experiment, incipient caries lesions had been formed in the specimen, as can be deduced from the value of $-0.376\,\mu mol\,Ca/mm^2$ mineral content at $t = 0$. The figures shows that the cycling scheme used and the presence of fluoride are parameters determining whether lesion progression or arrest occur. (With permission from [20])

Intraoral Studies

Even though the recently developed laboratory methods take into account many of the intraoral parameters, the caries mechanisms can never be completely and definitively unraveled from in vitro studies. Therefore, intraoral models were developed to study the caries formation in the mouth and to study enamel remineralization. Such studies involve the use of volunteers who usually wear dentures onto which enamel specimens are attached. This can be done in several ways [25]. One application is the use of partial dentures in which enamel specimens are placed in the buccal flange in the molar region. Specimens can be pretreated in the laboratory, e.g., to create incipient caries lesions, before they are placed in the mouth. Also, the specimens may be covered with gauze which would accumulate plaque formation [26].

The aims to conduct in situ studies have been threefold. First, intraoral models have been used to study the initiation and progression of caries condit-

Fig. 6. Results obtained for a group of 18 participants in an intraoral study to assess the degree of lesion repair when a placebo paste, a fluoride paste, and a fluoride + pyrophosphate (anti-tartar) paste were compared. The percentage of lesion repair (positive values) is given. In some individuals, lesions were found to further demineralize, as indicated by the negative repair values. Panellist initials are given at the x-axis, as are the group averages

placebo paste

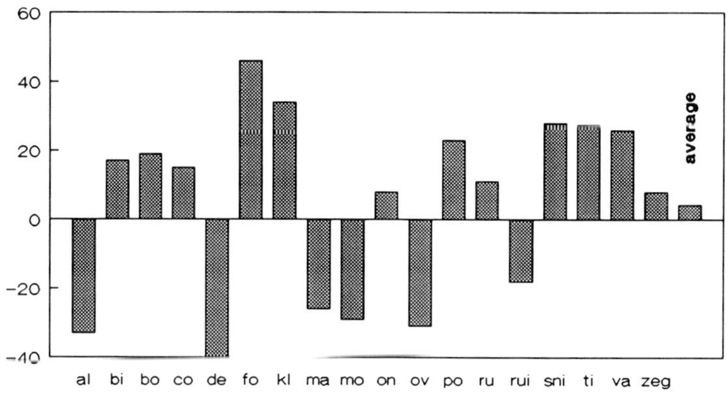

fluoride paste

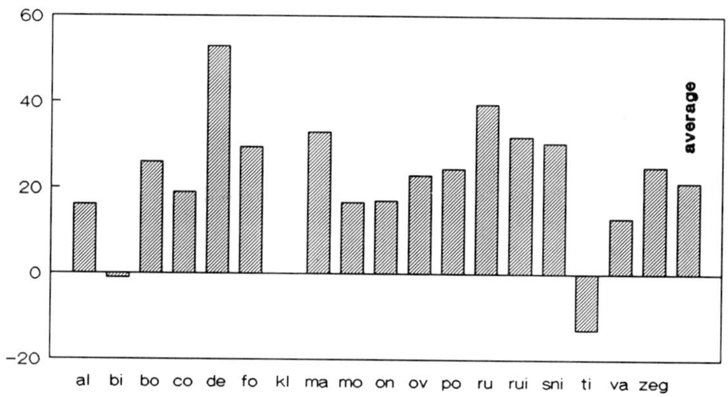

fluoride + pyrophosphate paste

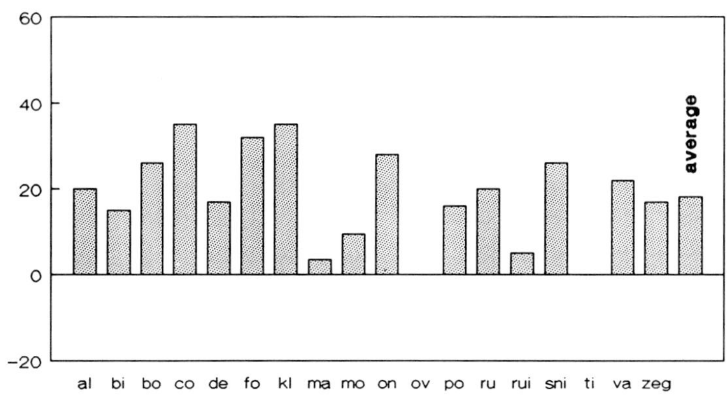

ions involving extraoral sucrose rinses with or without a caries preventive agent as additive. Second, studies have been conducted to determine the amount of fluoride uptake as a result of a given preventive scheme. Third, intraoral studies have been aimed at determining the remineralization of preformed lesions, in particular also studying the effect of caries preventive agents in their normal usage on this process. Unlike in the laboratory where conditions can be monitored continuously and can be made optimal for enamel remineralization in the mouth, the pH and the calcium and phosphate levels of the fluids in which the specimens are bathing vary continuously with time within each individual as well as between individuals. Therefore, it is not surprising that either caries progression or reversal may be found among individuals with the same experimental protocol. The same is even reported for individuals in subsequent trials [27]. In the mouth, a complicating factor is the presence of many organic constituents in saliva and plaque fluid which affect the process of mineral dissolution and precipitation. In saliva, many inhibitors have been identified which interact with the formation of calcium phosphates by lowering the driving force for precipitation due to binding with calcium. Other components bind strongly to mineral surfaces, resulting in inhibition of crystal growth. Therefore, in addition to the rates of mineral deposition and dissolution being determined by the mineral components of saliva and plaque fluid, an overriding effect may be the content and composition of saliva in terms of specifically binding proteins. In spite of their limitations, intraoral models give an insight into the magnitude and rate at which caries progression and lesion remineralization occur in individuals, and consequently these data enable researchers to fine-tune their laboratory models. A typical study of this type concerns a comparison of a fluoride and a placebo toothpaste with an anti-tartar paste. The aim of this study was whether the addition of anti-tartar ingredients resulted in an impaired anti-caries efficacy of the fluoride product. Eighteen panellists were asked to have enamel specimens with incipient caries lesions placed in their partial dentures during each of 3 7-week experiments. In the first leg of the study, a placebo paste was supplied to all participants. Then, in the next two legs of the study, the people were randomly assigned to one of the experimental groups which brushed their dentures with either a fluoride dentifrice or a fluoride anti-tartar dentifrice. At the end of each 7-week period, the specimens were removed and subjected to microradiographic analysis to determine the mineral content patterns. A comparison of the lesions before and after the in situ period indicated whether the experimental conditions had led to an enhancement of the caries defect or to a repair of the lesion. The result of this study is depicted in Fig. 6. From the data a number of conclusions can be drawn: irrespective of the type of dentifrice used there was a large inter-individual difference in intraoral response. For a placebo paste further demineralization was observed in some volunteers while remineralization occurred in others. With both types of fluoride dentifrices (with or without pyrophosphate as anti-tartar active) lesion repair was observed in all but one of the specimens, illustrating the fluoride enhancing effect on enamel remineralization.

In conclusion, intraoral studies show that the process of remineralization may contribute significantly to a slowing down, arrest, or even reversal of caries progression. Fluoride causes a shift in the demineralization-remineralization equilibrium in the mouth. Such effects can be observed even at the very low fluoride levels in the oral fluids such as those produced by fluoride supplementation schemes used in caries prevention.

References

1. Koulourides T, Feagin F, Pigman W (1965) Remineralisation of dental enamel by saliva in vitro. Ann NY Acad Sci 131:751–757
2. Backer Dirks O (1966) Posteruptive changes in dental enamel. J Dent Res 45: 503–511
3. Silverstone LM (1973) Structure of carious enamel, including the early lesion. In: Melcker AH, Zarb GA (eds) Dental enamel, development, structure and caries. Munksgaard, Copenhagen, pp 100–160
4. Gray JA (1966) Kinetics of enamel dissolution during formation of incipient caries-like lesions. Arch Oral Biol 11:397–422
5. Feagin FF (1971) Calcium, phosphate and fluoride depositions on enamel surfaces. Calcif Tissue Res 8:154–164
6. Feagin FF, Gonzalez M, Jeansonne BG (1972) Kinetic reactions of calcium phosphate and fluoride ions at the enamel surface-solution interface. Calcif Tissue Res 10:113–127
7. Cate JM ten, Arends J (1977) Remineralisation of artificial enamel lesions in vitro. Caries Res 11:277–286
8. Feagin FF, Jeansonne BG, (1973) Effective fluoride concentrations to promote apatite mineralisation at the enamel surface. Ala J Med Sci 10:107–114
9. Koulourides T, Cueto H, Pigman W, (1961) Rehardening of softened enamel surfaces on human teeth by solutions of calcium phosphates. Nature 189:226
10. Cate JM ten, Jongebloed WL, Arends J, (1981) Remineralisation of artificial enamel lesions in vitro, IV. Influence of fluorides and diphosphonates on short and long term remineralisation. Caries Res 15:60–69
11. Silverstone LM (1983) Remineralisation and enamel caries: New concepts. Dental Update 10:4
12. Cate JM ten, Arends J (1980) Remineralisation of artificial enamel lesions in vitro, III. A study of the deposition mechanism. Caries Res 14:351–358
13. Cate JM ten (1983) A model for enamel lesion remineralisation. In: Leach SA, Edgar WM (eds) Demineralisation and remineralisation of the teeth. IRL, Oxford, pp 129–144
14. Groeneveld A (1985) Longitudinal study of prevalence of enamel lesions in a fluoridated and non-fluoridated area. Community Dent Oral Epidemiol 13:159–163
15. Larsen MJ, Jensen SJ (1985) An X-ray diffraction and solubility study of equilibration of human enamel-powder suspensions in fluoride-containing buffer. Arch Oral Biol 30:471–475
16. Duff EJ (1976) Fluoride incorporation into powdered human enamel under conditions of fluctuating pH. Caries Res 10:234–240
17. Cate JM ten, Duijsters PPE (1983) Influence of fluoride in solution on tooth demineralisation. I Chemical data. Caries Res 17:193–199

18. Cate JM ten, Duijsters PPE (1983) Influence of fluoride in solution on tooth demineralisation. II. Microradiographic data. Caries Res 17:513–519
19. Margolis HC, Moreno EC, Murphy BJ (1986) Effect of low levels of fluoride in solution on enamel demineralization in vitro. J Dent Res 65:23–29
20. Cate JM ten, Duijsters PPE (1982) Alternating demineralisation and remineralisation of artificial enamel lesions. Caries Res 16:201–210
21. Featherstone JDB, O'Reilly MM, Shariati M, Brugler S (1986) Enhancement of remineralisation in vitro and in vivo. In: Leach, SA (ed) Factors relating to demineralisation and remineralisation of the teeth. IRL, Oxford, pp 23–35
22. Gerrard WA, Winter PJ (1986) Evaluation of toothpastes by their ability to assist rehardening of enamel in vitro. Caries Res 20:209–217
23. White DJ (1987) Reactivity of fluoride dentifrices with artificial caries. Caries Res 21:126–140
24. Almqvist H, Lagerlöf F, Angmar-Mansson B (1990) Automatic pH-cycling caries model applied on root hard tissue. Caries Res 24:1–5
25. Wefel JS (1990) Effects of fluoride on caries development and progression using intra-oral models. J Dent Res 69:626–633
26. Koulourides T, Phantumvanit P, Munksgaard EC, Housch T (1974) An intra-oral model used for studies of fluoride-incorporation in enamel. J Oral Path 3:185–195
27. Mellberg JR, Petrou ID, Deutchman M, Grote N (1988) The effects of 1% pyrophosphate and 0.02% sodium fluoride on artificial caries lesions in vivo. J Dent Res 67:1461–1465

Calcification in Octocorals

NORIMITSU WATABE[1] and RONI J. KINGSLEY[2]

Summary. The octocorals are colonial cnidarians, many of which contain calcareous spicules in the mesoglea and, in some species, in the axis as well. The octocorals are excellent organisms for probing the mechanisms of biomineralization and demineralization, and the processes of calcification of spicules have been investigated extensively in recent years, especially in *Leptogorgia virgulata*. In this review, octocoral spicule formation is discussed relative to (1) the characteristics of scleroblasts, the spicule-forming cells, (2) calcium transport by the colonies, (3) rate of calcification, (4) role of carbonic anhydrase, (5) hormonal control of calcification, and (6) organic matrices and their possible functions in spicule formation.

Key words: Octocorals — Spicules — Organic matrix — Ca-transport — Carbonic anhydrase

Introduction

The octocorals are colonial cnidarians which are comprised of animals of 6 orders: the Stolonifera (e.g., stolon corals, organ-pipe corals), Telestacea, Alcynoacea (soft corals), Coenothecalia (blue corals), Gorgonacea (e.g., sea whips, sea fans, and red corals), and Pennatulacea (e.g., sea pens and sea pansy). Many octocorals contain calcareous spicules in the mesoglea and also in the axis in some species. The spicule is generally spindle-shaped in form with various modifications characterizing different taxonomic categories [1]. The octocorals are excellent organisms for probing the mechanisms of biomineralization and demineralization. Thus, the processes of calcification of spicules have been investigated extensively in recent years, especially in the sea whip *Leptogorgia virgulata* (Fig. 1).

[1] Electron Microscopy Center, The University of South Carolina, Columbia, SC 29208, USA
[2] Department of Biology, University of Richmond, Richmond, VA 23173, USA

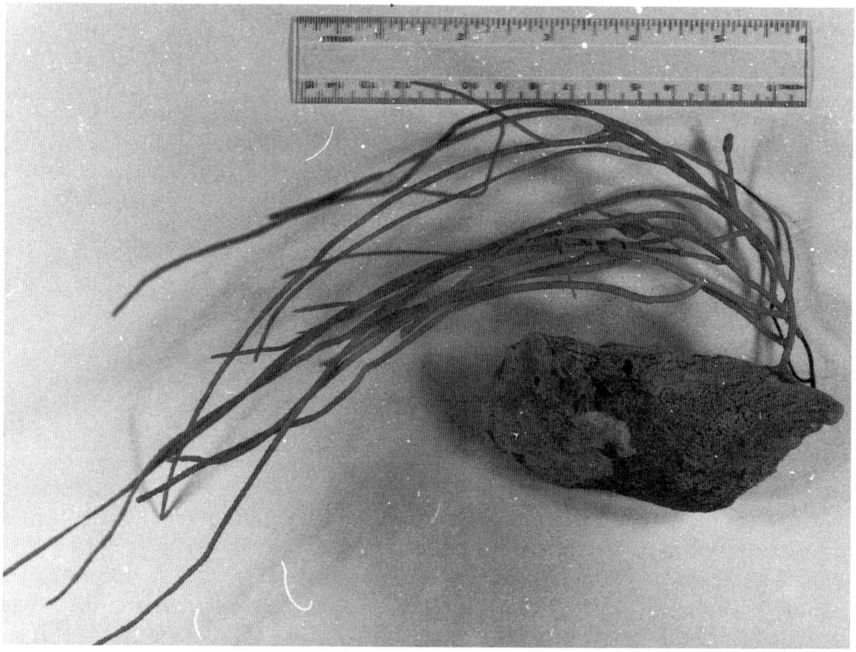

Fig. 1. Colonies of the gorgonian *Leptogorgia virgulata* attached to an oyster shell

Many octocoral spicules are composed of high magnesium calcite with 6–12 wt% magnesium carbonate [2]. The content of magnesium carbonate in all branch regions of the gorgonian is constant within a species [3]. The mature spicule of *Leptogorgia virgulata* is usually up to a few hundred μm long, and exhibits a more or less symmetrical spindle shape with many branches or small wart-like surface ornamentations [4] (Fig. 2). The spicule of the sea pansy *Renilla reniformis* is approximately 200 μm long and 25 μm wide, trilobed in cross sections, and characterized by the presence of grooves running parallel or obliquely to the morphological long axis of the spicule [5]. In both *Leptogorgia* and *Renilla*, the spicule is not of a single crystal but a polycrystalline aggregate of minute crystallites [4, 5]. In *Leptogorgia*, the surface ornamentation frequently shows aggregates of several rhombohedral calcite (Kingsley and Watabe, unpublished work). In cross sections, the crystallites are arranged in concentric layers, and organic matrix materials are present between the layers as well as within the crystals [4–7] (Fig. 3).

The Scleroblasts and Processes of Spicule Formation

The spicules are formed by the scleroblasts. The scleroblasts are the mesogleal cells derived from interstitial cells that have migrated inward from the epidermis [1]. In *Leptogorgia virgulata*, the preliminary phase of spicule formation

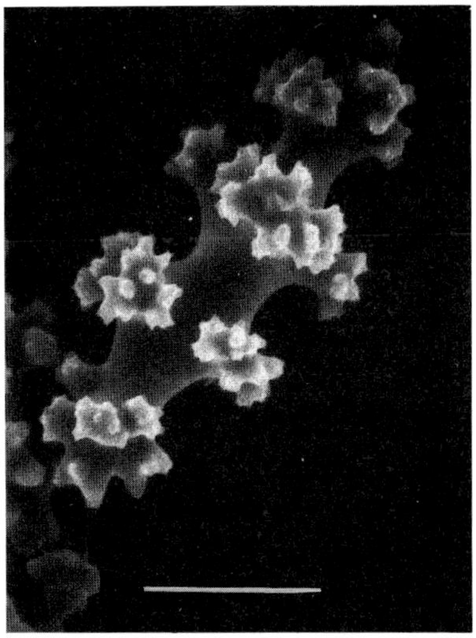

Fig. 2. Isolated mature spicule of *Leptogorgia virgulata*. SEM. *Bar* = 60 μm

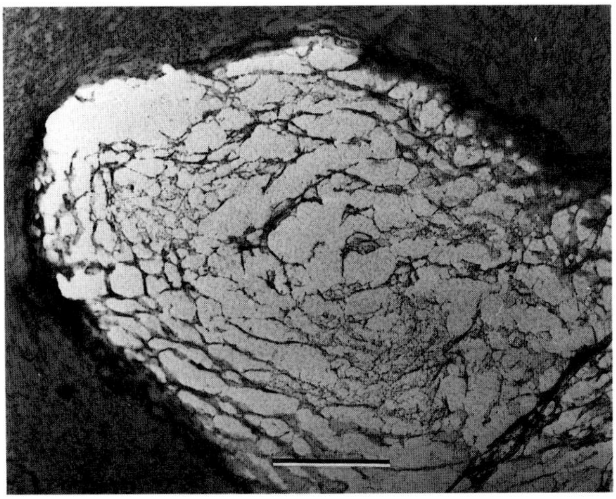

Fig. 3. Insoluble organic matrix. TEM of spicule section stained and decalcified with uranyl acetate. The matrix framework is more or less concentrically arranged. Perforations represent areas where calcite crystallites were present prior to decalcification. *Leptogorgia virgulata*. *Bar* = 1 μm

is the aggregation of the scleroblasts. Each cell contains a large nucleus, prominent Golgi apparatus, rough endoplasmic reticulum, free ribosomes, mitochondria, electron-dense bodies, vacuoles, and vesicles. A deposition of fibrous organic matrix within a vacuole, probably derived from the Golgi apparatus, is the first stage of spicule formation. The first crystals are laid down in close association with fibrous matrix materials in the vacuoles (Fig. 4).

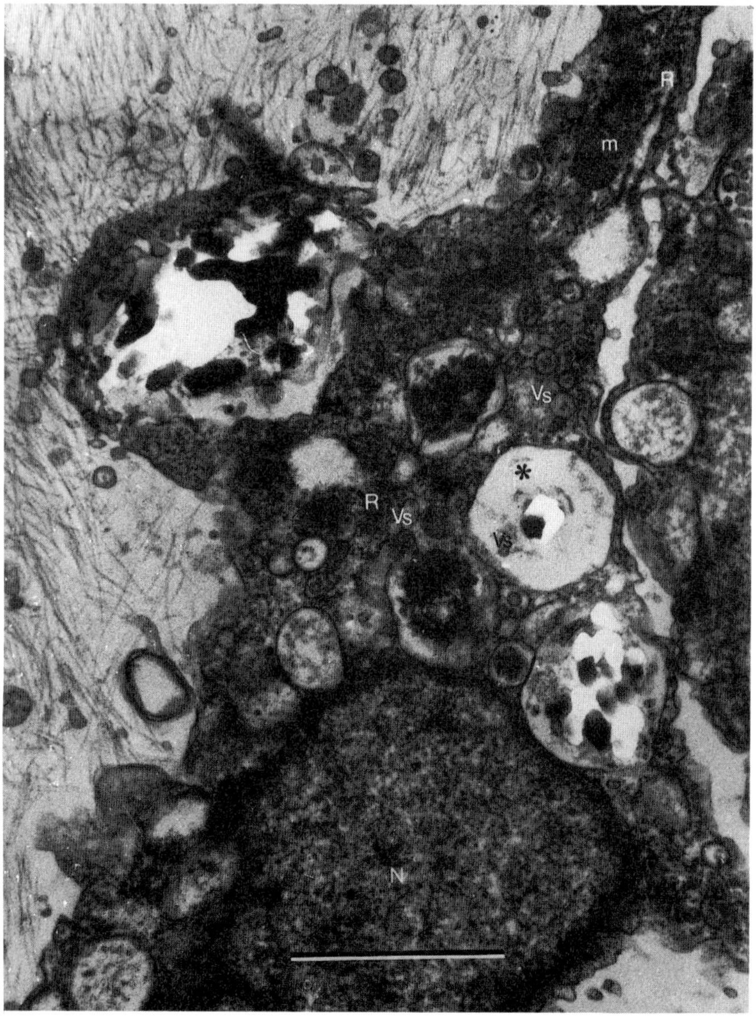

Fig. 4. TEM of early deposition of calcite crystals on organic matrix (*) within vacuoles of the scleroblast. *Leptogorgia virgulata*. Vs, Vesicles; *R*, rough endoplasmic reticulum; *N*, nucleus; m, mitochondrion. *Bar* = 1 μm. (With permission from [4])

The matrix and crystals continue to be deposited until the vacuole is almost completely occupied by the spicule. At or slightly before this stage, the scleroblasts dissociate from the aggregate and disperse. As the spicule grows, the vacuole enlarges and the nucleus becomes deformed and displaced toward the periphery of the scleroblast. The cytoplasm occupies a narrow space between the plasma membrane and the enlarged vacuole membrane. Wart-like structures become prominent at this stage. One of the most significant phenomena is the transition of the spicule from the intracellular to the extracellular entity.

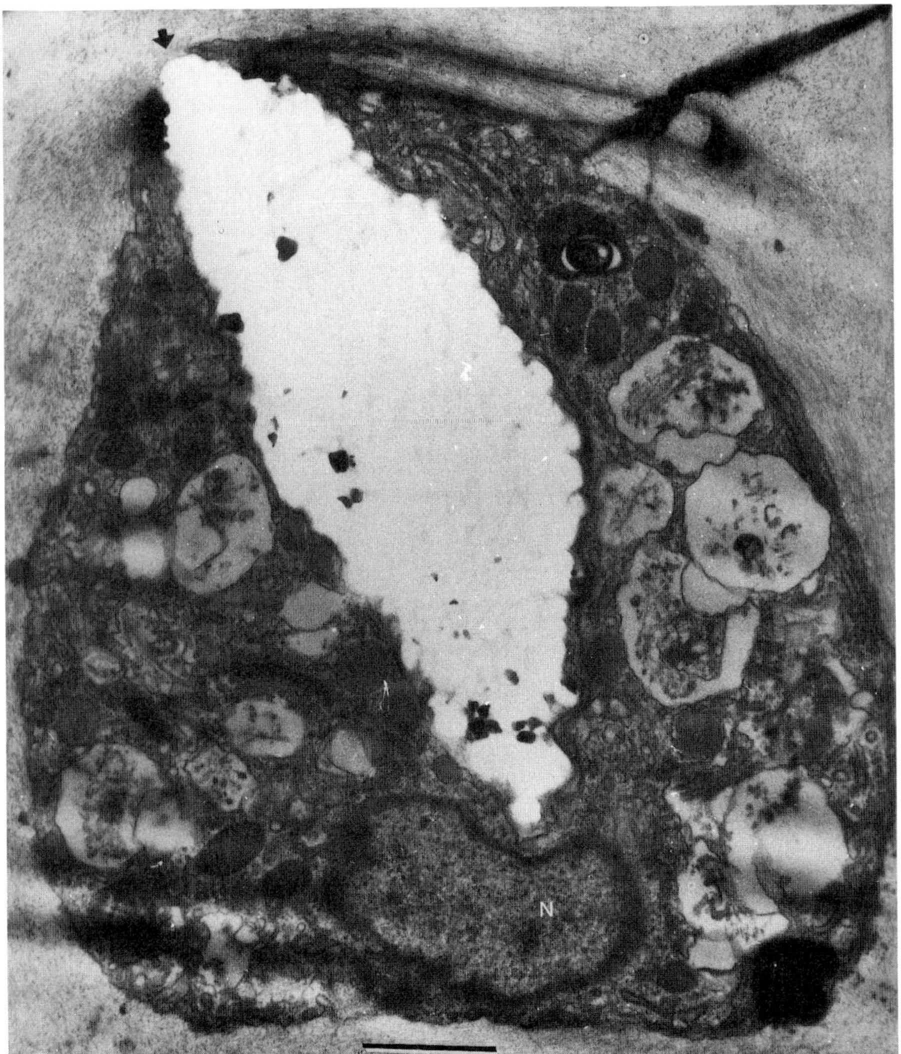

Fig. 5. TEM of a scleroblast of *Leptogorgia virgulata* showing maturing spicule (the *oblong space in the center*, the spicule crystals having been dislodged during sectioning) and a breach in the plasma membrane (*arrow*). *N*, Nucleus. *Bar* = 1 μm. (With permission from [4])

This is accomplished by a fusion of the spicule vacuole membrane with the plasma membrane of the scleroblast, followed by a breach formation in the scleroblast (Fig. 5). This causes the spicule to extend into the extracellular space. This extrusion has also been confirmed by cell culture of the scleroblast [8] (Fig. 6). Further growth of the spicule in the extracellular environment has

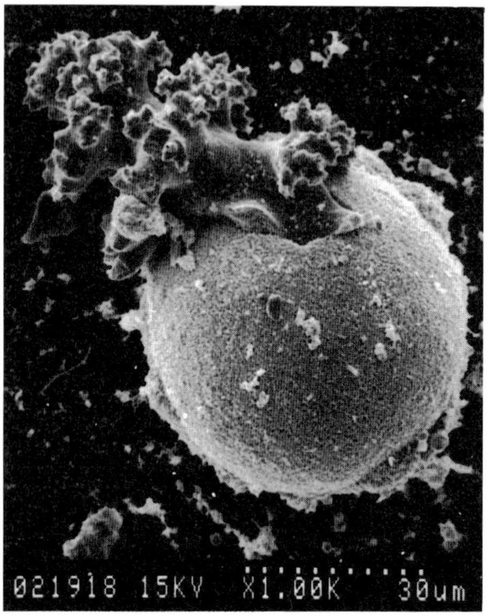

Fig. 6. Cultured scleroblast of *Leptogorgia virgulata* with mature spicule extruding from the cell. SEM. (With permission from [8])

not been confirmed, although it is possible that the growth occurs by the secretory activities of the original and/or secondary scleroblast [9]. In the gorgonian *Pseudoplexaura flagellosa*, scleroblast vesicles calcify and the crystals are released to the mesoglea [6]. Several of those primordial crystals integrate into a single mature spicule through the action of several secondary scleroblasts. How those primordial crystals assemble into a regularly shaped spicule is not clear. Further investigation is needed to clarify the additional growth of spicules in the extracellular environment.

Processes of formation of axial spicules have not been investigated in detail. In the primary polyp of the sea pen *Ptilosarcus gurneyi*, crystal development at an early stage of axis formation has been reported to occur in stylet cells in the endoderm [10]. However, in another sea pen *Veretillum cynomorium* the initial crystal-like deposits are not preformed within the epithelium surrounding the axis. They are presumed to be formed by precipitation from inorganic ions secreted into the general compartment of the axis [11]. Initial mineral deposits give rise to calcitic nodules which enlarge by the radial growth of several lobes. These nodules fuse and form the core of the axis (Fig. 7).

Calcium Transport for the Spicule Formation

Calcium transport for the spicule formation in *Leptogorgia virgulata* has been investigated by Kingsley and Watabe by ^{45}Ca pulse-chase experiments [12], ^{45}Ca light- and EM-autoradiography [13], and ATPases localization at the EM level [14]. The calcium movement from the seawater into the spicule vacuole and the axis occurs in several steps. First, calcium ions enter the outer epithe-

Fig. 7. Longitudinal fracture surface of a fully mineralized part of the axis of the sea pen *Veretillum cynomorium* after etching. Mineral columns (co) are radiating out from the axis core (c). SEM. *Bar* = 60 μm. (With permission from [11])

lium through diffusion (Fig. 8①). From the epithelium, these ions are actively transported out into the mesoglea (Fig. 8②) and also back to the seawater by the activity of Ca-ATPase localized at the epithelial plasma membranes. The exit of ions to the seawater is mediated largely through the polyp outer epithelium rather than the other outer epithelium. High activities of Ca-ATPase are found on the inner plasma membranes of the microvilli and lateral and basal plasma membranes of the polyp epithelium. From the mesoglea, some calcium ions diffuse into the scleroblasts. However, the majority of the calcium moves into the axis, which is the initial and primary site of calcium accumulation, through the axial epithelium (Fig. 8③④). The microvilli of the attachment cells (democytes) penetrate deep into the axis and inner membranes of those microvilli show very strong activities of Ca-ATPase. Approximately 30% of the calcium ions remain in the axis, probably being deposited as amorphous calcareous granules. The other 70% leave the axis and return to the mesoglea

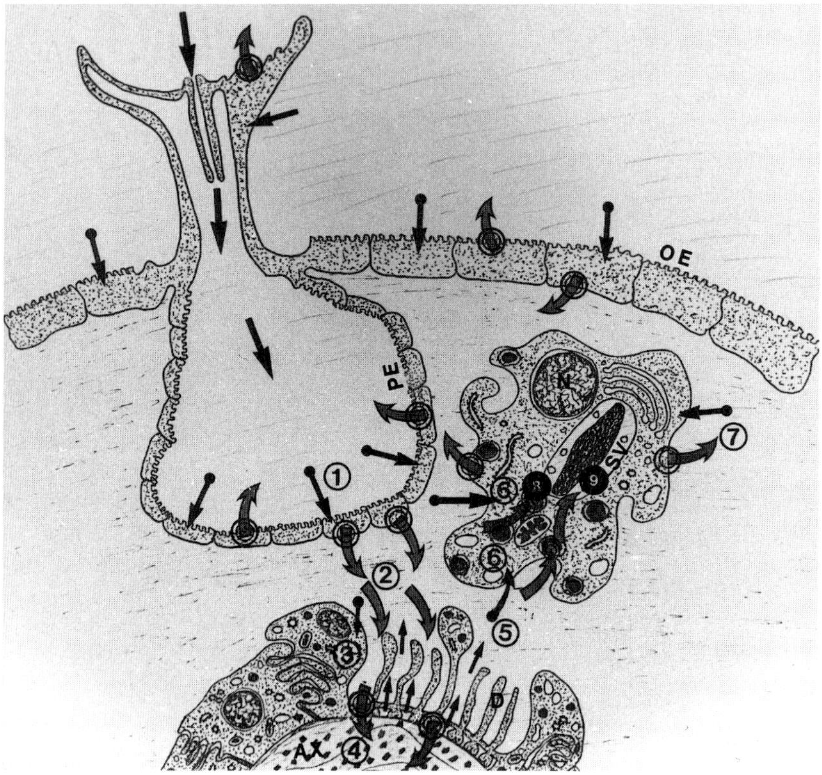

Fig. 8. A diagram depicting Ca^{2+} pathways from the environment into the spicule forming vacuole of the scleroblast of *Leptogorgia virgulata*. *Dark arrows* with *small dark circle* indicate diffusion. *Pale arrows* with *double circles* indicate active transport. *Other arrows* indicate directions of Ca^{2+} movement. *OE*, Outer epithelium; *PE*, polyp epithelium; *SV*, spicule vacuole; *N*, nucleus of the scleroblast; *AX*, axis; *D*, democyte

(Fig. 8⑤), and diffuse into the scleroblasts (Fig. 8⑥). They are further transported to the spicule-forming vacuole of the scleroblast via electron-dense bodies (Fig. 8⑧) to form the spicule (Fig. 8⑨). Portions of the Ca ions diffused into the scleroblasts are pumped back to the mesoglea (Fig. 8⑦). The plasma membrane of the scleroblasts and the limiting membrane of the electron-dense bodies also show ATPase activities. The autoradiographic study indicated that ^{45}Ca accumulates in the axis from 1 to 15 min after the 15-min incubation of the colony [13]. After 15 min, the label moves out of the axis. By 60 min, much of the label is in the spicules, reaching its maximum level at 120 min.

The calcium ions which leave the axis but do not enter the scleroblasts are released into the surrounding seawater. This backflow of calcium from the axis is a process also involving the active transport. This is different from the other gorgonian *Eunicella papillosa* and hermatypic corals in which the backflow is considered to be a "leakage" [15, 16]. In *Eunicella papillosa*, 60% to 70% of

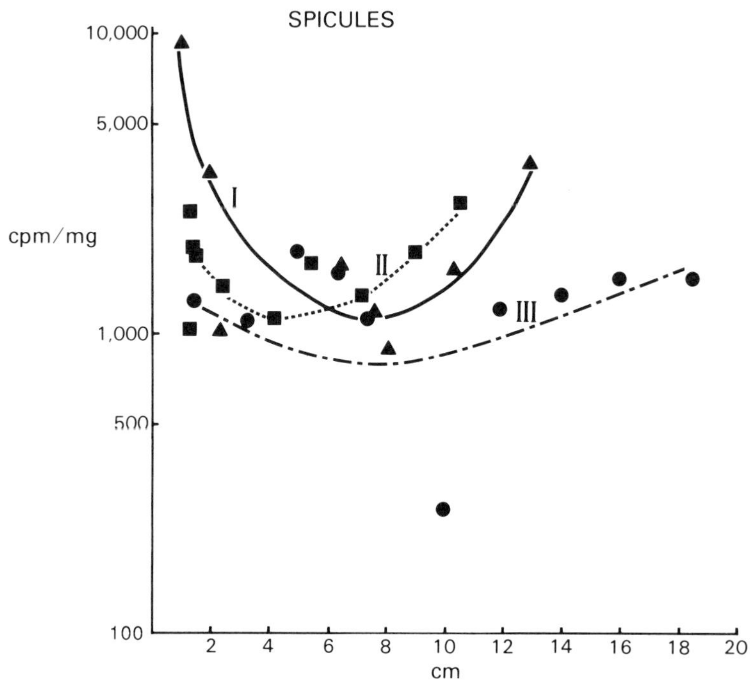

Fig. 9. ^{45}Ca uptake in spicules at different portions along length of branches of Colonies *I*, *II*, and *III*. *Ordinate*: Count · min^{-1} · mg^{-1} tissue. *Abscissa*: Distance from branch tip toward base. *Leptogorgia virgulata*. (With permission from [17])

the ^{45}Ca taken up by the animal remains in the tissue and is not deposited in the spicules [15]; 45% of the tracer which is taken up is returned to the water.

Rate of Calcification

In *Leptogorgia virgulata* spicule formation occurs throughout the animal colony and is directly related to the cellular activities of the branches [17]. In the branch tips, the degree of organic matrix formation and cell division are the highest, as indicated by the highest uptake of ^{3}H-aspartic acid and ^{3}H-thymidine. The rate of calcification is the highest in these regions and decreases to the mid-portion of colonies, followed by an increase toward the bases (Fig. 9). This latter rise is due to the increase in size and density of spicules which causes an increase in isotopic exchange. The rate of spicule formation is also the highest in the branch tips, but it decreases progressively in the lower branches in *Eunicella papillosa* [15].

Both spicule and axis weights progressively increase toward the basal portions in *Leptogorgia* [17] and *Eunicella* [3, 15]. Evidently, the branch increases in both length and diameter during growth with accompanying spicule forma-

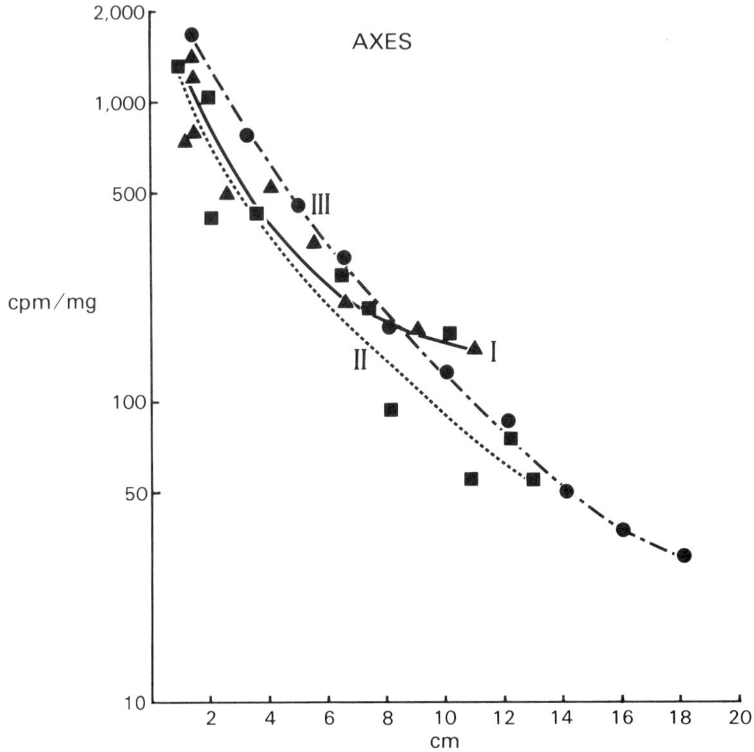

Fig. 10. ^{45}Ca uptake in axes at different portions along length of branches of Colonies *I*, *II*, and *III*. *Ordinate*: Counts · min^{-1} · mg^{-1} axis. *Abscissa*: Distance from branch tip toward base. *Leptogorgia virgulata*. (With permission from [17])

tion and axial growth. Thus, the spicule and axial weights become larger toward the base with increasing age of the branch, although the rate of calcification and cell division for both spicules and axes are lower at the base. Axes of apical regions show high rates of Ca uptake (Fig. 10) which indicates its active role in the calcification process.

Only few data are available on calcification rates of corals. In *Eunicella*, the average net calcification rate is 4.8 μg Ca mg $N^{-1} h^{-1}$ at 14°C [15]. The rates of ahermatypes range from 5.5 to 12.0 μg Ca mg $N^{-1} h^{-1}$ [18].

The Role of Carbonic Anhydrase in Spicule Formation

Carbonic anhydrase (CA) are associated with calcification in a number of invertebrates [19]. CA may affect calcification by catalyzing the conversion of CO_2 to HCO_3^- (equation 1), which would then be available for $CaCO_3$ formation (equation 2)

Table 1. ^{45}Ca uptake by control and Diamox-treated specimens. (From [19])

Sample	*Mean count min^{-1} mg^{-1} (n = 20)*	*SD*
Spicules		
10^{-6} M Diamox	764.57*	±78.37
10^{-4} M Diamox	760.03*	±75.87
Control	689.03	±89.32
Tissue		
10^{-6} M Diamox	596.03*	±326.88
10^{-4} M Diamox	830.70	±157.50
Control	841.47	±231.04
Axis		
10^{-6} M Diamox	2331.20*	±734.00
10^{-4} M Diamox	2126.50*	±850.90
Control	1615.00	±385.30

* Significantly different from control

$$CO_2 + H_2O \rightleftharpoons HCO_3^- + H^+. \tag{1}$$

$$Ca^{2+} + HCO_3^- \rightleftharpoons CaCO_3 + H^+. \tag{2}$$

CA may also catalyze the conversion of HCO_3^- to CO_2, the reverse reaction of Eq. (1), which would reduce the acidity of the calcifying milieu and provide a more conducive environment for $CaCO_3$ deposition. Another function assigned to CA is that of promoting decalcification. This enzyme may catalyze the reaction (1) to the right, producing protons and causing dissolution of calcareous structures.

The mean activity of CA in *Leptogorgia virgulata* is 2.7 nmol $\mu g^{-1} s^{-1}$, which is comparable to the values of crab gills and epidermis at intermolt and of some tissues of several molluscan species [18]. The activity in the polyps of the scleractinian coral *Pocillopora damicornis* is 0.162 nmol $\mu g^{-1} s^{-1}$ [20, 21]. Histochemically, intense CA activity is observed on the limiting membrane of the spicule-forming vacuole of *Leptogorgia* [19]. The membrane of the desmocyte microvilli, which is the closest cellular structure to the calcareous granules of the axis, also strongly exhibits this enzyme activity.

The function of CA in *Leptogorgia* spicule formation was investigated by a ^{45}Ca uptake experiment in the presence of the CA inhibitor Diamox [19]. In contrast to the results in other invertebrate calcification, treatment with Diamox causes an increase instead of a decrease in ^{45}Ca uptake in the spicules and axes (Table 1). This increase is interpreted to be related to the dynamics of HCO_3^- on the membranes adjacent to the calcifying structures. Assuming that HCO_3^- is not a limiting factor, as would be expected in a marine organism, CA may be involved in regulating this abundant ion in the spicule-forming vacuole and axis. CA may catalyze reaction (1) to the left and provide a means to regulate the amount of HCO_3^- available to react with Ca^{2+}. Thus, inhibition of CA would result in the increased availability of HCO_3^- for reaction with Ca^{2+}, thereby enhancing calcium carbonate precipitation.

It is possible that the CA in *Leptogorgia* is involved in ion regulation as in other aquatic invertebrates. Some CA, such as that on the plasma membrane of the scleroblasts, may catalyze the hydration of metabolic CO_2, generating H^+ and HCO_3^-. These ions could act as counter ions of Na^+ and Cl^-. The CA in this organism may also be involved in such functions as diffusion of CO_2 through membranes and maintenance of H^+ gradients across membranes [22]. The inhibition of CA involved in such processes may indirectly affect ^{45}Ca uptake. A priori, inhibition of such processes could potentially alter the physiological status of the organism, consequently altering the calcification process.

Hormonal Control of Calcification

Little is known concerning the hormonal control of calcification in octocorals. However, diiodotyrosine is found in gorgonians and iodine occurs in the axis [23]. Administration of thyroxine reduces the number of statoliths in the rhopalia of developing jellyfish [24]. Tissues of *Leptogorgia virgulata* stains positively for neurosecretory activity with paraldehyde fuchsin and lead hematoxylin. Immunofluorescence and immunogold labeling studies indicate that thyroxine or a thyroxine-like substance (TX) is present in the polyp vacuoles, gastrodermal tubes and nuclei, and electron dense bodies and spicules of scleroblasts of this animal. It is possible that TX is formed in the polyp and is transferred, by way of the gastrodermal tubes, to the scleroblasts, where it is involved in spicule formation (M. Corcoran and R.J. Kingsley, unpublished work).

Organic Matrix of Calcified Skeletons

Organic matrices of gorgonian spicules are approximately 1%–5% of the total weight of the spicules and contain lipids, carbohydrates, and proteins [25–27]. Quantitative data on those components are few. In *Pseudoplexaura flagellosa*, more than one-half of the weight of the matrix consists of lipids, about one-third is proteins, and the remainder is carbohydrates [7] (Table 2). In *Eugorgia ampla*, lipids are approximately 0.6% of the weight of spicules, or 60% of the organic matrix [26]. The data indicate that the lipids, followed by proteins are the major constituents of the gorgonian matrix.

The role of lipids in spicule formation is not well understood. However, long-chain fatty acids have been found to be responsible for the color of spicules and, accordingly, for the orange-yellow color of colonies of *Eugorgia ampla* [26]. One of these fatty acids is the "eugorgiaenoic acid" named by Fox et al. [26]. This is a yellowish, chemically unstable, non-aromatic, polyene-chained acid, which exhibits chemical and spectral properties suggestive of a

Table 2. Proximate analysis of *Pseudoplexaura* spicule matrix. (From [7])

Undecalcified spicules (n = 15)	*% Undecalcified spicule wt*	*% Total organic*	*% Insoluble organic*	*% Soluble organic*
Organic carbon	4.5 ± 0.4	100	88 ± 3.0	12.7 ± 2.0
Protein[a]	1.5 ± 0.2	33.3 ± 4.4	27 ± 4.6	4.3 ± 2.0
Carbohydrate[b]	0.4 ± 0.2	8.9 ± 0.4	0.5 ± 0.10	8.4 ± 1.0
Lipoidal pigment	2.6 ± 0.4	57.8 ± 8.9	55 ± 8.9	0

[a] Estimated by both amino acid analysis and fluorescence assay

[b] Estimated by phenol-sulfuric acid assay, using a glucose standard and by gas chromatography

chemical relationship with dihydrobixin. This pigment is conjugated to divalent cations, such as Ca and Mg in situ within the spicules.

Carbohydrates in the spicules of *Pseudoplexaura* include glucose and glucosamine which occupy nearly 60% of the total carbohydrate, followed by galactose and galactosamine (35%) and mannose (5%) [7]. Glucosamine or chitin is also present in *Leptogorgia virgulata* [28, 29]. Chitinous spicules are found in the sea pen, *Leioptilus guerneyi* [30].

The organic matrix of many invertebrates is known to consist of EDTA-soluble and -insoluble fractions. The insoluble matrix of the *Leptogorgia* spicules is usually fibrous, and the fibers are approximately 30 nm in diameter, displaying no periodic banding (Fig. 3). The autoradiographic investigation [31] using ^{3}H-aspartic acid as the tracer revealed that the protein moiety of the organic matrix is synthesized in the RER during the 5-min period following the 10-min incubation of the animal in the tracer. Within 5 min thereafter, the label moves from the RER to the Golgi complexes where the carbohydrate moiety is presumed to be synthesized. The label can be seen from 60 to 120 min in the spicules in the Golgi vacuole [31]. The entire process of the matrix synthesis and transport into the spicule takes approximately 2 h at 21°C in the summer.

Unlike vertebrate osseous and dental tissues made of calcium phosphates, invertebrate hard tissues of calcium carbonate rarely contain collagenous components in the insoluble matrices. The spicules of *Leptogorgia virgulata* are unique in that the major constituent of the insoluble matrix protein is a collagen without any periodic banding, and that the collagen content changes with the season [9]. As shown in Table 3 (columns A and C) the insoluble matrix of spicules collected in the summer months exhibits amino acid compositions typical of a collagen, i.e., having significant amounts of Hyp (Hydroxyproline), Hyl (Hydroxylysine) and 30%–33% Gly (Glycine). The composition is closer to that of the mesogleal collagen of sea anemone (Table 3, column I) than to the bovine bone collagen (Table 3, column J). However, the *Leptogorgia* spicule collagen contains less Hyp and more acidic amino acids than the sea anemone collagen. In fact, the *Leptogorgia* collagen is different in its amino acid composition from any other collagens reported in the cnidaria [29].

Table 3. Amino acids compositions of matrix protein fractions from isolated calcite spicules of *Leptogorgia virgulata*. (From [9])

	Residues per 1000 total residues									
	A TIM S.C.	B TIM 3/87	C TIM 7/87	D TIM 12/87	E Pepsin sol.	F Pepsin res.	G SM 12/87	H SM 7/87	I Sea anemone collagen[d]	J Bovine bone[e]
Hyp	74	8	82	31	80	11	1		103	98
Asp	93	132	76	130	84	153	516	577	73	45
Thr[a]	36	57	34	48	32	63	18	16	37	17
Ser[a]	43	57	42	53	43	78	16	13	41	34
Glu	109	87	94	92	112	76	34	32	97	74
Pro	66	67	65	59	64	50	17	19	67	123
Gly	295	158	332	225	335	161	210	185	339	337
Ala	83	78	82	76	77	67	130	108	65	109
Val	27	58	26	37	22	55	17	16	25	20
Cys[b]/2	2	16	1	10		17				
Met[c]	4	11	7	10	5	10				5
Ile	18	38	15	26	13	46	6	5	22	11
Leu	27	53	26	42	23	58	7	6	30	25
Tyr	8	24	8	20	6	32		2	3	4
Phe	9	31	9	22	7	35	4	4	8	13
His	3	17	4	11	3	11	2	1	1	4
Hyl	33	7	34	15	34	6			25	6
Lys	15	44	12	33	10	33	12	9	16	26
Arg	56	55	50	53	50	39	8	5	68	50

TIM, Total insoluble matrix; S.C., summer collection in South Carolina; *3/87, 7/87, 12/87*, dates of collection in North Carolina; *Pepsin sol.*, 7/87 collection, pepsin soluble material; *Pepsin res.*, 7/87 collection, pepsin insoluble material; *SM*, soluble matrix

[a] Uncorrected for hydrolysis

[b] Half Cys, sum of cysteic acid and cystine

[c] Sum of methionine sulphoxide and methionine

[d] [37]

[e] [38]

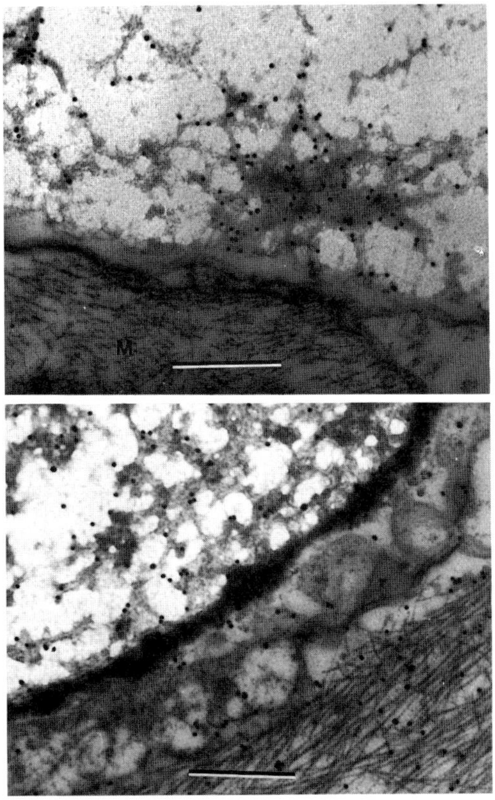

a

b

Fig. 11. a Immunogold labeling for the collagenous 140 kd protein in the summer spicule of *Leptogorgia virgulata*. The labels are on the framework of the insoluble matrix. Mesoglea fibers (*M*) are scarcely labeled. *Scale bar* = 1 μm. **b** Immunogold labeling for the 40 kd non-collagenous protein in the summer spicule of *Leptogorgia virgulata*. The labels are at the periphery of the insoluble matrix and on the mesoglea fibers. *Scale bar* = 1 μm

The collagenous nature of the spicule insoluble matrix has also been confirmed by the presence of the intermolecular reducible cross-link, dihydroxylysinonorleucine [9]. The collagen content decreases in the December samples, as indicated by the lower values of Hyp (31/1000), Hyl (15/1000), and Gly (225/1000) (Table 3, column D). The values further decrease in the March sample, showing very little of the collagenous characteristics (Table 3, column B). This composition is similar to that of the non-collagenous minor component of the summer insoluble matrix (Table 3, column F). These results are substantiated by SDS-PAGE and immunocytochemical findings [29] (Watabe N et al. (1992) unpublished material). SDS-PAGE patterns of the summer insoluble matrix consist of protein bands at 500 kd, 140 kd, 120 kd, and 40 kd, of which the 140 kd and probably 120 kd proteins are collagenous. The spring samples have only the 40 kd protein band, but are sometimes accompanied by very faint bands of 140 kd and120 kd.

In the summer spicules, the collagenous 140 kd protein is detected within the insoluble matrix framework by immunogold labeling with antibodies against this protein (Fig. 11.a). The mesogleal fibers surrounding the scleroblasts and spicules show very little labeling. On the other hand, the labels for non-

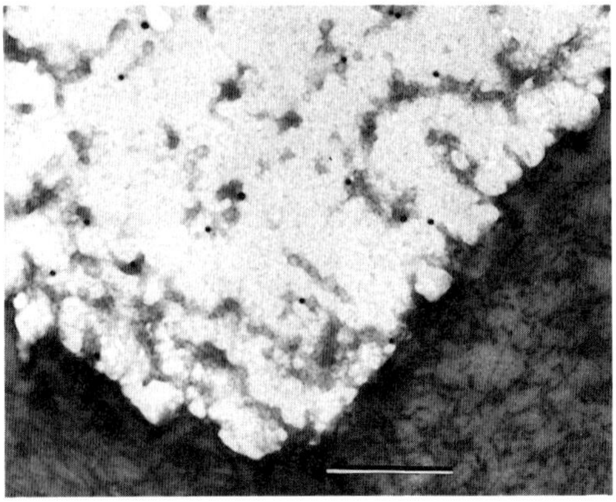

a

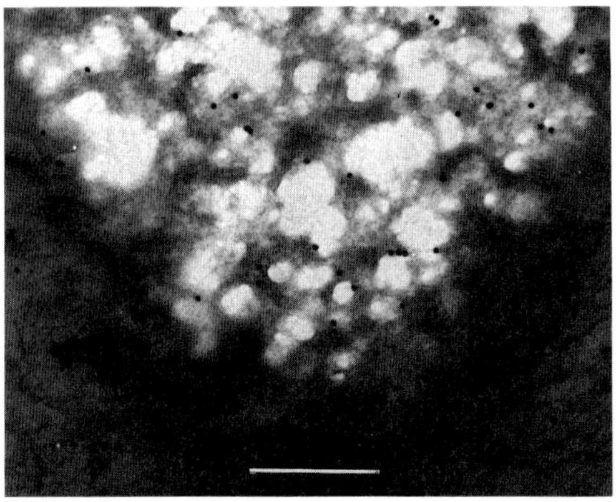

b

Fig. 12. **a** Matrix of the winter spicule of *Leptogorgia virgulata* showing few immunogold labels for the 140 kd collagenous protein. *Scale bar* = 1 μm. **b** Winter spicule matrix and mesoglea of *Leptogorgia virgulata* labeled with immunogold for the non-collagenous 40 kd protein. *Scale bar* = 1 μm

collagenous 40 kd insoluble matrix protein are at the periphery of the insoluble matrix framework, and the mesogleal fibers show a heavy concentration of immunogold which labels antibodies against this protein (Fig. 11.b). In the winter/spring spicules, both the insoluble matrix framework and the mesogleal fibers are scarcely labeled for the 140 kd collagen (Fig. 12.a), whereas the periphery of the insoluble matrix framework and the mesogleal fibers are labeled heavily for the non-collagenous 40 kd protein (Fig. 12.b). However, the 40 kd labels are not as heavy as those in the summer.

In contrast to the insoluble matrix, the biochemical characteristics of the soluble matrix do not change throughout the year. The matrix is very high in aspartate, approximately 500 to 650 residues per thousand residues, glycine, and alanine [25] (Table 3, columns G and H). A highly acidic calcium-binding glycoprotein is present in this matrix [28], and SDS-PAGE shows that the apparent molecular weight of this protein is 88,000 [29]. In addition, approximately 10 minor protein bands ranging from 12 kd to 100 kd were detected in 12% gels of both the summer and winter samples [29]. Similar to the labels for the 40 kd insoluble protein, the immunogold marker for the soluble 88 kd glycoprotein decorates the periphery of the insoluble matrix framework (Fig.

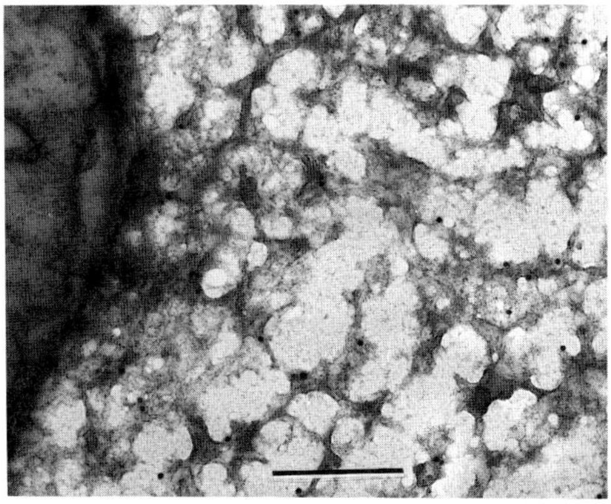

Fig. 13. Immunogold particles labeling the 88 kd soluble protein at the periphery of the insoluble matrix. *Leptogorgia virgulata*. *Scale bar* = 0.5 μm

13). The spatial relationship between the 88 kd soluble and 40 kd insoluble proteins is unknown.

The results of the biochemical and immunocytochemical investigations indicate that the content of the spicule collagen is very much reduced in the winter/spring animals, but that of the non-collagenous components is not. This implies that the collagen component within the spicules becomes degraded and digested during the fall/winter months, which is preceded by (partial) spicule decalcification. At present, no information is available concerning the process of degradation and digestion of this collagen. It may involve activation of some enzymes, such as a specific collagenase, by certain seasonal cues. Spicule decalcification may result from shifts in the activities of some enzymes, e.g., lysosomal enzymes, carbonic anhydrase (CA), etc. In fact, these enzymes have been cytochemically localized in the scleroblasts [4, 19]. By some seasonal cue, CA may be activated to catalyze the reaction $CO_2 + H_2O \rightarrow HCO_3^- + H^+$. The H^+ react with $CaCO_3$ and decalcify spicules ($CaCO_3 + H^+ = Ca^{2+} + HCO_3^-$). The activation may be hormone dependent. Lysosomal enzymes may react with spicules and dissolve crystals. The enzymes may also digest degraded collagen. It is known that lysosomal enzymes are affected by thyroid hormone [32], and substances which immunologically cross-react with thyroxine (T4) have been localized in *Leptogorgia* (see Hormonal Control of Calcification). Furthermore, an inhibition of T4 activities has been shown to affect the calcification in this organism (R.J. Kingsley and N. Watabe, unpublished work). Thus, it is conceivable that season-dependent activities of hormone or hormone-like substances and enzymes regulate collagenous (and non-collagenous) organic matrix formation and degradation as well as spicule calcification and decalcification in *Leptogorgia*.

Functions of the organic matrix in spicule formation are not clear. Acidic glycoprotein and/or proteoglycan macromolecules are generally considered to influence calcium carbonate crystal nucleation and growth in vitro. As for the spicule matrix, information is lacking concerning the molecular organization of various protein fractions. However, immunocytochemical evidence shows that the collagen is the backbone of the organic matrix framework of the summer spicule in *Leptogorgia virgulata*, and that the non-collagenous insoluble and soluble fractions are attached to it. The fact that the collagenous component occurs only in the summer when the spicule calcification is very active suggests that this spicule collagen is essential for the initial calcite nucleation. In fact, the insoluble matrix obtained after decalcification of spicules of *Leptogorgia* has been shown to recalcify in vitro [33]. Nevertheless, this does not necessarily mean that the collagenous insoluble matrix itself is the calcite nucleator in vivo. Although this matrix is calcifiable, it is covered by the soluble matrix and may not be able to attract Ca^{2+} ions. Recently, Crenshaw [34] and Gunthorpe et al. [35] presented a theory and evidence for the role of organic matrix in calcification in vitro. They have shown that polyanions, such as those of the soluble matrix, induce nucleation when immobilized on an insoluble substrate. No nucleation occurs when those polyanions are in solution. This scheme may be applicable to the *Leptogorgia* crystal nucleation. The soluble matrix attached to the collagenous insoluble matrix may be the nucleator and the collagenous matrix may regulate crystal growth. In *Veretillum cynomorium*, the axial rod contains calcite crystals through which collagen fibers penetrate, but the collagen fibers are never impregnated with the calcite. Ledger and Franc consider that the collagen is not responsible for the initial nucleation in this organism [11]. More data are necessary to determine the roles of the insoluble and soluble matrices in spicule formation in vivo.

Nothing is known concerning the morphogenesis of the spicules. The organic matrix (insoluble) deposited initially in the spicule vacuole of *Leptogorgia virgulata* does not assume the final spicule morphology. This implies that the final, species-specific and symmetrical shape of spicules (not of the individual crystals) is not likely to be determined by the initial insoluble matrix morphology. The form of the initial spicule is a rather smooth-surfaced rod without warts and decorations [4]. Thus, the basic form of the spicule may be the rod, which is presumably determined solely by the insoluble matrix substrate to which the soluble matrix is attached, or by combined effects of crystal growth and the substrate. The spicule continues to grow into a larger rod or spindle by further alternate deposition of the matrix and crystal layers. After attaining a certain size, symmetrical modification of the spicule form occurs, resulting in the elaborate and yet systematic decorations of the spicule. Watabe and Kingsley [36] proposed that the presence of regularly spaced inhibitors of crystal growth in the vacuole may be one of the factors controlling this morphogenesis. Patterns of distribution of those inhibitors along the vacuole membrane may be species specific. Biochemical, physiological, cytochemical, and ultrastructural analyses of those spicule-forming vacuoles and vacuole membranes may help to elucidate the morphogenesis of spicules as well as certain other biominerals.

Acknowledgements. All the authors' gorgonian research mentioned herein have been supported by the National Science Foundation grants #PCM-

8201389, DCB-8502689, DCB-880109, and partially by DIR-8805143, and by the funds of the Electron Microscopy Center, The University of South Carolina. We thank Betty Bynum and Robert Tye Whitaker for assistance in the preparation of the manuscript.

References

1. Bayer FM, Harding BO (1968) The free-living lower invertebrates. Macmillan, London, pp 229
2. Vinogradov AP (1953) The elementary chemical composition of marine organisms. Mem Sears Fdn mar Res No. 2, New Haven, pp 647
3. Velimerov B, Bohm EL (1976) Calcium and magnesium carbonate concentrations in different growth regions of gorgonians. Mar Biol 35:269–275
4. Kingsley RJ, Watabe N (1982) Ultrastructural investigation of spicule formation in the gorgonian *Leptogorgia virgulata* (Lamarck) (Coelenterata: Gorgonacea). Cell Tissue Res 223:325–334
5. Dunkelberger DG, Watabe N (1974) An ultrastructural study on spicule formation in the Pennatulid colony *Renilla reniformis*. Tissue Cell 6:573–586
6. Goldberg WM, Benyahu Y (1987) Spicule formation in the gorgonian coral *Pseudoplexaura flagellosa* 1. Demonstration of intracellular and extracellular growth and the effect of Ruthenium red during decalcification. Bull Mar Sci 40:287–303
7. Goldberg WM (1988) Chemistry, histochemistry and microscopy of the organic matrix of spicules from a gorgonian coral. Relationship to Alcian blue staining and calcium binding. Histochemistry 89:163–170
8. Kingsley RJ, Bernhardt AM, Wilbur KM, Watabe N (1987) Scleroblast cultures from the gorgonian *Leptogorgia virgulata* (Lamarck) (Coelenterata: Gorgonacea). In Vitro Cell Dev Biol 23:297–302
9. Kingsley RJ, Tsuzaki M, Watabe N, Mechanic GL (1990) Collagen in spicule organic matrix of the gorgonian *Leptogorgia virgulata*. Biol Bull 179:207–213
10. Chia FS, Crawford B (1977) Comparative fine structural studies of planulae and primary polyp of identical age of the sea pen, *Ptilosarcus gurneyi*. J Morphol 151:131–158
11. Ledger PW, Franc S (1978) Calcification of the collagenous axial skeleton of *Veretillum cynomorium* Pall. (Cnidaria: Pennatulacea). Cell Tissue Res 192: 249–266
12. Kingsley RJ, Watabe N (1984) Calcium uptake in the gorgonian *Leptogorgia virgulata*. The effects of ATPase inhibitors. Comp Biochem Physiol [A] 79:487–491
13. Kingsley RJ, Watabe N (1985) An autoradiographic study of calcium transport in spicule formation in the gorgonian *Leptogorgia virgulata* (Lamarck) (Coelenterata: Gorgonacea). Cell Tissue Res 239:305–310
14. Kingsley RJ, Watabe N (1985) Ca-ATPase localization and inhibition in the gorgonian *Leptogorgia virgulata* (Lamarck) (Coelenterata: Gorgonacea). J Exp Mar Biol Ecol 93:157–167
15. Velimirov B, King J (1979) Calcium uptake and net calcification rate in the octocoral *Eunicella papillosa*. Mar Biol 50:349–358
16. Barnes DJ, Crossland CJ (1972) Coral calcification: Sources of error in radioisotope techniques. Mar Biol 42:119–129

17. Kingsley RJ, Watabe N (1989) The dynamics of spicule calcification in whole colonies of the gorgonian *Leptogorgia virgulata* (Lamarck) (Coelenterata: Gorgonacea). J Exp Mar Biol Ecol 133:57–65
18. Goreau TF (1961) On the relation of calcification to primary productivity in reef-building organisms. In: Lenhoff HM, Loomis WF (eds) The biology of *Hydra*. University of Miami Press, Miami, pp 269–285
19. Kingsley RJ, Watabe N (1987) Role of carbonic anhydrase in calcification in the gorgonian *Leptogorgia virgulata*. J Exp Zool 241:171–180
20. Graham D, Smillie RM (1976) Carbonate dehydratase in marine organisms of the Great Barrier Reef. Aust J Plant Physiol 3:113–119
21. Sikes CS, Wheeler AP (1982) Carbonic anhydrase and carbon fixation in coccolithophorids. J Phycol 18:423–426
22. Jones WC, Ledger PW (1986) The effect of diamox and various concentrations of calcium on spicule secretion in the calcareous sponge *Sycon ciliatum*. Comp Biochem Physiol [A] 84:149–158
23. Gorbman AM, Clements M, O'Brien R (1959) Utilization of radioiodine by invertebrates, with special study of several annelida and mollusca. J Exp Zool 127:75–89
24. Spangenberg DB (1984) Effects of exogenous thyroxine on statolith synthesis and resorption in *Aurelia*. Am Zool 24:917–924
25. Kingsley RJ, Watabe N (1983) Analysis of proteinaceous components of the organic matrices of spicules from the gorgonian *Leptogorgia virgulata*. Comp Biochem Physiol [B] 76:443–447
26. Fox DL, Smith VE, Grigg RW, MacLeod WD (1969) Some structural and chemical studies of the microspicules in the fan-coral *Eugorgia ampla* Verrill. Comp Biochem Physiol 28:1103–1114
27. Silberberg MS, Ciereszko LS, Jacobson RA, Smith EC (1972) Evidence for a collagen-like protein within spicules of coelenterates. Comp Biochem Physiol [B] 43:323–332
28. Samata T, Kingsley RJ, Watabe N (1989) Ca-binding glycoprotein from the spicules of the octocoral *Leptogorgia virgulata*. Comp Biochem Physiol [B] 94:651–654
29. Watabe N, Oishi M, Kingsley RJ (1991) The organic matrix of spicules of the gorgonian *Leptogorgia virgulata*. In: Suga S, Nakahara H (eds) Mechanisms and phylogeny of mineralization in biological systems. Springer, Tokyo, pp 9–16
30. Shapeero W (1969) A positive chitosan test for spicules in the Anthozoan order, Pennatulacea. Pacif Sci 23:261–263
31. Kingsley RJ, Watabe N (1984) Synthesis and transport of the organic matrix of the spicules of the gorgonian *Leptogorgia virgulata* (Lamarck) (Coelenterata: Gorgonacea) An autoradiographic investigation. Cell Tissue Res 235:533–538
32. Gaton DD, Gaton E, Wolman M (1987) Are the effects of thyroid hormone on target organs mediated through lysosomal enzymes? Cell Mol Biol 33:619–624
33. Watabe N, Berhnardt AM, Kingsley RJ, Wilbur KM (1986) Recalcification of decalcified spicule matrices of the gorgonian *Leptogorgia virgulata* (Cnidaria: Anthozoa). Trans Am Microsc Soc 105:311–318
34. Crenshaw MA (1990) Mineral induction by immobilized polyanions. In: Suga S, Nakahara H (eds) Mechanisms and phylogeny of mineralization in biological systems. Springer, Tokyo, pp 101–105
35. Gunthorpe ME, Sikes CS, Wheeler AP (1990) Promotion and inhibition of calcium carbonate crystallization in vitro by matrix protein from blue crab exoskeleton. Biol Bull 179:191–200

36. Watabe N, Kingsley RJ (1990) Extra-, inter-, and intracellular mineralization in invertebrates and algae. In: Crick RE (ed) Origin, evolution, and modern aspects of biomineralization in plants and animals. Plenum, New York, pp 209–223
37. Norwack H, Nordwig A (1974) Sea-anemone collagen: Isolation and characterization of the cyanogen-bromide peptides. Eur J Biochem 45:333–342
38. Herring GM (1972) The organic matrix of bone. In: Bourne GH (ed) The biochemistry and physiology of bone structure, vol 1. Academic, New York, pp 127–189

Involvement of Mammalian Insulin and Insulin-like Peptides in Shell Growth and Shell Regeneration in Molluscs

Abu S.M. Saleuddin, Venkatalakshmi M. Sevala, Veeresh L. Sevala, Spencer T. Mukai, and Hamid R. Khan[1]

Summary. A brief review of the neuroendocrine control of shell formation/shell regeneration is presented. In *Helisoma duryi*, neurosecretory mediodorsal cells (MDC), which occur in 2 major groups in the two cerebral ganglia, are coupled via gap junctions, as revealed by electron microscopy and Lucifer yellow injections. The axons of MDC terminate in the two median lip nerves where neurosecretory granules are released by exocytosis. Extracellular electrical recordings from the median lip nerve, the neurohemal area of MDC, show that MDC are active in non-reproducing virgin snails and less active in reproducing snails. Many MDC are marked by porcine insulin antiserum, suggesting that they produce insulin-like peptide(s) (ILP). ILP and porcine insulin stimulate periostracal protein synthesis in vitro in the mantle. Immunoblotting experiments suggest the presence of two ILP immunoreactive bands in the hemolymph of *Helisoma* and one band in *Otala*. When a piece of shell is removed from the edge in *Helisoma*, shell regeneration is faster in fast growing snails than in slow growing ones. However, ILP titers are highest in slow growing snails and are generally lower at the peak of shell regeneration. ILP titers in the hemolymph fluctuate during a 24 h period, and are influenced by photoperiods. In *Helix aspersa*, microgram quantities of bovine insulin stimulate growth.

Key words: Insulin-like peptide—Shell growth—Molluscs—Radioimmunoassay—Shell regeneration

Introduction

The molluscan shell, which offers protection from predators and from extreme environmental fluctuations, such as temperature and humidity, is constructed on an uniform plan. Typically, it consists of an outermost organic layer, the periostracum, which overlies a series of calcareous layers made primarily of $CaCO_3$ crystals. These inorganic layers are supported by an organic matrix. While the shell structure of many molluscs is built on an uniform plan, it's

[1] Department of Biology, York University, North York (Toronto), Ontario, Canada M3J 1P3

mode of formation is not. In many molluscs, shell growth is incremental while in some molluscs a daily rhythm of shell deposition can be detected. This incremental growth of the shell whether it be sub-daily, daily, weekly, monthly, annually, or irregular can be influenced by various internal as well as external factors [1]. Among the latter category, tides, salinity, and photoperiod have been implicated. It is known that photoperiod not only influences the shell increment rate but also the thickness of a specific shell layer [2]. Environmental conditions, such as temperature, also influence the types of crystal morphs of $CaCO_3$ to be deposited [1]. Various control mechanisms of shell formation in molluscs have been described wherein each of these must involve a complex series of cellular responses by the mantle in first secreting the periostracum and then the calcareous shell. In order to deposit the inorganic shell material, the mantle is involved in ionic transport of Ca^{2+} and HCO_3^- to the extrapallial space whereby mechanism(s), still not quite well understood, form ordered layers of $CaCO_3$ crystals on an organic matrix. Although a role of the central nervous system (CNS) in shell formation has been speculated upon for some time, experimental evidence was not forthcoming until Dillaman et al. [3] showed that a group of neurosecretory cells in the visceral ganglion of the CNS in *Helisoma duryi* exhibited cyclical fluctuation of neurosecretory material in response to shell damage. Earlier, Lubet [4] showed that a factor from the cerebral ganglia in the prosobranch *Crepidula fornicata* stimulated body growth.

When a piece of shell is removed from a mollusc, a new piece is regenerated in the area of injury. The rate of shell repair is variable from species to species and is also dependent upon the site of shell injury. For example, shell repair rate is high if the injury is at the shell edge of *Helisoma* but the rate is slow if the area of injury is distant to the edge [5, 6]. By contrast, the repair is faster in *Helix* if the injury is made distant to the shell edge [7, 8]. Structurally, the repaired shell can be identical to the normal shell if the injury occurs at the edge [5] or it can be quite different, not only structurally but also in mineralogy, when the injury is made distant to the edge, as in *Helix* [7, 9]. Both regular incremental shell growth and shell regeneration involve regular and rapid turnover of $CaCO_3$ and proteins, and these systems are thus amenable to experimental analysis when involvement of the CNS is sought. To address the question whether or not the CNS exerts a direct influence on shell formation and/or shell regeneration, we refer to the findings of Dillaman et al. [3]. These authors first established the neuroendocrine centers in the CNS of *H. duryi*, and then studied the effect of shell damage on the synthetic activity of neurosecretory cells in the CNS. They reported that a group of neurosecretory cells in the visceral ganglion showed cyclical activities of paraldehyde fuchsin (PAF)-positive materials, suggesting that these cells are involved in the shell repair process. These cells were later called Type-1 cells [10] and Saleuddin et al. [11] suggested that they are involved in calcium mobilization, a prerequisite for the shell repair process. Several reports have shown that the injection of suboesophageal ganglion extract increases hemolymph calcium levels in a number of pulmonates [12, 13].

Another line of evidence for the involvement of the CNS in shell formation in *H. duryi* came from the study of Kunigelis and Saleuddin [2] who showed that brain extracts from fast growing snails, when injected into slow growing snails, caused faster shell growth in the treated group. Furthermore, injections of either supra- or suboesophageal fractions of the CNS had a more pronounced effect on shell growth than whole CNS [11].

The involvement of neurosecretory cells in body and shell growth in *Lymnaea stagnalis* was attributed to one kind of cell located in 4 groups in the cerebral ganglia [14]. These cells, termed light green cells (LGC) which are homologous to mediodorsal cells (MDC) in *H. duryi*, are under an inhibitory control of two lateral lobes located within the cerebral ganglia [15]. However the exact cellular origin of this inhibition is not known. This inhibitory role was extrapolated from the observation that the surgical removal of both lateral lobes caused giant growth in snails and that their reimplantation restored normal growth [15]. The lateral lobes of *H. duryi* also exert an inhibitory effect on its MDC [16]. An alternate explanation for this inhibition of growth would be that lateral lobes stimulate the endocrine dorsal bodies involved in female reproduction, and their removal would thus inhibit reproduction. Because there is an antagonistic relationship between reproduction and growth, growth (shell) will accelerate when the reproduction is inhibited [17]. In a series of papers, Dogterom et al. showed that the growth factor from LGC in *L. stagnali* increased uptake of tritiated tyrosine and ^{45}Ca from the medium by the mantle edge. They detected a calcium-binding protein in the mantle edge whose concentration was effected by the removal of LGC [18–20]. Neurosecretory cells in the CNS of several other pulmonates are known to be involved in growth [21].

The influence of the CNS on in vitro periostracum formation and calcium transport was studied by Kunigelis and Saleuddin [16, 22] who showed that the addition of the extract of CNS from a fast growing snail increased periostracum formation in the mantle from a slow growing snail. Furthermore, the putative brain factor increased incorporation of certain amino acids, such as glycine. They also showed that the periostracum formed in vitro incorporated ^{45}Ca from the medium, and that the CNS from a fast growing snail increased the incorporation by the mantle from a slow growing snail. Finally, they found that the dorsal epithelium of the mantle edge was important for ^{45}Ca incorporation, and that the lateral lobes exerted an inhibitory influence both on periostracum formation and ^{45}Ca incorporation.

The LGC once thought to produce somatostatin [23] in *L. stagnalis* have now been shown to produce molluscan insulin-like peptides (MIPs), one of which has been sequenced [24]. In situ hybridization with cDNA probes for MIPs showed that the transcription of MIPs occurs in LGC and canopy cells [21]. MIP has been shown to have significant homology with insect prothoracicotropic hormone (PTTH) and mammalian insulin [25]. The effect of MIP and vertebrate insulin on shell growth in *Lymnaea* has not yet been demonstrated.

MDC Produce Insulin-like Peptide (ILP) in *H. Duryi*

The distribution and the fine structure of MDC have been described [17, 26]. There are about 30 MDC in each cerebral ganglion lying close to the dorsal bodies (DB) (Figs. 1–3). As in *L. stagnalis*, the axons of the MDC in *Helisoma* terminate in the median lip nerves which comprise the neurohemal area of the MDC (Figs. 1, 3). When the cells are injected with Lucifer yellow, it can be seen that a few axons of MDC also traverse the cerebral commissure. It is not known where these axons terminate (Figs. 1, 3). In *L. stagnalis*, Benjamin et al. [27] suggested that, while MDC in one cerebral ganglion are mostly coupled, they are not coupled with the contralateral cells. However when Ca^{2+} was replaced by Ba^{2+} in the bathing solution, synchronous bursting activities were recorded from both median lip nerves, which suggested that all MDC/LGC are coupled [28]. In *Helisoma*, axons of MDC from one cerebral ganglion were seen to pass toward contralateral cells. This raises the possibility of coupling between the ipsilateral and contralateral cells. Although *Helisoma* is a hermaphroditic pulmonate, it reproduces by cross fertilization. Reproducing animals grow more slowly than virgin and castrated snails [17]. Extracellular electrical recordings from the median lip nerve (the neurohemal area of MDC) showed very little activity in randomly mating snails, whereas recordings from virgin and castrated snails showed increased frequency of spontaneous action potentials (Fig. 4). This finding suggests an increased release of neurosecretory material from the axons of MDC located in the median lip nerves of virgins but not mated snails, further suggesting that the MDC are involved in the control of shell growth. The MDC in *Helisoma* were found to be about 45 μm in diameter and were full of neurosecretory granules measuring 150 nm in diameter (Fig. 5). Gap junctions were detected between the cell bodies as well as between axon endings of MDC (Figs. 3, 6, 7). Axons of the MDC terminated at the median lip nerve and were also filled with neurosecretory granules (Fig. 8). The axons showed release of neurosecretory granular content in the connective tissue by single or multiple exocytosis (Figs. 9–12). The rate of exocytosis by these cells was higher in virgins than in mated snails and was also higher during the night. It is worth nothing that

---

Fig. 1. Whole mount of *Helisoma* brain in which 10% Lucifer yellow was microinjected into an MDC (*large arrow*); the dye was subsequently transferred into two other MDC (*small arrows*); the dye-filled axons of MDC are seen in cerebral ganglion (*CG*), cerebral commissure (*CM*), and median lip nerve (*MLN*). *VG*, Visceral ganglion. *Bar* = 100 μm

Fig. 2. Cross-section of the cerebral ganglion (*CG*) of *Helisoma* showing indirect immunofluorescent anti-insulin staining of MDC. *DB*, Dorsal bodies. *Bar* = 30 μm

Fig. 3. Diagrammatic illustration of the supraesophageal portion of the CNS of *Helisoma* showing the *MDC* and the location of *gap junctions*. Axons of MDC do travel to contralateral sides but their final destinations are not known. Not drawn to scale

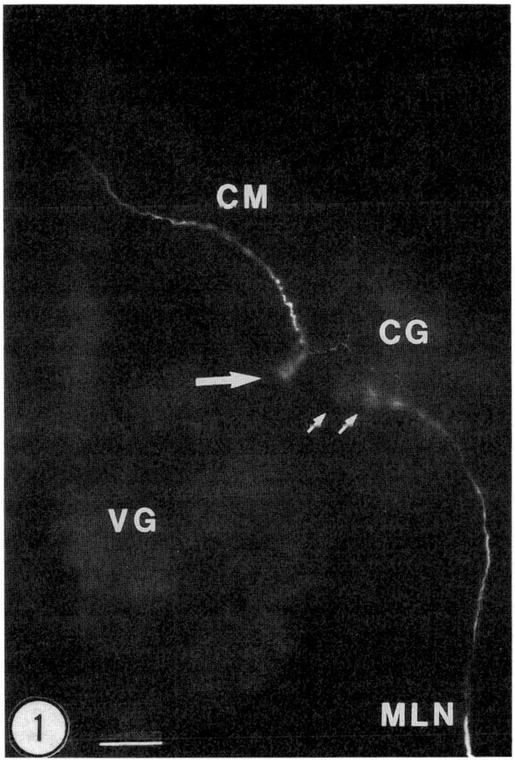
CM
CG
VG
MLN
1

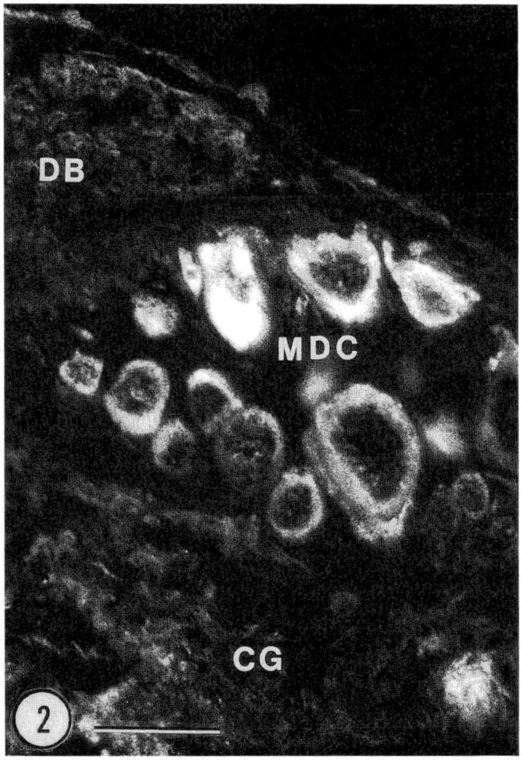
DB
MDC
CG
2

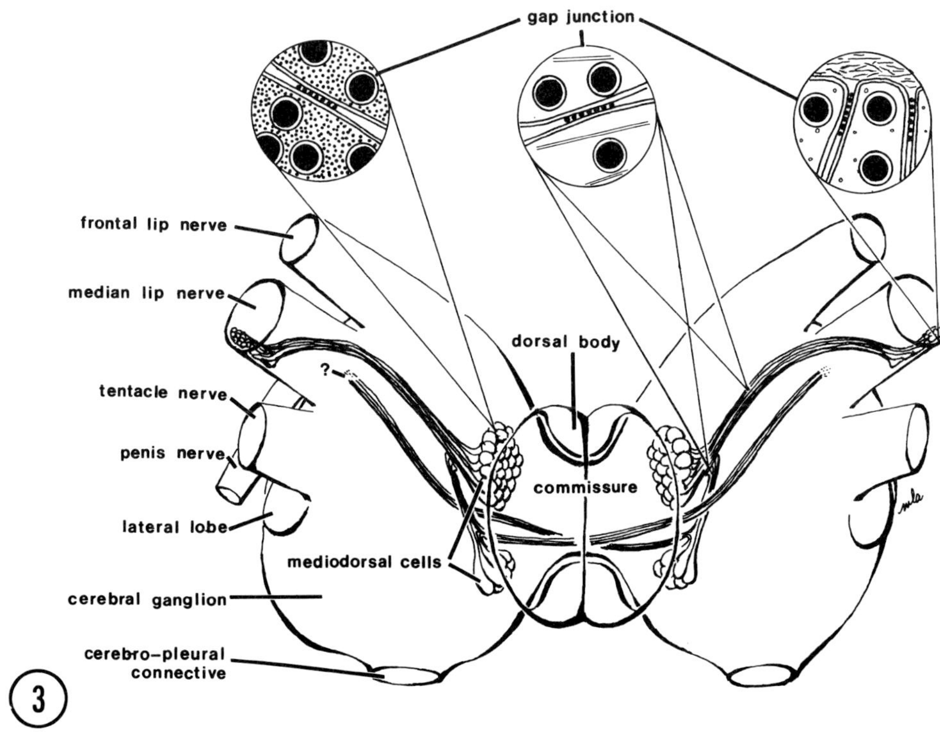
gap junction
frontal lip nerve
median lip nerve
dorsal body
tentacle nerve
penis nerve
commissure
lateral lobe
mediodorsal cells
cerebral ganglion
cerebro-pleural
connective
3

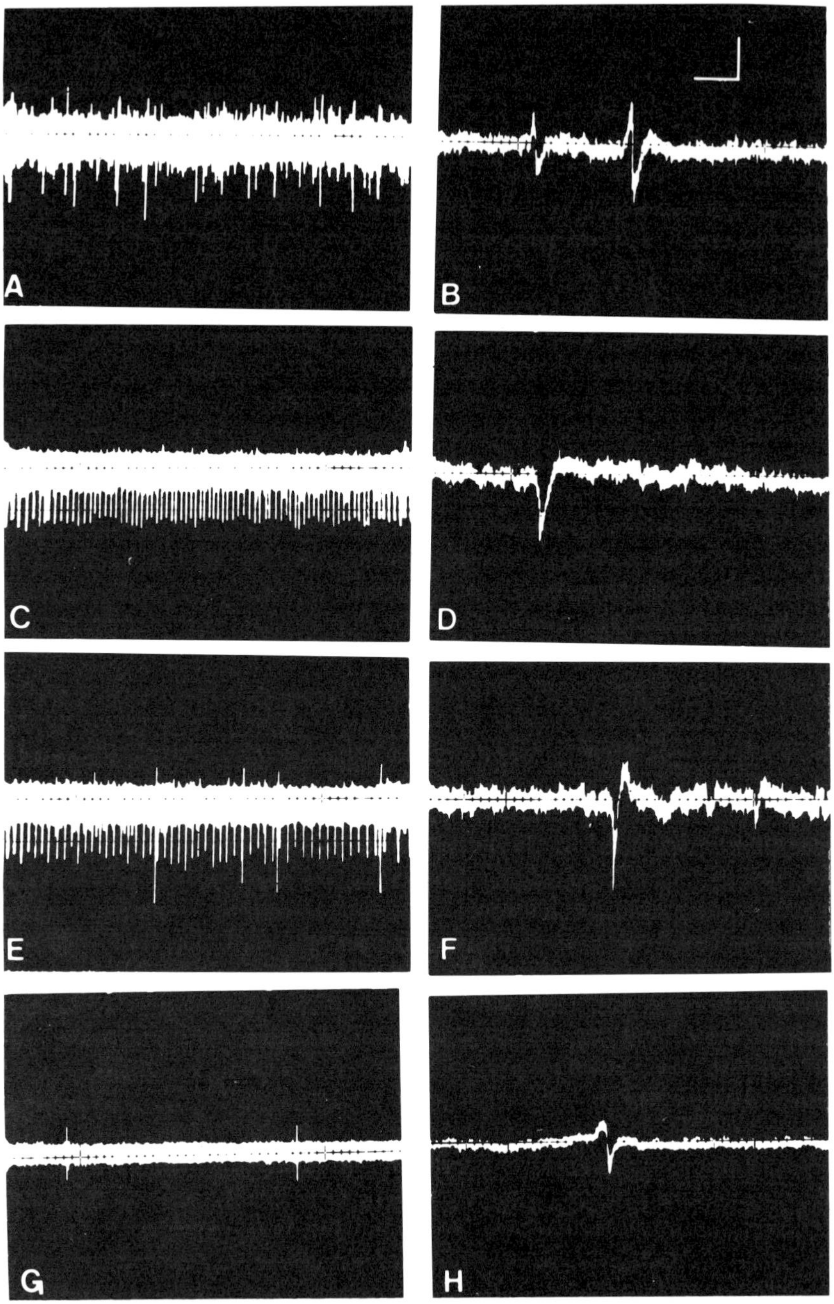

Fig. 4. Spontaneously occurring action potentials recorded extracellularly using suction electrodes from the median lip nerve of virgin (**A–D**) and castrated snails (**E–F**) (*Helisoma*). Little spontaneous activity was observed from the median lip nerve of randomly mating snails (**G–H**). The traces on the *right* are extensions of the time scale from the *left* traces. Scale mark: *left panel*, vertical: 100 µv, horizontal: 1s; *right panel*, vertical: 100 uv, horizontal 10 ms. (From [17] with permission)

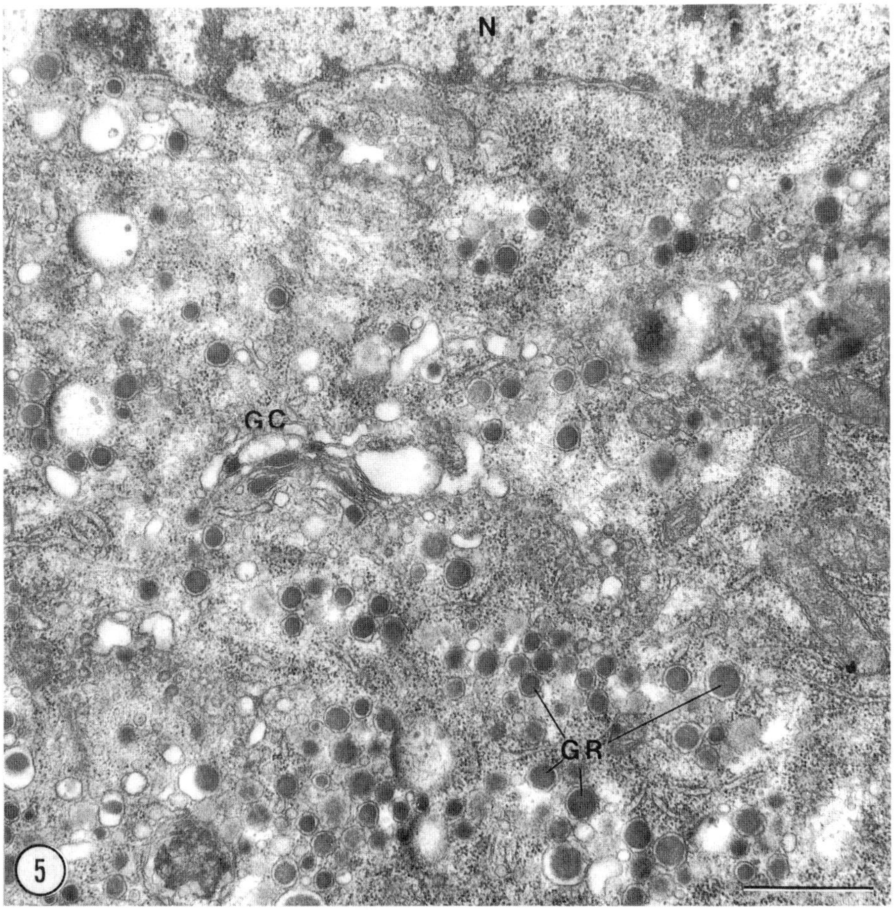

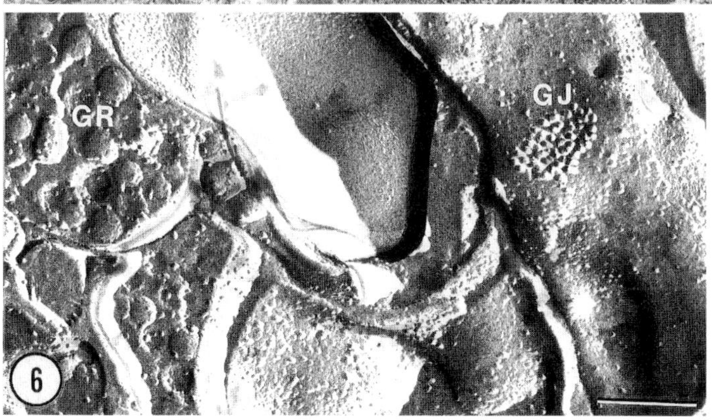

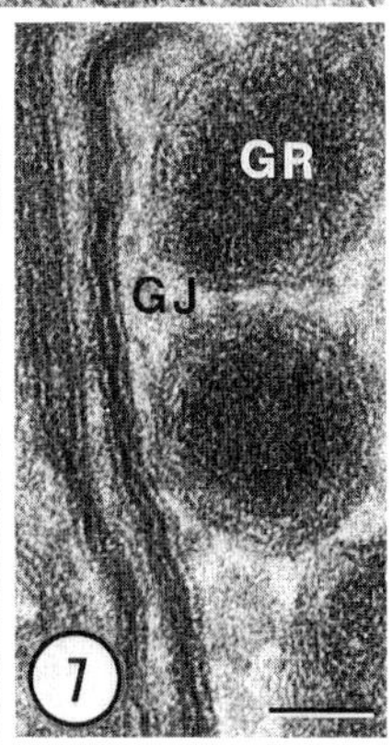

Fig. 5. Cross-section of MDC of a virgin snail showing Golgi complex (*GC*), elementary granules (*GR*), and nucleus. *Bar* = 0.5 μm

Fig. 6. Freeze-fracture replica of the basal portion of an MDC showing gap junctions (*GJ*) and elementary granules (*GR*). *Bar* = 0.2 μm

Fig. 7. *Helisoma*. Cross-section of MDC showing the gap junction (*GJ*) and elementary granules (*GR*). *Bar* = 0.02 μm

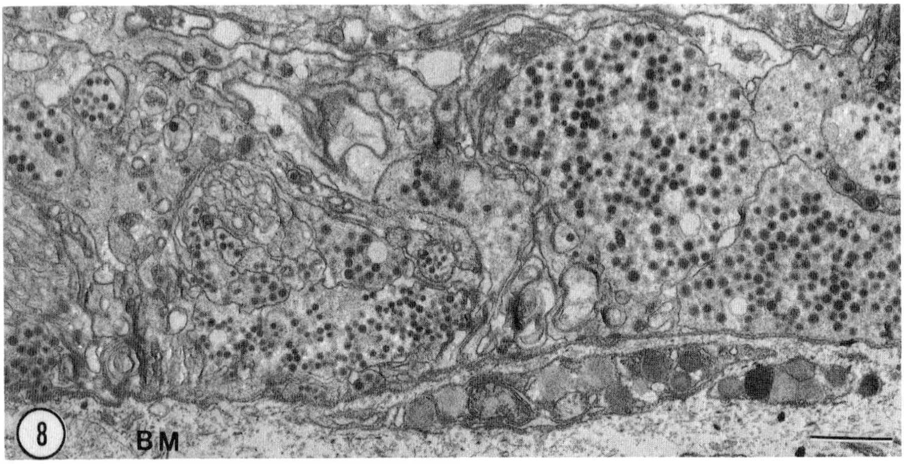

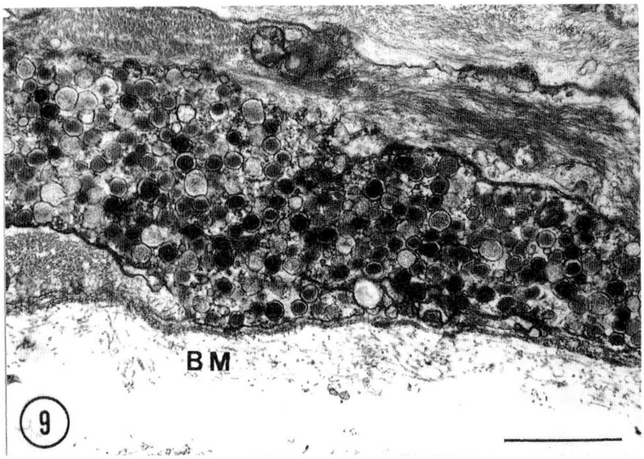

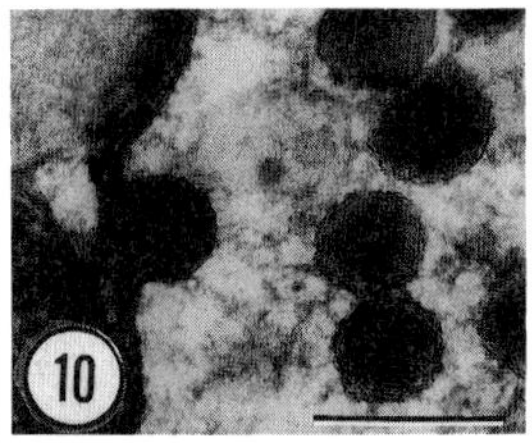

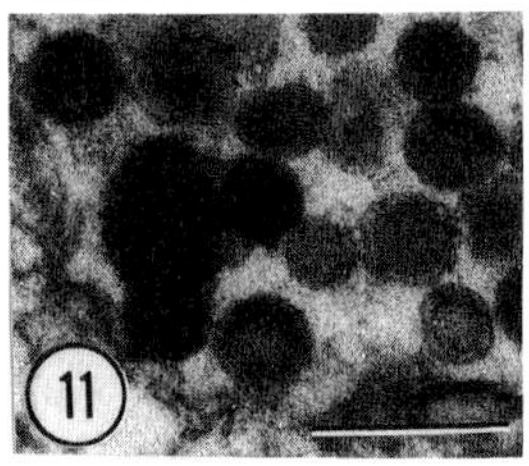

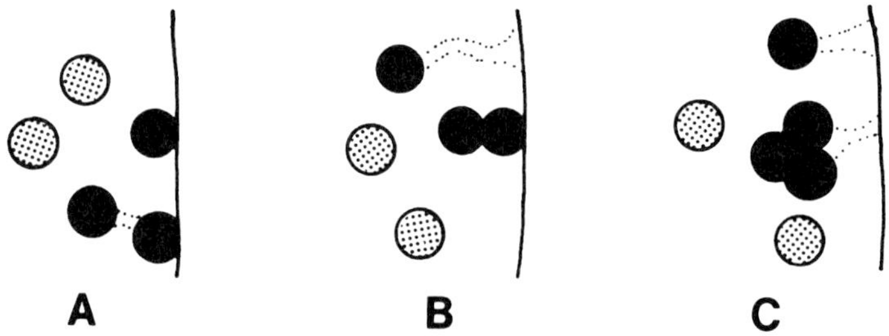

12

Fig. 13. The titer of hemolymph insulin-like peptide (ILP) in slow, medium, and fast growing snails (*Helisoma*)

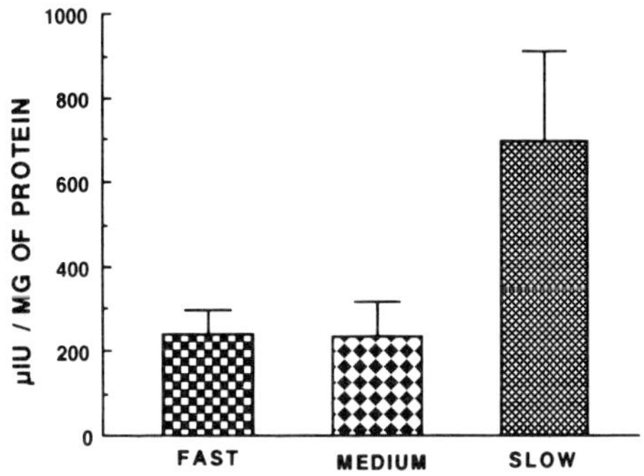

snails maintained in total darkness (00L.24D) grew significantly faster than those in 24L:00D and 12L:12D photoperiods [2, 11].

Using porcine and/or human insulin antiserum, we showed that the MDC as well as the median lip nerve in *Helisoma* contained insulin-like immunoreactive material (Fig. 2). Serial sectioning of the brain showed that not all cells were immunoreactive nor were they uniformly stained. Similar observations have been made on other pulmonates [29]. This may be partly explained in that bovine anti-insulin may not recognize the multiple forms of insulin-like material which are probably present in the population of MDC.

Detection of Insulin-like Peptide (ILP) by Radioimmunoassay (RIA) and Western Blotting in *H. Duryi*

The amount of ILP in the hemolymph of fast, medium, and slow growing snails was measured by RIA against human insulin. Surprisingly, the titer was highest in the slow growing snails (Fig. 13). This can be explained by the fact that the rate of use of ILP might be lowest in slow growing as opposed to fast growing snails, and thus its accumulation in the hemolymph in the former group. This

←

Fig. 8. *Helisoma.* Cross-section of median lip nerve showing the MDC axon terminals at their neurohemal area. *BM*, Basement membrane. *Bar* = 1.0 µm

Fig. 9. *Helisoma.* TARI-treated median lip nerve of a virgin snail in low magnification showing numerous electron dense exocytotically released granules. Note that the unreleased granules are less electron dense. *Bar* = 1.0 µm

Fig. 10. *Helisoma.* Higher magnification showing the excytotic release profile of a single granule. Note that the unreleased granules are less electron dense. *Bar* = 0.2 µm

Fig. 11. *Helisoma.* Release profile of multiple exocytosis. *Bar* = 0.2 µm

Fig. 12. *Helisoma.* Explanation of TARI-revealed exocytosis. **A** Exocytosis of single granules with or without attached plasma membrane. **B** Exocytosis of fused granules. **C** Exocytosis of many granules without attached plasma membrane. *Dotted line* represents membrane invagination or membrane caveolae above or below the plane of sections. Not to scale

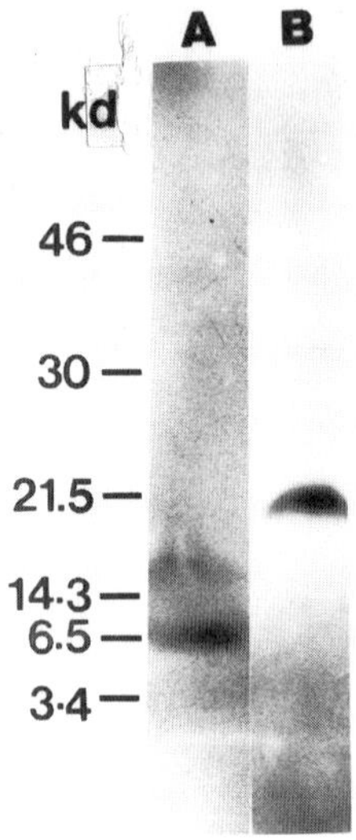

Fig. 14. Immunoblot of ILP in the hemolymph of *Helisoma duryi* (lane **A**) and *Otala lactea* (lane **B**). The samples were resolved by SDS-PAGE and then immunoblotted with bovine insulin antiserum. Note: two bands of 16 kd and 7 kd in *Helisoma* and one band of 21 kd in *Otala*

explanation relies on the premise that the rate of synthesis and release of ILP in slow and fast growing snails are identical. An alternative explanation would be that, at a higher concentration, ILP inhibits shell growth. Both high K^+ and 4- aminopyridine (4-AP) caused an increased release of ILP from MDC (H.R. Khan and A.S.M. Saleuddin, unpublished results). This release was maximum within 4 h after the addition of high K^+ and 4-AP. Similar results have been obtained with *L. stagnalis* [25]. After eletrophoresis in SDS PAGE followed by transfer to nitrocelluose, an immunoblotting technique revealed two insulin immunoreactive bands in the hemolymph of *Helisoma* but only one band in that of *Otala*. The peptides appeared to be 16 and 7 kd in *Helisoma* and 21 kd in *Otala lactea* (Fig. 14).

→

Fig. 15. Scanning micrograph of the mantle edge of *Helisoma* in vitro culture for 48 h. Note the newly formed periostracum (*P*) is protruding from the periostracel groove (*PG*). *VS*, Ventral surface; *MEG*, mantle edge gland. *Bar* = 100 μm

Fig. 16. Diagrammatic longitudinal section of the mantle edge of *Helisoma* showing the general histological features. Periostracum formation is initiated by periostracal cells (*PC*) but is thickened by the mantle edge gland (*MEG*). The shell part (*SH*) is formed by the dorsal epithelium (*DE*). The presumed sites of action of insulin-like peptides are shown by *lines*. *C*, Ciliary tufts; *CB*, receptor cell body; *CE*, ciliated epithelium; *DR*, dorsal receptors; *IN*, intrinsic neuron; *IND*, dorsal plexus intrinsic neuron; *MC*, mucous cell; *NV*, longitudinal branch of the pallial nerve; *PG*, periostracal groove; *RE*, receptor ending; *RP*, receptor process; *T*, tip of ventral mantle edge; *TZ*, transitional zone; *VS*, ventral surface. Not drawn to scale. Modified from [42]

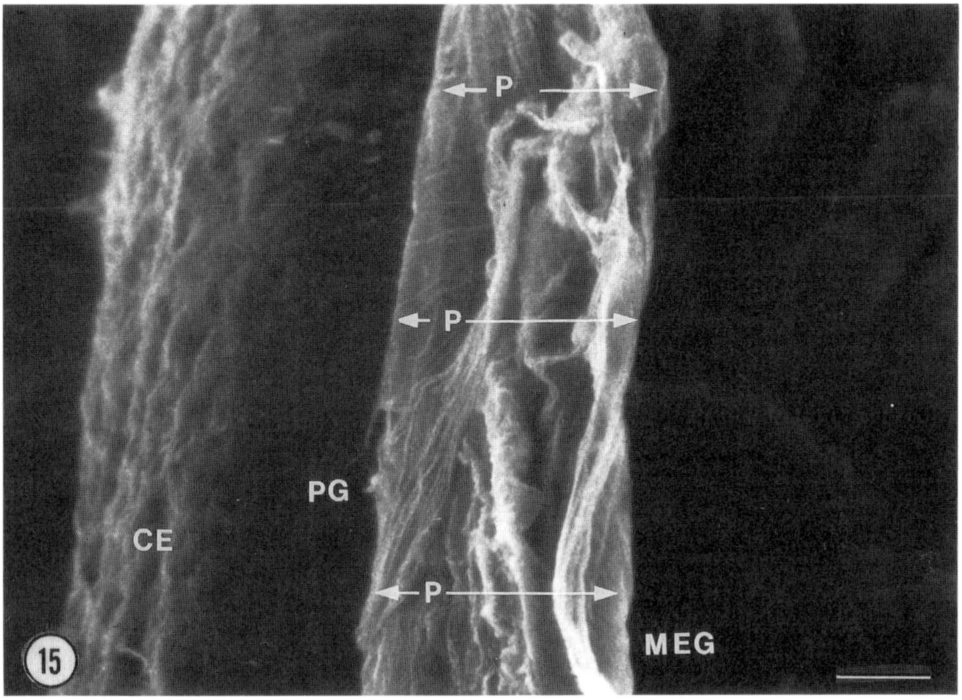
P
P
P
PG
CE
MEG
15

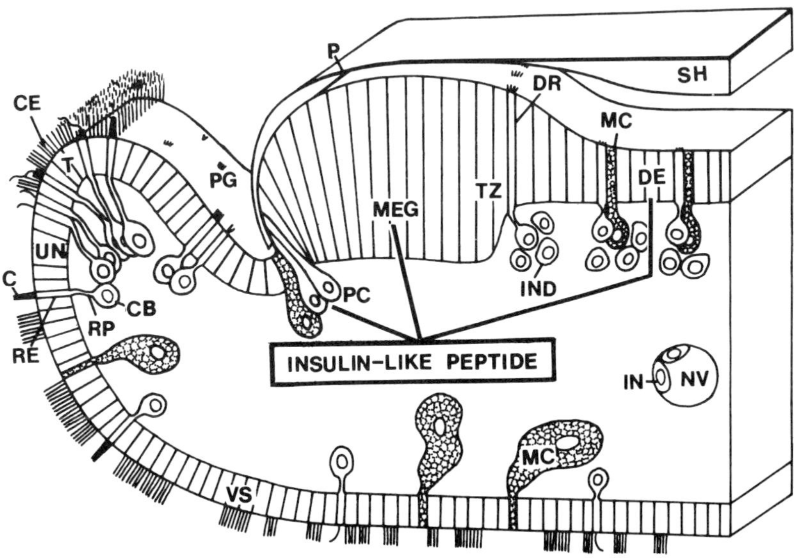
P
SH
DR
CE
MC
T
PG
DE
TZ
MEG
UN
C
CB
PC
IND
RP
RE
INSULIN-LIKE PEPTIDE
IN
NV
MC
VS

16

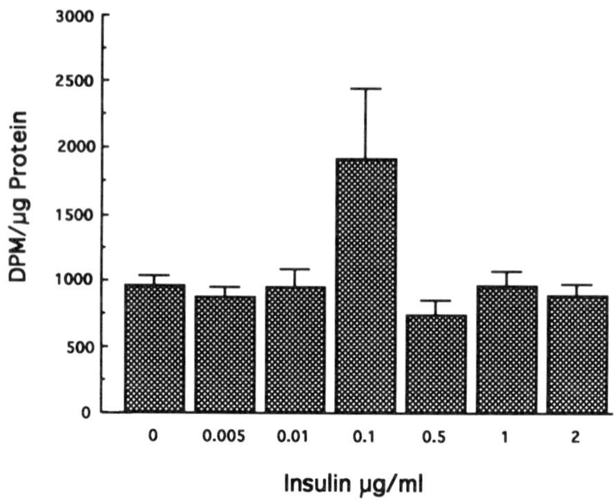

Fig. 17. Dose-response curve of porcine insulin on protein synthesis in the mantle of *Helisoma* in vitro

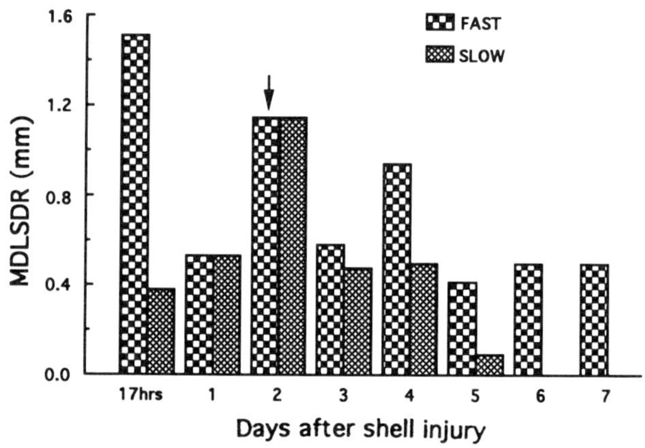

Fig. 18. Shell deposition rate (*MDLSDR*) in fast and slow growing snails (*Helisoma*) following shell injury at the edge. *Arrow* indicates the completion of shell repair

Effect of Porcine Insulin on Shell Growth and Shell Regeneration in *H. Duryi*

When a microgram of porcine insulin was injected into snails, the shell growth was stimulated. The significance of this observation is enhanced by the finding that the injection of porcine insulin antiserum inhibited shell growth [26]. One of the targets of ILP was the mantle which secretes the shell. When cultured in vitro, the mantle secreted the periostracum (Figs. 15, 16), and thus the effect of porcine insulin on periostracum synthesis could be assayed by measuring

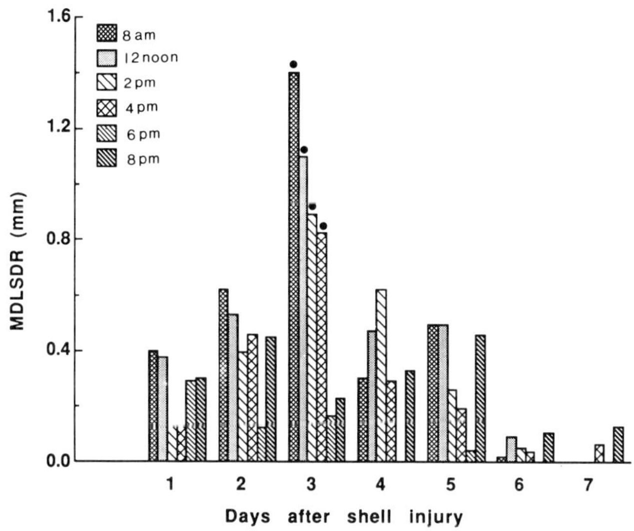

Fig. 19. Shell deposition rate (*MDLSDR*) in *Helisoma* following shell injury at the edge performed at different times of the day. *Solid circles* indicate resumption of shell growth at the edge

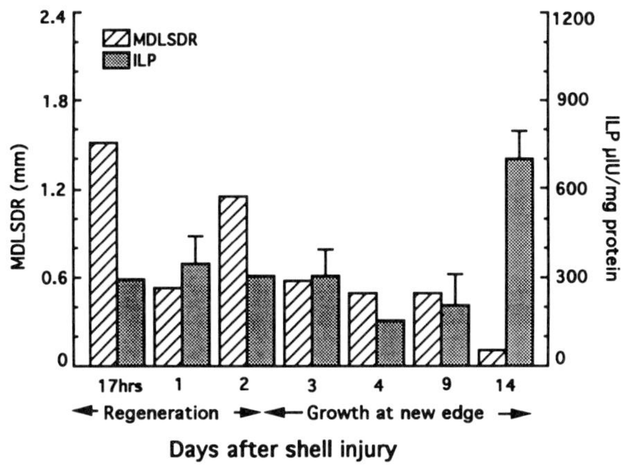

Fig. 20. Fluctuations in the hemolymph titers of ILP following shell injury at the edge in fast growing snails (*Helisoma*)

incorporation of tritiated amino acids into the periostracum in vitro [16, 30]. The mantle from a slow growing snail incorporated a significantly higher amount of amino acids in vitro when porcine insulin was present in the medium (Fig. 17). Similar results were also obtained with insulin-like growth factor II (IGF-II) (results not shown here).

In *Helisoma*, when a piece of shell was removed from the edge of the shell,

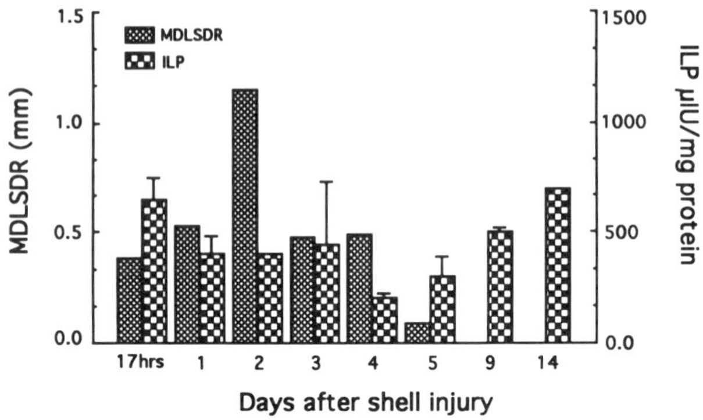

Fig. 21. Fluctuations in the hemolymph titers of ILP following shell injury at the edge in slow growing snails (*Helisoma*)

a new piece rapidly formed [2, 6], making this system suitable for designing experimental protocols for studying the hormonal titer during shell regeneration which involves rapid transport of organic and inorganic components. Normally, in fast growing snails, an injured area of 5 × 5 mm is repaired within 2 days, whereas in slow growing snails, it took nearly 5 days to complete the same repair process. (Fig. 18). In order to determine the best time to obtain the fastest regeneration, we damaged an uniform area of the shell every 2 hours, from 8 AM to 8 PM. Shell regeneration was best when injury was made between 8 AM and 4 PM. At 6 PM as well as at 8 PM, shell regeneration was slow (Fig. 19). Therefore, throughout the shell regeneration study, shells were damaged between 8 AM and 4 PM. The reason for this difference may become apparent when we will later discuss the ILP titer. We also found that the shell regeneration rate did fluctuate. Typically, in fast growing snails, the rate was high in the first 24 h, and the damaged area was fully repaired by day 2. Normal growth at the edge resumed by day 3 and continued at a steady rate until day 9 (Fig. 20). In slow growing snails, the duration of shell repair of the damaged area was longer, 5 days instead of 2 (Fig. 21). The peak rate of repair rate occurred at day 2, followed by a reduced rate until day 9, when there was no shell growth at all. Shell growth at the edge resumed at day 14 (results not shown). Having found differences in shell growth rates during shell regeneration in fast and slow growing snails, we wanted to see if ILP titers fluctuate during shell regeneration and, if they did indeed fluctuate, how did they correlate to shell repair rates in fast and slow growing snails? Figure 20 shows the ILP titer pattern during shell regeneration in fast growing snails over a 14-day period. During the first 24 h, when the repair rate was maximal, the ILP titer was low. At day 14, when the shell growth was low, the ILP titer was high. In slow growing snails, the general pattern of titer was similar to fast growing ones, in that the titer was high when the shell repair rate was slow and vice versa. Again, when the animals stopped growing, the titers were high (Fig. 21). It is worth emphasizing that ILP titers were higher in slow growing than in

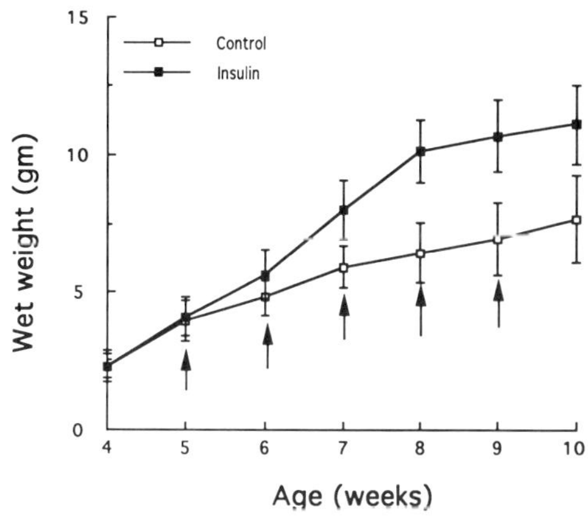

Fig. 22. The effect of bovine insulin on growth (weight in grams) in young *Helix aspersa maxima* during 6 weeks of study: 0.5 μg insulin/gram weight was dissolved in 10 μl 1% bovine serum albumin in saline. *Arrows* indicate when injections of insulin were made (Courtesy of A. Gomot and L. Gomot)

fast growing snails. The results we obtained from shell regeneration studies seem to correspond well with shell growth as far as the relationships of ILP titer of shell growth is concerned.

Is ILP Titer Controlled by Photoperiod in *H. Duryi*?

To study this, we first determined ILP titers every 2 hours over a 12L:12D period. Two major peaks were detected, one at 8:30 AM which followed the scoto phase, and the other at 6 PM immediately prior to the lights-off time (7:30 PM). The titer remained low during the scoto phase, with a small peak at 2 AM. When the snails were acclimatized in 24D:OL, two peaks were still present, but they had shifted by about 2 h, occurring at 6 AM and 4 PM. These peaks were also present when snails were raised in 24L:OOD, but the ILP titers decreased significantly. These experiments do not prove that ILP titer follows a circadian rhythm but rather suggest the possibility. We are now conducting further experiments to demonstrate conclusively the presence of a rhythm in ILP titer and that it is influenced by photoperiod. Such rhythm in hormone titers occurs in vertebrates and insects [31, 32]. As far as we are aware, this is the first report of the presence of a rhythm of ILP in molluscs. ILP could be involved in shell growth and/or shell regeneration by acting in concert on three main cellular components, the periostracal cells, mantle edge gland, and the dorsal epithelium, on the mantle edge (Fig. 16).

ILP in Other Molluscs

We have already mentioned that LGC and canopy cells in *L. stagnalis* transcribe molluscan insulin-like peptides whose molecular structure has been determined by recombinant DNA technology [24]. Previous to this finding, the presence of insulin-like material in molluscs was suggested by Thorpe and Duve [33]. By immunohistochemical techniques using anti-MIP-C, immunoreactive cells have been localized in neurons in the CNS of several molluscs [21]. In *Helix aspersa*, immunoreactive cells were detected in the cerebral ganglia and, in *Limax maximus*, similar cells were found in the pedal and pleural ganglia. In *Aplysia californica*, immunoreactive cells were found in all ganglia of the CNS [21]. Using anti-porcine insulin, we have localized immunoreactive cells in all ganglia of the CNS of *Otala lactea*; however, many were located in the mesocerebrum (H.R. Khan and A.S.M. Saleuddin, unpublished results). ILP has also been found in the hemolymph, digestive gland (hepatopancreas), and in brain extracts of *Otala lactea* and *H. aspersa*. The amounts of ILP in the hemolymph and digestive gland are 90 ± 20 uIU/mg protein and 1000 ± 50 uIU/mg protein, respectively, in *H. aspersa*, and 413 ± 97 uIU/mg protein and 2077 ± 56 uIU/mg protein, respectively, in *O. lactea*. Injections of microgram quantities of bovine insulin caused young *H. aspersa maxima* to grow significantly large compared to controls over a 6-week period (A. Gomot and L. Gomot, unpublished results) (Fig. 22). A few comments about the use of immunocytochemical data using mammalian insulin antiserum are necessary here. It is well accepted that positive reactions of cells/products to certain antibodies do not necessarily mean that the product(s) are identical to the structure of the antigen generating the probe. However, immunohistochemistry has been a popular technique to localize vertebrate-like peptides in invertebrates and thus a large volume of work has been published in recent years. In many of these publications, only the localization of vertebrate hormones in various tissues of invertebrates has been sought, without considering the possible functions of these peptides/hormones. Our rationale for localizing ILP is that a molecule of this type is involved in shell growth, as we have documented by in vitro and in vivo experiments in this study. We were at first surprised to find that porcine insulin and IGF-II stimulated the periostracum synthesis by the mantle of *Helisoma* in vitro and, furthermore, that the injection of anti-insulin into snails caused retardation in shell growth. The mammalian insulin receptor binding site involves residues 8, 9, and 10 in the A chain and residues 2, 3, and 4 in the B chain. These residues have proven to vary considerably in many invertebrate insulin-like peptides [25]. Of course it is possible that some other site on the porcine insulin molecule is responsible for binding to the *Helisoma* ILP receptor or that the *Helisoma* receptor binding site is by chance very similar to that of the vertebrate. It has been suggested that mammalian anti-insulin does not bind to molluscan insulin-like peptide (MIP) or to other invertebrate insulin-like peptides; however, we can detect ILP in the hemolymph, CNS, and digestive gland of *Helisoma*, *Helix*, and *Otala* with anti-human insulin ([26]). Perhaps the ILP in *Helisoma* is structurally more similar to mammalian insulin than is *Lymnaea* MIP. Alternatively, the receptors for ILP in *Helisoma* tissues have a wider receptivity than the more restrictive MIP receptors.

Our findings demonstrate the presence of an ILP in the neurosecretory MDC and that this peptide is involved in shell growth. We are now using recombinant DNA technology to elucidate the gene structure and expression of ILP in *Helisoma*.

Somatostatin-like Growth Factor in Pulmonate Molluscs

Schot et al. [23] reported that LGC in *L. Stagnalis* contained immunoreactive somatostatin-like material (SSLM) and, because these cells were implicated in the production of a growth hormone, SSLM was speculated to be the growth hormone in this snail. This report prompted Grimm-Jørgensen [34] to undertake a series of experiments to investigate the role of SSLM on shell growth in *L. stagnalis* and *Physa*. She found a significantly higher amount of SSLM in fast growing young snails than in slow growing older snails. By HPLC, it was shown that SSLM occured in the blood and that the SSLM originated from the CNS. Finally, by use of in vitro bioassay, she showed that a factor from the CNS stimulated the secretion of tyrosine-rich material by the mantle. However, when mammalian somatostatin was used, the synthesis and secretion of tyrosine-rich material was not altered. In *H. aspersa*, Marchand et al. [35] detected SSLM in the hemolymph, CNS, hepatopancreas (digestive gland), and mantle by immunocytochemistry and immunoassay. They observed fluctuations in the level of SSLM in the hemolymph and in other tissues following shell injury in young and adult snails. The following major points emerge from their studies. Except in the CNS, the level of SSLM is higher in young than old, shell regenerating snails. Regenerating snails from short days (8L:16D) have more SSLM than those kept in long days (18L:6D). In young, shell regenerating snails, the level of SSLM increases during the early phase (2–3 h) whereas it increases during the later part of shell regeneration (24–48 h) in adults. They also found that injections of nanogram quantities of mammalian somatostatin caused an increase not only in the volume of the periostracal gland but also in its secretory products in both adult and young snails. The SSLM was purified and partially sequenced and it was found to be quite different from mammalian somatostatin [36]. These studies [34–36] suggested that an SSLM is involved in shell formation and/or shell regeneration and thus it could be a growth hormone. Ebberink and Joosse [37] suggested that the growth hormone in *L. stagnalis* has a molecular weight of 1 kd and thus is in line with the molecular weight of mammalian somatostatin 14. They found the growth factor activities in the two median lip nerves which comprise the neurohemal area of LGC. Median lip nerves are also the sites from which DeWith et al. [38] isolated two natremic factors, each with a molecular weight of 4 kd, which are involved in Na^+ influx into the skin (mantle) of *L. stagnalis*. Joosse [39] suggested that the natremic factors are synthesized by LGC. Smit et al. [24] isolated and sequenced an insulin-like peptide from the peptides of the median lip nerves by using recombinant DNA techniques and termed it molluscan insulin-like peptide (MIP). In situ hybridization showed that LGC and canopy cells of the lateral lobes produce this peptide. So far, we have seen that LGC in *L. stagnalis* possibly produce a growth hormone (a SSLM), two natremic factors, and an MIP. In *H. duryi*, the MDC are homologous to LGC of *L. stagnalis*, but we prefer to call them MDC because of their location in the cerebral ganglia and because they do not stain green with the alcian yellow and alcian blue technique. These cells seem to produce an immunoreactive ILP. Various lines of studies in our laboratory indicated that this peptide acts as a shell growth hormone and/or perhaps as a general growth hormone. Both in vivo and in vitro experiments showed that mammalian insulin and insulin-like growth factors (IGF) stimulate periostracum synthesis as well as shell growth.

While the presence of MIP in LGC in *L. stagnalis* is conclusive, the occurrence of natremic factors and SSLM in these cells is not. Co-localization of these peptides in the small cell has not been shown. However, it is likely that among LGC, 3 distinct populations exist, each producing a category of peptide(s) mentioned above. It is tempting to suggest that MIP and SSLM-like producing LGC cells occur in close association, and that SSLM cells control MIP cells. Similar close associations have been found between B cells and immunoreactive somatostatin-14 cell in the pancreas of salmonids [40]. We are now working on co-localization of SSLM-like and ILP in MDC in *Helisoma*. Additionally, we are studying the effect of mammalian somatostatin on periostracal protein synthesis in the mantle in vitro in order to determine whether or not its effects in growth are direct.

Growth Factor in Non-Pulmonate Molluscs

As already mentioned, Lubet [4] was the first to show that the cerebral ganglia of the prosobranch *Crepidula fornicata* contained a factor which stimulated body growth. Le Gall [41], while working on the same prosobranch, performed a series of extirpation and reimplanation experiments to show that certain cells in the anterior parts of the cerebral ganglia were involved in body growth. Extirpation of these cells caused cessation of body growth. This could be partly corrected by implantation of cerebral ganglia. Furthermore, he reported that severing of nerves arising from the cerebral ganglia also caused retardation of body growth. He therefore concluded that both hormonal and nervous stimuli were necessary for body growth in *Crepidula*. It should be noted that he did not study shell growth per se.

Conclusions

The use of antisera against well characterized vertebrate hormones allowed the recognition and localization of related peptides in pulmonate molluscs. ILPs appear to be involved in the control of growth (shell) and shell regeneration. In addition, there seems to be a circadian rhythm of ILP in *H. duryi*. Although ILPs appear to be involved in the regulation of shell/body growth, at least one other peptide (somatostatin-like), originating from the same group of cells as the ILPs, has substantial effect on growth/shell in *Lymnaea*. While comparable studies in non-pulmonate molluscs are lacking to date, there is evidence of the presence of a growth factor in the CNS.

Acknowledgements. This work was supported by NSERC of Canada. We thank Professor L. Gomot of université de Besançon, France, for allowing us to use unpublished results. Thanks are also due to Dr. B.G. Loughton for help in preparing the manuscript and to Ms M-L. Ashton for drawing Fig. 3.

References

1. Lutz RA, Rhoads DC (1980) Growth patterns within the mollusan shell. In: Rhoads DC, Lutz RA (eds) Skeletal growth of aquatic organisms. Plenum, New York, PP 203–254
2. Kunigelis SC, Saleuddin ASM (1978) Regulation of shell growth in the pulmonate gastropod *Helisoma duryi*. Can J Zool 56:1975–1980
3. Dillaman RM, Saleuddin ASM, Jones GM (1976) Neurosecretion and shell regeneration in *Helisoma duryi* (Mollusca: Pulmonata). Can J Zool 54:1771–1778
4. Lubet P (1971) Influence des ganglion cérébroides sur la croissance de *Crepidula fornicata* Phil. (Mollusque Mésogastéropode). C R Acad Sci [III] 273:2309–2311
5. Wong V, Saleuddin ASM (1972) Fine structure of normal and regenerated shell of *Hetisoma duryi*. Can J Zool 50:1563–1568
6. Kunigelis SC, Saleuddin ASM (1983) Shell repair rates and carbonic anhydrase activity during shell repair in *Helisoma duryi* (Mollusca). Can J Zool 61:597–602
7. Saleuddin ASM, Wilbur KM (1969) Shell regeneration in *Helix pomatia*. Can J Zool 47:51–53
8. Saleuddin ASM, Chan W (1969) Shell regeneration in *Helix*: Shell matrix composition and crystal formation. Can J Zool 47:1107–1111
9. Saleuddin ASM (1970) Electron microscopic study of the mantle of normal and regenerating *Helix*. Can J Zool 48:409–416
10. Khan HR, Saleuddin ASM (1979) Osmotic regulation and osmotically induced changes in the neurosecretory cells of the pulmonate snail *Helisoma*. Can J Zool 57:1371–1383
11. Saleuddin ASM, Kunigelis SC, Khan HR, Jones GM (1980) Possible control mechanisms in mineralization in *Helisoma duryi* (Mollusca: Pulmonata). In: Omori M, Watabe N (eds). The mechanisms of biomineralization in animals and plants. Tokai University Press, Tokyo, pp 121–129
12. Subramanyan OV (1973) Neuroendocrine control of calcium levels in the blood of *Ariophanta semirugata*, a terrestrial pulmonate snail. Endocrinol Exp (Bratisl) 7:315–317
13. Burton RF (1977) Haemolymph calcium in Helix pomatia: Effects of EGTA, ganglion extracts, ecdysterone, cyclic AMP and ionophase A23187. Comp Biochem Physiol [C] 57:135–137
14. Geraerts WPM (1976a) Control of growth by the neurosecretory hormone of the light green cells in the freshwater snail *Lymnaea stagnalis*. Gen Comp Endocrinol 29:61–71

15. Geraerts WPM (1976b) The role of lateral lobes in the control of growth and reproduction in the hermaphrodite freshwater snail *Lymnaea stagnalis*. Gen Comp Endocrinol 29:97–108
16. Kunigelis SC, Saleuddin ASM (1984) Studies on the *in vitro* formation of periostracum in *Helisoma duryi*: The influence of the dorsal epithelium on calcium incorporation. Can J Zool 62:1177–1180
17. Saleuddin ASM, Mukai ST, Khan HR (1990) Hormonal control of reproduction and growth in the freshwater snail *Helisoma duryi* (Mollusca: Pulmonata). In: Loughton BG, Saleuddin ASM (eds) Neurobiology and endocrinology of selected invertebrates. Captus University Press, Toronto, pp 163–182
18. Dogterom AA, van Loenhout H, vander Schors RC (1979) The effect of growth hormone of *Lymnaea stagnalis* on shell formation. Gen Comp Endocrinol 39:63–68
19. Dogterom AA, Jentjens T (1980) The effect of growth hormone of the pond snail *Lumnaea stagnalis* on periostracum formation. Comp Biochem Physiol [A] 66: 687–690
20. Dogterom AA, Doderer A (1981) A hormone dependent calcium-binding protein in the mantle edge of the freshwater snail *Lymnaea stagnalis*. Calcif Tissue Res 33:505–508
21. Boer HH, van Minnen J, Kerkhoven RM (1991) Evolutionary aspects of morphology and function of peptidergic systems, with particular reference to the pond snail *Lymnaea* stagnalis. In: Florey E, Stefano GB (eds) Comparative aspects of neuropeptide function. Manchester University Press, Manchester, pp 154–171
22. Kunigelis SC, Saleuddin ASM (1985) Studies on the *in vitro* formation of periostracum in *Helisoma duryi*: The influence of the brain. J Comp Physiol 13:177–183
23. Schot LPC, Boer HH, Swaab DF, van Noorden S (1981) Immunocytochemical demonstration of peptidergic neurons in the central nervous system of the pond snail *Lymnaea stagnalis* with anti-sera raised to biologically active peptides of vertebrates. Cell Tissue Res 216:273–291
24. Smit AB, Vreugdenhil E, Ebberink RHM, Geraerts WPM, Klootwijk J, Joosse J (1988) Growth controlling molluscan neurons produce the precursor of an insulin-related peptide. Nature 331:535–538
25. Ebberink RHM, Smit AB, van Minnen J (1989) The insulin family: Evolution of structure and function in vertebrates and invertebrates. Biol Bull 177:176–182
26. Saleuddin ASM, Khan HR, Sevala M, Sevala V (1991) Hormonal control of confirmed shell growth in the snail *Helisoma duryi* (Mollusca: Gastropoda). In: Suga S, Nakahara H (eds) Mechanisms and phylogeny of mineralization in biological systems. Springer, Tokyo, pp 161–165
27. Benjamin PR, Swindale NV, Slade CT (1976) Electrophysiology of identified neurosecretory neurones in the pond snail *Lymnaea stagnalis* (L.). In: Salanki J (ed) Neurobiology of invertebrates. Akademiai Kiado, Budapest, pp 85–100
28. Vleiger TA de, Lodder JC (1983) Morphological and electrophysiological features of the growth producing light green cells of *Lymnaea stagnalis*. In: Lever J, Boer HH (eds) Molluscan neuroendocrinology. North-Holland, Amsterdam, pp 81–85
29. Minnen J van, Schallig H (1990) Demonstration of insulin-related substances in the central nervous system of pulmonates and *Aplysia californica*. Cell Tissue Res 260:381–386
30. Saleuddin ASM, Kunigelis SC (1984) Neuroendocrine control mechanisms in shell formation. Am Zool 24:911–916
31. Saunders DS (1979) Insect clocks. Pergamon, Oxford, pp 31–40

32. Menaker M, Binkley S (1981) Normal and endocrine control of circadian rhythms in the vertebrates. In: Aschoff J (ed) Handbook of behavioral neurobiology, vol. 4. Biological rhythms. Plenum, New York, pp 234–255
33. Thorpe A, Duve H (1984). Insulin and glucagon-like peptides in insects and molluscs. Mol Physiol 5:235–260
34. Grimm-Jørgensen Y (1983) Possible physiological roles of thyrotropin releasing hormone and a somatostatin-like peptide in gastropods. In: Lever J, Boer HH (eds) Molluscan neuroendocrinology. North-Holland, Amsterdam, pp 21–27
35. Marchand CR, Assaka L, Strosser MT (1989) Variations of somatostatin-like immunoreactivity in the circumesophageal ganglia, the hepatopancreas, the mantle edge and the hemolymph of shell-repairing snails (*Helix aspersa*). Gen Comp Endocrinol 73:59–68
36. Assaka L (1988) Etude biochimique d'une (ou de) substance(s) immunologiquement apparentée(s) ã la somatostatine-14 chez l'escargot *Helix aspersa*: Correlation avec la croissancé et lá réparation de la coquille. Thesis, Université de Besancon, France
37. Ebberink RHM, Joosse J (1985) Molecular properties of various snail peptides from brain and gut. Peptides 6 (Suppl):451–457
38. de With ND, Ebberink RHM, vander Schors RC (1987) Neurohormonal control of sodium metabolism in the pulmonate freshwater snail *Lumnaea stagnalis*. In: Boer HH, Geraerts WPM, Joosse J (eds) Neurobiology, molluscan models. North-Holland, Amsterdam, pp 357–359
39. Joosse J (1988) The hormones of molluscs. In: Laufer H, Downer RGH (eds) Endocrinology of selected invertebrate types. Alan R. Liss, New York, pp 89–140
40. Nozaki M, Miyata K, Oota Y, Gorbman A, Plisetskaya EM (1988) Different cellular distributions of two somatostatins in brain and pancreas of salmonids, and their associations with insulin and glucagon secreting cells. Gen Comp Endocrinol 69:267–280
41. Le Gall P (1980) Etude expérimentale de l'association en chaine et de son influence sur la croissance et la sexualité chez la crépidule *Crepidula fornicata*. Université de Caen, France, pp 1–251
42. Jones GM, Saleuddin ASM (1978) Cellular mechanisms of periostracum formation in *Physa* spp. (Mollusca: Pulmonata). Can J Zool 56:2299–2311

Phosphoproteins of Oyster (*Crassostrea virginica*) Shell Organic Matrix

A.P. WHEELER[1]

Summary. As in most biominerals, the oyster shell is a composite of inorganic crystals and an organic matrix. Its mineral is primarily calcitic calcium carbonate and its matrix is made up largely of highly acidic phosphoproteins. Among molluscs, this type of matrix may be confined to those which form the foliated microstructure. These phosphoproteins form a continuum of molecular sizes ranging from modest molecular weight soluble proteins to insoluble complexes. The fact that similar sequences or even peptides can be identified among the various size classes suggests that all the classes may have common precursors. The primary structure of the major reverse phase HPLC fraction is roughly analogous to the phosphoryns of tooth dentin, in that it contains domains enriched in aspartic acid, other domains enriched in phosphoserine, and a terminal hydrophobic domain. The proteins have high affinities for crystal nuclei and calcite crystals, but not calcium, at high ionic strength. From these activities, it is hypothesized that the insoluble complexes bind nuclei, initiating crystal or crystal layer growth, and that the soluble components adsorb to crystal surfaces and, thereby, both control crystal morphology and limit crystal growth.

Key words: Phosphoproteins—Matrix—Mineralization—Shell—Molluscs

Introduction

The calcium carbonate shells of molluscs are similar to most biominerals in that they contain a small percentage of organic matrix by weight. One of the main functions of the matrix is to produce a composite with structural characteristics far more resistant to stress than that of inorganic carbonate. Further, as is generally postulated for biominerals, it seems evident that the matrix serves to regulate the growth of the mineral phase [1], producing structures clearly different from those grown inorganically.

[1] Department of Biological Sciences, Clemson University, Clemson, SC 29634, USA

In many ways the molluscan shell appears ideal for studying the role of matrix in mineral formation. The biomineral is plentiful and readily prepared for study. Furthermore, the process of matrix mineralization occurs in an extracellular compartment (the extrapallial fluid) defined by the forming shell on one side and a tissue layer (outer mantle epithelium) on the other. Therefore, the mineralization environment itself is theoretically accessible.

One of the limitations of using the molluscan shell system as a model for mineralization is the diversity of shell microstructures produced by this group. Some structures have been studied more than others; for example, microscopic observation of mature and growing nacre abound and some data on the matrix of this structure have been accumulated [2]. However, even the studies of nacre is scattered over a wide number of species.

To partially avoid the dilemma of trying to provide an appropriately balanced treatment of the findings from so many systems, studies on the matrix from a single species, the oyster *Crassostrea virginica*, will be highlighted and used as a point of departure for discussing results obtained for other molluscan matrix or matrix analog systems. The oyster is perhaps no better a model than any other molluscan system and no doubt has many properties that are not general to other molluscan systems. In particular, the presence of significant quantities of highly phosphorylated proteins may be, among the bivalves, unique to the foliated microstructure which makes up the majority of the oyster shell [3]. Interestingly, these matrix proteins may be analogous to similar proteins found in calcium phosphate systems, especially the phosphophoryns of tooth dentin [4]. However, it should be noted that the oyster, in general, is a well-studied species. Further, there has recently been an accumulation of data on the oyster matrix in which the author has been actively involved. It is hoped that this contribution will serve to draw attention to this interesting class of proteins.

Shell and Matrix Structure

Typical of many molluscs, the oyster shell is constructed principally of two layers which are microstructurally distinct. These are a thin outer prismatic layer and an inner foliated (calcitostracum; pseudonacre) layer which makes up the majority of the shell. Much like nacre, the foliated microstructure is laminar, with each sheet (laminae; lamellae) of the layer being made up of flattened polygonal crystal tablets (folia) which grow together. In both cases, the appropriate vertical sections through several lamellae can yield a "brick wall" appearance (Fig. 2) and the *c*-axis of the individual crystals is found more or less along the thickness of the crystals and approximately perpendicular to the surfaces of the shell [5]. In contrast to nacre, the crystals of the foliated structure are more elongated and may be irregular and are composed of calcite rather than aragonite. In addition, the manner in which the crystals grow and the final faces expressed is no doubt different for the two microstructures.

Several classes of matrix have been identified in molluscan shell [5]. They include fibrous sheets which may be reticulated [6] and which lie between layers of crystals (interlamellar) and at the edges of crystals (intercrystalline) (Fig. 2). In addition, the crystals themselves may be divided into blocks by matrix [5]. In some systems, the matrix is layered—especially the interlamellar matrix—with the innermost (crystal-facing) layer of the interlamellar and intercrystallic matrix having been referred to by some as the envelope [7]. In the oyster, the interlamellar matrix appears poorly developed and the intercrystalline matrix is quite thin. However, intracystalline sheets are evident, dividing the crystals vertically [5].

In addition to the morphological designations of matrix, a class of matrix has been identified based on solubility when shell is dissolved with dilute acids or, more commonly, near-neutral solutions of EDTA [8]. Although, as expected, much of the matrix is often insoluble upon dissolution of the shell (insoluble matrix; IM), soluble proteins and glyoproteins have been released upon shell dissolution as well in every case to date. The soluble matrix (SM) proteins are often highly anionic, deriving their charge from the carboxylate groups of aspartic and, to a lesser extent, glutamic acid, phosphoserine, and sulfated carbohydrate residues [1]. In general, it is thought that these soluble proteins play a significant role in crystal growth regulation in situ, as these proteins are capable of affecting mineral growth in vitro (see below). One should bear in mind that, at this stage, no single biological function has been definitely assigned to any given matrix protein.

The origin of the SM is in doubt, but it may derive from envelope layers of the intercrystalline and interlamellar matrix. This conclusion comes from the fact that certain species appear to have only this layer [7] and that the matrix from these species is almost entirely soluble [9]. In addition, dissolution of pieces of prismatic shell by acid can ostensibly result in isolated envelopes which, when analyzed, contain highly anionic proteins [10]. However, other origins of SM have been mentioned, including the reticulum of the sheets [11] and the intracrystalline matrix. Evidence for the latter can be inferred from microscopic observations of dissolved or partially dissolved sections of shell, including that of the oyster [5], and from the observation that, compared to the insoluble matrix, the soluble fraction appears to be selectively protected from oxidation when powdered shell is exposed to oxidant [8].

Isolation and Fractionation of Oyster Phosphoproteins

As has been described for numerous other calcium biominerals, the primary mode of extraction of matrix proteins involves dissolution of mineral with EDTA at near neutral pH. A method for rapid dissolution has been developed [12] in which the shell is dissolved in dialysis bags that are rotated through large volumes of EDTA. The rotation serves to keep the powdered shell suspended within the bag, effecting a reduction in the time for dissolution. To date, no protease inhibitors or denaturants have been added to the dissolution media, as is common practice for extracting proteins from vertebrate calcium biomineral such as teeth or bones. In these latter cases, the opportunity for tissue contamination is much greater than for the shell, which during growth comes to lie entirely outside of the mineral-forming tissue. Accordingly, Weiner [13] reported that gel electrophoretic patterns of *Mercenaria* were unaffected by the inclusion of such agents in the extraction media. In contrast, Keith et al. [14] have shown different electrophoretic patterns of *Mytilus* shell extracts depending on whether or not protease inhibitors plus denaturing agents are included in the extraction media. The question of whether or not entities identified by electrophoresis constitute a significant fraction of the total protein extracted is not known (see below), making interpretation of these findings difficult.

As is normally the procedure for matrix extracts, the oyster protein is first fractionated into soluble and insoluble fractions, with the latter classified as that which forms into pellets upon centrifugation. The pellet from oyster has two distinct layers. The upper layer is a viscous, translucent material which is referred to as light insoluble matrix (LIM). The lower layer is dark and relatively particulate and has been termed dark insoluble matrix (DIM). The soluble matrix from the supernatant has been frationated by gel chromatography on Sephacryl 300, which resolves proteins having two classes of molecular weights, those greater than about 500 kd (SE I) and those with a mean around 50 kd (SE II). Using other gel chromatographic media, it has been determined that SE I can be further sub-fractionated with components having molecular weights ranging into the millions. In general, it appears that the soluble proteins form a continuum of molecular weights with a possible discrete class represented by SE II.

The soluble fractions can also be fractionated by ion exchange chromatography or reverse-phase HPLC. In DEAE-ion exchange, a pattern of elution for oyster SM [15] is roughly equivalent to that described for other molluscan species [9]. This technique has not proven especially useful in achieving better definition of fractions of matrix, in that large molecular weight fractions elute at zero ionic strength, probably because they are so large that only a small percentage of their ionic groups have access to the resin. Further, much, if not all, of SE II elutes near 0.3 M ionic strength with no distinct evidence of fractionation. This lack of fractionation is even more pronounced when various discrete reverse-phase HPLC fractions elute at the same ionic strength in step gradients (A.P. Wheeler, unpublished). However, more refined continuous gradients may result in subfractionation of SM.

Reverse-phase HPLC has resolved three fractions of oyster SE II termed RP-1, -2, and -3, with the first being the most hydrophilic [16]. Although RP-1 is typically collected as a near-void peak, attempts to further fractionate the peak led to the conclusion that it contains a discrete class of proteins [17]. When RP-2 was again subjected to chromatographic conditions identical to those used in the original fractionation, it partially separated into RP-1 and RP-3. This raises the question of whether or not there exists a relatively stable interaction between these latter two classes of proteins.

Wourms and Weiner [18] have also reported that two or three general reverse-phase classes of proteins form several species of molluscs, including a member of the same genus (*C. irredescens*). These fractions were obtained using a reverse-phase cartridge and step-wise elution. As with *C. virginica*, much of the protein elutes with the weakest buffer solvent and a second class elutes when the eluent is switched to a much stronger solvent (50% acetonitrile). A third fraction elutes with a step of dimethyl sulfoxide. Although it can be argued that these fractions roughly correspond to RP-1 through RP-3, the different solvent systems used in the two cases and possibly the different species studied make such a comparison tenuous at best (see below).

For several molluscan species, Weiner and co-workers have resolved numerous peaks from ion-exchange or reverse-phase cartridge fractions of soluble matrix using reverse-phase HPLC [13, 18, 19]. Since much of the total protein fractionationated in these studies appears at or near the onset of the gradient, it is difficult to assess the significance of the much smaller peaks which eluted later. Such a multitude of peaks was not observed for *C. virginica* when various gradients were used (K.W. Rusenko and A.P. Wheeler, unpublished data). However, in most of these latter cases, the sensitivity of the detector was adjusted to the scale of the major peaks and thus minor peaks, if any, could have escaped detection.

It should be mentioned that electrophoretic determination of protein constituents of soluble matrix from oyster has been hampered by the difficulty in staining the protein. Much like dentin phosphoproteins [20], these proteins are not reactive to Coomassie blue staining following standard acrylamide gel procedures. A lack of discrete Coomassie bands has been observed for soluble matrix from the calcite layers of *C. irredescens* and *C. gigas* [21], implying that similar types of proteins may be found in all the species of this genus. Discrete molecular weight components for other species have been identified using electrophoresis [14, 21], but the percent of the actual protein identified or entering the gel may be low. For the same reasons, caution should be applied to avoid overemphasizing the observation that SE II or RP-I occasionally produce dimeric bands around 50 kd when gels are stained using siliver techniques (A.P. Wheeler, unpublished data).

Composition of Oyster Fractions

The first observation that can be made regarding the oyster fractions is the similarity of amino acid composition among the components of the molecular weight fractions SE I and SE II and LIM (Table 1). In these cases, the aspartic acid, glycine, and serine make up the majority of the amino acids. In addition to a similarity of amino acid composition, all three fractions are highly phosphorylated, and, from β-elimination studies with base, it is clear that it is the serine groups which are phosphorylated. In summary, these are highly anionic proteins and, in this regard, typify in the extreme the soluble proteins extracted from diverse mineralized systems. For example, reverse-phase fractions with as high as 50% aspartic acid have been isolated from *C. irredescens* [18], and the envelope fraction isolated from prismatic layers of various bivalves was

Table 1. Composition of oyster shell matrix fractions[a]

Analysis	*LIM*	*SE I*	*SE II*	*RP-1*	*RP-3*	*DIM*
Amino acid[b]						
Asp	258	288	302	315	216	88
Thr	9	13	12	5	5	18
Ser	201	216	221	283	135	102
Glu	36	45	46	42	85	37
Pro	30	30	27	9	56	59
Gly	334	285	292	300	222	228
Ala	14	18	16	9	49	246
Cys	6	5	4			0
Val	14	11	11	2	23	35
Met	3	4	3			4
Ile	5	7	5	1	23	35
Leu	12	13	9	1	38	29
Dopa	0	0	0			8
Tyr	24	24	25	17	27	44
Phe	4	6	4	1	15	35
Lys	21	16	11	6	18	8
Arg	22	14	5	3	22	18
C/HP[c]	4.3	4.3	5.1	15.9	1.7	0.35
Carbohydrate	35.0	33.5	14.4	14.4	46.4	
Phosphate:[d]						
Total	109	129	140	153	49	
By OD_{240}	101		132	141	16	

[a] *LIM* is the light (translucent) insoluble matrix which forms the upper part of the pellet of oyster shell EDTA extracts upon centrifugation. *DIM* is dark insoluble matrix which is particulate and sediments under LIM. *SE I* and *SE II* are the principal molecular weight classes of soluble matrix obtained on Sephacryl S-300, with I containing material greater than about 500 kd and II a broad peak whose molecular weight is about 50 kd. *RP-1* and *3* are reverse-phase HPLC fractions of SE II, with RP-1 being the more hydrophilic and approx. 54 kd and RP-3 the more hydrophobic and about 70 kd. Data from [15–17]

[b] Residues per thousand

[c] Sum of the residues per thousand for the charged amino acids (asp, glu, lys, arg) divided by the sum of hydrophobic amino acids (pro, ala, val, ile, leu, phe)

[d] Phosphate as calculated from dehydroalanine detected at 240 nm during incubation of fractions in NaOH

reported to contain in excess of 70% aspartic acid [10]. In addition, some matrices have been reported as having sulfated carbohydrates [8].

Although there are relatively few studies which compare IM to SM, in cases where the comparison is clearly made, the IM tends to be somewhat lower in anionic amino acids than SM and have higher relative concentrations of certain hydrophobic amino acids [22]. The universeality of this observation is unknown; however, in LIM from oyster, a clear case for a dichotomy between SM and IM based on composition would be hard to make. In contrast, DIM is clearly distinctive from the other matrix fractions, having much lower relative aspartate levels and much higher relative levels of hydrophobic amino acids. In particular, alanine is a major constituent of DIM and dihydroxyphenylalanine (dopa) is present in this fraction (Table 1). It is possible that similar IM fractions exist in other species and simply have not been separated. For example, Travis et al. [6] analyzed IM from oyster and found compositions that could be described as intermediate between DIM and LIM.

The insolubility of IM and, for that matter, the molecular weight continuum of the SM most likely result from cross-linking between molecular species. However, no cross-links have been definitively described to date. The presence of phenoloxidase in shell extracts or sections of various species, including a member of the genus *Crassostrea* [23, 24], would suggest the involvement of dopa residues in the cross-linking of DIM.

LIM can be solubilized and the molecular weight of SE I reduced upon treatment with 0.1 N NaOH. This treatment results in fragments that have chromatographic properties similar to SM when passed over gel permeation [15] or reverse-phase HPLC media [17]. Whether the base is disrupting cross-links or fragmenting the backbone of the molecules is unknown. However, a linkage that would be especially base susceptible is a phosphodiester cross-link, a bond which has been described in other systems [25]. Given the significant levels of phosphorylation of both SE I and LIM, it is possible that at least a small percentage of the phosphates are involved in such a linkage.

Other potential cross-links have not been explored, and therefore cannot be excluded. For example, histidinoalanine has been described in both the phosphoprotein particles of molluscan mantle and extrapallial fluid [26], as well as, with lysinoalanine, in dentin phosphophoryn [27]. A histidinoalanine type of linkage is made less likely by the fact that histidine is present in very low quantities in the oyster phosphoproteins; however, lysinoalanine is a distinct possibility. It is possible that non-covalent interactions stabilize LIM or SE I. However, treatment of SE I with denaturants of non-covalent linkages, such as sodium dodecyl sulfate or cetylpyridinium chloride, does not result in reduction in size of this fraction.

Compositions of the reverse-phase fractions RP-1 and RP-3 indicate that there is actually some heterogeneity within the matrix phosphoproteins (Table 1). However, one must bear in mind that the heterogeneity may result from a relatively small percentage of the total protein in that RP-1 makes up greater than 90% of the SE II fractionated by reverse-phase. This fact is evident from the similarity of compositon of SE II and RP-1.

One of the most significant differences between RP-1 and RP-3 fractions is the relative concentration of phosphate and carbohydrate, with RP-3 being lower in the former and higher in the latter. Further, the phosphate of RP-1 appears to be more or less entirely linked via serine residues, as the total phosphate of the fraction agrees with that which can be determined as dehydroalanine following β-elimination. In contrast, using the same criteria, only a fraction of the phosphate of RP-3 appears to be O-linked via serine, suggesting that other O-linkages or N-linkages are possible.

Additional differences between the fractions are reflected by the amino acid composition of the two samples in that Asp, Ser, and Gly make up approximately 90% of the amino acids of RP-1 and less than 60% of RP-3. Further, as expected from their relative elution from reverse-phase HPLC, the fraction of hydrophobic residues in RP-3 is much higher than in RP-1. In comparing the "A" or most hydrophilic fraction isolated by Wourms and Weiner [18] from *C. irredescens* with RP-1 of *C. viriginica*, a significant disparity is observed in that although the "A" fraction is also 90% Asp + Ser + Gly, it is less than 10% Ser. The "C" fraction of Wourms and Weiner roughly corresponds to RP-3 in the composition of amino acids and in the fact that they both may be more glycosylated than their more hydrophilic counterparts. However, it is also clear that these more hydrophobic fractions isolated by the two methods differ significantly in detail. In short, it is apparent that RP-1 and RP-3 are unique fractions, and the extent to which they may be present in other species, especially those of the same genus or those having the foliated microstructure, will require extraction and fractionation of the matrix from various species using the same procedures as employed for *C. virginica*.

It should be noted that reverse-phase fractions that elute much like RP-1 and RP-3 have been obtained from preparations of base-solubilized LIM [17]. Predictably, their composition differs from the soluble matrix fractions in that base-induced β-elimination at serine will cause a loss of that amino acid. However, taking that loss into account, the residual amino acid distribution remains remarkably similar to the fractions from SM. This finding accentuates the overall homogeneity of the phosphoprotein complexes.

Sulfate, presumably as sulfated carbohydrate, has been identified as a significant component of both *Mercenaria mercenaria* and *Mytilus californianus* SM proteins, being 10% by weight of the former [8] and in a ratio of about 70 residues per 100 amino acids in the latter [28]. Simkiss [29], using the results of infrared analyses, suggested the presence of sulfate in matrix extracts of *C. virginica* as well. Chemical analysis of SE II resulted in only approximately 1% sulfate by weight (D.M. Swift, unpublished data). It is interesting to note that neither *Mytilus edulis* nor *M. mercenaria* SM is phosphorylated to any extent [3]. Given that all these species have different shell microstructures, the dichotomy of sulfated versus phosphorylated matrix raises the question of whether the type of anion found in matrix is an important factor in controlling the shell crystal architecture.

Primary Structure of Phosphoproteins

Previous to studies of the oyster matrix, much of the evidence regarding the primary sequence of molluscs was interpreted as suggesting that these proteins contain domains of $(\text{Asp-X})_n$ in which X is primarily Gly or Ser. These conclusions were based on the finding that Asp, Ser, and Gly were released from SM proteins during a mild acid hydrolysis [13, 30–32]. Although this technique is selective for peptide bonds on either side of aspartic acid residues, they are not entirely specific for the release of this amino acid [1, 16, 17]. Consequently, if the hydrolysis is continued for extended periods, as it was in the above-mentioned studies, then the accumulated release of other amino acids is expected even if they are not adjacent to aspartic acid.

To avoid this complication, time-course hydrolyses were performed on a number of oyster phosphoprotein fractions including LIM, RP-1, and RP-3. In every case, the rate of release of aspartic acid far exceeded that of Gly or Ser [16, 17]. Further, Gly and Ser release were linear, whereas the rate of release of Asp approached a plateau in 2–4 h. These data, along with those for the hydrolysis of model peptides, suggest that Asp-Asp bonds are quite common in the proteins, possibly existing as domains of polyaspartate. If domains of Asp-X are present, they are certainly in the minority in oyster proteins. Further, domains of Asp-X-Y, if present at all, are also in the minority, as there was no evidence of Gly-Ser or Ser-Gly dipeptides released during the hydrolysis. These dipeptides were identified in the hydrolysates of SM from the calcitic-prismatic layer of *Pinctada margaritifera* [32]. However, they may be microstructure-specific, as they were not identified in the nacreous layer of this species. It is interesting to note that the matrix of the calcitic-prismatic layer of *Mytilus californianus* contains the dipeptide Pro-Thr which is also specific to that layer [13].

Additional sequence information has been obtained for RP-1 using reverse-phase HPLC fractions of the residual material following mild acid hydrolysis [16, 17] and fragments produced by hydroxylamine cleavage [33]. In both cases, the fragments ranged dramatically in relative hydrophobicity, indicating distinct domains. One of the more hydrophilic fragments produced by acid hydrolysis, when subjected to automated Edman degradation, revealed an NH_2-terminal sequence composed almost entirely of serine. Because of the hydrophilicity of the fragment and the fact that phosphate is concentrated in the dialyzed residue over the native material, this fragment may represent a phosphoserine-enriched domain. Edman degradation of several of the hydroxylamine fragments revealed runs of Asp, confirming the findings from mild acid hydrolysis. Furthermore, in the fragments sequenced to date, $(Asp\text{-}X)_n$ appears to be virtually absent, confirming the results of mild acid hydrolysis.

The identification of hydrophobic fragments from RP-1 is interesting in that only a small percent of the protein is composed of hydrophobic amino acids. From hydrolysis studies using a variety of carboxypeptidases, it is clear that one of those domains is at the carboxy terminus of the intact protein [16, 17]. One of the polyaspartate domains is at the carboxy terminus as well. The only other molluscan matrix fractions for which this type of sequencing has been attempted is a blot-transferred electrophoretic fraction isolated from *Mytilus edulis* [14]. In this case, a hydrophobic domain was tentatively identified at the amino terminus.

The distribution of the glycine residues in oyster phosphoprotein is not yet clear. However, because glycine in dialyzed residues obtained from mild acid hydrolysis is in about the same molar proportion as in the native protein, it could be surmised that there are domains of polyglycine or at least domains highly enriched for the amino acid. These domains would provide a large

NH2-	[Aspn]x [(P)Sern]x [(Gly(P)Ser)n]$_x$? [Glyn]$_x$?	-[Aspn]-Hydrophobic-Lys-Gly-Tyr-Tyr-COOH

Fig. 1. Suggested amino acid domains of the oyster phosphoprotein, RP-1. The carboxy terminus was elucidated by the use of carboxypeptidases. The terminal four amino acid sequence is consistent with the results from hydrolysis of three enzymes. The hydrophobic domain appears to contain phe, leu, ile, val, and ala—all minor components of RP-1. The sequence of these amino acids within the domain is not known. The elucidation of an Asp_n domain near the carboxy terminus was elucidated with carboxypeptidase P. The existence of Asp_n domains in general follows from the rate of release of this amino acid during mild acid hydrolysis and from partial sequence analysis of hydroxylamine-cleaved fragments. The existence of $(P)Ser_n$ regions follows from compositional analysis of the residue of mild acid hydrolysis and from partial sequence of reverse phase fractions from these and hydroxylamine-cleaved RP-1. The possibility of $Gly(P)Ser_n$ is suggested but not confirmed by these studies. Gly_n domains are entirely speculative at this time and based on the composition of residues following hydrolysis. Sequence analysis and reverse-phase profiles of cleavage fragments also suggest that some mixed domains exist as well (not shown). Structures summarized from data in [16, 17, 33]

degree of rotational freedom to the protein. A summary of the putative domains for RP-1 identified to date is given in Fig. 1.

A comparison can be made between RP-1 and phosphophoryn, the aspartic acid-rich phosphoprotein from tooth dentin. This latter protein also appears to contain hydrophobic domains in the end regions of the molecule and both Asp and (P)Ser domains [34]. The difference is the presence of putative Asp-(P)Ser domains in the dentin protein which may be absent in RP-1. To some extent it appears that phosphoproteins from the two mineral systems may be structurally analogous. The structural analogy indicates that they may function in similar ways despite the fact that the minerals evolved along different lines and at different periods.

Functional Aspects of Phosphoproteins

If one accepts the hypothesis that matrix serves to regulate crystal growth, it then follows that the oyster phosphoproteins must be shown to interact with calcite, preferably in more or less specific ways. For example, it is now well established that immobilized matrix molecules can induce crystal growth from slightly supersaturated solutions. One technique used to demonstrate this phenomenon involves extraction of matrix in fixatives so that SM remains associated with IM [35]. Other approaches involve attaching SM proteins to

Table 2. Calcium, nuclei, and crystal association properties of oyster phosphoproteins and synthetic peptides[a]

Calcium binding:	
Low ionic strength	SE II—affinity K, 10^{+4}, $10^{+5}\,M^{-1}$
	Capacity, 1-2 µmol mg^{-1}
	LIM, polyasp—binding occurs
500 mM NaCl	SE II, SE I, polyasp-binding negligible

CaCO₃ Nucleation—order of inhibitory activity on weight basis in 500 mM NaCl[b]

$Asp_{20}PSer > (Asp(P)Ser)_{10} > RP\text{-}1 \cong Asp_{20} > RP\text{-}3$
$(AspSer)_{10} \cong Asp_{10} > (AspSerGly)_{10} \cong (Asp(P)SerGly)_{10}$

$CaCO_3$ Seed crystal-binding and growth inhibition in 500 mM NaCl:

Molar affinity:
$RP\text{-}1 \gg Asp_{19} \cong (AspSer)_{10} \cong (AspSerGly)_{10} > (Asp(P)SerGly)_{10}$

Molar capacity[c]
$Asp_{19} > (Asp(P)SerGly)_{10} > (AspSer)_{10} > (AspSerGly)_{10} \cong RP\text{-}1$

Max % crystal growth inhibition when crystal sties saturated for peptide[d]
$RP\text{-}1 = Asp_{19} = 100\% > (Asp(P)SerGly)_{10} > (AspSer)_{10} > (AspSerGly)_{10}$

[a] Data from [12, 41, 42] and A.P. Wheeler and A.S. Mount, unpublished results. *(P)Ser* designation means partial phosphorylation of Ser

[b] The relative capability to inhibit nucleation should reflect the affinity of peptides for ion clusters or nuclei. On a molar basis the inhibitory activity of RP-1 would exceed that of all peptides tested to date, with the possible exception of polyaspartates having terminal PSer residues

[c] The low molar capacity of RP-1 relative to its high affinity and capability for inducing 100% inhibition suggests that it can bind to more than one growth site

[d] RP-1 achieves 100% inhibition when sites are only 50% saturated, suggesting that attachment of RP-1 to one Asp_{19} site influences growth at adjacent sites. The maximal percent inhibition for other peptides is identical to the percent of Asp_{19} sites to which they can bind, suggesting that binding sites are equivalent to growth sites and more than one kind of growth site exists on crystals

polystyrene films [28] or gel beads [36]. The idea in each case is to simulate the relationship between the crystal active component (polyanionic SM) and the framework molecules (IM). An ideal matrix material that combines the anionic and insoluble characteristics is the LIM of oyster, a system that has not been studied in this regard.

The mechanism for the induction of mineral by insolubilized matrix is often explained in terms of calcium binding to the anionic substrate which results in the accumulation of layers of both cations and subsequently anions, and thus formation of crystal nuclei at that site [37]. This hypothesis is based on the observation that SM has a low affinity, high capacity calcium binding when in solution, especially at low ionic strength [1] (Table 2). Using soluble oyster phosphoproteins, Wheeler et al. [12] has demonstrated that most of the calcium-binding capacity of these proteins is lost at physiological ionic strength. Probably more significant to the theories of induction of crystal growth is the question to what extent the insoluble phosphoprotein LIM, which binds calcium at low ionic strength (A.P. Wheeler and A.S. Mount, unpublished data), retains its binding capacity at physiological strength. It should be mentioned that the effect of ionic strength on oyster soluble protein stands in contrast to observations on phosophophoryn in which significant calcium binding is retained at high ionic strength [38]. Because studies with model proteins suggest that repetitive carboxyl-phosphate groups may enhance calcium binding [39], it may be that the lack of Asp-PSer domains in oyster matrix but apparent presence in phosphophoryn may explain their different binding properties.

An alternate hypothesis to explain the effect of immobilized polyanions on induction of crystal growth involves the binding of nuclei rather than ions from metastable solutions [12]. This is a credible hypothesis in that soluble oyster phosphoproteins, on a molar basis, are as good as or better than other polyanions as inhibitors of calcite induction in bulk solutions [40] (Table 2) and thus probably have a high affinity for nuclei. In addition, unlike calcium-binding affinity, a high affinity for nuclei is retained at elevated ionic strength, as inhibitory assays are routinely run in 500 mM NaCl. Based on studies with model peptides, it is clear that this inhibitory activity is quite sensitive to primary structure. That is, peptides with runs of polyanions (devoid of spacer amino acids), much like those found in oyster phosphoproteins, are especially effective inhibitors. The addition of PSer termini to polyaspartate peptides dramatically increases their efficacy as inhibitors of calcite induction [41] (Table 2), indicating that phosphorylated domains other than Asp-(P)Ser calcium-binding domains may be excellent nuclei-binding agents.

The oyster soluble phosphoproteins are also excellent inhibitors of calcite crystal growth and effect this inhibition through high affinity binding to crystal surfaces [42] (Table 2). For the natural proteins, the runs of polyaspartate no doubt contribute to the overall efficacy of RP-1 inhibition in that synthetic polyaspartate is more effective as a crystal growth inhibitor on a mole basis than $(Ser\text{-}Asp)_n$ or $(Gly\text{-}Ser\text{-}Asp)_n$ peptides. The addition of phosphate to Ser in these model peptides increases their efficacy as an inhibitor as well, suggesting an importance for the PSer of oyster matrix in its interaction with growing calcitic crystals. This was confirmed in studies in which alkaline phosphatase dephosphorylated SE II fractions were shown to have significantly lower inhibitory activity than the native protein. In fact, the activity approached that of non-phosphorylated SM from other species [3].

Most data on crystal growth inhibition have been amassed from in vitro systems. However, it has been established that these oyster phosphoproteins can inhibit sea urchin embryo calcitic spicule growth in vivo [43]. Furthermore, it was determined that the protein actually bound to the crystals, a process that involved transport of the protein through cellular layers. This transport would not necessarily be expected of polyanions. Possibly one function of the hydrophobic tail of the phosphoproteins is to allow for recognition by and transport across the cells.

The relevance of inhibition by matrix proteins measured in vitro to crystal growth in situ is not known. At first glance, this activity may be related to simple limitation of crystal growth either in preformed compartments [7] or by periodic secretion of relevant layers [44]. Another role for inhibitory matrix proteins might involve control of crystal microstructure. In this process, both immobilized and soluble matrix components might participate. For example, it has been shown that *Mytilus* SM proteins adsorbed onto a surface can increase the frequency of development of unusual crystal faces and growth along the *c*-axis of the crystal [45]. As mentioned earlier, this axis is elaborated in many molluscan microstructures, including the oyster foliated system. Whether immobilized oyster SM or LIM can induce similar effects is not known. However, given that cooperation between the sulfate and carboxylate groups of the *Mytilus* proteins may be necessary for growth regulation [28], a similar effect may be produced by the phosphate and carboxylate groups of oyster matrix.

Interactions of matrix from solution with growing crystals may effect growth modifications as well, especially if specific matrix proteins adsorb to and inhibit growth at specific sites or faces, forcing growth in other areas. For example, it has been shown that the maximal percent inhibition of crystal growth induced by Asp-containing peptides and the number of binding sites on calcite for the peptides varies with their primary structure [42] (Table 2). Specifically, Asp_{19} effects 100% maximal inhibition and binds twice the number of sites as $(Ser\text{-}Asp)_{10}$, which effects only about 50% maximal inhibition. In this regard, RP-1 appears to be rather non-specific in that, like polyaspartate, it effects 100% inhibition of crystal growth but has a molar affinity for the crystal surface much greater than even polyaspartate. Because, on a molar basis, RP-1 binds to fewer sites than does Asp_{19}, its higher affinity for crystals may indicate that the polymer is binding to more than one Asp_{19}-type site. In fact, it has been shown via scanning tunneling microscopy that RP-1 may form a ring structure when adsorbed onto the surface of calcite [33]. The significance of this configuration is unclear; but, if these proteins do absorb in this fashion, it suggests that they may have the capacity to bind to multiple sites.

The significance of the above-mentioned observations regarding the relationship of RP-1 to the regulation of mineral growth is not known. However, the non-specific effect of such strongly anionic structures seems to carry over to the morphology of the crystals grown in their presence. This applies to both oyster phosphoproteins [46] and to sulfated SM proteins from *Mytilus* [47]. In both these cases, gross changes in crystal morphology occur but no faces atypical of inorganically grown crystals appear. In contrast, matrix proteins of lower anion content can induce specific morphologies and also be more effectively incorporated into crystals than are the highly anionic proteins [47]. In this regard, it might be speculated that RP-3 (or similar peptides), having a much lower anion content, might regulate crystal morphology and subsequently become part of the intracrystalline matrix. The role of the highly charged proteins (RP-1) may be relegated to that of limitation of crystal growth.

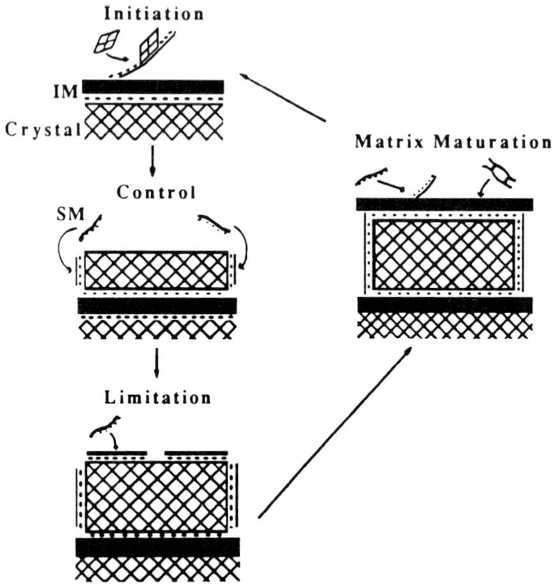

Fig. 2. A hypothetical scheme for the involvement of oyster matrix in crystal/crystal lamellae formation. *Initiation* The interlamellar matrix from the previous layer induces the next layer. Here, a polyanionic insoluble matrix such as LIM may be the induction factor. Nucleation is shown as binding nuclei rather than calcium since oyster matrix is a poor calcium binding agent at physiological ionic strength but apparently has a high affinity for nuclei under these conditions (Table 2). *Control* Soluble components may interact specifically with crystal sites, inhibiting growth in certain dimensions but allowing growth in others thus shaping the crystal. RP-3-type proteins, being less acidic, may have such selectivity and may be more easily overgrown, coming to reside as intracrystalline matrix. *Limitation* A highly charged, relatively non-specific inhibitor such as RP-1 may be involved in overall termination of growth. *Maturation* At this stage it is envisioned that structural elements (such as those cross-linked to form DIM) are added. In addition, the next layer of polyanion, which may include soluble elements, is cross-linked in some fashion to form a new LIM layer. Overall, a charged layer (envelope) comes to lie next to the crystal and other layers form between the envelopes. This process of matrix production is shown as being sequential with crystal growth. However, the matrix structure may be preformed, at least in part, and still function substantively as depicted

Conclusions

Despite their apparent complexity, the phosphoproteins that make up the majority of the oyster matrix may well consist of a fairly limited number of major classes of proteins. It is the interaction among these classes in time and space that no doubt controls the process of mineralization. A simple scheme of these processes is shown in Fig. 2. This scheme is not intended to be exhaustive or inclusive of all ideas or possibilities for matrix function. It does, however, reflect some of the organizing principles identified thus far. Furthermore, such a model indicates the need for continued study in several areas. For example, the interaction among the matrix components is far from clear. In addition to continued study on the interaction of components identified in extracts, knowing the form (or forms) of matrix molecules secreted by the mantle would help in the reconstruction of matrix molecular morphology. In turn, knowing how the matrix is structured comprehensively may help in the design of additional assays for its function. Continued analysis of the composition and primary sequences of matrix components is clearly a requisite to understanding both overall matrix structure and function. Combining these features of the oyster phosphoprotein-foliated calcite microstructure system would facilitate endeavors involving other shell structures in particular and mineralizing systems in general.

References

1. Wheeler AP, Sikes CS (1989) Matrix-crystal interactions in $CaCO_3$ biomineralization. In: Mann S, Webb J, Williams RJP (eds) Biomineralization: Chemical and biochemical perspectives. VCH, Weinheim, pp 95–131
2. Weiner S (1986) Organization of extracellularly mineralized tissues: A comparative study of biological crystal growth. CRC Crit Rev Biochem 20:365–408
3. Borbas JE, Wheeler AP, Sikes CS (1991) Molluscan shell matrix phosphoproteins: Correlation of degree of phosphorylation to shell mineral microstructure and to in vitro regulation of mineralization. J Exp Zool 258:1–13
4. Veis A (1989) Biochemical studies of vertebrate tooth mineralization. In: Mann S, Webb J, Williams RJP (eds) Biomineralization: Chemical and biochemical perspectives. VCH, Weinheim, pp 189–222
5. Watabe N (1965) Studies on shell formation. XI. Crystal-matrix relationships in the inner layers of mollusk shells. J Ultrastruct Mol Struct Res 12:351–370
6. Travis DF, Francois CJ, Bonar LC, Glimcher MJ (1967) Comparative studies of the organic matrices of invertebrate mineralized tissues. J Ultrastruct Mol Struct Res 18:519–550
7. Bevelander G, Nakahara H (1980) Compartment and envelope formation in the process of biological mineralization. In: Omori M, Watabe N (eds) Tokai University Press, Tokyo, pp 19–27
8. Crenshaw MA (1972) The soluble matrix from *Mercenaria mercenaria* shell. Biomineralization 6:6–11
9. Weiner S (1979) Aspartic acid-rich proteins: Major components of the soluble organic matrix of mollusk shells. Calcif Tissue Int 29:163–167
10. Nakahara H, Mitsuo K, Bevelander G (1980) Five structure and amino acid composition of the organic "envelope" in the prismatic layer of some bivalve shells. Venus (Jpn J Malacol) 39:167–177
11. Iwata K (1975) Ultrastructure of the chonchiolin matrices in molluscan nacreous layer. J Fac Sci, Hokkaido Univ, IV 17:173–229
12. Wheeler AP, Rusenko KW, George JW, Sikes CS (1987) Evaluation of calcium binding by molluscan shell organic matrix and its relevance to biomineralization. Comp Biochem Physiol [B] 87:953–960

13. Weiner S (1983) Mollusk shell formation: Isolation of two organic matrix proteins associated with calcite deposition in the bivalve *Mytilus californianus*. Biochemistry 22:4139–4145
14. Keith JA, Stockwell SA, Fletcher MR, Napier K, Donnley TS, Kaplan DL (1991) Characterization of *Mytilus edulis* matrix proteins and the influence of temperature on crystallization of $CaCO_3$ under monolayers. Chem Eng News, May 11, 1991, p 72
15. Wheeler AP, Rusenko KW, Sikes CS (1988) Regulation of in vitro and in vivo $CaCO_3$ crystallization by fractions of oyster soluble matrix. Mar Biol 98:71–80
16. Rusenko KW, Donachy JE, Wheeler AP (1991) Purification and characterization of a shell matrix phosphoprotein from the American oyster. In: Sikes CS, Wheeler AP (eds) Surface reactive peptides and polymers. ACS, Washington, pp 108–124
17. Rusenko KW (1988) Studies on the structure and function of shell matrix proteins from the American oyster, *Crassostrea virginica*. PhD dissertation, Clemson University, Clemson, South Carolina
18. Wourms D, Weiner S (1986) Mollusk shell organic matrix: Fourier transform infrared study of the acidic macromolecules. J Exp Zool 237:11–20
19. Weiner S (1982) Separation of acidic proteins from mineralized tissues by reversed phase high-performance liquid chromatography. J Chromatogr 245:148–154
20. Butler WT, Bhown M, Dimuzio T, Linde A (1981) Noncollagenous proteins of dentin. Isolation and partial characterization of rat dentin proteins and proteoglycans using a three-step preparative method. Coll Res 1:187–199
21. Weiner S, Lowenstam HA (1977) Discrete molecular weight components of the organic matrices of mollusc shells. J Exp Mar Biol Ecol 30:45–51
22. Meenakshi VR, Hare PE, Wilbur KM (1971) Amino acids of the organic matrix of the gastropod shells. Comp Biochem Physiol [B] 40:1037–1043
23. Samata T, Sanguansri P, Cazaux C, Hamm M, Engles J, Krampitz G (1980) Biochemical studies on components of mollusc shells. In: Omori M, Watabe N (eds) The mechanisms of biomineralization in animals and plants. Tokai University Press, Tokyo, pp 37–47
24. Gordon J, Carriker MR (1980) Sclerotized protein in the shell matrix of a bivalve mollusc. Mar Biol 57:251–260
25. Gustafson GL, Milner LA (1980) Occurrence of N-acetylglucosamine-1-phosphate in proteinase I from *Dictyostelium discoideum*. J Biol Chem 255:7208–7210
26. Marsh ME (1986) Histidinoalanine, a naturally occurring cross-link derived from phosphoserine and histidine residues in mineral-binding phosphoproteins. Biochemistry 25:2392–2396
27. Kuboki Y, Fujisawa R, Tsuzaki M, Liu CF, Sasaki S (1984) Presence of lysinoalanine and histidinoalanine in bovine dentin phosphoprotein. Calcif Tissue Int 36:126–128
28. Addadi L, Moradian J, Shay E, Maroudas NG, Weiner S (1987) A chemical model for the cooperation of sulfates and carboxylates in calcite crystal nucleation: Relevance to biomineralization. Proc Natl Acad Sci USA 84:2732–2736
29. Simkiss K (1965) The organic matrix of the oyster shell. Comp Biochem Physiol 16:427–435
30. Weiner S, Hood L (1975) Soluble protein of the organic matrix of mollusc shells: A potential template for shell formation. Science 190:987–989
31. Akiyama M (1980) Thermal alteration experiments of scallop shells in relation to the diagenetic change of fossil proteins. In: Omori M, Watabe N (eds) The mechanisms of biomineralization in animals and plants. Tokai University Press, Tokyo, pp 257–262

32. Weiner S (1981) Repeating amino acid sequences in the soluble protein of mollusc shell organic matrices: Their involvement in crystal formation. In: Veis A (ed) The chemistry and biology of mineralized connective tissues. Elsevier, North Holland, pp 517–521
33. Donachy JE (1991) Atomic force microscopy and sequence analysis of a matrix protein from calcium carbonate biomineral. Chem Eng News, Mar. 11, 1991, p 72
34. Veis A (1991) Phosphoproteins as mediators of biomineralization. In: Sikes CS, Wheeler AP (eds) Surface reactive peptides and polymers. ACS, Washington, pp 1–12
35. Greenfield E (1987) In vitro mineral induction by soluble matrix from molluscan shells. PhD dissertation, University of North Carolina, Chapel Hill, North Carolina
36. Crenshaw MA, Linde A, Lussi A (1988) Biologically formed composites. In: Aksay IA, McVay GL, Stoebe TG, Wager JG (eds) Atomic and molecular processing of electronic and ceramic materials. Materials Research Society, Pittsburg, pp 99–107
37. Greenfield EM, Wilson DT, Crenshaw MA (1984) Ionotropic nucleation of calcium carbonate by molluscan matrix. Am Zool 24:925–932
38. Stetler-Stevenson WG, Veis A (1987) Bovine dentin phosphophoryn: Calcium ion binding properties of a high molecular weight preparation. Calcif Tissue Int 40:97–102
39. Lee SL, Veis A (1980) Cooperativity in calcium ion binding to repetitive, carboxylate-serylphosphate polypeptides and the relationship of this property to dentin mineralization. J Peptide Protein Res 16:231–240
40. Sikes CS, Wheeler AP (1988) Regulators of biomineralization. CHEMTECH 18:620–626
41. Sikes CS, Yeung ML, Wheeler AP (1991) Inhibition of calcium carbonate and calcium phosphate crystallization by peptides enriched in aspartic acid and phosphoserine. In: Sikes CS, Wheeler AP (eds) Surface reactive peptides and polymers. ACS, Washington, pp 50–71
42. Wheeler AP, Low KC, Sikes CS (1991) $CaCO_3$ crystal-binding properties of peptides and their influence on crystal growth. In: Sikes CS, Wheeler AP (eds) Surface reactive peptides and polymers. ACS, Washington, pp 72–84
43. Sikes CS, Wheeler AP (1986) The organic matrix of oyster shell as a regulator of calcification in vivo. Biol Bull 170:494–505
44. Watabe N (1981) Crystal growth of calcium carbonate in the invertebrates. Prog Crystal Growth Character 4:99–147
45. Addadi L, Weiner S (1985) Interactions between acidic proteins and crystals: Steriochemical requirements in biomineralization. Proc Natl Acad Sci USA 82: 4110–4114
46. Wheeler AP, Sikes CS (1984) Regulation of carbonate calcification by organic matrix. Am Zool 24:933–944
47. Berman A, Addadi L, Weiner S (1988) Interactions of sea urchin skeleton macromolecules with growing calcite crystals—a study of intracrystalline proteins. Nature 331:546–548

Index